Lecture Notes in Computer Science

Edited by G. Goos, J. Hartmanis, and J. va...

Lecture Notes in Computer Science 2779
Edited by G. Goos, J. Hartmanis, and J. van Leeuwen

Springer
Berlin
Heidelberg
New York
Hong Kong
London
Milan
Paris
Tokyo

Dan Boneh (Ed.)

Advances in Cryptology – CRYPTO 2003

23rd Annual International Cryptology Conference
Santa Barbara, California, USA, August 17-21, 2003
Proceedings

Springer

Series Editors

Gerhard Goos, Karlsruhe University, Germany
Juris Hartmanis, Cornell University, NY, USA
Jan van Leeuwen, Utrecht University, The Netherlands

Volume Editor

Dan Boneh
Stanford University
Computer Science Department
Gates 475, Stanford, CA, 94305-9045, USA
E-mail: dabo@cs.stanford.edu

Cataloging-in-Publication Data applied for

A catalog record for this book is available from the Library of Congress.

Bibliographic information published by Die Deutsche Bibliothek
Die Deutsche Bibliothek lists this publication in the Deutsche Nationalbibliografie;
detailed bibliographic data is available in the Internet at <http://dnb.ddb.de>.

CR Subject Classification (1998): E.3, G.2.1, F.-2.1-2, D.4.6, K.6.5, C.2, J.1

ISSN 0302-9743
ISBN 3-540-40674-3 Springer-Verlag Berlin Heidelberg New York

Springer-Verlag Berlin Heidelberg New York
a member of BertelsmannSpringer Science+Business Media GmbH

http://www.springer.de

© International Association for Cryptologic Research 2003
Printed in Germany

Typesetting: Camera-ready by author, data conversion by PTP-Berlin GmbH
Printed on acid-free paper   SPIN: 10929063   06/3142   5 4 3 2 1 0

Preface

Crypto 2003, the 23rd Annual Crypto Conference, was sponsored by the International Association for Cryptologic Research (IACR) in cooperation with the IEEE Computer Society Technical Committee on Security and Privacy and the Computer Science Department of the University of California at Santa Barbara.

The conference received 169 submissions, of which the program committee selected 34 for presentation. These proceedings contain the revised versions of the 34 submissions that were presented at the conference. These revisions have not been checked for correctness, and the authors bear full responsibility for the contents of their papers. Submissions to the conference represent cutting-edge research in the cryptographic community worldwide and cover all areas of cryptography. Many high-quality works could not be accepted. These works will surely be published elsewhere.

The conference program included two invited lectures. Moni Naor spoke on cryptographic assumptions and challenges. Hugo Krawczyk spoke on the 'SIGn-and-MAc' approach to authenticated Diffie-Hellman and its use in the IKE protocols. The conference program also included the traditional rump session, chaired by Stuart Haber, featuring short, informal talks on late-breaking research news.

Assembling the conference program requires the help of many many people. To all those who pitched in, I am forever in your debt.

I would like to first thank the many researchers from all over the world who submitted their work to this conference. Without them, Crypto could not exist.

I thank Greg Rose, the general chair, for shielding me from innumerable logistical headaches, and showing great generosity in supporting my efforts.

Selecting from so many submissions is a daunting task. My deepest thanks go to the members of the program committee, for their knowledge, wisdom, and work ethic. We in turn relied heavily on the expertise of the many outside reviewers who assisted us in our deliberations. My thanks to all those listed on the pages below, and my thanks and apologies to any I have missed. Overall, the review process generated over 400 pages of reviews and discussions.

I thank Victor Shoup for hosting the program committee meeting in New York University and for his help with local arrangements. Thanks also to Tal Rabin, my favorite culinary guide, for organizing the postdeliberations dinner. I also thank my assistant, Lynda Harris, for her help in the PC meeting pre-arrangements.

I am grateful to Hovav Shacham for diligently maintaining the Web system, running both the submission server and the review server. Hovav patched security holes and added many features to both systems. I also thank the people who, by their past and continuing work, have contributed to the submission and review systems. Submissions were processed using a system based on software written by Chanathip Namprempre under the guidance of Mihir Bellare. The review process was administered using software written by Wim Moreau and Joris Claessens, developed under the guidance of Bart Preneel.

I thank the advisory board, Moti Yung and Matt Franklin, for teaching me my job. They promptly answered any questions and helped with more than one task.

Last, and more importantly, I'd like to thank my wife, Pei, for her patience, support, and love. I thank my new-born daughter, Naomi Boneh, who graciously waited to be born after the review process was completed.

June 2003

Dan Boneh
Program Chair
Crypto 2003

CRYPTO 2003

August 17–21, 2003, Santa Barbara, California, USA

Sponsored by the
International Association for Cryptologic Research (IACR)

in cooperation with
IEEE Computer Society Technical Committee on Security and Privacy,
Computer Science Department, University of California, Santa Barbara

General Chair
Greg Rose, Qualcomm Australia

Program Chair
Dan Boneh, Stanford University, USA

Program Committee

Mihir Bellare	U.C. San Diego, USA
Jan Camenisch	IBM Research, Zurich
Don Coppersmith	IBM Research, Watson, USA
Jean-Sebastien Coron	Gemplus Card International, France
Ronald Cramer	BRICS, Denmark
Antoine Joux	DCSSI Crypto Lab, France
Charanjit Jutla	IBM Research, Watson, USA
Jonathan Katz	University of Maryland, USA
Eyal Kushilevitz	Technion, Israel
Anna Lysyanskaya	Brown University, USA
Phil MacKenzie	Bell Labs, USA
Mitsuru Matsui	Mitsubishi Electric, Japan
Tatsuaki Okamoto	NTT, Japan
Rafail Ostrovsky	Telcordia Technologies, USA
Benny Pinkas	HP Labs, USA
Bart Preneel	Katholieke Universiteit Leuven, Belgium
Tal Rabin	IBM Research, Watson, USA
Kazue Sako	NEC, Japan
Victor Shoup	NYU, USA
Jessica Staddon	PARC, USA
Ramarathnam Venkatesan	Microsoft Research, USA
Michael Wiener	Canada

Advisory Members

Moti Yung (Crypto 2002 Program Chair)	Columbia University, USA
Matthew Franklin (Crypto 2004 Program Chair)	U.C. Davis, USA

External Reviewers

Masayuki Abe
Amos Beimel
Alexandra Boldyreva
Jesper Buus Nielsen
Christian Cachin
Ran Canetti
Matt Cary
Suresh Chari
Henry Cohn
Nicolas Courtois
Christophe De Canniere
David DiVincenzo
Yevgeniy Dodis
Pierre-Alain Fouque
Atsushi Fujioka
Eiichiro Fujisaki
Jun Furukawa
Rosario Gennaro
Philippe Golle
Stuart Haber
Shai Halevi
Helena Handschuh
Susan Hohenberger
Yuval Ishai
Mariusz Jakubowski
Rob Johnson
Mads Jurik
Aviad Kipnis
Lars Knudsen
Tadayoshi Kohno
Hugo Krawczyk

Ted Krovetz
Joe Lano
Gregor Leander
Arjen Lenstra
Matt Lepinski
Yehuda Lindell
Moses Liskov
Tal Malkin
Jean Marc Couveignes
Gwenaelle Martinet
Alexei Miasnikov
Daniele Micciancio
Kazuhiko Minematsu
Sara Miner
Michel Mitton
Brian Monahan
Frédéric Muller
David Naccache
Kobbi Nissim
Kaisa Nyberg
Satoshi Obana
Pascal Paillier
Adriana Palacio
Sarvar Patel
Jacques Patarin
Chris Peikert
Krzysztof Pietrzak
Jonathan Poritz
Michael Quisquater
Omer Reingold
Vincent Rijmen

Phillip Rogaway
Pankaj Rohatgi
Ludovic Rousseau
Atri Rudra
Taiichi Saitoh
Louis Salvail
Jasper Scholten
Hovav Shacham
Dan Simon
Nigel Smart
Diana Smetters
Martijn Stam
Doug Stinson
Reto Strobl
Koutarou Suzuki
Amnon Ta Shma
Yael Tauman
Stafford Tavares
Vanessa Teague
Isamu Teranishi
Yuki Tokunaga
Nikos Triandopoulos
Shigenori Uchiyama
Frédéric Valette
Bogdan Warinschi
Lawrence Washington
Ruizhong Wei
Steve Weis
Stefan Wolf
Yacov Yacobi
Go Yamamoto

Table of Contents

Public Key Cryptanalysis II

Universal Composability

Zero-Knowledge

Algebraic Geometry

Public Key Constructions

Invited Talk II

New Problems

Symmetric Key Constructions

New Models

Symmetric Key Cryptanalysis II

Factoring Large Numbers with the TWIRL Device

Adi Shamir and Eran Tromer

Department of Computer Science and Applied Mathematics
Weizmann Institute of Science, Rehovot 76100, Israel
{shamir,tromer}@wisdom.weizmann.ac.il

Abstract. The security of the RSA cryptosystem depends on the difficulty of factoring large integers. The best current factoring algorithm is the Number Field Sieve (NFS), and its most difficult part is the sieving step. In 1999 a large distributed computation involving hundreds of workstations working for many months managed to factor a 512-bit RSA key, but 1024-bit keys were believed to be safe for the next 15-20 years. In this paper we describe a new hardware implementation of the NFS sieving step (based on standard 0.13μm, 1GHz silicon VLSI technology) which is 3-4 orders of magnitude more cost effective than the best previously published designs (such as the optoelectronic TWINKLE and the mesh-based sieving). Based on a detailed analysis of all the critical components (but without an actual implementation), we believe that the NFS sieving step for 512-bit RSA keys can be completed in less than ten minutes by a $10K device. For 1024-bit RSA keys, analysis of the NFS parameters (backed by experimental data where possible) suggests that sieving step can be completed in less than a year by a $10M device. Coupled with recent results about the cost of the NFS matrix step, this raises some concerns about the security of this key size.

1 Introduction

The hardness of integer factorization is a central cryptographic assumption and forms the basis of several widely deployed cryptosystems. The best integer factorization algorithm known is the Number Field Sieve [12], which was successfully used to factor 512-bit and 530-bit RSA moduli [5,1]. However, it appears that a PC-based implementation of the NFS cannot practically scale much further, and specifically its cost for 1024-bit composites is prohibitive. Recently, the prospect of using custom hardware for the computationally expensive steps of the Number Field Sieve has gained much attention. While mesh-based circuits for the matrix step have rendered that step quite feasible for 1024-bit composites [3, 16], the situation is less clear concerning the sieving step. Several sieving devices have been proposed, including TWINKLE [19,15] and a mesh-based circuit [7], but apparently none of these can practically handle 1024-bit composites.

One lesson learned from Bernstein's mesh-based circuit for the matrix step [3] is that it is inefficient to have memory cells that are "simply sitting around,

D. Boneh (Ed.): CRYPTO 2003, LNCS 2729, pp. 1–26, 2003.

twiddling their thumbs" — if merely storing the input is expensive, we should utilize it efficiently by appropriate parallelization. We propose a new device that combines this intuition with the TWINKLE-like approach of exchanging time and space. Whereas TWINKLE tests sieve location one by one serially, the new device handles thousands of sieve locations in parallel at every clock cycle. In addition, it is smaller and easier to construct: for 512-bit composites we can fit 79 independent sieving devices on a 30cm single silicon wafer, whereas each TWINKLE device requires a full GaAs wafer. While our approach is related to [7], it scales better and avoids some thorny issues.

The main difficulty is how to use a single copy of the input (or a small number of copies) to solve many subproblems in parallel, without collisions or long propagation delays and while maintaining storage efficiency. We address this with a heterogeneous design that uses a variety of routing circuits and takes advantage of available technological tradeoffs. The resulting cost estimates suggest that for 1024-bit composites the sieving step may be surprisingly feasible.

Section 2 reviews the sieving problem and the TWINKLE device. Section 3 describes the new device, called TWIRL[1], and Section 4 provides preliminary cost estimates. Appendix A discusses additional design details and improvements. Appendix B specifies the assumptions used for the cost estimates, and Appendix C relates this work to previous ones.

2   Context

2.1   Sieving in the Number Field Sieve

Our proposed device implements the sieving substep of the NFS relation collection step, which in practice is the most expensive part of the NFS algorithm [16]. We begin by reviewing the sieving problem, in a greatly simplified form and after appropriate reductions.[2] See [12] for background on the Number Field Sieve.

The inputs of the sieving problem are $R \in \mathbb{Z}$ (*sieve line width*), $T > 0$ (*threshold*) and a set of pairs (p_i, r_i) where the p_i are the prime numbers smaller than some *factor base bound* B. There is, on average, one pair per such prime. Each pair (p_i, r_i) corresponds to an arithmetic progression $P_i = \{a : a \equiv r_i \pmod{p_i}\}$. We are interested in identifying the sieve locations $a \in \{0, \dots, R-1\}$ that are members of many progressions P_i with large p_i:

$$g(a) > T \quad \text{where} \quad g(a) = \sum_{i : a \in P_i} \log_h p_i$$

for some fixed h (possibly $h > 2$). It is permissible to have "small" errors in this threshold check; in particular, we round all logarithms to the nearest integer.

In the NFS relation collection step we have two types of sieves: *rational* and *algebraic*. Both are of the above form, but differ in their factor base bounds (B_R

---

[1] TWIRL stands for "The Weizmann Institute Relation Locator"

[2] The description matches both line sieving and lattice sieving. However, for lattice sieving we may wish to take a slightly different approach (cf. A.8).

vs. B_A), threshold T and basis of logarithm h. We need to handle H *sieve lines*, and for sieve line both sieves are performed, so there are $2H$ sieving instances overall. For each sieve line, each value a that passes the threshold in both sieves implies a *candidate*. Each candidate undergoes additional tests, for which it is beneficial to also know the set $\{i : a \in P_i\}$ (for each sieve separately). The most expensive part of these tests is *cofactor factorization*, which involves factoring medium-sized integers.[3] The candidates that pass the tests are called *relations*. The output of the relation collection step is the list of relations and their corresponding $\{i : a \in P_i\}$ sets. Our goal is to find a certain number of relations, and the parameters are chosen accordingly a priori.

2.2   TWINKLE

Since TWIRL follows the TWINKLE [19,15] approach of exchanging time and space compared to traditional NFS implementations, we briefly review TWIN-KLE (with considerable simplification). A TWINKLE device consists of a wafer containing numerous independent cells, each in charge of a single progression P_i. After initialization the device operates for R clock cycles, corresponding to the sieving range $\{0 \le a < R\}$. At clock cycle a, the cell in charge of the progression P_i emits the value $\log p_i$ iff $a \in P_i$. The values emitted at each clock cycle are summed, and if this sum exceeds the threshold T then the integer a is reported. This event is announced back to the cells, so that the i values of the pertaining P_i is also reported. The global summation is done using analog optics; clocking and feedback are done using digital optics; the rest is implemented by digital electronics. To support the optoelectronic operations, TWINKLE uses Gallium Arsenide wafers which are small, expensive and hard to manufacture compared to silicon wafers, which are readily available.

3   The New Device

3.1   Approach

We next describe the TWIRL device. The description in this section applies to the rational sieve; some changes will be made for the algebraic sieve (cf. A.6), since it needs to consider only a values that passed the rational sieve.

For the sake of concreteness we provide numerical examples for a plausible choice of parameters for 1024-bit composites.[4] This choice will be discussed in Sections 4 and B.2; it is not claimed to be optimal, and all costs should be taken as rough estimates. The concrete figures will be enclosed in double angular brackets: $\langle\!\langle x \rangle\!\rangle_R$ and $\langle\!\langle x \rangle\!\rangle_A$ indicate values for the algebraic and rational sieves respectively, and $\langle\!\langle x \rangle\!\rangle$ is applicable to both.

We wish to solve $H \langle\!\langle \approx 2.7 \cdot 10^8 \rangle\!\rangle$ pairs of instances of the sieving problem, each of which has sieving line width $R \langle\!\langle = 1.1 \cdot 10^{15} \rangle\!\rangle$ and smoothness bound

---

[3] We assume use of the "2+2 large primes" variant of the NFS [12,13].

[4] This choice differs considerably from that used in preliminary drafts of this paper.

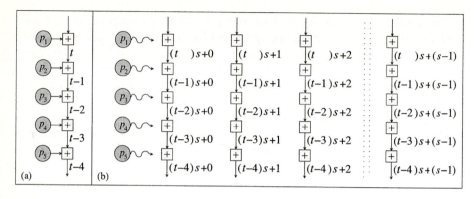

Fig. 1. Flow of sieve locations through the device in (a) a chain of adders and (b) TWIRL.

$B \langle\!\langle = 3.5 \cdot 10^9 \rangle\!\rangle_R \langle\!\langle = 2.6 \cdot 10^{10} \rangle\!\rangle_A$. Consider first a device that handles one sieve location per clock cycle, like TWINKLE, but does so using a pipelined systolic chain of electronic adders.[5] Such a device would consist of a long unidirectional bus, $\log_2 T \langle\!\langle = 10 \rangle\!\rangle$ bits wide, that connects millions of conditional adders in series. Each conditional adder is in charge of one progression P_i; when activated by an associated timer, it adds the value[6] $\lfloor \log p_i \rceil$ to the bus. At time t, the z-th adder handles sieve location $t - z$. The first value to appear at the end of the pipeline is $g(0)$, followed by $g(1), \dots, g(R)$, one per clock cycle. See Fig. 1(a).

We reduce the run time by a factor of $s \langle\!\langle = 4{,}096 \rangle\!\rangle_R \langle\!\langle = 32{,}768 \rangle\!\rangle_A$ by handling the sieving range $\{0, \dots, R - 1\}$ in chunks of length s, as follows. The bus is thickened by a factor of s to contain s logical lines of $\log_2 T$ bits each. As a first approximation (which will be altered later), we may think of it as follows: at time t, the z-th stage of the pipeline handles the sieve locations $(t - z)s + i$, $i \in \{0, \dots, s - 1\}$. The first values to appear at the end of the pipeline are $\{g(0), \dots, g(s - 1)\}$; they appear simultaneously, followed by successive disjoint groups of size s, one group per clock cycle. See Fig. 1(b).

Two main difficulties arise: the hardware has to work s times harder since time is compressed by a factor of s, and the additions of $\lfloor \log p_i \rceil$ corresponding to the same given progression P_i can occur at different lines of a thick pipeline. Our goal is to achieve this parallelism without simply duplicating all the counters and adders s times. We thus replace the simple TWINKLE-like cells by other units which we call *stations*. Each station handles a small portion of the progressions, and its interface consists of bus input, bus output, clock and some circuitry for loading the inputs. The stations are connected serially in a pipeline, and at the end of the bus (i.e., at the output of the last station) we place a threshold check unit that produces the device output.

An important observation is that the progressions have periods p_i in a very large range of sizes, and different sizes involve very different design tradeoffs. We

---

[5] This variant was considered in [15], but deemed inferior in that context.

[6] $\lfloor \log p_i \rceil$ denote the value $\log_h p_i$ for some fixed h, rounded to the nearest integer.

thus partition the progressions into three classes according to the size of their p_i values, and use a different station design for each class. In order of decreasing p_i value, the classes will be called *largish, smallish* and *tiny*.[7]

This heterogeneous approach leads to reasonable device sizes even for 1024-bit composites, despite the high parallelism: using standard VLSI technology, we can fit $\langle\!\langle 4 \rangle\!\rangle_{\mathrm{R}}$ rational-side TWIRLs into a single 30cm silicon wafer (whose manufacturing cost is about \$5,000 in high volumes; handling local manufacturing defects is discussed in A.9). Algebraic-side TWIRLs use higher parallelism, and we fit $\langle\!\langle 1 \rangle\!\rangle_{\mathrm{A}}$ of them into each wafer.

The following subsections describe the hardware used for each class of progressions. The preliminary cost estimates that appear later are based on a careful analysis of all the critical components of the design, but due to space limitations we omit the descriptions of many finer details. Some additional issues are discussed in Appendix A.

3.2   Largish Primes

Progressions whose p_i values are much larger than s emit $\lfloor \log p_i \rfloor$ values very seldom. For these largish primes $\langle\!\langle p_i > 5.2 \cdot 10^5 \rangle\!\rangle_{\mathrm{R}} \langle\!\langle p_i > 4.2 \cdot 10^6 \rangle\!\rangle_{\mathrm{A}}$, it is beneficial to use expensive logic circuitry that handles many progressions but allows very compact storage of each progression. The resultant architecture is shown in Fig. 2. Each progression is represented as a *progression triplet* that is stored in a memory bank, using compact DRAM storage. The progression triplets are periodically inspected and updated by special-purpose processors, which identify emissions that should occur in the "near future" and create corresponding *emission triplets*. The emission triplets are passed into *buffers* that merge the outputs of several processors, perform fine-tuning of the timing and create *delivery pairs*. The delivery pairs are passed to pipelined *delivery lines*, consisting of a chain of *delivery cells* which carry the delivery pairs to the appropriate bus line and add their $\lfloor \log p_i \rfloor$ contribution.

Scanning the progressions. The progressions are partitioned into many $\langle\!\langle 8{,}490 \rangle\!\rangle_{\mathrm{R}} \langle\!\langle 59{,}400 \rangle\!\rangle_{\mathrm{A}}$ DRAM banks, where each bank contains some d progression $\langle\!\langle 32 \leq d < 2.2 \cdot 10^5 \rangle\!\rangle_{\mathrm{R}} \langle\!\langle 32 \leq d < 2.0 \cdot 10^5 \rangle\!\rangle_{\mathrm{A}}$. A progression P_i is represented by a progression triplet of the form (p_i, ℓ_i, τ_i), where ℓ_i and τ_i characterize the next element $a_i \in P_i$ to be emitted (which is not stored explicitly) as follows. The value $\tau_i = \lfloor a_i/s \rfloor$ is the time when the next emission should be added to the bus, and $\ell_i = a_i \bmod s$ is the number of the corresponding bus line. A processor repeats the following operations, in a pipelined manner:[8]

---

[7] These are not to be confused with the "large" and "small" primes of the high-level NFS algorithm — all the primes with which we are concerned here are "small" (rather than "large" or in the range of "special-q").

[8] Additional logic related to reporting the sets $\{i : a \in P_i\}$ is described in Appendix A.7.

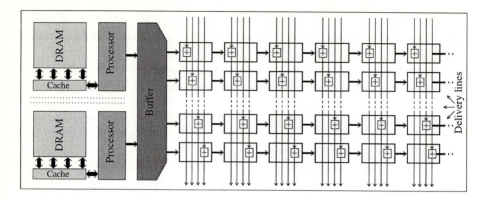

Fig. 2. Schematic structure of a largish station.

1. Read and erase the next state triplet (p_i, ℓ_i, τ_i) from memory.
2. Send an emission triplet $(\lfloor \log p_i \rfloor, \ell_i, \tau_i)$ to a buffer connected to the processor.
3. Compute $\ell' \leftarrow (\ell + p) \bmod s$ and $\tau_i' \leftarrow \tau_i + \lfloor p/s \rfloor + w$, where $w = 1$ if $\ell' < \ell$ and $w = 0$ otherwise.
4. Write the triplet (p_i, ℓ_i', τ_i') to memory, according to τ_i' (see below).

We wish the emission triplet $(\lfloor \log p_i \rfloor, \ell_i, \tau_i)$ to be created slightly before time τ_i (earlier creation would overload the buffers, while later creation would prevent this emission from being delivered on time). Thus, we need the processor to always read from memory some progression triplet that has an imminent emission. For large d, the simple approach of assigning each emission triplet to a fixed memory address and scanning the memory cyclically would be ineffective. It would be ideal to place the progression triplets in a priority queue indexed by τ_i, but it is not clear how to do so efficiently in a standard DRAM due to its passive nature and high latency. However, by taking advantage of the unique properties of the sieving problem we can get a good approximation, as follows.

Progression storage. The processor reads progression triplets from the memory in sequential cyclic order and at a constant rate ⟪of one triplet every 2 clock cycles⟫. If the value read is empty, the processor does nothing at that iteration. Otherwise, it updates the progression state as above and stores it at a different memory location — namely, one that will be read slightly before time τ_i'. In this way, after a short stabilization period the processor always reads triplets with imminent emissions. In order to have (with high probability) a free memory location within a short distance of any location, we increase the amount of memory ⟪by a factor of 2⟫; the progression is stored at the first unoccupied location, starting at the one that will be read at time τ_i' and going backwards cyclically.

If there is no empty location within ⟪64⟫ locations from the optimal designated address, the progression triplet is stored at an arbitrary location (or a dedicated overflow region) and restored to its proper place at some later stage;

when this happens we may miss a few emissions (depending on the implementation). This happens very seldom,[9] and it is permissible to miss a few candidates.

Autonomous circuitry inside the memory routes the progression triplet to the first unoccupied position preceeding the optimal one. To implement this efficiently we use a two-level memory hierarchy which is rendered possibly by the following observation. Consider a largish processor which is in charge of a set of d adjacent primes $\{p_{\min}, \dots, p_{\max}\}$. We set the size of the associated memory to $p_{\max}/s$ triplet-sized words, so that triplets with $p_i = p_{\max}$ are stored right before the current read location; triplets with smaller p_i are stored further back, in cyclic order. By the density of primes, $p_{\max} - p_{\min} \approx d \cdot \ln(p_{\max})$. Thus triplet values are always stored at an address that precedes the current read address by at most $d \cdot \ln(p_{\max})/s$, or slightly more due to congestions. Since $\ln(p_{\max}) \leq \ln(B)$ is much smaller than s, memory access always occurs at a small window that slides at a constant rate of one memory location every $\langle\!\langle 2 \rangle\!\rangle$ clock cycles. We may view the $\langle\!\langle 8{,}490 \rangle\!\rangle_{\mathrm{R}} \langle\!\langle 59{,}400 \rangle\!\rangle_{\mathrm{A}}$ memory banks as closed rings of various sizes, with an active window "twirling" around each ring at a constant linear velocity.

Each sliding window is handled by a fast SRAM-based cache. Occasionally, the window is shifted by writing the oldest cache block to DRAM and reading the next block from DRAM into the cache. Using an appropriate interface between the SRAM and DRAM banks (namely, read/write of full rows), this hides the high DRAM latency and achieves very high memory bandwidth. Also, this allows simpler and thus smaller DRAM.[10] Note that cache misses cannot occur. The only interface between the processor and memory are the operations "read next memory location" and "write triplet to first unoccupied memory location before the given address". The logic for the latter is implemented within the cache, using auxiliary per-triplet occupancy flags and some local pipelined circuitry.

Buffers. A buffer unit receives emission triplets from several processors in parallel, and sends delivery pairs to several delivery lines. Its task is to convert emission triplets into delivery pairs by merging them where appropriate, fine-tuning their timing and distributing them across the delivery lines: for each received emission triplet of the form $(\lfloor \log p_i \rfloor, \ell, \tau)$, the delivery pair $(\lfloor \log p_i \rfloor, \ell)$ should be sent to some delivery line (depending on ℓ) at time exactly τ.

Buffer units can be be realized as follows. First, all incoming emission triplets are placed in a parallelized priority queue indexed by τ, implemented as a small

---

[9] For instance, in simulations for primes close to $\langle\!\langle 20{,}000s \rangle\!\rangle_{\mathrm{R}}$, the distance between the first unoccupied location and the ideal location was smaller than $\langle\!\langle 64 \rangle\!\rangle_{\mathrm{R}}$ for all but $\langle\!\langle 5 \cdot 10^{-6} \rangle\!\rangle_{\mathrm{R}}$ of the iterations. The probability of a random integer $x \in \{1, \dots, x\}$ having k factors is about $(\log \log x)^{k-1}/(k-1)! \log x$. Since we are (implicitly) sieving over values of size about $x \approx \langle\!\langle 10^{64} \rangle\!\rangle_{\mathrm{R}} \langle\!\langle 10^{101} \rangle\!\rangle_{\mathrm{A}}$ which are "good" (i.e., semi-smooth) with probability $p \approx \langle\!\langle 6.8 \cdot 10^{-5} \rangle\!\rangle_{\mathrm{R}} \langle\!\langle 4.4 \cdot 10^{-9} \rangle\!\rangle_{\mathrm{A}}$, less than $10^{-15}/p$ of the good a's have more than 35 factors; the probability of missing other good a's is negligible.

[10] Most of the peripheral DRAM circuitry (including the refresh circuitry and column decoders) can be eliminated, and the row decoders can be replaced by smaller stateful circuitry. Thus, the DRAM bank can be smaller than standard designs. For the stations that handle the smaller primes in the "largish" range, we may increase the cache size to d and eliminate the DRAM.

mesh whose rows are continuously bubble-sorted and whose columns undergo random local shuffles. The elements in the last few rows are tested for τ matching the current time, and the matching ones are passed to a pipelined network that sorts them by ℓ, merges where needed and passes them to the appropriate delivery lines. Due to congestions some emissions may be late and thus discarded; since the inputs are essentially random, with appropriate choices of parameters this should happen seldom.

The size of the buffer depends on the typical number of time steps that an emission triplet is held until its release time τ (which is fairly small due to the design of the processors), and on the rate at which processors produce emission triplets 《about once per 4 clock cycles》.

Delivery lines. A delivery line receives delivery pairs of the form $(\lceil \log p_i \rceil, \ell)$ and adds each such pair to bus line ℓ exactly $\lfloor \ell/k \rfloor$ clock cycles after its receipt. It is implemented as a one-dimensional array of cells placed across the bus, where each cell is capable of containing one delivery pair. Here, the j-th cell compares the ℓ value of its delivery pair (if any) to the constant j. In case of equality, it adds $\lfloor \log p_i \rfloor$ to the bus line and discards the pair. Otherwise, it passes it to the next cell, as in a shift register.

Overall, there are $《2,100120》_{\mathrm{R}} 《14,900》_{\mathrm{A}}$ delivery lines in the largish stations, and they occupy a significant portion of the device. Appendix A.1 describes the use of interleaved carry-save adders to reduce their cost, and Appendix A.6 nearly eliminates them from the algebraic sieve.

Notes. In the description of the processors, DRAM and buffers, we took the τ values to be arbitrary integers designating clock cycles. Actually, it suffices to maintain these values modulo some integer 《2048》 that upper bounds the number of clock cycles from the time a progression triplet is read from memory to the time when it is evicted from the buffer. Thus, a progression occupies $\log_2 p_i + 《\log_2 2048》$ DRAM bits for the triplet, plus $\log_2 p_i$ bits for re-initialization (cf. A.4).

The amortized circuit area per largish progression is $\Theta(s^2(\log s)/p_i + \log s + \log p_i)$.[11] For fixed s this equals $\Theta(1/p_i + \log p_i)$, and indeed for large composites the overwhelming majority of progressions $《99.97\%》_{\mathrm{R}} 《99.98\%》_{\mathrm{A}}$ will be handled in this manner.

3.3   Smallish Primes

For progressions with p_i close to s, $《256 < p_i < 5.2 \cdot 10^5》_{\mathrm{R}} 《256 < p_i < 4.2 \cdot 10^6》_{\mathrm{A}}$, each processor can handle very few progressions because it can produce at most one emission triplet every 《2》 clock cycles. Thus, the amortized cost of the processor, memory control circuitry and buffers is very high. Moreover, such progression cause emissions so often that communicating their emissions to distant bus lines (which is necessary if the state of each progression is maintained

---

[11] The frequency of emissions is s/p_i, and each emission occupies some delivery cell for $\Theta(s)$ clock cycles. The last two terms are due to DRAM storage, and have very small constants.

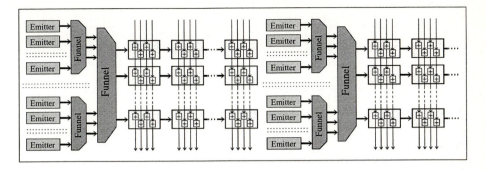

Fig. 3. Schematic structure of a smallish station.

at some single physical location) would involve enormous communication bandwidth. We thus introduce another station design, which differs in several ways from the largish stations (see Fig 3).

Emitters and funnels. The first change is to replace the combination of the processors, memory and buffers by other units. Delivery pairs are now created directly by *emitters*, which are small circuits that handle a single progression each (as in TWINKLE). An emitter maintains the state of the progression using internal registers, and occasionally emits delivery pairs of the form $(\lfloor \log p_i \rceil, \ell)$ which indicate that the value $\lfloor \log p_i \rceil$ should be added to the ℓ-th bus line some fixed time interval later. Appendix A.2 describes a compact emitters design.

Each emitter is continuously updating its internal counters, but it creates a delivery pair only once per roughly $\sqrt{p_i}$ (between $\langle\!\langle 8 \rangle\!\rangle_R$ and $\langle\!\langle 512 \rangle\!\rangle_R$ clock cycles — see below). It would be wasteful to connect each emitter to a dedicated delivery line. This is solved using *funnels*, which "compress" their sparse inputs as follows. A funnel has a large number of input lines, connected to the outputs of many adjacent emitters; we may think of it as receiving a sequence of one-dimensional arrays, most of whose elements are empty. The funnel outputs a sequence of much shorter arrays, whose non-empty elements are exactly the non-empty elements of the input array received a fixed number of clock cycle earlier. The funnel outputs are connected to the delivery lines. Appendix A.3 describes an implementation of funnels using modified shift registers.

Duplication. The other major change is duplication of the progression states, in order to move the sources of the delivery pairs closer to their destination and reduce the cross-bus communication bandwidth. Each progression is handled by $n_i \approx s/\sqrt{p_i}$ independent emitters[12] which are placed at regular intervals across the bus. Accordingly we fragment the delivery lines into segments that span $s/n_i \approx \sqrt{p_i}$ bus lines each. Each emitter is connected (via a funnel) to a different segment, and sends emissions to this segment every $p_i/sn_i \approx \sqrt{p}$ clock cycles. As emissions reach their destination quicker, we can decrease the total

---

[12] $\langle\!\langle n_i = s/2\sqrt{p_i} \rangle\!\rangle$ rounded to a power of 2 (cf. A.2), which is in the range $\langle\!\langle \{2, \dots, 128\} \rangle\!\rangle_R$.

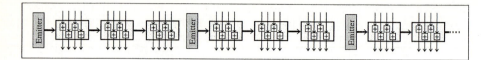

Fig. 4. Schematic structure of a tiny station, for a single progression.

number of delivery lines. Also, there is a corresponding decrease in the emission frequency of any specific emitter, which allows us to handle p_i close to (or even smaller than) s. Overall there are $\langle\!\langle 501 \rangle\!\rangle_R$ delivery lines in the smallish stations, broken into segments of various sizes.

Notes. Asymptotically the amortized circuit area per smallish progression is $\Theta((s/\sqrt{p_i}+1)(\log s+\log p_i))$. The term 1 is less innocuous than it appears — it hides a large constant (roughly the size of an emitter plus the amortized funnel size), which dominates the cost for large p_i.

3.4   Tiny Primes

For very small primes, the amortized cost of the duplicated emitters, and in particular the related funnels, becomes too high. On the other hand, such progressions cause several emissions at every clock cycle, so it is less important to amortize the cost of delivery lines over several progressions. This leads to a third station design for the tiny primes $\langle\!\langle p_i < 256 \rangle\!\rangle$. While there are few such progressions, their contributions are significant due to their very small periods.

Each tiny progression is handled independently, using a dedicated delivery line. The delivery line is partitioned into segments of size somewhat smaller than p_i,[13] and an emitter is placed at the input of each segment, without an intermediate funnel (see Fig 4). These emitters are a degenerate form of the ones used for smallish progressions (cf. A.2). Here we cannot interleave the adders in delivery cells as done in largish and smallish stations, but the carry-save adders are smaller since they only (conditionally) add the small constant $\lfloor\log p_i\rfloor$. Since the area occupied by each progression is dominated by the delivery lines, it is $\Theta(s)$ regardless of p_i.

Some additional design considerations are discussed in Appendix A.

4   Cost Estimates

Having outlined the design and specified the problem size, we next estimate the cost of a hypothetical TWIRL device using today's VLSI technology. The hardware parameters used are specified in Appendix B.1. While we tried to produce realistic figures, we stress that these estimates are quite rough and rely on many approximations and assumptions. They should only be taken to indicate

---

[13] The segment length is the largest power of 2 smaller than p_i (cf. A.2).

the order of magnitude of the true cost. We have not done any detailed VLSI design, let alone actual implementation.

4.1   Cost of Sieving for 1024-Bit Composites

We assume the following NFS parameters: $B_R = 3.5 \cdot 10^9$, $B_A = 2.6 \cdot 10^{10}$, $R = 1.1 \cdot 10^{15}$, $H \approx 2.7 \cdot 10^8$ (cf. B.2). We use the cascaded sieves variant of Appendix A.6.

For the rational side we set $s_R = 4,096$. One rational TWIRL device requires 15,960mm^2 of silicon wafer area, or 1/4 of a 30cm silicon wafer. Of this, 76% is occupied by the largish progressions (and specifically, 37% of the device is used for the DRAM banks), 21% is used by the smallish progressions and the rest (3%) is used by the tiny progressions. For the algebraic side we set $s_A = 32,768$. One algebraic TWIRL device requires 65,900mm^2 of silicon wafer area — a full wafer. Of this, 94% is occupied by the largish progressions (66% of the device is used for the DRAM banks) and 6% is used by the smallish progressions. Additional parameters of are mentioned throughout Section 3.

The devices are assembled in clusters that consist each of 8 rational TWIRLs and 1 algebraic TWIRL, where each rational TWIRL has a unidirectional link to the algebraic TWIRL over which it transmits 12 bits per clock cycle. A cluster occupies three wafers, and handles a full sieve line in R/s_A clock cycles, i.e., 33.4 seconds when clocked at 1GHz. The full sieving involves H sieve lines, which would require 194 years when using a single cluster (after the 33% saving of Appendix A.5.) At a cost of \$2.9M (assuming \$5,000 per wafer), we can build 194 independent TWIRL clusters that, when run in parallel, would complete the sieving task within 1 year.

After accounting for the cost of packaging, power supply and cooling systems, adding the cost of PCs for collecting the data and leaving a generous error margin,[14] it appears realistic that all the sieving required for factoring 1024-bit integers can be completed within 1 year by a device that cost \$10M to manufacture. In addition to this per-device cost, there would be an initial NRE cost on the order of \$20M (for design, simulation, mask creation, etc.).

4.2   Implications for 1024-Bit Composites

It has been often claimed that 1024-bit RSA keys are safe for the next 15 to 20 years, since both NFS relation collection and the NFS matrix step would be unfeasible (e.g., [4,21] and a NIST guideline draft [18]). Our evaluation suggests that sieving can be achieved within one year at a cost of \$10M (plus a one-time cost of \$20M), and recent works [16,8] indicate that for our NFS parameters the matrix can also be performed at comparable costs.

---

[14] It is a common rule of thumb to estimate the total cost as twice the silicon cost; to be conservative, we triple it.

With efficient custom hardware for both sieving and the matrix step, other subtasks in the NFS algorithm may emerge as bottlenecks.[15] Also, our estimates are hypothetical and rely on numerous approximations; the only way to learn the precise costs involved would be to perform a factorization experiment.

Our results do not imply that breaking 1024-bit RSA is within reach of individual hackers. However, it is difficult to identify any specific issue that may prevent a sufficiently motivated and well-funded organization from applying the Number Field Sieve to 1024-bit composites within the next few years. This should be taken into account by anyone planning to use a 1024-bit RSA key.

4.3   Cost of Sieving for 512-Bits Composites

Since several hardware designs [19,15,10,7] were proposed for the sieving of 512-bit composites, it would be instructive to obtain cost estimates for TWIRL with the same problem parameters. We assume the same parameters as in [15,7]: $B_R = B_A = 2^{24} \approx 1.7 \cdot 10^7$, $R = 1.8 \cdot 10^{10}$, $2H = 1.8 \cdot 10^6$. We set $s = 1,024$ and use the same cost estimation expressions that lead to the 1024-bit estimates.

A single TWIRL device would have a die size of about 800mm^2, 56% of which are occupied by largish progressions and most of the rest occupied by smallish progressions. It would process a sieve line in 0.018 seconds, and can complete the sieving task within 6 hours.

For these NFS parameters TWINKLE would require 1.8 seconds per sieve line, the FPGA-based design of [10] would require about 10 seconds and the mesh-based design of [7] would require 0.36 seconds. To provide a fair comparison to TWINKLE and [7], we should consider a single wafer full of TWIRL devices running in parallel. Since we can fit 79 of them, the effective time per sieve line is 0.00022 seconds.

Thus, in factoring 512-bit composites the basic TWIRL design is about 1,600 times more cost effective than the best previously published design [7], and 8,100 times more cost effective than TWINKLE. Adjusting the NFS parameters to take advantage of the cascaded-sieves variant (cf. A.6) would further increase this gap. However, even when using the basic variant, a single wafer of TWIRLs can complete the sieving for 512-bit composites in under 10 minutes.

4.4   Cost of Sieving for 768-Bits Composites

We assume the following NFS parameters: $B_R = 1 \cdot 10^8$, $B_A = 1 \cdot 10^9$, $R = 3.4 \cdot 10^{13}$, $H \approx 8.9 \cdot 10^6$ (cf. B.2). We use the cascaded sieves variant of Appendix A.6, with $s_R = 1,024$ and $s_A = 4,096$. For this choice, a rational sieve occupies 1,330mm^2 and an algebraic sieve occupies 4,430mm^2. A cluster consisting of 4 rational sieves and one algebraic sieve can process a sieve line in 8.3 seconds, and 6 independent clusters can fit on a single 30cm silicon wafer.

---

[15] Note that for our choice of parameters, the cofactor factorization is cheaper than the sieving (cf. Appendix A.7).

Thus, a single wafer of TWIRL clusters can complete the sieving task within 95 days. This wafer would cost about \$5,000 to manufacture — one tenth of the RSA-768 challenge prize [20].[16]

4.5 Larger Composites

For largish progressions, the amortized cost per progression is $\Theta(s^2(\log s)/p_i + \log s + \log p_i)$ with small constants (cf. 3.2). For smallish progressions, the amortized cost is $\Theta((s/\sqrt{p_i}+1)(\log s + \log p_i))$ with much larger constants (cf. 3.3). For a serial implementation (PC-based or TWINKLE), the cost per progression is clearly $\Omega(\log p_i)$. This means that asymptotically we can choose $s = \tilde{\Theta}(\sqrt{B})$ to get a speed advantage of $\tilde{\Theta}(\sqrt{B})$ over serial implementations, while maintaining the small constants. Indeed, we can keep increasing s essentially for free until the area of the largish processors, buffers and delivery lines becomes comparable to the area occupied by the DRAM that holds the progression triplets.

For some range of input sizes, it may be beneficial to reduce the amount of DRAM used for largish progressions by storing only the prime p_i, and computing the rest of the progression triplet values on-the-fly in the special-purpose processors (this requires computing the roots modulo p_i of the relevant NFS polynomial).

If the device would exceed the capacity of a single silicon wafer, then as long as the bus itself is narrower than a wafer, we can (with appropriate partitioning) keep each station fully contained in some wafer; the wafers are connected in a serial chain, with the bus passing through all of them.

5 Conclusion

We have presented a new design for a custom-built sieving device. The device consists of a thick pipeline that carries sieve locations through thrilling adventures, where they experience the addition of progression contributions in myriad different ways that are optimized for various scales of progression periods. In factoring 512-bit integers, the new device is 1,600 times faster than best previously published designs. For 1024-bit composites and appropriate choice of NFS parameters, the new device can complete the sieving task within 1 year at a cost of \$10M, thereby raising some concerns about the security of 1024-bit RSA keys.

Acknowledgments.   This work was inspired by Daniel J. Bernstein's insightful work on the NFS matrix step, and its adaptation to sieving by Willi Geiselmann and Rainer Steinwandt. We thank the latter for interesting discussions of their design and for suggesting an improvement to ours. We are indebted to Arjen K. Lenstra for many insightful discussions, and to Robert D. Silverman,

---

[16] Needless to say, this disregards an initial cost of about \$20M. This initial cost can be significantly reduced by using older technology, such as 0.25μm process, in exchange for some decrease in sieving throughput.

Andrew "bunnie" Huang and Michael Szydlo for valuable comments and suggestions. Early versions of [14] and the polynomial selection programs of Jens Franke and Thorsten Kleinjung were indispensable in obtaining refined estimates for the NFS parameters.

References

1. F. Bahr, J. Franke, T. Kleinjung, M. Lochter, M. Böhm, *RSA-160*, e-mail announcement, Apr. 2003, http://www.loria.fr/~zimmerma/records/rsa160
2. Daniel J. Bernstein, *How to find small factors of integers*, manuscript, 2000, http://cr.yp.to/papers.html
3. Daniel J. Bernstein, *Circuits for integer factorization: a proposal*, manuscript, 2001, http://cr.yp.to/papers.html
4. Richard P. Brent, *Recent progress and prospects for integer factorisation algorithms*, proc. COCOON 2000, LNCS **1858** 3–22, Springer-Verlag, 2000
5. S. Cavallar, B. Dodson, A.K. Lenstra, W. Lioen, P.L. Montgomery, B. Murphy, H.J.J. te Riele, et al., *Factorization of a 512-bit RSA modulus*, proc. Eurocrypt 2000, LNCS **1807** 1–17, Springer-Verlag, 2000
6. Don Coppersmith, *Modifications to the number field sieve*, Journal of Cryptology **6** 169–180, 1993
7. Willi Geiselmann, Rainer Steinwandt, *A dedicated sieving hardware*, proc. PKC 2003, LNCS **2567** 254–266, Springer-Verlag, 2002
8. Willi Geiselmann, Rainer Steinwandt, *Hardware to solve sparse systems of linear equations over GF(2)*, proc. CHES 2003, LNCS, Springer-Verlag, to be published.
9. International Technology Roadmap for Semiconductors 2001, http://public.itrs.net/
10. Hea Joung Kim, William H. Magione-Smith, *Factoring large numbers with programmable hardware*, proc. FPGA 2000, ACM, 2000
11. Robert Lambert, *Computational aspects of discrete logarithms*, Ph.D. Thesis, University of Waterloo, 1996
12. Arjen K. Lenstra, H.W. Lenstra, Jr., (eds.), *The development of the number field sieve*, Lecture Notes in Math. **1554**, Springer-Verlag, 1993
13. Arjen K. Lenstra, Bruce Dodson, *NFS with four large primes: an explosive experiment*, proc. Crypto '95, LNCS **963** 372–385, Springer-Verlag, 1995
14. Arjen K. Lenstra, Bruce Dodson, James Hughes, Paul Leyland, *Factoring estimates for 1024-bit RSA modulus*, to be published.
15. Arjen K. Lenstra, Adi Shamir, *Analysis and Optimization of the TWINKLE Factoring Device*, proc. Eurocrypt 2002, LNCS **1807** 35–52, Springer-Verlag, 2000
16. Arjen K. Lenstra, Adi Shamir, Jim Tomlinson, Eran Tromer, *Analysis of Bernstein's factorization circuit*, proc. Asiacrypt 2002, LNCS **2501** 1–26, Springer-Verlag, 2002
17. Brian Murphy, *Polynomial selection for the number field sieve integer factorization algorithm*, Ph. D. thesis, Australian National University, 1999
18. National Institute of Standards and Technology, *Key management guidelines, Part 1: General guidance (draft)*, Jan. 2003, http://csrc.nist.gov/CryptoToolkit/tkkeymgmt.html
19. Adi Shamir, *Factoring large numbers with the TWINKLE device (extended abstract)*, proc. CHES'99, LNCS **1717** 2–12, Springer-Verlag, 1999

20. RSA Security, *The new RSA factoring challenge*, web page, Jan. 2003,
    `http://www.rsasecurity.com/rsalabs/challenges/factoring/`
21. Robert D. Silverman, *A cost-based security analysis of symmetric and asymmetric key lengths*, Bulletin 13, RSA Security, 2000,
    `http://www.rsasecurity.com/rsalabs/bulletins/bulletin13.html`
22. Web page for this paper, `http://www.wisdom.weizmann.ac.il/~tromer/twirl`

A   Additional Design Considerations

A.1   Delivery Lines

The delivery lines are used by all station types to carry delivery pairs from their source (buffer, funnel or emitter) to their destination bus line. Their basic structure is described in Section 3.2. We now describe methods for implementing them efficiently.

Interleaving. Most of the time the cells in a delivery line act as shift registers, and their adders are unused. Thus, we can reduce the cost of adders and registers by interleaving. We use larger delivery cells that span $r \langle\!\langle= 4\rangle\!\rangle_{\mathrm{R}}$ adjacent bus lines, and contain an adder just for the q-th line among these, with q fixed throughout the delivery line and incremented cyclically in the subsequent delivery lines. As a bonus, we now put every r adjacent delivery lines in a single bus pipeline stage, so that it contains one adder per bus line. This reducing the number of bus pipelining registers by a factor of r throughout the largish stations.

Since the emission pairs traverse the delivery lines at a rate of r lines per clock cycle, we need to skew the space-time assignment of sieve locations so that as distance from the buffer to the bus line increases, the "age" $\lfloor a/s \rfloor$ of the sieve locations decreases. More explicitly: at time t, sieve location a is handled by the $\lfloor (a \bmod s)/r \rfloor$-th cell[17] of one of the r delivery lines at stage $t - \lfloor a/sr \rfloor - \lfloor (a \bmod s)/r \rfloor$ of the bus pipeline, if it exists.

In the largish stations, the buffer is entrusted with the role of sending delivery pairs to delivery lines that have an adder at the appropriate bus line; an improvement by a factor of 2 is achieved by placing the buffers at the middle of the bus, with the two halves of each delivery line directed outwards from the buffer. In the smallish and tiny stations we do not use interleaving.

Note that whenever we place pipelining registers on the bus, we must delay all downstream delivery lines connected to this buffer by a clock cycle. This can be done by adding pipeline stages at the beginning of these delivery lines.

Carry-save adders. Logically, each bus line carries a $\log_2 T \langle\!\langle= 10\rangle\!\rangle$-bit integer. These are encoded by a redundant representation, as a pair of $\log_2 T$-bit integers whose sum equals the sum of the $\lfloor \log p_i \rfloor$ contributions so far. The additions at the delivery cells are done using carry-save adders, which have inputs a,b,c and whose output is a representation of the sum of their inputs in the form of a pair e,f such that $e + f = a + b + c$. Carry-save adders are very compact and support a high

---

[17] After the change made in Appendix A.2 this becomes $\lfloor \mathrm{rev}(a \bmod s)/r \rfloor$, where $\mathrm{rev}(\cdot)$ denotes bit-reversal of $\log_2 s$-bit numbers and s,r are powers of 2.

clock rate, since they do not propagate carries across more than one bit position. Their main disadvantage is that it is inconvenient to perform other operations directly on the redundant representation, but in our application we only need to perform a long sequence of additions followed by a single comparison at the end. The extra bus wires due to the redundant representation can be accommodated using multiple metal layers of the silicon wafer.[18]

To prevent wrap-around due to overflow when the sum of contributions is much larger than T, we slightly alter the carry-save adders by making their most significant bits "sticky": once the MSBs of both values in the redundant representation become 1 (in which case the sum is at least T), further additions do not switch them back to 0.

A.2   Implementation of Emitters

The designs of smallish and tiny progressions (cf. 3.3, 3.4) included *emitter* elements. An emitter handles a single progression P_i, and its role is to emit the delivery pairs ($\lceil \log p_i \rceil, \ell$) addressed to a certain group G of adjacent lines, $\ell \in G$. This subsection describes our proposed emitter implementation. For context, we first describe some less efficient designs.

Straightforward implementations. One simple implementation would be to keep a $\lceil \log_2 p_i \rceil$-bit register and increment it by s modulo p_i every clock cycle. Whenever a wrap-around occurs (i.e., this progression causes an emission), compute ℓ and check if $\ell \in G$. Since the register must be updated within one clock cycle, this requires an expensive carry-lookahead adder. Moreover, if s and $|G|$ are chosen arbitrarily then calculating ℓ and testing whether $\ell \in G$ may also be expensive. Choosing s, $|G|$ as power of 2 reduces the costs somewhat.

A different approach would be to keep a counter that counts down the time to the next emission, as in [19], and another register that keeps track of ℓ. This has two variants. If the countdown is to the next emission of this triplet regardless of its destination bus line, then these events would occur very often and again require low-latency circuitry (also, this cannot handle $p_i < s$). If the countdown is to the next emission into G, we encounter the following problem: for any set G of bus lines corresponding to adjacent residues modulo s, the intervals at which P_i has emissions into G are irregular, and would require expensive circuitry to compute.

Line address bit reversal. To solve the last problem described above and use the second countdown-based approach, we note the following: the assignment of sieve locations to bus lines (within a clock cycle) can be done arbitrarily, but the partition of wires into groups G should be done according to physical proximity. Thus, we use the following trick. Choose $s = 2^\alpha$ and $|G| = 2^{\beta_i} \approx \sqrt{p_i}$ for some integers $\alpha \langle\!\langle = 12 \rangle\!\rangle_{\mathrm{R}} \langle\!\langle = 15 \rangle\!\rangle_{\mathrm{A}}$ and β_i. The residues modulo s are assigned to bus

---

lines with bit-reversed indices; that is, sieve locations congruent to w modulo s are handled by the bus line at physical location $\mathrm{rev}(w)$, where

$$w = \sum_{i=0}^{\alpha-1} c_i 2^i , \quad \mathrm{rev}(w) = \sum_{i=0}^{\alpha-1} c_{\alpha-1-i} 2^i \quad \text{for some } c_0, \dots, c_{\alpha-1} \in \{0,1\}$$

The j-th emitter of the progression P_i, $j \in \{0, \dots, 2^{\alpha-\beta_i}\}$, is in charge of the j-th group of 2^{β_i} bus lines. The advantage of this choice is the following.

Lemma 1. *For any fixed progression with $p_i > 2$, the emissions destined to any fixed group occur at regular time intervals of $T_i = \lfloor 2^{-\beta_i} p_i \rfloor$, up to an occasional delay of one clock cycle due to modulo s effects.*

Proof. Emissions into the j-th group correspond to sieve locations $a \in P_i$ that fulfill $\lfloor \mathrm{rev}(a \bmod s)/2^{\beta_i} \rfloor = j$, which is equivalent to $a \equiv c_j \pmod{2^{\alpha-\beta_i}}$ for some c_j. Since $a \in P_i$ means $a \equiv r_i \pmod{p_i}$ and p_i is coprime to $2^{\alpha-\beta_i}$, by the Chinese Remainder Theorem we get that the set of such sieve locations is exactly $P_{i,j} \equiv \{a : a \equiv c_{i,j} \pmod{2^{\alpha-\beta_i} p_i}\}$ for some $c_{i,j}$. Thus, a pair of consecutive $a_1, a_2 \in P_{i,j}$ fulfill $a_2 - a_1 = 2^{\alpha-\beta_i} p_i$. The time difference between the corresponding emissions is $\Delta = \lfloor a_2/s \rfloor - \lfloor a_1/s \rfloor$. If $(a_2 \bmod s) > (a_1 \bmod s)$ then $\Delta = \lfloor (a_2 - a_1)/s \rfloor = \lfloor 2^{\alpha-\beta_i} p_i/s \rfloor = T_i$. Otherwise, $\Delta = \lceil (a_2 - a_1)/s \rceil = T_i + 1$. $\square$

Note that $T_i \approx \sqrt{p_i}$, by the choice of β_i.

Emitter structure. In the smallish stations, each emitter consists of two counters, as follows.

- Counter A operates modulo $T_i = \lfloor 2^{-\beta_i} p_i \rfloor$ (typically $\langle\!\langle 7 \rangle\!\rangle_{\mathrm{R}} \langle\!\langle 5 \rangle\!\rangle_{\mathrm{A}}$ bits), and keeps track of the time until the next emission of this emitter. It is decremented by 1 (nearly) every clock cycle.
- Counter B operates modulo 2^{β_i} (typically $\langle\!\langle 10 \rangle\!\rangle_{\mathrm{R}} \langle\!\langle 15 \rangle\!\rangle_{\mathrm{A}}$ bits). It keeps track of the β_i most significant bits of the residue class modulo s of the sieve location corresponding to the next emission. It is incremented by $2^{\alpha-\beta_i} p_i \bmod 2^{\beta_i}$ whenever Counter A wraps around. Whenever Counter B wraps around, Counter A is suspended for one clock cycle (this corrects for the modulo s effect).

A delivery pair $(\lfloor \log p_i \rfloor, \ell)$ is emitted when Counter A wraps around, where $\lfloor \log p_i \rfloor$ is fixed for each emitter. The target bus line ℓ gets β_i of its bits from Counter B. The $\alpha - \beta_i$ least significant bits of ℓ are fixed for this emitter, and they are also fixed throughout the relevant segment of the delivery line so there is no need to transmit them explicitly.

The physical location of the emitter is near (or underneath) the group of bus lines to which it is attached. The counters and constants need to be set appropriately during device initialization. Note that if the device is custom-built for a specific factorization task then the circuit size can be reduced by hard-wiring many of these values[19]. The combined length of the counters is roughly

---

[19] For sieving the rational side of NFS, it suffices to fix the smoothness bounds. Similarly for the preprocessing stage of Coppersmith's Factorization Factory [6] .

$\log_2 p_i$ bits, and with appropriate adjustments they can be implemented using compact ripple adders[20] as in [15].

Emitters for tiny progressions. For tiny stations, we use a very similar design. The bus lines are again assigned to residues modulo s in bit-reversed order (indeed, it would be quite expensive to reorder them). This time we choose β_i such that $|G| = 2^{\beta_i}$ is the largest power of 2 that is smaller than p_i. This fixes $T_i = 1$, i.e., an emission occurs every one or two clock cycles. The emitter circuitry is identical to the above; note that Counter A has become zero-sized (i.e., a wire), which leaves a single counter of size $\beta_i \approx \log_2 p_i$ bits.

A.3   Implementation of Funnels

The smallish stations use *funnels* to compact the sparse outputs of emitters before they are passed to delivery lines (cf. 3.3). We implement these funnels as follows.

An n-to-m funnel $(n \gg m)$ consists of a matrix of n columns and m rows, where each cell contains registers for storing a single progression triplet. At every clock cycle inputs are fed directly into the top row, one input per column, scheduled such that the i-th element of the t-th input array is inserted into the i-th column at time $t + i$. At each clock cycle, all values are shifted horizontally one column to the right. Also, each value is shifted one row down if this would not overwrite another value. The t-th output array is read off the rightmost column at time $t + n$.

For any $m < n$ there is some probability of "overflow" (i.e., insertion of input value into a full column). Assuming that each input is non-empty with probability ν independently of the others $(\nu \approx 1/\sqrt{p_i}$; cf. 3.3), the probability that a non-empty input will be lost due to overflow is:

$$\sum_{k=m+1}^{n} \binom{n}{k} \nu^k (1 - \nu)^{n-k} (k - m)/k$$

We use funnels with $\langle\!\langle m = 5 \rangle\!\rangle_{\mathrm{R}}$ rows and $\langle\!\langle n \approx 1/\nu \rangle\!\rangle_{\mathrm{R}}$ columns. For this choice and within the range of smallish progressions, the above failure probability is less than 0.00011. This certainly suffices for our application.

The above funnels have a suboptimal compression ratio $n/m \langle\!\langle \approx 1/5\nu \rangle\!\rangle_{\mathrm{R}}$, i.e., the probability $\nu' \langle\!\langle \approx 1/5 \rangle\!\rangle_{\mathrm{R}}$ of a funnel output value being non-empty is still rather low. We thus feed these output into a second-level funnel $\langle\!\langle$with $m' = 35$, $n' = 14 \rangle\!\rangle_{\mathrm{R}}$, whose overflow probability is less than 0.00016, and whose cost is amortized over many progressions. The output of the second-level funnel is fed into the delivery lines. The combined compression ratio of the two funnel levels is suboptimal by a factor of $5 \cdot 14/34 = 2$, so the number of delivery lines is twice the naive optimum. We do not interleave the adders in the delivery lines as done for largish stations (cf. A.1), in order to avoid the overhead of directing delivery pairs to an appropriate delivery line.[21]

---

[20] This requires insertion of small delays and tweaking the constant values.

[21] Still, the number of adders can be reduced by attaching a single adder to several bus lines using multiplexers. This may impact the clock rate.

A.4 Initialization

The device initialization consists of loading the progression states and initial counter values into all stations, and loading instructions into the bus bypass re-routing switches (after mapping out the defects).

The progressions differ between sieving runs, but reloading the device would require significant time (in [19] this became a bottleneck). We can avoid this by noting, as in [7], that the instances of sieving problem that occur in the NFS are strongly related, and all that is needed is to increase each r_i value by some constant value $\tilde{r}_i$ after each run. The $\tilde{r}_i$ values can be stored compactly in DRAM using $\log_2 p_i$ bits per progression (this is included in our cost estimates) and the addition can be done efficiently using on-wafer special-purpose processors. Since the interval R/s between updates is very large, we don't need to dedicate significant resources to performing the update quickly. For lattice sieving the situation is somewhat different (cf. A.8).

A.5 Eliminating Sieve Locations

In the NFS relation collection, we are only interesting in sieve locations a on the b-th sieve line for which $\gcd(a',b) = 1$ where $a' = a - R/2$, as other locations yield duplicate relations. The latter are eliminated by the candidate testing, but the sieving work can be reduced by avoiding sieve locations with $c\,|\,a',b$ for very small c. All software-based sievers consider the case $2\,|\,a',b$ — this eliminates 25% of the sieve locations. In TWIRL we do the same: first we sieve normally over all the odd lines, $b \equiv 1 \,(\mathrm{mod}\ 2)$. Then we sieve over the even lines, and consider only odd a' values; since a progression with $p_i > 2$ hits every p_i-th odd sieve location, the only change required is in the initial values loaded into the memories and counters. Sieving of these odd lines takes half the time compared to even lines.

We also consider the case $3\,|\,a',b$, similarly to the above. Combining the two, we get four types of sieve runs: full-, half-, third- and sixth-length runs, for $b \bmod 6$ in $\{1,5\}$, $\{2,4\}$, $\{3\}$ and $\{0\}$ respectively. Overall, we get a 33% time reduction, essentially for free. It is not worthwhile to consider $c\,|\,a',b$ for $c > 3$.

A.6 Cascading the Sieves

Recall that the instances of the sieving problem come in pairs of *rational* and *algebraic* sieves, and we are interested in the a values that passed both sieves (cf. 2.1). However, the situation is not symmetric: $B_{\mathrm{R}} \langle\!\langle 2.6 \cdot 10^{10} \rangle\!\rangle_{\mathrm{A}}$ is much larger than $B_{\mathrm{R}} \langle\!\langle = 3.5 \cdot 10^9 \rangle\!\rangle_{\mathrm{R}}$.[22] Therefore the cost of the algebraic sieves would dominate the total cost when s is chosen optimally for each sieve type. Moreover, for 1024-bit composites and the parameters we consider (cf. Appendix B), we cannot make the algebraic-side s as large as we wish because this would exceed the capacity of a single silicon wafer. The following shows a way to address this.

---

[22] B_{A} and B_{R} are chosen as to produce a sufficient probability of semi-smoothness for the values over which we are (implicitly) sieving: circa $\langle\!\langle 10^{101} \rangle\!\rangle_{\mathrm{A}}$ vs. circa $\langle\!\langle 10^{64} \rangle\!\rangle_{\mathrm{R}}$.

Let s_R and s_A denote the s values of the rational and algebraic sieves respectively. The reason we cannot increase s_A and gain further "free" parallelism is that the bus becomes unmanageably wide and the delivery lines become numerous and long (their cost is $\Theta(s^2)$). However, the bus is designed to sieve s_A sieve locations per pipeline stage. If we first execute the rational sieve then most of these sieve locations can be ruled out in advance: all but a small fraction $\langle\!\langle 1.7 \cdot 10^{-4} \rangle\!\rangle$ of the sieve locations do not pass the threshold in the rational sieve,[23] and thus cannot form candidates regardless of their algebraic-side quality.

Accordingly, we make the following change in the design of algebraic sieves. Instead of a wide bus consisting of s_A lines that are permanently assigned to residues modulo s_A, we use a much narrower bus consisting of only $u \langle\!\langle = 32 \rangle\!\rangle_A$ lines, where each line contains a pair (C,L). $L = (a \bmod s_A)$ identifies the sieve location, and C is the sum of $\lfloor \log p_i \rfloor$ contributions to a so far. The sieve locations are still scanned in a pipelined manner at a rate of s_A locations per clock cycle, and all delivery pairs are generated as before at the respective units.

The delivery lines are different: instead of being long and "dumb", they are now short and "smart". When a delivery pair $(\lfloor \log p_i \rfloor, \ell)$ is generated, ℓ is compared to L for each of the u lines (at the respective pipeline stage) in a single clock cycle. If a match is found, $\lfloor \log p_i \rfloor$ is added to the C of that line. Otherwise (i.e., in the overwhelming majority of cases), the delivery pair is discarded.

At the head of the bus, we input pairs $(0, a \bmod s_A)$ for the sieve locations a that passed the rational sieve. To achieve this we wire the outputs of rational sieves to inputs of algebraic sieves, and operate them in a synchronized manner (with the necessary phase shift). Due to the mismatch in s values, we connect s_A/s_B rational sieves to each algebraic sieves. Each such cluster of s_A/s_B+1 sieving devices is jointly applied to one single sieve line at a time, in a synchronized manner. To divide the work between the multiple rational sieves, we use interleaving of sieve locations (similarly to the bit-reversal technique of A.2). Each rational-to-algebraic connection transmits at most one value of size $\log_2 s_R \langle\!\langle 12 \rangle\!\rangle$ bits per clock cycle (appropriate buffering is used to average away congestions).

This change greatly reduces the circuit area occupied by the bus wiring and delivery lines; for our choice of parameters, it becomes insignificant. Also, there is no longer need to duplicate emitters for smallish progressions (except when $p_i < s$). This allows us to use a large $s \langle\!\langle = 32{,}768 \rangle\!\rangle_A$ for the algebraic sieves, thereby reducing their cost to less than that of the rational sieve (cf. 4.1). Moreover, it lets us further increase B_A with little effect on cost, which (due to tradeoffs in the NFS parameter choice) reduces H and R.

A.7   Testing Candidates

Having computed approximations of the sum of logarithms $g(a)$ for each sieve location a, we need to identify the resulting candidates, compute the corresponding sets $\{i : a \in P_i\}$, and perform some additional tests (cf. 2.1). These are implemented as follows.

---

[23] Before the cofactor factorization. Slightly more when $\lfloor \log p_i \rfloor$ rounding is considered.

Identifying candidates. In each TWIRL device, at the end of the bus (i.e., downstream for all stations) we place an array of comparators, one per bus line, that identify a values for which $g(a) > T$. In the basic TWIRL design, we operate a pair of sieves (one rational and one algebraic) in unison: at each clock cycle, the sets of bus lines that passed the comparator threshold are communicated between the two devices, and their intersection (i.e., the candidates) are identified. In the cascaded sieves variant, only sieve locations that passed the threshold on the rational TWIRL are further processed by the algebraic TWIRL, and thus the candidates are exactly those sieve locations that passed the threshold in the algebraic TWIRL. The fraction of sieve locations that constitute candidates is very small $\langle\!\langle 2 \cdot 10^{-11} \rangle\!\rangle$.

Finding the corresponding progressions. For each candidate we need to compute the set $\{i : a \in P_i\}$, separately for the rational and algebraic sieves. From the context in the NFS algorithm it follows that the elements of this set for which p_i is relatively small can be found easily.[24] It thus appears sufficient to find the subset $\{i : a \in P_i, p_i \text{ is largish}\}$, which is accomplished by having largish stations remember the p_i values of recent progressions and report them upon request.

To implement this, we add two dedicated pipelined channels passing through all the processors in the largish stations. The *lines channel*, of width $\log_2 s$ bits, goes upstream (i.e., opposite to the flow of values in the bus) from the threshold comparators. The *divisors channel*, of width $\log_2 B$ bits, goes downstream. Both have a pipeline register after each processor, and both end up as outputs of the TWIRL device. To each largish processor we attach a *diary*, which is a cyclic list of $\log_2 B$-bit values. Every clock cycle, the processor writes a value to its diary: if the processor inserted an emission triplet $(\lfloor \log p_i \rfloor, \ell_i, \tau_i)$ into the buffer at this clock cycle, it writes the triple (p_i, ℓ_i, τ_i) to the diary; otherwise it writes a designated NULL value. When a candidate is identified at some bus line ℓ, the value ℓ is sent upstream through the lines channel. Whenever a processor sees an ℓ value on the lines channel, it inspects its diaries to see whether it made an emission that was added to bus line ℓ exactly z clock cycles ago, where z is the distance (in pipeline stages) from the processor's output into the buffer, through the bus and threshold comparators and back to the processor through the lines channel. This inspection is done by searching the $\langle\!\langle 64 \rangle\!\rangle$ diary entries preceeding the one written z clock cycles ago for a non-NULL value (p_i, ℓ_i) with $\ell_i = \ell$. If such a diary entry is found, the processor transmits p_i downstream via the divisors channel (with retry in case of collision). The probability of intermingling data belonging to different candidates is negligible, and even then we can recover (by appropriate divisibility tests).

In the cascaded sieves variant, the algebraic sieve records to diaries only those contributions that were not discarded at the delivery lines. The rational diaries are rather large ($\langle\!\langle 13{,}530 \rangle\!\rangle_{\mathrm{R}}$ entries) since they need to keep their entries a long time — the latency z includes passing through (at worst) all rational

---

[24] Namely, by finding the small factors of $F_j(a - R{,}b)$ where F_j is the relevant NFS polynomial and b is the line being sieved.

bus pipeline stages, all algebraic bus pipeline stages and then going upstream through all rational stations. However, these diaries can be implemented very efficiently as DRAM banks of a degenerate form with a fixed cyclic access order (similarly to the memory banks of the largish stations).

Testing candidates. Given the above information, the candidates have to be further processed to account for the various approximations and errors in sieving, and to account for the NFS "large primes" (cf. 2.1). The first steps (computing the values of the polynomials, dividing out small factors and the diary reports, and testing size and primality of remaining cofactors) can be effectively handled by special-purpose processors and pipelines, which are similar to the division pipeline of [7, Section 4] except that here we have far fewer candidates (cf. C).

Cofactor factorization. The candidates that survived the above steps (and whose cofactors were not prime or sufficiently small) undergo cofactor factorization. This involves factorization of one (and seldom two) integers of size at most $\langle\!\langle 1 \cdot 10^{24}\rangle\!\rangle$. Less than $\langle\!\langle 2 \cdot 10^{-11}\rangle\!\rangle$ of the sieve locations reach this stage (this takes $\lceil \log p_i\rceil$ rounding errors into consideration), and a modern general-purpose processor can handle each in less than 0.05 seconds. Thus, using dedicated hardware this can be performed at a small fraction of the cost of sieving. Also, certain algorithmic improvements may be applicable [2].

A.8 Lattice Sieving

The above is motivated by NFS line sieving, which has very large sieve line length R. Lattice sieving (i.e., "special-q") involves fewer sieving locations. However, lattice sieving has very short sieving lines (8192 in [5]), so the natural mapping to the lattice problem as defined here (i.e., lattice sieving by lines) leads to values of R that are too small.

We can adapt TWIRL to efficient lattice sieving as follows. Choose s equal to the width of the lattice sieving region (they are of comparable magnitude); a full lattice line is handled at each clock cycle, and R is the total number of points in the sieved lattice block. The definition (p_i, r_i) is different in this case — they are now related to the vectors used in lattice sieving by vectors (before they are lattice-reduced). The handling of modulo s wrap-around of progressions is now somewhat more complicated, and the emission calculation logic in all station types needs to be adapted. Note that the largish processors are essentially performing lattice sieving by vectors, as they are "throwing" values far into the "future", not to be seen again until their next emission event.

Re-initialization is needed only when the special-q lattices are changed (every $8192 \cdot 5000$ sieve locations in [5]), but is more expensive. Given the benefits of lattice sieving, it may be advantageous to use faster (but larger) re-initialization circuits and to increase the sieving regions (despite the lower yield); this requires further exploration.

A.9 Fault Tolerance

Due to its size, each TWIRL device is likely to have multiple local defects caused by imperfections in the VLSI process. To increase the yield of good devices, we make the following adaptations.

   If any component of a station is defective, we simply avoid using this station. Using a small number of spare stations of each type (with their constants stored in reloadable latches), we can handle the corresponding progressions.

   Since our device uses an addition pipeline, it is highly sensitive to faults in the bus lines or associated adders. To handle these, we can add a small number of spare line segments along the bus, and logically re-route portions of bus lines through the spare segments in order to bypass local faults. In this case, the special-purpose processors in largish stations can easily change the bus destination addresses (i.e., ℓ value of emission triplets) to account for re-routing. For smallish and tiny stations it appears harder to account for re-routing, so we just give up adding the corresponding $\lfloor \log p_i \rceil$ values; we may partially compensate by adding a small constant value to the re-routed bus lines. Since the sieving step is intended only as a fairly crude (though highly effective) filter, a few false-positives or false-negatives are acceptable.

B Parameters for Cost Estimates

B.1 Hardware

The hardware parameters used are those given in [16] (which are consistent with [9]): standard 30cm silicon wafers with 0.13μm process technology, at an assumed cost of \$5,000 per wafer. For 1024-bit and 768-bit composites we will use DRAM-type wafers, which we assume to have a transistor density of 2.8 μm^2 per transistor (averaged over the logic area) and DRAM density of 0.2μm^2 per bit (averaged over the area of DRAM banks). For 512-bit composites we will use logic-type wafers, with transistor density of 2.38μm^2 per transistor and DRAM density of 0.7μm^2 per bit. The clock rate is 1GHz clock rate, which appears realistic with judicious pipelining of the processors.

   We have derived rough estimates for all major components of the design; this required additional analysis, assumptions and simulation of the algorithms. Here are some highlights, for 1024-bit composites with the choice of parameters specified throughout Section 3. A typical largish special-purpose processor is assumed to require the area of $\langle\!\langle 96,400 \rangle\!\rangle_{\mathrm{R}}$ logic-density transistors (including the amortized buffer area and the small amount of cache memory, about $\langle\!\langle 14\mathrm{Kbit} \rangle\!\rangle_{\mathrm{R}}$, that is independent of p_i). A typical emitter is assumed to require $\langle\!\langle 2,037 \rangle\!\rangle_{\mathrm{R}}$ transistors in a smallish station (including the amortized costs of funnels), and $\langle\!\langle 522 \rangle\!\rangle_{\mathrm{R}}$ in a tiny station. Delivery cells are assumed to require $\langle\!\langle 530 \rangle\!\rangle_{\mathrm{R}}$ transistors with interleaving (i.e., in largish stations) and $\langle\!\langle 1220 \rangle\!\rangle_{\mathrm{R}}$ without interleaving (i.e., in smallish and tiny stations). We assume that the memory system of Section 3.2 requires $\langle\!\langle 2.5 \rangle\!\rangle$ times more area per useful bit than standard DRAM, due to the required slack and and area of the cache. We assume that bus wires don't require

Table 1. Sieving parameters.

Parameter	Meaning	1024-bit	768-bit	512-bit
R	Width of sieve line	$1.1 \cdot 10^{15}$	$3.4 \cdot 10^{13}$	$1.8 \cdot 10^{10}$
H	Number of sieve lines	$2.7 \cdot 10^{8}$	$8.9 \cdot 10^{6}$	$9.0 \cdot 10^{5}$
B_R	Rational smoothness bound	$3.5 \cdot 10^{9}$	$1 \cdot 10^{8}$	$1.7 \cdot 10^{7}$
B_A	Algebraic smoothness bound	$2.6 \cdot 10^{10}$	$1 \cdot 10^{9}$	$1.7 \cdot 10^{7}$

wafer area apart from their pipelining registers, due to the availability of multiple metal layers. We take the cross-bus density of bus wires to be ⟨⟨0.5⟩⟩ bits per μm, possibly achieved by using multiple metal layers.

Note that since the device contains many interconnected units of non-uniform size, designing an efficient layout (which we have not done) is a non-trivial task. However, the number of different unit types is very small compared to designs that are commonly handled by the VLSI industry, and there is considerable room for variations. The mostly systolic design also enables the creation devices larger than the reticle size, using multiple steps of a single (or very few) mask set.

Using a fault-tolerant design (cf. A.9), the yield can made very high and functional testing can be done at a low cost after assembly. Also, the acceptable probability of undetected errors is much higher than that of most VLSI designs.

B.2 Sieving Parameters

To predict the cost of sieving, we need to estimate the relevant NFS parameters (R, H, B_R, B_A). The values we used are summarized in Table 1. The parameters for 512-bit composites are the same as those postulated for TWINKLE [15] and appear conservative compared to actual experiments [5].

To obtain reasonably accurate predictions for larger composites, we followed the approach of [14]; namely, we generated concrete pairs of NFS polynomials for the RSA-1024 and RSA-768 challenge composites [20] and estimated their relations yield. The search for NFS polynomials was done using programs written by Jens Franke and Thorsten Kleinjung (with minor adaptations). For our 1024-bit estimates we picked the following pair of polynomials, which have a common integer root modulo the RSA-1024 composite:

$$
\begin{aligned}
f(x) = \quad & 1719304894236345143401011418080x^5 \\
& - 6991973488866605861074074186043634471x^4 \\
& + 2708603048356953289405097425785134664952131 4x^3 \\
& + 4693758405266857450288679183553655227741024235 9042x^2 \\
& - 1010702948425721113717814588506968458777068995453945 01384x \\
& - 226669159394909405786175246770453711891289098997165603 98434136 \\
g(x) = \quad & 93877230837026306984571367477027x \\
& - 37934895496425027513691045755639637174211483324451628365
\end{aligned}
$$

Subsequent analysis of relations yield was done by integrating the relevant smoothness probability functions [11] over the sieving region. Successful factorization requires finding sufficiently many *cycles* among the relations, and for two large primes per side (as we assumed) it is currently unknown how to predict the number of cycles from the number of relations, but we verified that the numbers appear "reasonable" compared to current experience with smaller composites. The 768-bit parameters were derived similarly. More details are available in a dedicated web page [22] and in [14].

Note that finding better polynomials will reduce the cost of sieving. Indeed, our algebraic-side polynomial is of degree 5 (due to a limitation of the programs we used), while there are theoretical and empirical reasons to believe that polynomials of somewhat higher degree can have significantly higher yield.

C   Relation to Previous Works

TWINKLE. As is evident from the presentation, the new device shares with TWINKLE the property of time-space reversal compared to traditional sieving. TWIRL is obviously faster than TWINKLE, as two have comparable clock rates but the latter checks one sieve location per clock cycle whereas the former checks thousands. None the less, TWIRL is smaller than TWINKLE — this is due to the efficient parallelization and the use of compact DRAM storage for the largish progressions (it so happens that DRAM cannot be efficiently implemented on GaAs wafers, which are used by TWINKLE). We may consider using TWINKLE-like optical analog adders instead of electronic adder pipelines, but constructing a separate optical adder for each residue class modulo s would entail practical difficulties, and does not appear worthwhile as there are far fewer values to sum.

FPGA-based serial sieving. Kim and Mangione-Smith [10] describe a sieving device using off-the-shelf parts that may be only 6 times slower than TWINKLE. It uses classical sieving, without time-memory reversal. The speedup follows from increased memory bandwidth – there are several FPGA chips and each is connected to multiple SRAM chips. As presented this implementation does not rival the speed or cost of TWIRL. Moreover, since it is tied to a specific hardware platform, it is unclear how it scales to larger parallelism and larger sieving problems.

Low-memory sieving circuits. Bernstein [3] proposes to completely replace sieving by memory-efficient smoothness testing methods, such as the Elliptic Curve Method of factorization. This reduces the asymptotic time × space cost of the matrix step from $y^{3+o(1)}$ to $y^{2+o(1)}$, where y is subexponential in the length of the integer being factored and depends on the choice of NFS parameters. By comparison, TWIRL has a throughput cost of $y^{2.5+o(1)}$, because the speedup factor grows as the square root of the number of progressions (cf. 4.5). However, these asymptotic figures hide significant factors; based on current experience, for 1024-bit composites it appears unlikely that memory-efficient smoothness testing would rival the practical performance of traditional sieving, let alone that of TWIRL, in spite of its superior asymptotic complexity.

Mesh-based sieving. While [3] deals primarily with the NFS matrix step, it does mention "sieving via Schimmler's algorithm" and notes that its cost would be $L^{2.5+o(1)}$ (like TWIRL's). Geiselmann and Steinwandt [7] follow this approach and give a detailed design for a mesh-based sieving circuit. Compared to previous sieving devices, both [7] and TWIRL achieve a speedup factor of $\tilde{\Theta}(\sqrt{B})$.[25] However, there are significant differences in scalability and cost: TWIRL is 1,600 times more efficient for 512-bit composites, and ever more so for bigger composites or when using the cascaded sieves variant (cf. 4.3, A.6).

One reason is as follows. The mesh-based sorting of [7] is effective in terms of *latency*, which is why it was appropriate for the Bernstein's matrix-step device [3] where the input to each invocation depended on the output of the previous one. However, for sieving we care only about *throughput*. Disregarding latency leads to smaller circuits and higher clock rates. For example, TWIRL's delivery lines perform trivial one-dimensional unidirectional routing of values of size $\langle\!\langle 12+10\rangle\!\rangle_R$ bits, as opposed to complicated two-dimensional mesh sorting of progression states of size $\langle\!\langle 2 \cdot 31.7\rangle\!\rangle_R$.[26] For the algebraic sieves the situation is even more extreme (cf. A.6).

In the design of [7], the state of each progression is duplicated $\lceil\tilde{\Theta}(B/p_i)\rceil$ times (compared to $\lceil\tilde{\Theta}(\sqrt{B/p_i})\rceil$ in TWIRL) or handled by other means; this greatly increases the cost. For the primary set of design parameters suggested in [7] for factoring 512-bit numbers, 75% of the mesh is occupied by duplicated values even though all primes smaller than 2^{17} are handled by other means: a separate division pipeline that tests potential candidates identified by the mesh, using over 12,000 expensive integer division units. Moreover, this assumes that the sums of $\lfloor\log p_i\rfloor$ contributions from the progressions with $p_i > 2^{17}$ are sufficiently correlated with smoothness under all progressions; it is unclear whether this assumption scales.

TWIRL's handling of largish primes using DRAM storage greatly reduces the size of the circuit when implemented using current VLSI technology (90 DRAM bits vs. about 2500 transistors in [7]).

If the device must span multiple wafers, the inter-wafer bandwidth requirements of our design are much lower than that of [7] (as long as the bus is narrower than a wafer), and there is no algorithmic difficulty in handling the long latency of cross-wafer lines. Moreover, connecting wafers in a chain may be easier than connecting them in a 2D mesh, especially in regard to cooling and faults.

---

[25] Possibly less for [7] — an asymptotic analysis is lacking, especially in regard to the handling of small primes.

[26] The authors of [7] have suggested (in private communication) a variant of their device that routes emissions instead of sorting states, analogously to [16]. Still, mesh routing is more expensive than pipelined delivery lines.

New Partial Key Exposure Attacks on RSA

Johannes Blömer and Alexander May

Faculty of Computer Science, Electrical Engineering and Mathematics
Paderborn University
33102 Paderborn, Germany
{bloemer,alexx}@uni-paderborn.de

Abstract. In 1998, Boneh, Durfee and Frankel [4] presented several attacks on RSA when an adversary knows a fraction of the secret key bits. The motivation for these so-called partial key exposure attacks mainly arises from the study of side-channel attacks on RSA. With side channel attacks an adversary gets either most significant or least significant bits of the secret key. The polynomial time algorithms given in [4] only work provided that the public key e is smaller than $N^{\frac{1}{2}}$. It was raised as an open question whether there are polynomial time attacks beyond this bound. We answer this open question in the present work both in the case of most and least significant bits. Our algorithms make use of Coppersmith's heuristic method for solving modular multivariate polynomial equations [8]. For known most significant bits, we provide an algorithm that works for public exponents e in the interval $[N^{\frac{1}{2}}, N^{0.725}]$. Surprisingly, we get an even stronger result for known least significant bits: An algorithm that works for all $e < N^{\frac{7}{8}}$.

We also provide partial key exposure attacks on fast RSA-variants that use Chinese Remaindering in the decryption process (e.g. [20,21]). These fast variants are interesting for time-critical applications like smart-cards which in turn are highly vulnerable to side-channel attacks. The new attacks are provable. We show that for small public exponent RSA half of the bits of $d_p = d \bmod p-1$ suffice to find the factorization of N in polynomial time. This amount is only a quarter of the bits of N and therefore the method belongs to the strongest known partial key exposure attacks.

Keywords: RSA, known bits, lattice reduction, Coppersmith's method

1 Introduction

Let (N, e) be an RSA public key with $N = pq$, where p and q are of equal bit-size. The secret key d satisfies $ed = 1 \bmod \phi(N)$.

In 1998, Boneh, Durfee and Frankel [4] introduced the following question: How many bits of d does an adversary need to know in order to factor the modulus N? In addition to its theoretical impact on understanding the complexity of the RSA-function, this is an important practical question arising from the intensive study of side-channel attacks on RSA in cryptography (e.g. fault attacks, timing attacks, power analysis, see for instance [6,15,16]).

D. Boneh (Ed.): CRYPTO 2003, LNCS 2729, pp. 27–43, 2003.

In many scenarios, an attacker using a side-channel attack either succeeds to obtain the most significant bits (MSBs) or the least significant bits (LSBs) of d in *consecutive order*. Whether he gets MSBs or LSBs depends on the different ways of computing an exponentiation with d during the decryption process. Therefore in this work, we just focus on the case where an adversary knows either MSBs or LSBs of d and we ignore attacks where an adversary has to know both sorts of bits or intermediate bits.

Cases have been reported in the literature [9] where side-channel attacks are able to reveal a fraction of the secret key bits, but fail to reveal the entire key. For instance it is often the case that an attacker gets the next bit of d under the conditional probability that his hypothesis of the previous bits is correct. Hence, it gets harder and harder for him to make a correct guess with a certain probability. This makes it essential to know how many bits of d suffice to discover the whole secret information.

Boneh, Durfee and Frankel [4] were the first that presented polynomial time algorithms when an attacker knows only a fraction of the bits. In the case of known least significant bits, they showed that for low public exponent RSA (e.g. $e = \text{poly}(\log N)$) a quarter of the bits of d are sufficient to find the factorization of N. Their method makes use of a well-known theorem due to Coppersmith [8]: Given half of the bits of p, the factorization of N can be found in polynomial time.

Considering known MSBs, Boneh, Durfee and Frankel presented an algorithm that works for all $e < N^{\frac{1}{2}}$, again using Coppersmith's theorem. However it remained an open question in [4] whether there are polynomial time algorithms that find the factorization of N for *values of e substantially larger than $N^{\frac{1}{2}}$* given only a subset of the secret key bits.

In this work, we answer this question both in the case of known MSBs and of known LSBs.

MSBs of d Known:
We present a method that works for all public exponents e in the interval $[N^{\frac{1}{2}}, N^{0.725}]$. The number of bits of d that have to be known increases with e. Let us provide some examples of the required bits: For $e = N^{0.5}$ one has to know half of the MSBs of d, for $e = N^{0.55}$ a 0.71-fraction suffices whereas for $e = N^{0.6}$ a fraction of 0.81 is needed to factor N.

In contrast to Boneh, Durfee and Frankel we do not use Coppersmith's result for known bits of p. Instead we directly apply Coppersmith's method for finding roots of modular multivariate polynomial equations [8]. This method has many applications in cryptography. Since it is a heuristic in the multivariate case, our result is heuristic as well. However, in various other applications of Coppersmith's method (see [1,3,10,14]) a systematic failure of the multivariate heuristic has never been reported. Hence the heuristic is widely believed to work perfectly in practice. We also provide various experiments that confirm the reliability: None of our experiments failed to yield the factorization of N.

In Figure 1 we illustrate our result for MSBs. The size of the fraction of the bits that is needed in our attack is plotted as a function of the size of the public exponent e. We express the size of e in terms of the size of N (i.e. we use $\log_N(e)$). For a comparison with previous results, we also include in our graphs the results of Boneh, Durfee and Frankel. The marked regions in Figure 1 are the feasible regions for the various approaches.

Note that the area belonging to BDF2 requires that the factorization of e is known. The result BDF3 is not explicitly mentioned as a polynomial time algorithm in [4], but can be easily derived from a method stated by the same authors in [5]: The upper $\log_N(e)$ bits of d immediately yield half of the MSBs of d and the attacker can use the remaining quarter of bits to factor N.

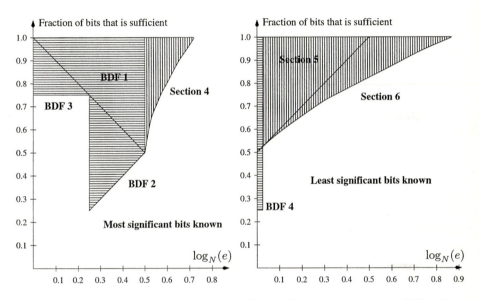

Fig. 1. The results for known MSBs of d **Fig. 2.** The results for known LSBs of d

LSBs of d Known:

We start by proving a result for all but a negligible fraction of the public exponents $e < N^{\frac{1}{2}}$. Previously, only polynomial time algorithms for e of the order $\mathrm{poly}(\log N)$ were known [4]. Our approach uses a 3-dimensional lattice to find the factorization of N using a single lattice basis reduction, whereas the method in [4] requires about e lattice reductions. We tested our attack with the frequently used RSA-exponent $e = 2^{16} + 1$. Our algorithm is faster than the method in [4] but requires more bits of d.

Interestingly, our approach makes use of the linear independence of two sufficiently short vectors in the lattice and we do not need to apply Coppersmith's heuristic in this case. This makes our method rigorous and at the same time introduces a new method to solve modular multivariate polynomial equations of a special form. Therefore we believe that our approach is of independent interest.

Next, we generalize the 3-dimensional approach to multi-dimensional lattices. This improves the bound up to all $e < N^{\frac{7}{8}}$, which is the largest bound for e in partial key exposure attacks that is known up to now. Unfortunately, since our attack relies on Coppersmith's method for modular multivariate polynomial equations, it becomes heuristic. But again in our experiments, we could not find a single failure of the multivariate heuristic. The results are illustrated in Figure 2 in the same fashion as before.

We raise the question whether it is possible to derive results for all keys $e < \phi(N)$. In the light of our new results, this bound does not seem to be out of reach. Maybe a modification of our lattices could already suffice (e.g. using non-triangular lattice bases), but at the moment this is an open question.

Known Bits in CRT-Variants:

We present results on known bits of $d_p = d \bmod p - 1$ (and symmetrically on $d_q = d \bmod q - 1$). The value d_p is used in fast Chinese Remainder variants of the decryption process. This includes the well-known Quisquater-Couvreur method [21]. With suitable modifications, the attack applies also to other fast RSA-variants like for instance Takagi's scheme [20], which uses a modulus of the form $p^k q$.

These fast variants of RSA are especially interesting for time-critical applications. Therefore they are frequently used on smart-cards. On the other hand, it is well-known that smart-cards are highly vulnerable to different sorts of side-channel attacks. Hence it is of important practical interest to study the complexity of partial key exposure attacks for CRT-variants.

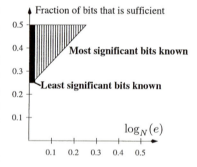

Fig. 3. LSBs/MSBs of d_p.

We provide provable attacks for both cases: LSBs and MSBs. Interestingly, in our proofs we use a less known variant of a result of Coppersmith [8] that is due to Howgrave-Graham. Coppersmith showed that an approximation of p up to an additive error of $N^{\frac{1}{4}}$ yields the factorization of N. Howgrave-Graham [13] observed that an approximation of kp for some (unknown) k with the same error bound already suffices.

We prove that for low public exponents e (i.e. $e = \mathrm{poly}(\log N)$), half of the LSBs of d_p always suffice to factor N. Therefore the attack is a threat to RSA-implementations with the commonly used public exponent $e = 2^{16} + 1$. Note that half of the bits of d_p is only an amount of a quarter of the bits of N and therefore the result is as strong as the best known partial key exposure attacks.

In the case of known MSBs of d_p, we present an algorithm that even works for all $e < N^{\frac{1}{4}}$ in polynomial time. Again for low public exponent RSA, it requires only half of the MSBs of d_p in order to factor N. The results are illustrated in Figure 3.

Detailed Overview:

We briefly overview all known polynomial time partial key exposure attack by giving the precise functions of the bits that have to be known. Let $\alpha = \log_N(e)$ denote the size of e in terms of N. In Figure 4, the upper half of the table states the results for known MSBs whereas the lower half is dedicated to the results for known LSBs. The attacks for known bits of d_p are stated in the last lines of each half.

	$\alpha = \log_N(e)$	Fraction of bits that is needed	Restriction/Comment
BDF [4]	$[\frac{1}{4}, \frac{1}{2}]$	α	e prime/known fact.
BDF [4]	$[0, \frac{1}{2}]$	$1 - \alpha$	$\frac{d}{\phi(N)} = \Omega(1)$
Section 4	$[\frac{1}{2}, \frac{\sqrt{6}-1}{2}]$	$\frac{1}{8}\left(3 + 2\alpha + \sqrt{36\alpha^2 + 12\alpha - 15}\right)$	heuristic
BDF [5]	$[0, \frac{1}{2}]$	$\frac{3}{4}$	$\frac{d}{\phi(N)}, \frac{|p-q|}{\sqrt{N}} = \Omega(1)$
Section 2	$[0, \frac{1}{4}]$	$\frac{1}{4} + \alpha$	bits of d_p
BDF [5]	$\mathcal{O}(\log_N \log N)$	$\frac{1}{4}$	$N = 3 \bmod 4$
Section 5	$[0, \frac{1}{2}]$	$\frac{1}{2} + \alpha$	all but $O(N^{\alpha-\epsilon})$ e's
Section 6	$[0, \frac{7}{8}]$	$\frac{1}{6} + \frac{1}{3}\sqrt{1 + 6\alpha}$	heuristic
Section 2	$\mathcal{O}(\log_N \log N)$	$\frac{1}{4}$	bits of d_p

Fig. 4. Detailed summary of the results

The paper is organized as follows: In Section 2, we present our methods for the CRT-variants. Here we use lattice reduction methods only as a black-box. In order to give the more elaborate results for partial key exposure attacks with large public exponent, we have to define some lattice notation in Section 3. The method for MSBs is presented in Section 4, the LSB-attacks are given in Section 5 and 6.

2   Known MSBs/LSBs and Chinese Remaindering

Throughout this work we will consider RSA-public keys (N, e) with $N = pq$, where p and q are of equal bit-size. Therefore $p, q \leq 2\sqrt{N}$. Furthermore, we assume wlog that $p \leq q$ which implies $p \leq \sqrt{N}$ and

$$p + q \leq 3\sqrt{N}.$$

The secret exponent d corresponding to (N, e) satisfies the equality $ed = 1 \bmod \phi(N)$, where $\phi(N)$ is the Euler totient function.

We will often talk of known most or least significant bits (MSBs/LSBs) of d, but we want to point out that this should only be understood as a helpful simplification to explain our results in the context of side-channel attacks. To be more precise, when we talk of k known LSBs of d, then in fact we only need to know integers d_0, M such that $d_0 = d \bmod M$, where $M \geq 2^k$. Thus, $M = 2^k$ is only the special case where we really know the bits. Analogously, in the case of known MSBs: We do not really need to know the MSBs but only an approximation $\tilde{d}$ of d such that $|d - \tilde{d}|$ can be suitably upper-bounded.

In order to speed up the decryption/signing process, it is common practice to use the values $d_p = d \bmod p - 1$ and $d_q = d \bmod q - 1$. To sign m, one computes $m^{d_p} \bmod p$ and $m^{d_q} \bmod q$ and combines the results using the Chinese Remainder Theorem (CRT).

These fast RSA-variants are especially interesting for time-critical applications like smart-cards, which are highly vulnerable to side-channel attacks. However, it has never been studied how many bits of d_p (or symmetrically of d_q) suffice in order to find the factorization of N. We present two provable results for RSA-variants with CRT in this section.

Both of our proofs use the following variation of a well-known theorem of Coppersmith [8] that is due to Howgrave-Graham. Coppersmith showed how to factor N given half of the MSBs of p. Howgrave-Graham [13] observed that this holds in more general form for the MSBs of multiples of p.

Theorem 1 (Howgrave-Graham) *Let $N = pq$ be an RSA-modulus and k be an unknown integer which is not a multiple of q. Given an approximation of kp with additive error at most $N^{\frac{1}{4}}$, the factorization of N can be found in polynomial time.*

First, we consider the case of known LBSs of d_p. We show that whenever the public exponent e is of size poly$(\log N)$, then half of the lower bits of d_p are sufficient to find the factorization of N in polynomial time.

Theorem 2 *Let (N, e) be an RSA public key with $N = pq$ and secret key d. Let $d_p = d \bmod p - 1$. Given d_0, M with $d_0 = d_p \bmod M$ and*

$$M \geq N^{\frac{1}{4}}.$$

Then the factorization of N can be found in time $e \cdot poly(\log N)$.

Proof: We know that

$$ed_p - 1 = k(p - 1)$$

for some $k \in \mathbb{N}$. Since $d_p < p - 1$, we know that $k = \frac{ed_p - 1}{p - 1} < e$. Let us write $d_p = d_1 M + d_0$, where $d_1 < \frac{d_p}{M} < \frac{p}{M} \leq N^{\frac{1}{4}}$. We can rewrite our equation as

$$ed_0 + k - 1 = kp - eMd_1.$$

Let E be the inverse of eM modulo N, e.g. there exist a $c \in \mathbb{N}$ such that $E \cdot eM = 1 + cN$ (if E does not exist, we obtain the factorization of N). Multiplying the above equation by E yields

$$E(ed_0 + k - 1) = (Ek - cqd_1)p - d_1.$$

The only unknown parameter on the left hand side of the equation is k. We make a brute force search for k in the interval $[1, e)$. The correct guess of k gives us a multiple of p up to an additive error $d_1 < N^{\frac{1}{4}}$. Thus, when the algorithm of Theorem 1 is applied to the correct guess of k, we obtain the factorization of N. Note that q divides the term $Ek - cqd_1$ iff q divides k which is easily testable (q cannot divide k in the case $e < q$). This concludes the proof of the theorem. $\boxed{}$

In our second approach, we consider the case when MSBs of d_p are known.

Theorem 3 *Let (N, e) be an RSA public key with secret key d and $e = N^\alpha$ for some $a \in [0, \frac{1}{4}]$. Furthermore, let $d_p = d \bmod p - 1$. Given $\tilde{d}$ with*

$$|d_p - \tilde{d}| \leq N^{\frac{1}{4} - \alpha}.$$

Then N can be factored in polynomial time.

Proof: We start again by looking at the equation $ed_p - 1 = k(p - 1)$. Since $d_p < p-1$, we know that $k < N^\alpha$, which implies that q cannot divide k. Compute $\tilde{p} = e\tilde{d} - 1$. Now, $\tilde{p}$ is an approximation of kp up to an additive error of at most

$$|\tilde{p} - kp| = |e(\tilde{d} - d_p) - k| \leq N^{\frac{1}{4}} + N^\alpha \leq 2N^{\frac{1}{4}}.$$

Thus, either $\tilde{p} + N^{\frac{1}{4}}$ or $\tilde{p} - N^{\frac{1}{4}}$ is an approximation of kp with error at most $N^{\frac{1}{4}}$. Applying the algorithm of Theorem 1 to both values yields the factorization of N. $\boxed{}$

3   Preliminaries on Lattices

Since our partial key exposure attacks for large public exponents use polynomial arithmetic, we introduce some helpful notations. Let $f(x, y) = \sum_{i,j} a_{i,j} x^i y^j$ be a bivariate polynomial with coefficients $a_{i,j} \in \mathbb{Z}$. All terms $x^i y^j$ with non-zero coefficients are called monomials. The coefficient vector of f is defined by the vector of the coefficients $a_{i,j}$. We define the norm of f as the Euclidean norm of the coefficient vector: $\|f\|^2 = \sum_{i,j} a_{i,j}^2$. The definitions for trivariate polynomials are analogous. In the following, we state a few basic facts about lattices and lattice basis reduction and refer to the textbooks [7,11,18] for an introduction to the theory of lattices.

Let $v_1, \ldots, v_n \in \mathbb{R}^n$ be linearly independent vectors. A lattice L spanned by $\{v_1, \ldots, v_n\}$ is the set of all integer linear combinations of $v_1, \ldots, v_n$. We call n the dimension of L, which we denote by $\dim(L)$.

The set $B = \{v_1, \ldots, v_n\}$ is called a basis of L, the $(n \times n)$-matrix consisting of the row vectors $v_1, \ldots, v_n$ is called basis matrix. A basis of L can be transformed into another basis by applying an unimodular transformation to the basis matrix. The determinant $\det(L)$ is the absolute value of the determinant of a basis matrix.

The famous L^3-lattice reduction algorithm of Lenstra, Lenstra and Lovász [17] can be used to approximate a shortest vector.

Theorem 4 (Lenstra, Lenstra, Lovász) *Let $L \in \mathbb{Z}^n$ be a lattice spanned by $\{v_1, \ldots, v_n\}$. The L^3-algorithm outputs in polynomial time a reduced lattice basis $\{v_1', \ldots, v_n'\}$ with*

$$\|v_i'\| \leq 2^{\frac{n(n-1)+(i-1)(i-2)}{4(n-i+1)}} \det(L)^{\frac{1}{n-i+1}} \quad for \ i = 1, \ldots, n.$$

This theorem can easily be proven using [7], Theorem 2.6.2.

In Sections 4 and 6, we will use a heuristic of Coppersmith [8] for multivariate modular polynomial equations. This heuristic has proven to be very useful in many attacks (see [1,3,10,14]). We made various experiments for our approaches and the methods never failed to reveal the desired factorization of N. Therefore, we make the following assumption which refers to the only heuristic part in our computations of Section 4 and 6.

Assumption 5 *The resultant computations for the multivariate polynomials constructed in our approaches yield non-zero polynomials.*

4   MSBs Known: A Method for $e \in [N^{\frac{1}{2}}, N^{0.725}]$

In this section, we present an attack on RSA for public exponents e in the interval $[N^{\frac{1}{2}}, N^{\frac{\sqrt{6}-1}{2}}]$ given most significant bits of d. This answers an open question of Boneh, Durfee and Frankel [4] whether there are partial key exposure attacks in the case of known MSBs beyond the bound $e = \sqrt{N}$. Our approach makes use of Coppersmith's method for modular polynomial equations in the trivariate case.

Theorem 6 *Under Assumption 5, for every $\epsilon > 0$ there exists an integer N_0 such that for every $N > N_0$ the following holds:*
Let (N, e) be an RSA public key, where $\alpha = \log_N(e)$ is in the range $[\frac{1}{2}, \frac{\sqrt{6}-1}{2}]$. Given an approximation $\tilde{d}$ of d with

$$|d - \tilde{d}| \leq N^{\frac{1}{8}\left(5 - 2\alpha - \sqrt{36\alpha^2 + 12\alpha - 15}\right) - \epsilon}.$$

Then N can be factored in time polynomial in $\log N$.

Before we start to prove Theorem 6, in Figure 5 we provide some experimental results to give an idea of the amount of bits that is needed in our partial key exposure attack. The experiments also confirm the reliability of the multivariate heuristic and support our Assumption 5.

Define $\delta = \frac{1}{8}\left(5 - 2\alpha - \sqrt{36\alpha^2 + 12\alpha - 15}\right) - \epsilon$. Then a fraction of $1 - \delta$ of the MSBs of d is required (asymptotically) for the new attack. For $\alpha = 0.55$ this is a 0.710-fraction and for $\alpha = 0.6$ we require a 0.809-fraction. Note that these theoretical bounds hold as N and the lattice dimension go to infinity. All of our experiments were carried out on a 500-MHz workstation using Shoup's NTL [19].

N	e	known MSBs	Lattice parameters	L^3-time
1000 bit	600 bit	955 bit	$m = t = 1,\ \dim(L) = 7$	1 sec
1000 bit	550 bit	855 bit	$m = t = 1,\ \dim(L) = 7$	1 sec
1000 bit	600 bit	905 bit	$m = t = 2,\ \dim(L) = 19$	40 sec
1000 bit	550 bit	810 bit	$m = t = 2,\ \dim(L) = 19$	40 sec
1000 bit	600 bit	880 bit	$m = t = 3,\ \dim(L) = 50$	57 min
1000 bit	550 bit	785 bit	$m = t = 3,\ \dim(L) = 50$	72 min

Fig. 5. Experimental results for known MSBs

Proof (Theorem 6). : We start by looking at the public key equation

$$ed - 1 = k\phi(N), \quad \text{where } k \in \mathbb{Z}. \tag{1}$$

Boneh, Durfee and Frankel [4] observed that a suitable fraction of the MSBs of d yields the parameter k. The main drawback of the methods presented in [4] is that they all require that k is known exactly. This restricts the methods' usability to public exponents $e \leq \sqrt{N}$.

Now let us relax this restriction and look at the case where one obtains only an approximation $\tilde{k}$ of k. Let $\tilde{k} = \frac{e\tilde{d}-1}{N+1}$, then

$$|k - \tilde{k}| = \left| \frac{ed - 1}{\phi(N)} - \frac{e\tilde{d} - 1}{N + 1} \right|$$

$$= \left| \frac{(ed - 1)(N + 1) - (e\tilde{d} - 1)(N + 1 - (p + q))}{\phi(N)(N + 1)} \right|$$

$$\leq \left| \frac{e(d - \tilde{d})}{\phi(N)} \right| + \left| \frac{(p + q)(e\tilde{d} - 1)}{\phi(N)(N + 1)} \right| \leq \frac{e}{\phi(N)}(N^\delta + 3N^{-\frac{1}{2}}\tilde{d})$$

We claim that the hard case is the one where the term $N^{-\frac{1}{2}}\tilde{d}$ dominates N^δ. Let us first assume the opposite, i.e. $N^\delta > N^{-\frac{1}{2}}\tilde{d}$. In this case, $|k - \tilde{k}|$ can be bounded by $N^{\alpha+\delta-1}$, where we neglect low order terms. Hence whenever $\alpha + \delta - 1 \le 0$, then k can be determined exactly. Note that the condition in Theorem 6 implies the desired inequality $\delta \le 1 - \alpha$.

But if k is known, we can compute $p+q = N+1-k^{-1}$ mod e. On the other hand $e \ge N^{\frac{1}{2}}$ and therefore we get $p + q$ over the integers and not modulo e. This leads to the factorization of N.

Hence, we assume in the following that $N^{-\frac{1}{2}}\tilde{d} \ge N^\delta$. In this case, we can bound $|k - \tilde{k}|$ by $4N^{\alpha-\frac{1}{2}}$.

Now, let us define $d_0 = d - \tilde{d}$ and $k_0 = k - \tilde{k}$. Then, we can reformulate equation (1) as

$$e(\tilde{d} + d_0) - 1 = (\tilde{k} + k_0)\phi(N).$$

This can also be written as

$$ed_0 + (\tilde{k} + k_0)(p+q-1) + e\tilde{d} - 1 = (\tilde{k} + k_0)N. \tag{2}$$

Equation (2) gives us a trivariate polynomial

$$f_N(x,y,z) = ex + (\tilde{k} + y)z + e\tilde{d} - 1$$

with the root $(x_0, y_0, z_0) = (d_0, k_0, p+q-1)$ modulo N. Define the upper bounds $X = N^\delta$, $Y = 4N^{\alpha-\frac{1}{2}}$ and $Z = 3N^{\frac{1}{2}}$. Then, we have $x_0 \le X$, $y_0 \le Y$ and $z_0 \le Z$.

Now we use Coppersmith's method [8] in order to construct from $f_N(x,y,z)$ a polynomial $f(x,y,z)$ with the same root (x_0, y_0, z_0) over $\mathbb{Z}$ (and not just modulo N). The following theorem due to Howgrave-Graham [12] is a convenient reformulation of Coppersmith's method.

Theorem 7 (Howgrave-Graham) *Let $f(x,y,z)$ be a polynomial that is a sum of at most ω monomials. Suppose that*

(1) $f(x_0, y_0, z_0) = 0$ mod N^m, where $|x_0| \le X$, $|y_0| \le Y$ and $|z_0| \le Z$
(2) $\|f(xX, yY, zZ)\| < \frac{N^m}{\sqrt{\omega}}$.

Then $f(x_0, y_0, z_0) = 0$ holds over the integers.

Next, we construct polynomials that all satisfy condition *(1)* of Howgrave-Graham's Theorem. Thus, every integer linear combination of these polynomials also satisfies the first condition. We search among these linear combinations for a polynomial f that satisfies condition *(2)*. This will be done using the L^3-lattice reduction algorithm.

Let us start by defining the following polynomials $g_{i,j}(x,y,z)$ and $h_{i,j}(x,y,z)$ for some fixed integers m and t:

$$g_{i,j,k} = x^{j-k}z^k N^i f_N^{m-i} \quad \text{for } i = 0,\ldots,m; \ j = 0,\ldots,i; \ k = 0,\ldots,j$$
$$h_{i,j,k} = x^j y^k N^i f_N^{m-i} \quad \text{for } i = 0,\ldots,m; \ j = 0,\ldots,i; \ k = 1,\ldots,t$$

The parameter t has to be optimized as a function of m.

One can build a lattice $L(m)$ by using the coefficient vectors of the polynomials $g_{i,j,k}(xX, yY, zZ)$ and $h_{i,j,k}(xX, yY, zZ)$ as basis vectors for a basis $B(m)$ of $L(m)$. The following lemma shows, that the L^3-algorithm always finds at least three different vectors in $L(m)$ that satisfy condition (2) of Howgrave-Graham's Theorem. The proof makes use of Theorem 4.

Lemma 8 *Let* $X = N^\delta$, $Y = N^{\alpha-\frac{1}{2}}$ *and* $Z = N^{\frac{1}{2}}$. *Then one can find three linearly independent vectors in* $L(m)$ *with norm smaller than* $\dfrac{N^m}{\sqrt{\dim L(m)}}$ *using the* L^3*-algorithm.*

Proof: Let $n = \dim L(M)$ denote the lattice dimension. We want to find a reduced basis of $L(m)$ with three basis vectors smaller than $\frac{N^m}{\sqrt{n}}$. Applying Theorem 4, we know that for an L^3-reduced basis $\{v_1', v_2', \ldots, v_n'\}$

$$\|v_1'\| \leq \|v_2'\| \leq \|v_3'\| \leq 2^{\frac{n(n-1)+2}{4(n-2)}} \det L(M)^{\frac{1}{n-2}}.$$

Since we need $\|v_3'\| < \frac{N^m}{\sqrt{n}}$, we have to satisfy the condition

$$\det(L) < cN^{m(n-2)},$$

where $c = 2^{-\frac{n(n-1)+2}{4}} n^{-\frac{n-2}{2}}$ does not depend on N and therefore contributes to the error term ϵ.

Most of the following computations are straightforward but tedious. So we only sketch the rest of the proof. Let $t = \tau m$, then the determinant of $L(M)$ is

$$\det L(M) = \left(N^{8\tau+3} X^{4\tau+1} Y^{6\tau^2+4\tau+1} Z^{4\tau+2}\right)^{\frac{1}{24}m^4(1+o(1))}.$$

Using the bounds $X = N^\delta$, $Y = 4N^{\alpha-\frac{1}{2}}$ and $Z = 3N^{\frac{1}{2}}$ we obtain

$$\det L(M) = N^{\frac{1}{24}m^4(3\tau^2(2\alpha-1)+4\tau(\delta+\alpha+2)+\delta+\alpha+\frac{7}{2})(1+o(1))}.$$

An easy calculation shows that $n = \frac{1}{24}m^3(12\tau + 4)(1 + o(1))$. Neglecting low order terms, our condition simplifies to

$$3\tau^2(2\alpha - 1) + 4\tau(\delta + \alpha - 1) + \delta + \alpha - \frac{1}{2} < 0.$$

The left hand side is minimized for the choice $\tau = \frac{2}{3}\frac{1-\delta-\alpha}{2\alpha-1}$. Plugging this value in, we obtain the desired condition

$$\delta \leq \frac{1}{8}\left(5 - 2\alpha - \sqrt{36\alpha^2 + 12\alpha - 15}\right),$$

which concludes the proof. $\square$

Combining Theorem 7 and Lemma 8, from the three vectors with norm smaller than $\dfrac{N^m}{\sqrt{\dim L(m)}}$ we obtain three polynomials $f_1(x, y, z)$, $f_2(x, y, z)$ and $f_3(x, y, z)$

with the common root (x_0, y_0, z_0). Our goal is to extract the value $z_0 = p+q-1$. The equation $N = pq$ together with the number z_0 yields the factorization of N. Therefore, we take the resultants $\mathrm{res}_x(f_1, f_2)$ and $\mathrm{res}_x(f_1, f_3)$ with respect to x. The resulting polynomials g_1 and g_2 are bivariate polynomials in y and z. In order to remove the unknown y, we compute the resultant $\mathrm{res}_y(g_1, g_2)$ which is an univariate polynomial in z. The root z_0 most be among the roots of this polynomial. Thus, if $\mathrm{res}_y(g_1, g_2)$ is not the zero polynomial (Assumption 5) then z_0 can be found by standard root finding algorithms. This concludes the proof of Theorem 6.

5   LSBs Known: A Provable Method for $e < N^{\frac{1}{2}}$

In this section, we present a provable attack on RSA with public key $e < N^{\frac{1}{2}}$, where we know $d_0 = d \bmod M$ for some modulus M. For instance assume that an attacker succeeds to get the lower k bits of d, then $M = 2^k$.

In the following we show that whenever M is sufficiently large then N can be factored in polynomial time for all but a negligible fraction of choices for e.

Theorem 9 *Let N be an RSA-modulus and let $0 < \alpha, \epsilon < \frac{1}{2}$. For all but a $\mathcal{O}(\frac{1}{N^\epsilon})$-fraction of the public exponents e in the interval $[3, N^\alpha]$ the following holds: Let d be the secret key. Given d_0, M satisfying $d = d_0 \bmod M$ with*

$$N^{\alpha+\frac{1}{2}+\epsilon} \leq M \leq 2N^{\alpha+\frac{1}{2}+\epsilon}.$$

Then the factorization of N can be found in polynomial time.

Before we prove the theorem, we want to give some experimental results. We tested our algorithm with the commonly used public exponent $e = 2^{16} + 1$ and varying 1000-bit moduli N, where we knew 525 LSBs of d. Note that in comparison to the Boneh-Durfee-Frankel-approach for LSBs, we need about twice as many bits but in their method one has to run a lattice reduction about e times. The running time of our algorithm is about 1 second on a 500 MHz workstation. In 100 experiments, the algorithm never failed to yield the factorization of N.

Proof (Theorem 9). We start by looking at the RSA key equation $ed-1 = k\phi(N)$. Let us write $d = d_1 M + d_0$, where d_1 is the unknown part of d. Then

$$ed_1 M + k(p+q-1) - 1 + ed_0 = kN. \tag{3}$$

Equation (3) in turn gives us a bivariate polynomial

$$f_N(x, y) = eMx + y + ed_0$$

with a root $(x_0, y_0) = (d_1, k(p+q-1) - 1)$ modulo N. In order to bound y_0 notice that

$$k = \frac{ed-1}{\phi(N)} < e\frac{d}{\phi(N)} < e \leq N^\alpha.$$

Since $d_1 \leq \frac{N}{M}$, we can set the bounds $X = N^{\frac{1}{2}-\alpha-\epsilon}$ and $Y = 3N^{\frac{1}{2}+\alpha}$ satisfying $x_0 \leq X$ and $y_0 \leq Y$.

As in Section 4, we want to transform our polynomial $f_N(x, y)$ into a polynomial $f(x, y)$ with the root (x_0, y_0) over the integers. Therefore, we apply Howgrave-Graham's Theorem (Theorem 7) in the bivariate case. For this purpose we take the auxiliary polynomials N and Nx which are both the zero polynomial modulo N. Thus, every integer linear combination $f = a_0 N + a_1 N x + a_2 f_N(x, y)$ has the root (x_0, y_0) modulo N.

According to the second condition of Howgrave-Graham's Theorem we have to look for an integer linear combination f satisfying $\|f(xX, yY)\| \leq \frac{N}{\sqrt{3}}$. Thus, we search for a suitably small vector in the lattice L given by the span of the row vectors of the following (3×3)-lattice base

$$
B = \begin{bmatrix} N & & \\ & NX & \\ ed_0 & eMX & Y \end{bmatrix}.
$$

Now, our goal is to find two linearly independent vectors $(a_0, a_1, a_2)B$ and $(b_0, b_1, b_2)B$ both having norm smaller than $\frac{N}{\sqrt{3}}$. Since L has dimension 3, we can compute two shortest linearly independent vectors in L in polynomial time using an algorithm of Blömer [2]. In practice, the L^3-algorithm will suffice.

Assume we can find two linearly independent vectors with norm smaller than $\frac{N}{\sqrt{3}}$. Then we obtain from Theorem 7 the following two equations

$$
a_0 N + a_1 N x_0 + a_2 f_N(x_0, y_0) = 0 \quad \text{and}
$$
$$
b_0 N + b_1 N x_0 + b_2 f_N(x_0, y_0) = 0.
$$

From equation (3) we know that $f(x_0, y_0) = kN$. Hence, our equations simplify to the linear system

$$
\begin{aligned}
a_1 x_0 + a_2 k &= -a_0 \\
b_1 x_0 + b_2 k &= -b_0
\end{aligned}
\tag{4}
$$

If $(a_0, a_1, a_2), (b_0, b_1, b_2) \in \mathbb{Z}^3$ are linearly independent and satisfy (4), then the 2-dimensional vectors $(a_1, a_2), (b_1, b_2)$ are also linearly independent. But this implies that we can determine x_0, k as the unique solution of the linear system. Afterwards, we can derive y_0 by $y_0 = kN - eMx_0 - ed_0$. Therefore, $\frac{y_0+1}{k} = p+q-1$ gives us the necessary term to factor the modulus N.

It remains to show that L contains indeed two linearly independent vectors with norm smaller than $\frac{N}{\sqrt{3}}$. The following lemma proves that this is satisfied for most choices of e using a counting argument.

Lemma 10 *Given N, α, ϵ and M as defined in Theorem 9. Then for all but $\mathcal{O}(N^{\alpha-\epsilon})$ choices of e in the interval $[3, N^\alpha]$ the following holds: Let $X = N^{\frac{1}{2}-\alpha-\epsilon}$ and $Y = 3N^{\frac{1}{2}+\alpha}$. Then the lattice L contains two linearly independent vectors with norm less than $\frac{N}{\sqrt{3}}$.*

Proof: In terms of lattice theory, we have to show that for most of the choices of e the second successive minima λ_2 of L is strictly less than $\frac{N}{\sqrt{3}}$. By Minkowski's second theorem we know that for any 3-dimensional lattice L and its successive minima $\lambda_1, \lambda_2, \lambda_3$

$$\lambda_1 \lambda_2 \lambda_3 \leq 2 \det(L).$$

In our case $\det(L) = N^2 XY$. Hence for all e such that $\lambda_1 > 6XY$, we get $\lambda_2 < \frac{N}{\sqrt{3}}$ and we are done.

Now assume $\lambda_1 \leq 6XY$. Hence, we can find coefficients $c_0, c_1, c_2 \in \mathbb{Z}$ such that $\|(c_0, c_1, c_2)B\| < 6XY$. This implies

$$|c_2| \leq 6X$$

$$\left| \frac{c_1}{c_2} + \frac{eM}{N} \right| \leq \frac{6Y}{c_2 N}$$

Using $XY \leq 3N^{1-\epsilon}$, the second inequality implies

$$\left| \frac{c_1}{c_2} + \frac{eM}{N} \right| \leq \frac{18}{c_2 X N^\epsilon} \tag{5}$$

Next we bound the number of e's in $[3, N^\alpha]$ that can satisfy (5) for some ratio $\frac{c_1}{c_2}$.

Since $\frac{eM}{N}$ is positive, without loss of generality we can assume that $c_1 < 0$ and $c_2 > 0$. Now we make the following series of observations.

- The difference between any two numbers of the form $\frac{eM}{N}$ is at least $\frac{M}{N} \geq N^{\alpha - \frac{1}{2} + \epsilon}$.
- If (5) is true for some ratio $\frac{c_1}{c_2}$ and some e then $\frac{eM}{N}$ must lie in the interval $\left[\frac{c_1}{c_2} - \frac{18}{c_2 X N^\epsilon}, \frac{c_1}{c_2} + \frac{18}{c_2 X N^\epsilon} \right]$.
- Combining the first two observations we conclude that for a fixed ratio $\frac{c_1}{c_2}$ there are at most $\frac{36}{c_2 X N^{\alpha - \frac{1}{2} + 2\epsilon}}$ public keys e such that (5) is satisfied.
- Since $e \leq N^\alpha$ and $M \leq 2N^{\alpha + \frac{1}{2} + \epsilon}$, we get $\frac{eM}{N} \leq 2N^{2\alpha - \frac{1}{2} + \epsilon}$. Consider a fixed but arbitrary c_2. Then (5) is satisfied for some c_1 and some public key e only if $c_1 \in [-2N^{2\alpha - \frac{1}{2} + \epsilon} c_2, -1]$.
- The previous two observations imply that for fixed c_2 the number of e's satisfying (5) is bounded by $\frac{72 N^{\alpha - \epsilon}}{X}$.
- The previous observation and $c_2 \leq 6X$ imply, that the number of public keys e for which (5) is satisfied for some ratio $\frac{c_1}{c_2}$ is bounded by $432 N^{\alpha - \epsilon}$.

The last observation concludes the proof of Lemma 10. $\qquad \square$

6 LSBs Known: A Method for All e with $e < N^{\frac{7}{8}}$

In this section, we improve the approach of Section 5 by taking multi-dimensional lattices. In contrast to Section 5 our results are not rigorous. As in Section 4 they rely on Coppersmith's heuristic for multivariate modular equations. However, the results are even stronger: We obtain an attack for all $e < N^{\frac{7}{8}}$.

Theorem 11 *Under Assumption 5, for every $\epsilon > 0$ there exists N_0 such that for every $N \geq N_0$ the following holds:*
Let (N, e) be an RSA public key with $\alpha = \log_N(e) \leq \frac{7}{8}$. Let d be the secret key. Given d_0, M satisfying $d = d_0 \bmod M$ with

$$M \geq N^{\frac{1}{6}+\frac{1}{3}\sqrt{1+6\alpha}+\epsilon}.$$

Then N can be factored in polynomial time.

Before we start with the proof of Theorem 11, in Figure 6 we provide some experimental results to give an idea of the number of bits that are needed in our partial key exposure attack. We fixed a bit-size of 1000 for the modulus N and used varying sizes of 300, 400 and 500 bits for e. Theorem 11 states that we need to know at least 725, 782 and 834 LSBs of d, respectively.

N	e	**known LSBs**	Lattice parameters	L^3-time
1000 bit	300 bit	**805 bit**	$m = 1, t = 0,\ \dim(L) = 3$	1 sec
1000 bit	300 bit	**765 bit**	$m = 7, t = 1,\ \dim(L) = 44$	405 min
1000 bit	400 bit	**880 bit**	$m = 3, t = 1,\ \dim(L) = 14$	40 sec
1000 bit	400 bit	**840 bit**	$m = 6, t = 1,\ \dim(L) = 35$	196 min
1000 bit	500 bit	**920 bit**	$m = 4, t = 1,\ \dim(L) = 20$	7 min
1000 bit	500 bit	**890 bit**	$m = 8, t = 2,\ \dim(L) = 63$	50 hours

Fig. 6. Experimental results for known LSBs

Proof (Theorem 11). We start by looking at the equation $ed - 1 = k\phi(N)$. As in Section 5, we write $d = d_1 M + d_0$. This gives us the equation

$$k(N - (p + q - 1)) - ed_0 + 1 = eMd_1. \tag{6}$$

From (6) we obtain the bivariate polynomial

$$f_{eM}(y, z) = y(N - z) - ed_0 + 1$$

with the root $(y_0, z_0) = (k, p + q - 1)$ modulo eM. Analogous to Section 5 we can derive the bounds $Y = N^\alpha$ and $Z = 3N^{\frac{1}{2}}$ satisfying $y_0 \le Y$ and $z_0 \le Z$.

Fix some integers m and t. Define the polynomials

$$g_{i,j} = y^j (eM)^i f_{eM}^{m-i} \quad \text{for } i = 0, \ldots, m; \ j = 0, \ldots, i$$
$$h_{i,j} = z^j (eM)^i f_{eM}^{m-i} \quad \text{for } i = 0, \ldots, m; \ j = 1, \ldots, t.$$

The parameter t has to be optimized as a function of m.

Since all the polynomials have a term $(eM)^i f_{eM}^{m-i}$, all integer linear combinations of the polynomials have the root (y_0, z_0) modulo $(eM)^m$, i.e. they satisfy the first condition of Howgrave-Graham's theorem (in the bivariate case). Let $L(m)$ be the lattice defined by the basis $B(m)$, where the coefficient vectors of $g_{i,j}(yY, zZ)$ and $h_{i,j}(yY, zZ)$ are the basis vectors of $B(m)$ (with the same parameter choices of i and j as before).

In order to fulfill the second condition in Howgrave-Graham's theorem, we have to find vectors in $L(m)$ with norm less than $\frac{(eM)^m}{\sqrt{\dim L(m)}}$. The following lemma states that one can always find two such sufficiently short vectors in $L(m)$ using the L^3-algorithm.

Lemma 12 *Let e, M be as defined in Theorem 11. Suppose $Y = N^\alpha$ and $Z = 3N^{\frac{1}{2}}$. Then the L^3-algorithm finds at least two vectors in $L(M)$ with norm smaller than $\frac{(eM)^m}{\sqrt{\dim L(m)}}$.*

Proof. Since the proof is analogous to the proof of Lemma 8, we omit it.

Combining Theorem 7 and Lemma 12, we obtain two polynomials $f_1(y, z)$, $f_2(y, z)$ with the common root (y_0, z_0) over the integers. By Assumption 5, the resultant $\text{res}_y(f_1, f_2)$ is non-zero such that we can find $z_0 = p + q - 1$ using standard root finding algorithms. This gives us the factorization of N.

Acknowledgement. We want to thank Jean-Pierre Seifert for suggesting to look at partial key exposure attacks on CRT-variants of RSA.

References

1. D. Bleichenbacher, "On the Security of the KMOV public key cryptosystem", Advances in Cryptology – Crypto '97, Lecture Notes in Computer Science vol. 1294. Springer-Verlag, pp. 235–248, 1997
2. J. Blömer, "Closest vectors, successive minima, and dual HKZ-bases of lattices", Proc. of 17th ICALP, Lecture Notes in Computer Science 1853, pp. 248–259, 2000.
3. D. Boneh, G. Durfee, "Cryptanalysis of RSA with private key d less than $N^{0.292}$", IEEE Trans. on Information Theory vol. 46(4), 2000
4. D. Boneh, G. Durfee, Y. Frankel, "An attack on RSA given a small fraction of the private key bits", Advances in Cryptology - AsiaCrypt '98, Lecture Notes in Computer Science vol. 1514, Springer-Verlag, pp. 25–34, 1998

5. D. Boneh, G. Durfee, Y. Frankel, "Exposing an RSA Private Key Given a Small Fraction of its Bits", Full version of the work from Asiacrypt'98, available at `http://crypto.stanford.edu/~dabo/abstracts/bits_of_d.html`, 1998

6. D. Boneh, R. DeMillo, R. Lipton, "On the importance of checking cryptographic protocols for faults", Advances in Cryptology - Eurocrypt'97, Lecture Notes in Computer Science vol. 1233, Springer-Verlag, pp. 37–51, 1997.

7. H. Cohen, "A Course in Computational Algebraic Number Theory", Springer-Verlag, 1996

8. D. Coppersmith, "Small Solutions to Polynomial Equations, and Low Exponent RSA Vulnerabilities", Journal of Cryptology 10(4), 1997

9. J. F. Dhem, F. Koeune, P. A. Leroux, P. Mestre, J. J. Quisquater, and J. L. Willems, "A practical implementation of the timing attack", In Proc. of CARDIS 98 – Third smart card research and advanced application conference, 1998

10. G. Durfee, P. Nguyen, "Cryptanalysis of the RSA Schemes with Short Secret Exponent from Asiacrypt '99", Advances in Cryptology – Asiacrypt 2000, Lecture Notes in Computer Science vol. 1976, Springer, pp. 14–29, 2000

11. M. Gruber, C.G. Lekkerkerker, "Geometry of Numbers", North-Holland, 1987

12. N. Howgrave-Graham, "Finding small roots of univariate modular equations revisited", Proc. of Cryptography and Coding, Lecture Notes in Computer Science 1355, Springer-Verlag, 1997

13. N. Howgrave-Graham, "Approximate Integer Common Divisors", CaLC 2001, Lecture Notes in Computer Science vol. 2146, pp. 51–66, 2001

14. C. Jutla, "On finding small solutions of modular multivariate polynomial equations", Advances in Cryptology - Eurocrypt '98, Lecture Notes in Computer Science vol. 1403, pp. 158–170, 1998

15. P. Kocher, "Timing attacks on implementations of Diffie-Hellman, RSA, DSS and other systems", Advances in Cryptology - Crypto '96, Lecture Notes in Computer Science vol. 1109, pp. 104–113, 1996

16. P. Kocher, J. Jaffe and B. Jun, "Differential power analysis", Advances in Cryptology – CRYPTO '99, Lecture Notes in Computer Science vol. 1666, pp. 388–397, 1999

17. A. Lenstra, H. Lenstra and L. Lovász, "Factoring polynomials with rational coefficients", Mathematische Annalen, 1982

18. L. Lovász, "An Algorithmic Theory of Numbers, Graphs and Convexity", Conference Series in Applied Mathematics, SIAM, 1986

19. V. Shoup, NTL: A Library for doing Number Theory, online available at http://www.shoup.net/ntl/index.html

20. T. Takagi, "Fast RSA-Type Cryptosystem Modulo $p^k q$", Advances in Cryptology – Crypto '98, Lecture Notes in Computer Science vol. 1462, pp. 318–326, 1998

21. J.-J. Quisquater, C. Couvreur, "Fast decipherment algorithm for RSA public-key cryptosystem", Electronic Letters 18, pp. 905–907, 1982

Algebraic Cryptanalysis of Hidden Field Equation (HFE) Cryptosystems Using Gröbner Bases

Jean-Charles Faugère[1] and Antoine Joux[2]

[1] Projet SPACES LIP6/LORIA CNRS/UPMC/INRIA
[2] DCSSI/Crypto Lab, 51 Bd de Latour-Maubourg, 75700 PARIS 07 SP

Abstract. In this paper, we review and explain the existing algebraic cryptanalysis of multivariate cryptosystems from the hidden field equation (HFE) family. These cryptanalysis break cryptosystems in the HFE family by solving multivariate systems of equations. In this paper we present a new and efficient attack of this cryptosystem based on fast algorithms for computing Gröbner basis. In particular it was was possible to break the first HFE challenge (80 bits) in only two days of CPU time by using the new algorithm F5 implemented in C.

From a theoretical point of view we study the algebraic properties of the equations produced by instance of the HFE cryptosystems and show why they yield systems of equations easier to solve than random systems of quadratic equations of the same sizes. Moreover we are able to bound the maximal degree occuring in the Gröbner basis computation.

As a consequence, we gain a deeper understanding of the algebraic cryptanalysis against these cryptosystems. We use this understanding to devise a specific algorithm based on sparse linear algebra. In general, we conclude that the cryptanalysis of HFE can be performed in polynomial time. We also revisit the security estimates for existing schemes in the HFE family.

1 Introduction

The idea of using multivariate quadratic equations as a basis for building public key cryptosystems appeared with the Matsumoto-Imai cryptosystem in [18]. This system was broken by Patarin in [20]. Shortly after, Patarin proposed to repair it and thus devised the hidden field equation (HFE) cryptosystem in [22]. While the basic idea of HFE is simple, several variations are proposed in [22]. As of today, the security of HFE, especially with variations is not well understood. Some systems have been attacked, others seem to resist. However, it is hard to draw the boundary between secure and insecure schemes in the HFE family.

Among the known attacks, we can roughly distinguish two classes. The first class consists of specific attacks which focus on one particular variation and breaks it due to specific properties. Typical examples are the attack of Kipnis and Shamir against Oil and Vinegar [14] and the attack by Gilbert and Minier [13] against the first version of the NESSIE proposal Sflash. The second class of

D. Boneh (Ed.): CRYPTO 2003, LNCS 2729, pp. 44–60, 2003.

attack consists of general purpose algorithms that solve multivariate system of equations. In this class, we find the relinearization technique of [15], the XL algorithm [8] and also the Gröbner basis [1,2,3] approach.

The relinearization technique of Kipnis and Shamir [15] is well suited to the basic HFE schemes without variations. The first step is to remark that some part of the secret key is a solution of some overdefined quadratic system of equations, with ϵm^2 equations involving m variables. The relinearization technique can even solve random systems as long as they are "sufficiently" overdefined. In fact, the success of the relinearization cryptanalysis against HFE is evaluated by estimating its complexity and success rate against random systems of the same size.

Gröbner basis is a well established and general method for solving polynomial system of equations. Of course the efficiency of this attack depends strongly on the algorithm used for computing the Gröbner basis: the first HFE challenge broken in section 3.4 is completely out of reach of the Buchberger algorithm (the historical algorithm for computing Gröbner bases). With this second class of attacks, the main difficulty is, given a set of parameters from an HFE cryptosystem, to determine the attacks' complexity.

In this paper, we show that the weakness of the systems of equations coming from HFE instances can be *explained* by the algebraic properties of the secret key. From this study we are able to predict the maximal degree occuring in the Gröbner basis computation. This accounts well for the experimental results: we made a series of computer simulations on real size HFE problems (up to 160 bits) so that we can establish precisely the complexity of the Gröbner attack and we compare with the theoretical bounds.

We also devise a specific algorithm based on efficient sparse linear algebra algorithms over $\mathbb{F}_2$ such as Block Lanczos. Finally, in section 5.4 and 5.1 we give concrete security estimates for existing cryptosystems in the HFE family.

2   General Description of Multivariate Cryptosystems

The basic idea of multivariate cryptosystems from the HFE family is to build the secret key on a polynomial S in one unknown x over some finite field (often $\mathbb{F}_{2^n}$). Clearly, such a polynomial can be easily evaluated, moreover, under reasonable hypothesis, it can also be "inverted" quite efficiently. By inverting, we mean finding any solution to the equation $S(x) = y$, when such a solution exists. The secret transformations (decryption and/or signature) are based on this efficient inversion. Of course, in order to build a cryptosystem, the polynomial S must be presented as a public transformation which hides the original structure and prevents inversion. This is done by viewing the finite field $\mathbb{F}_{2^n}$ as a vector space over $\mathbb{F}_2$ and by choosing two linear transformation of this vector space L_1 and L_2. Then the public transformation is $L_2 \circ S \circ L_1$.

By enforcing a simple constraint on the choice of S, this transformation can be be expressed as a set of quadratic polynomial in n unknowns over $\mathbb{F}_2$. The constraint on S is that all the monomials occuring in S should be either constant, of degree 2^t (for some t) or of degree $2^{t_1} + 2^{t_2}$. Indeed, if $x_1, \ldots, x_n$ are the coordinates of x in some basis of $\mathbb{F}_{2^n}$, we can express each coordinates of S over $\mathbb{F}_2$ as a polynomial in $x_1, \ldots, x_n$. In this expression into multivariate polynomials, the constant monomial from S is translated into constants, the monomials of degree 2^t are translated into linear terms and the monomials of degree $2^{t_1} + 2^{t_2}$ into quadratic terms. This is due to the simple fact that the Frobenius map on $\mathbb{F}_{2^n}$, i.e., squaring, is a linear transform over $\mathbb{F}_2$ of the vector space $\mathbb{F}_{2^n}$.

From this basic idea, we obtain a trapdoor one-way function. Indeed, the public transformation expressed in multivariate form as:

$$y = (p_1(x_1, \cdots, x_n), \cdots, p_n(x_1, \cdots, x_n)),$$

is easy to compute and seems hard to invert, while the secret expression in univariate form is easy to invert when S is correctly chosen. In cryptosystems based on HFE, S is usually a low degree polynomial, which can be inverted by using a general purpose root finding algorithm over $\mathbb{F}_{2^n}$. In some systems, such as Sflash, we encounter an alternative construction, quite similar to RSA, where S consists of a single monomial of degree $2^t + 1$, for some t. In that case, to find a $2^t + 1$-th root of a finite field element y, we proceed as for RSA decryption, i.e., we raise y to the power d, where d is the inverse of $2^t + 1$ modulo $2^n - 1$. However, in that case, the basic system reduces to the Matsumoto-Imai cryptosystem [18] which is not secure (see [20]) and thus, it should be strengthened by removing some public equations. This strengthening operation which can also be applied when S is a low degree polynomial is discussed in section 4.1. In the sequel, we focus on the case where the degree D of S is small. In this case, the performance of the cryptosystem is directly related to that of the root solving algorithm that inverts S in $\mathbb{F}_{2^n}$. To illustrate this performance, the following table reports the time needed to find one such solution using NTL[25] (PC PIII 1 Ghz):

(n, D)	(80,129)	(80,257)	(80,513)	(128,129)	(128,257)	(128,513)
NTL (CPU time)	0.6 sec	2.5 sec	6.4 sec	1.25 sec	3.1 sec	9.05 sec

From these data, we conclude that for the sake of speed, the degree D of the secret polynomial must remain quite small, say at most 512. For example, the recommended values for the first HFE challenge are $n = 80$ and $D = 96$. For this reason, throughout the rest of the paper, we focus on the study on HFE cryptosystems under the assumption that D remains fixed as n grows.

3   Gröbner Basis Cryptanalysis

We refer to [5,9] for basic definitions. Let k be a field and $R = k[x_1, \ldots, x_n]$ the ring of multivariate polynomials. To a system of equations

$$f_1(x_1, \ldots, x_n) = \cdots = f_m(x_1, \ldots, x_n) = 0$$

we associate the ideal I generated by $f_1, \ldots, f_m$. A Gröbner basis is a generating system of I that behaves nicely with respect to some order on monomials. We recall that a monomial in $x_1, \ldots, x_n$ is a term in $x_1, \ldots, x_n$ together a coefficient. Let $<$ denotes, an admissible ordering on the monomials in $x_1, \ldots, x_n$. A frequentely encountered example is the lexicographical ordering which is defined as $x_1^{\alpha_1} \cdots x_n^{\alpha_n} < x_1^{\beta_1} \cdots x_n^{\beta_n}$ iff $\alpha_i = \beta_i$ for $i = 1, \ldots, k$ and $\alpha_k < \beta_k$ for some k. Then for each polynomial f in R we can define its leading term $LT(f)$ (resp. its leading monomial $LM(f)$) to be the biggest term (resp. monomial) with respect to $<$.

Definition 1. *G a finite set of elements of I is a Gröbner basis of $(f_1, \ldots, f_m)$ wrt $<$ if for all $f \in I$ there exists $g \in G$ such that $LT(g)$ divides $LT(f)$.*

Let K be a field containing k, we can define the set of solutions in K which is the algebraic variety:

$$V_K = \{(z_1, \ldots, z_n) \in K \quad |f_i(z_1, \ldots, z_n) = 0 \; i = 1, \ldots, m\}$$

which is in fact the set of roots of the system of equations. Gröbner bases can be used in various situation (for instance when the number of solution is infinite of for computing real solutions). In the case of HFE we want to compute solutions of algebraic systems in $\mathbb{F}_2$. The following proposition tell us how to use Gröbner bases in order to *solve* a system over $\mathbb{F}_2$:

Proposition 1. *The Gröbner basis of $[f_1, \ldots, f_m, x_1^2 - x_1, \ldots, x_n^2 - x_n]$, in $\mathbb{F}_2[x_1, \ldots, x_n]$, describe all the solutions of $V_{\mathbb{F}_2}$. Particular useful cases are:*

i) $V_{\mathbb{F}_2} = \emptyset$ (no solution) iff $G = [1]$.
2) $V_{\mathbb{F}_2}$ has exactly one solution iff $G = [x_1 - a_1, \ldots, x_n - a_n]$ where $a_i \in \mathbb{F}_2$. Then $(a_1, \ldots, a_n)$ is the solution in $\mathbb{F}_2$ of the algebaric system.

This proposition tell us that we have to add the "field equations" $x_i^2 = x_i$ to the list of equations that we want to solve. Consequently we have to compute a Gröbner basis of $m + n$ polynomials and n variables. In fact, the more equations you have the more able you are to compute a Gröbner basis.

3.1   Useful Properties of Gröbner Bases

Another order on monomials is the Degree Reverse lexicographical order or (**DRL** order). This order is less intuitive than the lexicographical order but it has been shown that the DRL ordering is the most efficient, in general, for computating Gröbner bases.

$x_1^{\alpha_1} \cdots x_n^{\alpha_n} >_{\text{DRL}} x_1^{\beta_1} \cdots x_n^{\beta_n}$ iff $deg(x^\alpha) = \sum_{i=1}^n \alpha_i > deg(x^\beta)$ or $deg(x^\alpha) > deg(x^\beta)$ and, in $\alpha - \beta \in \mathbb{Z}^n$, the right-most nonzero entry is negative.

We have seen in proposition 1 that Gröbner bases are useful to solve a system but they can also be used to discover low degree relations:

Proposition 2. *If G is a Gröbner basis of an ideal I for $<_{\text{DRL}}$ then G contains all the (independent) equations in I of lowest total degree.*

By computing Gröbner bases it is even possible to find *all* the algebraic relations among $f_1, \ldots f_m$ (see [9] page 338 for a precise definition of the ideal of relations).

Proposition 3. *([9] page 340)*
Fix a monomial order in $k[x_1, \ldots, x_n, y_1, \ldots, y_m]$ where any monomial involving one of the $x_1, \ldots, x_n$ is greater than all monomials in $k[y_1, \ldots, y_m]$ (lexicographical ordering for instance) and let G be the Gröbner basis for this ordering. Then $G \cap k[y_1, \ldots, y_m]$ describe all the relations among $f_1, \ldots, f_m$.

By combining proposition 2 and 3 we can find the lowest degree relations among the f_i. This will enable us to give predict the computation of a Gröbner basis in the multivariate world from another (easy) Gröbner basis computation in the univariate world (see section 5.1).

3.2   Algorithms for Computing Gröbner Bases

Note that definition 1 does *not depend on a particular algorithm*. In the section, we quickly review existing algorithms and especially the recent improvements in the field. Historically the first algorithm for computing Gröbner basis was presented by Buchberger [1,2,3]. The Buchberger algorithm is a very practical algorithm and it is implemented in all Computer Algebra Systems (a non exhaustive list of efficient implementations is: Magma, Cocoa, Singular, Macaulay, Gb, ...); section 3.4 contains a comparison between them for the HFE problem. More recently, more efficient algorithms for computing Gröbner bases have been proposed. The first one F_4 [10] reduces the computation to a (sparse) linear algebra problem (from a theoretical point of view the link between solving algebraic systems and linear algebra is very old, e.g., see [17,16]). More precisely the algorithm F_4 incrementally construct matrices in degree 2, 3, ... D:

$$A_D = \begin{array}{c} m_1 \times f_{i_1} \\ m_2 \times f_{i_2} \\ m_3 \times f_{i_3} \\ \ldots \end{array} \overset{\textit{momoms degree} \ \leq \ D \ \textit{in} \ x_1, \ldots, x_n}{\left(\begin{array}{cccc} & & \ldots & \\ & & \ldots & \\ & & \ldots & \\ & & \ldots & \end{array} \right)},$$

where $m_1, m_2, \ldots$ are monomials such that the total degree of $m_j f_{i_j}$ is less than D. The next step in the algorithm is to compute a row echelon of A_D using linear algebra techniques. It must be emphasized that:

- The rows of A_D is a *small subset* of all the possible rows $\{mf_i \text{ s.t. } m \text{ any monomial } deg(m) \leq D - deg(f_i)\}$.
- The rows $m_j f_{i_j}$ in the matrix are not necessarily obtained from the initial set of equations $f_1, \ldots, f_m$. Thus we can have $i_j > m$, in that case f_{i_j} was produced in a previous step of the algorithm (in degree $D' < D$).

The main drawback of this algorithm is that very often the matrix A_D does not have full rank. A new algorithm called F_5 that avoids this drawback have been proposed in [12]. This new algorithm replaces A_D by a full rank matrix, which is thus minimal (if the input system is regular see [12]). For the special case of $\mathbb{F}_2$ we use a dedicated version of this algorithm (called $F_5/2$) that also takes into account the action of the Frobenius on the solutions. Of course, the implementation of the linear algebra is also dedicated to $\mathbb{F}_2$. In the current implementation of this algorithm $F_5/2$ in the program FGb (written in C) we use a naive implementation of Gauss algorithm. The resulting algorithm/implementation is much more efficient than any other program for solving algebraic equations. We expect that by using sophisticated techniques described in section 5.2 the efficiency of such algorithm could be significantly improved.

From a complexity point of view the important parameters are: D the maximal degree occurring in the computation and N_D, the size of the matrix A_D. The overall complexity is N_D^ω where $2 \leq \omega \leq 3$ is the exponent of the linear algebra algorithm.

3.3 Complexity of Gröbner Bases

The complexity of Gröbner bases algorithms has been studied in a huge number of papers. Over $\mathbb{F}_2$, when the "field equations" $x_i^2 - x_i$ are added, the set of solutions have a simple geometry. All the ideals are radicals (no multiple roots), zero dimensional (finite number of solutions). In fact it is easy to prove:

Proposition 4. *The maximal degree D of the polynomials occurring in the computation of a Gröbner basis including field equations $x_i^2 = x_i$ is less than n. The complexity of the whole computation is bounded by a polynomial in 2^n.*

Remark. Note that this result is only a rough upper bound. This must be compared with the complexity of the exhaustive search $\mathcal{O}(n2^n)$. In practice, however, efficient algorithms for computing Gröbner bases behave much better than in the worst case.

A crucial point in the cryptanalysis of HFE is the ability to distinguish a "random" (or generic) algebraic system from an algebraic system coming from HFE. We will establish in section 3 that this can be done by computing Gröbner bases and comparing the maximal degree occurring in these computations. As a consequence we have to describe theoretically the behavior of such a computation. This study is beyond the scope of this paper and is the subject of another paper [4] from which we extract the asymptotic behavior of the maximal degree occurring in the computation is: $d = \max$ total degree $\approx \frac{n}{11.114...}$

From this result we know that computing Gröbner bases of random systems is simply exponential; consequently, in practice, it is impossible to solve a system of n equations of degree 2 in n variables when n is big (say $n \geq 80$).

3.4 First HFE Challenge Is Broken

The first HFE Challenge was proposed in [23] with a (symbolic) prize of 500\$. This correspond to a HFE($d = 96, n = 80$) problem and can be downloaded from

[21]. For this problem, the exhaustive search attack require $\geq 2^{80}$ operations, hence is not feasible.

We have computed a Gröbner basis of this system with the algorithm $F_5/2$. As explained in section 3.2, the most time consuming operation is the linear algebra: for this example we have solved a 307126×1667009 matrix over $\mathbb{F}_2$. The total running time was 187892 sec ($\approx$ 2 days and 4 hours) on an HP workstation with an alpha EV68 processor at 1000 Mhz and 4Go bytes of RAM. Some care had to be taken for the memory management since the size of the process was 7.65 Giga bytes.

For this algebraic system [21], we found that there were is fact *four solutions*. Encoded as numbers by $X = \sum_{i=1}^{80} x_i 2^{i-1}$ the solutions are 64431800523905114-0554718, 93434489004594109861521 4, 1022677713629028761203046 and 103704608265180114959 4670. It must be emphasized that this computation is far beyond reach of all the other implementations and algorithms for computing Gröbner basis as is made clear by the table 1. Because 80 equations of degree 2 was a previously untractable problem, this Gröbner basis computation represents a breakthrough in research on polynomial system solving.

Table 1. Comparison of various algorithms and implementations (PC PIII 1 Ghz)

Algorithm	Buchberger				F_4	F_5
System	Maple	Magma	Macaulay	Singular	FGb	FGb
CPU time < **10m**	12	17	18	19	22	**35**
CPU time < **2h**	14	19	20	21	28	**45**

4   Explanation of the Algebraic Cryptanalysis

In the cryptosystems described in section 2, the basic problem is to find solutions of equations of the form $y = L_2 \circ S \circ L_1(x)$, hidden in multivariate expressions. In section 3, we saw that general purpose Gröbner basis can be used to solve this problem. However, in general, computing such a Gröbner basis of a system of equations is quite difficult. We saw in section 3.4 that instances of HFE can be solved much faster than random systems of equations of the same size (in fact the complexity is only polynomial).

In order to explain this qualitative difference, we first need to show that the structure of the equations in an instance of HFE implies a relatively small upper bound on the degree m of the polynomials which occur during the Gröbner basis computation. In fact, we show in the sequel, that this bound depends on the degree of the secret polynomial S but not on the size of the field $\mathbb{F}_{2^n}$. On the other hand, with random systems, the degree of the intermediate polynomials strongly depends on n. Computing an exact bound on m is difficult, since the detailed behavior of general purpose Gröbner basis algorithms is quite complicated. Thus, we derive our bound by analyzing the basic linear algebra approach described in section 3. With this algorithm, we search for linear polynomials in the variables x_i in the space of linear combination of multiples of the public polynomials p_i. When the degree of the intermediate polynomials is m, the degree of the multipliers is $m - 2$. First, remark that a solution of $y = L_2 \circ S \circ L_1(x)$ is also a solution of $L_2^{-1}(y) = S \circ L_1(x)$ and that the degree of the multiples that need to be considered is thus independent of the linear transformations L_1 and L_2. As a consequence, it suffices to prove our statement for the equations directly derived from the secret equation $S - L_2^{-1}(y) = 0$. This can be done by working in the single variable description of S. Without loss of generality and to simplify the exposition, we assume the constant term is incorporated into S and look for a solution of $S = 0$. In that case, the multivariate equations associated to S can be expressed in term of S and its Frobenius transforms S^2, S^4, ... Moreover, low degree multiples of the multivariate equations can be obtained by multiplying S and its Frobenius transforms by monomials of the form x^d, where d is a sum of a small number of powers of two. More precisely, multiplying the multivariate equations by a monomial of degree t can be performed by multiplying S and its transforms by a polynomial whose monomials are of the form x^d, where the value of d is a sum of t (different) powers of two. Thus, it suffices to consider expressions of the form:

$$\sum_i \sum_j c_{i,j} x^j S^i, \tag{1}$$

where i is a power of two, j a sum of at most m powers of two, and $c_{i,j}$ are arbitrary elements in the field $\mathbb{F}_{2^n}$. Such an expression is linear as a multivariate polynomial, if and only if, it can be written as a polynomial in x whose monomials are either constant or x^{2^t} for some value of t.

   With these remarks in mind, the original problem can be reformulated as follows: Find a linear combination of polynomials $x^j S^i$ (i power of two and j sum of at most $m-2$ powers of two), where all monomials x^d, where d is the sum of two or more powers of two, cancel out. Whenever such a combination exists it can, of course, be found through linear algebra. When, as in multivariate cryptosystems, the degree of S is bounded by D, we limit the degree[1] of the polynomials $x^j S^i$ we are considering by some bound H. Then, we count these polynomials, and denotes their number by N_P. We also count the number N_M of different monomials x^d, where d is a sum of at least two or more powers of two, appearing in all these polynomials. When N_P is larger than N_M, we have more polynomials to combine than terms to cancel out. Thus, unless some degeneracy occurs, the linear combination we want must exists. Moreover, the condition of degeneracy can be expressed algebraically by writing down some minor determinants that must vanish when a degeneracy is encountered. Thus,

---

[1] When the degree of $x^j S^i$ is a power of two, the higher degree monomial correspond to a linear multivariate polynomial, as a consequence it can be ignored. In that case, we can slightly improve our bounds by replacing the degree of $x^j S^i$, by the degree of its second monomial. The datum provided in table 2 accounts for this fact.

unless the (randomly chosen) coefficients of S satisfy these algebraic equations, we can guarantee the success of the attack. Clearly, the proportion of bad secret polynomials is exponentially vanishing in n. As a consequence, given D and m, it suffices to evaluate the numbers N_P and N_M for various values of H to determine whether Gröbner based cryptanalysis can attack an instance of HFE while considering intermediate polynomials of degree no higher than m. We do not explain any further how to solve this combinatorial problem. However, we now state some results and give some values N_P and N_M for many proposed parameters of HFE systems. It should be recalled that this results are upper bounds and than in practice, the attack may succeed even better.

The degree of the secret polynomial S is at least 3, otherwise, the public equations are entirely linear. For S of degree 3 and 4, it suffices to take $m = 3$, i.e., to multiply the public polynomial by the x_i variables and search for a linear combination. For degrees from 5 to 15, it suffices to take $m = 4$. From 16 to 127, $m = 5$ suffices. From 128 to 1280, $m = 6$ is enough. Finally, $m = 7$ works from degree 1281 up to degree 20480. These bounds really match the experimental complexity of the simple linear algebra approach except for the boundary cases of degree 16 and 128. On the other hand, more sophisticated Gröbner basis algorithms work even better.

The detailed results giving the relation between the degree of S and m, up to degree $D = 4096$, are shown in table 2. Note that some values for the degree of S are missing from this table, this is due to the simple fact that the degree of S is either a power of two or a sum of two such powers. Table 2 also gives the maximal number Max_{Eq} of independent equations that may be obtained by pushing the value of H even further.

4.1   Extension of the Attack to HFE Variations

In order to increase the security of HFE, several variations can be introduced. In particular, four simple variations are proposed in [22]. These variations are:

1. Remove some (multivariate) equations from the public key. Usually denoted by HFE-.
2. Add (useless) random equations to the public key. Usually denoted by HFE+.
3. Add extra variables with values in $\mathbb{F}_2$, usually denoted by HFEv. More precisely, the secret polynomial S is now a function of these extra parameters. However, the multivariate expression should remain quadratic. This is ensured by replacing the coefficient of "linear" monomials of the form x^{2^t} by an affine polynomial in the extra variables, similarly the constant term is replaced by a degree 2 polynomial in the extra variables.
4. Remove some variables by fixing their values. Usually denoted by HFEf.

All these variations can be combined in cryptosystems. The only restrictions are that variations 1 and 3 (HFE- and HFEv) are more suited to signature schemes and that variations 2 and 4 (HFE+ and HFEf) are more suited to encryption schemes. For example, when applying variation 1 to a signature scheme,

Table 2. Upper bound on the efficiency of Gröbner cryptanalysis

Degree D	m	H	N_P	N_M	Max_{Eq}	Degree D	m	H	N_P	N_M	Max_{Eq}
3	3	3	2	1	2	4	3	6	4	3	1
5	4	10	7	6	16	6	4	14	12	10	14
8	4	20	16	15	12	9	4	27	23	22	10
10	4	30	26	25	9	12	4	57	49	48	7
16	5	44	39	38	248	17	5	51	46	45	230
18	5	54	49	48	219	20	5	60	55	54	210
24	5	76	67	66	200	32	5	120	103	102	162
33	5	142	127	126	153	34	5	150	135	134	149
36	5	169	154	152	142	40	5	192	171	169	132
48	5	242	209	208	119	64	5	548	403	402	88
65	5	601	452	451	80	66	5	650	483	481	79
68	5	664	491	490	76	72	5	724	528	527	71
80	5	936	620	619	64	96	5	1856	1000	998	52
128	6	576	514	513	3763	129	6	584	527	526	3701
130	6	592	536	535	3671	132	6	616	558	557	3633
136	6	659	596	595	3550	144	6	689	620	619	3431
160	6	800	703	701	3285	192	6	977	829	828	3163
256	6	1665	1285	1284	2516	257	6	2060	1484	1482	2486
258	6	2070	1497	1495	2467	260	6	2090	1517	1515	2447
264	6	2147	1572	1571	2417	272	6	2320	1709	1708	2363
288	6	2470	1825	1824	2262	320	6	2828	2032	2031	2143
384	6	3633	2407	2406	1988	512	6	6528	3645	3643	1444
513	6	8337	4212	4211	1420	514	6	8408	4273	4272	1410
516	6	8784	4537	4536	1402	520	6	8896	4606	4605	1388
528	6	9234	4734	4733	1366	544	6	9520	4906	4904	1332
576	6	10378	5180	5179	1261	640	6	11792	5636	5633	1157
768	6	15616	6444	6443	1042	1024	6	74056	17376	17375	573
1025	6	82128	18487	18486	500	1026	6	82656	18689	18688	499
1028	6	83328	18858	18857	495	1032	6	84560	19069	19068	489
1040	6	85520	19139	19138	483	1056	6	89120	19438	19437	462
1088	6	152960	26249	26248	404	1152	6	180480	28270	28268	312
1280	6	246144	31102	31101	138						
1536	7	12805	8698	8695	41733	2048	7	19072	11702	11699	38779
2049	7	20487	12285	12284	38709	2050	7	20537	12441	12439	38688
2052	7	20563	12486	12484	38672	2056	7	20588	12510	12509	38648
2064	7	20652	12567	12566	38615	2080	7	20784	12669	12668	38571
2112	7	21072	12859	12857	38435	2176	7	21766	13211	13210	37882
2304	7	23072	13711	13709	36761	2560	7	25880	14484	14483	32707
3072	7	32512	16370	16368	31803	4096	7	71952	29494	29492	28394

we can remove a large number of equations from the public key, it will not prevent signature verification. On the other hand, with an encryption scheme, we can only delete a small number of public equations. Indeed, before performing decryption, it is necessary to guess the values of all deleted equations.

In order to evaluate the impact of the variations on the security of HFE, we need to compute the new values of m (the maximal degree of multiplier monomials) when each variation is applied. The simplest cases are variations 2 and 4, which preserve m. Indeed, adding random polynomials as in variation 2 increases the number of columns in the matrix M but does not change the kernel we are looking for (since the number of lines in M is much larger than the number of columns). Similarly, fixing the value of some variables as in variation 4 collides some lines together, but leave the kernel unchanged.

The action of variation 1 (or HFE-) is more complicated. When a single equation is removed, the secret key view of the problem is changed. Instead of having an equation $S(x) = y$, we now have $S(x) = y_0$ or $S(x) = y_1$ depending on the value of the removed equation. These two possibility can be merged into the following equation (over $\mathbb{F}_{2^n}$):

$$(S + y_0)(S + y_1) = 0 \text{ or } S^2 + (y_0 + y_1)S + y_0 y_1 = 0.$$

Since this new equation is obtained by combining the Frobenius of S and S, it is again quadratic in the multivariate view. As a consequence, removing a single equation corresponds (at worse) to a doubling of the degree of S. Thus, the an upper bound on the value of m can once again be deduced from table 2. On the other hand, it should be noted that when some public equations are removed, the cryptanalysis can be somewhat improved. Indeed, if t equations are removed, then the system is underdetermined. Thus, we may guess the value of any set of t variables and with good probability still find a solution. Clearly, this slightly reduces the size of the linear algebra, since any line involving one of the t variables is either removed (when the variable value is 0) or merge with another line (corresponding to the monomial without this variable). With more sophisticated Gröbner basis algorithms the speed-up is quite noticeable.

The most complicated variation to deal with is variation 3 (or HFEv). In that case, in addition to the main variable x in $\mathbb{F}_{2^n}$, we have extra variables v_1, ..., v_s over $\mathbb{F}_2$. We assume here that s is small, which must be the case with encryption schemes and is usually true with concrete instances of signature scheme (typically, $s = 4$). In that case, any equation computed from S which is "linear" in the x_i may contains additional quadratic terms of the form $x_i v_j$ or $v_i v_j$. The problem for the cryptanalyst is that, after the outward linear transform, this simple structure is hidden. Luckily, there are in fact, many kernel equations coming from the low degree expression in x, and not a single one. With enough equations, we can carry the linear algebra step further on and remove the $x_i v_j$ and $v_i v_j$ terms. Indeed, we need to cancel the coefficients in v_1, ..., v_s before each "linear" term of the form x^{2^t}, we also need to cancel the coefficients of the $v_i v_j$. The total number of extra multivariate equations needed is $ns + s(s-1)/2$. Thus, for fixed s, it suffices to have $s+1$ linear equation in the univariate world to ensure that the attack again succeeds, without any change in the value of m.

5   Performance Optimizations

5.1   Optimal Degree with Gröbner Basis Algorithms

We have collected a lot of experimental data by running thousand of HFE systems for various $D \leq 1024$ and $n \leq 160$ (see [11]). In graph 3, the maximal degree m_{F_5} occurring in the Gröbner basis computation of an algebraic system coming from HFE (resp. from a random system as described in table 1 section 3.3) is plotted. This graph clearly shows that HFE algebraic systems do not behave as random systems since the maximal degree occurring in the computation does not depend on n the size of the field. The second point, is that the maximal degree m_{F_5} is always less than the upper theoretical bound m found in

Table 3. Small dots correspond to a computer simulation.

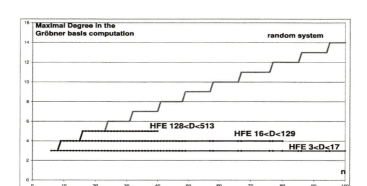

section 4; more precisely when $D < 513$, m_{F_5} is in fact equal to $m - 1$ (except for $D = 16$ or 128) and when it is big enough $m_{F_5} = m$.

In section 4 we have derived an upper bound m by searching *linear* polynomials as a linear combination of "low degree" multiples of $S(x)$. By using again Gröbner bases and proposition 2 we can derive lower bound of complexity for any given instance of HFE. We search now *low degree* polynomials in the univariate world and not only linear relations. Let m^- be the minimal degree for which we have such a low degree relation. Of course, $m^- \le m_{F_5} \le m$. Another meaningful interpretation of m^- is that in this degree we can distinguish between an instance of HFE and a random system.

Since, we have to express that the degree of a monomial x^j where $j = 2^{i_1} + \cdots + 2^{i_k}$ is $w(j) = k$ (so that the degree is not j). First, we introduce new variables X_i corresponding to x^{2^i} together with the relations $X_i^2 = X_{i+1}$. Next, we define a "low degree relation" by saying that the monomials of highest degree vanish. To reach this goal, we introduce a new homogenization variable h so that all the variables X_i are bigger than h and we transform the equations into homogeneous equations; if we find in the ideal generated by these equations a polynomial f such that $f = hf'$ (in other words h is a factor of f) then we obtain a low degree relation. For instance if $S(x) = x^3 + a\,x^2$, where $a \in \mathbb{F}_{2^n}$ and $c \in \mathbb{F}_{2^n}$ the ciphertext, we have the algebraic system of degree 2:

$$[X_0 X_1 + a X_1 h - (c^3 + a\,c^2)\,h^2, X_0^2 - X_1 h, X_1^2 - X_2 h]$$

We compute a Gröbner basis of this system and we find the following equation in degree $m^- = 3$:

$$h^2 \left(\left(c^3 + ac^2 \right) X_0 + a^2 X_1 - X_2 - ahc^2 \left(c + a \right) \right)$$

Note that for this simple example we have in fact a linear relation. In general, we have to compute a Gröbner basis of an algebraic system of degree 2 similarly to equations (1):

$$S^{2^j}(X_0, X_1, \dots, h) - S(c)^{2^j} h^2, X_j^2 - X_{j+1} h, \quad j = 0, 1, \dots \qquad (2)$$

Hence we have a very efficient method (the Gröbner basis computation takes only several seconds) to predict the exact behavior of the Gröbner basis computation for an explicit instance of HFE. This could be used to detect weak secret key instances. We see that when D is big enough $m = m_{F_5} = m^-$. For practical

Table 4. Comparison of various bounds m

Degree D	m	m_{F_5}	m^-
D=3 or D=4	3	3	3
$5 \leq D \leq 12$	4	3	3
D=16	5	3	3
$17 \leq D \leq 96$	5	4	4
D=128	6	4	4
$129 \leq D \leq 512$	6	5	5
$513 \leq D \leq 1024$	6	6	5
$1025 \leq D \leq 1280$	6	6	6

value of D we can establish precisely the experimental complexity of the Gröbner attack (see [11] for details) as follows:

Table 5. Experimental complexity of the Gröbner basis attack.

degree of $S(x)$	$d \leq 16$	$17 \leq d \leq 128$	$129 \leq d \leq 512$
Gröbner complexity	$\mathcal{O}(n^6)$	$\mathcal{O}(n^8)$	$\mathcal{O}(n^{10})$

5.2   With Fast Linear Algebra Techniques

In this section, we show how fast sparse linear algebra techniques can be used to improve the asymptotic complexity of our attacks. Thanks to the analysis from section 4 and the implied bounds on m, we know that we can find linear relations among all linear combinations of the original multivariate polynomials multiplied by all monomials of degree up to $m - 2$. In order to compute these relations, we form a matrix M representing all these polynomials. Each column in the matrix correspond to one of the polynomials, each line is associated to one of the monomials. Each entry is the coefficient in $\mathbb{F}_2$ of the monomial corresponding to the entry's line in the polynomial corresponding to the entry's column. Then, we form a truncated copy M' of M by removing the lines corresponding to the constant and linear monomials. Then, any vector k in the kernel of M', yields a "linear" polynomial L_k in x. The expression of this polynomial in term of the elementary polynomials and their multiples can be read directly from k. The value of L_k is obtained by multiplying the matrix M by k and by reading the values of the lines deleted from M'.

Thus, in order to solve the original HFE cryptosystem, it suffices to compute many independent vectors from the kernel of M'. We further remark that M' is relatively sparse, indeed, each line contains as many entries as the initial quadratic equations. With n variables and n polynomials, there are $O(n^2)$ quadratic monomials and $O(n^{m+1})$ columns in M' and the proportion of non-zero entries is small. Thus, the right approach is to use algorithms that take advantage of this sparsity. Over $\mathbb{F}_2$, we can for example use the block Wiedeman or block Lanczos algorithms (see [6,19]). Their complexity to find $O(n)$ elements in the kernel of M' in term of n-bit word operations is the product of the total number of non-zero entries in the matrix by its rank. Approximating the rank by the smallest dimension (i.e., by the number of columns $O(n^{m-1})$), and by counting $O(n^2)$ non-zero entry per column, we find a total complexity of $O(n^{m-1}n^{m-1}n^2) = O(n^{2m})$. We can slightly reduce this complexity by removing most of the extra lines in M'. Indeed, since M' has $O(n^m)$ lines and $O(n^{m-1})$ columns, we can form a truncated copy M'' of M' by randomly removing lines in M' until the number of lines left is almost equal to the number of columns. Any vector in the kernel of M' is of course in the kernel of M''. On the other hand, a vector from the kernel of M'' is, with high probability, in the kernel of M'. If we keep 50 more lines than columns, the probability to find a bad vector in the kernel of M'' and not in the kernel of M' is about 2^{-50}. Assuming that the number of non-zero entry decreases linearly during this randomized truncation, the complexity of the cryptanalysis is lowered to $O(n^{2m-1})$ operations of n-bit words.

5.3   Incremental Linear Algebra and Almost "Secret Key" Recovery

In fact, this cryptanalysis can be further improved by taking into account some specific property of M. The first (and probably most important) remark is that when decrypting many times with the same HFE system the matrix M is almost the same. More precisely, in the basic quadratic system of equation derived from HFE, all the non-constant terms are identical from one resolution to the next. After multiplication by monomials, we find that the lines of the M that correspond to monomials of total degree m and $m-1$ are left unchanged. As a consequence, the contribution of the columns corresponding to original equations multiplied by a monomial of degree $m-2$ or $m-3$ to the kernel elements is left unchanged. This implies that the linear algebra can be split into two parts. We first find the kernel on the submatrix consisting of the lines and columns described above. This is done once for each HFE public key. Once, the kernel vectors on this submatrix are found, we add back the deleted lines and evaluate their values for each of these vectors. In the second part of the linear algebra, which should be performed for each individual decryption, we solve a much smaller system which involve the previously deleted lines and columns. More precisely, we want, for each kernel element from the first phase, to cancel its contribution to the deleted lines. Since the linear algebra in the second phase involves a much smaller matrix, we have almost recovered an equivalent of the secret key. Indeed, after a first expensive step, we can now decrypt quite efficiently.

Looking further into the specific properties of the matrix M we find that the linear algebra can be improved again. Indeed, assume that we want to find the projection of the kernel of M' on a subset of the column vectors. A possible technique is to form a submatrix with these columns and all the lines which are equal to zero on all the other columns. If the number of lines is greater than the number of columns, we probably find the right result. If the number of lines is too small, we find a larger (probably useless) subspace than the kernel. Luckily for us, we can find subsets of the column vectors with enough corresponding lines. One possible construction is to choose a subset S_k of k variables among $x_1, \ldots x_n$ and form the subset of the columns corresponding to the original equations multiplied by monomials of degree $m - 2$ formed of the product of $m - 2$ variables from S_k. Clearly, the lines corresponding to the product of m variables from S_k are zero outside of the chosen subset. The number of chosen columns is $n \cdot \binom{m-2}{k}$, while the number of lines is $\binom{m}{k}$. From an asymptotic point of view, k is thus of the order of $\sqrt{n}$. This reduces the dimension of the linear system to $O(k^m)$, moreover, the number of non-zero entries per line is now $O(k^2)$. Thus the complexity of the initial step of the attack to $O(k^{2m+2})$ or $O(n^{m+1})$ operations on n-bit words. However, we now need to add the remaining columns and find their contribution. In all generality, it is cumbersome to do this efficiently. Yet, for concrete examples, this is easily done (see section 5.4).

Implementation considerations. When implementing the above approach with sparse linear algebra technique, the memory complexity can also be greatly reduced. Indeed, the matrices we are considering have a very regular structure and can be generated by adding together copies of the much smaller matrix that encodes the original public equation. In the copied matrices, the coefficient are moved around through multiplication by some monomial. In sparse linear algebra algorithms, it is important to efficient compute matrix by vector products (for the matrix and its transpose). To save memory, we do not precompute the expanded matrix, but only store the small one. Then the matrix-vector multiplication can be performed by following the structure of the large matrix. For each multiplier monomial, it suffices to extract a small subvector, to multiply it by the small matrix and to add the result in some subvector of the result. Moreover, the result matrix-vector multiplication algorithm is respectful of the memory cache structure. As a consequence, it runs faster than the ordinary sparse matrix multiplication that uses the expanded matrix.

5.4   Application to Quartz

The Quartz cryptosystem [24] is a signature scheme from the HFE family with variation 1 and 3 applied. Each signature computation requires four HFE inversions. The underlying finite field is $\mathbb{F}_{2^{103}}$ and the degree of the secret polynomial is 129. Variation 1 is applied by deleting 3 public equations, variation 3 adds 4 extra variables. Looking up in table 2 at degree $2^3 * 129 = 1032$, we find that $m = 6$. With incremental linear algebra, the starting point is to multiply the 99 public equations by all monomials of degree 4 in 60 variables. This yields a system of dimension around 48 millions and approximately 1800 entries per lines. The estimated complexity of the initial linear algebra step is 2^{62}. While out of range, this is much lower than the estimated security bound of 2^{80} triple DES announced in [24] and [7].

6   Conclusion

We have presented a very efficient attack on the basic HFE cryptosystem based on Gröbner bases computation. It is not only a theoretical attack with a good complexity but also a very practical method since our implementation was able to break the first HFE challenge (80 bits). The results presented in this paper shed a new light on the security of many schemes in the HFE family. The main result is that when the degree D of the secret polynomial is fixed, the cryptanalysis of an HFE system requires polynomial time in the number of variables. Of course, if D and n are large enough, the cryptanalysis may still be out of practical reach. Yet, this shows the need for a re-evaluation of the security of HFE based cryptosystems.

Acknowledgements. We would like to thank the J.-F. Michon and D. Augot for their programs. The first named author is indebted to the LIP6 for its partial support of this work (Alpha DS25).

References

1. B. Buchberger. *Ein Algorithmus zum Auffinden der Basiselemente des Restklassenringes nach einem nulldimensionalen Polynomideal.* PhD thesis, Innsbruck, 1965.
2. B. Buchberger. An Algorithmical Criterion for the Solvability of Algebraic Systems. *Aequationes Mathematicae*, 4(3):374–383, 1970. (German).
3. B. Buchberger. A Criterion for Detecting Unnecessary Reductions in the Construction of Gröbner Basis. In *Proc. EUROSAM 79*, volume 72 of *Lect. Notes in Comp. Sci.*, pages 3–21. Springer Verlag, 1979.
4. M. Bardet, J.-C. Faugère, and B. Salvy. Complexity of Gröbner basis computation for regular, overdetermined. in preparation, 2003.
5. T. Becker and V. Weispfenning. *Groebner Bases, a Computationnal Approach to Commutative Algebra.* Graduate Texts in Mathematics. Springer-Verlag, 1993.
6. D. Coppersmith. Solving homogeneous linear equations over GF(2) via block Wiedemann algorithm. *Math. Comp*, 62:333–350, 1994.
7. N. Courtois. The security of Hidden Field Equations (HFE). In *Cryptographers' Track RSA Conference*, volume 2020 of *Lectures Notes in Computer Science*, pages 266–281, 2001.
8. N. Courtois, A. Shamir, J. Patarin, and A. Klimov. Efficient algorithms for solving overdefined systems of multivariate polynomial equations. In *Advances in Cryptology – Eurocrypt'2000*, volume 1807 of *Lectures Notes in Computer Science*, pages 392–407. Springer Verlag, 2000.

9. D. Cox, J. Little, and D. O'Shea. *Using Algebraic Geometry*. Springer Verlag, New York, 1998.
10. J.-C. Faugère. A new efficient algorithm for computing Gröbner bases (F4). *Journal of Pure and Applied Algebra*, 139(1–3):61–88, June 1999.
11. J.-C. Faugère. Algebraic cryptanalysis of HFE using Gröbner bases. Technical Report 4738, INRIA, 2003.
12. J.-C. Faugère. A new efficient algorithm for computing Gröbner bases without reduction to zero F5. In T. Mora, editor, *Proceedings of ISSAC*, pages 75–83. ACM Press, July 2002.
13. H. Gilbert and M. Minier. Cryptanalysis of SFLASH. In L. Knudsen, editor, *Advances in Cryptology – Eurocrypt'2002*, volume 2332 of *LNCS*, pages 288–298. Springer, 2002.
14. A. Kipnis and A. Shamir. Cryptanalysis of the oil and vinegar signature scheme. In H. Krawczyk, editor, *Advances in Cryptology – Crypto'98*, volume 1462 of *LNCS*, pages 257–266. Springer Verlag, 1998.
15. A. Kipnis and A. Shamir. Cryptanalysis of the HFE Public Key Cryptosystem by R elinearization. In M. Wiener, editor, *Advances in Cryptology – Crypto'99*, volume 1666 of *LNCS*, pages 19–30. Springer Verlag, 1999.
16. D. Lazard. Gaussian Elimination and Resolution of Systems of Algebraic Equations. In *Proc. EUROCAL 83*, volume 162 of *Lect. Notes in Comp. Sci*, pages 146–157, 1983.
17. F. S. Macaulay. *The algebraic theory of modular systems.*, volume xxxi of *Cambridge Mathematical Library*. Cambridge University Press, 1916.
18. T. Matsumoto and H. Imai. Public quadratic polynomial-tuples for efficient signature-veri fication and message-encryption. In C. Günther, editor, *Advances in Cryptology – EuroCrypt '88*, pages 419–454, Berlin, 1988. Springer-Verlag. Lecture Notes in Computer Science Volume 330.
19. P. L. Montgomery. A block Lanczos algorithm for finding dependencies over GF(2). In L. Guillou and J.-J. Quisquater, editors, *Advances in Cryptology - EuroCrypt '95*, pages 106–120, Berlin, 1995. Springer-Verlag. Lecture Notes in Computer Science Volume 921.
20. J. Patarin. Cryptanalysis of the Matsumoto and Imai public key scheme o f Eurocrypt '88. In D. Coppersmith, editor, *Advances in Cryptology – Crypto '95*, pages 248–261, Berlin, 1995. Springer-Verlag. Lecture Notes in Computer Science Volume 963.
21. J. Patarin. *HFE first challenge*, 1996. http://www.minrank.org/challenge1.txt.
22. J. Patarin. Hidden fields equations (HFE) and isomorphisms of polynomials (IP): two new families of asymmetric algorithms. In U. Maurer, editor, *Advances in Cryptology – EuroCrypt '96*, pages 33–48, Berlin, 1996. Springer-Verlag. Lecture Notes in Computer Science Volume 1070.
23. J. Patarin. Hidden Fields Equations (HFE) and Isomorphisms of Polynomials (IP): Two New Families of Asymmetric Algorithms. Extended version, 1996.
24. J. Patarin, L. Goubin, and N. Courtois. Quartz: An Asymetric Signature Scheme for Short Signatu res on PC, submission to NESSIE, 2000.
25. V. Shoup. *NTL 5.3.1, a Library for doing Number Theory*, 2003. http://www.shoup.net/ntl.

On Constructing Locally Computable Extractors and Cryptosystems in the Bounded Storage Model*

Salil P. Vadhan**

Harvard University, Cambridge, MA, `salil@eecs.harvard.edu`

Abstract. We consider the problem of constructing randomness extractors that are *locally computable*; that is, read only a small number of bits from their input. As recently shown by Lu (*CRYPTO '02*), locally computable extractors directly yield secure private-key cryptosystems in Maurer's bounded storage model (*J. Cryptology*, 1992).

We suggest a general "sample-then-extract" approach to constructing locally computable extractors. Plugging in known sampler and extractor constructions, we obtain locally computable extractors, and hence cryptosystems in the bounded storage model, whose parameters improve upon previous constructions and come quite close to the lower bounds.

The correctness of this approach follows from a fundamental lemma of Nisan and Zuckerman (*J. Computer and System Sciences,* 1996), which states that sampling bits from a weak random source roughly preserves the min-entropy rate. We also present a refinement of this lemma, showing that the min-entropy rate is preserved up to an arbitrarily small additive loss, whereas the original lemma loses a logarithmic factor.

1   Introduction

Maurer's *bounded storage model* for private-key cryptography [2] has been the subject of much recent activity. In this model, one assumes that there is public, high-rate source of randomness and that all parties have limited storage so that they cannot record all the randomness from the source. Remarkably, this quite plausible model makes it possible to construct private-key cryptosystems that are information-theoretically secure and require no unproven complexity assumptions (in contrast to most of modern cryptography). Intuitively, a shared secret key can be used by legitimate parties to randomly select bits from the random source about which the adversary has little information (due to the bound on its storage). With some further processing, the legitimate parties can convert these unpredictable bits into ones which the adversary cannot distinguish from truly random (in an information-theoretic sense), and hence they can safely be used for cryptographic purposes, e.g. as a one-time pad for encryption.

---

* A preliminary version of this paper has appeared on the Cryptology e-print archive [1] and a full version will appear in the Journal of Cryptology.
** Supported by NSF grants CCR-0205423 and CCR-0133096.

A sequence of works [2,3,4,5,6,7,8] has given increasingly secure and efficient protocols in this model. In particular, the works of Aumann, Ding, and Rabin [5, 6] showed that protocols in this model have the novel property of "everlasting security" — the security is preserved even if the key is reused an exponential number of times and is subsequently revealed to the adversary.

Recently, Lu [8] showed that work in this model can be cast nicely in the framework of *randomness extractors*. Extractors, introduced by Nisan and Zuckerman [9], are procedures for extracting almost-uniform bits from sources of biased and correlated bits. These powerful tools have been the subject of intense study, and have found many applications to a wide variety of topics in the theory of computation. (See the surveys [10,11].) One of the first applications, in the original paper of Nisan and Zuckerman, was to construct pseudorandom generators for space-bounded computation. Thus, they seem a natural tool to use in the bounded storage model, and indeed Lu [8] showed that any extractor yields secure private-key cryptosystems in the bounded storage model. However, the efficiency considerations of the bounded storage model require a nonstandard property from extractors — namely that they are *locally computable*;[1] that is, they can be computed by reading only a few bits from the random source. Lu constructed locally computable extractors by first constructing locally computable error-correcting codes, and then plugging them into the specific extractor construction of Trevisan [13].

In this paper, we suggest a general "sample-then-extract" approach to constructing locally computable extractors: use essentially any randomness-efficient "sampler" to select bits from the source and then apply essentially any extractor to the selected bits. Plugging in known sampler and extractor constructions, we obtain locally computable extractors, and hence cryptosystems in the bounded storage model, whose parameters improve upon previous constructions and come quite close to the lower bounds.

The correctness of this approach follows directly from a fundamental lemma of Nisan and Zuckerman [9]. Roughly speaking, the lemma states that a random of sample of bits from a string of high min-entropy[2] also has high min-entropy. We also present a refinement of this lemma, showing that the min-entropy rate is preserved up to an arbitrarily small additive loss, whereas the original lemma loses a logarithmic factor. This improvement is not necessary for the sample-then-extract approach to work, but increases its efficiency. Together with some of our techniques for constructing samplers, it has also played a role in the recent explicit construction of extractors that are "optimal up to constant factors" [14].

In retrospect, several previous cryptosystems in the bounded storage model, such as [3] and [5], can be viewed as special cases of the sample-then-extract

---

[1] This terminology was suggested by Yan Zong Ding and we prefer it to the terminology "on-line extractors," which was used (with different meanings) in [12,8]. The issue of "local computation" versus "on-line computation" is discussed in more detail in Section 3.

[2] Like Shannon entropy, the min-entropy of a probability distribution X is a measure of the number of bits of "randomness" in X. A formal definition is given in Section 3.

approach, with particular choices for the extractor and sampler. By abstracting the properties needed from the underlying tools, we are able to use state-of-the-art extractors and samplers, and thereby obtain our improvements.

2   Preliminaries

Except where otherwise noted, we refer to random variables taking values in discrete sets. We generally use capital letters for random variables and lower-case letters for specific values, as in $\Pr[X = x]$. If S is a set, then $x \overset{\text{R}}{\leftarrow} S$ indicates that x is selected uniformly from S. For a random variable A and an event E, we write $A|_E$ to mean A conditioned on E.

The *statistical difference* (or variation distance) between two random variables X, Y taking values in a universe $\mathcal{U}$ is defined to be

$$\Delta(X, Y) \overset{\text{def}}{=} \max_{S \subseteq \mathcal{U}} \left| \Pr[X \in S] - \Pr[Y \in S] \right| = \frac{1}{2} \sum_{x \in \mathcal{U}} \left| \Pr[X = x] - \Pr[Y = x] \right|.$$

We say X and Y are ε-*close* if $\Delta(X, Y) \leq \varepsilon$.

The *min-entropy* of X is $\mathrm{H}_\infty(X) \overset{\text{def}}{=} \min_x \log(1/\Pr[X = x])$. (All logarithms in this paper are base 2.) Intuitively, min-entropy measures randomness in the "worst case," whereas standard (Shannon) entropy measures the randomness in X "on average." X is called a k-*source* if $\mathrm{H}_\infty(X) \geq k$, i.e. for all x, $\Pr[X = x] \leq 2^{-k}$.

3   The Bounded Storage Model

The Random Source. The original model of Maurer [2] envisioned the random source as a high-rate stream of perfectly random bits being broadcast from some natural or artificial source of randomness. However, since it may difficult to obtain perfectly random bits from a physical source, particularly at a high rate, we feel it is important to investigate the minimal conditions on the random source under which this type of cryptography can be performed. As noted in [8, 7], the existing constructions still work even if we only assume that the source has sufficient "entropy". Below we formalize this observation, taking particular note of the kind of independence that is needed when the cryptosystem is used many times.

We model the random source as a sequence of random variables $X_1, X_2, \ldots$, each distributed over $\{0, 1\}^n$, where X_t is the state of the source at time period t. To model a random source which is a high-rate "stream" of bits, the X_t's can be thought of as a partition of the stream into contiguous substrings of length n. However, one may also consider random sources that are not a stream, but rather a (natural or artificial) "oracle" of length n, which changes over time and can be probed at positions of one's choice. In both cases, n should be thought of as very large, greater than the storage capacity of the adversary (and the legitimate parties).

To obtain the original model of a perfectly random stream, each the X_t's can be taken to be uniform on $\{0,1\}^t$ and independent of each other. Here we wish to allow biases and correlations in the source, only assuming that each X_t has sufficient randomness, as measured by min-entropy (as advocated in [15,16]). That is, we will require each X_t to be an αn-source for some $\alpha > 0$. Using a worst-case measure like min-entropy rather than Shannon entropy is important because we want security to hold with high probability and not just "on average". (The results will also apply for random sources that are statistically close to having high min-entropy, such as those of high Renyi entropy.)

For our cryptosystems, we will actually need to require that the random source has high min-entropy conditioned on the future.

Definition 1. *A sequence of random variables $X_1, X_2, \ldots$, each distributed over $\{0,1\}^n$ is a* reverse block source *of entropy rate α if for every $t \in \mathbb{N}$ and every $x_{t+1}, x_{t+2}, \ldots$, the random variable $X_t|_{X_{t+1}=x_{t+1}, X_{t+2}=x_{t+2}, \ldots}$ is an αn-source.*

As the terminology suggests, this is the same as the Chor-Goldreich [15] notion of a *block source*, but "backwards" in time. Intuitively, it means that X_t possesses αn bits of randomness that will be "forgotten" at the next time step. This is somewhat less natural than the standard notion of a (forward) block source, but it still may be a reasonable model for some physical sources of randomness that are not perfectly random.[3] Below we will see why some condition of this form (high entropy *conditioned on the future*) is necessary for the cryptography. In the special case $\alpha = 1$, Definition 1 is equivalent to requiring that X_t's are uniform and independent, so in this case the issue of reversal is moot.

Cryptosystems. Here, as in previous works, we focus on the task of using a shared key to extract pseudorandom bits from the source. These pseudorandom bits can then be used for private communication or message authentication. A *pseudorandom extraction scheme in the bounded storage model* is a function $\text{Ext} : \{0,1\}^n \times \{0,1\}^d \to \{0,1\}^m$ (typically with $d, m \ll n$). Such a scheme is to be used as follows. Two parties share a key $K \in \{0,1\}^d$. At time t, they apply $\text{Ext}(\cdot, K)$ to the random source X_t to obtain m pseudorandom bits, given by $Y_t = \text{Ext}(X_t, K)$. At time $t+1$ (or later), Y_t will be pseudorandom to the adversary (if the scheme is secure), and hence can be used by the legitimate parties as a shared random string for any purpose (e.g. as a one-time pad for encryption). The pseudorandomness of Y_t will rely on the fact that, at time $t+1$ and later, X_t is no longer accessible to the adversary. More generally, we need X_t to be unpredictable from future states of the random source, as captured by our notion of a reverse block source. Note that even if Y_t will only be used exactly at time $t+1$, we still need X_t to have high min-entropy given the entire future, because the adversary can store Y_t.

We now formally define security for a pseudorandom extraction scheme. Let βn be the bound on the storage of the adversary A, and denote by

---

[3] The consideration of such sources raises interesting philosophical questions: does the universe keep a perfect record of the past? If not, then reverse block sources seem plausible.

$S_t \in \{0,1\}^{\beta n}$ the state of the adversary at time t. For a sequence of random variables $Z_1, Z_2, \ldots$, we will use the shorthand $Z_{[a,b]} = (Z_a, Z_{a+1}, \ldots, Z_b)$, $Z_{[a,\infty)} = (Z_a, Z_{a+1}, \ldots)$. Following the usual paradigm for pseudorandomness, we consider the adversary's ability to distinguish two experiments — the "real" one, in which the extraction scheme is used, and an "ideal" one, in which truly random bits are used. Let A be an arbitrary function representing the way the adversary updates its storage and attempts to distinguish the two experiments at the end.

Real Experiment: Let $X_1, X_2, \ldots$ be the random source, and let $K \xleftarrow{\text{R}} \{0,1\}^d$. For all t, let $Y_t = \text{Ext}(X_t, K)$ be the extracted bits. Let $S_0 = 0^{\beta n}$, and for $t = 1, \ldots, T$, let $S_t = A(Y_{[1,t-1]}, S_{t-1}, X_t)$. Output $A(Y_{[1,T]}, S_T, X_{[T+1,\infty)}, K) \in \{0,1\}$.

Ideal Experiment: Let $X_1, X_2, \ldots$ be the random source, and let $K \xleftarrow{\text{R}} \{0,1\}^d$. For all t, let $Y_t \xleftarrow{\text{R}} \{0,1\}^m$. Let $S_0 = 0^{\beta n}$, and for $t = 1, \ldots, T$, let $S_t = A(Y_{[1,t-1]}, S_{t-1}, X_t)$. Output $A(Y_{[1,T]}, S_T, X_{[T+1,\infty)}, K) \in \{0,1\}$.

Note that at each time step we give the adversary access to all the past Y_i's "for free" (i.e. with no cost in the storage bound), and in the last time step, we give the adversary the adversary access to all future X_i's and the key K. The benefits of doing this are explained below.

Definition 2. *We call* $\text{Ext} : \{0,1\}^n \times \{0,1\}^d \to \{0,1\}^m$ ε*-secure for storage rate β and entropy rate α if for every reverse block source (X_t) of entropy rate α, every adversary A with storage bound βn, and every $T > 0$, A distinguishes between the real and ideal experiments with advantage at most $T \cdot \varepsilon$.*

Remarks:

- In the real experiment, we give the extracted strings Y_t explicitly to the adversary, as is typical in definitions of pseudorandomness. However, when they are used in a cryptographic application (e.g. as one-time pads), they of course will not be given explicitly to the adversary. The string Y_{t-1} extracted at time $t-1$ is not given to the adversary (i.e. is not used in the application) until time t. As mentioned above, this is crucial for security.
- The definition would be interesting even if $Y_1, \ldots, Y_{t-2}$ were not given to the adversary at time t, (i.e. $S_t = A(Y_{t-1}, S_{t-1}, X_t)$), and if K and $X_{T+2}, X_{T+3}, \ldots$ were not given to the adversary at the end. Giving all the previous Y_i's implies that it is "safe" to use Y_i at any time period after i (rather than exactly at time $i+1$). Giving the adversary all subsequent X_i's at the end is to guarantee that the security does not deteriorate if the adversary waits and watches the source for some future time periods. Giving the adversary the key at the end means that even if the secret key is compromised, earlier transactions remain secure. This is the remarkable property of "everlasting security" noticed and highlighted by [5].
- We require that the security degrade only linearly with the number of times the same key is reused.

– No constraint is put on the computational power of the adversary except for the storage bound of βn (as captured by $S_t \in \{0,1\}^{\beta n}$). This means that the distributions of $(Y_{[1,T-1]}, S_T, X_{[T+1,\infty)}, K)$ in the real and ideal experiments are actually close in a statistical sense — they must have statistical difference at most $T \cdot \varepsilon$.

– The definition is impossible to meet unless $\alpha > \beta$: If $\alpha \le \beta$, we can take each X_t to have its first αn bits uniform and the rest fixed to zero. Then an adversary with βn storage can entirely record X_t, and thus can compute Y_t once K is revealed. (Even if K is not revealed, in this example the bounded-storage model still clearly provides no advantage over the standard private-key setting, and hence is subject to the usual limitations on information-theoretic security [17].)

As usual, the above definition implies that to design a cryptosystem (e.g. private-key encryption or message authentication) one need only prove its security in the ideal experiment, where the two parties effectively share an infinite sequence of random strings $Y_1, Y_2, \ldots$. Security in the bounded storage model immediately follows if these random Y_i's are then replaced with ones produced by a secure pseudorandom extraction scheme.

Efficiency Considerations. In addition to security, it is important for the extraction scheme to be efficient. In the usual spirit of cryptography, we would like the honest parties to need much smaller resources than the adversary is allowed. In this case, that means we would like the computation of Ext to require much less *space* than the adversary's storage bound of βn. Note that the honest parties will have to store the entire extracted key $Y_t \in \{0,1\}^m$ during time t (when it is not yet safe to use), so reducing their space to m is the best we can hope for (and since we envision $m \ll n$, this is still very useful). However, since n is typically envisioned to be huge, it is preferable to reduce not just the space for Ext to much less than n, but also the time spent.[4] Thus, we adopt as our efficiency measure the *number of bits read from the source*. Of course, once these bits are read, it is important that the actual computation of Ext is efficient with respect to both time and space. In our constructions (and all previous constructions), Ext can be computed in polynomial time and polylogarithmic work space (indeed even in NC). Thus the total storage required by the legitimate parties is dominated by the number of bits read from the source.

The following (proven in the full version) shows that the number of bits read from the source must be linear in m, and must grow when the difference between the entropy rate and storage rate goes to zero.

Proposition 3. *If* Ext $: \{0,1\}^n \times \{0,1\}^d \to \{0,1\}^m$ *is an ε-secure pseudorandom extraction scheme for storage rate β and entropy rate α, then* Ext$(\cdot, K)$ *depends on at least* $(1 - \varepsilon - 2^{-m}) \cdot (1/(\alpha - \beta)) \cdot m$ *bits of its input (on average, taken over K).*

---

[4] If having Ext computable in small space with one pass through the random source is considered sufficiently efficient, then the work of Bar-Yossef et al. [12] is also applicable here. See Section 4.

Another common complexity measure in cryptography is the key length, which should be minimized. Figure 1 describes the performance of previous schemes and our new constructions with respect to these two complexity measures.[5] With respect to both measures, our constructions are within a constant factor of optimal. In fact, for the number of bits read, this constant factor can be made arbitrarily close to 1.

reference	key length	# bits read	restrictions
[3]	$O(\log n)$	$O(m/\varepsilon^2)$	interactive
[4]	$O(\log n \cdot \log(1/\varepsilon))$	$O(m \cdot \log(1/\varepsilon))$	$\alpha = 1, \beta < 1/m$
[5]	$O(m \cdot \log n \cdot \log(1/\varepsilon))$	$O(m \cdot \log(1/\varepsilon))$	
[6]	$O(\log n \cdot \log(1/\varepsilon))$	$O(m \cdot \log(1/\varepsilon))$	$\alpha = 1, \beta < 1/\log m$
[7]	$O(\log n \cdot \log(1/\varepsilon))$	$O(m \cdot \log(1/\varepsilon))$	
[8]	$O(m \cdot (\log n + \log(1/\varepsilon)))$	$O(m \cdot \log(1/\varepsilon))$	
[8]	$O((\log^2(n/\varepsilon)/\log n))$	$O(m \cdot \log(1/\varepsilon))$	$m \leq n^{1-\Omega(1)}$
here	$O(\log n + \log(1/\varepsilon))$	$O(m + \log(1/\varepsilon))$	$\varepsilon > \exp(-n/2^{O(\log^* n)})$

Fig. 1. Comparison of pseudorandom extraction schemes in the bounded storage model. Parameters are for ε-secure schemes Ext : $\{0,1\}^n \times \{0,1\}^d \to \{0,1\}^m$, for constant storage rate β and entropy rate α, where $\alpha > \beta$. We only list the parameters for the case that the number of bits read from the source is $o(n)$, as n is assumed to be huge and infeasible.

We now touch upon a couple of additional efficiency considerations. First, if the random source is indeed a high-rate stream (as opposed to an "oracle source"), it is important that the positions to be read from the source can be computed offline (from just the key) and sorted so that they can be quickly read in order as the stream goes by. This is the case for our scheme and previous ones.

Second, one can hope to reduce the space of the legitimate parties to exactly $m+d$ (i.e., the length of the extracted string plus the key). That is, even though the schemes read more than m bits from the source, the actual computation of the m-bit extracted string can be done "in place" as the bits from the source are read. This property holds for most of the previous constructions, as each bit of the extracted string is a parity of $O(\log(1/\varepsilon))$ bits of the source. Our construction does not seem to have this property in general (though specific instantiations may); each bit of the output can be a function of the entire $O(m + \log(1/\varepsilon))$

---

[5] Most of the previous schemes were explicitly analyzed only for the case of a perfectly random source, i.e. $\alpha = 1$, but the proofs actually also work for weak random sources provided $\alpha > \beta$ (except where otherwise noted) [8]. Also note that the schemes with key length greater than m do not follow trivially from the one-time pad, because the same key can be used many times.

bits read from the source. Still the space used by our scheme is $O(m+\log(1/\varepsilon))$, only a constant factor larger than optimal.

4   Locally Computable Extractors

In this section, we define extractors and locally computable extractors, and recall Lu's result [8] about their applicability to the bounded storage model. Then we discuss averaging samplers and describe how using them to sample bits from a random source preserves the min-entropy rate, via a lemma of Nisan and Zuckerman [9] which we refine. Finally, we give our general sample-then-extract construction which combines any extractor and any averaging sampler to yield a locally computable extractor.

An extractor is a procedure for extracting almost-uniform bits from any random source of sufficient min-entropy. This is not possible to do deterministically, but it is possible using a short *seed* of truly random bits, as captured in the following definition of Nisan and Zuckerman.

Definition 4 ([9]). Ext $: \{0,1\}^n \times \{0,1\}^d \rightarrow \{0,1\}^m$ *is a* strong[6] (k,ε)-extractor *if for every k-source X, the distribution $U_d \circ \mathrm{Ext}(X, U_d)$ is ε-close to $U_d \times U_m$.*

The goal in constructing extractors is to minimize the seed length d and maximize the output length m. We will be precise about the parameters in later sections, but, for reference, an "optimal" extractor has a seed length of $d = O(\log n + \log(1/\varepsilon))$ and an output length of $m = k - O(\log(1/\varepsilon))$, i.e. almost all of the min-entropy is extracted using a seed of logarithmic length.

Recently, Lu proved that any extractor yields secure cryptosystems in the bounded storage model:

Theorem 5 (implicit in [8]). *If* Ext $: \{0,1\}^n \times \{0,1\}^d \rightarrow \{0,1\}^m$ *is a strong* $(\delta n - \log(1/\varepsilon), \varepsilon)$-extractor, then for any $\beta > 0$, Ext *is an* 2ε-secure pseudorandom extraction scheme for storage rate β and entropy rate $\beta + \delta$.

However, as noted by Lu, for a satisfactory solution to cryptography in the bounded storage model, the extractor should only read only a few bits from the source.

Definition 6. Ext $: \{0,1\}^n \times \{0,1\}^d \rightarrow \{0,1\}^m$ *is t-locally computable (or t-local) if for every $r \in \{0,1\}^d$, $\mathrm{Ext}(x, r)$ depends on only t bits of x, where the bit locations are determined by r.*

Thus, in addition to the usual goals of minimizing d and maximizing m, we also wish to minimize t. Bar-Yossef, Reingold, Shaltiel, and Trevisan [12] studied a related notion of *on-line extractors*, which are required to be computable in

---

[6] A standard (i.e. non-strong) extractor requires only that $\mathrm{Ext}(X, U_d)$ is ε-close to uniform.

small space in one pass. They show that space approximately m is necessary and sufficient to evaluate extractors with output length m. Since the small-space requirement is weaker than being locally computable, their lower bound also applies here.[7] But a stronger lower bound for locally computable extractors can be obtained by combining Proposition 3 and Theorem 5 with $\beta = 0$, or by mimicking the proof of Proposition 3 directly for t-local extractors to obtain the following slightly better bound:

Proposition 7. *If* $\mathrm{Ext} : \{0,1\}^n \times \{0,1\}^d \to \{0,1\}^m$ *is a t-local strong $(\delta n, \varepsilon)$-extractor, then* $t \geq (1 - \varepsilon - 2^{-m}) \cdot (1/\delta) \cdot m$.

Lu [8] observed that the encryption schemes of Aumann, Ding, and Rabin [4, 5,6] can be viewed as locally computable extractors, albeit with long seeds. He constructed locally computable extractors with shorter seeds based on Trevisan's extractor [13]. The construction of Dziembowski and Maurer [7] is also a locally computable extractor. The parameters of these constructions can be deduced from Figure 1.

Our construction of locally computable extractors is based on a fundamental lemma of Nisan and Zuckerman [9], which says that if one samples a random subset of bits from a weak random source, the min-entropy rate of the source is (nearly) preserved. More precisely, if $X \in \{0,1\}^n$ is a δn-source and $X_S \in \{0,1\}^t$ is the projection of X onto a *random* set $S \subset [n]$ of t positions, then, with high probability, X_S is ε-close to a $\delta' t$-source, for some δ' depending on δ. Thus, to obtain a locally computable extractor, we can simply apply a (standard) extractor to X_S, and thereby output roughly $\delta' t$ almost-uniform bits. That is, part of the seed of the locally computable extractor will be used to select S, and the remainder as the seed for applying the extractor to X_S.

However, choosing a completely random set S of positions is expensive in the seed length, requiring approximately $|S| \cdot \log n$ random bits. (This gives a result analogous to [5], because $|S| \geq m$.) To save on randomness, Nisan and Zuckerman [9] showed that S could be sampled in a randomness-efficient manner, using k-wise independence and/or random walks on expander graphs. More generally, their proof only requires that w.h.p. S has large intersection with any subset of $[n]$ of a certain density (cf., [18]). In order to achieve improved performance, we will impose a slightly stronger requirement on the sampling method: for any $[0,1]$-valued function, w.h.p. its average on S should approximate its average on $[n]$. Such sampling procedures are known as *averaging* (or *oblivious*) *samplers*, and have been studied extensively [19,20,21,22]. Our definition differs slightly from the standard definition, to allow us to obtain some savings in parameters (discussed later).

Definition 8. *A function* $\mathrm{Samp} : \{0,1\}^r \to [n]^t$ *is a* (μ, θ, γ) *averaging sampler if for every function* $f : [n] \to [0,1]$ *with average value* $\frac{1}{n} \sum_i f(i) \geq \mu$, *it holds that*

---

[7] This is because their space lower bounds apply also to a nonuniform branching program model of computation, where the space is always at most the number of bits read from the input.

$$\Pr_{(i_1,\ldots,i_t) \xleftarrow{R} \mathrm{Samp}(U_r)} \left[\frac{1}{t} \sum_{j=1}^{t} f(i_j) < \mu - \theta \right] \leq \gamma.$$

Samp *has* distinct samples *if for every* $x \in \{0,1\}^r$, *the samples produced by* Samp(x) *are all distinct.*

That is, for any function f whose average value is at least μ, with high probability (i.e., at least $1 - \gamma$) the sampler selects a sample of positions on which the average value of f is not much smaller than μ. The goal in constructing averaging samplers is usually to simultaneously minimize the randomness r and sample complexity t. We will be precise about the parameters in later sections, but, for reference, an "optimal" averaging sampler uses only $t = O(\log(1/\gamma))$ samples and $r = O(\log n + \log(1/\gamma))$ random bits (for constant μ, θ).

In contrast to most applications of samplers, we will not necessarily be interested in minimizing the sample complexity. Ideally, we prefer samplers where the number of distinct samples can be chosen anywhere in the interval $[t_0, n]$, where t_0 is the minimum possible sample complexity. (Note that without the requirement of distinct samples, the number of samples can be trivially increased by repeating each sample several times.) Another atypical aspect of our definition is that we make the parameter μ explicit. Averaging samplers are usually required to give an approximation within additive error θ regardless of the average value of f, but being explicit about μ will allow us to obtain some savings in the parameters.

Using averaging samplers (rather than just samplers that intersect large sets) together with an idea from [23] allows us to obtain a slight improvement to the Nisan–Zuckerman lemma. Specifically, Nisan and Zuckerman show that sampling bits from a source of min-entropy rate δ yields a source of min-entropy rate $\Omega(\delta/\log(1/\delta))$; our method can yield min-entropy rate $\delta - \tau$ for any desired τ.

For a string $x \in \{0,1\}^n$ and a sequence $s = (i_1,\ldots,i_t) \in [n]^t$, define $x_s \in \{0,1\}^t$ to be the string $x_{i_1} x_{i_2} \cdots x_{i_t}$. Recall that for a pair of jointly distributed random variables (A, B), we write $B|_{A=a}$ for B conditioned on the event $A = a$.

Lemma 9 (refining [9]). *Suppose* Samp $: \{0,1\}^r \to [n]^t$ *is an* (μ, θ, γ) *averaging sampler with distinct samples for* $\mu = (\delta - 2\tau)/\log(1/\tau)$ *and* $\theta = \tau/\log(1/\tau)$. *Then for every* δn-*source* X *on* $\{0,1\}^n$, *the random variable* $(U_r, X_{\mathrm{Samp}(U_r)})$ *is* $(\gamma + 2^{-\Omega(\tau n)})$-*close to* (A, B) *where* A *is uniform on* $\{0,1\}^r$ *and for every* $a \in \{0,1\}^r$,[8] *the random variable* $B|_{A=a}$ *is* $(\delta - 3\tau)t$-*source.*

The above lemma is where we use the fact that the sampler has distinct samples. Clearly, sampling the same bits of X many times cannot increase the min-entropy of the output, whereas the above lemma guarantees that the min-entropy grows linearly with t, the number of samples.

---

[8] Intuitively, the reason we can guarantee that B has high min-entropy conditioned on *every* value of A, is that the "bad" values of A are absorbed in the $\gamma + 2^{-\Omega(\tau n)}$ statistical difference.

An alternative method to extract a shorter string from a weak random source while preserving the min-entropy rate up to a constant factor was given by Reingold, Shaltiel, and Wigderson [18], as a subroutine in their improved extractor construction. However, the string produced by their method consists of bits of an encoding of the source in an error-correcting code rather than bits of the source itself, and hence is not good for constructing locally computable extractors (which was not their goal). As pointed out to us by Chi-Jen Lu and Omer Reingold, Lemma 9 eliminates the need for error-correcting codes in [18].

The proof of Lemma 9 is deferred to the full version. Given the lemma, it follows that combining an averaging sampler and an extractor yields a locally computable extractor.

Theorem 10 (sample-then-extract). *Suppose that* Samp $: \{0,1\}^r \to [n]^t$ *is an* (μ, θ, γ) *averaging sampler with distinct samples for* $\mu = (\delta - 2\tau)/\log(1/\tau)$ *and* $\theta = \tau/\log(1/\tau)$. *and* Ext $: \{0,1\}^t \times \{0,1\}^d \to \{0,1\}^m$ *is a strong* $((\delta - 3\tau)t, \varepsilon)$ *extractor. Define* Ext$' : \{0,1\}^n \times \{0,1\}^{r+d} \to \{0,1\}^m$ *by*

$$\text{Ext}'(x, (y_1, y_2)) = \text{Ext}(x_{\text{Samp}(y_1)}, y_2).$$

Then Ext$'$ *is a t-local strong* $(\delta n, \varepsilon + \gamma + 2^{-\Omega(\tau n)})$ *extractor.*

Proof. For every (y_1, y_2), Ext$'(x, (y_1, y_2))$ only reads the t bits of x selected by Samp(y_1), so Ext$'$ is indeed t-local. We now argue that it is a $(\delta n, \varepsilon + \gamma + 2^{-\Omega(\tau n)})$ extractor. Let X be any δn-source. We need to prove that the random variable $Z = (U_r, U_d, \text{Ext}'(X, (U_r, U_d))) = (U_r, U_d, \text{Ext}(X_{\text{Samp}(U_r)}, U_d))$ is close to uniform. By Lemma 9, $(U_r, X_{\text{Samp}(U_r)})$ is $(\gamma + 2^{-\Omega(\tau n)})$-close to (A, B) where A is uniform on $\{0,1\}^r$ and $B|_{A=a}$ is a $(\delta - 3\tau)t$-source for every a. This implies that Z is $(\gamma + 2^{-\Omega(\tau n)})$-close to $(A, U_d, \text{Ext}(B, U_d))$. Since Ext is a strong $((\delta - 3\tau)t, \varepsilon)$ extractor, $(U_d, \text{Ext}(B|_{A=a}, U_d))$ is ε-close to $U_d \times U_m$ for all a. This implies that $(A, U_d, \text{Ext}(B, U_d))$ is ε-close to $A \times U_d \times U_m = U_r \times U_d \times U_m$. By the triangle inequality, Z is $(\varepsilon + \gamma + 2^{-\Omega(\tau n)})$-close to $U_r \times U_d \times U_m$. $\square$

For intuition about the parameters, consider the case when $\delta > 0$ is an arbitrary constant, $\tau = \delta/6$, and $\gamma = \varepsilon$. Then using "optimal" averaging samplers and extractors will give a locally computable extractor with seed length $r + d = O(\log n + \log(1/\varepsilon))$ and output length $m = \Omega(\delta t) - O(\log(1/\varepsilon))$. This matches, up to constant factors, the seed length of an optimal extractor (local or not) and the optimal relationship between the output length and the number of bits read from the source.

We stress that the above refinement to the Nisan–Zuckerman lemma is not necessary to achieve these parameters (or those in Figure 1). Those parameters can be obtained by applying the sample-then-extract method with the original lemma and sampler of Nisan and Zuckerman [9] together with the extractor of Zuckerman [21]. The advantage provided by the refined lemma lies in the hidden constant in the number of bits read from the source. Specifically, by taking $\tau \to 0$, the ratio between m and t approaches δ, which is essentially optimal by Proposition 7.

5   Non-explicit Constructions

In this section, we describe the locally computable extractors obtained by using truly optimal extractors and samplers in Theorem 10. This does not give efficient constructions of locally computable extractors, because optimal extractors and samplers are only known by nonconstructive applications of the Probabilistic Method. However, it shows what Theorem 10 will yield as one discovers constructions which approach the optimal bounds. In fact, the explicit constructions known are already quite close, and (as we will see in Section 6) match the optimal bounds within constant factors for the range of parameters most relevant to the bounded storage model.

5.1   The Extractor

The Probabilistic Method yields extractors with the following expressions for the seed length d and output length m, both of which are tight up to additive constants [24].

Lemma 11 (nonconstructive extractors (cf., [21,24])). *For every n, $k \leq n$, $\varepsilon > 0$, there exists a strong (k, ε)-extractor* Ext $: \{0,1\}^n \times \{0,1\}^d \to \{0,1\}^m$ *with $d = \log(n - k) + 2\log(1/\varepsilon) + O(1)$, $m = k - 2\log(1/\varepsilon) - O(1)$.*

5.2   The Sampler

Similarly, the following lemma states the averaging samplers implied by the Probabilistic Method. There are matching lower bounds for both the randomness complexity and the sample complexity [20] (except for the dependence on μ, which was not considered there). The proof of the lemma (given in the full version) follows the argument implicit in [21], with the modifications that it makes the dependence on μ explicit and guarantees distinct samples.

Lemma 12 (nonconstructive samplers). *For every $n \in \mathbb{N}$, $1/2 > \mu > \theta > 0$, $\gamma > 0$, there is a (μ, θ, γ) averaging sampler* Samp $: \{0,1\}^r \to [n]^t$ *that uses*

- *t distinct samples for any $t \in [t_0, n]$, where $t_0 = O\left(\frac{\mu}{\theta^2} \cdot \log \frac{1}{\gamma}\right)$.*
- *$r = \log(n/t) + \log(1/\gamma) + 2\log(\mu/\theta) + \log\log(1/\mu) + O(1)$ random bits.*

5.3   The Local Extractor

Plugging the above two lemmas into Theorem 10, we obtain the following.

Theorem 13 (nonconstructive local extractors). *For every $n \in \mathbb{N}$, $\delta > 0$, $\varepsilon > 0$, and $m \leq \delta n/2 - 2\log(1/\varepsilon) - O(1)$, there is a t-local strong $(\delta n, \varepsilon)$ extractor* Ext $: \{0,1\}^n \times \{0,1\}^d \to \{0,1\}^m$ *with*

- *$d = \log n + 3\log(1/\varepsilon) + \log\log(1/\delta) + O(1)$.*
- *$t = O\left(\frac{m + \log(1/\delta) \cdot \log(1/\varepsilon)}{\delta}\right)$.*

In fact, the hidden constant for the m/δ term in the expression for t can be made arbitrarily close to the optimal value of 1 (by paying a price in the other hidden constants). We defer the exact expressions to the full version.

6   Explicit Constructions

In the previous section, we showed that very good locally computable extractors exist, but for applications we need *explicit* constructions. For an extractor Ext : $\{0,1\}^n \times \{0,1\}^d \to \{0,1\}^m$ or a sampler Samp : $\{0,1\}^r \to [n]^t$, explicit means that it is computable in polynomial time and polylogarithmic work-space with respect to its input+output lengths (i.e., $n+d+m$ for an extractor and $r+t\log n$ for a sampler). For a t-local extractor Ext : $\{0,1\}^n \times \{0,1\}^d \to \{0,1\}^m$, we give it oracle access to its first input and view the input length as $\log n + t + d$ ($\log n$ to specify the length of the oracle, and t as the number of bits actually read).

There are many explicit constructions of averaging samplers and extractors in the literature and thus a variety of local extractors can be obtained by plugging these into Theorem 10. We do not attempt to describe all possible combinations here, but rather describe one that seems particularly relevant to cryptography in the bounded storage model. We recall the following features of this application:

- The local extractor should work for sources of min-entropy $(\alpha - \beta)n - \log(1/\varepsilon)$, which is $\Omega(n)$ for most natural settings of the parameters. (Recall that α is the entropy rate of the random source and β is the storage rate of the adversary.) That is, we can concentrate on constant min-entropy rate.
- Optimizing the number of bits read from the source seems to be at least as important as the seed length of the extractor.
- The error ε of the extractor will typically be very small, as this corresponds to the "security" of the scheme.
- We are not concerned with extracting all of the entropy from the source, since we anyhow will only be reading a small fraction of the source.

6.1   The Extractor

With the above criteria in mind, the most natural extractor to use (in Theorem 10) is Zuckerman's extractor for constant entropy rate [21]:

Lemma 14 ([21]). *For every constant $\delta > 0$, every $n \in \mathbb{N}$, and every $\varepsilon > \exp(-n/2^{O(\log^* n)})$, there is an explicit strong $(\delta n, \varepsilon)$-extractor Ext : $\{0,1\}^n \times \{0,1\}^d \to \{0,1\}^m$ with $d = O(\log n + \log(1/\varepsilon))$ and $m = \delta n/2$.*

6.2   The Sampler

For the averaging sampler, the well-known sampler based on random walks on expander graphs provides good parameters for this application. Indeed, its randomness and sample complexities are both optimal to within a constant factor when μ and θ are constant (and the minimal sample complexity is used). However, we cannot apply it directly because it does not guarantee distinct samples, and we do not necessarily want to minimize the number of samples. Nisan and Zuckerman [9] presented some methods for getting around these difficulties, but their analysis does not directly apply here since we impose a stronger requirement on the sampler. (As mentioned earlier, we could use their sampler with their version of Lemma 9, at the price of worse constant factors in the number of bits read from the source.) Thus we introduce some new techniques to deal with these issues.

The following gives a modification of the expander sampler which guarantees distinct samples.

Lemma 15 (modified expander sampler). *For every* $0 < \theta < \mu < 1$, $\gamma > 0$, *and* $n \in \mathbb{N}$, *there is an explicit* (μ, θ, γ) *averaging sampler* $\mathrm{Samp} : \{0,1\}^r \to [n]^t$ *that uses*

- *t distinct samples for any $t \in [t_0, n]$, where $t_0 = O\left(\frac{1}{\theta^2} \cdot \log(1/\gamma)\right)$, and*
- *$r = \log n + O(t \cdot \log(1/\theta))$ random bits.*

The main idea in the proof is to show that a "short" random walk on a "good" expander is unlikely to have "many" repeats. Specifically, in a random walk of length t on an n-vertex expander of normalized second eigenvalue λ, the expected fraction of repeated vertices is at most $t/n + O(\lambda)$. We prove this by the "trace method," which expresses repeat probabilities in terms of the trace of powers of the adjacency matrix. Given this bound on expected repeats, we obtain our sampler by taking several random walks on the expander (dependently, using another expander!) until we obtain a walk with a small fraction of repeats, from which we can safely discard the repeats without substantially affecting the sampler's estimate. Details are given in the full version.

A drawback of the expander sampler is that the randomness increases with the number of samples, whereas in the optimal sampler of Lemma 12 the randomness actually decreases with the number of samples. To fix this, we use the following lemma, which shows that the number of (distinct) samples can be increased at no cost.

Lemma 16. *Suppose there is an explicit* (μ, θ, γ) *averaging sampler* $\mathrm{Samp} : \{0,1\}^r \to [n]^t$ *with distinct samples. Then for every* $m \in \mathbb{N}$, *there is an explicit* (μ, θ, γ) *averaging sampler* $\mathrm{Samp}' : \{0,1\}^r \to [m \cdot n]^{m \cdot t}$ *with distinct samples.*

In fact, there is a gain as the sample complexity increases, because the randomness complexity depends only on the original value of n, rather than $n' = m \cdot n$. This simple observation about samplers (employed in conjunction with Lemma 9) has played a role in the recent construction of extractors that are "optimal up to constant factors" [14].

To apply this lemma to construct a sampler with given values of n, t, μ, θ, and γ, it is best to start with a sampler $\mathrm{Samp}_0 : \{0,1\}^{r_0} \to [n_0]^{t_0}$ using the minimal sample complexity $t_0 = t_0(\theta, \mu, \gamma) < t$ and domain size $n_0 = n \cdot (t_0/t)$. For example, for constant μ and θ, the sampler of Lemma 15 will give $t_0 = O(\log(1/\gamma))$ and $r_0 = \log n_0 + O(\log(1/\gamma)) = \log(n/t) + O(\log(1/\gamma))$. Then setting $m = t/t_0$, Lemma 16 gives a sampler for domain size $n_0 \cdot m = n$, using $t_0 \cdot m = t$ distinct samples, and r_0 random bits. This is how we obtain our final sampler, stated in the next lemma.

Lemma 17. *For every* $n \in \mathbb{N}$, $1 > \mu > \theta > 0$, $\gamma > 0$, *there is a* (μ, θ, γ) *averaging sampler* Samp $: \{0,1\}^r \to [n]^t$ *that uses*

- *t distinct samples for any* $t \in [t_0, n]$, *where* $t_0 = O\left(\frac{1}{\theta^2} \cdot \log \frac{1}{\gamma}\right)$, *and*
- $r = \log(n/t) + \log(1/\gamma) \cdot \operatorname{poly}(1/\theta)$ *random bits.*

Unfortunately, when t/t_0 and $n \cdot (t_0/t)$ are not integers, some care is needed to deal with the rounding issues in the argument given above. The tedious details are given in the full version.

6.3 The Local Extractor

Analogously to Theorem 13, we plug Lemmas 14 and 17 into Theorem 10 to obtain:

Theorem 18 (explicit local extractors). *For every constant* $\delta > 0$, $n \in \mathbb{N}$, $\varepsilon > \exp(-n/2^{O(\log^* n)})$, $m \le \delta n/4$, *there is an explicit t-local strong* $(\delta n, \varepsilon)$ *extractor* Ext $: \{0,1\}^n \times \{0,1\}^d \to \{0,1\}^m$ *with*

- $d = O(\log n + \log(1/\varepsilon))$.
- $t = O(m + \log(1/\varepsilon))$.

The above construction does generalize to the case of subconstant δ. The expression for t is actually $t = O(m/\delta + \log(1/\varepsilon)/\delta^2)$, which is not too bad compared with non-explicit construction of Theorem 13. In fact, as in Theorem 13, the hidden constant in the m/δ term can be made arbitrarily close to the optimal value of 1. The seed length d is $d = O((\log n + \log(1/\varepsilon)) \cdot \operatorname{poly}(1/\delta))$, but here the multiplicative dependence on $\operatorname{poly}(1/\delta)$ is much worse than the additive dependence on $\log\log(1/\delta)$ in Theorem 13. This is due to both underlying components — the extractor (Lemma 14) and the averaging sampler (Lemma 17). The dependence on δ in the extractor component can be made logarithmic by using one of the many known explicit extractors for subconstant min-entropy rate. For the averaging sampler, too, there are constructions whose randomness complexity is within a constant factor of optimal [21]. However, these constructions have a sample complexity that is only polynomial in the optimal bound, resulting in a t-local extractor with $t \ge \operatorname{poly}(\log(1/\gamma), 1/\delta)$. It is an interesting open problem, posed in [22], to construct averaging samplers whose sample and randomness complexities are both within a constant factor of optimal. (Without the "averaging" constraint, there are samplers which achieve this [25,26,22].)

6.4 Previous Constructions

Some previous constructions of cryptosystems in the bounded storage model can be understood using our approach, namely Theorem 10 together with Theorem 5 (of [8]). For example, the cryptosystem of Cachin and Maurer [3] amounts to using pairwise independence for both the averaging sampler and the extractor. (The fact that pairwise independence yields a sampler follows from Chebychev's

Inequality [27], and that it yields an extractor is the Leftover Hash Lemma of [28].) Actually, in the description in [3], the seed for the extractor is chosen at the time of encryption and sent in an additional interactive step. But it follows from this analysis that it actually can be incorporated in the secret key, so interaction is not necessary.

Our approach also yields an alternative proof of security for the ADR cryptosystem [4,5]. Consider the sampler which simply chooses a random t-subset of $[n]$ for $t = O(\log(1/\varepsilon))$ and the extractor $\text{Ext} : \{0,1\}^t \times \{0,1\}^t \to \{0,1\}$ defined by $\text{Ext}(x,r) = x \cdot r \bmod 2$. The correctness of the sampler follows from Chernoff-type bounds, and the correctness of the extractor from the Leftover Hash Lemma [28]. Combining these via Theorem 10 yields a locally computable extractor which simply outputs the parity of a random subset of $O(\log(1/\varepsilon))$ bits from the source. This is essentially the same as the ADR cryptosystem, except that the size of the subset is chosen according to a binomial distribution rather than fixed. However, the security of the original ADR cryptosystem follows, because subsets of size exactly $t/2$ are a nonnegligible fraction $(\Omega(1/\sqrt{t}))$ of the binomial distribution. To extract m bits, one can apply this extractor m times with independent seeds, as done in [5].

Acknowledgments. Noga Alon, Omer Reingold, and Amnon Ta-Shma also thought of similar approaches to this problem, and I am grateful to them for providing me with several helpful suggestions. I am also grateful to Yan Zong Ding and Chi-Jen Lu for pointing out errors in an earlier version of this paper. I thank Oded Goldreich for his encouragement and many helpful comments on the presentation. I thank Yan Zong Ding, Dick Lipton, and Michael Rabin for a number of illuminating discussions about the bounded storage model. Finally, I thank the anonymous CRYPTO and J. Cryptology reviewers for several helpful suggestions.

References

1. Vadhan, S.P.: On constructing locally computable extractors and cryptosystems in the bounded storage model. Cryptology ePrint Archive, 2002/162 (2002)
2. Maurer, U.: Conditionally-perfect secrecy and a provably-secure randomized cipher. J. Cryptology **5** (1992) 53–66
3. Cachin, C., Maurer, U.: Unconditional security against memory-bounded adversaries. In: CRYPTO '97. Spring LNCS 1294 (1997) 292–306
4. Aumann, Y., Rabin, M.O.: Information theoretically secure communication in the limited storage space model. In: CRYPTO '99. Springer LNCS 1666 (1999) 65–79
5. Aumann, Y., Ding, Y.Z., Rabin, M.O.: Everlasting security in the bounded storage model. IEEE Trans. Information Theory **48** (2002) 1668–1680
6. Ding, Y.Z., Rabin, M.O.: Hyper-encryption and everlasting security (extended abstract). In: 19th STACS. Springer LNCS 2285 (2002) 1–26
7. Dziembowski, S., Maurer, U.: Tight security proofs for the bounded-storage model. In: 34th STOC. (2002) 341–350 See also preliminary journal version, entitled "Optimal Randomizer Efficiency in the Bounded-Storage Model," Dec. 2002.

8. Lu, C.J.: Hyper-encryption against space-bounded adversaries from on-line strong extractors. In: CRYPTO '02. Springer LNCS 2442 (2002) 257–271
9. Nisan, N., Zuckerman, D.: Randomness is linear in space. J. Computer & System Sci. **52** (1996) 43–52
10. Nisan, N., Ta-Shma, A.: Extracting randomness: A survey and new constructions. J. Computer & System Sci. **58** (1999) 148–173
11. Shaltiel, R.: Recent developments in explicit constructions of extractors. Bull. EATCS **77** (2002) 67–95
12. Bar-Yossef, Z., Reingold, O., Shaltiel, R., Trevisan, L.: Streaming computation of combinatorial objects. In: 17th CCC. (2002) 165–174
13. Trevisan, L.: Extractors and pseudorandom generators. JACM **48** (2001) 860–879
14. Lu, C.J., Reingold, O., Vadhan, S., Wigderson, A.: Extractors: Optimal up to constant factors. In: 35th STOC. (2003)
15. Chor, B., Goldreich, O.: Unbiased bits from sources of weak randomness and probabilistic communication complexity. SIAM J. Computing **17** (1988) 230–261
16. Zuckerman, D.: Simulating BPP using a general weak random source. Algorithmica **16** (1996) 367–391
17. Shannon, C.E.: Communication theory of secrecy systems. Bell System Technical Journal **28** (1949) 656–715
18. Reingold, O., Shaltiel, R., Wigderson, A.: Extracting randomness via repeated condensing. In: 41st FOCS. (2000)
19. Bellare, M., Rompel, J.: Randomness-efficient oblivious sampling. In: 35th FOCS. (1994) 276–287
20. Canetti, R., Even, G., Goldreich, O.: Lower bounds for sampling algorithms for estimating the average. Information Processing Letters **53** (1995) 17–25
21. Zuckerman, D.: Randomness-optimal oblivious sampling. Random Struct. & Alg. **11** (1997) 345–367
22. Goldreich, O.: A sample of samplers: A computational perspective on sampling. Technical Report TR97-020, ECCC (1997)
23. Ta-Shma, A.: Almost optimal dispersers. Combinatorica **22** (2002) 123–145
24. Radhakrishnan, J., Ta-Shma, A.: Bounds for dispersers, extractors, and depth-two superconcentrators. SIAM J. Discrete Math. **13** (2000) 2–24 (electronic)
25. Bellare, M., Goldreich, O., Goldwasser, S.: Randomness in interactive proofs. Computational Complexity **3** (1993) 319–354
26. Goldreich, O., Wigderson, A.: Tiny families of functions with random properties: A quality-size trade-off for hashing. Random Struct. & Alg. **11** (1997) 315–343
27. Chor, B., Goldreich, O.: On the power of two-point based sampling. J. Complexity **5** (1989) 96–106
28. Håstad, J., Impagliazzo, R., Levin, L.A., Luby, M.: A pseudorandom generator from any one-way function. SIAM J. Comput. **28** (1999) 1364–1396

Unconditional Authenticity and Privacy from an Arbitrarily Weak Secret

Renato Renner[1] and Stefan Wolf[2]

[1] Department of Computer Science, ETH Zürich, Switzerland.
`renner@inf.ethz.ch`
[2] Département d'Informatique et R.O., Université de Montréal, Canada.
`wolf@iro.umontreal.ca`

Abstract. Unconditional cryptographic security cannot be generated simply from scratch, but must be based on some given primitive to start with (such as, most typically, a private key). Whether or not this implies that such a high level of security is necessarily impractical depends on how weak these basic primitives can be, and how realistic it is therefore to realize or find them in—classical or quantum—reality. A natural way of minimizing the required resources for information-theoretic security is to reduce the *length* of the private key. In this paper, we focus on the *level of its secrecy* instead and show that even if the communication channel is completely insecure, a shared string of which an arbitrarily large fraction is known to the adversary can be used for achieving fundamental cryptographic goals such as message authentication and encryption. More precisely, we give protocols—using such a weakly secret key—allowing for both the exchange of authenticated messages and the extraction of the key's entire amount of privacy into a shorter virtually secret key. Our schemes, which are highly interactive, show the power of two-way communication in this context: Under the given conditions, the same objectives cannot be achieved by one-way communication only.

Keywords. Information-theoretic security, authentication, privacy amplification, extractors, quantum key agreement.

1 Information-Theoretic Security and Its Price

1.1 Unconditional Authentication and Privacy Amplification with an Arbitrarily Weak Key by Completely Insecure Communication

The main advantage of *information-theoretic*—as opposed to *computational*—cryptographic security is the fact that it can be based on a mathematical proof which does not depend on any assumption on the hardness of certain computational tasks nor on an adversary's computing power or memory space. An important *disadvantage* of such unconditional security, on the other hand, is often perceived to be its impracticality. At the origin of this belief stands Shannon's

D. Boneh (Ed.): CRYPTO 2003, LNCS 2729, pp. 78–95, 2003.

famous result [22] stating that the perfectly secret transmission of a message over a public channel requires a private key of, roughly speaking, the same length.

In the present paper, we take a step towards making unconditional security more practical by showing that such a private key can be generated, by communication over a *completely insecure channel*, from an *arbitrarily weakly secret key*. One of the main ingredients of our protocol is a new interactive method for unconditionally secure message authentication requiring only a weak secret key as well. No such method has previously been known which works when the adversary knows more than half of the partial secret.

The problem of extracting a highly secret from a longer, partly compromised key—so-called *privacy amplification* [4], [3]—has been studied intensively since it is the final step of any information-theoretic key-agreement protocol based on classical or quantum correlations (e.g., quantum key agreement [2]). It is a direct consequence of our result that the assumption—usually made in the context of privacy amplification—that the communication channel is authentic can simply be dropped: Privacy amplification by communication over a *completely insecure* channel is possible even with arbitrarily weakly secret strings, and the length of the extractable private key is asymptotically the same as in the case of an authenticated channel or, equivalently, an only passive adversary. Previous results were pointing into another direction [14], [24], [17].

1.2   Towards Making Unconditional Security Practical

The main motivation for this work is to relax the conditions under which unconditional cryptographic security can be achieved. Our results should be seen in the context of a number of more or less recent steps, taken by various authors, towards making unconditionally provable security more practical by reducing the requirements for achieving it. For instance, techniques and protocols have been proposed allowing for generating provably secret keys from noisy channels [26], [6], weakly correlated classical information by public [13], [3] and even unauthenticated [15], [16], [17] communication, or from quantum channels [2]; reducing the required key size for authentication [23], [10] as well as encryption [21]; basing cryptographic tasks on keys from weak random sources [7]; or realizing information-theoretic secrecy against memory-bounded yet otherwise unlimited adversaries [12], [1], [8], to mention just a few.

1.3   Determining the Cryptographic Value of an Arbitrarily Weak Secret: The Power of Interaction

In the setting where two parties initially share some key in order to achieve cryptographic goals, two natural quantities to be minimized are the *length* and the *level of privacy* of this key. In this paper we address the question what the cryptographic value of a shared string is about which the adversary has almost complete information if we also assume her to have perfect read and write access to the communication channel. Our results show that such a key is useful both for achieving authenticity (Section 2) and privacy (Section 3) in this scenario.

We consider the following model. Two parties Alice and Bob both know an n-bit string S about which an adversary Eve has some partial information U. We also assume her to be able to read, modify, or delete any message sent over the communication channel connecting the legitimate partners. Ultimately, the goal of Alice and Bob is the exchange of a message M in a both authentic and confidential way. We achieve this in two steps which are described and analyzed in detail in Sections 2 and 3, respectively. First, we give an interactive protocol that allows, using the weak secret S, for the authentication of short messages. Second, we use this authentication technique as a building block of a protocol for distilling, from S, a *highly* secret string S'' the length of which is, roughly, the *min-entropy*[1] of S given Eve's knowledge $U = u$; this string can then be used for private key cryptography, in particular encryption and message authentication.

We describe our results in more detail. Let $0 < t \leq 1$ be an arbitrary constant, and assume that tn is a lower bound on the min-entropy of (the n-bit string) S from Eve's viewpoint (i.e., conditioned on $U = u$). Then Protocol AUTH of Section 2 allows for authenticating an l-bit message M, where $l = tn/s$ for a security parameter s: The probability of a successful active attack is of order $2^{-\Omega(s)}$. The protocol is computationally very efficient and uses $\Theta(l)$ rounds of communication. Note that all previously described protocols—interactive or one-way—for authentication with a partially secret key work only under the assumption that the key is more than "half secret" [24], [17].

Protocol AUTH can be used as a building block of a protocol allowing for distilling a short but highly secret key S'' from the initial string S, using communication over a completely insecure channel only. Let again tn be the min-entropy of the shared n-bit string S in Eve's view. Then Protocol PA of Section 3 allows Alice and Bob to generate a common string S'' of length $(1-o(1))tn$ about which Eve only has an exponentially (in n) small amount of information. In contrast to privacy amplification over an authenticated channel, privacy amplification secure against active adversaries has so far been known possible only for keys offering a relatively high secrecy level initially (at least two thirds of the key should be unknown to the adversary), and the length of the extractable secret was only a small fraction of the key's entropy [14], [24], [17]. It was speculated that this might be the price which has to be paid for the missing authenticity of the channel. Protocol PA shows that this is not so: Privacy amplification secure against active adversaries is equally powerful as against passive adversaries with respect to the condition on the initial string as well as to the size of the extractable secret.

Our results can alternatively be interpreted as realizing encryption and authentication using *private keys generated by weak random sources*—instead of *highly compromised keys*—as studied in [18], [7]. In [18] it was shown that weakly random keys from certain sources with substantial min-entropy do not allow for

---

[1] The *min-entropy* $H_\infty(X)$ of a random variable X with range $\mathcal{X}$ is simply the negative logarithm of the maximal probability occurring: $H_\infty(X) := -\log(\max_{x \in \mathcal{X}} P_X(x))$. We have $0 \leq H_\infty(X) \leq H(X) \leq \log|\mathcal{X}|$ for all random variables X. All logarithms, here and in the rest of the paper, are binary.

information-theoretically secure (one-way) encryption; in [7], it was proven that a weakly random key allows for (one-way) authentication only if its min-entropy exceeds half its length. Therefore, the results of [18] and [7] suggest that not all private keys with substantial randomness—i.e., min-entropy—are useful for basic cryptographic tasks. This is true, however, only in the one-way communication model: Our results add to this picture by showing that if *two-way communication* is allowed (and perfect randomness is available locally), then keys from *all* sources with non-negligible min-entropy allow for *both* authentication and encryption.

In [10], it has been shown that the use of interaction in authentication allows for dramatically reducing the *length* of the used (private) key. Our results underline the power of two-way communication, suggested by that result, in this context: Interaction alternatively allows for strongly relaxing the condition on the *degree of privacy* of the used key.

2   Authentication with an Arbitrarily Weak Key

2.1   Intuition and Building Blocks

In standard (one-way) authentication, the message to be authenticated is sent together with a so-called *authenticator*, i.e., an additional string depending on that message and the secret key. These methods fail as soon as the adversary has substantial knowledge about the key (more precisely, half the knowledge in terms of min-entropy [7]) since, very roughly speaking, this knowledge could consist of the correct authenticator for one or several messages. A possible way of overcoming this problem is to use a challenge-response protocol. In [24] (see also [17]), for instance, it was proposed that one party, the *sender*, sends the message as a *challenge*, the reception of which is confirmed by the other, the *receiver*, by sending *back* an authenticator. In this case the person in the middle Eve *has* to find the correct authenticator of a message which is *not of her choice*, even in the case of a substitution attack. As shown in [24], [17], one advantage of this scheme is that the authenticator can be short and thus leaks only a small amount of additional information about the key. On the other hand, however, its security could be shown only under the assumption that the adversary knows less than half the key; the same condition that characterizes the possibility of one-way authentication [7]. The reason is the attack where Eve uses the receiver of the message as an oracle and gets the correct response to a challenge of her choice (where she can make this choice adaptively after having seen the challenge for which she has to generate the correct response). In summary, such an interactive authentication method, where *the challenge is identical with the message* to be authenticated, may be preferable to one-way authentication in certain cases [24], [17], but *cannot*, in order to resist adaptive substitution attacks, tolerate Eve to have more knowledge about the "private" key—namely roughly half of it—than simple one-way authentication.

In Section 2.2 we propose a new protocol solving this problem by, roughly speaking, *preventing adaptive substitution attacks completely*. The main idea is to encode the message differently: The message bits do not determine the challenge strings (which are just random), but rather *which of them will be answered.*

Let us first have a look at how the authenticator should depend on the key and the message. Since we want the adversary to be able to compute the correct authenticator to only very few messages unless she knows the entire key, a natural way is to interpret the key as a polynomial, and let the authenticator be its evaluation at a point determined by the challenge. (This idea was already used in previous protocols of this type [24], [17].) Lemma 1 states that when this function is used, then even an adversary who knows almost the entire key cannot correctly respond to a random challenge except with small probability. A similar result was shown in [24], [17] with respect to Rényi entropy H_2.

Lemma 1. *Let n, k, and a be positive integers such that $n = k \cdot a$ holds, and let, for $x \in \{0,1\}^k$, $f_x : \{0,1\}^n \to \{0,1\}^k$ be the function $f_x(s) := \sum_{i=0}^{a-1} s_i x^i$. Here, the strings $s_i \in \{0,1\}^k$ are defined by $s = (s_0, s_1, \ldots, s_{a-1})$, and the k-bit strings s_i, x, and $f_x(s)$ are interpreted as elements of $GF(2^k)$ with respect to a fixed basis of $GF(2^k)$ over $GF(2)$. Let now S be a random variable with range $\mathcal{S} \subseteq \{0,1\}^n$ and distribution P_S such that when given $x \in \{0,1\}^k$ chosen according the uniform distribution, the probability that $f_x(S)$ can be guessed correctly is α. Then we have $\max_{s \in \mathcal{S}} P_S(s) \geq (\alpha - a/2^k)^a$ or, equivalently, $\alpha \leq 2^{-H_\infty(S)/a} + a/2^k$.*

Proof. We can assume that the guessing strategy is deterministic, i.e., only depends on x. For $s \in \mathcal{S}$, let α_s be the number of x for which $f_x(s)$ is guessed correctly, divided by 2^k. Then we have $\alpha = E_S[\alpha_s]$. The probability that $f_{x_i}(S)$ is guessed correctly simultaneously for a randomly chosen $x_1, \ldots, x_a \in GF(2^k)$ is lower bounded by

$$E_S \left[\prod_{i=0}^{a-1} \left(\alpha_s - \frac{i}{2^k} \right) \right] \geq E_S \left[\left(\alpha_s - \frac{a}{2^k} \right)^a \right] \geq \left(E_S \left[\alpha_s - \frac{a}{2^k} \right] \right)^a = \left(\alpha - a/2^k \right)^a . \quad (1)$$

(The second inequality of (1) is Jensen's inequality [9].) Therefore, there must exist a particular a-tuple $x_1, \ldots, x_a$ such that the values $f_{x_i}(S)$ are simultaneously guessed correctly with probability at least $(\alpha - a/2^k)^a$. On the other hand, S is uniquely determined by these $f_{x_i}(S)$ since $f_x(s)$ is a polynomial in x of degree at most $a - 1$ with coefficients $s_0, \ldots, s_{a-1}$. Hence there must exist a value $s \in \mathcal{S}$ with probability $P_S(s)$ at least (1), and this concludes the proof. □

During the execution of Protocol AUTH and Protocol PA, the adversary observes a number of messages that depend on the key and hence leak information about it. An important argument in the analysis of these protocols is an upper bound on the effect of such information on the min-entropy of the key (from the adversary's viewpoint). Roughly speaking, the min-entropy does not, except with small probability, decrease much more than by the number of physical bits observed. Results similar to Lemma 2 were proven in [5], [24], [17].

Lemma 2. *Let S, V, and W be discrete random variables with ranges $\mathcal{S}$, $\mathcal{V}$, and $\mathcal{W}$, respectively, such that S and V are independent, and let $b \geq 0$. Then $\text{Prob}_{VW}[H_\infty(S|V = v, W = w) \geq H_\infty(S) - \log |\mathcal{W}| - b] \geq 1 - 2^{-b}$.*

Proof. We have $\text{Prob}\,[P_{W|V}(w,v) < 2^{-b}/|\mathcal{W}|] < 2^{-b}$ (where $P_{W|V}$ stands for the conditional distribution of W given V), which implies that $P_{S|VW}(s,v,w) = P_{SVW}(s,v,w)/P_{VW}(v,w) = P_S(s) \cdot P_V(v) \cdot P_{W|SV}(w,s,v)/(P_V(v) \cdot P_{W|V}(w,v)) \leq P_S(s)/P_{W|V}(w,v) \leq P_S(s) \cdot |\mathcal{W}| \cdot 2^b$ holds with probability at least $1 - 2^{-b}$ over V and W. The statement now follows by maximizing over $s \in \mathcal{S}$ and by taking negative logarithms. $\qquad\square$

Lemma 3 finally gives a bound on the min-entropy of substrings in terms of the min-entropy of the full string [14]. It follows from the fact that every r-bit string s' corresponds to exactly 2^{n-r} n-bit strings s.

Lemma 3. *Let S be a random variable with range $\mathcal{S} \in \{0,1\}^n$, and let S' be an r-bit substring of S. Then we have $H_\infty(S') \geq H_\infty(S) - (n - r)$.*

2.2 The Authentication Protocol and Its Analysis

We now give Protocol AUTH. Let s be a string of length n and k a divisor of n. For $s \in \{0,1\}^n$ and $x \in \{0,1\}^k$, let $f_x(s)$ be defined as in Lemma 1. Finally, $b = (b_1,\ldots,b_r)$ are the bits to be authenticated; the bits are authenticated separately, one after another.

——————— **Protocol AUTH (Authentication)** ———————

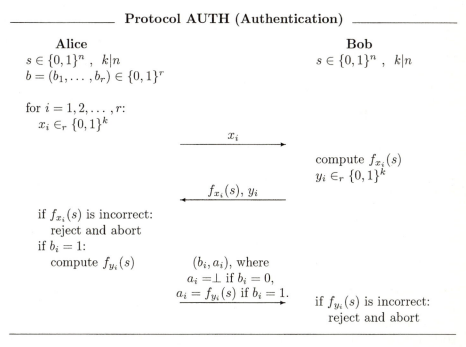

Let us first discuss some properties of Protocol AUTH intuitively. Note first that since the values x_i and y_i, which are chosen randomly, are independent from the message bit b_i, a person-in-the-middle attack in which x_i or y_i is substituted by another value is of no use for changing a message bit: only the fact *whether*

a response was given or not is important. If no such response is given by the legitimate party, then it is, according to Lemma 1, in any case difficult for the adversary to generate one, provided the min-entropy of the key is still large enough. This is why it is hard for the adversary to flip a bit from 0 to 1.

Since, on the other hand, an active adversary can simply delete a given response, it is trivial to flip a message bit 1 into a 0. Furthermore, message bits 0 can always be generated towards Bob without participation of Alice at all. We will take care of these problems later and transform the "semi-authentication protocol" into a complete authentication allowing for no undetected modifications of the message at all.

Let us, however, first make precise what Protocol AUTH, as given above, achieves. Lemma 4 states that the two above-mentioned types of undetected modifications of the message are the only ones possible unless Eve is able to generate a random challenge's response by herself. Note that statement 2 in Lemma 4 is a formalization of the fact that the string Bob receives can be obtained from the string sent by Alice by changing 1s into 0s and generating 0s from scratch.

Lemma 4. *Assume that Alice and Bob execute Protocol AUTH in the presence of an adversary Eve, that Alice has not aborted the protocol and has, so far, authenticated a string $b = (b_1, \ldots, b_j)$. Assume further that either Bob has rejected and aborted, or that in his view, a protocol round has just been completed (i.e., that the last message he received was a message bit together with the authentication string if the bit was 1) and that the string sent and authenticated up to this point is, still in Bob's view, $b' = (b'_1, \ldots, b'_{j'})$.*

Then if Eve has been passive, Bob has not rejected and $b' = b$ holds. If on the other hand Eve is active, at least one of the following three statements is true.

1. *Bob has rejected and aborted.*
2. *There exists an injective monotonically increasing function $g : \{1, \ldots, j\} \to \{1, \ldots, j'\}$ such that for all $1 \leq i \leq j'$, $b'_i = 1$ implies both $i \in \mathrm{Im}(g)$ and $b_{g^{-1}(i)} = 1$. (Note that this implies, in particular, $j' \geq j$ and $w_H(b') \leq w_H(b)$, where w_H denotes the Hamming weight.)*
3. *Eve has successfully computed and sent $f_z(s)$ for a value $z \in \{0, 1\}^k$ that she received from Alice or Bob without having received another message in-between. (In this case, we say that Eve was able to answer a random challenge without help.)*

Proof. Assume first that there is no active adversary, i.e., that no message sent has been modified or deleted. Then, clearly, Bob is accepting and $b' = b$ holds.

Let us suppose that Eve is (potentially) an active adversary. We prove the statement by induction over j. Let first $j = 0$, i.e., Alice has not sent (nor received) any message. Assume that Bob is accepting and has received the string $b' = (b'_1, \ldots, b'_{j'})$. We have to show $b'_i = 0$ for all $1 \leq i \leq j'$ unless Eve was able to answer a challenge without help. Assume $b'_i = 1$ for some i. Then Bob's challenge y_i must have been correctly answered (by $f_{y_i}(s)$) without Bob sending any other message between sending y_i and receiving the response. Since also Alice has not sent any message so far, Eve must have generated $f_{y_i}(s)$ without help.

Suppose now that the statement is true for $j \geq 0$; we prove its validity for $j + 1$. Assume that a protocol round has been concluded in Bob's view, and that 1. and 3. are not true. Let $(b'_1, \ldots, b'_{j''})$ be the string authenticated so far in Bob's view. Just before Bob's receiving of x_{j+1} (or a possibly modified value x'_{j+1}) and sending of $f_{x_{j+1}}(s)$ (or of the value $f_{x'_{j+1}}(s)$ that will allow Eve to determine $f_{x_{j+1}}(s)$—and since 3. is wrong she *must* have received such a message), a protocol round had been concluded in his view and the message received up to that point was an initial substring of b', i.e., $(b'_1, \ldots, b'_{j'})$ for some $j' \leq j''$. At that point, Alice had authenticated the string $(b_1, \ldots, b_j)$. By the induction hypothesis, and since 1. and 3. do not hold, there exists $g : \{0,1\}^j \to \{0,1\}^{j'}$ with the required properties. For establishing the statement for $j+1$ (i.e., proving that g can be extended to $\{1, \ldots, j+1\}$), we have to show two facts. First, $j'' > j'$ must hold, and secondly, we must have $w_H((b_{j'+1}, \ldots, b_{j''})) \leq 1$, where equality implies $b_{j+1} = 1$.

Since Alice has received $f_{x_{j+1}}(s)$, and since this cannot have been generated by Eve without help, Bob must have sent at least one message after Alice's sending of x_{j+1}. Thus, because Bob is still accepting, we have $j'' > j'$. On the other hand, for every value $i \in \{j'+1, \ldots, j''\}$ with $b'_i = 1$, Bob must have received $f_{y_i}(s)$ correctly after his challenge y_i. Since Eve has not computed this value without help (3. is untrue), Alice must have sent a value $f_{y'_{j+1}}$ between Bob's sending and receiving of y_i and $f_{y_i}(s)$, respectively. This implies both $w_H((b_{j'+1}, \ldots, b_{j''})) \leq 1$—since Alice has sent at most one such value during what was a single protocol round in her view—and that in case of equality, Alice must have authenticated the bit $b_{j+1} = 1$ in the last step. This concludes the induction step and the proof.   □

Clearly, Protocol AUTH cannot be used *directly* for the authentication of messages $(b_1, \ldots, b_r)$ by the following three reasons. First of all, it is, for an active adversary, easy to flip a bit from 1 to 0 without being detected, or to insert a 0 at any point. Secondly, Eve can block all messages sent after some point without Bob realizing that he only received part of the message. (In this case, Alice, but not Bob, would realize the attack, reject, and abort.) Finally, Bob can be used as an oracle for finding out the entire key: Eve simply impersonates Alice and authenticates a sufficient number of 0s to Bob.

The third problem can be solved by limiting the length of the message; this limit L must be chosen such that even $2L$ values $f_z(s)$ (where $2L$ different values for z can be chosen by Eve) do not reveal the entire key, but leave sufficient uncertainty in terms of min-entropy to guarantee the security of the protocol.

In order to get rid of the first two problems, we restrict the set of possible messages (of even length r): a string $b = (b_1, \ldots, b_r)$ is a valid message only if it is *balanced*, i.e., if half the bits are 0s and the other half are 1s, and if every initial substring $(b_1, \ldots, b_i)$, $i < r$, is *"underweight"*: the number of 1s is strictly less than $i/2$. If, given that the sent string b satisfies these conditions,

Bob accepts the outcome[2] only if the received string b' is balanced, then he is prevented from erroneously accepting in case Eve performs one of the described attacks, and $b' = b$ must hold. Fortunately, the given restriction on the strings to be authenticated only reduces the effective message length insignificantly, as Lemma 5 shows. It follows from a well-known result on random walks (see for example [9]), and from Stirling's formula.

Lemma 5. *Let r be an even integer and let $z(r)$ be the number of r-bit strings $b = (b_1, \ldots, b_r)$ satisfying $w_H((b_1, \ldots, b_r)) = r/2$ and $w_H((b_1, \ldots, b_i)) < i/2$ for all $1 \leq i < r$. Then we have $z(r) = \binom{r}{r/2}/(2(r-1)) = \Theta(2^r/r^{3/2})$, hence $\log(z(r)) = (1 - o(1))r$.*

Let us assume from now on that Protocol AUTH is used in the way described above: Bob rejects and aborts when given more than L challenges (where L is an additional protocol parameter to be properly chosen), and he accepts the outcome only if the received message is balanced. We will prove that with this modification, Protocol AUTH is a secure authentication protocol.

Note that since the protocol uses two-way communication, also Alice can detect an active attack, reject, and abort the protocol. Unfortunately, it is not possible to achieve agreement of Alice's and Bob's acceptance states in every case in the presence of an active adversary (who can, for instance, delete the final message sent). However, our protocol *does* achieve that whenever Bob accepts, then so does Alice, and Bob has received the correct string (except with small probability). This means that every active attack detected by Alice is automatically also perceived by Bob; the final decision whether the authentication succeeded is hence up to the receiver—just as in one-way authentication.

Theorem 1 makes the security of Protocol AUTH precise. It is only due to simplicity that the result is stated asymptotically. The protocol is useful already for short strings. The proof of Theorem 1 explicitly shows all the involved constants that are neglected in the asymptotic notation.

Theorem 1. *Let S be a random variable, with range $\mathcal{S} \subseteq \{0,1\}^n$, known to Alice and Bob, and let U summarize an adversary Eve's entire knowledge about S. Assume $H_\infty(S|U = u) \geq tn$ for the particular value $u \in \mathcal{U}$ known to Eve, where $0 < t \leq 1$ is a constant. Let now $k < tn/7$ be of order[3] $k = \omega(\log n)$. Then, for some $l = (1 - o(1))(tn - k)/3k$, Protocol AUTH can be used to authenticate, by communication over a completely insecure channel, a message m of at most l bits sent from Alice to Bob. More precisely, the following holds: If Eve is passive, then Alice and Bob accept the outcome of the protocol and Bob receives the correct message. If Eve is active, then, with probability $1 - 2^{-\Omega(k)}$, either Bob rejects and aborts the protocol, or Alice and Bob both accept and Bob receives the correct message m.*

---

[2] We say that a party *accepts the outcome* of a protocol if he has not rejected and aborted and the execution of the protocol is, or could be, finished from his point of view.

[3] Here, $f = \omega(g)$ stands for $f/g \to \infty$.

Remark. Note that k can be freely chosen subject to the conditions $k < tn/7$ and $k = \omega(\log n)$. Since the success probability of an active attack is bounded as $2^{-\Omega(k)}$, choosing a greater k is more secure; on the other hand, the smaller k is, the longer can the authenticated message be.

Proof of Theorem 1. We first observe that we can assume n to be a multiple of k if we replace, at the same time, the entropy condition by $H_\infty(S|U = u) > tn - k > 6k$. The reason is that Alice and Bob can cut at most $k - 1$ bits at the end of S, reducing the min-entropy by less than k according to Lemma 3.

Let now L be the greatest even integer such that $L \le (tn - k)/3k$ holds, and let $l := \lfloor \log z(L) \rfloor = (1 - o(1))(tn - k)/3k$. Here, z is the function defined in Lemma 5; the maximum message length l is hence chosen such that there exists a one-to-one mapping from $\{0,1\}^l$ to the set of L-bit strings satisfying the conditions of Lemma 5. If the length of the *actual* message m to be authenticated is shorter, then the number r of bits b_i in Protocol AUTH is smaller as well. Let in the following $b = (b_1, \dots, b_r)$ be the r-bit string (where $r \le L$ holds) which satisfies the conditions of Lemma 5 and corresponds to the message m.

Assume that Alice and Bob execute Protocol AUTH with respect to the key S, the parameter k, maximal message length L, and the string b. Let first Eve be passive. Then, clearly, Alice and Bob accept the outcome of the protocol and Bob receives the correct message, as sent by Alice.

Let now Eve be a possibly active adversary. Since neither Alice nor Bob generate and send responses for more than L challenges during the execution of the protocol (L is the maximum possible length of the string b; if Bob, for instance, is challenged for more than L times, he will conclude that there is an active attack, reject, and abort), we have at every point in the protocol that, for any $a \ge 0$,

$$\text{Prob}_C[H_\infty(S|U = u, C = c) \ge tn - k - 2Lk - 2La] \ge 1 - 2L2^{-a} \qquad (2)$$

holds, where C stands for the collection of all messages sent by Alice and Bob so far. Inequality (2) follows from $2L$-fold application of Lemma 2: At most $2L$ times, Eve has observed a string $f_x(s)$ (or $f_y(s)$) where X and S are, given Eve's entire knowledge at this point, independent. (It is important to see that the latter is true even if x is chosen by Eve herself, depending on all her knowledge. When applying Lemma 2 here, the distribution P_S in the statement of the lemma has to be replaced by $P_{S|U=u,C'=c'}$, where $U = u$ and $C' = c'$ summarize this knowledge.)

Since the total number of challenges generated by Alice or Bob in Protocol AUTH with maximal message length L is also upper bounded by $2L$, the probability $\text{Prob}[\mathcal{A}]$ of the event $\mathcal{A}$ that Eve can correctly answer one of them without help is, according to Lemma 1, inequality (2), and the union bound, at most

$$\text{Prob}[\mathcal{A}] \le 2L\left(2^{-(tn-k-2L(k+a))/(n/k)} + \frac{n/k}{2^k}\right) + 2L2^{-a}$$

for any $a \ge 0$; the choice $a := (tn - k)/(12L)$ leads to

$$\text{Prob}\,[\mathcal{A}] \;\leq\; 2\,\frac{tn-k}{3k}\left(2^{-(tn-k)/(6n/k)} + \frac{n/k}{2^k} + 2^{-(tn-k)/(12L)}\right)$$
$$= \; O((n/k)^2)\cdot 2^{-\Omega(k)} \;=\; 2^{-\Omega(k)}\,, \tag{3}$$

where the first "equality" in (3) holds because of $k < tn/7$, and the second one since k is of order $\omega(\log n)$.

Let us assume that $\mathcal{A}$ does not occur, and that Bob accepts the outcome of the protocol. Let $(b_1, \dots, b_j)$, for $j \leq r$, be the bits authenticated in Alice's view. According to Lemma 4, the string $b' = (b'_1, \dots, b'_{j'})$ that Bob receives can be obtained from $(b_1, \dots, b_j)$ by inserting 0s and flipping 1s to 0s. Then, $w_H(b') = j'/2$—a necessary condition for Bob to accept—implies $w_H((b_1, \dots, b_j)) \geq j'/2 \geq j/2$. Since b satisfies the conditions of Lemma 5, we have $w_H((b_1, \dots, b_j)) = j/2$ and $j = r$ (i.e., Alice has sent the entire message already and hence accepts the outcome), as well as $j' = j = r$ and $b' = (b'_1, \dots, b'_{j'}) = (b_1, \dots, b_j) = (b_1, \dots, b_r) = b$: Bob receives the correct string b and message m. $\qquad\square$

3   Confidentiality from an Arbitrarily Weak Key

3.1   Privacy Amplification and Extractors

Privacy amplification means extracting a weakly secret string's randomness as seen from the adversary's viewpoint, and has been shown possible under the assumption that either the communication channel is authentic [4], [3], [5], or that the initial key's privacy level is already high [14], [24], [25], [17]. Here, we show that these two restrictions can be dropped simultaneously.

It was shown in [3] that—in the authentic-channel model—*universal hashing* is a good technique for privacy amplification, allowing for extracting virtually all the so-called *Rényi entropy* into a highly secret key. Another randomness-extraction technique, which has attracted a lot of attention recently in the context of derandomization of probabilistic algorithms, are *extractors*, which allow, by using only very few additional truly random bits, for extracting a weakly random source's complete *min-entropy*. The fact that extractors can distill only the min-entropy—instead of the Rényi entropy, which is a priori up to two times greater—is insignificant according to a recent unpublished result [11] stating that for every distribution P, there exists a distribution P' such that the variational distance between P and P' (both with range $\mathcal{X}$), defined as $d(P, P') := (\sum_{x\in\mathcal{X}} |P(x) - P'(x)|)/2$, is small and $H_\infty(P') \approx H_2(P)$ holds. This means that universal hashing is ultimately nothing else than a particular extractor—one which is, however, very inefficient with respect to the number of additional randomness required.

Using extractors, we show in Sections 3.2 and 3.3 that privacy amplification over an unauthenticated public channel can be (almost) as powerful—both with respect to the conditions for the *possibility in principle* and to the *length* of the resulting key—as over an authentic channel. When used for privacy amplification over a public channel, extractors should be *strong*, meaning that the output's distribution is close to uniform *even when given the truly random bits*. The existence of such extractors, distilling virtually all the source's min-entropy, was proven in [19], [20]. Theorem 2 is a direct consequence of these results.

Theorem 2. *For integers $D \leq n$ and a real number ε of order $\Omega(2^{-n/\log n})$ there exist $r = O(((\log n)^2 + \log(1/\varepsilon))\log D)$, $m = D - 2\log(1/\varepsilon) - O(1)$, and a function $E : \{0,1\}^n \times \{0,1\}^r \to \{0,1\}^m$ with the following property. If X is a random variable with range $\mathcal{X} \subseteq \{0,1\}^n$ and such that $H_\infty(X) \geq D$ holds, and if U_r stands for a random variable independent of X and with uniform distribution over $\{0,1\}^r$, then we have $d\left(P_{(E(X,U_r),U_r)}, P_{U_{m+r}}\right) \leq \varepsilon$, where $P_{U_{m+r}}$ is the uniform distribution over $\{0,1\}^{m+r}$.*

A function E having the properties given in Theorem 2 is called a *strong (D,ε)-extractor*.

Lemma 6 below justifies the use of strong extractors for privacy amplification (both in the passive- and active-adversary cases, and with respect to our new simplified notion of security of privacy amplification given below): The extractor's output is, with high probability, equal to a perfectly uniformly distributed "ideal" key independent of the random bits: The adversary has no information at all about this ideal key.

Lemma 6. *Let E be a function as in Theorem 2 with parameters n, D, r, m, and ε, let S be a random variable with $H_\infty(S) \geq D$, and let $R = U_r$ be the random variable corresponding to a uniformly distributed r-bit string independent of S. Let $S' := E(S, R)$. Then there exists a uniformly distributed m-bit random variable S'_{id} which is independent of R and such that $\mathrm{Prob}\left[S' = S'_{id}\right] \geq 1 - \varepsilon$ holds.*

Proof. For every value r_0 that R can take, there exists a uniformly distributed string $S'_{id}(r_0)$ with $\mathrm{Prob}\left[E(S, r_0) \neq S'_{id}(r_0)\right] = d(P_{E(S,r_0)}, P_{U_m})$ (again, P_{U_m} is the uniform distribution over $\{0,1\}^m$).

Let S'_{id} be the random variable defined by all the $S'_{id}(r_0)$. The statement now follows from $d\left(P_{(E(S,R),R)}, P_{U_{m+r}}\right) = E_R\left[d\left(P_{E(S,R)}, P_{U_m}\right)\right]$, which is true since R is uniform, and from $\mathrm{Prob}\left[E(S,R) \neq S'_{id}\right] = E_R[\mathrm{Prob}\left[E(S,R) \neq S'_{id}(R)\right]]$. □

3.2   The Idea Behind Protocol PA

Given Protocol AUTH of Section 2 and the extractors described in Section 3.1, it seems obvious how to achieve privacy amplification over an unauthenticated channel: Alice chooses the extractor's random input bits and sends them, using Protocol AUTH, to Bob. However, this solution has a conceptual error. Eve can, knowing the bits $b_1, \ldots, b_i$ already sent, perform active attacks—replacing x_{i+1} or y_{i+1} by values x'_{i+1} and y'_{i+1} of her choice and therefore learning $f_{x'_{i+1}}(S)$ and $f_{y'_{i+1}}(S)$—and obtain information about S that *depends on the bits b_j*; in other words, the extractor's second input would in this case not be independent of S from Eve's viewpoint, and Theorem 2 would not apply.

On the other hand, however, the string S *must* be used *both* for authentication *and* as the input for privacy amplification. This dilemma can be resolved by a two-step protocol: First, an extractor is used to generate, from S, a short key S' about which all the information—depending on the extractor bits or not—revealed during Protocol AUTH gives Eve less than half the total information (except with small probability). In a second step, the random bits actually used to apply privacy amplification on S are authenticated with the key S'. The crucial point here is that the information Eve learns about S during this second authentication is at most S', since "the rest" of S is not used at all. Such a two-step approach hence allows for controlling the information Eve obtains during the two authentication phases, even when she is carrying out active attacks.

For the second authentication, a key—namely S'—about which Eve knows less than half (in terms of min-entropy) can be used. In this case, a simpler and more efficient method than Protocol AUTH can be applied, namely *strongly universal (SU-) hashing* [23]. A result similar to Lemma 7, but with respect to Rényi entropy H_2, was proven in [14].

Lemma 7. *Let n be an even integer. Assume that Alice and Bob both know the value taken by a random variable S with range $\mathcal{S} \subseteq \{0,1\}^n$ and conditional min-entropy $H_\infty(S|U=u) \geq n/2+R$, where $U = u$ summarizes an adversary's entire knowledge about S. Assume further that Alice and Bob use S to authenticate an $n/2$-bit message M, where M and S are independent, given $U = u$, with strongly universal hashing, i.e., with the authenticator $A = MS_1 + S_2$, where S_1 and S_2 are the first and second halves of S, and where M, S_1, S_2, and A are interpreted as elements of $GF(2^{n/2})$ with respect to a fixed basis of $GF(2^{n/2})$ over $GF(2)$. Then Bob always accepts and receives the correct message if the adversary is passive, and in general we have with probability $1-2^{-\Omega(R)}$ that Bob either rejects or receives the correct message M, even if the adversary has full control over the communication channel.*

Proof. If Eve is passive, then, clearly, Bob accepts and receives the correct message. In general, the probability of a successful *impersonation* attack is at most the maximum probability of a subset of $2^{n/2}$ keys, i.e., at most $2^{n/2} \cdot 2^{-H_\infty(S|U=u)} \leq 2^{-R}$. Let now (m, a) be the correctly authenticated message observed by Eve in a *substitution attack*. According to Lemma 2, applied to the distribution $P_{S|U=u}$, this pair is with probability $\geq 1 - 2^{-R/2}$ such that $H_\infty(S|M = m, A = a, U = u) \geq R/2$ holds. (Note that M is independent of S, given $U = u$.) Because generating another correct pair (m', a'), for $m' \neq m$, is equivalent to guessing S and because of the union bound, the success probability of this attack is upper bounded by $2^{-R/2} + 2^{-R/2} = 2^{-(R/2-1)}$. $\square$

3.3 Asymptotically Optimal Privacy Amplification by Insecure Communication

We are now ready to give Protocol PA, and to prove our second main result. In the following, s is the n-bit key known to Alice and Bob about which Eve's information $U = u$ is limited by $H_\infty(S|U = u) \geq tn$, where t is an arbitrary constant with $0 < t \leq 1$. E_1 and E_2 are suitably chosen extractors.

───────────── **Protocol PA (Privacy Amplification)** ─────────────

Alice		**Bob**
$s \in \{0,1\}^n$		$s \in \{0,1\}^n$

$r \in_r \{0,1\}^{\Theta(\sqrt{n})}$

$$\xrightarrow{\quad \text{AUTH}(r) \text{ with } k = \Theta(\sqrt{n}/\log n) \quad}$$

$s' = E_1(s,r)$		$s' = E_1(s,r)$
$\text{len}(s') = l' = \Theta(n/\log n)$		
$s' = (s_1', s_2')$		
$r' \in_r \{0,1\}^{l'/2}$		
$a = s_1' r' + s_2'$		

$$\xrightarrow{\quad r', a \quad}$$

if a is incorrect:
reject
if a is correct:

$s'' = E_2(s,r')$		$s'' = E_2(s,r')$
$\text{len}(s'') = (1 - o(1))tn$		

In the first protocol phase, Alice authenticates the string r with Protocol AUTH (with parameter k).

Before stating the main result, we make a few remarks on the security achieved by Protocol PA. Note first that privacy amplification cannot be *guaranteed* to work in every case if Eve is assumed to have full control over the communication channel; the best one can hope for is that a possible active attack is detected.

A natural security definition would require that, with high probability, any party accepting the outcome of the protocol indeed has a highly secret key, and that if both parties accept, their keys are identical. (As already mentioned in the context of Protocol AUTH, one cannot demand for similar acceptance decisions of Alice and Bob.) Our protocol achieves even more than that: *If Bob accepts, then everything went well* (in particular, Alice also accepts).

The condition concerning the privacy of the resulting key stated and proven below is, although equivalent to, somewhat simpler than security definitions used previously in this context: Instead of giving a condition on the Shannon entropy of the resulting key, we consider privacy amplification successful if the generated key is, except with small probability, *equal to a perfectly secret ideal key*. This allows for simplifying the security proofs.

Theorem 3. *Assume that Alice and Bob know an n-bit string S satisfying $H_\infty(S|U = u) \geq tn$ for arbitrary $0 < t \leq 1$, where $U = u$ summarizes an* *adversary Eve's entire knowledge about S. Then Protocol PA allows, for suitable choices of the parameters and the extractors E_1 and E_2, for privacy amplification by communication over a completely insecure channel, distilling an arbitrarily large fraction of the min-entropy of S, given $U = u$, into a virtually secret string. More precisely, Protocol PA satisfies the following two conditions.*

1. *If Eve is* passive, *then Alice and Bob accept the outcome of the protocol and end up with the same string S'' of length $l'' = (1 - o(1))tn$ with the following property: There exists a string S''_{id} of the same length such that for all possible protocol communications $C = c$, $P_{S''_{id}|C=c,U=u}$ is the uniform distribution over the set of l''-bit strings, and such that $\text{Prob}\,[S'' = S''_{id}] = 1 - 2^{-\Omega(n/(\log n)^2)}$ holds.*

2. *If Eve is* active, *then the probability that either Bob rejects (and Alice either rejects as well or accepts and has computed a key S'' satisfying 1.), or that both Alice and Bob accept and that all the conditions of 1. hold, is of order $1 - 2^{-\Omega(\sqrt{n}/\log n)}$.*

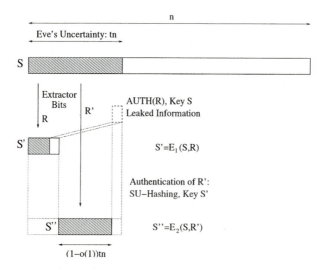

Fig. 1. Privacy amplification over an unauthenticated channel with Protocol PA. The privacy of S is extracted in two steps. The short key S' is more than half secret and used for authenticating the extractor bits for S''. This second key is the output of the protocol and highly secret although the information Eve obtains in the authentication depends on the extractor bits.

Proof. Note first that if Eve is passive, both parties accept the outcome and compute the same string S'', the secrecy of which, as stated in 1., follows from the subsequent analysis of the general case.

Let us hence assume that Eve is a possibly active adversary. Let $\varepsilon > 0$ be of order $2^{-\Theta(\sqrt{n}/\log n)}$. According to Theorem 2, there exist $r = \Theta(\sqrt{n})$, $k = \Theta(\sqrt{n}/\log n)$, and $l' = 7rk = o(n)$ as well as a strong extractor $E_1 : \{0,1\}^n \times \{0,1\}^r \to \{0,1\}^{l'}$ extracting l' bits out of S (distributed according to

$P_{S|U=u}$) with "error probability" ε. (Note that in fact, l' could be chosen much larger, namely almost tn, according to Theorem 2.)

The r randomly chosen bits R are now sent and authenticated using Protocol AUTH with parameter k. According to Theorem 1, the probability of a successful active attack to this authentication is $2^{-\Theta(k)} = 2^{-\Theta(\sqrt{n}/\log n)}$.

Let $S' = E_1(S, R)$ be the extractor's output. Because of Lemma 6, there exists an l'-bit string S'_{id} that is uniformly distributed conditioned on $U = u$, independent of the bits R (the second part of the extractor's input), and such that $\text{Prob}\,[S' = S'_{id}] \geq 1 - \varepsilon = 1 - 2^{-\Omega(\sqrt{n}/\log n)}$ holds.

Let C be all the messages sent by Alice and Bob during the execution of Protocol AUTH. Since every party sends at most $l = (1 + o(1))r$ messages (of length k) of the form $f_{x'_i}(s)$ or $f_{y'_i}(s)$ (and since the respective challenges x'_i and y'_i are, even if generated by Eve, independent of S, given Eve's knowledge about S at this moment), Lemma 2—applied $2l$ times—implies that

$$H_\infty(S'_{id}|C = c, U = u) \geq l' - 2lk - lk = l'/2 + \Omega(rk) \qquad (4)$$

holds with probability at least $1 - 2l2^{-lk} = 1 - 2^{-\Omega(n/\log n)}$. The "equality" in (4) is true because l' has been defined to be equal to $7rk$, and because of $l = (1 + o(1))r$.

According to Lemma 7, the success probability of an active attack on the second authentication, using strongly universal hashing with the key S', is hence of order

$$2^{-\Omega(n/\log n)} + 2^{-\Omega(rk)} + 2^{-\Omega(\sqrt{n}/\log n)} = 2^{-\Omega(\sqrt{n}/\log n)} \ . \qquad (5)$$

(The first term in (5) is the probability that (4) does not hold, the second term is the attack success probability if the key S'_{id} would be used and given that (4) holds, and the third term is the probability that the actually used key S' differs from S'_{id}. The bound (5) then follows from the union bound.)

Let us now look at the remaining min-entropy of S, given all the communication Eve has observed. Note first that the last authentication reveals information about S to Eve that depends on the random bits R' sent in this step. This dependence is a potential problem since R' must be chosen completely independently from S given Eve's knowledge and is, with respect to authentication with Protocol AUTH, the reason for the "two-step" nature of Protocol PA. However, under the (pessimistic) assumption that Eve learns the *entire* key S', she cannot obtain *any additional* information about S, in particular no information depending on R', since the rest of S is not used at all in this authentication. In other words, if we assume Bob to announce S' to Eve after the second authentication (what, of course, he does not actually have to do), then R' is independent of S given Eve's total knowledge.

We now have that $H_\infty(S|C = c, S' = s', U = u) \geq tn - \Theta(rk)$ holds with probability $1 - 2^{-\Omega(rk)} = 1 - 2^{-\Omega(n/\log n)}$, as above for S'. Because of Theorem 2, there exists a strong extractor $E_2 : \{0,1\}^n \times \{0,1\}^{r'} \to \{0,1\}^{l''}$ with parameters $r' \leq l'/2 = \Theta(n/\log n)$ (note that $l'/2$ is the possible message length in the last authentication), $\varepsilon' = 2^{-\Theta(n/(\log n)^2)}$, and $l'' = tn - \Theta(rk) - 2\log(1/\varepsilon') =$

$tn - o_1(n) - o_2(n) = (1 - o(1))tn$. The extractor's output $S'' = E_2(S, R')$ satisfies, according to Lemma 6, the following condition. There exists an l''-bit string S''_{id} such that $P_{S''_{id}|C=c,U=u}$ is the uniform distribution (where C is the entire protocol communication) and $\text{Prob}\,[S'' = S''_{id}] \geq 1 - \varepsilon' = 1 - 2^{-\Omega(n/(\log n)^2)}$ holds. The final statement now follows from the union bound.

4   Concluding Remarks

We have shown that two parties who are connected by a communication channel under full adversarial control and who share a key that is arbitrarily weakly secret can not only exchange authenticated messages, but also generate an unconditionally secret key. The given protocols for achieving this are computationally very efficient for the legitimate parties; they require two-way communication, where the number of rounds is of order $O(r)$ for the authentication protocol (if r is the length of the message to be authenticated) and $O(\sqrt{n})$ for privacy amplification of a weak n-bit secret. Clearly, the extracted highly secret key can then be used for all sorts of cryptographic tasks. The fact that unconditional security can be achieved even under assumptions as weak as that shows that this—most desirable—type of security might be more practical than generally assumed.

It is a natural question in this context whether such protocols can be given which even tolerate Alice's and Bob's initial strings to differ in a certain fraction of the positions (and how large this fraction can be). A positive answer to that would be useful in the context of quantum key agreement, for instance, since the usually-made assumption that the classical channel—used for the processing of the raw key—is authenticated, or that Alice and Bob share a short secret key already initially, could be dropped.

Acknowledgments. The authors thank Ueli Maurer for many interesting discussions, and Ronald Cramer as well as two anonymous reviewers for their helpful comments. The first author was supported by the Swiss National Science Foundation (SNF), and the second author by Canada's NSERC.

References

1. Y. Aumann, Y. Z. Ding, and M. O. Rabin, Everlasting security in the bounded storage model, *IEEE Trans. on Information Theory*, Vol. 48, pp. 1668–1680, 2002.
2. C. H. Bennett and G. Brassard, Quantum cryptography: public key distribution and coin tossing, *Proceedings of the IEEE International Conference on Computers, Systems and Signal Processing*, pp. 175–179, 1984.
3. C. H. Bennett, G. Brassard, C. Crépeau, and U. M. Maurer, Generalized privacy amplification, *IEEE Trans. on Information Theory*, Vol. 41, No. 6, pp. 1915–1923, 1995.
4. C. H. Bennett, G. Brassard, and J.-M. Robert, Privacy amplification by public discussion, *SIAM Journal on Computing*, Vol. 17, pp. 210–229, 1988.
5. C. Cachin, *Entropy measures and unconditional security in cryptography*, Ph. D. Thesis, ETH Zürich, Hartung-Gorre Verlag, Konstanz, 1997.

6. I. Csiszár and J. Körner, Broadcast channels with confidential messages, *IEEE Trans. on Information Theory*, Vol. 24, pp. 339–348, 1978.
7. Y. Dodis and J. Spencer, On the (non)universality of the one-time pad, *Proceedings of FOCS 2002*, 2002.
8. S. Dziembowski and U. M. Maurer, Tight security proofs for the bounded-storage model, *Proceedings of STOC 2002*, pp. 341–350, 2002.
9. W. Feller, *An introduction to probability theory and its applications*, 3rd edition, Vol. 1, Wiley International, 1968.
10. P. Gemmell and M. Naor, Codes for interactive authentication, *Advances in Cryptology – CRYPTO '93*, LNCS, Vol. 773, pp. 355–367, Springer-Verlag, 1993.
11. T. Holenstein, U. M. Maurer, and R. Renner, personal communication.
12. U. M. Maurer, Conditionally-perfect secrecy and a provably-secure randomized cipher, *Journal of Cryptology*, Vol. 5, No. 1, pp. 53–66, 1992.
13. U. M. Maurer, Secret key agreement by public discussion from common information, *IEEE Trans. on Information Theory*, Vol. 39, No. 3, pp. 733–742, 1993.
14. U. M. Maurer and S. Wolf, Privacy amplification secure against active adversaries, *Advances in Cryptology – CRYPTO '97*, LNCS, Vol. 1294, pp. 307–321, Springer-Verlag, 1997.
15. U. M. Maurer and S. Wolf, Secret-key agreement over unauthenticated public channels – Part I: Definitions and a completeness result, *IEEE Trans. on Information Theory*, Vol. 49, No. 4, pp. 822–831, 2003.
16. U. M. Maurer and S. Wolf, Secret-key agreement over unauthenticated public channels – Part II: The simulatability condition, *IEEE Trans. on Information Theory*, Vol. 49, No. 4, pp. 832–838, 2003.
17. U. M. Maurer and S. Wolf, Secret-key agreement over unauthenticated public channels – Part III: Privacy amplification, *IEEE Trans. on Information Theory*, Vol. 49, No. 4, pp. 839–851, 2003.
18. J. L. McInnes and B. Pinkas, On the impossibility of private key cryptography with weakly random keys, *Advances in Cryptology – CRYPTO '90*, LNCS, Vol. 537, pp. 421–436, Springer-Verlag, 1990.
19. R. Raz, O. Reingold, and S. Vadhan, Extracting all the randomness and reducing the error in Trevisan's extractors, *Proceedings of STOC '99*, pp. 149–158, 1999.
20. R. Raz, O. Reingold, and S. Vadhan, Error reduction for extractors, *Proceedings of FOCS '99*, pp. 191–201, 1999.
21. A. Russell and H. Wang, How to fool an unbounded adversary with a short key, *Advances in Cryptology – EUROCRYPT 2002*, LNCS, Vol. 2332, pp. 133–148, Springer-Verlag, 2002.
22. C. E. Shannon, Communication theory of secrecy systems, *Bell System Technical Journal*, Vol. 28, pp. 656–715, 1949.
23. D. R. Stinson, Universal hashing and authentication codes, *Advances in Cryptology – CRYPTO '91*, LNCS, Vol. 576, pp. 74–85, Springer-Verlag, 1992.
24. S. Wolf, Strong security against active attacks in information-theoretic secret-key agreement, *Advances in Cryptology – ASIACRYPT '98*, LNCS, Vol. 1514, pp. 405–419, Springer-Verlag, 1998.
25. S. Wolf, *Information-theoretically and computationally secure key agreement in cryptography*, ETH dissertation No. 13138, ETH Zürich, 1999.
26. A. D. Wyner, The wire-tap channel, *Bell System Technical Journal*, Vol. 54, No. 8, pp. 1355–1387, 1975.

On Cryptographic Assumptions and Challenges

Moni Naor*

Weizmann Institute of Science
Rehovot 76100, Israel
naor@wisdom.weizmann.ac.il

Abstract. We deal with computational assumptions needed in order to design secure cryptographic schemes. We suggest a classification of such assumptions based on the complexity of falsifying them (in case they happen not to be true) by creating a challenge (competition) to their validity. As an outcome of this classification we propose several open problems regarding cryptographic tasks that currently do not have a good challenge of that sort. The most outstanding one is the design of an efficient block ciphers.

1 The Main Dilemma

Alice and Bob are veteran cryptographers (see Diffie [15] for their history; apparently RSA [38] is their first cooperation). One day, while Bob is sitting in his office his colleague Alice enters and says: "I have designed a new signature scheme. It has an 120 bits long public key and the signatures are 160 bits long". That's fascinating, says Bob, but what computational assumption is it based on? Well, says Alice, it is based on a new trapdoor permutation f_k and a new hash function h and the assumption that after given f_k (but not the trapdoor information) and many pairs of the form $(m_i, f_k^{-1}(h(m_i)))$ it is still hard to come up with a new pair $(m, f_k^{-1}(h(m)))$. So what should be Bob's response? There are several issues regarding these assumptions. For instance, given that the trapdoor permutation and hash function are "new" can they be trusted? But a more bothersome point is whether the assumption Alice uses is really weaker than the assumption "this signature scheme is secure".

So how can we differentiate between the strengths of assumptions and avoid circularity in our arguments? This is the main question we would like to address in this paper and talk and, in particular, to allow answering such doubts and thoughts in a somewhat quantitative manner. We suggest several categories of computational assumptions. Given an assumption it should then be possible to put it into a weaker or stronger class and evaluate it according to this classification. Furthermore, this classification raises many interesting open problems regarding the possibility of basing schemes on assumptions from a weaker class than is currently known.

Designing secure systems is a complex task, even more so than most other design tasks, as there are some malicious entities trying to interfere with the operation of the system. Furthermore, demonstrating that the design is a 'good' one is also not so simple.

* Incumbent of the Judith Kleeman Professorial Chair. Research supported in part by a grant from the Israel Science Foundation.

First, of course, one should define what 'good' means. The last twenty years have seen great progress in defining what a good systems is in the area of Foundations of Cryptography (see Goldreich [24] for a more formal treatment and Bellare and Goldwasser [8] for a more concrete one). For many cryptographic tasks (e.g. signature schemes) we have precise definitions of what a good system means i.e. what the attack is and what it means to break the system. In particular:

- The power of the adversary in terms of access to the system is specified (in the context of signature schemes, whether it is an adaptive message attack or not).
- What constitutes failure of the system (in the context of signature schemes, existential forgery means that the adversary comes up with a single message and a valid signature on it that was not explicitly signed by the signer).

For almost any interesting cryptographic task we need to make computational hardness assumptions, i.e. that certain computational tasks require a lot of resources e.g. time and memory[1]. Furthermore, given the state-of-the-art in Complexity Theory this hardness must be assumed rather than proved.

We deal with hardness assumptions where the adversary is assumed to be limited in its computation time and may succeed with a low probability. The assumptions are stated formally using three parameters n, t and ϵ as follows . Let n be the instance size (which can be thought of as the security parameter), t the upper bound on the time available to the adversary and ϵ the probability of success. We assume that there is some underlying distribution on instances (this could be for instance the choice of a key in signature scheme). For an assumption to be 'hard' we require that for instances of size n no adversary whose run time is bounded by t can succeed (where the notion of succeeds depends on the problem type) with probability better than ϵ. The probability of success is taken over both the distribution of the instance and the coin flips of the adversary. The upper bound on the time t and upper bound on success probability ϵ are a function n. Note that this is a concrete parameters treatment, rather than an asymptotic one, but it is more a matter of style rather than substance. We could have specified our classes in an asymptotic manner.

In general when we are proving that a System X is secure based on an Assumption A, then what we show is that either (i) X is secure or (ii) Assumption A is false. However, the problem with this sort of conditional statement is that it might be hard to demonstrate that A is false (especially since it's collapse is relatively rear), so we don't have good guidelines to decide whether to use the scheme or not.

A proof of security is interesting if (but not necessarily only if) a system whose security requirement is (apparently) not efficiently falsifiable is based on an assumption that is efficiently falsifiable.

Paper plan: In Section 1.1 we list four assumptions representing the various types of assumptions we deal with in this paper. In Section 2 we present our main thesis regarding assumption and challenges. Section 3 we discuss common cryptographic tasks

---

[1] Notable exceptions to tasks that must rely on the adversary's computational limitation are encryption with a one-time pad, authentication with a one-time key and multi-party computation where a majority of the players are honest and private lines are available.

and where they naturally reside. In Section 4 we provide a collection of open problems that our methodology raises. Finally, in Section 5 we provide some caveats regarding our classification.

1.1 Examples of Assumptions

We now describe four assumptions that will be used to demonstrate the new concept introduced in the paper. The assumptions we use involve factoring, RSA and Discrete log type. The first two are well known while the last two are more esoteric. Let

$$Z^{(2)}(n) = \{N = pq | p, q \text{ are } n\text{-bit primes}\}.$$

Factoring: Consider the assumption that factoring is hard: Let the input be an N chosen uniformly at random from $Z^{(2)}(n)$ (denoted $N \in_R Z^{(2)}(n)$) and the computational task the adversary $\mathcal{A}$ is faced with is to find n-bit primes p and q such that $N = p \cdot q$. The (n, t, ϵ) hardness of factoring assumption is: no t-time algorithm $\mathcal{A}$ satisfies:

$$\Pr[\mathcal{A}(N) = (p, q) \text{ where } p \cdot q = N)] > \epsilon.$$

The probability is over the choice of N and the coin flips of $\mathcal{A}$. Note that it is easy to verify a solution, and since we have good primality algorithms (either probabilistic or deterministic) we can even check that N is indeed in $Z^{(2)}(n)$.

The RSA Assumption: The input distribution is a random $N \in_R Z^{(2)}(n)$, e that is relatively prime to $\phi(N)$ and $s \in_R Z_N^*$. The computational task the adversary $\mathcal{A}$ is faced with is to find a such that $a^e = s \bmod N$. The (n, t, ϵ) hardness of RSA assumption is: no t-time algorithm $\mathcal{A}$ satisfies:

$$\Pr[\mathcal{A}(N, e, s) = a \text{ where } a^e = s \bmod N] > \epsilon.$$

Difference RSA assumption: The difference RSA assumption deals with the hardness of finding two RSA preimages such that the difference of their images is a given quantity D, even when many examples with this property are given. The input distribution is a random $N \in_R Z^{(2)}(n)$, e (an RSA exponent) and $D \in_R Z_N^*$. The adversary $\mathcal{A}$ has access to sequence of $m - 1$ pairs (x_i, y_i) s.t. $x_i^e - y_i^e = D \bmod N$ where $\mathcal{A}$ chooses x_i. The adversary should find new (x_m, y_m) (not in $((x_1, y_1), (x_2, y_2) \ldots (x_{m-1}, y_{m-1}))$ such that $X_m^e - Y_m^e = D \bmod N$.

The (n, t, ϵ) Difference RSA assumption: *no* t-time algorithm $\mathcal{A}$ that is given (N, e, D) as above and may access is given input pairs with the first value being his choice satisfies:

$$\Pr[\mathcal{A}(N, e) = (x_m, y_m) \text{ s.t. } x_m^e - x_m^e = D \bmod N \text{ and } (x_m, y_m) \text{ is new}] > \epsilon.$$

Knowledge of Exponent: [26] Let Q and P be primes such that $Q|P-1$, and g a generator of a subgroup of size $Q \bmod P$. The input distribution is a random $h = g^x \bmod P$ for $x \in_R [Q]$ (i.e. h is a random element in the group generated by g; the exponent x is secret). The adversary tries to come with h_1 and h_2 where $h_1 = g^z \bmod P$ and $h_2 = h^z \bmod P$

for some $z \in [Q]$. The point of the assumption is that the only concievable way of doing so is to first pick z and then exponentiate g and h with z; therefore any algorithm that outputs such a pair really 'knows' z. The assumption is for any t-time algorithm $\mathcal{A}$ there is a t'-time algorithm $\mathcal{A}'$ s.t:

$$| \sum_z \Pr[\mathcal{A}(g, h) = (g^z, h^z)] - \sum_z \Pr[\mathcal{A}'(g, h) = (z, g^z, h^z)]| < \epsilon$$

Note that this is essentially a proof of knowledge, but unlike the usual definition of proof of knowledge there is no black-box knowledge extractor here (since there is no interaction), but rather the assumption is that it is always possible to output the same distribution as the original one but with z as well.

An observant reader can detect dependencies between these assumptions. Specifically, RSA assumes Factoring and Difference-RSA assumes RSA. But are there more fundamental differences? these assumptions will be used to exemplify the various categories we introduce.

2   Thesis

In order to be able to evaluate whether an assumption is true or not it must be falsifiable, i.e. there must be a (constructive) way to demonstrate that it is false, *if this is the case*. Furthermore, the complexity of checking the refutation of the assumption is of interest and should be a major consideration in how acceptable the assumption is.

We are mostly interested in falsification by challenge, where a public challenge is advertised. If the assumption is false, then it is possible to solve the challenge (and the solution can be verified) in time related to the time it takes to break the assumption. Falsification by challenge embodies a fundamental principle that underlies the scientific investigation of cryptography: that public evaluation and discussion of cryptographic schemes enhances their security and is essential for the acceptance of any scheme[2].

We define the complexity of falsifying an assumption from the perspective of the protocol designer or the one who should evaluate how good is the design. Therefore we will be interested in

- How difficult it is to generate a random challenge and in particular what is the size of the challenge (which should be small).
- How difficult it is to verify that a proposed solution to the challenge is indeed a correct one. Ideally it should involve computing a simple function or even simply comparing the result to a precomputed solution.
- Is the verification of the solution public, i.e. simply a function of the public challenge and the solution or is secret information needed, i.e. the challenge is generated together with some additional input that can help check the solution. The former is preferable of course, since the verification is public and there is no need for a trusted party (that would leak the solution...).

---

[2] Although this principle can be traced back to Kerckhoffs in 1883 [29], it has been executed in earnest only since the 1970's.

An important entity in our world view is the *falsifier*. The role of the falsifier it to solve the challenges published by the protocol designer (or someone wishing to use it). In other words, what the adversary is to the assumption the falsifier is to the challenge. There are two main differences between them, one is that an assumption might be interactive (though none of those in Section 1.1 are interactive) whereas a challenge is by definition not, but is of the "send-and-forget" type. Another difference is that if an assumption is false, then the probability of success might be rather small, just over ϵ, whereas a falsifier needs a much better probability of success. The protocol designer would like to encourage would be falsifiers (or falsifier designers, keep in mind that the acceptance of a protocol is a social process), so the probability of success should be high, close to 1, (if the assumption A is false).

2.1 The Categories of Falsification

In this section we present our categories of falsification via a challenge. As mentioned above, a major consideration is the size of the challenge d and the run time of the verifier V that should verify the solution y to the challenge d. A challenge is specified by D_n a distribution of challenges of size n. In general given a challenge d chosen according to D_n there should be at least one solution y such that V accepts y (as a solution to d). One additional parameter we need is δ, which is an upper bound on the failure of the challenge, i.e. even if the original assumption A is false, there might be a chance (bounded by δ) that the challenge will not be solvable.

So what we want is a one-to-one correspondence between the adversary for breaking A and the falsifier for the challenge. If there is a an adversary that runs in time t and succeeds with probability better than ϵ in breaking A, then there should be a falsifier that runs in time t' which is some (fixed) polynomial in t and $1/\epsilon$ and succeeds with high probability. This assures us that instead of looking for ways to break A (where the reward is ϵ) we may as well look for ways for solving the challenge. Similarly, we would like the challenge to be non-trivial, i.e. if it can be solved with probability γ, then the original assumption A should also be solvable, with probability which is at least polynomial in γ and $1/n$.

We present three categories of assumptions which we call *efficiently falsifiable, falsifiable* and *somewhat falsifiable*. The differences between them are based mainly on the time to verify the solution. Since the process we have in mind regarding the challenges does not assume that the one who suggests the falsifier and the protocol designer belong to the same body or entity, it is important that the interaction between them be as simple as possible. Therefore ideas such as simulating the falsifier (running his program) should be used only as a last resort.

An adversary $\mathcal{A}$ to Assumption A with parameters (n, t, ϵ) means that $\mathcal{A}$ working in time at most t can break the assumption on instances of size n with probability better than ϵ.

Definition 1 (Efficiently falsifiable (EF)). *An (n, t, ϵ) assumption A is* efficiently falsifiable *if there is a distribution on challenges D_n and a verification procedure $V : \{0,1\}^* \times \{0,1\}^* \mapsto \{accept, reject\}$ such that*

- *Sampling from D_n can be done efficiently, in time polynomial in $n, \log 1/\epsilon$ and $\log 1/\delta$.*
- *The run time of V is polynomial in n, $\log 1/\epsilon$ and $\log 1/\delta$.*
- *If the assumption A is false then there exists a falsifier B that for a $d \in_R D_n$ finds a y for which $V(d, y)$ is 'accept' with probability at least $1 - \delta$ and the run time of B is t' which is polynomial in the run-time of A as well as $\log 1/\epsilon$ and $\log 1/\delta$ and n.*
- *If there is a falsifier B that solves random challenges $d \in_R D_n$ in time t with probability at least γ, then there is a (t', ϵ') adversary A for breaking the original assumption A where t' and ϵ' are polynomially related to t and γ.*

Which assumptions are efficiently falsifiable? An interesting class of assumptions is those that are based on a random self reducible problem, i.e. given any instance it is possible to generate a random instance (or instances) whose solution would yield one for the original. One such problem is discrete log (and related problems such as Diffie-Hellman). Similarly the Ajtai and Dwork [1,2] reductions also yield efficiently falsifiable assumptions. The example of factoring is a little less straightforward and is quite instructive:

Factoring is efficiently falsifiable (EF): we first attempt to show that factoring is EF by choosing the challenge distribution D_n as random $N \in_R Z^{(2)}(n)$. The desired solution is p and q such that $N = p \cdot q$. However, there is a problem: suppose that there are a fraction of 2ϵ of $Z^{(2)}(n)$ that are very easy to factor (denote them by Z'. Then the assumption that factoring is (t, ϵ)-hard is false. However the probability that the challenge will be chosen from Z' is small, 2ϵ, whereas the probability that a falsifier should succeed in solving the challenge should be large, $1 - \delta$. So the problem is that the bad set is too sparse and we need 'easiness' amplification (which is related to the much used hardness amplification). The idea is to allow sampling roughly $1/\epsilon$ samples from $Z^{(2)}(n)$. First note that $Z^{(2)}(n)$ is not so sparse with respect to all $2n$-bit numbers. So if we sample polynomially many $2n$-bit numbers we are bound to hit $Z^{(2)}(n)$ many times. We employ standard technique for amplification using pair-wise independent hashing. Let H be a family of pairwise independent hash functions where for $h \in H$ we have $h : \{0,1\}^n \mapsto \{0,1\}^{n-2\log n + 2\log \epsilon}$. One additional property we need from H is that it will be easy to invert h on a given point c (standard construction enjoy this property). The challenge d is specified by (h, c) where $h \in_R H$ and $c \in_R \{0,1\}^{n-\log n + 2\log \epsilon}$. A solution is p and q such that $h(p \cdot q) = c$ and p and q are n-bit primes. This gives us a fixed δ, so to get any δ we repeat it $\log 1/\delta$ times and the challenge is to solve one instance.

The claim is that if more there is an adversary A that succeeds in factoring with probability better than ϵ, then there is a falsifier that succeeds with probability $1 - \delta$. The falsifier simply runs A on all $N \in h^{-1}(c)$ and checks the results for primality.

When relaxing the notion of 'efficiently falsifiable' to 'falsifiable' we change the requirement that verification time will be proportional to (polynomial in) to that of polynomial in $1/\epsilon$. In particular the challenge size can be much larger now.

Definition 2 (Falsifiable). *An (n, t, ϵ) assumption A is falsifiable if everything is as in Definition 1, except that the run time of sampling D_n and V may depend on $1/\epsilon$ (rather than $\log 1/\epsilon$).*

RSA is falsifiable: The RSA problem is similar in nature to factoring. Therefore it is simple to make it into a falsifiable assumption: sample $1/\epsilon$ times from $Z^{(2)}(n)$ and choose a random prime to generate a pair (N_i, e_i). Choose a random s and publish the results as a challenge. A solution is any y such that there exits an i such that $y^{e_i} = s \bmod N_i$. However, there does not seem to be a simple way of making the RSA Assumption efficiently falsifiable, since we don't know how to obliviously sample from the domain of the function. The reason it did not hurt with the factoring example is that the solution filtered out those that were not of the proper form.

Definition 3. *An (n, t, ϵ) assumption A is* somewhat falsifiable *if everything is as in Definition 1 accept that the run time of V may depend on $1/\epsilon$ (rather than $\log 1/\epsilon$) and the run time of $\mathcal{B}$. In particular this means that V may simulate $\mathcal{B}$.*

Difference RSA is somewhat falsifiable: When considering the problem of Difference RSA, the problem in making it falsifiable is that the amount of information the attacker is allowed to obtain is large and is not bounded by a fixed polynomial in n and $1/\epsilon$. Furthermore the adversary can choose it dynamically. So there does not seem to be a publishable challenge. Instead what the verifier can do is to simply run the program of the supposed falsifier (each time with a self chosen $N \in_R Z^{(2)}(n)$, so that it is possible to answer the query). This should be executed $1/\epsilon$ times and if in at least one of them the program succeeds in finding a different pair than was given to it we conclude that the Difference RSA assumption is false.

None of the above: finally, there seem to be assumption that do not fall into any of the categories mentioned above. These includes assumptions where it is not so clear whether the adversary has won or not. This include the *Knowledge of an exponent* problem, since given a program that outputs consistent pairs (where both h_1 and h_2 have the same exponent) it is not clear that it is hard to extract the exponent from it. What seems to be the problem in making this assumption even somewhat falsifiable is the order of quantifiers in the specification (for all program there exists an extract).

3   Some Cryptographic Tasks and Their Classification

We now list a few cryptographic tasks and discuss in what class the assumption "this scheme is a secure implementation of the task" lies (at least to the best of our knowledge, since we do not how to demonstrate that an assumption is not in a weaker class than what we know how to show).

One-way permutations: given a function $f : \{0,1\}^n \mapsto \{0,1\}^n$ is it a way permutation, i.e. no t time algorithm can find x for a random $y = f(x)$ with probability better than ϵ. We claim that the assumption "f is a one-way permutation" is efficiently falsifiable. This is true for similar reasons as factoring: choose a set of challenges specified as a range of a function. Let H be a family of pairwise independent hash functions, for $h \in H$ we have $h : \{0,1\}^n \mapsto 0, 1^{n-\log n + 2\log \epsilon + \log \delta}$. The challenge is specified by $h \in_R H$ and $c \in \{0,1\}^{n-\log n + 2\log \epsilon + \log \delta}$. A solution is x such that $f(x) = y$ and $h(y) = c$.

Note that we are not checking the permutation part, i.e. that f is indeed $1-1$ and length preserving (this is an unconditional property and could be proved irrespective of the developments in Complexity Theory.

Regarding an assumption of the form "f is a one-way function" (as opposed to a permutation), then we do not know whether it is necessarily an efficiently falsifiable assumption, since it is not clear how to sample from the domain of the function f (without explicitly computing the values). Therefore the best we can say is that it is a falsifiable assumption by generating $1/\epsilon$ values $y_i = f(x_i)$ and the challenge is to solve at least one of them.

Pseudo-random Sequences: We say that a function $G : \{0,1\}^m \mapsto \{0,1\}^n$ is a cryptographically strong pseudo-random sequence generator if it passes all polynomial time statistical tests. More concretely that it is (t,ϵ)-pseudo-random if no test $\mathcal{A}$ running in time t can distinguish outputs of G from random string of length n with advantage greater than ϵ. We do not know how to make the assumption "G is a pseudo-random sequence generator" into an efficiently falsifiable one. However, we can show that it is a falsifiable one. The idea is to publish a bunch of pairs of strings (random, output of G) in random order and the challenge is to distinguish correctly for many (more than half) of them between the truly random part and the $G(x)$ part. More specifically, the challenge consists of "many" (polynomial in $1/\epsilon$ and $\log 1/\delta$) pairs $\{(x_i, y_i)\} \in \{0,1\}^{2n}$, where one element in the pair is random and one pseudo-random and the solution should identify correctly $1/2 + \epsilon/2$ of the pairs. Note that here we are assuming that V has some secret information - the correct ordering of the pairs (which is in violation of Definition 1).

Pseudo-random functions, Block Ciphers and MACs: We now deal with three types of keyed functions where during a learning phase the adversary can query them at many points of its choice and then has to pass a test related to a point it chooses but one that has not been queried before. In the case of a pseudo-random function [22] the adversary has to distinguish between a value of a given point and a truly random string, in the case of a MAC or an unpredictable function (the terminology is from [34]) it has to guess its value on the point. For block copers, or equivalently pseudo-random permutation [30, 33] the function is a permutation on the inputs and the adversary can ask encryption or decryption queries (to go forward or backwards in the permutation). It then has to distinguish the given value from random. Given the interactive nature of the queries of the adversary and given that the adversary can choose both the queries and the test point, there does not seem to be a way to make a challenge that is only of size proportional to $\log 1/\epsilon$ and n, so the best we know how to do is simulate the adversary (and then we can effectively test whether it has succeeded). This makes all assumptions of the form "F_k is a family of (pseudo-random functions|pseudo-random permutations| unpredictable functions)" into somewhat falsifiable ones.

On the other hand there are constructions of pseudo-random functions that are efficiently falsifiable, for instance the Goldreich, Goldwasser and Micali construction [22] when the underlying pseudo-random generator is based on a one-way permutation, or the constructions of Naor, Reingold and Rosen based on factoring and Decisional Diffie-Hellman [32,35]. Similarly for pseudo-random permutations (block-ciphers).

Signatures Schemes: The acceptable notion of a good signature scheme is that of an existentially unforgeable under an adaptive message attack [23]. A signature scheme satisfies this notion if no t-time adversary has probability of success better than ϵ of breaking the scheme where:

- The access to the system consists of adaptively choosing messages m_i and receiving valid signature s_i for $q - 1$ rounds ($q \leq t$).
- The adversary tries to find a pair (m_q, s_q) where $m_q \notin \{m_1, m_2, \ldots m_{q-1}\}$ and s_q is a valid signature on s_q.

As far as we know the requirement that a signatures scheme be existentially unforgeable under an adaptive chosen message attack is not efficiently falsifiable of even just plain falsifiable, for the same reason as pseudo-random functions are not: the attacker can adaptively choose both the queries and the test. Again it is a somewhat falsifiable assumption, since by simulating the adversary it is possible to check whether it has succeeded or not.

Encryption Schemes: the definitions of a good encryption scheme come in many shapes and sizes, depending on whether the scheme is shared or public key, the type of attack (known plaintext, chosen plaintext, chosen ciphertext of various shades) and what it means to break the system (semantic security, non-malleability). (See [16] and [5] for more details.) All these requirement are somewhat falsifiable, since it is possible to simulate the adversary and check whether it succeeds or not and hence evaluate the probability of success. On the other hand they do not seem to be in any weaker class. There is one exception which is falsifiable: when the scheme is public key, the messages are encrypted bit-by-bit, the attack is chosen plaintext (the weakest one relevant in this context) and the security requirement is semantic security.

Zero-knowledge: the requirement that a proof system be zero knowledge is that for any behavior by the verifier denoted by V^* there should be a simulator S that outputs a simulated conversation between the prover and the verifier. The distributions on the conversations between $(\mathcal{P}, V^*)$ and the outputs of S should be indistinguishable. This requirement does not seem to be even somewhat falsifiable due to the order of the quantifiers: suppose that a falsifier comes up with a seemingly bad V^*, how do we (the protocol designers) know that there is not simulator for V^*. This is especially acute in case of non block-box simulation [4].

In contrast, the property of Witness Indistinguishable is somewhat falsifiable.

4   Challenging Problems

One of the main points of this paper is that the classification just proposed yields many open problems. For any cryptographic task the question is whether it is possible to base it on an assumption from a weaker class than where the tasks lies "naturally" (i.e. where the assumption "this system implements the task in a secure manner" lies). One can view many of the significant and interesting results in cryptography this way, basing a system on an assumption from a weaker class; e.g. any cryptographic scheme based on factoring (and there are many examples of such) is based on an efficiently falsifiable assumption, no matter where it naturally lies.

We now list a few such problems. In most of the cases there is an implementation based on random oracles [27,6], i.e. the analysis is based on the assumption that a certain function behaves as a random function. Once the random function is substituted with a concrete one we may phrase an assumption on what is needed from the function. However, usually it is often not simpler than "this system is secure."

Efficient Block ciphers: common block ciphers such as DES or AES (Reijndael) are several orders of magnitude faster than schemes based on the hardness of factoring or any other scheme that can be based on an efficiently falsifiable assumption. On the other hand they do not seem to yield any assumption that is simpler than "this scheme is secure". So the most outstanding problem our methodology raises is whether it is possible to come up with an efficient proposal to a block cipher, competitive with the state-of-the-art, together with a meaningful succinct challenge , i.e. a block cipher based on an efficiently falsifiable assumption.

Signatures: while there have been many proposals of signatures schemes based on efficiently falsifiable or just falsifiable assumption, they are mostly less efficient than ones based on random oracles [7], but some of them give the latter a run for their money [14,21]. However, there have been several parameters where no non random oracles constructions are known:

Short signatures: The schemes with the shortest signature construction known are based on Weil Pairing and random oracles, and get to less than 200 bits per signature [10]. These sort of succinct signatures are important when the signature is carried over a very low bandwidth channel, e.g. a human saying it over the phone. So the question is whether there exists an efficiently refutable assumption allowing such short signatures.

Low overhead signatures: There are methods that allow signatures with a short additional string, mostly by using the signatures itself to store the string (signatures with recovery). This is important to lower the overall bandwidth overhead of sending a signature. These schemes are based on RSA or Rabin plus a random oracle [7]. The question is whether it is possible to get to this overhead with overhead signatures which are based on a falsifiable assumption.

Discrete log signatures: To the best of our knowledge no discrete log type signature scheme (El Gamal, Schnorr) has been shown secure without the help of random oracles.

Low exponent verification: RSA or Rabin provide signatures with extremely efficient verification. Is it possible to obtain such schemes based on a falsifiable assumption?

Identity Based Encryption: In 1984 Adi Shamir [40] suggested the notion of an identity based encryption (IBE), where a short public key determined implicitly the public keys of all participants (in conjunction with their 'identity') and a center that know a secret can provide them with the corresponding secret key. One of the more interesting developments in the last two or three years has been the first proposals for IBEs by Boneh and Franklin [9] and Cocks [13]. While the scheme of [9] is provided with a rigorous analysis, as far as we can tell one does not yield an efficient falsifiable assumption from it, so the open question is how to obtain a scheme based on such an assumption.

Blind Signatures and anonymity: A blind signature is one where the signer does not know what it is signing yet it is impossible to generate more signed messages than the signer provided. This notion, proposed by Chaum [11] has many application from untraceable electronic money [12] to fairness in two party computation [36]. It has been analyzed in the random oracle model [37] and there is a highly interactive (between the signer and signee). There is also construction with no random oracles based on secure function evaluation [28]. So the question is whether a two round signature scheme (request and response) is possible based on efficiently falsifiable assumption. This is also true for other anonymity preserving tools such as deniable group signature [39] (based on RSA and random oracles vs. the interactive version given in [31] (based on any good encryption scheme).

Three round zero knowledge: One of the outstanding open problems regarding zero-knowledge proofs is whether it is possible to obtain a three round (move) protocol with low probability of error. Hada and Tanaka [26] introduced the assumption of "Knowledge of Exponent" in order to resolve this issue, but as far as we know this assumption is not even somewhat falsifiable. So the question is whether it is possible to obtain such a protocol with an efficiently falsifiable assumption. See [19,20] for recent work on the subject.

Incompressible functions: Let $f : \{0,1\}^n \mapsto \{0,1\}^m$ be an expanding function. Suppose that one party Alice knows x, and wants the other party Bob to learn $f(x)$ *without revealing more information than necessary about* x. Alice could of course simply send $f(x)$ to Bob, but the goal is to have a low communication protocol. A function f is called *incompressible* if any $o(m)$ bit message enabling the computation of $f(x)$ reveals x. This property was introduced by [17] in the context of preventing the illicit distribution of keys for broadcast encryption. They also conjectured that the function $f(x) = g^x \bmod P \circ g^{x^2} \bmod P \circ \ldots \circ g^{x^\ell} \bmod P$ is incompressible. However, as far as we can tell the assumption that a function f is incompressible is not even somewhat falsifiable. So the open problem is to come up with a candidate for an incompressible function where the assumption that it is incompressible is efficiently falsifiable.

Moderately hard functions: Moderately hard functions, where the time it takes to evaluate them is neither too small nor too large have many applications in cryptography and computer security, from abuse prevention (See [18]) to zero-knowledge [19]. The question is whether it is possible to incorporate assumptions regarding moderately hard functions into this framework.

Physical security: We have dealt so far with the usual mathematical abstract of a distributed system and ignored to a large degree the physical realization, however, as we have seen in the last few years often times "physical" attacks and issues such as: power consumption, tamper resistance, fault Resistance and timing make the implementation very vulnerable (see [3]). So ideally we would like to base security on simple assumption that can be falsified similarly to computational assumption.

This is true also for humans. It is often said that the human user is the weakest link in the security of a system. So again we would like to incorporate human ability considerations into the security design in a rigorous manner and in particular make falsifiable assumptions about human user's ability and prove the security of a system based on these assumptions?

5 Caveats

While we believe that the classification we have presented is widely applicable and can clarify many issues, it is not a panacea and many important issues are not handled by it.

Implication not preserved by classification: If Assumption A implies Assumption B and A is efficiently falsifiable we cannot conclude that B is efficiently falsifiable as well. The reason is that we cannot simply use the challenges for A to check B, since a falsifier for A is not necessarily one for B (recall that we should be able to use it to break A).

Classification does not highlight all major results in cryptography: One of the most outstanding results in Foundation of Cryptography is the HILL (Håstad, Impagliazzo, Levin and Luby) [25] stating that it is possible to construct a pseudo-random sequence generator from any one-way function. On the other hand, according to our classification both the assumption and the outcome are in the class of falsifiable tasks (though the latter is there only when the verifier has secret information). So such a major result is not highlighted by our classification. This true for many general results, since there is no particular payoff for a general result. In particular the issue may be that falsifiability provides only a partial order between assumptions (given the we have only four categories then clearly this is not a linear scale).

History of investigating an assumption is not accounted for. A major reason to adopt an assumption is the history of computational attempts to resolve it. This is completely ignored in our framework, but on the other hand our framework yields the opportunity to create such a history for an assumption or an area.

Do not verify run time if the one taking up the challenge has more computational resources than the protocol designer assumes then he may simply break the challenge "in brute force" (without violating the formal statement). But this can be seen as violating another implicit assumption "that t is an infeasible amount of total computational time available to the attacker."

Acknowledgements. I have had many interesting conversation with my colleagues regarding the issues outlined in the paper. In particular I would like to thank Dan Boneh, Cynthia Dwork, Benny Pinkas and Steven Rudich. I thank Dalit Naor for many comments regarding the writeup.

References

1. M. Ajtai, *Generating Hard Instances of Lattice Problems*, Proceedings of the 27th ACM Symposium on Theory of Computing, 1996, 99–108.
2. M. Ajtai and C. Dwork, *A Public-Key Cryptosystem with Worst-Case/Average-Case Equivalence*, Proceedings of the 28th ACM Symposium on Theory of Computing, 1997, pp. 284–293.
3. R. Anderson, *Security Engineering: A guide to Building Dependeable Distributed Systems*, Wiley, 2001.
4. B. Barak, *How to Go Beyond The Black-Box Simulation Barrier*, Proc. of the 42nd IEEE Symposium on the Foundation of Computer Science, 2001.
5. M. Bellare, A. Desai, D. Pointcheval and P. Rogaway. *Relations among notions of security for public-key encryption schemes,* Advances in Cryptology – CRYPTO'98, Lecture Notes in Computer Science, vol. 1462, Springer, 1998, pp. 26–45.
6. M. Bellare and P. Rogaway, *Random oracles are practical: A paradigm for designing efficient protocols Authors*, ACM Conference on Computer and Communications Security, 1993, pp. 62–73.
7. M. Bellare and P. Rogaway, *The Exact Security of Digital Signatures - HOw to Sign with RSA and Rabin*, Advances in Cryptology – Eurocrypt 1996: pp. 399–416
8. S. Goldwasser and M. Bellare, Lecture Notes on Cryptography, 1996, available `http://philby.ucsd.edu/cryptolib/BOOKS/gb.html`
9. D. Boneh and M. Franklin, *Identity Based Encryption from the Weil Pairing*, Advances in Cryptology – CRYPTO 2001, Lecture Notes in Computer Science 2139, Springer, 2001, pp. 213–229.
10. D. Boneh, B. Lynn and H. Shacham, *Short signatures from the Weil pairing*, Advances in Cryptology – ASIACRYPT 2001, Lecture Notes in Computer Science, Springer, pp. 514–532.
11. D. Chaum. *Blind signatures for untraceable payments*, Advances in Cryptology - Proceedings of Crypto 82, Plenum Press, New York and London, pp. 199–203. 1983
12. D. Chaum, A. Fiat and M. Naor, *Untraceable Electronic Cash* Advances in Cryptology – CRYPTO 1988, ecture Notes in Computer Science, Springer, pp. 319–327.
13. C. Cocks, *An identity based encryption scheme based on quadratic residues*, Cryptography and Coding, Lecture Notes in Computer Science 2260, Springer, 2001, pp. 360–363.
14. R. Cramer and V. Shoup, *Signature Schemes Based on the Strong RSA Assumption*, ACM Conference on Computer and Communications Security, 1999, pp. 46–51.
15. W. Diffie, *The First Ten Years of Public Key Cryptography*, in *Contemporary Cryptography The Science of Information Integrity*, edited by G. J. Simmons, IEEE Press, 1992.
16. D. Dolev, C. Dwork and M. Naor, *Non-malleable Cryptography*, Siam J. on Computing, vol 30, 2000, pp. 391–437.
17. C. Dwork, J. Lotspiech and M. Naor, *Digital Signets: Self-Enforcing Protection of Digital Information,* Proceedings of the 27th ACM Symposium on Theory of Computing, 1996, pp. 489–498.
18. C. Dwork and M. Naor, *Pricing via Processing, Or, Combatting Junk Mail*, Advances in Cryptology – CRYPTO'92, Lecture Notes in Computer Science No. 740, Springer, 1993, pp. 139–147.
19. C. Dwork and M. Naor, *Zaps and their Applications*, Proceedings of the IEEE 41st Annual Symposium on the Foundations of Computer Science, 2000, pp. 283–293.
20. C. Dwork and L. Stockmeyer,
21. R. Gennaro, S. Halevi and T. Rabin, *Secure Hash-and-Sign Signatures Without the Random Oracle*, Advances in Cryptology – Eurocrypt'99 Proceeding, Lecture Notes in Computer Science, Springer, 1999, pp. 123–139

22. O. Goldreich, S. Goldwasser and S. Micali, *How to Construct Random Functions,* JACM 33(4), 1986, pp. 792–807.
23. S. Goldwasser, S. Micali and R. Rivest, *A secure digital signature scheme*, SIAM J. on Computing 17, 1988, pp. 281–308.
24. O. Goldreich, *On the Foundations of Modern Cryptography,* Advances in Cryptology – Crypto'97 Proceeding, Lecture Notes in Computer Science, Springer, 1997. See `http://www.wisdom.weizmann.ac.il/~oded/foc.html`
25. J. Håstad, R. Impagliazzo, L. A. Levin and M. Luby, *A Pseudorandom Generator from any One-way Function*, SIAM J. Computing, 28(4), 1999, pp. 1364–1396.
26. S. Hada and T. Tanaka, *On the Existence of 3-Round Zero-Knowledge Protocols,* Advances in Cryptology – Crypto'98 Proceeding, Lecture Notes in Computer Science No. 1462, Springer, 1998, pp. 408–423.
27. R. Impagliazzo and S. Rudich, *Limits on the Provable Consequences of One-Way Permutations*, STOC 1988.
28. A. Juels, M. Luby and R. Ostrovsky *Security of Blind Digital Signatures*, Advances in Cryptology – Crypto'97 Proceeding, Lecture Notes in Computer Science, pp. 150–164
29. A. Kerckhoffs, *La Cryptographie Militaire*, Journal des Sciences Militaires, 1883, pp. 5–38. Available at `http://www.cl.cam.ac.uk/usres/fapp2/kerckhoffs/`.
30. M. Luby and C. Rackoff, *How to construct pseudorandom permutations and pseudorandom functions*, SIAM J. Computing, vol. 17, 1988, pp. 373–386.
31. M. Naor, *Deniable Ring Authentication*, Advances in Cryptology – CRYPTO 2002 Proceeding, Lecture Notes in Computer Science 2442, Springer, 2002, pp. 481–498.
32. M. Naor and O. Reingold, *Number-theoretic constructions of efficient pseudo-random functions*, Proceedings of the IEEE 38th Symposium on the Foundations of Computer Science, Oct. 1997, pp. 458–467.
33. M. Naor and O. Reingold, *On the Construction of Pseudo-Random Permutations: Luby-Rackoff Revisited*, Journal of Cryptology, vol 12, 1999, pp. 29–66.
34. M. Naor and O. Reingold, *From Unpredictability to Indistinguishability: A Simple Construction of Pseudo-Random Functions from MACs*, Advances in Cryptology – Crypto'98 Proceeding, Lecture Notes in Computer Science No. 1462, Springer, 1998, pp. 267–282.
35. M. Naor, O. Reingold and A. Rosen. *Pseudo-Random Functions and Factoring.* Siam J. on Computing Vol. 31 (5), pp. 1383–1404, 2002.
36. B. Pinkas, *Fair Secure Two-Party*, Advances in Cryptology - Eurocrypt'2003, Lecture Notes in Computer Science, Springer, 2003, pp. 87–105.
37. D. Pointcheval, J. Stern, *Security Arguments for Digital Signatures and Blind Signatures*, J. of Cryptology 13(3): pp. 361–396, 2000.
38. R. L. Rivest, A. Shamir, and L. M. Adleman, *A method for obtaining digital signature and public key cryptosystems*, Communications of the ACM, vol. 21, Feb 1978, pp. 120–126.
39. R. L. Rivest, A. Shamir, and Y. Tauman, *How to Leak A Secret*, Advances in Cryptology - ASIACRYPT 2001, Lecture Notes in Computer Science, Vol. 2248, Springer, pp. 552–565.
40. A. Shamir, *Identity-Based Cryptosystems and Signature Schemes*, Advances in Cryptology - CRYPTO'84, Lecture Notes in Computer Science, Springer, 1985, pp. 47–53.

Scalable Protocols for Authenticated Group Key Exchange

Jonathan Katz[1] and Moti Yung[2]

[1] Dept. of Computer Science, University of Maryland, College Park, MD
jkatz@cs.umd.edu
[2] Dept. of Computer Science, Columbia University, New York, NY
moti@cs.columbia.edu

Abstract. We consider the fundamental problem of authenticated group key exchange among n parties within a larger and insecure public network. A number of solutions to this problem have been proposed; however, all provably-secure solutions thus far are not scalable and, in particular, require n rounds. Our main contribution is the first *scalable* protocol for this problem along with a rigorous proof of security in the standard model under the DDH assumption; our protocol uses a constant number of rounds and requires only $O(1)$ modular exponentiations per user (for key derivation). Toward this goal and of independent interest, we first present a scalable compiler that transforms any group key-exchange protocol secure against a passive eavesdropper to an *authenticated* protocol which is secure against an active adversary who controls all communication in the network. This compiler adds only one round and $O(1)$ communication (per user) to the original scheme. We then prove secure — against a passive adversary — a variant of the two-round group key-exchange protocol of Burmester and Desmedt. Applying our compiler to this protocol results in a provably-secure three-round protocol for *authenticated* group key exchange which also achieves forward secrecy.

1 Introduction

Protocols for authenticated key exchange (AKE) allow a group of parties within a larger and completely insecure public network to establish a common secret key (a *session key*) and furthermore to be guaranteed that they are indeed sharing this key with *each other* (i.e., with their intended partners). Protocols for securely achieving AKE are fundamental to much of modern cryptography. For one, they are crucial for allowing symmetric-key cryptography to be used for encryption/authentication of data among parties who have no alternate "out-of-band" mechanism for agreeing upon a common key. Furthermore, they are instrumental for constructing "secure channels" on top of which higher-level protocols can be designed, analyzed, and implemented in a modular manner. Thus, a detailed understanding of AKE — especially the design of provably-secure protocols for achieving it — is critical.

The case of 2-party AKE has been extensively investigated (e.g., [19,8,20,6, 25,4,30,15,16,17] and others) and is fairly well-understood; furthermore, a variety

D. Boneh (Ed.): CRYPTO 2003, LNCS 2729, pp. 110–125, 2003.

of efficient and provably-secure protocols for 2-party AKE are known. Less attention has been given to the important case of *group* AKE where a session key is to be established among n parties; we survey relevant previous work in the sections that follow. Group AKE protocols are essential for applications such as secure video- or tele-conferencing, and also for collaborative (peer-to-peer) applications which are likely to involve a large number of users. The recent foundational papers of Bresson, et al. [13,11,12] (building on [6,7,5]) were the first to present a formal model of security for group AKE and the first to give rigorous proofs of security for particular protocols. These represent an important initial step, yet much work remains to be done to improve the efficiency and scalability of existing solutions.

1.1 Our Contributions

We may summarize the prior "state-of-the-art" for group AKE as follows (see Section 1.2 for a more detailed discussion of previous work):

 - The best-known provably-secure solutions in the standard model are those of [13,11,12], building on [31]. These protocols do not scale well: to establish a key among n participants, they require n rounds and additionally require (for some players) $O(n)$ modular exponentiations and $O(n)$ communication.
 - Subsequent to the present work, a constant-round protocol for group AKE has been proven secure *in the random oracle model* [10]. Unfortunately, this protocol does *not* achieve forward secrecy (an explicit attack is known [10]). The protocol is also not symmetric; furthermore, the initiator of the protocol must perform $O(n)$ encryptions and send $O(n)$ communication.

Our main result is the first *constant-round* and *fully-scalable* protocol for group AKE which is provably-secure in the standard model. Security is proved (in the same security model used in previous work [13,11,12,10]) via reduction to the decisional Diffie-Hellman (DDH) assumption. The protocol also achieves forward secrecy [20] in the sense that exposure of principals' long-term secret keys does not compromise the security of previous session keys.[1] Our 3-round protocol remains practical even for large groups: it requires only $O(1)$ communication, 3 modular exponentiations, and $O(n)$ signature verifications per user.

The difficulty of analyzing protocols for group AKE has seemingly hindered the development of practical and provably-secure solutions, and has led to the proposal of some protocols which were later found to be flawed (see, e.g., the attacks given in [29,10]). To manage this complexity, we take a modular approach which greatly simplifies the design and analysis of group AKE protocols and should therefore prove useful for future work. Specifically, we show a *compiler* that transforms any group key-exchange protocol secure against a passive eavesdropper to one secure against a stronger (and more realistic) *active* adversary who controls all communication in the network. If the original protocol achieves

---

[1] We of course also require that exposure of (multiple) session keys does not compromise the security of unexposed session keys; see the formal model in Section 2.

forward secrecy, the compiled protocol does too. Adapting work of Burmester and Desmedt [14], we then present a 2-round group key-exchange protocol and rigorously prove its security — against a passive adversary — under the DDH assumption.[2] Applying our compiler to this protocol gives our main result.

We note two additional and immediate applications of the compiler presented here. First, the compiler may be applied to the group key-exchange protocols of [31] to yield a group AKE protocol similar to that of [13] but with a much simpler security proof which holds for groups of polynomial size (the proof given in [13] holds only for groups of *constant* size). Second, we may compile the 1-round, 3-party key-exchange protocol of Joux [23] to obtain a simple, 3-party AKE protocol requiring 2 rounds. The simplicity of the resulting security proof in these cases makes a modular approach of this sort compelling, especially when this approach is compared to the largely *ad hoc* methods which are often used when analyzing group AKE protocols (as in, e.g., [1,26,27]).

1.2   Previous Work

Group key exchange. A number of works have considered the problem of extending the 2-party Diffie-Hellman protocol [19] to the multi-party setting. Most well-known among these are perhaps the works of Ingemarsson, et al. [22], Burmester and Desmedt [14], and Steiner, et al. [31]. These works all assume a passive (eavesdropping) adversary, and only [31] provides a rigorous proof of security (but see footnote 2).

Authenticated protocols are designed to be secure against the stronger class of adversaries who —in addition to eavesdropping — control all communication in the network (cf. Section 2). A number of protocols for authenticated group key exchange have been suggested [24,9,2,3,32]; unfortunately, none of these works present rigorous security proofs and thus confidence in these protocols is limited. Indeed, attacks on some of these protocols have been presented [29], emphasizing the need for rigorous proofs in a well-defined model. Tzeng and Tzeng [33] prove security of a group AKE protocol using a non-standard adversarial model; an explicit attack on their protocol has recently been identified [10].

Provably-secure protocols. As mentioned earlier, only recently have Bresson, et al. [13,11,12] given the first formal model of security and the first provably-secure protocols for the group AKE setting. Their security model builds on earlier work of Bellare and Rogaway in the 2-party setting [6,7] as extended by Bellare, et al. [5] to handle (among other things) forward secrecy.

The provably-secure protocols of Bresson, et al. [13,11,12] are based on the protocols of Steiner, et al. [31], and require n rounds to establish a key among a group of n users. The initial work [13] deals with the static case, and shows a

---

[2] Because no proof of security appears in [14], the Burmester-Desmedt protocol has been considered "heuristic" and not provably-secure (see, e.g., [13,10]). Subsequent to our work we became aware that a proof of security for a variant of the Burmester-Desmedt protocol (in a weaker model than that considered here) appears in the pre-proceedings of Eurocrypt '94 [18]. See Section 4 for further discussion.

protocol which is secure (and achieves forward secrecy) under the DDH assumption.[3] Unfortunately, the given proof of security applies only when n is *constant*; in contrast, the proofs given here allow $n = \mathsf{poly}(k)$.

Later work [11,12] focuses on the dynamic case where users join or leave and the session key must be updated whenever this occurs. Although we do not explicitly address this case, note that dynamic group membership can be handled efficiently — when using a constant-round protocol — by running the group AKE protocol from scratch among members of the new group. For the protocol given here, the complexity of this approach is roughly equivalent[4] to the Join and Remove protocols of [11,12]. Yet, handling dynamic membership even more efficiently remains an interesting topic for future research.

More recently (in work subsequent to ours), a constant-round group AKE protocol with a security proof in the random oracle model has been shown [10]. The given protocol does not provide forward secrecy; in fact (as noted by the authors) an *explicit attack* is possible when long-term keys are exposed. Finally, the protocol is not symmetric but instead requires a "group leader" to perform $O(n)$ encryptions and send $O(n)$ communication each time a group key is established.

Compilers for key-exchange protocols. A modular approach such as ours has previously been used in the design and analysis of key-exchange protocols. Mayer and Yung [28] give a compiler which converts any 2-party protocol into a centralized (non-contributory) group protocol; their compiler invokes the original protocol $O(n)$ times, however, and is therefore not scalable. In work with similar motivation as our own, Bellare, et al. [4] show a compiler which converts unauthenticated protocols into authenticated protocols in the 2-party setting. Their compiler was not intended for the group setting and does not scale as well as ours; extending [4] to the group setting gives a compiler which triples the number of rounds and furthermore requires n signature computations/verifications and an $O(n)$ increase in communication per player per round. In contrast, the compiler presented here adds only a *single* round and introduces an overhead of 1 signature computation, n signature verifications, and $O(1)$ communication per player per round. (In fact, the compiler introduced here is slightly more efficient than that of [4] even in the 2-party case.)

1.3 Outline

In Section 2, we review the security model of Bresson, et al. [13]. We present our compiler in Section 3 and a two-round protocol secure against passive adversaries in Section 4. Applying our compiler to this protocol gives our main result: an efficient, fully-scalable, and constant-round group AKE protocol.

---

[3] The given reduction is in the random oracle model using the CDH assumption but they mention that the protocol can be proven secure in the standard model under the DDH assumption.

[4] For example, the Join algorithm of [11,12] requires 2 rounds when one party joins and $O(n)$ rounds when n parties join; running our group AKE protocol from scratch requires only 3 rounds regardless of the number of parties who have joined.

2   The Model and Preliminaries

Our security model is the standard one of Bresson, et al. [13] which builds on prior work from the 2-party setting [6,7,5] and which has been widely used to analyze group key-exchange protocols (e.g., [11,12,10]). We explicitly define notions of security for both passive and active adversaries; this will be necessary for stating and proving meaningful results about our compiler in Section 3.

Participants and initialization.   We assume for simplicity a fixed, polynomial-size set $\mathcal{P} = \{U_1, \ldots, U_\ell\}$ of potential participants. Any subset of $\mathcal{P}$ may decide at any point to establish a session key, and we do not assume that these subsets are always the same size or always include the same participants. Before the protocol is run for the first time, an initialization phase occurs during which each participant $U \in \mathcal{P}$ runs an algorithm $\mathcal{G}(1^k)$ to generate public/private keys (PK_U, SK_U). Each player U stores SK_U, and the vector $\langle PK_i \rangle_{1 \leq i \leq |\mathcal{P}|}$ is known by all participants (and is also known by the adversary).

Adversarial model. In the real world, a protocol determines how principals behave in response to signals from their environment. In the model, these signals are sent by the adversary. Each principal can execute the protocol multiple times with different partners; this is modeled by allowing each principal an unlimited number of *instances* with which to execute the protocol. We denote instance i of user U as Π_U^i. A given instance may be used only once. Each instance Π_U^i has associated with it the variables state_U^i, term_U^i, acc_U^i, used_U^i, sid_U^i, pid_U^i, and sk_U^i; the last of these is the *session key* whose computation is the goal of the protocol, while the function of the remaining variables is as in [5].

The adversary is assumed to have complete control over all communication in the network. An adversary's interaction with the principals in the network (more specifically, with the various instances) is modeled by the following *oracles*:

- Send(U, i, M) — This sends message M to instance Π_U^i, and outputs the reply generated by this instance. We allow the adversary to prompt the unused instance Π_U^i to initiate the protocol with partners $U_2, \ldots, U_n$ by calling Send$(U, i, \langle U_2, \ldots, U_n \rangle)$.
- Execute$(U_1, \ldots, U_n)$ — This executes the protocol between unused instances of players $U_1, \ldots, U_n \in \mathcal{P}$ and outputs the transcript of the execution. The number of group members and their identities are chosen by the adversary.
- Reveal(U, i) — This outputs session key sk_U^i.
- Corrupt(U) — This outputs the long-term secret key SK_U of player U.
- Test(U, i) — This query is allowed only once, at any time during the adversary's execution. A random bit b is generated; if $b = 1$ the adversary is given sk_U^i, and if $b = 0$ the adversary is given a random session key.

A *passive adversary* is given access to the Execute, Reveal, Corrupt, and Test oracles, while an *active adversary* is additionally given access to the Send oracle. (Even though the Execute oracle can be simulated via repeated calls to the Send oracle, allowing the adversary access to the Execute oracle allows for a tighter definition of forward secrecy.)

Partnering. Partnering is defined via *session IDs* and *partner IDs*. The session ID for instance Π_U^i (denoted sid_U^i) is a protocol-specified function of all communication sent and received by Π_U^i; for our purposes, we will simply set sid_U^i equal to the concatenation of all messages sent and received by Π_U^i during the course of its execution. The partner ID for instance Π_U^i (denoted pid_U^i) consists of the identities of the players in the group with whom Π_U^i intends to establish a session key, including U itself; note that these identities are always clear from the initial call to the Send or Execute oracles. We say instances Π_U^i and $\Pi_{U'}^j$ are *partnered* iff (1) $\mathsf{pid}_U^i = \mathsf{pid}_{U'}^j$, and (2) $\mathsf{sid}_U^i = \mathsf{sid}_{U'}^j$. Our definition of partnering is much simpler than that of [13] since, in our protocols, all messages are sent to all other members of the group taking part in the protocol.

Correctness. Of course, we wish to rule out "useless" protocols from consideration. In the standard way, we require that for all U, U', i, j such that $\mathsf{sid}_U^i = \mathsf{sid}_{U'}^j$, $\mathsf{pid}_U^i = \mathsf{pid}_{U'}^j$, and $\mathsf{acc}_U^i = \mathsf{acc}_{U'}^j = \text{TRUE}$ it is the case that $\mathsf{sk}_U^i = \mathsf{sk}_{U'}^j \neq \text{NULL}$.

Freshness. Following [5,13], we define a notion of *freshness* appropriate for the goal of forward secrecy. An instance Π_U^i is *fresh* unless one of the following is true: (1) at some point, the adversary queried $\mathsf{Reveal}(U, i)$ or $\mathsf{Reveal}(U', j)$ where Π_U^i and $\Pi_{U'}^j$ are partnered; or (2) a Corrupt query was asked before a query of the form $\mathsf{Send}(U, i, *)$.

Definitions of security. We say event Succ occurs if the adversary queries the Test oracle on a fresh instance and correctly guesses the bit b used by the Test oracle in answering this query. The advantage of an adversary $\mathcal{A}$ in attacking protocol P is defined as $\mathsf{Adv}_{\mathcal{A},P}(k) \stackrel{\text{def}}{=} |2 \cdot \Pr[\mathsf{Succ}] - 1|$. We say protocol P is a *secure group key exchange (KE) protocol* if it is secure against a passive adversary; that is, for any PPT passive adversary $\mathcal{A}$ it is the case that $\mathsf{Adv}_{\mathcal{A},P}(k)$ is negligible. We say protocol P is a *secure authenticated group key exchange (AKE) protocol* if it is secure against an active adversary; that is, for any PPT active adversary $\mathcal{A}$ it is the case that $\mathsf{Adv}_{\mathcal{A},P}(k)$ is negligible.

To enable a concrete security analysis, we define $\mathsf{Adv}_P^{\mathsf{KE-fs}}(t, q_{\mathsf{ex}})$ to be the maximum advantage of any passive adversary attacking P, running in time t and making q_{ex} calls to the Execute oracle. Similarly, we define $\mathsf{Adv}_P^{\mathsf{AKE-fs}}(t, q_{\mathsf{ex}}, q_{\mathsf{s}})$ to be the maximum advantage of any active adversary attacking P, running in time t and making q_{ex} calls to the Execute oracle and q_{s} calls to the Send oracle.

Protocols without forward secrecy. Throughout this paper we will be concerned primarily with protocols achieving forward secrecy; the definitions above already incorporate this requirement since the adversary has access to the Corrupt oracle in each case. However, our compiler may also be applied to KE protocols which do not achieve forward secrecy (cf. Theorem 2). For completeness, we define $\mathsf{Adv}_P^{\mathsf{KE}}(t, q_{\mathsf{ex}})$ and $\mathsf{Adv}_P^{\mathsf{AKE}}(t, q_{\mathsf{ex}}, q_{\mathsf{s}})$ in a manner completely analogous to the above, with the exception that the adversary in each case no longer has access to the Corrupt oracle.

Authentication. We do not define any notion of explicit authentication or, equivalently, confirmation that the other members of the group have computed the common key. Indeed, our protocols do not explicitly provide such confirmation. However, explicit authentication in our protocols can be achieved at little additional cost. Previous work (e.g., [13, Sec. 7]) shows how to achieve explicit authentication for any secure group AKE protocol using one additional round and minimal extra computation. (Although [13] use the random oracle model, their techniques can be extended to the standard model by replacing the random oracle with a pseudorandom function.) Applying their transformation to our final protocol will result in a constant-round group AKE protocol with explicit authentication.

2.1   Notes on the Definition

Although the above definition is standard for the analysis of group key-exchange protocols — it is the definition used, e.g., in [13,11,10] — there are a number of concerns that it does *not* address. For one, it does not offer any protection against malicious insiders, or users who do not honestly follow the protocol. Similarly, the definition is not intended to ensure any form of "agreement" and thus secure protocols for group AKE do not contradict known impossibility results for asynchronous distributed computing (e.g., [21]). (Actually, since the public-key model is assumed here, many of these impossibility results do not apply.) Finally, the definition inherently does not protect against "denial of service" attacks, and cannot prevent the adversary from causing an honest instance to "hang" indefinitely; this is simply because the model allows the adversary to refuse to deliver messages to any instance.

Some of these concerns can be addressed — at least partially — within the model above. For example, to achieve confirmation that all intended participants have computed the (correct, matching) session key following execution of a protocol, we may augment *any* group AKE protocol in the following way: after computing key sk, each player U_i computes $x_i = F_{sk}(U_i)$, signs x_i, broadcasts x_i and the corresponding signature, and computes the "actual" session key $sk' = F_{sk}(\perp)$ (here, F represents a pseudorandom function and "$\perp$" represents some distinguished string); other players check the validity of these values in the obvious way.[5] Although this does not provide agreement (since an adversary can refuse to deliver messages to some of the participants), it *does* prevent a corrupted user from sending different messages to different parties, thereby causing them to generate and use non-matching keys.

Addressing the other concerns mentioned above represents an interesting direction for future work.

3   A Scalable Compiler for Group AKE Protocols

We show here a compiler transforming any secure group KE protocol P to a secure group AKE protocol P'. Without loss of generality, we assume the following about P: (1) Each message sent by an instance Π_U^i during execution of

---

[5] This is slightly different from the approach of [13, Sec. 7] in that we require a signature on the broadcast value x_i.

P includes the sender's identity U as well as a sequence number which begins at 1 and is incremented each time Π_U^i sends a message (in other words, the jth message sent by an instance Π_U^i has the form $U|j|m$); (2) every message of the protocol is sent — via point-to-point links — to every member of the group taking part in the execution of the protocol (that is, Π_U^i sends each message to all users in pid_U^i). For simplicity, we refer to this as "broadcasting a message" but stress that we do *not* assume a broadcast channel and, in particular, an active adversary or a corrupted user can deliver different messages to different members of the group. Note that any secure group KE protocol $\tilde{P}$ can be readily converted to a secure group KE protocol P in which the above assumptions hold (recall, security of a KE protocol is with respect to a passive adversary only).

Let $\Sigma = (\text{Gen}, \text{Sign}, \text{Vrfy})$ be a signature scheme which is strongly unforgeable under adaptive chosen message attack (where "strong" means that an adversary is also unable to forge a new signature for a previously-signed message), and let $\text{Succ}_\Sigma(t)$ denote the maximum advantage of any adversary running in time t in forging a new message/signature pair. We furthermore assume that the signature length is independent of the length of the message signed; this is easy to achieve by hashing the message (using a collision-resistant hash function) before signing. Given P as above, our compiler constructs protocol P' as follows:

1. During the initialization phase, each party $U \in \mathcal{P}$ generates the verification/signing keys (PK_U', SK_U') by running $\text{Gen}(1^k)$. This is in addition to any keys (PK_U, SK_U) needed as part of the initialization phase for P.
2. Let $U_1, \ldots, U_n$ be the identities (in lexicographic order) of users wishing to establish a common key, and let $\mathcal{U} = U_1|\cdots|U_n$. Each user U_i begins by choosing a random nonce $r_i \in \{0,1\}^k$ and broadcasting $U_i|0|r_i$ (note we assign this message the sequence number "0"). After receiving the initial broadcast message from all other parties, each instance stores $\mathcal{U}$ and $r_1|\cdots|r_n$ as part of its state information.
3. The members of the group now execute P with the following changes:
   - Whenever instance Π_U^i is supposed to broadcast $U|j|m$ as part of protocol P, the instance instead signs $j|m|\mathcal{U}|r_1|\cdots|r_n$ using SK_U' to obtain signature σ, and then broadcasts $U|j|m|\sigma$.
   - When instance Π_U^i receives message $V|j|m|\sigma$, it checks that: (1) $V \in \text{pid}_U^i$, (2) j is the next expected sequence number for messages from V, and (3) (using PK_V') σ is a correct signature of V on $j|m|\mathcal{U}|r_1|\cdots|r_n$. If any of these are untrue, Π_U^i aborts the protocol and sets $\text{acc}_U^i = \text{FALSE}$ and $\text{sk}_U^i = \text{NULL}$. Otherwise, Π_U^i continues as it would in P upon receiving message $V|j|m$.
4. Each non-aborted instance computes the session key as in P.

Theorem 1. *If P is a secure group KE protocol achieving forward secrecy, then P' given by the above compiler is a secure group AKE protocol achieving forward secrecy. Namely:*

$$\text{Adv}_{P'}^{\text{AKE-fs}}(t, q_{\text{ex}}, q_{\text{s}}) \leq (q_{\text{ex}} + q_{\text{s}}) \cdot \text{Adv}_P^{\text{KE-fs}}(t, 1) + |\mathcal{P}| \cdot \text{Succ}_\Sigma(t) + \frac{q_{\text{s}}^2 + 2q_{\text{ex}}q_{\text{s}} + |\mathcal{P}|q_{\text{ex}}^2}{2^{k+1}}.$$

Proof. Given an *active* adversary $\mathcal{A}'$ attacking P', we will construct a *passive* adversary $\mathcal{A}$ attacking P where $\mathcal{A}$ makes only a single Execute query; relating the success probabilities of $\mathcal{A}'$ and $\mathcal{A}$ gives the stated result.

Before describing $\mathcal{A}$, we first define events Forge and Repeat and bound their probabilities of occurrence. Let Forge be the event that $\mathcal{A}'$ outputs a new, valid message/signature pair with respect to the public key PK'_U of some user $U \in \mathcal{P}$ *before* querying Corrupt(U), and let Pr[Forge] denote $\text{Pr}_{\mathcal{A}',P'}[\text{Forge}]$ for brevity. Using $\mathcal{A}'$, we may construct an algorithm $\mathcal{F}$ that forges a signature with respect to signature scheme Σ as follows: given a public key PK, algorithm $\mathcal{F}$ chooses a random $U \in \mathcal{P}$, sets $PK'_U = PK$, and honestly generates all other public/private keys for the system. $\mathcal{F}$ simulates the oracle queries of $\mathcal{A}'$ in the natural way (accessing its signing oracle when necessary); this results in a perfect simulation unless $\mathcal{A}'$ queries Corrupt(U). If this occurs, $\mathcal{F}$ simply aborts. Otherwise, if $\mathcal{A}'$ ever outputs a new, valid message/signature pair with respect to $PK'_U = PK$, then $\mathcal{F}$ outputs this pair as its forgery. The success probability of $\mathcal{F}$ is exactly $\frac{\text{Pr[Forge]}}{|\mathcal{P}|}$; this immediately implies that

$$\text{Pr[Forge]} \leq |\mathcal{P}| \cdot \text{Succ}_{\Sigma}(t).$$

Let Repeat be the event that a nonce is used twice by a particular user; i.e., that there exists a user $U \in \mathcal{P}$ and i,j ($i \neq j$) such that the nonce used by instance Π_U^i is equal to the nonce used by instance Π_U^j. A straightforward "birthday problem" calculation shows that $\text{Pr[Repeat]} \leq \frac{|\mathcal{P}|(q_{\text{ex}}+q_{\text{s}})^2}{2^{k+1}}$, since each user $U \in \mathcal{P}$ chooses at most $(q_{\text{ex}} + q_{\text{s}})$ nonces from $\{0,1\}^k$. A more careful analysis (omitted in the present abstract) in fact shows that

$$\text{Pr[Repeat]} \leq \frac{q_{\text{s}}^2 + 2q_{\text{ex}}q_{\text{s}} + |\mathcal{P}|q_{\text{ex}}^2}{2^{k+1}}.$$

We now construct our passive adversary $\mathcal{A}$ attacking protocol P. Recall that as part of the initial setup, adversary $\mathcal{A}$ is given public keys $\{PK_U\}_{U \in \mathcal{P}}$ if any are defined as part of protocol P. We first have $\mathcal{A}$ obtain all secret keys $\{SK_U\}_{U \in \mathcal{P}}$ using multiple Corrupt queries. Next, $\mathcal{A}$ runs Gen(1^k) to generate keys (PK'_U, SK'_U) for each $U \in \mathcal{P}$; the set of public keys $\{PK'_U, PK_U\}_{U \in \mathcal{P}}$ is then given to $\mathcal{A}'$. We now have $\mathcal{A}$ run $\mathcal{A}'$, simulating the oracle queries of $\mathcal{A}'$ as described below.

Before describing the details, we provide a high-level overview. Let $Q = q_{\text{ex}} + q_{\text{s}}$ denote the total number of Execute and Send queries made by $\mathcal{A}'$. Intuitively, $\mathcal{A}$ chooses an $\alpha \in \{1, \ldots, Q\}$ representing a guess as to which Send/Execute query of $\mathcal{A}'$ activates the instance for which $\mathcal{A}'$ will ask its Test query. For the αth such query of $\mathcal{A}'$, we will have $\mathcal{A}$ respond by making an Execute query, obtaining a transcript of an execution of P, modifying this transcript to obtain a valid transcript for P', and then returning an appropriate response to $\mathcal{A}'$. (We also need to ensure that this provides $\mathcal{A}'$ with a consistent view; these details are discussed below.) For all other (unrelated) Send/Execute queries of $\mathcal{A}'$, we have $\mathcal{A}$ respond by directly running protocol P'; note that $\mathcal{A}$ can do this since it has the secret keys for all players. $\mathcal{A}$ aborts and outputs a random bit if it determines that its guess α was incorrect, or if events Forge or Repeat occur. Otherwise, $\mathcal{A}$ outputs whatever bit is output by $\mathcal{A}'$. We now describe the simulation of the oracle queries of $\mathcal{A}'$ in detail.

Execute queries. If an Execute query is *not* the αth Send/Execute query of $\mathcal{A}'$, then $\mathcal{A}$ simply generates on its own a transcript of an execution of P' and returns this to $\mathcal{A}'$ (as noted above, $\mathcal{A}$ can do this since it knows all the secret keys for all players). If an Execute query *is* the αth Send/Execute query of $\mathcal{A}'$, let the query be $\text{Execute}(U_1, \ldots U_n)$ and let $\mathcal{U} = U_1 | \cdots | U_n$. Adversary $\mathcal{A}$ sends the same query to its Execute oracle and receives in return a transcript T of an execution of P. To simulate a transcript T' of an execution of P', $\mathcal{A}$ first chooses random $r_1, \ldots, r_n \in \{0,1\}^k$. The initial messages of T' are set to $\{U_i | 0 | r_i\}_{1 \le i \le n}$. Then, for each message $U | j | m$ in transcript T, $\mathcal{A}$ computes $\sigma \leftarrow \text{Sign}_{SK'_U}(j | m | \mathcal{U} | r_1 | \cdots | r_n)$ and places $U | j | m | \sigma$ in T'. When done, the complete transcript T' is given to $\mathcal{A}'$.

Send queries. If a Send query is not the αth Send/Execute query of $\mathcal{A}'$, the intuition is to have $\mathcal{A}$ simulate by itself the actions of this instance and thus generate the appropriate responses for $\mathcal{A}'$. On the other hand, if a Send query is the αth Send/Execute query of $\mathcal{A}'$, then $\mathcal{A}$ should obtain a transcript T from its Execute oracle and generate responses for $\mathcal{A}'$ by modifying T using the signature keys $\{SK'_U\}_{U \in \mathcal{P}}$. The actual simulation is slightly more difficult, since we need to ensure consistency in the view of $\mathcal{A}'$.

Consider an arbitrary instance Π_U^ℓ. Denote the initial Send query to this instance (i.e., protocol initiation) by Send_0; this query always has the form $\text{Send}_0(U, \ell, \langle U_1, \ldots, U_n \rangle)$ for some n. We set $\mathcal{U}_U^\ell = U | U_1 | \cdots | U_n$, where we assume without loss of generality that these are in lexicographic order. We denote the second Send query to the instance by Send_1; this query always has the form $\text{Send}(U, \ell, U_1 | 0 | r_1, \ldots, U_n | 0 | r_n)$. After a Send_1 query, we may set $\mathcal{R}_U^\ell = r_U^\ell | r_1 | \cdots | r_n$, where r_U^ℓ is the nonce generated by instance Π_U^ℓ. To aid the simulation, $\mathcal{A}$ will maintain a list Nonces whose function will become clear below.

If a Send query is *not* the αth Send/Execute query of $\mathcal{A}'$, then:

- On query $\text{Send}_0(U, \ell, *)$, $\mathcal{A}$ simply chooses a random nonce r_U^ℓ and replies to $\mathcal{A}'$ with $U | 0 | r_U^\ell$. Note that $\mathcal{U}_U^\ell$ is now defined.
- If the query is not a Send_0 query and has the form $\text{Send}(U, \ell, M)$, then $\mathcal{R}_U^\ell$ is defined (either by the present query or by some previous query). $\mathcal{A}$ looks in Nonces for an entry of the form $(\mathcal{U}_U^\ell | \mathcal{R}_U^\ell, c)$. There are two cases to consider:
  - If such an entry exists and $c = 1$ then $\mathcal{A}$ has already queried its Execute oracle and received in return a transcript T. First, $\mathcal{A}$ verifies correctness of the current incoming message(s) as in the description of the compiler (and aborts execution of Π_U^ℓ if verification fails). $\mathcal{A}$ then finds the appropriate message $U | j | m$ in T, computes $\sigma \leftarrow \text{Sign}_{SK'_U}(j | m | \mathcal{U}_U^\ell | \mathcal{R}_U^\ell)$, and replies to $\mathcal{A}'$ with $U | j | m | \sigma$.
  - If no such entry exists, $\mathcal{A}$ stores $(\mathcal{U}_U^\ell | \mathcal{R}_U^\ell, 0)$ in Nonces. In this case or if the entry exists and $c = 0$, then $\mathcal{A}$ simulates on its own the actions of this instance ($\mathcal{A}$ can do this since it knows all relevant secret keys).

If a Send query *is* the αth Send/Execute query of $\mathcal{A}'$, then:

- If the query is *not* a Send_1 query, then $\mathcal{A}$ aborts (and outputs a random bit) since its guess α was incorrect.
- If a Corrupt query has previously been made by $\mathcal{A}'$, then $\mathcal{A}$ aborts. The current instance is no longer fresh, and therefore the guess α is incorrect.
- If the query is a Send_1 query to instance Π_U^ℓ, then $\mathcal{U}_U^\ell$ and $\mathcal{R}_U^\ell$ are now both defined. $\mathcal{A}$ looks in Nonces for an entry of the form $(\mathcal{U}_U^\ell|\mathcal{R}_U^\ell, 0)$. If such an entry exists, then $\mathcal{A}$ aborts (and outputs a random bit). Otherwise, $\mathcal{A}$ stores $(\mathcal{U}_U^\ell|\mathcal{R}_U^\ell, 1)$ in Nonces. Next, $\mathcal{A}$ queries $\text{Execute}(\mathcal{U}_U^\ell)$, obtains in return a transcript T (which is stored for later use), and finds the message $U|1|m$ in T. The signature $\sigma \leftarrow \text{Sign}_{SK_U'}(1|m|\mathcal{U}_U^\ell|\mathcal{R}_U^\ell)$ is computed, and the message $U|1|m|\sigma$ is returned to $\mathcal{A}'$.

Corrupt queries. On query $\text{Corrupt}(U)$, $\mathcal{A}$ returns (SK_U, SK_U') (recall that $\mathcal{A}$ has obtained SK_U already, and knows SK_U' since it was generated by $\mathcal{A}$).

Reveal queries. When $\mathcal{A}'$ queries $\text{Reveal}(U, i)$ for a terminated instance it must be the case that $\mathcal{U}_U^\ell$ and $\mathcal{R}_U^\ell$ are both defined. $\mathcal{A}$ locates the entry $(\mathcal{U}_U^\ell|\mathcal{R}_U^\ell, c)$ in Nonces and aborts (and outputs a random bit) if $c = 1$ since its guess α was incorrect. Otherwise, $c = 0$ implies that this instance was simulated by $\mathcal{A}$ itself; thus $\mathcal{A}$ can compute the appropriate key sk_U^ℓ and returns this key to $\mathcal{A}'$.

Test queries. When $\mathcal{A}'$ queries $\text{Test}(U, i)$ for a terminated instance it must be the case that $\mathcal{U}_U^\ell$ and $\mathcal{R}_U^\ell$ are both defined. $\mathcal{A}$ finds the entry $(\mathcal{U}_U^\ell|\mathcal{R}_U^\ell, c)$ in Nonces and aborts (and outputs a random bit) if $c = 0$ since its guess α was incorrect. Otherwise, $c = 1$ implies that this instance corresponds to an instance for which $\mathcal{A}$ had asked its single Execute query. So, $\mathcal{A}$ asks its own Test query for any such instance (it does not matter which, since they are all partnered and all hold the same key) and returns the result to $\mathcal{A}'$.

Let Guess denote the event that $\mathcal{A}$ correctly guesses α. We claim that as long as Guess and $\overline{\text{Forge}}$ and $\overline{\text{Repeat}}$ occur, the above simulation is perfect. Indeed, assuming Guess occurs the only difference between the simulation and a real execution of $\mathcal{A}'$ occurs for those instances Π_U^ℓ for which $(\mathcal{U}_U^\ell|\mathcal{R}_U^\ell, 1) \in$ Nonces. Here, the simulation is perfect unless $\mathcal{A}'$ forges a signature or can "splice in" a message from a different execution. However, neither of these events can happen as long as neither Forge nor Repeat occur.

Letting $\text{Good} \stackrel{\text{def}}{=} \overline{\text{Forge}} \wedge \overline{\text{Repeat}}$ and $\text{Bad} \stackrel{\text{def}}{=} \overline{\text{Good}}$, and recalling that $Q = q_{\text{ex}} + q_s$ denotes the total number of Send/Execute queries asked by $\mathcal{A}'$, a straightforward probability calculation shows that:

$$2 \cdot \left| \Pr_{\mathcal{A}, P}[\text{Succ}] - \tfrac{1}{2} \right|$$

$$= 2 \cdot \left| \Pr_{\mathcal{A}', P'}[\text{Succ} \wedge \text{Guess} \wedge \text{Good}] + \tfrac{1}{2} \Pr_{\mathcal{A}', P'}[\overline{\text{Guess}} \vee \text{Bad}] - \tfrac{1}{2} \right|$$

$$= 2 \cdot \left| \tfrac{1}{Q} \Pr_{\mathcal{A}', P'}[\text{Succ} \wedge \text{Good}] + \tfrac{1}{2} \Pr_{\mathcal{A}', P'}[\text{Bad}] \right.$$
$$\left. + \tfrac{1}{2} \Pr[\overline{\text{Guess}}|\overline{\text{Bad}}] \Pr_{\mathcal{A}', P'}[\overline{\text{Bad}}] - \tfrac{1}{2} \right|$$

$$= 2 \cdot \left| \tfrac{1}{Q} \Pr\nolimits_{\mathcal{A}',P'}[\mathsf{Succ}] - \tfrac{1}{Q} \Pr\nolimits_{\mathcal{A}',P'}[\mathsf{Succ} \wedge \mathsf{Bad}] + \tfrac{1}{2} \Pr\nolimits_{\mathcal{A}',P'}[\mathsf{Bad}] \right.$$

$$\left. + \tfrac{1}{2}(\tfrac{Q-1}{Q})(1 - \Pr\nolimits_{\mathcal{A}',P'}[\mathsf{Bad}]) - \tfrac{1}{2} \right|$$

$$\geq \tfrac{1}{Q} \cdot |2 \cdot \Pr\nolimits_{\mathcal{A}',P'}[\mathsf{Succ}] - 1| - \tfrac{1}{Q} |2 \cdot \Pr\nolimits_{\mathcal{A}',P'}[\mathsf{Succ} \wedge \mathsf{Bad}] - \Pr\nolimits_{\mathcal{A}',P'}[\mathsf{Bad}]|.$$

Since $2 \left| \Pr_{\mathcal{A},P}[\mathsf{Succ}] - \tfrac{1}{2} \right| \leq \mathsf{Adv}_P^{\mathsf{KE-fs}}(t,1)$ by assumption, we obtain:

$$\mathsf{Adv}_{P'}^{\mathsf{AKE-fs}}(t, q_{\mathrm{ex}}, q_{\mathrm{s}}) \leq Q \cdot \mathsf{Adv}_P^{\mathsf{KE-fs}}(t,1) + \Pr\nolimits_{\mathcal{A}',P'}[\mathsf{Forge}] + \Pr[\mathsf{Repeat}],$$

which immediately yields the statement of the theorem. ∎

We remark that the above theorem is a generic result that applies to the invocation of the compiler on an *arbitrary* group KE protocol P. For specific protocols, a better exact security analysis may be obtainable. Furthermore, the compiler above may also be applied to KE protocols that do not achieve forward secrecy. In this case, we obtain the following tighter security reduction.

Theorem 2. *If P is a secure group KE protocol (without forward secrecy), then P' given by the above compiler is a secure group AKE protocol (without forward secrecy). Namely:*

$$\mathsf{Adv}_{P'}^{\mathsf{AKE}}(t, q_{\mathrm{ex}}, q_{\mathrm{s}}) \leq \mathsf{Adv}_P^{\mathsf{KE}}(t, q_{\mathrm{ex}} + q_{\mathrm{s}}) + |\mathcal{P}| \cdot \mathsf{Succ}_\Sigma(t) + \frac{q_{\mathrm{s}}^2 + 2q_{\mathrm{ex}}q_{\mathrm{s}} + |\mathcal{P}|q_{\mathrm{ex}}^2}{2^{k+1}}.$$

The proof is largely similar to that of Theorem 1, and will appear in the full version of this paper.

4   A Constant-Round Group KE Protocol

Let $\mathbb{G}$ be any finite cyclic group of prime order q (e.g., letting p, q be prime such that $p = \beta q + 1$ we may let $\mathbb{G}$ be the subgroup of order q in $\mathbb{Z}_p^*$), and let g be an arbitrary generator of $\mathbb{G}$. We define $\mathsf{Adv}_{\mathbb{G}}^{\mathsf{ddh}}(t)$ as the maximum value, over all adversaries A running in time at most t, of:

$$\left| \Pr[x, y \leftarrow \mathbb{Z}_q : A(g, g^x, g^y, g^{xy}) = 1] - \Pr[x, y, z \leftarrow \mathbb{Z}_q : A(g, g^x, g^y, g^z) = 1] \right|.$$

Informally, we say the *DDH assumption holds in* $\mathbb{G}$ if $\mathsf{Adv}_{\mathbb{G}}^{\mathsf{ddh}}(t)$ is "small" for "reasonable" values of t. We now describe an efficient, two-round group KE protocol whose security is based on the DDH assumption in $\mathbb{G}$. Applying the compiler of the previous section to this protocol immediately yields an efficient, three-round group AKE protocol.

The protocol presented here is essentially the protocol of Burmester and Desmedt [14], except we assume that $\mathbb{G}$ is a finite, cyclic group of prime order in which the DDH assumption holds. Our work was originally motivated by the fact that no proof of security appears in the proceedings version of [14]; furthermore, subsequent work in this area (e.g., [13,10]) implied that the Burmester-Desmedt protocol was "heuristic" and had not been proven secure. (Indeed, presumably for this reason the group AKE protocols of [13,11,12] are based on the

$O(n)$-round group KE protocol of Steiner, et al. [31] rather than the Burmester-Desmedt protocol.) Subsequent to our work, however, we became aware that a proof of security for a variant of the Burmester-Desmedt protocol appears in the pre-proceedings of Eurocrypt '94 [18].[6] Even so, we note the following:

- The given proof shows only that an adversary cannot compute the *entire* session key; in contrast to our work, it says nothing about whether the key is indistinguishable from random. On the other hand, for this reason their proof uses only the weaker CDH assumption.
- A proof of security is given only for an *even* number of participants n. A modified, asymmetric protocol (which is slightly less efficient) is introduced and proven secure for the case of n odd.
- Finally, the previously-given proof of security makes no effort to optimize the concrete security of the reduction (since this issue was not generally considered at that time).

As required by the compiler of the previous section, our protocol ensures that players send every message to all members of the group via point-to-point links; although we refer to this as "broadcasting" we stress that no broadcast channel is assumed (in any case, the distinction is moot since we are dealing here with a passive adversary). In our protocol P, no public keys are required but for simplicity we assume a group $\mathbb{G}$ and generator $g \in \mathbb{G}$ have been fixed in advance and are known to all parties in the network. Note that this assumption can be avoided at the expense of an additional round in which the first player simply generates and broadcasts these values (that this is secure is clear from the fact that we are now considering a *passive* adversary). When n players $U_1, \ldots, U_n$ wish to generate a session key, they proceed as follows (the indices are taken modulo n so that player U_0 is U_n and player U_{n+1} is U_1):

Round 1. Each player U_i chooses a random $r_i \in \mathbb{Z}_q$ and broadcasts $z_i = g^{r_i}$.
Round 2. Each player U_i broadcasts $X_i = (z_{i+1}/z_{i-1})^{r_i}$.
Key computation. Each player U_i computes their session key as:

$$K_i = (z_{i-1})^{nr_i} \cdot X_i^{n-1} \cdot X_{i+1}^{n-2} \cdots X_{i-2}.$$

(It may be easily verified that all users compute the same key $g^{r_1 r_2 + r_2 r_3 + \cdots + r_n r_1}$.)

We do not explicitly include sender identities and sequence numbers as required by the compiler of the previous section; however, as discussed there, it is easy to modify the protocol to include this information. Note that each user only computes three (full-length) exponentiations since $n \ll q$ in practice.

Theorem 3. *Protocol P is a secure group KE protocol achieving forward secrecy. Namely:*

$$\mathsf{Adv}_P^{\mathsf{KE-fs}}(t, q_{\mathrm{ex}}) \leq 4 \cdot \mathsf{Adv}_{\mathbb{G}}^{\mathsf{ddh}}(t).$$

---

[6] We are happy to publicize this, especially since it appears to have been unknown to many others in the cryptographic community as well!

Proof. Let $\varepsilon(t) \stackrel{\text{def}}{=} \mathsf{Adv}_{\mathbb{G}}^{\mathsf{ddh}}(t)$. We provide here a proof for the case of an adversary making only a single Execute query, and show the weaker result that $\mathsf{Adv}_P^{\mathsf{KE-fs}}(t, 1) \le 2|\mathcal{P}|\varepsilon(t)$. Note that this is sufficient for the purposes of applying Theorem 1, and also immediately yields (via a standard hybrid argument) that $\mathsf{Adv}_P^{\mathsf{KE-fs}}(t, q_{\mathsf{ex}}) \le 2q_{\mathsf{ex}}|\mathcal{P}|\varepsilon(t)$. A proof of the stronger result stated in the theorem can be obtained using random self-reducibility properties of the DDH problem (following [30]), and will appear in the full version.

Since there are no public keys in the protocol, we may ignore Corrupt queries. Assume an adversary $\mathcal{A}$ making a single query $\mathsf{Execute}(U_1, \ldots, U_n)$ (we stress that the number of parties n is chosen by the adversary; however, since the protocol is symmetric and there are no public keys the identities of the parties are unimportant). The distribution of the transcript T and the resulting session key sk is given by:

$$\mathsf{Real} \stackrel{\text{def}}{=} \left\{ \begin{array}{c} r_1, \ldots, r_n \leftarrow \mathbb{Z}_q; \ z_1 = g^{r_1}, \ z_2 = g^{r_2}, \ \ldots, \ z_n = g^{r_n} \\ X_1 = \frac{g^{r_2 r_1}}{g^{r_n r_1}}, \ X_2 = \frac{g^{r_3 r_2}}{g^{r_1 r_2}}, \ \ldots, \ X_n = \frac{g^{r_1 r_n}}{g^{r_{n-1} r_n}} \\ \mathsf{T} = (z_1, \ldots, z_n, X_1, \ldots, X_n) \\ \mathsf{sk} = (g^{r_1 r_2})^n \cdot (X_2)^{n-1} \cdots X_n \end{array} : (\mathsf{T}, \mathsf{sk}) \right\}.$$

Consider the following modified distribution:

$$\mathsf{Fake}_1 \stackrel{\text{def}}{=} \left\{ \begin{array}{c} w_{1,2}, r_1, \ldots, r_n \leftarrow \mathbb{Z}_q; \ z_1 = g^{r_1}, \ z_2 = g^{r_2}, \ \ldots, \ z_n = g^{r_n} \\ X_1 = \frac{g^{w_{1,2}}}{(g^{r_1})^{r_n}}, \ X_2 = \frac{(g^{r_2})^{r_3}}{g^{w_{1,2}}}, \ \ldots, \ X_n = \frac{(g^{r_1})^{r_n}}{g^{r_{n-1} r_n}} \\ \mathsf{T} = (z_1, \ldots, z_n, X_1, \ldots, X_n) \\ \mathsf{sk} = (g^{w_{1,2}})^n \cdot (X_2)^{n-1} \cdots X_n \end{array} : (\mathsf{T}, \mathsf{sk}) \right\}.$$

A standard argument shows that for any algorithm $\mathcal{A}'$ running in time t we have:

$$|\Pr[(\mathsf{T}, \mathsf{sk}) \leftarrow \mathsf{Real} : \mathcal{A}'(\mathsf{T}, \mathsf{sk}) = 1] - \Pr[(\mathsf{T}, \mathsf{sk}) \leftarrow \mathsf{Fake}_1 : \mathcal{A}'(\mathsf{T}, \mathsf{sk}) = 1]| \le \varepsilon(t).$$

We next make the following additional modification:

$$\mathsf{Fake}_2 \stackrel{\text{def}}{=} \left\{ \begin{array}{c} w_{1,2}, w_{2,3}, r_1, \ldots, r_n \leftarrow \mathbb{Z}_q; \ z_1 = g^{r_1}, \ z_2 = g^{r_2}, \ \ldots, \ z_n = g^{r_n} \\ X_1 = \frac{g^{w_{1,2}}}{g^{r_1 r_n}}, \ X_2 = \frac{g^{w_{2,3}}}{g^{w_{1,2}}}, \ \ldots, \ X_n = \frac{g^{r_1 r_n}}{g^{r_{n-1} r_n}} \\ \mathsf{T} = (z_1, \ldots, z_n, X_1, \ldots, X_n) \\ \mathsf{sk} = (g^{w_{1,2}})^n \cdot (X_2)^{n-1} \cdots X_n \end{array} : (\mathsf{T}, \mathsf{sk}) \right\},$$

where, again, a standard argument shows that:

$$|\Pr[(\mathsf{T}, \mathsf{sk}) \leftarrow \mathsf{Fake}_1 : \mathcal{A}'(\mathsf{T}, \mathsf{sk}) = 1] - \Pr[(\mathsf{T}, \mathsf{sk}) \leftarrow \mathsf{Fake}_2 : \mathcal{A}'(\mathsf{T}, \mathsf{sk}) = 1]| \le \varepsilon(t).$$

Continuing in this way, we obtain the distribution:

$$\mathsf{Fake}_n \stackrel{\text{def}}{=} \left\{ \begin{array}{c} w_{1,2}, w_{2,3}, \ldots, w_{n-1,n}, w_{n,1}, r_1, \ldots, r_n \leftarrow \mathbb{Z}_q \\ X_1 = \frac{g^{w_{1,2}}}{g^{w_{n,1}}}, \ X_2 = \frac{g^{w_{2,3}}}{g^{w_{1,2}}}, \ \ldots, \ X_n = \frac{g^{w_{n,1}}}{g^{w_{n-1,n}}} \\ \mathsf{T} = (g^{r_1}, \ldots, g^{r_n}, X_1, \ldots, X_n) \\ \mathsf{sk} = (g^{w_{1,2}})^n \cdot (X_2)^{n-1} \cdots X_n \end{array} : (\mathsf{T}, \mathsf{sk}) \right\},$$

such that, for any $\mathcal{A}'$ running in time t we have (via standard hybrid argument):

$$|\Pr[(\mathsf{T},\mathsf{sk}) \leftarrow \mathsf{Real} : \mathcal{A}'(\mathsf{T},\mathsf{sk}) = 1] - \Pr[(\mathsf{T},\mathsf{sk}) \leftarrow \mathsf{Fake}_n : \mathcal{A}'(\mathsf{T},\mathsf{sk}) = 1]|$$
$$\leq n \cdot \varepsilon(t). \quad (1)$$

In experiment Fake_n, the values $w_{1,2}, \ldots, w_{n,1}$ are constrained by T according to the following n equations

$$\log_g X_1 = w_{1,2} - w_{n,1}$$
$$\vdots$$
$$\log_g X_n = w_{n,1} - w_{n-1,n},$$

of which only $n-1$ of these are linearly independent. Furthermore, sk may be expressed as $g^{w_{1,2}+w_{2,3}+\cdots+w_{n,1}}$; equivalently, we have

$$\log_g \mathsf{sk} = w_{1,2} + w_{2,3} + \cdots + w_{n,1}.$$

Since this final equation is linearly independent from the set of equations above, the value of sk is independent of T. This implies that, for any adversary $\mathcal{A}$:

$$\Pr[(\mathsf{T},\mathsf{sk}_0) \leftarrow \mathsf{Fake}_n; \mathsf{sk}_1 \leftarrow \mathbb{G}; b \leftarrow \{0,1\} : \mathcal{A}(\mathsf{T},\mathsf{sk}_b) = b] = 1/2,$$

which — combined with Equation (1) and the fact that $n \leq |\mathcal{P}|$ — yields the desired result $\mathsf{Adv}_P^{\mathsf{KE-fs}}(t,1) \leq 2|\mathcal{P}|\varepsilon(t)$.

References

1. S.S. Al-Riyami and K.G. Paterson. Tripartite Authenticated Key Agreement Protocols from Pairings. Available at http://eprint.iacr.org/2002/035/.
2. G. Ateniese, M. Steiner, and G. Tsudik. Authenticated Group Key Agreement and Friends. ACM CCCS '98.
3. G. Ateniese, M. Steiner, and G. Tsudik. New Multi-Party Authentication Services and Key Agreement Protocols. *IEEE Journal on Selected Areas in Communications*, 18(4): 628–639 (2000).
4. M. Bellare, R. Canetti, and H. Krawczyk. A Modular Approach to the Design and Analysis of Authentication and Key Exchange Protocols. STOC '98.
5. M. Bellare, D. Pointcheval, and P. Rogaway. Authenticated Key Exchange Secure Against Dictionary Attacks. Eurocrypt 2000.
6. M. Bellare and P. Rogaway. Entity Authentication and Key Distribution. Crypto '93.
7. M. Bellare and P. Rogaway. Provably-Secure Session Key Distribution: the Three Party Case. STOC '95.
8. R. Bird, I. Gopal, A. Herzberg, P. Janson, S. Kutten, R. Molva, and M. Yung. Systematic Design of Two-Party Authentication Protocols. *IEEE J. on Selected Areas in Communications*, 11(5): 679–693 (1993). A preliminary version appeared in Crypto '91.
9. C. Boyd. On Key Agreement and Conference Key Agreement. ACISP '97.

10. C. Boyd and J.M.G. Nieto. Round-Optimal Contributory Conference Key Agreement. PKC 2003.
11. E. Bresson, O. Chevassut, and D. Pointcheval. Provably Authenticated Group Diffie-Hellman Key Exchange — The Dynamic Case. Asiacrypt 2001.
12. E. Bresson, O. Chevassut, and D. Pointcheval. Dynamic Group Diffie-Hellman Key Exchange under Standard Assumptions. Eurocrypt 2002.
13. E. Bresson, O. Chevassut, D. Pointcheval, and J.-J. Quisquater. Provably Authenticated Group Diffie-Hellman Key Exchange. ACM CCCS 2001.
14. M. Burmester and Y. Desmedt. A Secure and Efficient Conference Key Distribution System. Eurocrypt '94.
15. R. Canetti and H. Krawczyk. Key-Exchange Protocols and Their Use for Building Secure Channels. Eurocrypt 2001.
16. R. Canetti and H. Krawczyk. Universally Composable Notions of Key Exchange and Secure Channels. Eurocrypt 2002.
17. R. Canetti and H. Krawczyk. Security Analysis of IKE's Signature-Based Key-Exchange Protocol. Crypto 2002.
18. Y. Desmedt. Personal communication (including a copy of the pre-proceedings version of [14]), March 2003.
19. W. Diffie and M. Hellman. New Directions in Cryptography. *IEEE Transactions on Information Theory*, 22(6): 644–654 (1976).
20. W. Diffie, P. van Oorschot, and M. Wiener. Authentication and Authenticated Key Exchanges. *Designs, Codes, and Cryptography*, 2(2): 107–125 (1992).
21. M. Fischer, N. Lynch, and M. Patterson. Impossibility of Distributed Consensus with One Faulty Process. *J. ACM* 32(2): 374–382 (1985).
22. I. Ingemarsson, D.T. Tang, and C.K. Wong. A Conference Key Distribution System. *IEEE Transactions on Information Theory*, 28(5): 714–720 (1982).
23. A. Joux. A One Round Protocol for Tripartite Diffie Hellman. ANTS 2000.
24. M. Just and S. Vaudenay. Authenticated Multi-Party Key Agreement. Asiacrypt '96.
25. H. Krawczyk. SKEME: A Versatile Secure Key-Exchange Mechanism for the Internet. *Proceedings of the Internet Society Symposium on Network and Distributed System Security*, Feb. 1996, pp. 114–127.
26. H.-K. Lee, H.-S. Lee, and Y.-R. Lee. Multi-Party Authenticated Key Agreement Protocols from Multilinear Forms. Available at http://eprint.iacr.org/2002/166/.
27. H.-K. Lee, H.-S. Lee, and Y.-R. Lee. An Authenticated Group Key Agreement Protocol on Braid groups. Available at http://eprint.iacr.org/2003/018/.
28. A. Mayer and M. Yung. Secure Protocol Transformation via "Expansion": From Two-Party to Groups. ACM CCCS '99.
29. O. Pereira and J.-J. Quisquater. A Security Analysis of the Cliques Protocol Suites. *IEEE Computer Security Foundations Workshop*, June 2001.
30. V. Shoup. On Formal Models for Secure Key Exchange. Draft, 1999. Available at http://eprint.iacr.org/1999/012.
31. M. Steiner, G. Tsudik, and M. Waidner. Key Agreement in Dynamic Peer Groups. *IEEE Trans. on Parallel and Distributed Systems* 11(8): 769–780 (2000). A preliminary version appeared in ACM CCCS '96.
32. W.-G. Tzeng. A Practical and Secure Fault-Tolerant Conference Key Agreement Protocol. PKC 2000.
33. W.-G. Tzeng and Z.-J. Tzeng. Round Efficient Conference Key Agreement Protocols with Provable Security. Asiacrypt 2000.

Practical Verifiable Encryption and Decryption of Discrete Logarithms

Jan Camenisch[1] and Victor Shoup[2]

[1] IBM Zürich Research Lab jca@zurich.ibm.com
[2] New York University shoup@cs.nyu.edu

Abstract. This paper addresses the problem of designing practical protocols for proving properties about encrypted data. To this end, it presents a variant of the new public key encryption of Cramer and Shoup based on Paillier's decision composite residuosity assumption, along with efficient protocols for verifiable encryption and decryption of discrete logarithms (and more generally, of representations with respect to multiple bases). This is the first verifiable encryption system that provides chosen ciphertext security and avoids inefficient cut-and-choose proofs. The presented protocols have numerous applications, including key escrow, optimistic fair exchange, publicly verifiable secret and signature sharing, universally composable commitments, group signatures, and confirmer signatures.

1 Introduction

This paper concerns itself with the general problem of *proving properties about encrypted data*. In the case of public-key encryption, which is the setting in which we are interested here, there are two parties who are in a position to prove some property to another party about an encrypted message — namely, the party who created the ciphertext, and the party who holds the secret key. A protocol in which the encryptor is the prover is a *verifiable encryption* protocol, while a protocol in which the prover is the decryptor is a *verifiable decryption* protocol.

For example, suppose a party T has a public key/secret key pair $(\mathsf{PK}, \mathsf{SK})$ for a public key encryption scheme. Party A might encrypt, using T's public PK, a secret message m that satisfies a publicly-defined property θ, and give the resulting ciphertext ψ to another party B. The latter party might demand that A prove that ψ is an encryption of a message satisfying property θ. Ideally, the proof should be "zero knowledge," so that no unnecessary information about m is leaked to B as part of the proof. Another party B' might obtain the ciphertext ψ, and may request that T prove or disprove that ψ decrypts under SK to a message m satisfying a publicly-defined property θ'; a special case of this would be the situation where T simply gives m to B, and proves to B that the decryption was performed correctly. Again, ideally, the proof should be "zero knowledge."

Now, if one expects to obtain reasonably practical protocols for this problem, it seems necessary to restrict the type of properties that protocols should work with. In this paper, we consider only properties related to the discrete logarithm

D. Boneh (Ed.): CRYPTO 2003, LNCS 2729, pp. 126–144, 2003.
© International Association for Cryptologic Research 2003

problem. The message m encrypted by A above is the discrete logarithm of an element δ with respect to a base γ, and A proves to B that ψ is an encryption $\log_\gamma \delta$ under T's public key PK. Here, the common inputs to A and B in the proof protocol are PK, ψ, δ, and γ. Similarly, when a party B' presents ψ to T for decryption, T may state and prove whether or nor ψ decrypts to $\log_\gamma \delta$, or alternatively, T may give the decryption of ψ to B', and simply prove that the decryption was performed correctly. We also consider the obvious generalizations from discrete logarithms to representations with respect to several bases — i.e., proving that a ciphertext is an encryption of $(m_1, \ldots , m_k)$ such that $\delta = \gamma_1^{m_1} \cdots \gamma_k^{m_k}$.

Although the restriction to properties related to the discrete logarithm problem may seem excessive, it turns out (as we discuss in some detail below) that protocols for proving such properties have many useful applications in cryptography, including key escrow, optimistic fair exchange, publicly verifiable secret and signature sharing, universally composable commitments, group signatures, and confirmer signatures. One reason why this restriction is not really so excessive is because in the past few years, efficient protocols for proving numerous properties about committed values — using Pedersen's commitment scheme [Ped92] and generalizations to groups of unknown order — have been developed (c.f., [FO97, DF02,Bou00]); by using our scheme for verifiable encryption of a representation (i.e., an opening of a commitment), we immediately get corresponding protocols for proving properties about encrypted values.

The contribution of this paper is to present and analyze an efficient public-key encryption scheme, together with a suite of proof protocols for the properties related to the discrete logarithm problem outlined above. The encryption scheme is a variant of the new public key encryption of Cramer and Shoup based on Paillier's decision composite residuosity assumption, suitably modified so as to support our proof protocols. The proof protocols are all of the usual, three move "Σ-protocol" type, satisfying the usual, and very strong conditions of special honest verifier zero knowledge and special soundness. We note that any such protocol can be easily and efficiently converted into a "real" zero knowledge protocol using well known techniques, e.g., [Dam00]. Our system for verifiable encryption of discrete logarithms is the first one that provides chosen ciphertext security and avoids inefficient cut-and-choose proofs. It is also the first practical system for verifiable decryption of discrete logarithms.

Although our protocols do not rely on the random oracle heuristic, we hasten to point out that even allowing this heuristic, our protocols are much more efficient than previously known protocols for these problems.

1.1   Applications

In this section, we outline some of the numerous applications of verifiable encryption and decryption of discrete logarithms and representations. For all of them our protocols, used together with the existing solutions, yields more efficient solutions or adds security to chosen ciphertext attacks.

Key escrow. Party A may encrypt its own secret key for an asymmetric cryptographic primitive under the public key of a trusted third party T, and present to a second party B the ciphertext ψ and a proof that ψ is indeed an encryption of it's secret key. This problem area has attracted a good deal of attention, with specific schemes being proposed in [Sta96,BG96,YY98,ASW00,PS00].

Now, if A's secret key is, say, a key for a discrete log based scheme, such as Schnorr or DSS signatures or ElGamal encryption, we can use our verifiable encryption protocol directly. We note that for this and other applications, it is important to be able to bind some public data, called a *label*, to the ciphertext at both encryption and decryption time. In this application, user A would attach a label to ψ that indicates the conditions under which ψ should be decrypted, e.g., A's identity and perhaps and expiration date. The definition of chosen ciphertext security ensures that decrypting a ciphertext under any label different from the label used to create the ciphertext reveals no information about the original encrypted message.

Even though T is "trusted," it might be nice to minimize the trust we need to place in T. To this end, verifiable decryption comes in handy — we can force T to prove that it performed the decryption operation correctly. Of course, this does not prevent T from misbehaving in other ways, such as divulging a secret key to an unauthorized party.

If A's secret key is for a factoring based scheme, one can still use our protocol for verifiable encryption of a representation. One can use Pedersen's commitment scheme to commit to some quantity related to the secret key, and then use an appropriate protocol to prove that the committed value is indeed the right one, together with our protocol to prove that the encryption contains an opening of the commitment. The quantity committed to could be the factorization of an RSA modulus, the decryption exponent of an RSA scheme, or an appropriate root in a Guillou-Quisquater scheme — there are (not too terribly inefficient) protocols for proving that a committed value is of such a form [FO97,CM99a, DF02,PS00,Bou00].

Optimistic fair exchange. Two parties A and B want to exchange some valuable digital data (e.g., signatures on a contract, e-cash), but in a fair way: either each party obtains the other's data, or neither party does. One way to do this is by employing a trusted third party T, but, for the sake of efficiency, with T only involved in crisis situations. One approach to this problem is to have both parties verifiably encrypt to each other their data under T's public key, and only then do they reveal their data to each other — if one party backs out unexpectedly, the other can go to T to obtain the required data. The general problem of optimistic fair exchange has been extensively studied, c.f., [ASW97, BDM98,BP90,Mic,ASW00], while the solution using verifiable encryption was studied in detail in [ASW00].

Our scheme for verifiable encryption may be used directly to efficiently implement the fair exchange of Schnorr or DSS signatures. As outlined in [ASW00], if the public key of the Schnorr signature scheme consists of the base γ and the group element $\alpha = \gamma^x$, and A has a signature on a message m of the form (β, c, s), where $\beta = \gamma^r$, $c = H(\beta, m)$, $s = r + xc \bmod \rho$, and ρ is the group size, then A gives to B the triple (β, c, δ), where $\delta = \gamma^s$, along with an encryption ψ of s under T's public key, and proves to B that ψ is an encryption of $\log_\gamma \delta$. In addition to checking the proof that ψ is a correct encryption of $\log_\gamma \delta$, B also checks that $\delta = \beta \gamma^c$; with these checks, B can be sure that if the need arises, ψ can be decrypted so as to obtain a signature on m. As argued in [ASW00], this technique of reducing a signature to a discrete logarithm does not make it any easier for anyone to forge a signature. Moreover, as discussed in [ASW00], similar techniques can be used to facilitate the fair exchange of other items, such as electronic cash.

As in the escrow application, the label mechanism plays a crucial role here, helping to enforce the logic of the exchange protocol, and a verifiable decryption protocol may be used to hold T's feet to the fire.

Publicly verifiable secret sharing and signature sharing. Stadler [Sta96] introduced the notion of *publicly verifiable secret sharing*. Here, one party, the dealer, shares a secret with several proxies $P_1, \ldots, P_n$, in such a way that a third party (other than the dealer and the proxies) can verify that the sharing was done correctly. This can be done quite simply by sharing the secret using Shamir's secret sharing scheme: the dealer encrypts P_i's share under P_i's public key, and gives to the third party commitments to these shares, along with commitments to the coefficients of the blinding polynomial, and all of the ciphertexts, and proves to to the third party that the ciphertexts encrypt openings of the commitments to the shares. Since the openings to the commitments are just discrete logarithms, verifiable encryption of discrete logarithms is just the right tool.

Using the notion discussed above above for reducing a signature to a discrete logarithm, one can easily implement a (publicly) verifiable *signature* sharing scheme [FR95,CG98] for Schnorr and DSS signatures.

These two applications of verifiable encryption were discussed in [CD00].

Universally composable commitments. The notion of *universally composable (UC) commitments*, introduced by Canetti and Fischlin [CF01], is a very strong notion of security for a commitment scheme. It basically says that commitments in the real world acts like commitments in an ideal world in which, when a party A commits to a value x to a party B, A presents x to an idealized trusted party T (that does not exist in the real world), and when A opens the commitment, T gives x to B. In the ideal world, no information about x is revealed to B prior to opening, and A is forced to fix the value committed to when the commitment protocol runs.

This notion of security is so strong, in fact, that it can only be realized in the *common reference string (CRS)* model, where all parties have access to a string that was generated by a trusted party according to some prescribed distribution. In the CRS model, the simulator S in the ideal world is given the privilege of generating the common reference string, and so S may know some "side information" related to the common reference string that is not available to anyone in the real world.

Verifiable encryption of a representation may be used to implement UC commitments in the CRS model, as follows. The CRS consists of a public key for the encryption scheme, along with bases γ_1 and γ_2 for some suitable group. When A commits a value x to B, he creates a Pedersen commitment $C = \gamma_1^x \gamma_2^r$, and an encryption ψ of the representation (x, r) of C with respect to (γ_1, γ_2). A then gives (C, ψ) to B, and proves to B that ψ indeed decrypts to a representation of C. In order to satisfy the definition of security for UC commitments, and in particular, to prevent "man in the middle attacks," a label containing A's identity should be attached to ψ.

The reason this is secure is that the simulator S in the CRS model knows the secret key to the encryption scheme, which allows him to "extract" values committed by corrupted parties, and S knows the discrete logarithm of γ_2 with respect to γ_1, which allows him to "equivocate" values committed by honest parties. The proof that ψ is an encryption of a representation C ensures that the value extracted by the simulator at commitment time agrees with the value revealed at opening time.

The details of this construction and security proof are the subject of a forthcoming paper.

Confirmer signatures. In a confirmer signature scheme, a notion introduced in [Cha94], a party A creates an "opaque signature" ψ on a message m, which cannot be verified by any other party except a designated trusted third party T, who may either confirm or deny the validity of the signature to another party B. Under appropriate circumstances, T may also *convert* ψ into an ordinary signature, which may then be verified by anybody. Additionally, the party A may prove the validity of an opaque signature ψ to a party B, at the time that A creates and gives ψ to B. As described in [CM00], one may implement confirmer signatures as follows: A creates an ordinary signature σ on m, and encrypts σ under T's public key. Using verifiable encryption, A may prove to B that the resulting ciphertext ψ indeed encrypts a valid signature on m, and using verifiable decryption, T may confirm or deny the validity of ψ, or alternatively, just decrypt ψ, thus converting it to the ordinary signature σ. To implement this idea for Schnorr signatures, one again uses the idea outlined in above for reducing signatures to discrete logarithms. The details of all this are the subject of a forthcoming paper.

Group signatures and anonymous credentials. In a group signature scheme (see [ACJT00,KP98,CD00]), when a user joins a group (whose membership is controlled by a special party, called the *group manager*), the user may sign messages on behalf of the group, without revealing his individual identity; however, under appropriate circumstances, the identity of the individual who actually signed a particular message may be revealed (using a special party, called the *anonymity revocation manager*, which may be distinct from the group manager).

Without going into too many details, verifiable encryption may be used in the following way as a component in such a system. When a group member signs a message, he encrypts enough information under the public key of the anonymity revocation manager, so that later, if the identity of the signer needs to be revealed, this information can be decrypted. To prove that this information correctly identifies the signer, he makes a Pedersen commitment to this information, proves that the committed value identifies the user, encrypts the opening of the commitment, and proves that the ciphertext decrypts to an opening of the commitment. To turn this into a signature scheme, one must use the Fiat-Shamir heuristic [FS87] to make it non-interactive (the interactive version is called an *identity escrow* scheme).

Although one can implement group signatures without it, by using verifiable encryption, one can build a more modular system, in which the group manager and anonymity manager are separate entities with independently generated public keys. As pointed out in [KP97,CM99b,ASW00] such *separability* in system design is highly desirable in practice. Verifiable decryption can be used both to ensure the correct behavior of the anonymity revocation manager (preventing it from "framing" innocent users), and to allow even more fine-grained control of anonymity revocation: instead of simply revealing the identity of a particular signer, the anonymity revocation manager can state (and prove) whether or not a particular signature was generated by a particular user.

Credential systems [Cha85,CL01] are a generalization of group signatures that allow users to show credentials to various organizations, and obtain new credentials, without revealing their identity, except through the use of an anonymity revocation manager. Verifiable encryption can be used as a component in such systems in a manner similar to that described above for group signatures. In fact, our verifiable encryption scheme is used in a prototype credential system developed at IBM called *IDEMIX* [CVH02].

1.2   Previous Work and Further Discussion

In all applications mentioned in §1.1, it is essential that the underlying encryption scheme provide security against chosen ciphertext attacks. As pointed out in [ASW00], the earlier work on verifiable encryption in [Sta96,BG96,YY98] overlooked this fact, as does [PS00].

Our encryption scheme and proof protocols are quite efficient. In particular, the proof protocols are conventional "Σ-protocols," rather than the generally more expensive "cut and choose" protocols, such as in [Sta96,BG96,YY98, ASW00], that have been previously designed for the problem of verifiable encryption. Moreover, our verifiable encryption scheme actually produces a proof that a given ciphertext is correct, as opposed to the paradigm followed in [Sta96, BG96,YY98,ASW00], which intertwines the process of encrypting and proving, so that the entire transcript of the proof must be retained by the verifier in lieu of a (short) ciphertext. Additionally, the combined encrypting/proving paradigm makes it much harder to incorporate any type of verifiable decryption protocol.

Our verifiable decryption protocols are the first practical schemes of their kind.

Unlike, e.g., the schemes in [Sta96,YY98], we do not require that all users of the system work with the same algebraic group — in our system, there are no "double decker" discrete logarithms, and the encryption keys may be used with any group or groups, provided certain reasonable size restrictions are met.

To give the reader a rough idea of the complexity of of our protocols, consider a setting in which the discrete logarithms being encrypted are with respect to an element of order ρ, where ρ is, say, around $\ell' \approx 160$ bits. For such a ρ, it suffices to work with a modulus n of around $\ell \approx 1024$ bits for the Paillier encryption scheme. Counting just squarings, which are all that matter asymptotically, and ignoring lower order terms, the encryption algorithm takes 3ℓ squarings mod n^2, and the decryption algorithm takes 5ℓ squarings mod n^2. For the verifiable encryption protocol, the prover performs 2ℓ squarings mod n, 3ℓ squarings mod n^2, and ℓ' squarings in the underlying group; the verifier performs 3ℓ squarings mod n^2, ℓ squarings mod n, and ℓ' squarings in the group. The verifiable decryption protocols are several times slower than this. For representations with respect to several bases, the complexity of the encryption and decryption algorithms, and the corresponding proof protocols, grows linearly in the number of bases, as one would expect.

Our decryption procedure can be implemented as a *threshold decryption protocol*. This allows one to minimize the trust placed in the decryptor, and in some applications this may be a preferable alternative to verifiable decryption.

2   Preliminaries

2.1   Notation

For a real number a, $\lfloor a \rfloor$ denotes the largest integer $b \leq a$, $\lceil a \rceil$ the smallest integer $b \geq a$, and $\lceil a \rfloor$ the largest integer $b \leq a + 1/2$. For positive real numbers a and b, $[a]$ denotes the set $\{0, \dots, \lfloor a \rfloor - 1\}$ and $[a, b]$ the set $\{\lfloor a \rfloor, \dots, \lfloor b \rfloor\}$, and $[-a, b]$ the set $\{-\lfloor a \rfloor, \dots, \lfloor b \rfloor\}$.

Let a, b, and c be integers, with $b > 0$. Then $c = a \bmod b$ denotes $a - \lfloor a/b \rfloor b$ (and we have $0 \leq c < b$), and $c = a \operatorname{rem} b$ denotes $a - \lceil a/b \rfloor b$ (and we have $-b/2 \leq c < b/2$).

2.2 Σ-Protocols

A Σ-protocol [Cra96] is a protocol between a prover and a verifier, where y is their common input and x is the prover's additional input, which consists of three moves: in the first move the prover sends the verifier a "commitment" message t, in the second move the verifier sends the prover a random "challenge" message c, and in the third move the prover sends the verifier a "response" message s.

Such a protocol is *special honest verifier zero knowledge* if there exists a simulator that, on input (y, c), outputs (t, s) such that the distribution of the triple (t, c, s) is is indistinguishable from that of an actual conversation, conditioned on the event that the verifier's challenge is c. This property implies (ordinary) honest verifier zero knowledge, and also allows the protocol to be easily and efficiently transformed into one that satisfies much stronger notions of zero knowledge.

Such a protocol is said to satisfy the *special soundness condition with respect to a property* θ if it is computationally infeasible to find two valid conversations (t, c, s) and (t, c', s'), with $c \neq c'$, unless the input y satisfies θ. Via standard rewinding arguments, this notion of soundness implies the more general notion of computational soundness.

We use notation introduced by Camenisch and Stadler [CS97] for the various proofs of relations among discrete logarithms. For instance,

$$PK\{(a, b, c) : y = g^a h^b \ \wedge \ \mathfrak{y} = \mathfrak{g}^a \mathfrak{h}^c \ \wedge \ (u \leq a \leq v)\}$$

denotes a *"zero-knowledge* Proof of Knowledge of integers a, b, and c such that $y = g^a h^b$, $\mathfrak{y} = \mathfrak{g}^a \mathfrak{h}^c$, and $u \leq a \leq v$ holds," where $y, g, h, \mathfrak{y}, \mathfrak{g}$, and $\mathfrak{h}$ are elements of some groups $G = \langle g \rangle = \langle h \rangle$ and $\mathfrak{G} = \langle \mathfrak{g} \rangle = \langle \mathfrak{h} \rangle$. The convention is that the elements listed in the round brackets denote quantities the knowledge of which is being proved (and are in general not known to the verifier), while all other parameters are known to the verifier. Using this notation, a proof-protocol can be described by just pointing out its aim while hiding all details.

2.3   Secure Public-Key Encryption

We need the notion of a public-key encryption scheme secure against chosen ciphertext attacks [RS92] that supports *labels* [Sho01]. A label is an arbitrary bit string that is input to the encryption and decryption algorithms, specifying the "context" in which the encryption or decryption operation is to take place. The definition of security for such a scheme is the same as the one without labels except that now the adversary is given a target ciphertext ψ^* and a target label L^* and is then allowed to submit any queries (ψ, L) subject to $(\psi, L) \neq (\psi^*, L^*)$.

3   The Encryption Scheme

3.1   Background

Let p, q, p', and q' be distinct odd primes with $p = 2p' + 1$ and $q = 2q' + 1$, and where p' and q' are both ℓ bits in length. Let $n = pq$ and $n' = p'q'$. Consider the group $\mathbb{Z}_{n^2}^*$ and the subgroup $\mathbf{P}$ of $\mathbb{Z}_{n^2}^*$ consisting of all nth powers of elements in $\mathbb{Z}_{n^2}^*$.

Paillier's Decision Composite Residuosity (DCR) assumption [Pai99] is that given only n, it is hard to distinguish random elements of $\mathbb{Z}_{n^2}^*$ from random elements of $\mathbf{P}$.

We can decompose $\mathbb{Z}_{n^2}^*$ as an internal direct product $\mathbb{Z}_{n^2}^* = \mathbf{G}_n \cdot \mathbf{G}_{n'} \cdot \mathbf{G}_2 \cdot \mathbf{T}$, where each group $\mathbf{G}_\tau$ is a cyclic group of order τ, and $\mathbf{T}$ is the subgroup of $\mathbb{Z}_{n^2}^*$ generated by $(-1 \bmod n^2)$. This decomposition is unique, except for the choice of $\mathbf{G}_2$ (there are two possible choices). For any $x \in \mathbb{Z}_{n^2}^*$, we can express x uniquely as $x = x(\mathbf{G}_n)x(\mathbf{G}_{n'})x(\mathbf{G}_2)x(\mathbf{T})$, where for each $\mathbf{G}_\tau$, $x(\mathbf{G}_\tau) \in \mathbf{G}_\tau$, and $x(\mathbf{T}) \in \mathbf{T}$.

Note that the element $h = (1+n \bmod n^2) \in \mathbb{Z}_{n^2}^*$ has order n, i.e., it generates $\mathbf{G}_n$, and that $h^a = (1 + an \bmod n^2)$ for $0 \le a < n$. Observe that $\mathbf{P} = \mathbf{G}_{n'}\mathbf{G}_2\mathbf{T}$.

3.2   The Scheme

Let ℓ be a system parameter. The scheme makes use of a keyed hash scheme $\mathcal{H}$ that uses a key hk, chosen at random from some key space; the resulting hash function $\mathcal{H}_{\text{hk}}(\cdot)$ maps a triple (u, e, L) to a number in the set $[2^\ell]$. We shall assume that $\mathcal{H}$ is collision resistant, i.e., given a randomly chosen hash key hk, it is computationally infeasible to find two triples $(u, e, L) \ne (u', e', L')$ such that $\mathcal{H}_{\text{hk}}(u, e, L) = \mathcal{H}_{\text{hk}}(u', e', L')$.

Let abs : $\mathbb{Z}_{n^2}^* \to \mathbb{Z}_{n^2}^*$ map $(a \bmod n^2)$, where $a \in [n^2]$, to $(n^2 - a \bmod n^2)$ if $a > n^2/2$, and to $(a \bmod n^2)$, otherwise. Note that $v^2 = (\text{abs}(v))^2$ holds for all $v \in \mathbb{Z}_{n^2}^*$. We now describe the key generation, encryption, and decryption algorithms of the encryption scheme.

Key Generation. Select two random ℓ-bit Sophie Germain primes p' and q', with $p' \ne q'$, and compute $p := (2p'+1)$, $q := (2q'+1)$, $n := pq$, and $n' := p'q'$. Choose random $x_1, x_2, x_3 \in_R [n^2/4]$, choose a random $g' \in_R \mathbb{Z}_{n^2}^*$, compute $g := (g')^{2n}$, $y_1 := g^{x_1}$, $y_2 := g^{x_2}$, and $y_3 := g^{x_3}$. Also, generate a hash key hk from the key space of the hash scheme $\mathcal{H}$. The public key is (hk, n, g, y_1, y_2, y_3). The secret key is (hk, n, x_1, x_2, x_3).

In the rest of the paper, let $h = (1 + n \bmod n^2) \in \mathbb{Z}_{n^2}^*$, which as discussed above, is an element of order n.

Encryption. To encrypt a message $m \in [n]$ with label $L \in \{0, 1\}^*$ under a public key as above, choose a random $r \in_R [n/4]$ and computes the ciphertext (u, e, v) as follows.

$$u := g^r, \qquad e := y_1^r h^m, \qquad \text{and} \qquad v := \text{abs}\left((y_2 y_3^{\mathcal{H}_{\text{hk}}(u,e,L)})^r\right).$$

Decryption. To decrypt a ciphertext $(u, e, v) \in \mathbb{Z}_{n^2}^* \times \mathbb{Z}_{n^2}^* \times \mathbb{Z}_{n^2}^*$ with label L under a secret key as above, first check that $\text{abs}(v) = v$ and $u^{2(x_2 + \mathcal{H}_{\text{hk}}(u,e,L)x_3)} = v^2$. If this does not hold, then output reject and halt. Next, let $t = 2^{-1} \bmod n$, and compute $\hat{m} := (e/u^{x_1})^{2t}$. If $\hat{m}$ is of the form h^m for some $m \in [n]$, then output m; otherwise, output reject.

This scheme differs from the DCR-based schemes presented in [CS01], because in our situation, special attention must be paid to the treatment of elements of order 2 in the $\mathbb{Z}_{n^2}^*$, as these can cause some trouble for the proof systems we discuss in the next sections. Because of these differences, the above encryption scheme does not exactly fit into the general framework of [CS01], even though the basic ideas are the same. We therefore analyze the security of the scheme starting from first principles, rather than trying to modify their framework.

We remark on one of the more peculiar aspects of the scheme, namely, the role of the abs($\cdot$) function in the encryption and decryption algorithms. If one left this out, i.e., replaced abs($\cdot$) by the identity function, then the scheme would be malleable, as (u, e, v) is an encryption of some message m with label L, then so is $(u, e, -v)$. This particular type of malleability [ADR02,Sho01] is in fact rather "benign," and would be acceptable in most applications. However, we prefer to achieve non-malleability in the strictest sense, and because this comes at a marginal cost, we do so.

Theorem 1. *The above scheme is secure against adaptive chosen ciphertext attack provided the DCR assumption holds, and provided $\mathcal{H}$ is collision resistant.*

We refer to the full version of the paper [CS02] for the proof of Theorem 1.

Our scheme can easily be transformed to provide threshold decryption, where it comes in handy that the knowledge of the factorization of n is not required for decryption. This allows one to reduce the trust assumption for the TTP. This can be done either along the lines in [SG98], which requires a random oracle security argument, or along the lines in [CG99], which does not require that argument, but for which the decryption protocol is less efficient.

4   Verifiable Encryption

4.1   Definitions

At a high level, a verifiable encryption scheme for a binary relation $\mathcal{R}$ is a protocol that allows a prover to convince a verifier that a ciphertext ψ is an encryption under a given public key PK and label L of a value w such that $(\delta, w) \in \mathcal{R}$ for a given δ. Here, the common input to the prover and the verifier consists of PK, L, ψ, and δ, and the prover has as additional input the "witness" w and the random bit string that was used to create ψ. We shall require that the protocol is a Σ-protocol that is special honest verifier zero knowledge, and that satisfies the special soundness condition for the property described above.

We refer the reader to the full version of the paper [CS02] for a more detailed definition, but we briefly mention a few subtle points that apply here, as well as in other definitions in this paper: (1) our notion of security is computational, even to the extent that the we quantify "computationally" (rather than universally) over the common input to the prover and verifier in the definition of honest verifier zero knowledge and special soundness; (2) we assume that the public key/secret key pair for the encryption scheme is generated by a trusted party using the appropriate key generation algorithm; (3) in defining soundness, we only require that the proof convinces the verifier that plaintext can be easily transformed into a witness using some scheme-specific *reconstruction routine.*

4.2   The Protocol

Let $(\mathsf{hk}, n, g, y_1, y_2, y_3)$ be a public key of the encryption scheme provided in §3. Recall that the message space associated with this public key is $[n]$.

Let Γ be a cyclic group of order ρ generated by γ. We assume that γ and ρ are publicly known, and that ρ is prime. Let $W = [\rho]$ and $\Delta = \Gamma$, and let $\mathcal{R} = \{(w, \delta) \in W \times \Delta : \gamma^w = \delta\}$. The "discrete logarithm" relation $\mathcal{R}$ is the relation with respect to which we want to verifiably encrypt.

We shall of course require that $n > \rho$ (in fact, we will make a stronger requirement). The reconstruction routine will map a plaintext $m \in [n]$ to the integer $(m \operatorname{rem} n) \bmod \rho$, i.e., it computes the balanced remainder of m modulo n, and then computes the least non-negative remainder of this modulo ρ.

Setup. Our protocol requires the auxiliary parameters $\mathfrak{n}$, which must the product of two safe $(\mathfrak{l}+1)$-bit primes $\mathfrak{p} = 2\mathfrak{p}' + 1$ and $\mathfrak{q} = 2\mathfrak{q}' + 1$, and $\mathfrak{g}$ and $\mathfrak{h}$, which are two generators of $\mathfrak{G}_{\mathfrak{n}'} \subset \mathbb{Z}_\mathfrak{n}^*$, where $\mathfrak{n}' = \mathfrak{p}'\mathfrak{q}'$; $\mathfrak{G}_{\mathfrak{n}'}$ is the subgroup of $\mathbb{Z}_\mathfrak{n}^*$ of order $\mathfrak{n}'$, and $\mathfrak{l}$ is an additional system parameter.

One may view $\mathfrak{n}$, $\mathfrak{g}$, and $\mathfrak{h}$ as additional components of the public key of the encryption scheme, or as system parameters generated by a trusted party. Depending on the setting, we may simply put $\mathfrak{n} := n$ and $\mathfrak{g} := g$. In any event, the prover should not be privy to the factorization of $\mathfrak{n}$.

Let k and k' be further system parameters, where 2^{-k} and $2^{-k'}$ are negligible ($\{0,1\}^k$ is the "challenge space" of the verifier and k' controls the quality of the zero-knowledge property). We require that $2^k < \min\{p', q', \mathfrak{p}', \mathfrak{q}', \rho\}$ holds. Finally, we require that $\rho < n2^{-k-k'-3}$ holds, i.e., that $\log_\gamma \delta$ "comfortably fits into an encryption."

The protocol. The common input of the prover and verifier is: the public key $(\mathsf{hk}, n, g, y_1, y_2, y_3)$, the augmented public key $(\mathfrak{n}, \mathfrak{g}, \mathfrak{h})$, a group element (δ), a ciphertext (u, e, v), and a label L. The prover has additional inputs $m = \log_\gamma \delta$ and $r \in_R [n/4]$ such that $u = g^r$, $e = y_1^r h^m$, and $v = \operatorname{abs}((y_2 y_3^{\mathcal{H}_{\mathsf{hk}}(u,e,L)})^r)$.

1. The prover chooses a random $s \in_R [\mathfrak{n}/4]$ and computes $\mathfrak{k} := \mathfrak{g}^m \mathfrak{h}^s$. The prover sends $\mathfrak{k}$ to the verifier.
2. Then the prover and verifier engage in the following protocol.
   a) The prover chooses random $r' \in_R [-n2^{k+k'-2}, n2^{k+k'-2}]$, $s' \in_R [-\mathfrak{n}2^{k+k'-2}, \mathfrak{n}2^{k+k'-2}]$, and $m' \in_R [-\rho2^{k+k'}, \rho2^{k+k'}]$. The prover computes $u' := g^{r'}$, $e' := y_1^{r'} h^{m'}$, $v' := (y_2 y_3^{\mathcal{H}_{\mathsf{hk}}(u,e,L)})^{r'}$, $\delta' := \gamma^{m'}$, and $\mathfrak{k}' := \mathfrak{g}^{m'} \mathfrak{h}^{s'}$. The prover sends u', e', v', δ', and $\mathfrak{k}'$ to the verifier.
   b) The verifier chooses a random challenge $c \in_R \{0,1\}^k$ and sends c to the prover.
   c) The prover replies with $\tilde{r} := r' - cr$, $\tilde{s} := s' - cs$, and $\tilde{m} := m' - cm$ (computed in $\mathbb{Z}$).
   d) The verifier checks whether the relations $u'^2 = u^{2c} g^{2\tilde{r}}$, $e'^2 = e^{2c} y_1^{2\tilde{r}} h^{2\tilde{m}}$, $v'^2 = v^{2c}(y_2 y_3^{\mathcal{H}_{\mathsf{hk}}(u,e,L)})^{2\tilde{r}}$, $\delta' = \delta^c \gamma^{\tilde{m}}$, $\mathfrak{k}' = \mathfrak{k}^c \mathfrak{g}^{\tilde{m}} \mathfrak{h}^{\tilde{s}}$, and $-n/4 < \tilde{m} < n/4$ hold. If any of them does not hold, the verifier stops and outputs 0.
3. If $v = \operatorname{abs} v$ the verifier outputs 1; otherwise she outputs 0.

Using notation from [CS97] we denote the sub-protocol of step 2 as

$$PK\{(r,m,s): u^2 = g^{2r} \ \wedge \ e^2 = y_1^{2r}h^{2m} \ \wedge \ v^2 = (y_2y_3^{\mathcal{H}_{\mathsf{hk}}(u,e,L)})^{2r} \ \wedge$$
$$\delta = \gamma^m \ \wedge \ \mathfrak{k} = \mathfrak{g}^m\mathfrak{h}^s \ \wedge \ -n/2 < m < n/2\} \ .$$

Theorem 2. *Under the strong RSA assumption, the above system is a verifiable encryption scheme.*

We refer to the full version of the paper [CS02] for the proof of Theorem 2.

4.3   Extensions

It is straightforward to extend the above verifiable encryption scheme to a verifiable encryption scheme that encrypts a representation of a group element with respect to several bases. Further, all of these protocols can be easily adapted to the case where the order of the group Γ is not known, i.e., a subgroup of of $\mathbb{Z}_N^*$ for an RSA-modulus N, provided the order is not divisible by any small primes.

5   Proving the Inequality of Discrete Logarithms

Our protocol for verifiable decryption (below) requires that one party proves to another party whether or not two discrete logarithms are equal, where one of the discrete logarithms might *not be known* to the prover (that is, in the case the discrete logarithms are not equal). There are well-known, efficient, special honest-verifier zero-knowledge proof systems for proving that two discrete logarithms are equal (see [CP93]), so we focus on the problem of proving that two discrete logarithms are unequal. We discuss an efficient protocol for this problem separately as it is of independent interest and as the algebraic setting here is simpler than the one in the next section.

Let $G = \langle g \rangle$ be a group of prime order q. The prover and verifier have common inputs $g, h, y, z \in G$, where g and h are generators for G, and $\log_g y \neq \log_h z$. The prover has the additional input $x = \log_g y$. The prover and verifier then engage in the following protocol.

1. The prover chooses $r \in_R \mathbb{Z}_q$, computes the auxiliary commitment $C = (h^x/z)^r$, and sends C to the verifier.
2. The prover executes the protocol denoted $PK\{(\alpha, \beta) : C = h^\alpha \left(\frac{1}{z}\right)^\beta \ \wedge \ 1 = g^\alpha \left(\frac{1}{y}\right)^\beta\}$ with the verifier.
3. The verifier accepts if it accepts in step 2, and if $C \neq 1$; otherwise, the verifier rejects.

Theorem 3. *The above protocol is a special honest-verifier proof system for proving that satisfies the special soundness condition for the property $\log_g y \neq \log_h z$.*

We refer to the full version of this paper [CS02] for the proof of Theorem 3.

Let us discuss related work. Independently of our work, Bresson and Stern [BS02] provide a protocol to prove that two discrete logarithms are not equal that is similar to ours. However, their protocol is about a factor of two less efficient than ours and is only computationally sound. We finally note that the (efficient) protocol proposed by Michels and Stadler [MS98] to prove whether or not two discrete logarithms are equal is *not* zero-knowledge because it reveals the value h^x.

6 Verifiable Decryption

In this section we provide a protocol that allows the decryptor to prove that she decrypted correctly. In particular, we provide a protocol that allows the decryptor to prove whether or not a given ciphertext decrypts to a given plaintext. We then extend the protocol to one for proving whether or not a given ciphertext decrypts to the discrete logarithm of a given group element.

6.1 Definition of Verifiable Decryption

At a high level, a verifiable decryption scheme for a binary relation $\mathcal{R}$ is a protocol that allows a prover to convince a verifier *whether or not* a ciphertext ψ is an encryption under a given public key PK and label L of a value w such that $(\delta, w) \in \mathcal{R}$ for a given δ. Here, the common input to the prover and the verifier consists of PK, L, ψ, and δ, and the prover has as additional input the "witness" w and the *secret key* SK *corresponding to* PK. We shall require that the protocol is a Σ-protocol that is special honest verifier zero knowledge, and that satisfies the special soundness condition for the property described above.

We refer the reader to the full version of the paper [CS02] for a more detailed definition, but that as for verifiable encryption, the statement being proved (or disproved) is whether the plaintext reconstructs to a witness using the specified reconstruction routine. We also point out that since the prover tells whether or not the given condition holds, the zero-knowledge simulator must be given this one bit of information as well.

6.2 Verifiable Decryption of a Matching Plaintext

We give a protocol for the decryptor to prove whether or not a ciphertext (u, e, v) decrypts to a message m with label L, i.e., using this protocol she can show that she did correctly decrypt. This is a special case of verifiable decryption in which the relation $\mathcal{R}$ is equality and the reconstruction routine is the identity function.

For our encryption scheme in §3, this proof corresponds to proving whether or not the two equations

$$u^{2(x_2 + \mathcal{H}_{hk}(u,e,L)x_3)}/v^2 = 1 \qquad \text{and} \qquad (e/u^{x_1})^2/h^{2m} = 1 \qquad (1)$$

hold (assuming that the public test $\text{abs}(v) = v$ is satisfied). If the ciphertext is invalid, one or both of the two statements do not hold. If the ciphertext is valid but decrypts to another message, the first statements holds but the second one does not.

Proving that both of these equations hold is a fairly straightforward application of known techniques.

To prove that at least one of the equations does not hold, we can use the "proof of partial knowledge" technique of [CDS94], combined with the technique developed in §5. However, because in the present setting the group has non-prime order we can not prove the relationship among the secrets in the same way as in §5 and, more importantly, the resulting protocol would not be zero-knowledge. The former problem can be solved using an auxiliary group $\mathfrak{G}_{n'} \subset \mathbb{Z}_n^*$ as we did in §4. We consider the latter problem. Depending on the values of the secret keys x_1, x_2, and x_3, the left hand sides of the equations (1), and thus the auxiliary commitments to be provided in the protocol, lie in different (sub-)groups, i.e., in $\mathbf{G}_n$, $\mathbf{G}_{n'}$, or $\mathbf{G}_n\mathbf{G}_{n'}$. As the simulator does to know the values of $x_1, \dots, x_3$, it can not simulate these auxiliary commitments. We solve this problem using the fact that for all elements $a \in \mathbf{G}_n\mathbf{G}_{n'}$ we have $a \neq 1 \Leftrightarrow (a^n \in \mathbf{G}_{n'} \wedge a^n \neq 1) \vee (a \in \mathbf{G}_n \wedge a \neq 1)$. Thus, to prove that (at least) one of the equations (1) does not hold, we prove that either

$$\left(\frac{u^{2(x_2+\mathcal{H}_{\mathsf{hk}}(u,e,L)x_3)}}{v^2}\right)^n \neq 1 \tag{2}$$

or

$$\left(\frac{u^{2(x_2+\mathcal{H}_{\mathsf{hk}}(u,e,L)x_3)}}{v^2}\right)^n = 1 \qquad \text{and} \qquad \frac{u^{2(x_2+\mathcal{H}_{\mathsf{hk}}(u,e,L)x_3)}}{v^2} \neq 1 \tag{3}$$

or

$$\left(\frac{(e/u^{x_1})^2}{h^{2m}}\right)^n = (e/u^{x_1})^{2n} \neq 1 \tag{4}$$

or

$$\left(\frac{(e/u^{x_1})^2}{h^{2m}}\right)^n = 1 \qquad \text{and} \qquad \frac{(e/u^{x_1})^2}{h^{2m}} \neq 1 \tag{5}$$

holds. Now, whenever one of the four cases applies it is always well defined in which group the left-hand sides of the inequalities lie and we can apply the ideas underlying the protocol in Section 5. We remark that the case where the Statements (2-4) are false but the Statement (5) is true corresponds to the case, where the ciphertexts is a valid encryption of a message different from m.

We are now ready to describe the protocol between the decryptor and a verifier. Their common input is $((\mathsf{hk}, n, g, y_1, y_2, y_3), (\mathfrak{n}, \mathfrak{g}, \mathfrak{h}), (u, e, v), m, L)$ and the additional input to the decryptor is (x_1, x_2, x_3). The triple $(\mathfrak{n}, \mathfrak{g}, \mathfrak{h})$ is an auxiliary parameter as in §4.2. (As we assume here that n is generated by a trusted party as well, i.e., that the decryptor is not provided with n's factorization; also, n and $\mathfrak{n}$ could be identical.) In the following description we assume that all the messages the prover sends to the verifier prior to the execution of one of the possible PK protocols will in fact be bundled with the first message of that PK protocol. Here we provide the proof-protocols only by high-level notation; deriving the actual protocols is easily derived from it.

1. If $m \notin [n]$ or the ciphertext is malformed, (e.g., if $v \neq \text{abs}(v)$), the verifier outputs -1, and the protocol stops.
2. If (u, e, v) is a valid ciphertext with label L and decrypts to m, the decryptor sends 1 to the verifier, and then engages in the protocol denoted

$$PK\{(x_1, x_2, x_3) : y_1 = g^{x_1} \wedge y_2 = g^{x_2} \wedge y_3 = g^{x_3} \wedge$$
$$v^2 = u^{2x_2} u^{2\mathcal{H}_{hk}(u,e,L)x_3} \wedge e^2/h^{2m} = u^{2x_1}\}$$

with the verifier.
3. If (u, e, v) is an invalid ciphertext w.r.t. the label L or decrypts to some message different from m, then the decryptor sends -1 to the verifier. They proceed as follows.
   a) The decryptor chooses $a_1 \in_R [n/4]$, $a_2 \in_R [n^2/4]$, $a_3 \in_R [n/4]$, and $a_4 \in_R [n^2/4]$, along with $b_1, b_2, b_3, b_3 \in_R [\mathfrak{n}/4]$. She then computes $\mathfrak{C}_1 := \mathfrak{g}^{a_1} \mathfrak{h}^{b_1}$, $\mathfrak{C}_2 := \mathfrak{g}^{a_2} \mathfrak{h}^{b_2}$, $\mathfrak{C}_3 := \mathfrak{g}^{a_3} \mathfrak{h}^{b_3}$, and $\mathfrak{C}_4 := \mathfrak{g}^{a_4} \mathfrak{h}^{b_4}$. She chooses $C_1 \in_R \mathbf{G}_{n'}$, $C_2 \in_R \mathbf{G}_n$, $C_3 \in_R \mathbf{G}_{n'}$, and $C_4 \in_R \mathbf{G}_n$. Furthermore,

   if $u^{2n(x_2+\mathcal{H}_{hk}(u,e,L)x_3)} \neq v^{2n}$, she sets $\quad C_1 := (u^{x_2+\mathcal{H}_{hk}(u,e,L)x_3}/v)^{2na_1}$,

   else if $u^{2(x_2+\mathcal{H}_{hk}(u,e,L)x_3)} \neq v^2$, she sets $\quad C_2 := (u^{x_2+\mathcal{H}_{hk}(u,e,L)x_3}/v)^{2a_2}$,

   else if $(u^{x_1}/e)^2 \notin \langle h \rangle$, she sets $\quad C_3 := (u^{x_1}/e)^{2na_3}$,

   else $(u^{x_1}/e)^2 \neq h^{2m}$, and she sets $\quad C_4 := (u^{x_1} h^m/e)^{2a_4}$.

   The decryptor sends C_1, C_2, C_3, C_4, $\mathfrak{C}_1$, $\mathfrak{C}_2$, $\mathfrak{C}_3$, and $\mathfrak{C}_4$ to the verifier.
   b) The decryptor and the verifier carry out the protocol denoted

$$PK\Big\{(x_1, x_2, x_3, a_1, \ldots, a_4, b_1, \ldots, b_4, r_1, \ldots, r_4 \, s_1, \ldots, s_4) :$$

$$\Big[y_1 = g^{x_1} \wedge y_2 = g^{x_2} \wedge y_3 = g^{x_3} \wedge$$

$$C_1 = u^{2nr_1} \Big(\frac{1}{v}\Big)^{2na_1} \wedge \mathfrak{C}_1 = \mathfrak{g}^{a_1} \mathfrak{h}^{b_1} \wedge 1 = \Big(\frac{1}{\mathfrak{C}_1}\Big)^{x_2} \Big(\frac{1}{\mathfrak{C}_1}\Big)^{\mathcal{H}_{hk}(u,e,L)x_3} \mathfrak{g}^{r_1} \mathfrak{h}^{s_1}\Big]$$

$$\vee \Big[y_1 = g^{x_1} \wedge y_2 = g^{x_2} \wedge y_3 = g^{x_3} \wedge$$

$$C_2 = u^{2r_2} \Big(\frac{1}{v}\Big)^{a_2} \wedge \mathfrak{C}_2 = \mathfrak{g}^{a_2} \mathfrak{h}^{b_2} \wedge 1 = \Big(\frac{1}{\mathfrak{C}_2}\Big)^{x_2} \Big(\frac{1}{\mathfrak{C}_2}\Big)^{\mathcal{H}_{hk}(u,e,L)x_3} \mathfrak{g}^{r_2} \mathfrak{h}^{s_2}\Big]$$

$$\vee \Big[y_1 = g^{x_1} \wedge y_2 = g^{x_2} \wedge y_3 = g^{x_3} \wedge$$

$$C_3 = u^{2nr_3} \Big(\frac{1}{e}\Big)^{2na_3} \wedge \mathfrak{C}_3 = \mathfrak{g}^{a_3} \mathfrak{h}^{b_3} \wedge 1 = \Big(\frac{1}{\mathfrak{C}_3}\Big)^{x_1} \mathfrak{g}^{r_3} \mathfrak{h}^{s_3}\Big]$$

$$\vee \Big[y_1 = g^{x_1} \wedge y_2 = g^{x_2} \wedge y_3 = g^{x_3} \wedge$$

$$C_4 = u^{2r_4} \Big(\frac{h^m}{e}\Big)^{2a_4} \wedge \mathfrak{C}_4 = \mathfrak{g}^{a_4} \mathfrak{h}^{b_4} \wedge 1 = \Big(\frac{1}{\mathfrak{C}_4}\Big)^{x_1} \mathfrak{g}^{r_4} \mathfrak{h}^{s_4}\Big]\Big\} ,$$

where $r_1, \ldots, r_4, s_1, \ldots, s_4$ are temporary secrets (i.e., $r_1 = a_1(x_2 + \mathcal{H}_{hk}(u, e, L)x_3)$, $s_1 = b_1(x_2 + \mathcal{H}_{hk}(u, e, L)x_3)$, $r_2 = a_2(x_2 + \mathcal{H}_{hk}(u, e, L)x_3)$,

$s_2 = b_2(x_2 + \mathcal{H}_{\mathsf{hk}}(u, e, L)x_3)$, $r_3 = x_1 a_3$, $s_3 = x_1 b_3$, $r_4 = x_1 a_4$, $s_4 = x_1 b_4$,
(all computed in $\mathbb{Z}$)). (To derive the actual protocol one has to apply
the techniques by Cramer et al. [CDS94] for realizing the $\vee$'s.)
   c) The verifier checks that $C_1^2 \neq 1$, $C_2^2 \neq 1$, $C_3^2 \neq 1$, and $C_4^2 \neq 1$.

The computational load of the prover and the verifier is about one to four
times the load in the protocol for verifiable encryption described in §4.2 (depending on whether step 2 or step 3 gets carried out).

Theorem 4. *Assuming factoring is hard, the above scheme is a verifiable de-*
cryption scheme (for matching plaintexts).

We refer to the full version of this paper [CS02] for the proof of Theorem 4.

6.3   Verifiable Decryption of a Discrete Logarithm

We now describe how the protocol provided in the previous section can be modified to obtain a protocol for verifiable decryption of a discrete logarithm. The
setting and notation are as in §4.2; in particular, we make use of the same relation
$\mathcal{R}$ and the same reconstruction routine.
   We need to modify the protocol from the previous section only for the cases
where the ciphertext is valid. That is, instead of proving that the ciphertext
decrypts (or does not decrypt) to a given message, the decryptor now has to prove
that it decrypts (or does not decrypt) to a value m such that $(m \operatorname{rem} n) \equiv \log_\gamma \delta$
$(\bmod \rho)$. This corresponds to proving whether or not the three equations

$$u^{2(x_2 + \mathcal{H}_{\mathsf{hk}}(u,e,L)x_3)}/v^2 = 1 \ , \quad \left(\frac{e}{u^{x_1}}\right)^{2n} = 1 \ , \quad \text{and} \quad \delta = \gamma^{(\log_{h^2}(e/u^{x_1})^2 \operatorname{rem} n)} \quad (6)$$

hold. Note that $\log_{h^2}(e/u^{x_1})^2$ exist if and only if $(e/u^{x_1})^{2n} = 1$. The first two
statements of (6) can be handled as in §4.2. The last one can be handled by
proving knowledge of a secret, say m, that (1) equals the encrypted message
modulo n, (2) equals (or doesn't equal) $\log_\gamma \delta$ modulo q, and (3) lies in the
interval $[-(n-1)/2, (n-1)/2]$. The first two properties can be proved under
the strong RSA assumption using additional parameters $(\mathfrak{n}, \mathfrak{g}, \mathfrak{h})$ as in §4.2. We
discuss proving the last one. Different from the interval-proof used for verifiable
encryption, this interval-proof needs to be *exact*, i.e., if we allowed for the same
sloppiness, then the prover could for instance add a multiple of n to m and then
show that (u, e, v) does not (or does) decrypt to $\log_\gamma \delta$.
   Boudot [Bou00] presents several protocols to prove that in integer m lies exactly in an interval $[a, b]$. One protocol uses the fact that $x \in [a, b]$ is equivalent
to $b - x \geq 0$ and $x - a \geq 0$ and that one can show that an integer is positive
by proving knowledge of four values the squares of which sum up to the considered integer (in $\mathbb{Z}$), again under the strong RSA assumption using additional
parameters $(\mathfrak{n}, \mathfrak{g}, \mathfrak{h})$. Lagrange proved the an integer can always be represented as
four squares and Rabin and Shallit [RS86] provide an efficient algorithm to find
these squares. We note that in our case the interval is symmetric and it therefore
suffices to prove that $((n-1)/2)^2 - m^2 \geq 0$ holds, which is more efficient.

With these observations one can obtain a protocol for verifiable decryption of a discrete logarithm from the protocol presented in §4.2. For lack of space, we refer the reader to the full version of this paper [CS02] for the details. We also note that it is straightforward to adapt this protocol to verifiably decrypt representations with respect to several bases. One can also "mix and match," proving whether or not ψ decrypts to a representation, one or more components of which match specified values.

References

[ACJT00]   G. Ateniese, J. Camenisch, M. Joye, and G. Tsudik, *A practical and prov-ably secure coalition-resistant group signature scheme*, Advances in Cryptology — CRYPTO 2000, LNCS, vol. 1880, Springer Verlag, 2000, pp. 255–270.

[ADR02]   J. H. An, Y. Dodis, and T. Rabin, *On the security of joint signature and encryption*, Advances in Cryptology: EUROCRYPT 2002, LNCS, vol. 2332, Springer, 2002, pp. 83–107.

[ASW97]   N. Asokan, M. Schunter, and M. Waidner, *Optimistic protocols for fair exchange*, 4th ACM Conference on Computer and Communication Security, 1997, pp. 6–17.

[ASW00]   N. Asokan, V. Shoup, and M. Waidner, *Optimistic fair exchange of digital signatures*, IEEE Journal on Selected Areas in Communications **18** (2000), no. 4, 591–610.

[BDM98]   F. Bao, R. Deng, and W. Mao, *Efficient and practical fair exchange protocols with off-line TTP*, IEEE Symposium on Security and Privacy, IEEE Computer Society Press, 1998, pp. 77–85.

[BG96]   M. Bellare and S. Goldwasser, *Encapsulated key escrow*, Preprint, 1996.

[Bou00]   F. Boudot, *Efficient proofs that a committed number lies in an interval*, Advances in Cryptology — EUROCRYPT 2000, LNCS, vol. 1807, Springer Verlag, 2000, pp. 431–444.

[BP90]   H. Bürk and A. Pfitzmann, *Digital payment systems enabling security and unobservability*, Computer & Security **9** (1990), no. 8, 715–721.

[BS02]   E. Bresson and J. Stern, *Proofs of knowledge for non-monotone discrete-log formulae and applications*, Information Security (ISC 2002), LNCS, vol. 2433, Springer Verlag, 2002, pp. 272–288.

[CD00]   J. Camenisch and I. Damgård, *Verifiable encryption, group encryption, and their applications to group signatures and signature sharing schemes*, Advances in Cryptology — ASIACRYPT 2000, LNCS, vol. 1976, Springer Verlag, 2000, pp. 331–345.

[CDS94]   R. Cramer, I. Damgård, and B. Schoenmakers, *Proofs of partial knowledge and simplified design of witness hiding protocols*, Advances in Cryptology — CRYPTO '94, LNCS, vol. 839, Springer Verlag, 1994, pp. 174–187.

[CF01]   R. Canetti and M. Fischlin, *Universally composable commitments*, Advances in Cryptology — CRYPTO 2001, LNCS, vol. 2139, Springer Verlag, 2001, pp. 19–40.

[CG98]   D. Catalano and R. Gennaro, *New efficient and secure protocols for veri-fiable signature sharing and other applications*, Advances in Cryptology — CRYPTO '98 (Berlin), LNCS, vol. 1642, Springer Verlag, 1998, pp. 105–120.

[CG99]   R. Canetti and S. Goldwasser, *An efficient threshold public key cryptosystem secure against adaptive chosen ciphertext attack*, Advances in Cryptology — EUROCRYPT '99, LNCS, vol. 1592, Springer Verlag, 1999, pp. 90–106.

[Cha85]   D. Chaum, *Security without identification: Transaction systems to make big brother obsolete*, Communications of the ACM **28** (1985), no. 10, 1030–1044.

[Cha94]   D. Chaum, *Designated confirmer signatures*, Advances in Cryptology — EUROCRYPT '94, LNCS, vol. 950, Springer Verlag Berlin, 1994, pp. 86–91.

[CL01]   J. Camenisch and A. Lysyanskaya, *Efficient non-transferable anonymous multi-show credential system with optional anonymity revocation*, Advances in Cryptology — EUROCRYPT 2001, LNCS, vol. 2045, Springer Verlag, 2001, pp. 93–118.

[CM99a]   J. Camenisch and M. Michels, *Proving in zero-knowledge that a number n is the product of two safe primes*, Advances in Cryptology — EUROCRYPT '99, LNCS, vol. 1592, Springer Verlag, 1999, pp. 107–122.

[CM99b]   J. Camenisch and M. Michels, *Separability and efficiency for generic group signature schemes*, Advances in Cryptology — CRYPTO '99, LNCS, vol. 1666, Springer Verlag, 1999, pp. 413–430.

[CM00]   J. Camenisch and M. Michels, *Confirmer signature schemes secure against adaptive adversaries*, Advances in Cryptology — EUROCRYPT 2000, LNCS, vol. 1807, Springer Verlag, 2000, pp. 243–258.

[CP93]   D. Chaum and T. P. Pedersen, *Wallet databases with observers*, Advances in Cryptology — CRYPTO '92, LNCS, vol. 740, Springer-Verlag, 1993, pp. 89–105.

[Cra96]   R. Cramer, *Modular design of secure yet practical cryptographic protocols*, Ph.D. thesis, University of Amsterdam, 1996.

[CS97]   J. Camenisch and M. Stadler, *Efficient group signature schemes for large groups*, Advances in Cryptology — CRYPTO '97, LNCS, vol. 1296, Springer Verlag, 1997, pp. 410–424.

[CS01]   R. Cramer and V. Shoup, *Universal hash proofs and a paradigm for adaptive chosen ciphertext secure public-key encryption*, http://eprint.iacr.org/2001/108, 2001.

[CS02]   J. Camenisch and V. Shoup, *Practical verifiable encryption and decryption of discrete logarithms*, http://eprint.iacr.org/2002/161, 2002.

[CVH02]   J. Camenisch and E. Van Herreweghen, *Design and implementation of the idemix anonymous credential system*, Proc. 9th ACM Conference on Computer and Communications Security, 2002.

[Dam00]   I. Damgård, *Efficient concurrent zero-knowledge in the auxiliary string model*, Advances in Cryptology — EUROCRYPT 2000, LNCS, vol. 1807, Springer Verlag, 2000, pp. 431–444.

[DF02]   I. Damgård and E. Fujisaki, *An integer commitment scheme based on groups with hidden order*, Advances in Cryptology — ASIACRYPT 2002', LNCS, vol. 2501, 2002.

[FO97]   E. Fujisaki and T. Okamoto, *Statistical zero knowledge protocols to prove modular polynomial relations*, Advances in Cryptology — CRYPTO '97, LNCS, vol. 1294, Springer Verlag, 1997, pp. 16–30.

[FR95]   M. Franklin and M. Reiter, *Verifiable signature sharing*, Advances in Cryptology — EUROCRYPT '95, LNCS, vol. 921, Springer Verlag, 1995, pp. 50–63.

[FS87]   A. Fiat and A. Shamir, *How to prove yourself: Practical solution to identi-fication and signature problems*, Advances in Cryptology — CRYPTO '86, LNCS, vol. 263, Springer Verlag, 1987, pp. 186–194.

[KP97]   J. Kilian and E. Petrank, *Identity escrow*, Theory of Cryptography Library, Record Nr. 97-11, http://theory.lcs.mit.edu/~tcryptol, August 1997.

[KP98]   J. Kilian and E. Petrank, *Identity escrow*, Advances in Cryptology — CRYPTO '98 (Berlin), LNCS, vol. 1642, Springer Verlag, 1998, pp. 169–185.

[Mic]   S. Micali, *Efficient certificate revocation and certified e-mail with transpar-ent post offices*, Presentation at the 1997 RSA Security Conference.

[MS98]   M. Michels and M. Stadler, *Generic constructions for secure and efficient confirmer signature schemes*, Advances in Cryptology — EUROCRYPT '98, LNCS, vol. 1403, Springer Verlag, 1998, pp. 406–421.

[Pai99]   P. Paillier, *Public-key cryptosystems based on composite residuosity classes*, Advances in Cryptology — EUROCRYPT '99, LNCS, vol. 1592, Springer Verlag, 1999, pp. 223–239.

[Ped92]   T. P. Pedersen, *Non-interactive and information-theoretic secure verifiable secret sharing*, Advances in Cryptology – CRYPTO '91, LNCS, vol. 576, Springer Verlag, 1992, pp. 129–140.

[PS00]   G. Poupard and J. Stern, *Fair encryption of RSA keys*, Advances in Cryptology: EUROCRYPT 2000, LNCS, vol. 1087, Springer Verlag, 2000, pp. 173–190.

[RS86]   M. O. Rabin and J. O. Shallit, *Randomized algorithms in number theory*, Communications on Pure and Applied Mathematics **39** (1986), 239–256.

[RS92]   C. Rackoff and D. R. Simon, *Non-interactive zero-knowledge proof of knowl-edge and chosen ciphertext attack*, Advances in Cryptology: CRYPTO '91, LNCS, vol. 576, Springer, 1992, pp. 433–444.

[SG98]   V. Shoup and R. Gennaro, *Securing threshold cryptosystems against chosen ciphertext attack*, Advances in Cryptology: EUROCRYPT '98, LNCS, vol. 1403, Springer, 1998.

[Sho01]   V. Shoup, *A proposal for an ISO standard for public key encryption*, http://eprint.iacr.org/2001/112, 2001.

[Sta96]   M. Stadler, *Publicly verifiable secret sharing*, Advances in Cryptology — EUROCRYPT '96, LNCS, vol. 1070, Springer Verlag, 1996, pp. 191–199.

[YY98]   A. Young and M. Young, *Auto-recoverable auto-certifiable cryptosystems.*, Advances in Cryptology — EUROCRYPT '98, LNCS, vol. 1403, Springer Verlag, 1998, pp. 17–31.

Extending Oblivious Transfers Efficiently

Yuval Ishai[1], Joe Kilian[2], Kobbi Nissim[2*], and Erez Petrank[1**]

[1] Department of Computer Science, Technion – Israel Institute of Technology,
Haifa 32000, Israel. {yuvali|erez}@cs.technion.ac.il
[2] NEC Laboratories America, 4 Independence Way, Princeton, NJ 08550, USA.
{joe|kobbi}@nec-labs.com

Abstract. We consider the problem of extending oblivious transfers: Given a small number of oblivious transfers "for free," can one implement a large number of oblivious transfers? Beaver has shown how to extend oblivious transfers given a one-way function. However, this protocol is inefficient in practice, in part due to its non-black-box use of the underlying one-way function.
We give efficient protocols for extending oblivious transfers in the random oracle model. We also put forward a new cryptographic primitive which can be used to instantiate the random oracle in our constructions. Our methods suggest particularly fast heuristics for oblivious transfer that may be useful in a wide range of applications.

1 Introduction

Is it possible to base oblivious transfer on one-way functions? Partial answers to this question were given by Impagliazzo and Rudich [22] and Beaver [1]. Impagliazzo and Rudich [22] showed that a *black-box* reduction from oblivious transfer to a one-way function (or a one-way permutation) would imply P$\neq$NP. They gave an oracle that combines a random function and a PSPACE oracle and proved that relative to this oracle one-way functions exist, but secret-key agreement is impossible. In other words, even an *idealized* one-way function (a random oracle) is insufficient for constructing secret-key agreement and hence oblivious transfer. A number of papers have continued this line of research and drew the limits of black-box reductions in cryptography, mapping the separations between the power of cryptographic primitives in relativized worlds [34,15,16, 25,14].

It is not known whether a *non-black-box* reduction from oblivious transfer to one-way functions exists. Impagliazzo and Rudich's result strongly suggests that with the current knowledge in complexity theory we cannot base oblivious transfer on one-way functions. However, a remarkable theorem of Beaver [1] shows that a 'second-best' alternative *is* achievable – one-way functions are sufficient to *extend* a few oblivious transfers into many, i.e. it is possible to implement a large number of oblivious transfers given just a small number of oblivious transfers:

---

* Work partially done while the second author was at DIMACS, Rutgers University, 96 Frelinghuysen Road Piscataway, NJ 08854, USA.
** This research was supported by the E. AND J. BISHOP RESEARCH FUND.

D. Boneh (Ed.): CRYPTO 2003, LNCS 2729, pp. 145–161, 2003.

Theorem 1 ([1]). *Let k be a computational security parameter. If one-way functions exist, then for any constant $c > 1$ there exists a protocol for reducing k^c oblivious transfers to k oblivious transfers.*

Interestingly, Beaver's reduction is inherently non-black-box with respect to the one-way function it uses.

The results of Impagliazzo and Rudich and Beaver are motivated by both theory and practice. From a theoretical point of view, one is interested in the weakest assumptions needed for oblivious transfer, and the type of reductions employed. From a practical point of view, oblivious transfer protocols are based on public-key primitives, which seem to require more structure than private-key primitives, and in practice are more expensive to implement. Alternative physical or multi-party implementations of oblivious transfer are also comparatively expensive. Thus, it is highly desirable to obtain methods for implementing oblivious transfers with (amortized) cost comparable to that of *private-key* primitives.

Beaver's protocol shows how to implement most of one's oblivious transfers using simple primitives. Thus, if public-key primitives simply did not exist, one could implement a few oblivious transfers using, for example, multi-party computation, and then use one-way functions for the rest.

Unfortunately, Beaver's protocol appears to be inefficient in practice. In particular it requires that operations be performed for every gate of a circuit computing, among other things, a pseudo-random generator. Consequently, the protocol requires work at least quadratic in the circuit complexity of the pseudo-random generator.

1.1 Oblivious Transfer

Oblivious transfer (OT) [32,10,6,23] is a ubiquitous cryptographic primitive that may be used to implement a wide variety of other cryptographic protocols, including secret key exchange, contract signing [10], and secure function evaluation [36,19,20,23].

Oblivious transfer is a two-party protocol between a *sender* and a *receiver*. Several flavors of OT have been considered and shown equivalent [9,6]. In the most useful type of OT, often denoted $\binom{2}{1}$-OT [10] (for the single-bit version) or ANDOS [6] (for the multi-bit version), the sender holds a pair of strings[1] and the receiver holds a selection bit. At the end of the protocol the receiver should learn just the selected string, and the sender should not gain any new information. Moreover, it is required by default that the above properties hold even if the sender or the receiver maliciously deviate from the protocol. This notion of OT can be conveniently formalized within the more general framework

---

[1] In a typical application of OT these strings are used as key material, in which case their length should be equal to a cryptographic security parameter. Most direct implementations of OT (cf. [28]) in fact realize OT with these parameters. Moreover, OT of arbitrarily long strings can be reduced to such OT by making a simple use of a pseudo-random generator (see Appendix B). Hence, the following discussion is quite insensitive to the length of the strings being transferred.

of secure computation (see Section 2). In fact, OT is a *complete* primitive for general secure computation [20,23].

Efficiency is particularly crucial for oblivious transfer, due to its massive usage in secure protocols. For instance, general protocols for secure computation (e.g., [36,19,23]) require at least one OT invocation per input bit of the function being evaluated. This is also the case for more specialized or practically-oriented protocols (e.g., [30,29,17,26,12,27]), where oblivious transfers typically form the efficiency bottleneck.

1.2   Our Results

In light of the state of affairs described above, it is quite possible that one cannot extend oblivious transfers in a black-box manner. It was also possible, in analogy to [22,1], that even a random oracle cannot be used to extend oblivious transfers.We show that this is *not* the case. Specifically, we show the following black-box analogue of Theorem 1:

Theorem 2 (Main Theorem). *Let k be a computational security parameter. For any constant $c > 1$, there exists a protocol in the random oracle model for reducing k^c oblivious transfers to k oblivious transfers. Alternatively, the random oracle can be replaced by a black-box use of a correlation robust hash function, as defined in Definition 1, Section 5.*

We note that our result for the random oracle model is actually stronger. For any $\epsilon > 0$, we can reduce $2^{k^{1-\epsilon}}$ oblivious transfers to k oblivious transfers of k-bit strings. This reduction is essentially tight, in the sense that the negative result of [22] can be extended to rule out a similar reduction of $2^{\Omega(k)}$ oblivious transfers to k oblivious transfers (assuming that the adversary is allowed polynomial time in the number of oblivious transfers).

CONTRIBUTION PERSPECTIVE. It is instructive to draw an analogy between the problem of extending oblivious transfers and that of extending *public-key encryption*. To encrypt a *long* message (or multiple messages) using a public-key encryption scheme, it suffices to encrypt a *short* secret key κ for some private-key scheme, and then encrypt the long message using the private-key scheme with key κ. Thus, public-key encryption can be readily extended via a black-box use of a (far less costly) private-key primitive. The existence of such an efficient black-box reduction has a significant impact on our everyday use of encryption. The aim of the current work is to establish a similar result within the richer domain of secure computations.

EFFICIENCY. Our basic protocol is extremely efficient, and requires each party to make a small constant number of calls to the random oracle for each OT being performed. All other costs (on top of the initial seed of k OTs) are negligible. This basic protocol, however, is insecure against a malicious receiver who may deviate from the protocol's instructions. To obtain a fully secure protocol, we employ a cut-and-choose technique. This modification increases the cost of the basic protocol by a factor of σ, where σ is a statistical security parameter.

Specifically, any "cheating" attempt of a malicious receiver will be detected by the sender, except with probability $2^{-\Omega(\sigma)}$. Thus, in scenarios where a penalty is associated with being caught cheating, this (rather modest) cut-and-choose overhead can be almost entirely eliminated. We also discuss some optimized variants of our protocols, in particular ones that are tailored to an "on-line" setting where the number of desired oblivious transfers is not known in advance.

RELATED WORK. There is a vast literature on implementing OT in various settings and under various assumptions. Since OT implies key exchange, all these OT protocols require the use of expensive public-key operations. The problem of *amortizing* the cost of multiple oblivious transfers has been considered by Naor and Pinkas [28]. Their result is in a sense complementary to ours: while the savings achieved are modest, the amortization "kicks in" quite early. Thus, their techniques can be used for reducing the cost of the seed of k oblivious transfers required by our protocols.

1.3   On the Use of a Random Oracle

We describe and analyze our protocols in the random oracle model. Such an analysis is used in cryptography for suggesting the feasibility of protocols, and for partially analyzing protocols (in particular, practical ones). Instead of making purely heuristic arguments, one considers an idealized hash function, and proves rigorous results in this "nearby" version of the problem. The latter approach is advocated e.g. by Bellare and Rogaway [3]. They suggest to analyze protocols with hash functions modeled by an idealized version, implemented by some magic black box. For example, instead of trying to analyze a protocol that uses a specific hash function $h(x) : \{0,1\}^{2k} \to \{0,1\}^k$, one analyzes the system that uses a random function $H : \{0,1\}^{2k} \to \{0,1\}^k$, chosen uniformly from the space of all such functions.

All parties being considered, both legitimate parties and adversaries, are given access to H as a black box (or oracle). The protocol can instruct the parties to make queries to H. The adversary attacking the protocol is allowed to make arbitrary queries to H, at unit cost, but is bounded in how many queries it is allowed to make (if only by a bound on its running time). Using the complete lack of structure of H and the fact that the adversary can explore only a tiny portion of its inputs allows us to rigorously analyze idealized versions of systems that in vanilla form resist all analysis. The heuristic leap of faith is that the real world matches the idealized world insofar as security is concerned, i.e., that the hash function h is sufficiently "structureless" that the system remains secure if it uses h in place of H.

This approach has been applied to many practical systems. To mention just a few, it was used for optimal asymmetric encryption (OAEP) [5,33], for replacing interaction in the Fiat-Shamir signature scheme [13], and to justify hashing methodologies [4,35], a method for basing cryptosystems on one-way permutations [11], for analyzing the DESX construction [24], as well as RSA-based signature schemes with hashing [3]. In practical implementations, highly efficient hash functions such as SHA1 or RC5 are typically used to instantiate the random oracle.

There is no rigorous methodology allowing one to safely replace the hash function with a cryptographic instantiation. Indeed, Canetti, Goldreich and Halevi [8] give an explicit example of a cryptographic protocol that is secure using an ideal random function H, but is insecure for any polynomial-time computable instantiation of H. Their counterexample stems from the difference between an efficiently computable function and a random oracle: The efficient function has a small circuit whereas the random oracle does not. Further research along these lines appears in [31,21,2].

We would like to stress, however, that these concerns can be at least partially dismissed in the context of the current work. First, in our security proofs we treat the random oracle as being *non-programmable* [31], i.e., we do not allow the simulators used in these proofs to fake the answers received from the oracle. This version of the random oracle model is more conservative, in that it rules out some of the above counterexamples. More importantly, we provide an *alternative* to the random oracle model by suggesting a concrete and seemingly natural cryptographic primitive that can be used to instantiate our protocols. Specifically, we require an explicit $h : \{0,1\}^* \to \{0,1\}$ so that for a random and independent choice of (polynomially many) strings $s, t_1, \ldots, t_m \in \{0,1\}^k$, the joint distribution $(h(s \oplus t_1), \ldots, h(s \oplus t_m), t_1, \ldots, t_m)$ is pseudo-random. Since this is a *simple* property enjoyed by a random function, any evidence that hash functions such as SHA1 or RC5 violate this property could be considered a valid attack against these functions.

1.4 Road Map

Section 2 contains some preliminaries and basic tools. We give the basic version of our protocol in Section 3. This protocol is insecure against a malicious receiver. A modified version of the basic protocol that achieves full security is described in Section 4. In Section 5 we describe a primitive that captures a property of the random oracle that is sufficient for our constructions. We conclude with some open problems.

2   Preliminaries

2.1   Secure Two-Party Computation

It is convenient to define our problem within the more general framework of secure two-party computation. In the following we assume the reader's familiarity with standard simulation-based definitions of secure computation from the literature. For self-containment, the necessary definitions are sketched in Appendix A.

THE USE OF A RANDOM ORACLE. When augmenting the standard definition of two-party computation with a random oracle H, we assume that the oracle is picked at random in both the real function evaluation process involving the adversary and the ideal process involving the simulator. Furthermore, we require that the simulation be indistinguishable from the real execution even if the distinguisher is allowed to adaptively make polynomially many queries to the *same H* that was used in the process (real or ideal) generating its input. This is especially significant when simulating a semi-honest (i.e., passively corrupted) party, since this party by definition cannot make any additional calls to H. However, even when simulating a malicious (i.e., actively corrupted) party, who can make the additional calls himself, this requirement is important for ensuring that the transcripts provided by the simulator are indeed consistent with the instance of H to which it had access. This is referred to as the *non-programmable* random oracle model.

REDUCTIONS. It is often convenient to design secure protocols in a modular way: first design a high level protocol assuming an idealized implementation of a lower level primitive, and then substitute a secure implementation of this primitive. If the high level protocol securely realizes a functionality f and the lower level primitive a functionality g, the high level protocol may be viewed as a secure *reduction* from f to g. Such reductions can be formalized using a *hybrid* model, where the parties in the high level protocol are allowed to invoke g, i.e., a trusted party to which they can securely send inputs and receive the corresponding outputs. In our case, we will be interested in reductions where f implements "many" oblivious transfers and g "few" oblivious transfers. Moreover, we will usually restrict our attention to reductions making use of a *single* invocation to g. Appropriate composition theorems, e.g. from [7], guarantee that the call to g can be replaced by any secure[2] protocol realizing g, without violating the security of the high level protocol. Moreover, these theorems relativize to the random oracle model. Thus, it suffices to formulate and prove our reductions using the above hybrid model.

BLACK-BOX REDUCTIONS. The above framework for reductions *automatically* guarantees that the high-level protocol for f make a black-box use of g, in the sense that both the protocol and its security proof do not depend on the implementation of g. Thus, all our reductions in the random oracle model are fully black box. For comparison, Beaver's reduction [1] is clearly non-black-box with respect to the one-way function (or PRG) on which it relies. We refer the reader to [15,16] for a more thorough and general exposition to black-box reductions in cryptography.

2.2   Oblivious Transfer

An OT protocol can be defined as a secure two-party protocol between a sender S and a receiver R realizing the OT functionality (see Appendix A) . This definition implies the properties mentioned in the introduction: the sender cannot learn the receiver's selection, and the receiver cannot learn more than one of the two strings held by the sender. It is interesting to note that this definition imposes additional desirable features. For instance, a malicious sender must "know" a

---

[2] Here one should use an appropriate notion of security, e.g., the one from [7].

pair of string to which the receiver's selection effectively applies.[3] This may be important for using OT in the context of other protocols.

The most general OT primitive we consider, denoted OT_ℓ^m, realizes m (independent) oblivious transfers of ℓ-bit strings. That is, OT_ℓ^m represents the following functionality:

Inputs: S holds m pairs $(x_{j,0}, x_{j,1})$, $1 \leq j \leq m$, where each $x_{j,b}$ is an ℓ-bit
    string. R holds m selection bits $\mathbf{r} = (r_1, \ldots, r_m)$.
Outputs: R outputs x_{j,r_j} for $1 \leq j \leq m$. S has no output.

The task of extending oblivious transfers can be defined as that of reducing OT_ℓ^m to OT_k^k, where k is a security parameter and $m > k$. (For simplicity, one may think of ℓ as being equal to k.) That is, we would like to implement m oblivious transfers of ℓ-bit strings by using k oblivious transfers of k-bit strings. However, it will be more convenient to reduce OT_ℓ^m to OT_m^k. The latter primitive, in turn, can be very easily reduced to OT_k^k by generating $2m$ pseudo-random bits.[4] This reduction proceeds by performing OT of k pairs of independent seeds, and then sending the k pairs of strings masked with the outputs of the generator on the corresponding seeds. See Appendix B for a formal description.

We note that using known reductions, OT_k^k can be further reduced to k invocations of OT_k^1 (the type of OT typically realized by direct implementations) or $O(k^2)$ invocations of OT_1^1 [6]. Thus, any secure implementation for the most basic OT variant can be used as a black box to securely realize our high level OT_ℓ^m protocol. We prefer to use OT_k^k as our cryptographic primitive for extending oblivious transfers, due to the possibility for a more efficient direct implementation of this primitive than via k separate applications of OT_k^1 [28].

2.3   Notation

We use capital letters to denote matrices and small bold letters to denote vectors. We denote the jth row of a matrix M by $\mathbf{m}_j$ and its ith column by $\mathbf{m}^i$. The notation $b \cdot \mathbf{v}$, where b is a bit and $\mathbf{v}$ is a binary vector, should be interpreted in the natural way: it evaluates to 0 if $b = 0$ and to $\mathbf{v}$ if $b = 1$.

3   Extending OT with a Semi-honest Receiver

In this section we describe our basic protocol, reducing OT_ℓ^m to OT_m^k. As noted above, this implies a reduction to OT_k^k with a very small additional cost. The security of this protocol holds as long as the receiver is semi-honest. We will later modify the protocol to handle malicious receivers. The protocol is described in Fig. 1.

---

[3] This follows from the fact that a malicious sender's output should be simulated *jointly* with the honest receiver's output.
[4] A pseudo-random generator can be easily implemented in the random oracle model without additional assumptions.

INPUT OF S: m pairs $(x_{j,0}, x_{j,1})$ of ℓ-bit strings, $1 \leq j \leq m$.
INPUT OF R: m selection bits $\mathbf{r} = (r_1, \ldots, r_m)$.
COMMON INPUT: a security parameter k.
ORACLE: a random oracle $H : [m] \times \{0,1\}^k \to \{0,1\}^\ell$.
CRYPTOGRAPHIC PRIMITIVE: An ideal OT_m^k primitive.

1. S initializes a random vector $\mathbf{s} \in \{0,1\}^k$ and R a random $m \times k$ bit matrix T.
2. The parties invoke the OT_m^k primitive, where S acts as a receiver with input $\mathbf{s}$ and R as a sender with inputs $(\mathbf{t}^i, \mathbf{r} \oplus \mathbf{t}^i)$, $1 \leq i \leq k$.
3. Let Q denote the $m \times k$ matrix of values received by S. (Note that $\mathbf{q}^i = (s_i \cdot \mathbf{r}) \oplus \mathbf{t}^i$ and $\mathbf{q}_j = (r_j \cdot \mathbf{s}) \oplus \mathbf{t}_j$.) For $1 \leq j \leq m$, S sends $(y_{j,0}, y_{j,1})$ where $y_{j,0} = x_{j,0} \oplus H(j, \mathbf{q}_j)$ and $y_{j,1} = x_{j,1} \oplus H(j, \mathbf{q}_j \oplus \mathbf{s})$.
4. For $1 \leq j \leq m$, R outputs $z_j = y_{j,r_j} \oplus H(j, \mathbf{t}_j)$.

Fig. 1. Extending oblivious transfers with a semi-honest receiver.

It is easy to verify that the protocol's outputs are correct (i.e., $z_j = x_{j,r_j}$) when both parties follow the protocol.

EFFICIENCY. The protocol makes a single call to OT_m^k. In addition, each party evaluates at most $2m$ times (an implementation of) a random oracle mapping $k + \log m$ bits to ℓ bits. All other costs are negligible. The cost of OT_m^k is no more than k times the cost of OT_k^1 (the type of OT realized by most direct implementations) plus a generation of $2m$ pseudo-random bits. In terms of round complexity, the protocol requires a single message from S to R in addition to the round complexity of the OT_m^k implementation.

3.1   Security

We prove that the protocol is secure against a malicious sender and against a semi-honest receiver. More precisely, we show a *perfect* simulator for any malicious sender S^* and a *statistical* simulator for R. In the latter case, the output of the ideal process involving the simulator is indistinguishable from that of the real process even if the distinguisher is allowed to adaptively make $2^k/k^{\omega(1)}$ additional calls to H.

We let TP denote the trusted party for the OT_ℓ^m functionality in the ideal function evaluation process.

Simulating S^*. It is easy to argue that the output of an arbitrary S^* can be perfectly simulated. Indeed, all S^* views throughout the protocol is a k-tuple of uniformly random and independent vectors, received from the OT_m^k primitive in Step 2. This guarantees that the receiver's selections remain perfectly private. However, as discussed above, the OT definition we are using is stronger in the sense that it requires to simulate the output of the malicious sender jointly with the output of the honest receiver. Such a simulator for a malicious sender S^* may proceed as follows:

- Run S^* with a uniformly chosen random input ρ. Let $\mathbf{s}^*$ be the input S^* sends to the OT_m^k primitive in Step 2. Generate a random $m \times k$ matrix Q, and feed S^* with the columns of Q as the reply from the OT_m^k primitive.
- Let $(y_{j,0}^*, y_{j,1}^*)$ be the messages sent by S^* in Step 3. Call TP with inputs $x_{j,0}^* = y_{j,0}^* \oplus H(j, \mathbf{q}_j)$ and $x_{j,1}^* = y_{j,1}^* \oplus H(j, \mathbf{q}_j \oplus \mathbf{s}^*)$, $1 \le j \le m$.
- Output whatever S^* outputs.

CORRECTNESS: It is easy to verify that the joint distribution of $\rho, \mathbf{s}^*, Q$, the values $(y_{j,0}^*, y_{j,1}^*)$ and all values of H queried by S^* in the ideal process is identical to the corresponding distribution in the real process. It remains to show that, conditioned on all the above values, the receiver's outputs x_{j,r_j}^* in the ideal process are distributed identically to these outputs in the real process. This follows from the way the output of R is defined in the Step 4 of the protocol and from the fact that (in the real process) $\mathbf{t}_j^* = \mathbf{q}_j \oplus (r_j \cdot \mathbf{s}^*)$. Note that the above simulation remains perfect even if the distinguisher makes an arbitrary number of calls to H. Thus we have:

Claim. The protocol is perfectly secure with respect to an arbitrary sender.

Simulating R. The semi-honest receiver R can be simulated as follows:

- Call TP with input $\mathbf{r}$. Let z_j denote the jth output received from TP.
- Run the protocol between R and S, substituting the values z_j for the known inputs x_{j,r_j} of S and the default value 0^ℓ for the unknown inputs $x_{j,1-r_j}$. Output the entire view of R.

It is clear from the descriptions of the protocol and the simulator that, conditioned on the event $\mathbf{s} \ne 0$, the simulated view is distributed *identically* to the receiver's view in the real process. Indeed, if $\mathbf{s} \ne 0$, the values of H used for masking the unknown inputs $x_{j,1-r_j}$ are uniformly random and independent of the receiver's view and of each other. Thus, the simulator's output is 2^{-k}-close to the real view. However, to make a meaningful security statement in the random oracle model, we must also allow the distinguisher to (adaptively) make additional calls to H, where the answers to these calls are provided by the same H that was used in the process (real or ideal) generating the distinguisher's input.

Now, if the distinguisher can "guess" the oracle query used in the real process to mask some secret $x_{j,1-r_j}$ which R is not supposed to learn, then it clearly wins (i.e., it can easily distinguish between any two values for this unknown secret). On the other hand, as long as it does not guess such a critical query, the masks remain random and independent given its view, and so indistinguishability is maintained. The crucial observation is that from the distinguisher's point of view, each of these m offending queries is (individually) distributed uniformly at random over a domain of size 2^k. This follows from the fact that the distinguisher has no information about s as long as it makes no offending query. Hence, the distinguisher can only win the above game with negligible probability. This is formalized by the following lemma.

Lemma 1. *Any distinguisher D which makes at most t calls to H can have at most a $(t+1) \cdot 2^{-k}$-advantage in distinguishing between the output of the real process and that of the ideal process.*

Proof. Define the *extended* real (resp., ideal) process to be the real (resp., ideal) process followed by the invocation of D on the output of the process. The output of the extended process includes the output of the original process along with the transcript of the oracle calls made by D. For each of the extended processes, define an *offending query* to be a call to H on input $(j, \mathbf{t}_j \oplus \mathbf{s})$ for some $1 \leq j \leq m$, and define B to be the (bad) event that an offending query is ever made by either R or D. It is easy to verify that, as long as no offending query is made, the outputs of the two extended processes are perfectly indistinguishable. Thus, the event B has the same probability in both extended processes, and the outputs of the two extended processes are identically distributed conditioned on B not occurring. It remains to show that $\Pr[B] \leq (t+1) \cdot 2^{-k}$. This, in turn, follows by noting that: (1) R makes an offending query only if $\mathbf{s} = 0$, and (2) as long as no offending query is made, D's view is completely independent of the value of $\mathbf{s}$. □

Thus, we have:

Claim. As long as $m = 2^{o(k)}$, the protocol is statistically secure with respect to a semi-honest receiver and a polynomial-time distinguisher having access to the random oracle.

4   A Fully Secure Protocol

In this section we describe a variant of the protocol from Section 3 whose security also holds against a malicious receiver.

We begin by observing that the previous protocol is indeed insecure against a malicious R^*. Consider the following strategy for R^*. In Step 2, it chooses the ith pair of vectors sent to the OT_m^k primitive so that they differ only in their ith position.[5] For simplicity, assume without loss of generality that the ith bit of the first vector in the ith pair is 0. The above strategy guarantees that for $1 \leq j \leq k$ the vector $\mathbf{q}_j$ contains the bit s_j in its jth position and values known to R^* in the remaining positions. It follows that given an a-priori knowledge of $x_{j,0}$ the receiver can recover s_j by making two calls to H. This, in turn, implies that given the values of $x_{1,0}, \ldots, x_{k,0}$, the receiver can recover s and learn all $2m$ inputs of S.[6]

To foil this kind of attack, it suffices to ensure that the pairs of vectors sent by the receiver to the OT_m^k primitive are well-formed, i.e., correspond to some valid choice of $\mathbf{r}$ and T. Indeed, the simulator of R from the previous section can be readily modified so that it simulates any "well-behaved" R' satisfying this requirement. To deal with an arbitrarily malicious R^* we employ a cut-and-choose technique. Let σ denote a statistical security parameter. The players

---

[5] It is in fact possible for R^* to use this strategy in a completely undetectable way. Specifically, it can ensure that the m vectors received by S are uniformly random and independent.

[6] The security definition implies that the ideal process can properly simulate the real process given *any distribution* on the inputs. The above attack rules out proper simulation for an input distribution which fixes $x_{1,0}, \ldots, x_{k,0}$ and picks the remaining inputs uniformly at random.

engage in σ (parallel) executions of the previous protocol, where all inputs to these executions are picked randomly and independently of the actual inputs. Next, the sender challenges the receiver to reveal its private values for a random subset of $\sigma/2$ executions, and aborts if an inconsistency is found. This ensures S that except with $2^{-\Omega(\sigma)}$ probability, the remaining $\sigma/2$ executions contain at least one "good" execution where the receiver was well-behaved in the above sense. Finally, the remaining executions are combined as follows. Based on its actual selection bits, the receiver sends a correction bit for each of its $m\sigma/2$ random selections in the remaining executions, telling S whether or not to swap the corresponding pair of inputs. For each of its actual secrets $x_{j,b}$, the sender sends the exclusive-or of this secret with the $\sigma/2$ (random) inputs of the remaining executions which correspond to $x_{j,b}$ after performing the swaps indicated by the receiver. Having aligned all of the selected masks with the selected secrets, the receiver can now easily recover each selected secret x_{j,r_j}. This protocol is formally described in Fig. 2.

Note that the above protocol does not give a malicious S^* any advantage in guessing the inputs of R. Moreover, except with $2^{-\Omega(\sigma)}$ failure probability, security against a malicious R^* reduces to security against a well-behaved R'. A more detailed security analysis will be provided in the full version.

EFFICIENCY. The modification described above increases the communication and time complexity of the original protocol by a factor of σ. The probability of R^* getting away with cheating is $2^{-\Omega(\sigma)}$.[7] In terms of round complexity, the protocol as described above adds a constant number of rounds to the original protocol.

OPTIMIZATIONS. We did not attempt to optimize the exact round complexity. A careful implementation, using commitments, can entirely eliminate the overhead to the round complexity of the basic protocol. The length of the OT seed can be shortened from σk to $O(k)$ via the use of a modified cut-and-choose approach, at the expense of a comparable increase in the length of the input to H (and while achieving roughly the same level of security). Further details are omitted from this abstract.

AN ON-LINE VARIANT. In our default setting, we assume that the number m of desired OT is known in advance. However, in some scenarios it may be desirable to allow, following the initial seed of OT, an efficient generation of additional OT on the fly. While it is easy to obtain such an on-line variant for the basic protocol from Figure 1, this is not as obvious for the fully secure protocol. We note, however, that the modified cut-and-choose technique mentioned above can also be used to obtain an on-line variant of the fully secure protocol.

5   On Instantiating the Random Oracle

In this section we define an explicit primitive which can be used to replace the random oracle in our constructions.

---

[7] In particular, the distance between the output of the simulator for R^* and the output of the real process increases by at most $2^{-\Omega(\sigma)}$.

INPUT OF S: m pairs $(x_{j,0}, x_{j,1})$ of ℓ-bit strings, $1 \le j \le m$.
INPUT OF R: m selection bits $\mathbf{r} = (r_1, \ldots, r_m)$.
COMMON INPUT: security parameters k, σ.
ORACLE: a random oracle $H : [\sigma] \times [m] \times \{0,1\}^k \to \{0,1\}^\ell$.
CRYPTOGRAPHIC PRIMITIVE: An ideal $\text{OT}_m^{\sigma k}$ primitive.

1. For $1 \le p \le \sigma$, S initializes a random vector $\mathbf{s}^{(p)} \in \{0,1\}^k$ and m random pairs of ℓ-bit strings $(x_{j,0}^{(p)}, x_{j,1}^{(p)})$, $1 \le j \le m$. For $1 \le p \le \sigma$, R initializes a random vector $\mathbf{r}^{(p)} \in \{0,1\}^m$ and a random $m \times k$ bit matrix $T^{(p)}$.
2. The parties invoke the $\text{OT}_m^{\sigma k}$ primitive, where S acts as a receiver with inputs $\mathbf{s}^{(p)}$, $1 \le p \le \sigma$, and R as a sender with inputs $(\mathbf{t}^{(p),i}, \mathbf{r}^{(p)} \oplus \mathbf{t}^{(p),i})$, $1 \le p \le \sigma$, $1 \le i \le k$. Let $Q^{(p)}$ denote the pth $m \times k$ matrix of values received by S.
3. S picks a random subset $P \subset [\sigma]$ of size $\sigma/2$, and challenges R to reveal all values $\mathbf{r}^{(p)}$ and $T^{(p)}$ with $p \in P$. If the reply of R is not fully consistent with the values received in Step 2, S aborts.
4. For $1 \le p \le \sigma$ and $1 \le j \le m$, S sends $(y_{j,0}^{(p)}, y_{j,1}^{(p)})$ where

$$y_{j,b}^{(p)} = x_{j,b}^{(p)} \oplus H(p, j, \mathbf{q}_j^{(p)} \oplus b \cdot \mathbf{s}).$$

5. For all $1 \le j \le m$ and $p \notin P$, R sends a correction bit $c_j^{(p)} = r_j \oplus r_j^{(p)}$.
6. For $1 \le j \le m$ and $b \in \{0,1\}$, S sends

$$w_{j,b} = x_{j,b} \oplus \bigoplus_{p \notin P} x_{j, b \oplus c_j^{(p)}}^{(p)}.$$

7. For $1 \le j \le m$, R outputs

$$z_j = w_{j,r_j} \oplus \bigoplus_{p \notin P} (y_{j,r_j^{(p)}}^{(p)} \oplus H(p, j, \mathbf{t}_j^{(p)})).$$

Fig. 2. A fully secure protocol for extending oblivious transfers

We say that $h : \{0,1\}^k \to \{0,1\}$ is *correlation robust* if for a random and independent choice of (polynomially many) strings $s, t_1, \ldots, t_m \in \{0,1\}^k$, the joint distribution $(h(t_1 \oplus s), \ldots, h(t_m \oplus s))$ is pseudo-random *given* $t_1, \ldots, t_m$. More precisely:

Definition 1 (Correlation robustness). *An efficiently computable function $h : \{0,1\}^* \to \{0,1\}$ is said to be* correlation robust *if for any polynomials $p(\cdot), q(\cdot)$ there exists a negligible function $\epsilon(\cdot)$ such that the following holds. For any positive integer k, circuit C of size $p(k)$, and $m \le q(k)$*

$$\left| \Pr[C(t_1, \ldots, t_m, h(t_1 \oplus s), \ldots, h(t_m \oplus s)) = 1] - \Pr[C(U_{(k+1)m}) = 1] \right| \le \epsilon(k),$$

where the probability is over the uniform and independent choices of $s, t_1, \ldots, t_m$ from $\{0,1\}^k$.

A correlation robust h can be used to obtain a (weak) pseudo-random function family defined by $f_s(t) = h(d \oplus t)$, and hence also a one-way function.

However, we do not know whether correlation robustness can be based on one-way functions.

We now briefly discuss the application of correlation robustness to our problem. Consider the basic protocol from Fig. 1 restricted to $\ell = 1$, i.e., implementing *bit* oblivious transfers. Moreover, suppose that the first argument to the random oracle is omitted in this protocol (i.e., $H(\mathbf{u})$ is used instead of $H(j, \mathbf{u})$).[8] From the receiver's point of view, the value masking a forbidden input $x_{j,1-r_j}$ is $H(\mathbf{s} \oplus \mathbf{t}_j)$, where $\mathbf{s}$ is a uniformly random and *secret* k-bit string and $\mathbf{t}_j$ is a known k-bit string. The use of a correlation robust H suffices to ensure that the above m masks are *jointly* pseudo-random given the receiver's view.

It is also possible to modify the fully secure variant of this protocol so that its security can be based on correlation robustness. Such a modification requires the sender to randomize each evaluation point of H so as to prevent a malicious receiver from biasing the offsets t_j. Thus, we have:

Theorem 3. *Let k denote a computational security parameter. For any constant $c > 1$ it is possible to reduce k^c oblivious transfers to k oblivious transfers by making only a black-box use of a correlation robust function.*

6   Open Problems

We have shown how to extend oblivious transfers in an efficient black-box manner given a *random* function, or given a *correlation robust* function. The question whether it is possible to extend oblivious transfers in a black-box manner using a *one-way* function is still open. A negative answer (in the line of [22]) would further dilute our view of negative black-box reduction results as impossibility results.

A related question which may be of independent interest is that of better understanding our notion of correlation robustness. Its definition (while simple) appears to be somewhat arbitrary, and it is not clear whether similar variants, which still suffice for extending OT, are equivalent to our default notion. The questions of finding a "minimal" useful notion of correlation robustness and studying the relation between our notion and standard cryptographic primitives remain to be further studied.

Acknowledgments. We thank Moni Naor and Omer Reingold for useful comments.

---

[8] It is not hard to verify that as long as the receiver is honest, the protocol remains secure. The inclusion of j in the input to the random oracle slightly simplifies the analysis and is useful towards realizing the fully secure variant of this protocol.

References

1. Donald Beaver, *Correlated Pseudorandomness and the Complexity of Private Computations*, STOC 1996: 479–488
2. M. Bellare, A. Boldyreva and A. Palacio, *A Separation between the Random-Oracle Model and the Standard Model for a Hybrid Encryption Problem*, Electronic Colloquium on Computational Complexity (ECCC), 2003.
3. M. Bellare and P. Rogaway, *Random Oracles are Practical: a Paradigm for Designing Efficient Protocols*, Proc. of the 1st ACM Conference on Computer and Communications Security, pages 62–73, 1993. ACM press.
4. M. Bellare, J. Kilian and P. Rogaway, *The security of cipher block chaining*, Advances in Cryptology – CRYPTO '94, Lecture Notes in Computer Science, vol. 839, Springer-Verlag, 1994, pp. 341–358.
5. M. Bellare and P. Rogaway, *Optimal asymmetric encryption*, Crypto '94, pages 91–111, 1994.
6. G. Brassard, C. Crépeau, and J.-M. Robert, *All-or-nothing disclosure of secrets*, Crypto '86, pp. 234–238, 1987.
7. R. Canetti, *Security and composition of multiparty cryptographic protocols*, J. of Cryptology, 13(1), 2000.
8. R. Canetti, G. Goldreich and S. Halevi, *The Random Oracle Methodology, Revisited (preliminary version)*, STOC: ACM Symposium on Theory of Computing, 1998.
9. C. Crépeau, *Equivalence between two flavors of oblivious transfers*, Crypto '87, p. 350–354.
10. S. Even, O. Goldreich, and A. Lempel. A randomized protocol for signing contracts. *C. ACM*, 28:637–647, 1985.
11. S. Even and Y. Mansour, *A construction of a cipher from a single pseudorandom permutation*, Journal of Cryptology, vol. 10, no. 3, 151–162 (Summer 1997). Earlier version in Advances in Cryptology — ASIACRYPT '91. Lecture Notes in Computer Science, vol. 739, 210–224, Springer-Verlag (1992).
12. J. Feigenbaum, Y. Ishai, T. Malkin, K. Nissim, M. Strauss and Rebecca N. Wright, *Secure Multiparty Computation of Approximations*, ICALP 2001: 927–938
13. A. Fiat, A. Shamir. *How to Prove Yourself: Practical Solutions to Identification and Signature Problems*, Advances in Cryptology – CRYPTO '86, Lecture Notes in Computer Science, vol. 263, Springer-Verlag, 1986, pp. 186–194.
14. R. Gennaro and L. Trevisan, *Lower Bounds on the Efficiency of Generic Cryptographic Constructions*, IEEE Symposium on Foundations of Computer Science, 305–313, (2000)
15. Y. Gertner, S. Kannan, T. Malkin, O. Reingold and M. Viswanathan, *The Relationship between Public Key Encryption and Oblivious Transfer*, Proc. of the 41st Annual Symposium on Foundations of Computer Science (FOCS '00), 2000.
16. Y. Gertner, T. Malkin and O. Reingold, *On the Impossibility of Basing Trapdoor Functions on Trapdoor Predicates* Proc. of the 42st Annual Symposium on Foundations of Computer Science (FOCS '01) 2001.
17. N. Gilboa, *Two Party RSA Key Generation*, Proc. of CRYPTO 1999, pp. 116–129.
18. O. Goldreich, *Secure multi-party computation*, Available at http://philby.ucsb.edu/cryptolib/BOOKS, February 1999.
19. O. Goldreich, S. Micali and A. Wigderson, *Proofs that Yield Nothing but Their Validity and a Methodology of Cryptographic Protocol Design*, Proc. of the 27th FOCS, 1986, 174–187.
20. O. Goldreich and R. Vainish, *How to Solve Any Protocol problem – an Efficiency Improvement,* In proceedings, Advances in Cryptology: CRYPTO '87, 73–86, Springer (1988).

21. S. Goldwasser and Y. Tauman. *On the (In)security of the Fiat-Shamir Paradigm*, Electronic Colloquium on Computational Complexity (ECCC), 2003.
22. Impagliazzo, R. and S. Rudich, *Limits on the provable consequences of one-way permutations*, Proceedings of 21st Annual ACM Symposium on the Theory of Computing, 1989, pp. 44–61.
23. J. Kilian, *Founding Cryptography on Oblivious Transfer*, Proc of the 20th STOC, ACM, 1988, 20–29
24. J. Kilian and P. Rogaway, *How to protect DES against exhaustive key search*, Proceedings of Crypto '96, August 1996.
25. J.H. Kim, D.R. Simon, and P. Tetali. *Limits on the efficiency of one-way permutations- based hash functions*, Proceedings of the 40th IEEE Symposium on Foundations of Computer Science, pages 535–542, 1999.
26. Y. Lindell and B. Pinkas, *Privacy Preserving Data Mining*, Journal of Cryptology 15(3): 177–206 (2002)
27. M. Naor and K. Nissim, *Communication preserving protocols for secure function evaluation*, STOC 2001: 590–599.
28. M. Naor and B. Pinkas, *Efficient oblivious transfer protocols*, SODA 2001.
29. M. Naor and B. Pinkas, *Oblivious Transfer and Polynomial Evaluation*, STOC: ACM Symposium on Theory of Computing (STOC), (1999).
30. M. Naor, B. Pinkas, and R. Sumner, *Privacy preserving auctions and mechanism design*, ACM Conference on Electronic Commerce (1999), pp. 129–139.
31. J. Nielsen, *Separating random oracle proofs from complexity theoretic proofs: The non-committing encryption case*, Crypto 2002, pp. 111–126.
32. M. O. Rabin. How to exchange secrets by oblivious transfer. Technical Report TR-81, Harvard Aiken Computation Laboratory, 1981.
33. V. Shoup, *OAEP Reconsidered*, Proc. of Crypto '01, pp. 239–259.
34. D. Simon, *Finding Collisions on a One-Way Street: Can Secure Hash Functions Be Based on General Assumptions?*, Proc. of EUROCRYPT '98, pp. 334–345.
35. E. Petrank and C. Rackoff, *Message Authentication of Unknown Variable Length Data*, Journal of Cryptology, Vol. 13, No. 3, pp. 315–338, 2000.
36. A. Yao, *Protocols for Secure Computations (Extended Abstract)* Proc. of FOCS 1982, 160–164.

A   Secure Two-Party Computation

In this section we sketch the necessary definitions of secure two-party computation.[9] We refer the reader to, e.g., [7,18] for more details.

A secure two-party computation task is specified by a two-party *functionality*, i.e., a function mapping a pair of inputs to a pair of outputs. A protocol is said to *securely* realize the given functionality if an adversary attacking a party in a real-life execution of the protocol can achieve no more than it could have achieved by attacking the same party in an ideal implementation which makes use of a trusted party. In particular, the adversary should not learn more about the input and output of the uncorrupted party than it must inevitably be able to learn. This is formalized by defining a *real process* and an *ideal process*, and requiring that the interaction of the adversary with the real process can be *simulated* in the ideal process.

---

[9] The definitions we provide here apply to a simple stand-alone model assuming a non-adaptive adversary. However, the security of our reductions should hold for essentially any notion of security, including ones that support composition.

The real process. In the real process, the two parties execute the given protocol on their respective inputs and a common security parameter k. A probabilistic polynomial-time *adversary*, who may corrupt one of the parties and observe all of its internal data, intervenes with the protocol's execution. In the default case of an *active* adversary, the adversary has full control over the messages sent by the corrupted party. In this case we say that the corrupted party is *malicious*. Also of interest is the case of a *passive* adversary, who may only try to deduce information by performing computations on observed data, but otherwise follows the protocol's instructions.[10] Such a corrupted party is referred to as being *semi-honest*, or *honest but curious*. At the end of the interaction, the adversary may output an arbitrary function of its view. The *output of the real process* (on the the given pair of initial inputs) is defined as the random variable containing the *concatenation* of the adversary's output and the output of the uncorrupted party.

The ideal process. In the ideal process, an incorruptible trusted party is employed for computing the given functionality. That is, the "protocol" in the ideal process instructs each party to send its input to the trusted party, who computes the functionality and sends to each party its output. The interaction of the adversary with the ideal process and the output of the ideal process are defined analogously to the above definitions for the real process. The adversary attacking the ideal process is referred to as a *simulator*.

Security. A protocol is said to securely realize the given functionality if, for any (efficient) adversary attacking the real process, there exists an (efficient) simulator attacking the same party in the ideal process, such that on any pair of inputs, the outputs of the two processes are indistinguishable. The security is said to be perfect, statistical, or computational according to the type of indistinguishability achieved. In the latter two cases, indistinguishability is defined with respect to the security parameter k. The protocol is said to be secure against a semi-honest (resp., malicious) receiver, if the above security requirement holds with respect to a passive (resp., active) adversary corrupting the receiver. A similar definition applies to the sender. We note that in all cases the protocol is required to compute the given functionality if no party is corrupted.

B   Reducing OT_m^n to OT_k^n

Oblivious transfer of long strings can be efficiently reduced to oblivious transfer of shorter strings using any pseudo-random generator. The reduction is formally described in Fig. 3.

The security of this reduction is straightforward to prove. Its complexity overhead is dominated by the cost of generating $2m$ pseudo-random bits. By combining Step 3 with Step 2, the protocol's round complexity can be made the

---

[10] A passive adversary may model situations where an attacker (e.g., a computer virus) does not want to be exposed by disrupting the normal behavior of the attacked machine.

INPUT OF S: n pairs of m-bit strings $(x_{i,0}, x_{i,1})$, $1 \leq i \leq n$.
INPUT OF R: n selection bits $\mathbf{r} = (r_1, \ldots, r_n)$.
COMMON INPUT: a security parameter k.
ORACLE: A PRG $G : \{0,1\}^k \to \{0,1\}^m$.
CRYPTOGRAPHIC PRIMITIVE: An ideal OT_k^n primitive.

1. S initializes n pairs of random k-bit seeds $(s_{i,0}, s_{i,1})$.
2. The parties invoke the OT_k^n primitive, where S acts as a sender with inputs $(s_{i,0}, s_{i,1})$, $1 \leq i \leq n$, and R as a receiver with input $\mathbf{r}$.
3. For $1 \leq i \leq n$, S sends $(y_{i,0}, y_{i,1})$, where $y_{i,b} = x_{i,b} \oplus G(s_{i,b})$.
4. For $1 \leq i \leq n$, R outputs $z_i = y_{i,r_i} \oplus G(s_{i,r_i})$.

Fig. 3. Reducing OT_m^n to OT_k^n

same as that of OT_k^n. Finally, we note that the use of the PRG G can be emulated by standard means using calls to the random oracle H.

Algebraic Attacks on Combiners with Memory

Frederik Armknecht and Matthias Krause

Theoretische Informatik
Universität Mannheim
68131 Mannheim, Germany
{Armknecht,Krause}@th.informatik.uni-mannheim.de

Abstract. Recently, algebraic attacks were proposed to attack several cryptosystems, e.g. AES, LILI-128 and Toyocrypt. This paper extends the use of algebraic attacks to combiners with memory. A (k, l)-combiner consists of k parallel linear feedback shift registers (LFSRs), and the nonlinear filtering is done via a finite automaton with k input bits and l memory bits. It is shown that for (k, l)-combiners, nontrivial canceling relations of degree at most $\lceil k(l+1)/2 \rceil$ exist. This makes algebraic attacks possible. Also, a general method is presented to check for such relations with an even lower degree. This allows to show the invulnerability of certain (k, l)-combiners against this kind of algebraic attacks. On the other hand, this can also be used as a tool to find improved algebraic attacks.

Inspired by this method, the E_0 keystream generator from the Bluetooth standard is analyzed. As it turns out, a secret key can be recovered by solving a system of linear equations with $2^{23.07}$ unknowns. To our knowledge, this is the best published attack on the E_0 keystream generator yet.

1 Introduction

Stream ciphers are designed for online encryption of secret plaintext bitstreams $E = (e_1, e_2, \cdots)$ which have to pass an insecure channel. Depending on a given secret information $x^* \in \{0, 1\}^n$, the stream cipher produces a keystream $Z(x^*) = (z_1, z_2, \cdots)$ which is bitwise XORed with E. Knowing x^*, the decryption can be performed by using the same rule. It is common to evaluate the security of a stream cipher relative to the pessimistic scenario that an attacker has access not only to the encrypted bitstream, but even to a sufficiently long piece of keystream. Thus, the cryptanalysis problem of a given stream cipher consists in computing the secret information x^* from a sufficiently long prefix of $Z(x^*)$.

We call a stream cipher LFSR-based, if it consists of a certain number k of linear feedback shift registers (LFSRs) and an additional device, called the nonlinear combiner, which transforms the internal linear bitstream, produced by the LFSRs, into a nonlinear output keystream. Because of the simplicity of LFSRs and the excellent statistical properties of bitstreams produced by well-chosen LFSRs, LFSR-based stream ciphers are widely used in practice. A lot of different nontrivial approaches to the cryptanalysis of LFSR-based stream ciphers (fast correlation attacks, backtracking attacks, time-space tradeoffs, BDD-based

D. Boneh (Ed.): CRYPTO 2003, LNCS 2729, pp. 162–175, 2003.

attacks etc.) were discussed in the relevant literature, and a lot of corresponding design criterions (correlation immunity, large period and linear complexity, good local statistics etc.) for such stream ciphers were developed (see, e.g., Rueppel (1991)).

A (k, l)-combiner consists of k LFSRs and a finite Mealy automaton with k input bits, one output bit and l memory bits. Let n be the sum of the lengths of the k LFSRs. Starting from a secret initial assignment $x^* \in \{0, 1\}^n$, the LFSRs produce an internal linear bitstream $L(x^*)$, built by blocks x^t of k parallel bits for each clock t. Starting from a secret initial assignment $c^1 \in \{0, 1\}^l$ to the memory bits, in each clock t the automaton produces the t-th keystream bit z_t corresponding to x^t and c^t and changes the inner state to c^{t+1} (see figure 1). The secret information is given by x^* and c^1. Numerous ciphers of this type are used in practice. Note, e.g., that the E_0 keystream generator used in the Bluetooth wireless LAN system (see Bluetooth SIG (2001)) is a $(4, 4)$-combiner.

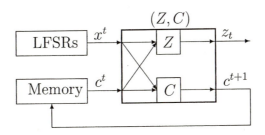

Fig. 1. A (k, l)-combiner

The aim of this paper is to analyze the security of (k, l)-combiners with respect to algebraic attacks, a new method for attacking stream and block ciphers. Algebraic attacks exist against AES and Serpent (Courtois and Pieprzyk (2002)) and Toyocrypt (Courtois (2002)). Related algebraic attacks were used to attack the HFE public key cryptosystem (Courtois (2001), cf. also Kipnis and Shamir (1999)).

Courtois and Meier (2003) discussed algebraic attacks on general LFSR-based stream ciphers and presented the best known attacks on Toyocrypt and LILI-128 so far. Very recently, Courtois introduced fast algebraic attacks on LFSR-based stream ciphers, an improved version of the algebraic attacks (Courtois (2003)).

An algebraic attack is based on a nontrivial low degree relation p for r clocks, i.e. a relation which holds for any sequence of r consecutive bits of the keystream and the corresponding kr internal bits. Given such a relation p of small degree d and a sufficiently long piece of a keystream $Z(x^*, c^1)$, p can be used to produce an overdefined system of T nonlinear equations in the initial bits of the LFSRs, which can be thought of as system of linear equations in the monomials of length at most d. If T is large enough then we get a unique solution which is induced by x^*, and from which x^* can be derived in a straightforward way.

Obviously, a higher value of d increases the running time significantly. Consequently, the nonexistence of nontrivial low degree relations is an important design criterion for (k, l)-combiners. One contribution of this paper is to provide an algorithm *FindRelation* which computes for a given (k, l)-combiner, represented by its automaton, and given d and r the set of all nontrivial degree d relation for r clocks (Section 3). One consequence is that nontrivial relations of degree $\lceil k(l + 1)/2 \rceil$ relations for $l + 1$ clocks (Theorem 1) cannot be avoided. Note that the running time is only polynomial in n if k and l are supposed to be constant. Hence, for each (k, l)-combiner exists a value n', such that the algebraic attack is more efficient than exhaustive search if $n \geq n'$.

E.g., this general bound implies a nontrivial degree 10 relation for 5 clocks for the E_0 generator, which yields, for $n = 128$, an algebraic attack of running time 2^{141}, which is much worse than exhaustive key-search. The algebraic attack would be better than exhaustive search if $n \geq 142$. Surprisingly, a nontrivial degree-4 relation for 4 clocks (Section 4) exists. This implies an algebraic attack of running time around $2^{67.58}$ and represents a serious weakness of this stream cipher. On the other hand, by using our method we can prove the nonexistence of nontrivial relations of degree smaller than 4, at least for 4 and 5 clocks. In the following section 2, we give basic definitions on boolean functions, LFSRs, and some notions around algebraic attacks.

2   Basics

2.1   Boolean Functions and $GF(2)$-Polynomials

In the following, we consider for all $k \geq 1$ the set B_k of k-ary boolean functions $f : \{0, 1\}^k \longrightarrow \{0, 1\}$ as a 2^k-dimensional vector space over the field $GF(2)$. It is a well known fact that each $f \in B_k$ has a unique representation as $GF(2)$-polynomial

$$p(x_1, \cdots, x_k) = \bigoplus_{\alpha \in \{0,1\}^k} a_\alpha m_\alpha, \tag{1}$$

where for all $\alpha \in \{0, 1\}^k$ the monomial m_α is defined as $m_\alpha = \Pi_{i, \alpha_i = 1} x_i$, and $a_\alpha \in GF(2)$. Let us denote $|\alpha| = |\{i, \alpha_i = 1\}|$ for all $\alpha \in \{0, 1\}^k$. The degree $\deg(p)$ of the polynomial p is defined as $\max\{|\alpha|, a_\alpha = 1\}$. For all $f \in B_k$ we denote by $\deg(f)$ the degree of the unique $GF(2)$-polynomial for f. Given a set $B \subseteq B_k$ we denote by $\mathcal{H}(B)$ the set of all linear combinations of functions from B. Note that the set of all k-ary boolean functions of degree at most d equals $\mathcal{H}(\mathcal{M}(k, d))$, where $\mathcal{M}(k, d) = \{m_\alpha, \alpha \in \{0, 1\}^k, |\alpha| \leq d\}$. The crucial computational problem here is $FindNullspace(B, X)$, where $B \subseteq B_k$ and $X \subseteq \{0, 1\}^k$ for some $k \geq 1$, which consists in the computation of all $h \in \mathcal{H}(B)$ for which $h(x) = 0$ for all $x \in X$. Clearly, all $h \in \mathcal{H}(B)$ can be represented as $h = \sum_{b \in B} a(h)_b b$, and the set of all coefficient vectors $a(h) \in GF(2)^B$ solving $FindNullspace(B, X)$ equals the set of solutions of the system

$$\sum_{b \in B} a(h)_b b(x) = 0, \quad \text{for all } x \in X, \tag{2}$$

of $GF(2)$-linear equations.

As usual, we call a boolean function $f \in B_k$ to be an implicant of another boolean function $g \in B_k$ if $f(x) = 1$ implies $g(x) = 1$ for all inputs $x \in \{0,1\}^k$.

2.2 LFSRs and (k, l)-Combiners

Let $k > 0$ and $l \geq 0$ be integers. A (k, l)-combiner $\mathcal{C} = (Z, C)$ consists of k linear feedback shift registers (LFSRs) $L_1, \cdots, L_k$ and a finite Mealy automaton which is defined by an output function $Z : \{0,1\}^k \times \{0,1\}^l \longrightarrow \{0,1\}$ and a feedback function $C : \{0,1\}^k \times \{0,1\}^l \longrightarrow \{0,1\}^l$. In this paper, we assume that the following reasonable condition holds: For each $c \in \{0,1\}^l$ exist $x, x' \in \{0,1\}^k$ with $Z(x, c) = 0$ and $Z(x', c) = 1$. Notice that all known (k, l)-combiners used in cryptosystems are of this kind.

For each $i, 1 \leq i \leq k$, LFSR L_i is defined by its length $n(i)$ and a generator polynomial $L_i = (L_{i,1}, \cdots, L_{i,n(i)}) \in GF(2)^{n(i)}$. Let $n = n(1) + \cdots + n(k)$. It is common to suppose that the generator polynomials of the LFSRs are public.

Given an initial assignment $x_i^* = (x_{i,1}^*, \cdots, x_{i,n(i)}^*) \in \{0,1\}^{n(i)}$ to each LFSR L_i, $1 \leq i \leq k$, the LFSRs compute at each clock t a block $x^t = (x_1^t, \cdots, x_k^t)$ of internal bits, where for each $i, 1 \leq i \leq k$, it holds $x_i^t = x_{t,i}^*$ if $t \leq n(i)$, and

$$x_i^t = L_{i,1} x_i^{t-1} \oplus L_{i,2} x_i^{t-2} \oplus \cdots \oplus L_{i,n(i)} x_i^{t-n(i)} \tag{3}$$

if $t > n(i)$. The bitstream $L(x^*) = (x^1, x^2, \cdots)$ is called the internal linear bitstream generated on the initial assignment $x^* = (x_1^*, \cdots, x_k^*)$. Note that for all $t \geq 0$, the $GF(2)$-linear mapping $L^t : GF(2)^n \longrightarrow GF(2)^n$ which assigns to x^* the t-th block x^t of the corresponding linear bitstream can be efficiently computed from the generator polynomials.

Given such an internal bitstream $x = (x^1, x^2, \cdots)$ and an initial assignment $c^1 \in \{0,1\}^l$ to the memory bits, the corresponding output bitstream $(Z, C)(x, c^1) = (z_1, z_2, \cdots)$ is defined according to

$$z^t = Z(x^t, c^t) \quad \text{and} \quad c^{t+1} = C(x^t, c^t), \tag{4}$$

for all $t \geq 1$. For all $r \geq 1$ let us denote by $(Z, C)^r(x^1, \cdots, x^r, c^1)$ the first r output bits of the keystream generated according to x and c^1.

Given the combiner $\mathcal{C} = (Z, C)$, the cryptanalysis problem consists in discovering the secret initial assignment $x^* \in \{0,1\}^n$ to the LFSRs and the secret initial assignment $c^1 \in \{0,1\}^l$ to the memory bits from a sufficiently long prefix of the output keystream $(Z, C)(L(x^*), c^1)$. Our results are motivated by an approach due to Courtois and Pieprzyk (2002) to this problem, which consists in performing a so-called algebraic attack, and which is based on finding nontrivial low-degree relations which hold for any sequence of r consecutive output bits and the corresponding kr bits of the internal bitstream, for some $r \geq 1$. Let us now give an outline of this kind of attack.

2.3   Nontrivial Relations and Algebraic Attacks

We use the same denotations as in the previous subsection.

Definition 1. *Let $r \geq 1$ and $z \in \{0,1\}^r$. A non-zero $GF(2)$-polynomial p in kr variables is called a z-relation for C if $p(x) = 0$ holds for all sequences $x = (x^1, x^2, \cdots, x^r) \in (\{0,1\}^k)^r$ of r consecutive blocks of the internal bitstream which have the property that $(Z,C)^r(x,c) = z$ for some initial assignments $c \in \{0,1\}^l$ to the memory bits.*

Let us suppose that C has a z-relation p of degree d for some $r \geq 1$. Fix arbitrary assignments $x^* \in \{0,1\}^n$ to the LFSRs and $c^1 \in \{0,1\}^l$ to the memory bits. Suppose that we have a sufficiently long prefix of the corresponding output bitstream $z^* = (Z,C)(L(x^*), c^1)$ and denote by $T(z)$ the set of all clocks t, for which $(z^*_t, \cdots, z^*_{t+(r-1)}) = z$. By the definitions, it holds for all $t \in T(z)$ that

$$P_t(x^*) := p(L^t(x^*), \cdots, L^{t+(r-1)}(x^*)) = 0. \tag{5}$$

P_t is a $GF(2)$-polynomial of degree d in n variables which can be efficiently computed. Consequently, the system

$$P_t(x_1, \cdots, x_n) = 0, \ t \in T(z) \tag{6}$$

of nonlinear equations can be considered as a system of linear equations in the unknowns $\{m_\alpha(x), \alpha \in \{0,1\}^n, |\alpha| \leq d\}$. If $|T(z)|$ is large enough then this system of linear equations has the unique solution $\{m_\alpha(x^*), |\alpha| \leq d\}$, from which the secret x^* can be easily derived. Obviously, $|T(z)|$ has to be at least $M(n,p)$. Here, $M(n,p)$ denotes the set of all monomials in $x_1, \cdots, x_n$ which can occur in a $GF(2)$-polynomial contained as equation in the system (6). Observe that $\Phi(n,p) := \sum_{i=0}^{d} \binom{n}{i}$ is a trivial upper bound for $|M(n,p)|$. Note that the minimum number of keystream bits which has to be available can be reduced if we know several degree-d z-relations for different strings z. In any case, it follows that the existence of low-degree z-relations implies a serious weakness of (k,l)-combiners. These attacks are called algebraic attacks. In Courtois and Meier (2003), the authors discuss algebraic attacks against combiners without memory. In this paper, we extend these attacks to combiners with memory.

3   On Constructing Nontrivial Relations

In this section, we show that for any (k,l)-combiner C, $r \geq k(l+1)$, and $d \geq \lceil k(l+1)/2 \rceil$, the existence of z-relations of degree d for some $z \in \{0,1\}^r$ cannot be avoided. Moreover, we present an algorithm which allows to construct all z-relations of degree at most d for any given r, d. Note that this solves an open problem stated, e.g., by Courtois (2003). This algorithm can be used for estimating the vulnerability of given (k,l)-combiners with respect to algebraic attacks (known from Courtois and Meier (2003)).

We first illustrate the problem of constructing nontrivial relations by means of some special cases. Let as before $\mathcal{C} = (Z, C)$ denote a (k, l)-combiner with output function Z and feedback function C. If $l = 0$, the construction of canceling relations for one clock is straightforward, as

$$Z(x_1^t, \cdots, x_k^t) \oplus z_t \quad , t \geq 0 \tag{7}$$

is always fulfilled. By arguments which will be given below this implies the existence of relations of degree at most $\lceil k/2 \rceil$.

Another tractable case is if $l = 1$ and the output function Z is linear in the feedback bit, i.e., $Z(x, c) = Z'(x) \oplus c$. Then the relation

$$z_2 = Z(x^2, C(x^1, z_1 \oplus x^1)) \tag{8}$$

is always true, which gives z-relations for all $z \in \{0, 1\}^2$. If $l \geq 1$ and the output function is nonlinear, the situation becomes more complicated as, via the feedback function C, z_t depends nonlinearly on $x^1, x^2, \cdots, x^t$ for all $t \geq 0$. One attempt for constructing nontrivial relations could be to consider the relation

$$\bigwedge_{c \in \{0,1\}^l} \left(Z(x^t, c) \oplus z_t \right), \tag{9}$$

which obviously gives 0 for all pairs of input and output streams generated via $\mathcal{C}$. The problem here is that this relation can become trivial. This is especially true if Z is linear in at least one memory bit, as is the case for the E_0 generator. We use a more systematic approach and show the following result.

Theorem 1. *Let $k \geq 1$, $l \geq 1$ and a (k, l)-combiner $\mathcal{C} = (Z, C)$ be arbitrarily fixed. Then for each $r > l$ there is a z-relation of degree $\lceil (k(l+1)/2 \rceil$ for $\mathcal{C}$ for some $z \in \{0, 1\}^r$.*

For the proof of this theorem we need some more technical definitions.

Definition 2. *(i)   For all $r \geq 1$, $z \in \{0, 1\}^r$, and $x = (x^1, \cdots, x^r) \in \left(\{0, 1\}^k\right)^r$, x is called z-critical for $\mathcal{C}$ if $(Z, C)^r(x, c) \neq z$ for all $c \in \{0, 1\}^l$. We denote by $Crit_{\mathcal{C}}(z)$ the set of all $x \in \left(\{0, 1\}^k\right)^r$ which are z-critical for $\mathcal{C}$, and by $NCrit_{\mathcal{C}}(z)$ the set of all x which are not.*

(ii)   The pair $(x, z) \in \left(\{0, 1\}^k\right)^r \times \{0, 1\}^r$ is called r-critical for $\mathcal{C}$ if x is z-critical for $\mathcal{C}$. We denote by $Crit_{\mathcal{C}}(r)$ the set of all r-critical $(x, z) \in \left(\{0, 1\}^k\right)^r \times \{0, 1\}^r$ and by $NCrit_{\mathcal{C}}(r)$ the set of all (x, r) which are not. Especially, we have $Crit_{\mathcal{C}}(r) \dot\cup NCrit_{\mathcal{C}}(r) = \{0, 1\}^{kr} \times \{0, 1\}^r$.

(iii)   For all $r \geq 1$ we denote by $\chi(\mathcal{C})_r : \left(\{0, 1\}^k\right)^r \times \{0, 1\}^r \longrightarrow \{0, 1\}$ the critical function of $\mathcal{C}$, which is defined as $\chi(\mathcal{C})_r(x, z) = 1$ iff (x, z) is r-critical for $\mathcal{C}$. For all $z \in \{0, 1\}^r$ we denote by $\chi(\mathcal{C})_r^z$ the subfunction $\chi(\mathcal{C})_r(\cdot, z)$ which outputs 1 on $x \in \left(\{0, 1\}^k\right)^r$ iff x is z-critical.

Observe that for all $r \geq 1$ and $z \in \{0, 1\}^r$, a nontrivial $GF(2)$-polynomial p in kr variables is a z-relation of $\mathcal{C}$ iff it outputs 0 for all $x \in NCrit_{\mathcal{C}}(z)$ and outputs 1 for at least one $x \in Crit_{\mathcal{C}}(z)$. This implies

Lemma 1. *For all $r \geq 1$ and $z \in \{0,1\}^r$ there is a z-relation for $\mathcal{C}$ iff $Crit_{\mathcal{C}}(z) \neq \emptyset$. If $Crit_{\mathcal{C}}(z) \neq \emptyset$ then $p : (\{0,1\}^k)^r \longrightarrow \{0,1\}$ is a z-relation for $\mathcal{C}$ if and only if it is a nontrivial implicant of $\chi(\mathcal{C})_r^z$.*

For each non-critical pair $(x,z) \in (\{0,1\}^k)^r \times \{0,1\}^r$ there exists at least one $c \in \{0,1\}^l$ such that $z = (Z,C)^r(x,c)$. Evidently, the number of non-critical pairs cannot exceed $2^{kr} \cdot 2^l$. We obtain

Lemma 2. *For all $r \geq 1$ it holds that $|NCrit_{\mathcal{C}}(r)| \leq 2^{kr+l}$.*

For $r = l + 1$, we have

$$|NCrit_{\mathcal{C}}(l+1)| \leq 2^{k \cdot (l+1)+l} < 2^{k \cdot (l+1)+l+1} = |Crit_{\mathcal{C}}(l+1)| + |NCrit_{\mathcal{C}}(l+1)| \quad (10)$$

Therefore, $|Crit_{\mathcal{C}}(r)| \neq 0$ and there is some $z \in \{0,1\}^r$ such that $|Crit_{\mathcal{C}}(z)| \neq \emptyset$.

For all $d \geq 0$ let us denote by $\mathcal{M}(kr,d)$ the set of all monomials over the kr variables $x^1, \cdots, x^r$ of length at most d. We derived

Lemma 3. *For all $r \geq 1$ and $z \in \{0,1\}^r$ the set of all z-relations for $\mathcal{C}$ equals the set of non-zero solutions of $FindNullspace(\mathcal{M}(kr,d), NCrit_{\mathcal{C}}(z))$.*

Lemma 4. *For each $r \geq 0$ and $z \in \{0,1\}^r$ the set $NCrit_{\mathcal{C}}(z)$ is not empty.*

Proof. We show this proposition by complete induction. As said in the beginning, we consider only combiners for which the following condition is true:

$$\forall c \in \{0,1\}^l \; \exists x, x' \in \{0,1\}^k : Z(x,c) = 0 \text{ and } Z(x',c) = 1 \quad (11)$$

This assures the proposition for $r = 1$. Let the proposition be true for some r. Choose $z = (z_1, \ldots, z_{r+1}) \in \{0,1\}^{r+1}$ arbitrarily. Then $NCrit_{\mathcal{C}}((z_1, \ldots, z_r)) \neq \emptyset$ by assumption. Let $x = (x^1, \ldots, x^r) \in NCrit_{\mathcal{C}}((z_1, \ldots, z_r))$. Then there exists a $c^1 \in \{0,1\}^l$ with $(Z,C)^r(x^1, \ldots, x^r, c^1) = (z_1, \ldots, z_r)$. By (11) we know that there is a least one $x^{r+1} \in \{0,1\}^k$ with $Z(x^{r+1}, c^{r+1}) = z_{r+1}$. Therefore, $(Z,C)^r(x^1, \ldots, x^{r+1}, c^1) = (z_1, \ldots, z_{r+1})$ and $(x^1, \ldots, x^{r+1}) \in NCrit_{\mathcal{C}}(z)$.

For showing the degree bound observe that if $|\mathcal{M}(kr,d)| = \Phi(kr,d)$ is greater than $|NCrit_{\mathcal{C}}(z)|$ then $FindNullspace(\mathcal{M}(kr,d), NCrit_{\mathcal{C}}(z))$ has a nontrivial solution. It suffices to prove the degree bound for $r = l+1$. Lemma 2 implies that $|NCrit_{\mathcal{C}}(l+1)| \leq 2^{k(l+1)+l} = \frac{1}{2}2^{k(l+1)+l+1}$, i.e., at most one half of all possible pairs (x,z) are not $(l+1)$-critical. Consequently, there exists at least one $z \in \{0,1\}^{l+1}$ for which at most half of all possible x are z-critical, i.e., $|NCrit_{\mathcal{C}}(z)| \leq \frac{1}{2}2^{k(l+1)}$. On the other hand, by lemma 4 we know that $|NCrit_{\mathcal{C}}(z)| > 0$. Using the fact that $\Phi(N, \lceil N/2 \rceil) > \frac{1}{2}2^N$ for all $N \geq 2$, we obtain the theorem.

We derived the following algorithm for the problem $FindRelation(Z,C,z,d)$ of computing all z-relations p of degree at most d for a given (k,l)-combiner $\mathcal{C} = (Z,C)$.

1 Compute $Crit_{\mathcal{C}}(z)$ and $NCrit_{\mathcal{C}}(z)$.
2 If $Crit_{\mathcal{C}}(z) \neq \emptyset$ then solve $FindNullspace(\mathcal{M}(kr,d), NCrit_{\mathcal{C}}(z))$.

Note that the computation of $Crit_C(z)$ and $NCrit_C(z)$ can be done in an elegant way by using an ordered binary decision diagram (OBDD) of size at most $(kr + r)2^{k+l+1}$ for $\chi(C)_r$ (see, e.g., Krause (2002) for the details). Step 2 requires to solve a system of $GF(2)$-linear equations with $\mathcal{M}(kr, d)$ unknowns and at most 2^{kr+l} linear equations.

4   Analyzing the E_0 Keystream Generator

In this section, we apply our results to the E_0 keystream generator. The E_0 keystream generator is part of the Bluetooth encryption system, used for wireless communication (see, e.g., Bluetooth SIG (2001)). It is a $(4, 4)$-combiner. Applying our results yields the existence of a nontrivial 5-relation of degree 10. The number of monomials is $T \leq \Phi(n, 10)$. Therefore, the secret key can be recovered by solving a system of linear equations in T unknowns. The fastest practial algorithm we are aware of to solve a system of linear equations is the algorithm by Strassen (1969). It requires about $7 \cdot T^{\log_2 7}$ operations. Our attack is more efficient than exhaustive search, if the following inequality holds:

$$2^n > 7 \cdot (\Phi(n, 10))^{\log_2 7} . \tag{12}$$

This is the case for $n \geq 142$. Notice, that in the Bluetooth encryption system the length of the secret key is $n = 128$.

If the E_0 keystream generator were optimally resistant against algebraic attacks, no canceling relations for $r < 5$ or $d < 10$ should exist. Surprisingly, for $d = 4$ and $r = 4$ such a relation can be found. In this case, it is even possible to show the existence directly.

Let us first recall the definitions of the keystream generator. The keystream generator consists of $k = 4$ regularly clocked LFSRs and $l = 4$ memory bits. With each clock, an output bit z_t is produced depending on the outputs $x^t = (x_1^t, x_2^t, x_3^t, x_4^t)$ of the four LFSRs and the four memory bits $c^t = (q^t, p^t, q^{t-1}, p^{t-1})$. Then, the next memory bits $c^{t+1} = (q^{t+1}, p^{t+1}, q^t, p^t)$ are calculated and so on. We see that the memory bits q^t and p^t are used in both clocks t and $t+1$. Let $\pi_s(t)$ be the symmetric $GF(2)$-polynomial over $x_1^t, x_2^t, x_3^t, x_4^t$ which consists of the sum of all monomials of length $s \leq 4$. Then the output bit z_t and the memory bits are computed by the following equations

$$z_t = \pi_1(t) \oplus p^t \tag{13}$$
$$c^{t+1} = (q^{t+1}, p^{t+1}, q^t, p^t) \tag{14}$$
$$= (\mathcal{S}_1^{t+1} \oplus q^t \oplus p^{t-1}, \mathcal{S}_0^{t+1} \oplus p^t \oplus q^{t-1} \oplus p^{t-1}, q^t, p^t), \tag{15}$$

where

$$\mathcal{S}_{t+1} = (\mathcal{S}_{t+1}^1, \mathcal{S}_{t+1}^0) = \left\lfloor \frac{x_1^t + x_2^t + x_3^t + x_4^t + 2 \cdot q^t + p^t}{2} \right\rfloor . \tag{16}$$

The values for c^1 and the contents of the LFSRs must be set before the start, the other values will then be calculated. Obviously, the value of q^{t+1} depends only on x^t, q^t, p^t and p^{t-1} and the value of p^{t+1} on x^t, q^t, q^{t-1}, p^t and p^{t-1}. The calculations of q^{t+1} and p^{t+1} are done via the following equations (see appendix A for details)

$$q^{t+1} = \pi_4(t) \oplus \pi_3(t)p^t \oplus \pi_2(t)q^t \oplus \pi_1(t)p^t q^t \oplus q^t \oplus p^{t-1} \qquad (17)$$
$$p^{t+1} = \pi_2(t) \oplus \pi_1(t)p^t \oplus q^t \oplus q^{t-1} \oplus p^{t-1} \oplus p^t \qquad (18)$$

If we define the following additional variables

$$a(t) = \pi_4(t) \oplus \pi_3(t)p^t \oplus p^{t-1}$$
$$b(t) = \pi_2(t) \oplus \pi_1(t)p^t \oplus 1,$$

equations (17) and (18) can be rewritten to

$$q^{t+1} = a(t) \oplus b(t)q^t \qquad (19)$$
$$p^{t+1} = b(t) \oplus 1 \oplus p^{t-1} \oplus p^t \oplus q^t \oplus q^{t-1}. \qquad (20)$$

By multiplying (19) with $b(t)$ we get another equation

$$0 = b(t)(a(t) \oplus q^t \oplus q^{t+1}). \qquad (21)$$

Equation (20) is equivalent to

$$q^t \oplus q^{t-1} = b(t) \oplus 1 \oplus p^{t-1} \oplus p^t \oplus p^{t+1}. \qquad (22)$$

Now we insert (22) into (21) with index $t+1$ instead of t and get

$$0 = b(t)\left(a(t) \oplus b(t+1) \oplus 1 \oplus p^t \oplus p^{t+1} \oplus p^{t+2}\right).$$

Using (13), we eliminate all memory bits in the equation and get the following equation which holds for every clock t:

$$\begin{aligned}
0 = \ &1 \oplus z_{t-1} \oplus z_t \oplus z_{t+1} \oplus z_{t+2} \\
&\oplus \pi_1(t) \cdot (z_t z_{t+2} \oplus z_t z_{t+1} \oplus z_t z_{t-1} \oplus z_{t-1} \oplus z_{t+1} \oplus z_{t+2} \oplus 1) \\
&\oplus \pi_2(t) \cdot (1 \oplus z_{t-1} \oplus z_t \oplus z_{t+1} \oplus z_{t+2}) \oplus \pi_3(t)z_t \oplus \pi_4(t) \\
&\oplus \pi_1(t-1) \oplus \pi_1(t-1)\pi_1(t)(1 \oplus z_t) \oplus \pi_1(t-1)\pi_2(t) \\
&\oplus \pi_1(t+1)z_{t+1} \oplus \pi_1(t+1)\pi_1(t)z_{t+1}(1 \oplus z_t) \oplus \pi_1(t+1)\pi_2(t)z_{t+1} \\
&\oplus \pi_2(t+1) \oplus \pi_2(t+1)\pi_1(t)(1 \oplus z_t) \oplus \pi_2(t+1)\pi_2(t) \\
&\oplus \pi_1(t+2) \oplus \pi_1(t+2)\pi_1(t)(1 \oplus z_t) \oplus \pi_1(t+2)\pi_2(t)
\end{aligned}$$

This gives a nontrivial degree-4 z-relation p for 4 clocks for any $z \in \{0,1\}^4$. The number $M(128, p)$ of monomials occuring in the corresponding system of nonlinear equations (see subsection 2.2) can not exceed $\Phi(128, 4) \approx 2^{23.39}$. In fact, if we look closely at p, we can see that not all monomials of $\mathcal{M}(kr, d)$ occur. Thus, we have $M(128, p) \leq T := 8,824,350 \approx 2^{23.07}$ (see appendix B for details).

With each clock t, we get a new equation in the bits of the secret key. If we have at least $M(128, p)$ linearly independent equations, x^* can be recovered by solving the system of linear equations. Using Strassen's algorithm, the secret key can be recovered with work $\leq 7 \cdot T^{\log_2 7} \approx 2^{67.58}$. The memory complexity is more or less the size of the matrix which is about $2^{46.14}$.

Obviously, to get enough linearly independent equations, we have to clock at least $M(128, p)$ times. The question is whether we have to clock more often. Until now, there is no satisfying answer to this question. Our assumption is that approximately T clocks should be enough, meaning that about $2^{23.07}$ key stream bits would be sufficient to mount the attack. We did some simulations for the same cryptosystem but with shorter LFSRs. The results can be seen in Table 1. Each time, the initial values of the LFSRs were successfully reconstructed. In all cases the number of clocks needed to reconstruct the secret key was close (or even equal) to $T + 3$.[1]

Table 1. Algebraic attacks on smaller E_0 crypto systems

$n(1), n(2), n(3), n(4)$	Initial Values	Feedback Taps	T	Clocks
1, 2, 3, 5	1 11 011 11110	1 11 101 10100	477	483
1, 2, 3, 5	1 10 101 01101	1 11 101 10100	477	481
1, 2, 3, 5,	1 01 010 01001	1 11 101 11011	477	480
1, 2, 3, 5	1 11 111 01111	1 11 101 11110	477	483
1, 2, 3, 5,	1 01 010 10100	1 11 110 11011	477	484
2, 3, 5, 7	10 010 11110 1100110	11 110 11101 1000100	2643	2647
2, 3, 5, 7	11 101 01101 0010011	11 101 10100 1101010	2643	2649
2, 3, 5, 7	10 100 10001 0010001	11 110 11110 1111000	2643	2647

Of course, a lower degree d would decrease the value of T and therefore allow a better attack. Applying our algorithm showed the non-existence of nontrivial relations of degree $d = 3$ for $r = 4$ and $r = 5$. Nevertheless, lower degree relations for $r > 5$ may exist.

It is important to mention that in the Bluetooth encryption system the secret key is changed after 2745 clocks. Therefore, we will never get enough equations in pratice. Note that the best published attack against the E_0 was proposed by Krause (2002) with time and memory effort of $\approx 2^{77}$, given only 128 known key stream bits. The attack by Fluhrer and Lucks (2001) needs about 2^{73} operations if 2^{43} bits are available. The memory needed is very small: about 10638 bits.

Recently, Courtois developed an improved version of algebraic attacks: fast algebraic attacks (Courtois (2003)). They allow an even better attack on the E_0 keystream generator. The estimation is that about 2^{49} operations are enough.

---

[1] As we need 4 succesive clocks to produce one equation the number of clocks needed is at least $T + 3$

We want to point out one remarkable fact. The output function was chosen to be linear in one memory bit to achieve maximum correlation immunity. The same attribute made it possible to eliminate the same memory bit in our relation. This may be a hint that some tradeoff between correlation immunity of the output function and resistance against algebraic attacks exists.

5   Discussion

We have seen that for all (k, l)-combiners, nontrivial relations of degree at most $\lceil k(l+1)/2 \rceil$ exist. This fact extends the attacks described by Courtois and Meier (2003) to combiners with memory. In consequence, each combiner is vulnerable against algebraic attacks if the length of the secret key n is large enough. E.g., for the E_0 keystream generator this is the case for $n \geq 142$. A (k, l)-combiner should be designed in such a way that an algebraic attack never becomes faster than exhaustive key-search. For this purpose, it should be checked if the automaton induces nontrivial degree-d relations for critical values of d. This can be done by applying the algorithm *FindRelation* presented in this paper, at least for a reasonable set of clocks.

The analysis of the E_0 generator shows that it may be dangerous to use a linear output function, since this may help replacing the memory bits and deriving nontrivial low-degree relations. It turns out that a nontrivial relation of degree 4 exists. This makes it possible to recover the secret key by solving a system of linear equations in at most $2^{23.07}$ unknowns.

Algebraic attacks work successfully only for LFSR-based stream ciphers which are oblivious in the sense that the attacker always knows which bit of the keystream depends on which bits of the internal bitstream. It would be interesting to know if similar attacks can also be applied to non-oblivious ciphers like the A5 generator or the shrinking generator.

Acknowledgment. The authors would like to thank Nicolas Courtois, Erik Zenner, Stefan Lucks and some unknown referees for helpful comments and discussions.

References

1. Bluetooth SIG, *Specification of the Bluetooth system*, Version 1.1, 1 February 22, 2001, available at http://www.bluetooth.com/.
2. Nicolas Courtois: *Higher Order Correlation Attacks, XL Algorithm and Cryptanalysis of Toyocrypt*, 5th International Conference on Information Security and Cryptology: ICISC 2002, November 2002, Seoul, Korea, Springer LNCS 2587. An updated version is available at http://eprint.iacr.org/2002/087.
3. Nicolas Courtois and Willi Meier: *Algebraic Attacks on Stream Ciphers with Linear Feedback*, Proceedings of Eurocrypt 2003, Warsaw, Poland, Springer LNCS 2656.
4. Nicolas Courtois: *Fast Algebraic Attacks on Stream Ciphers with Linear Feedback*, these proceedings.

5. Nicolas Courtois, Josef Pieprzyk: *Cryptanalysis of Block Ciphers with Overdefined Systems of Equations*, Proceedings of Asiacrypt '02, Springer LNCS 2501, 2002, pp. 267–287.
6. Nicolas Courtois: *Personal communication*, 2003
7. Scott R. Fluhrer, Stefan Lucks: *Analysis of the E_0 Encryption System*, Proceedings of Selected Areas of Cryptography '01, Springer LNCS 2259, 2001, pp. 38–48.
8. Matthias Krause: *BDD-Based Cryptanalysis of Keystream Generators*; Proceedings of Eurocrypt '02, Springer LNCS 2332, 2002, pp. 222–237.
9. Rainer A. Rueppel: *Stream Ciphers*; Contemporary Cryptology: The Science of Information Integrity. G. Simmons ed., IEEE Press New York, 1991.
10. Adi Shamir, Aviad Kipnis: *Cryptanalysis of the HFE Public Key Cryptosystem*; Proceedings of Crypto '99, Springer LNCS 1666, 1999, pp. 19–30.
11. Adi Shamir, Jacques Patarin, Nicolas Courtois, Alexander Klimov: *Efficient Algorithms for Solving Overdefined Systems of Multivariate Polynomial Equations*, Proceedings of Eurocrypt '00, Springer LNCS 1807, pp. 392–407.
12. Volker Strassen: *Gaussian Elimination is Not Optimal*; Numerische Mathematik, vol 13, pp 354–356, 1969.

A The Equations for q_{t+1} and p_{t+1}

In this section we prove the correctness of equations (17) resp. (18) for q^{t+1} resp. p^{t+1}. Let us recall the equation for c^{t+1}

$$c^{t+1} = (q^{t+1}, p^{t+1}, q^t, p^t) \tag{23}$$
$$= (\mathcal{S}_1^{t+1} \oplus q^t \oplus p^{t-1}, \mathcal{S}_0^{t+1} \oplus p^t \oplus q^{t-1} \oplus p^{t-1}, q^t, p^t) \tag{24}$$

where

$$\mathcal{S}_{t+1} = (\mathcal{S}_{t+1}^1, \mathcal{S}_{t+1}^0) = \left\lfloor \frac{x_1^t + x_2^t + x_3^t + x_4^t + 2 \cdot q^t + p^t}{2} \right\rfloor \tag{25}$$

Let f_0 resp. f_1 be the two boolean functions for which the equations

$$\mathcal{S}_{t+1}^i = f_i(x_1^t, x_2^t, x_3^t, x_4^t, q^t, p^t) \tag{26}$$

hold for $i \in \{0,1\}$. f_0 and f_1 can be found with the help of computers. If we write down f_0 and f_1 in algebraic normal form, we get

$$f_1 = \pi_4(t) \oplus \pi_3(t)p^t \oplus \pi_2(t)q^t \oplus \pi_1(t)p^t q^t \tag{27}$$
$$f_0 = \pi_2(t) \oplus \pi_1(t)p^t \oplus q^t \tag{28}$$

See section 4 for the definition of $\pi_k(t)$. In table 2, f_0 and f_1 are evaluated for all possible inputs and compared with $\mathcal{S}_{t+1}$. It is easy to see that f_0 and f_1 fulfill the requirements. Together with (24), we get the following expressions for q^{t+1} and p^{t+1}

$$q^{t+1} = \mathcal{S}_{t+1}^1 \oplus q^t \oplus p^{t-1} \tag{29}$$
$$= \pi_4(t) \oplus \pi_3(t)p^t \oplus \pi_2(t)q^t \oplus \pi_1(t)p^t q^t \oplus q^t \oplus p^{t-1} \tag{30}$$
$$p^{t+1} = \mathcal{S}_{t+1}^0 \oplus p^t \oplus q^{t-1} \oplus p^{t-1} \tag{31}$$
$$= \pi_2(t) \oplus \pi_1(t)p^t \oplus q^t \oplus q^{t-1} \oplus p^t \oplus p^{t-1} \tag{32}$$

Table 2. f_0 and f_1 evaluated for all possible inputs and compared with S_{t+1}

a_t	b_t	c_t	d_t	Q_t	P_t	S_{t+1}	f_1	f_0	a_t	b_t	c_t	d_t	Q_t	P_t	S_{t+1}	f_1	f_0
0	0	0	0	0	0	0	0	0	1	0	0	0	0	0	0	0	0
0	0	0	0	0	1	0	0	0	1	0	0	0	0	1	1	0	1
0	0	0	0	1	0	1	0	1	1	0	0	0	1	0	1	0	1
0	0	0	0	1	1	1	0	1	1	0	0	0	1	1	2	1	0
0	0	0	1	0	0	0	0	0	1	0	0	1	0	0	1	0	1
0	0	0	1	0	1	1	0	1	1	0	0	1	0	1	1	0	1
0	0	0	1	1	0	1	0	1	1	0	0	1	1	0	2	1	0
0	0	0	1	1	1	2	1	0	1	0	0	1	1	1	2	1	0
0	0	1	0	0	0	0	0	0	1	0	1	0	0	0	1	0	1
0	0	1	0	0	1	1	0	1	1	0	1	0	0	1	1	0	1
0	0	1	0	1	0	1	0	1	1	0	1	0	1	0	2	1	0
0	0	1	0	1	1	2	1	0	1	0	1	0	1	1	2	1	0
0	0	1	1	0	0	1	0	1	1	0	1	1	0	0	1	0	1
0	0	1	1	0	1	1	0	1	1	0	1	1	0	1	2	1	0
0	0	1	1	1	0	2	1	0	1	0	1	1	1	0	2	1	0
0	0	1	1	1	1	2	1	0	1	0	1	1	1	1	3	1	1
0	1	0	0	0	0	0	0	0	1	1	0	0	0	0	1	0	1
0	1	0	0	0	1	1	0	1	1	1	0	0	0	1	1	0	1
0	1	0	0	1	0	1	0	1	1	1	0	0	1	0	2	1	0
0	1	0	0	1	1	2	1	0	1	1	0	0	1	1	2	1	0
0	1	0	1	0	0	1	0	1	1	1	0	1	0	0	1	0	1
0	1	0	1	0	1	1	0	1	1	1	0	1	0	1	2	1	0
0	1	0	1	1	0	2	1	0	1	1	0	1	1	0	2	1	0
0	1	0	1	1	1	2	1	0	1	1	0	1	1	1	3	1	1
0	1	1	0	0	0	1	0	1	1	1	1	0	0	0	1	0	1
0	1	1	0	0	1	1	0	1	1	1	1	0	0	1	2	1	0
0	1	1	0	1	0	2	1	0	1	1	1	0	1	0	2	1	0
0	1	1	0	1	1	2	1	0	1	1	1	0	1	1	3	1	1
0	1	1	1	0	0	1	0	1	1	1	1	1	0	0	2	1	0
0	1	1	1	0	1	2	1	0	1	1	1	1	0	1	2	1	0
0	1	1	1	1	0	2	1	0	1	1	1	1	1	0	3	1	1
0	1	1	1	1	1	3	1	1	1	1	1	1	1	1	3	1	1

B The Number of Terms

In this section, we estimate the maximum number T of different monomials in the algebraic attack against the E_0 crypto system. With each clock t the following equation is produced

$$0 = 1 \oplus z_{t-1} \oplus z_t \oplus z_{t+1} \oplus z_{t+2}$$
$$\oplus \pi_1(t) \cdot (z_t z_{t+2} \oplus z_t z_{t+1} \oplus z_t z_{t-1} \oplus z_{t-1} \oplus z_{t+1} \oplus z_{t+2} \oplus 1)$$
$$\oplus \pi_2(t) \cdot (1 \oplus z_{t-1} \oplus z_t \oplus z_{t+1} \oplus z_{t+2}) \oplus \pi_3(t) z_t \oplus \pi_4(t)$$
$$\oplus \pi_1(t-1) \oplus \pi_1(t-1)\pi_1(t)(1 \oplus z_t) \oplus \pi_1(t-1)\pi_2(t)$$
$$\oplus \pi_1(t+1)z_{t+1} \oplus \pi_1(t+1)\pi_1(t)z_{t+1}(1 \oplus z_t) \oplus \pi_1(t+1)\pi_2(t)z_{t+1}$$

$$\oplus \pi_2(t+1) \oplus \pi_2(t+1)\pi_1(t)(1 \oplus z_t) \oplus \pi_2(t+1)\pi_2(t)$$
$$\oplus \pi_1(t+2) \oplus \pi_1(t+2)\pi_1(t)(1 \oplus z_t) \oplus \pi_1(t+2)\pi_2(t).$$

As we can see, every occurring term has to be one of the following types

$$a, b, c, d, ab, ac, ad, bc, bd, cd, abc, acd, abd, bcd, abcd, aa'bc, aa'cd, aa'bd,$$
$$bb'ac, bb'cd, bb'ad, cc'ab, cc'ad, cc'bd, dd'ab, dd'ac, dd'bc, aa'bb', aa'cc',$$
$$aa'dd', bb'cc', bb'dd', cc'dd', aa'b, aa'c, aa'd, bb'a, bb'c, bb'd, cc'a, cc'b,$$
$$cc'd, dd'a, dd'b, dd'c, aa', bb', cc', dd'$$

Here, $a, a' \in \{x_{1,1}^*, \ldots, x_{1,n_1}^*\}$ with $a \neq a'$, etc. In table 3 the number of possible terms for each type is presented depending on the length n_1, n_2, n_3, and n_4 of the four LFSRs. In addition, we give for each type one product in which it can occur. Note that some terms may occur in other products too[2]. Of course, these types have to be counted only once. The sum is the number of possible terms T. In E_0, the lengths are $n_1 = 25$, $n_2 = 31$, $n_3 = 33$ and $n_4 = 39$, so $T = 8,824,350$, which is approximately $2^{23.07}$.

Table 3. All possible terms and their number depending on n_i

type	occur in	number
a,b,c,d	$\pi_1(t)$	$n_1 + n_2 + n_3 + n_4$
ab, ac, ad, bc, bd, cd	$\pi_2(t)$	$n_1(n_2 + n_3 + n_4) + n_2(n_3 + n_4) + n_3 n_4$
abc, acd, abd, bcd	$\pi_3(t)$	$n_1(n_2 n_3 + n_2 n_4 + n_3 n_4) + n_2 n_3 n_4$
$abcd$	$\pi_4(t)$	$n_1 n_2 n_3 n_4$
aa', bb', cc', dd'	$\pi_1(t) \cdot \pi_1(t')$	$\sum_{i=1}^{4} \frac{1}{2} n_i(n_i - 1)$
$aa'b, aa'c, aa'd$	$\pi_1(t) \cdot \pi_2(t')$	$\frac{1}{2} n_1(n_1 - 1)(n_2 + n_3 + n_4)$
$bb'a, bb'c, bb'd$	$\pi_1(t) \cdot \pi_2(t')$	$\frac{1}{2} n_2(n_2 - 1)(n_1 + n_3 + n_4)$
$cc'a, cc'b, cc'd$	$\pi_1(t) \cdot \pi_2(t')$	$\frac{1}{2} n_3(n_3 - 1)(n_1 + n_2 + n_4)$
$dd'a, dd'b, dd'c$	$\pi_1(t) \cdot \pi_2(t')$	$\frac{1}{2} n_4(n_4 - 1)(n_1 + n_2 + n_3)$
$aa'bc, aa'cd, aa'bd$	$\pi_2(t) \cdot \pi_2(t')$	$\frac{1}{2} n_1(n_1 - 1)(n_2 n_3 + n_2 n_4 + n_3 n_4)$
$bb'ac, bb'cd, bb'ad$	$\pi_2(t) \cdot \pi_2(t')$	$\frac{1}{2} n_2(n_2 - 1)(n_1 n_3 + n_1 n_4 + n_3 n_4)$
$cc'ab, cc'ad, cc'bd$	$\pi_2(t) \cdot \pi_2(t')$	$\frac{1}{2} n_3(n_3 - 1)(n_1 n_2 + n_1 n_4 + n_2 n_4)$
$dd'ab, dd'ac, dd'bc$	$\pi_2(t) \cdot \pi_2(t')$	$\frac{1}{2} n_4(n_4 - 1)(n_1 n_2 + n_1 n_3 + n_2 n_3)$
$aa'bb', aa'cc', aa'dd'$	$\pi_2(t) \cdot \pi_2(t')$	$\frac{1}{2} n_1(n_1 - 1)\left(\sum_{i=2}^{4} \frac{1}{2} n_i(n_i - 1)\right)$
$bb'cc', bb'dd'$	$\pi_2(t) \cdot \pi_2(t')$	$\frac{1}{4} n_2(n_2 - 1)[n_3(n_3 - 1) + n_4(n_4 - 1)]$
$cc'dd'$	$\pi_2(t) \cdot \pi_2(t')$	$\frac{1}{4} n_3(n_3 - 1) n_4(n_4 - 1)$

---

[2] For example, a term of type abc can occur in $\pi_1(t)\pi_2(t')$ and in $\pi_2(t)\pi_2(t')$

Fast Algebraic Attacks on Stream Ciphers with Linear Feedback

Nicolas T. Courtois

Cryptography Research, Schlumberger Smart Cards, 36-38 rue de la Princesse,
BP 45, F-78430 Louveciennes Cedex, France, courtois@minrank.org

Abstract. Many popular stream ciphers apply a filter/combiner to the state of one or several LFSRs. Algebraic attacks on such ciphers [10,11] are possible, if there is a multivariate relation involving the key/state bits and the output bits. Recent papers by Courtois, Meier, Krause and Armknecht [1,2,10,11] show that such relations exist for several well known constructions of stream ciphers immune to all previously known attacks. In particular, they allow to break two ciphers using LFSRs and completely "well designed" Boolean functions: Toyocrypt and LILI-128, see [10,11]. Surprisingly, similar algebraic attacks exist also for the stateful combiner construction used in Bluetooth keystream generator E0 [1]. More generally, in [2] it is proven that they can break in polynomial time, any combiner with a fixed number of inputs and a fixed number of memory bits.

In this paper we present a method that allows to substantially reduce the complexity of all these attacks. We show that when the known keystream bits are consecutive, an important part of the equations will have a recursive structure, and this allows to partially replace the usual sub-cubic Gaussian algorithms for eliminating the monomials, by a much faster, essentially linear, version of the Berlekamp-Massey algorithm. The new method gives the fastest attack proposed so far for Toyocrypt, LILI-128 and the keystream generator that is used in E0 cipher. Moreover we present two new fast general algebraic attacks for stream ciphers using Boolean functions, applicable when the degree and/or the number of inputs is not too big.

Keywords: Algebraic attacks, stream ciphers, multivariate equations, nonlinear filters, Boolean functions, combiners with memory, LFSR synthesis, Berlekamp-Massey algorithm, Toyocrypt, Cryptrec, LILI-128, Nessie, E0, Bluetooth.

1 Introduction

In this paper we study stream ciphers with linear feedback. In such ciphers there is a linear component, and a stateful or stateless nonlinear combiner that produces the output, given the state of the first part. For stateless combiners – using a Boolean function, most of the general attacks known are correlation attacks, see for example [20,16,15,9]. In order to resist such attacks, many authors focused

D. Boneh (Ed.): CRYPTO 2003, LNCS 2729, pp. 176–194, 2003.

on proposing Boolean functions that will have no good linear approximation and that will be correlation immune with regard to a subset of several input bits, see for example [9]. Unfortunately there is a tradeoff between these two properties. One of the proposed remedies is to use a stateful combiner, as for example in the Bluetooth cipher E0 [6].

Recently the scope of application of the correlation attacks have been extended to consider higher degree correlation attacks with respect to non-linear low degree multivariate functions, or in other words, allowing to exploit low degree approximations [10]. The paper [10], proposes a novel algebraic approach to the cryptanalysis of stream ciphers. It will reduce the problem of key recovery, to solving an overdefined system of algebraic equations (i.e. many equations). Following [10] and [11], all stream ciphers with linear feedback are potentially vulnerable to algebraic attacks. If for one state we are able, by some method, to deduce from the output bits, only one multivariate equation of low degree in the state bits, then the same can (probably) be done for many other states. Each equation remains also linear with respect to any other state, and given many keystream bits, we inevitably obtain a very overdefined system of equations. Then we may apply the XL algorithm from Eurocrypt 2000 [25], adapted for this purpose in [10], or the simple linearization as in [11], to efficiently solve the system.

In the paper [11], the scope of algebraic attacks is substantially extended, by showing new non-trivial methods to obtain low degree equations, that are not low degree approximations. The method to reduce the degree of the equations, is analogous to the method proposed by Courtois and Pieprzyk to attack some block ciphers [13], and the basic idea goes back to [12] and [23]. Instead of considering outputs as functions of inputs, one should rather study multivariate relations between the input and output bits. They turn out to have a substantially lower degree. In this paper we take this idea further, and consider more general equations that include potentially many output bits, instead of one (or few) considered in [11,1]. Then we propose a new fast method to find and exploit such equations, mainly due to an application of a non-trivial asymptotically fast algorithm. This method allows to obtain equations that cannot be obtained by any other known method (due to their size), and to get much faster algebraic attacks.

The paper is organized as follows: in Section 2 we give a general view of algebraic attacks on stream ciphers. In Sections 2.2, 2.4 and 3 we discuss in details the types of equations that will be used in our attacks. In Section 4 we design our pre-computation attack and in Section 5 we show how to speed-up this attack. Then in Section 6 we apply it to Toyocrypt, LILI-128 and E0. In Section 7.1 we present a new general fast attack on ciphers with Boolean functions of low degree. Finally in Section 7.2 we apply our fast method to substantially speed up the general attack on stream ciphers using (arbitrary) Boolean functions with a small number of inputs from [11].

2   Algebraic Attacks on Stream Ciphers

For simplicity we restrict to binary stream ciphers defined over $GF(2)$. We consider only synchronous stream ciphers [21], in which there is a state $s \in GF(2)^n$. At each clock t the state s is updated by a "connection function" $s \mapsto L(s)$ that is assumed to be linear over $GF(2)$. Then a combiner f is applied to s, to produce the output bit $b_t = f(s)$. In principle also, we consider only regularly clocked stream ciphers. However this condition can sometimes be relaxed, see the attacks on LILI-128 described in [11]. Then the successive keystream bits $b_0, b_1, \ldots$ are XORed with the plaintext.

We assume that L and f are public, and only the state is secret. In many cases we assume that f is stateless, i.e. a Boolean function. Then our description covers "filter generators", in which the state of a single LFSR is transformed by a Boolean function, and also not less popular "nonlinear function generators", in which outputs of several LFSRs are combined by a Boolean function [21], and also many other known constructions. We will also apply our fast algebraic attacks to the case when the combiner f is not a Boolean function, and contains memory bits, as for example in E0. Then if if the number of memory bits is not too big, efficient algebraic attacks will still exist.

2.1   The General Framework of the Attack

The problem we want to solve is the following: find the initial state given some keystream bits. In principle, it is a known plaintext attack. However in some cases, ciphertext-only attacks are possible: For example if the plaintext is written in the Latin alphabet, and does not use too many special characters, we expect all the bytes to have their most significant bit equal to 0. Then all our attacks will work, only multiplying the amount of ciphertext needed by 8. Indeed this is equivalent to knowing all consecutive keystream bits for the same cipher, in which we replaced L by its eight-fold composition (L^8).

The goal of our attacks is to recover the initial state $(k_0, \ldots, k_{n-1})$ from some m consecutive keystream bits $b_0 \ldots b_{m-1}$, by solving multivariate equations. (Unlike in [11], our fast algebraic attacks do require consecutive bits.)

There are many different closely related approaches to algebraic attacks on stream ciphers and they differ mainly by the types of equations they use. This usually determines the methods used to generate, and to solve these equations.

Attacks Based on Direct Equations Following [10,11]

This approach applies only for stateless combiners. The problem of cryptanalysis of such a stream cipher is described as follows in [11].

Let $(k_0, \ldots, k_{n-1})$ be the initial state. Then the output of the cipher (i.e. the keystream) gives the following system of equations:

$$\begin{cases} b_0 = f \quad (k_0, \ldots, k_{n-1}) \\ b_1 = f\left(L\left(k_0, \ldots, k_{n-1}\right)\right) \\ b_2 = f\left(L^2(k_0, \ldots, k_{n-1})\right) \\ \quad \vdots \end{cases}$$

In [10] such equations are solved directly. In [11] they are first multiplied by well chosen multivariate polynomials, which decreases their degree, and then solved. In this paper we describe a more advanced method of generating (better) derived equations.

2.2   More General Attacks Using "ad hoc" Equations

Following [11], multivariate equations/relations that relate (only) key bits and output bits, may exist, for very different reasons, for (potentially) any cipher:

$$\begin{cases} 0 = \alpha + \sum \beta_i k_i + \sum \gamma_i b_i + \sum \delta_{ij} k_i b_j + \sum \epsilon_{ijk} k_i b_j b_k + \\ \quad \vdots \end{cases}$$

In this paper we restrict ourselves to equations that are true with probability 1.
Types of Equations: Following the notations introduced in Section 5.1. of [12], the equations given above may be called of type $1 \cup k \cup b \cup kb \cup kb^2$. This notation is obvious to understand, for example 1 denotes the presence of a constant monomial and kb^2 of all monomials of type $k_i b_j b_k$. Following another convention used in [12], we use capital letters to denote types of equations that also contain lower degree monomials, for example $K^2 = k^2 \cup k \cup 1$ and $B = b \cup 1$. Therefore the equations given above can also be said of type $KB \cup kb^2$ but they are not of type KB^2 because they do not contain monomials in $b_i b_j$ (a.k.a. monomials of type b^2).
Using ad hoc Such Equations: Our attacks will proceed in three steps:

- **Step A1.** Pre-computation stage: Find these equations.
- **Step A2.** Given some keystream, substitute the b_i in the equations to get an overdefined system of multivariate equations in the k_i.
- **Step A3.** Solve this (very overdefined) system of equations. Given sufficiently many keystream bits, we apply the simple linearization technique [25,11], which consists of adding one new variable for each monomial that appears in the system, and then solves a big linear system. If less keystream is available, one should use a version of the XL algorithm [25] for equations of any small degree over $GF(2)$, described and studied in [10].

2.3   Remarks on ad hoc Equations

The idea of ad-hoc equations follows closely the idea called scenario S5 in Section 7 of [11]. The word emphasises their unexpected character: they may be found be clever elimination by hand as in [1,2] or by an indirect unexpectedly fast method as in the present paper. The idea itself can be seen as a higher-degree generalisation of the concept of "augmented function" proposed by Anderson in [3] and recently exploited in [18]. It is also, yet another application in cryptography of looking for multivariate relations of low degree, main method for attacking numerous multivariate asymmetric schemes [12,23], and recently proposed also for attacking block ciphers [13].

Obviously, if such equations involving, say, T monomials exist, they can be found in time T^ω, with $\omega < 3$ being the exponent of the Gaussian reduction. The important contribution of this paper is to show that in some cases such equations can be found in a time linear or even sub-linear[1] in T.

A different view of "ad-hoc" equations is adopted in [2]. Instead of considering the equations with no limitation in the degree in the b_i, and study their degree when we substitute for b_i their respective values, it is possible to look at the equations in the state/key bits that are true each when several consecutive outputs are fixed to some fixed values. This approach used in [2] allows to study equations that contain much less monomials, which allows to find them faster, yet cannot be applied to systems in which the number of outputs is very big, as in the present paper.

Another way of looking at "ad-hoc" equations will be to consider that the stream cipher having one output, is in fact using several output functions $f_i \stackrel{def}{=} f \circ L^i$. Then "ad-hoc" equations boil down to look for algebraic combinations of type $\sum_i f_i g_i$ that would be of unusually low degree, exactly as in Section 2 of [12]. Then, if the g_i are also of low degree, the attack will exploit, for $j = 0, 1, 2, \ldots$, the following equation of low degree:

$$\sum_i f_i(L^j(s)) \cdot g_i(s) = \sum_i b_{i+j} \cdot g_i(s).$$

This approach is much more powerful than the S3 attack scenario [11]. Such equations may (and will) exist, without a low degree product fg to exist. They may (and do) exist even for stateful combiners f [1,2].

2.4   Equations Used in the S3 Attack [11] and Additional Properties

Let f be a Boolean function. We assume that the multivariate polynomial f has some multiple fg of low degree, with g being some non-zero multivariate polynomial. Let $deg(fg) = d$ and let d be not too big, then following [11], efficient attacks exist. For each known keystream bit at position t, we obtain a concrete value of $b_t = f(s)$ and this gives the following equation:

$$f(s) \cdot g(s) = b_t \cdot g(s),$$

which, for the current $s = L^t(k_0, \ldots, k_{n-1})$, rewrites as:

$$f\left(L^t(k_0, \ldots, k_{n-1})\right) \cdot g\left(L^t(k_0, \ldots, k_{n-1})\right) = b_t \cdot g\left(L^t(k_0, \ldots, k_{n-1})\right).$$

---

[1] It will be essentially linear in the number of equations involved in elimination, denoted later by L. This, assuming that the equations are written in a compressed form, can indeed be sub-linear in their size T.

If the degree of g is also $\leq d$, this equation can be used for any value of b_t, otherwise it can still be used when $b_t = 0$, i.e. for half of the time. We get one multivariate equation for each one or about two keystream bits. Each of these equations will be of the same degree and we inevitably obtain a very overdefined system of equations (number of equations $\gg n$), that can be efficiently solved, see [10,11,25].

Important Remark: Following [11], the chief advantage of the equations explained above may be of very low degree, without f being of low degree. In this paper we explore an additional property of these equations. Let e the degree of g. In [11], no distinction is made[2] between d and e. In this paper, on the contrary, we will exploit equations in which $e < d$, and show that for the same n and d, a smaller e leads allows to compute another type of "ad-hoc" equations, that will only be of degree e in the k_i, leading to much faster attacks than in [11].

More generally, in the next section we define a subclass of "ad-hoc" equations (including the example above with $e < d$) that will be subsequently used in the next sections to compute efficiently other (much better) "ad-hoc" equations.

3   The Double-Decker Equations

Definition 3.0.1 (Double-Decker Equations). For any $e < d$, we call "double-decker equations" with degrees (d, e, f), any set of multivariate equations of type $K^d \cup K^e B^f$, with $d, e, f \in \mathbb{N}$. In other words, equations with a maximum degree d in the monomials containing only the k_i, and the maximum degree e in the k_i among the monomials being divisible by one of the b_j.

The Double-decker equations will be used to find even better "ad-hoc" equations of type $K^e B^f$, with $f \in \mathbb{N}$, and finally used in our attacks. These equations cannot be obtained directly due to their size, and will be found indirectly in two major steps that will be described later:

- **Step A1.1.** Find some "double-decker equations" with degrees (d, e, f).
- **Step A1.2.** Then eliminate (at least) all the monomials of types $k^{e+1} \ldots k^d$, leaving only monomials of type $K^e B^f$ with a maximum degree e in the k_i.

3.1   Step A1.1. – Find the Basic Double-Decker Equations

The exact method to find the "double-decker equations" we will use later varies from one cipher to another.

Toyocrypt: Toyocrypt is a stream cipher, submitted to the Japanese government call for cryptographic primitives Cryptrec and accepted to the second phase. An impractical attack on Toyocrypt has been proposed [22] and it has

---

[2] In the extended version of the paper [11], two versions of the main attack are studied, called S3a and S3b, that simply require that $e \leq d$. The complexity of the attack does not depend on e, only on $\binom{n}{d}$. For another proposed version S3c, successful attacks could be possible even when e is very big.

been rejected by Cryptrec. A description can be found in [22]. From Section 5.1. of [11] we know that there are 2 equations of type $f(s) \cdot g(s) = b_t \cdot g(s)$ in which $e = deg(g) = 1$ and $d = deg(fg) = 3$. For example, anyone can verify that when $g(s) = (s_{23} - 1)$, then in $f(s)g(s)$, all the terms of degrees ≥ 4 cancel out and what remains is of degree 3.

LILI-128: LILI-128 is a stream cipher, that was submitted to the European evaluation effort Nessie, and was subsequently rejected due to attacks [27]. A description can be found in [26]. From Section 5.1. of [11] we know that there are 4 equations of type $f(s) \cdot g(s) = b_t \cdot g(s)$ in which $e = deg(g) = 2$ and $d = deg(fg) = 4$. For example, when $g(s) = s_{44}s_{80}$, anyone can verify that $f(s)g(s)$ is equal to the following multivariate polynomial of degree 4:

$$f(s) \cdot s_{44}s_{80} = s_{44}s_{80} \left(s_1 s_{65} + s_3 s_{30} + s_7 s_{30} + s_{12} s_{65} + s_0 + s_7 + s_{12} + s_{20} \right).$$

E0: E0 is the keystream generator used in the Bluetooth wireless interface [6]. We will look for some equations of the following type:

$$h(s) = \sum_i b_i \cdot g_{i\,0}(s) + \sum_i b_i b_j \cdot g_{i\,j}(s),$$

in which $e = \max(deg(g_{i\,j})) < d = deg(h)$. They may be called of type $K^d \cup B^2 K^e$ following notation of Section 2.2 or [12]. Such an equation of type $K^4 \cup B^2 K^3$, combining only 4 successive states, and eliminating all the state bits has been found by careful study of the cipher and successive elimination done by hand in [1,2]. In this equation we have $d = 4$ and $e = 3$. Our simulations confirmed that this equation exists, and is always true. We also found that it was unique (when combining only 4 consecutive states).

3.2   Summary: The Equations That Will Be Used in Our Attacks

For all the three ciphers Toyocrypt, LILI-128 and E0 there are "double-decker equations" (i.e. multivariate equations of type $K^d \cup K^e B^f$), with the following degrees:

Table 1. The equations that will be used in our attacks

stream cipher	the degrees			equation type	number of equations	successive b_i per equation
	d	e	f			
Toyocrypt	3	1	1	$K^3 \cup bk^1$	2	1
LILI-128	4	2	1	$K^4 \cup bK^2$	4	1
E0	4	3	2	$K^4 \cup B^2 K^3$	1	4

Important Remark: Nothing proves that there are no better equations. For any pair (d, e), even very small, when the number of b_i used grows, the simulations are becoming quickly impractical and cannot detect all equations that would lead to a (fast) algebraic attack.

4   The Pre-computation Attack on Stream Ciphers

This idea has been suggested to us by Philip Hawkes. Starting from the above "double-decker equations" with degrees (d, e, f), we will be able to compute "ad-hoc" equations of degree only e, by eliminating the monomials of type K^d, as defined in Step A1.2. First, it will be done by Gaussian elimination, then by a much faster method.

4.1   The General (Slow) Pre-computation Algebraic Attack

We assume that for a given stream cipher, we have a system of "double-decker equations" being of type $K^d \cup K^e B^f$, with f being small, for example in all our attacks $f \leq 2$. We also assume that the size of these initial equations is a small constant $\mathcal{O}(1)$ (for example, they are sparse or/and use a small subset of state bits). We will write the equations in the following form

$$\mathrm{Left}(L^t(k)) = \mathrm{Right}(L^t(k), b),$$

and we will put all the monomials of type K^d on the left side, and all the other monomials of type $K^e B^f$ on the right side. For monomials of type K^e we may place them indifferently on one or the other side. Then we do the following:

1. We assume that we have at least $R = \binom{n}{d} + \binom{n}{e} \approx \binom{n}{d}$ equations, this can be achieved given about at most[3] about $\binom{n}{d}$ keystream bits.
2. We will do a complete Gaussian elimination of all monomials that appear on the left sides, i.e. of all monomials of degree up to d in the k_i. This is possible because R is chosen to exceed the number of monomials. This gives at least one linear combination α of the left-hand sides that is 0:

$$0 = \sum_t \alpha_t \cdot \mathrm{Left}(L^t(k)).$$

   Since $R = \binom{n}{d} + \binom{n}{e}$, we are able to produce at least $\binom{n}{e}$ such linearly independent linear combinations. Each of them involves about $\binom{n}{d}$ left sides.
3. The complexity of this first step is about $\binom{n}{d}^\omega$ to find several solutions α (we need about $\binom{n}{e}$ solutions, and $\binom{n}{e} \ll \binom{n}{d}$).
4. We apply these linear combinations to the right sides. We get a system of equations of type $B^f K^e$.

$$0 = \sum_t \alpha_t \cdot \mathrm{Right}(L^t(k)).$$

   For example, for Toyocrypt or LILI-128, we get:

$$0 = \sum_t \alpha_t \cdot b_t \cdot g\left(L^t(k)\right).$$

---

[3] As a matter of fact, here for many ciphers there will be several, say M, equations that exist for each keystream bit. Then given about $\binom{n}{d}/M$ keystream bits, we still get about $\binom{n}{d}$ equations as required. Unfortunately, for the fast attack described in Section 5 below, we only know how to use one of them and will assume $M = 1$.

Similarly for E0 we get equations of the type $B^2 K^2$ using up to $\binom{n}{4}$ consecutive b_i.

5. Now we obtain a pre-computed information (a sort of trapdoor) that consists of $\binom{n}{e}$ equations. Each of these equations is of size $\mathcal{O}(\binom{n}{d})$, this is because we assume that the size of the initial equations is a small constant $\mathcal{O}(1)$. This trapdoor information allows, given a sequence b_i at the previously determined positions t that were in the sequence, to compute the secret key k by solving a system of equations of degree e instead of d.

6. Given a sequence b_t, we substitute it to all the equations. This step about takes $\mathcal{O}(\binom{n}{d} \cdot \binom{n}{e})$ computations.

7. Then we have a system of $\binom{n}{e}$ equations of degree e, that can be solved by linearization.

8. The complexity of the last linearization step is about $\binom{n}{e}^{\omega}$.

Remark: If all the equations are not linearly independent, the attack still works. Simulations done in the extended version of [11] suggest that the number of linearly dependent equations should always be negligible. Moreover, it is possible to see that even if we had to produce, say 10 times more equations, it would not greatly increase the complexity of the attack.

4.2   Summary: The General (Slow) Pre-computation Attack

To summarize we have a pre-computation attack that, given $\binom{n}{d}$ initial equations, allows to compute a trapdoor information, that later for any values of b_t obtained from the cipher, will allow to compute the key using only roughly $\binom{n}{d} \cdot \binom{n}{e} + \binom{n}{e}^{\omega}$ operations. For now, when applied with equations from Section 3.2, it does not improve on the attacks from [11] and [1,2], except that with the same complexity we may do a pre-computation once, and then break the cipher again and again with a much lower complexity.

5   The Fast Pre-computation Attack

In this (fast) attack there are three additional requirements (or limitations):

1. The equation used has to be exactly the same for each keystream bit, modulo a variable change resulting from the fact that $s = L^t$. For example in the previous attack we could use, for each keystream bit b_t, one or several equations of type $f(L^t(k)) \cdot g(L^t(k)) = b_t \cdot g(L^t(k))$ with different linearly independent functions g. In this attack we may only use one g that has always to be the same.

2. The slow pre-computation attack given above, as all attacks given in [11], will work given any subset of keystream bits. The improved (fast) attack will require consecutive keystream bits. (The attack will also work if they are taken at regular intervals. This amounts simply to taking a different linear feedback function L and without loss of generality we will assume that all the keystream bits are consecutive.)

3. Moreover we assume that the linear feedback L is non-singular, in a sense that the sequence $k, L(k), L^2(k), \ldots$ is always periodic, and we will also assume that this sequence has only one single cycle. This is true for all known stream ciphers with linear feedback, in particular when they are based on one or several maximum-length LFSRs.

When all the equations are of the form Equation$(L^t(k))$ and if all the keystream bits are consecutive, then the system has a very regular recursive structure:

$$\begin{cases} \text{Left } (k_0, \ldots, k_{n-1}) = \text{Right } (k_0, \ldots, k_{n-1}; b_0 \ldots b_{m-1}) \\ \quad \vdots \\ \text{Left } (L^t(k_0, \ldots, k_{n-1})) = \text{Right } (L^t(k_0, \ldots, k_{n-1}); b_t \ldots b_{m+t-1}) \\ \quad \vdots \end{cases}$$

Now we are going (for some time) to ignore the right sides of these equations. Their left sides do not depend on the b_i, they are just multivariate polynomials of type K^d. If we consider at least $\binom{n}{d}$ consecutive equations, a linear dependency α must exist. This linear dependency does not depend on the outputs b_i. Let S be the size of the smallest such linear dependency α. We have $S \leq \binom{n}{d}$.

One dependency is enough. Due to the recursive structure of the equations, this dependency α can be applied at any place. Indeed we have:

$$\forall k \quad 0 = \sum_{t=0}^{S-1} \alpha_t \cdot \text{Left } \left(L^t(k) \right) \quad \Rightarrow \quad \forall k \; \forall i \quad 0 = \sum_{t=0}^{S-1} \alpha_t \cdot \text{Left } \left(S^{t+i}(k) \right).$$

Moreover, since we assumed that the sequence $k, L(k), L^2(k), \ldots$ has one single cycle, this dependency α does not depend on the secret key of the cipher, and is the same for all k.

In the previous (slow) pre-computation attack we had to find $\binom{n}{e}$ linear dependencies α. Here we find one α and re-use it on $\binom{n}{e} \ll S$ successive windows of $S \leq \binom{n}{d}$ equations.

5.1   Finding α Faster – LFSR Synthesis

We have shown a (well known) fact, that for every initial k, the values of the left sides of our equations can be obtained from some LFSR of length at most S and defined by α. We can also do the reverse: recover α from the sequence (LFSR synthesis). It can be done given $2S$ bits of the sequence, see [19,21]. For this we choose a random key k', (α does not depend on k), and we will compute $2S$ output bits of this LFSR $c_t = \text{Left } (L^t(k'))$ for $t = 0 \ldots 2S - 1$. Then we apply the well known Berlekamp-Massey algorithm [19,21] to find the connection polynomial of this LFSR that will be essentially α. Done in this way, both these steps (computing the sequence c_i and the LFSR synthesis) will take

$\mathcal{O}(S^2)$ operations.[4] However both can be improved to become essentially linear in S.

Finding α Faster:

◇ First of all we observe that all the $L^i(k')$ can be computed in time about $\mathcal{O}(Sn^2)$ and even in $\mathcal{O}(Sn)$ if L is composed only of a combination of LF-SRs and MLFSRs. Then, since $n \ll S \approx \binom{n}{d}$, the time is indeed essentially linear in S. Moreover, in many interesting cases in practice, the equations Left $(S^t(k'))$ are sparse or/and have a known structure that allows to compute them very fast, in a time that can be assumed constant, being less than n and therefore much smaller than S. It is the case for Toyocrypt, LILI-128 and E0. We exploit here the fact that these ciphers have been designed to be very fast.

◇ To recover α takes $\mathcal{O}(S^2)$ computations using Berlekamp-Massey Algorithm, but it will take only $\mathcal{O}(S \log(S))$ operations using improved asymptotically fast versions of the Berlekamp-Massey Algorithm, see [5,14,7]. However we do not know how fast are these algorithms for the concrete values of S used in this paper.

How to Use α: We recall that the same linear dependency will be used $\binom{n}{e}$ times:

$$\forall i \quad 0 = \sum_{t=i}^{S+i-1} \alpha_t \cdot \text{Right} \left(S^t(k); b_t \ldots b_{m+t-1} \right).$$

Since $Right(t)$ does depend on many b_i, in general a sliding subset, and the output sequence b_i is not likely to have any periodic structure, all these equations are expected to be very different for different i. From simulations on a simpler but similar algebraic attack on Toyocrypt done in [11], we also expect that only a negligible number of these equations will be redundant (i.e. linearly dependent).

5.2   Summary – The Fast Pre-computation Attack

1. Given about $m = \binom{n}{d} + \binom{n}{e} \approx \binom{n}{d}$ consecutive keystream bits.
2. We compute the linear dependency α using an improved version of the Berlekamp-Massey Algorithm. This step is expected to be essentially linear in S and take at most $\mathcal{O}(S \log(S) + Sn)$ steps with $S = \binom{n}{d}$. This α is our pre-computed information (or the trapdoor). It allows, given a sequence b_i of consecutive keystream bits, to recover the secret key k by solving a system of equations of degree only $e < d$.
3. The second step takes (as before) about $\mathcal{O}(\binom{n}{d} \cdot \binom{n}{e}) + \binom{n}{e}^\omega$ operations.

Memory Requirements: In this attack, the most memory-consuming operation will be to store the $\binom{n}{e}$ equations of size at most $\binom{n}{e}$ (we only store them after substituting the b_i by the output bits). It will be at most $\binom{n}{e}^2$ bits.

---

[4] A quadratic time is already much faster than doing this by Gaussian elimination in $\mathcal{O}(S^\omega)$, and only this would already give the fastest attack known on all the three stream ciphers studied in this paper.

6 Application to Toyocrypt, LILI-128, and E0

In this paper, write $\mathcal{O}(2^{20})$ to say that the complexity is at most $C \cdot 2^{20}$ for some constant C, yet for simplicity we did not evaluate the exact value of C. All our results given in such form should be regarded as rough (and optimistic) approximations.

6.1 Application to Toyocrypt

For Toyocrypt we have $n = 128$, $d = 3$, $e = 1$. We obtain the following attack:
◇ With a pre-computation step in $\mathcal{O}(2^{23})$.
◇ Given $2^{18.4}$ consecutive keystream bits.
◇ The secret key can be computed in about $\mathcal{O}(2^{20})$ CPU clocks and with about 2^{14} bits of memory.

6.2 Application to LILI-128

In the second component of LILI-128 we have $n = 89$, $d = 4$, $e = 2$. The sequence will only become regularly clocked if we clock the first LFSR of LILI-128 $2^{39} - 1$ steps at a time, see [11]. We get the following attack:
◇ With a pre-computation step in $\mathcal{O}(2^{26})$.
◇ Given $2^{21.3+39} \approx 2^{60}$ consecutive keystream bits.
◇ The state of the second component can be computed in about $\mathcal{O}(2^{31})$ CPU clocks and with about 2^{24} bits of memory.
◇ Once the initial state of the second LFSR is recovered, the state of the first LFSR can be found easily in less than about 2^{20} (many thanks to Philip Hawkes for remarking this). Indeed, once the second LFSR is known, we can predict any number of consecutive bits a_i of the second component (without decimation), then we may guess several (for example 20) consecutive output bits c_i of the first component, which determines a subsequence of the a_i that should be equal to the observed output sequence fragment b_i. Our choice will be confirmed only for on average 1 or 2 strings of 20 bits of c_i. For each of these (very few) cases, we will find the remaining $39 - 20 = 19$ bits by the exhaustive search.

6.3 Application to E0

For E0 we have $n = 128$, $d = 4$, $e = 3$. Then we obtain:
◇ With a pre-computation step in $\mathcal{O}(2^{28})$.
◇ Given $2^{23.4}$ consecutive keystream bits.
◇ The secret key can be computed in about $\mathcal{O}(2^{49})$ CPU clocks and with about 2^{37} bits of memory.

Note: In the real-life implementation of Bluetooth cipher, at most about $2745 \approx 2^{11}$ bits can be obtained, see [17,6]. However this attack shows that the design of E0 is not (cryptographically) very good. It is possible that even a real-life application of E0 will be broken by our attack, if some other equations with a smaller d are found. This possibility cannot at all be excluded, the computer simulations only allow to explore equations that combine a few (e.g. up to 10) output bits. Better equations may exist when more bits are combined, and may be found by a clever elimination as in [1,2].

6.4 Summary of the Results

The fast algebraic attack gives the best attack known on three well known stream ciphers:

Table 2. The results of this paper vs. the best previous attacks

cryptosystem	Toyocrypt			LILI-128				E0	
n	128	128	128	89	89	89	89	128	128
d		3	3	4	4		4	4	4
e			1				2		2
Attack	[22]	[11]	new	[11]	[11]	[27]	new	[1,2]	new
Data	2^{48}	2^{18}	2^{18}	2^{18}	2^{57}	2^{46}	2^{60}	2^{24}	2^{24}
Memory	2^{48}	2^{37}	2^{14}	2^{43}	2^{43}	2^{51}	2^{24}	2^{48}	2^{37}
Pre-computation	2^{80}		$O(2^{23})$				2^{56}	$O(2^{26})$		$O(2^{28})$
Attack Cxty	2^{64}	2^{49}	$O(2^{20})$	2^{96}	2^{57}	2^{56}	$O(2^{31})$	2^{68}	$O(2^{49})$

7 Fast General Attacks on Stream Ciphers Using Boolean Functions

In [11] it is shown that for any cipher with linear feedback and a non-linear stateless filtering function that uses only a small subset of state bits, for example k bits out of n state bits, the key can be recovered in essentially $\binom{n}{k/2}^{\omega}$ operations, which at most about $n^{k\frac{\omega}{2}} \leq n^{1.19k}$ using the fairly theoretical result from [8]. In practice, if we consider the Strassen's algorithm, we get rather $n^{1.4k}$. This attack works given any subset of $\binom{n}{k/2}$ keystream bits.

In this paper we show that, given less than $\binom{n}{k}$ keystream bits, that (now) have to be consecutive, we may recover the key using only essentially n^k computations instead of $n^{1.4k}$. In the two following subsections we will show two separate attacks of this type that both will be better than $n^{k\omega/2}$.

The first attack assumes that the Boolean function of the cipher is constructed in such a way that it can be computed in constant time (for example using a table, the usual case, otherwise the cipher is not practical !). Given $\binom{n}{d}$ consecutive keystream bits, the key is recovered in $n^{d+\mathcal{O}(1)}$ computations, with $d \leq k$ being the degree of the Boolean function used. This attack can be sometimes much faster than n^k, as frequently we have $d < k$.

We also present a fast version of the general attack in $n^{k\frac{\omega}{2}}$ from [11], that for ciphers using only a small subset of output bits k, (k is a small constant, e.g. $k = 10$), will require (substantially) less than $\binom{n}{k}$ consecutive keystream bits, and about $n^{k+\mathcal{O}(1)}$ computations.

7.1   Fast General Attack on Stream Ciphers Using Boolean Functions of Small Degree

Theorem 7.1.1 (New General Attack on Stream Ciphers with Linear Feedback). If the degree of the Boolean function f is d, and its output can be computed in time $\mathcal{O}(1)$, then given $\binom{n}{k}$ consecutive keystream bits, one can recover the key using only $\mathcal{O}(n^{d+2})$ operations.

All this is achieved immediately by a straightforward application of our fast algebraic attack. The cipher is first described by the following set of equations:

$$f\left(L^t(k_0, \ldots, k_{n-1})\right) = b_t.$$

We will split this equation into two $0 = f(s) + b_t = \text{Left}(s) + \text{Right}(s, b_t)$ with:

- $\text{Right}(s, b_t)$ containing b_t and the linear part of f.
- $\text{Left}(s)$ being exactly the part of degrees $2 \ldots d$ of f.

It is easy to see that, from the assumption that the computation of f is done in constant time, the time to compute $Left()$ will be in $\mathcal{O}(n)$, knowing that it differs from f only by the linear part. Then we get the following set of equations for some m consecutive bits $t = 0 \ldots m - 1$:

$$\text{Left}\left(L^t(k_0, \ldots, k_{n-1})\right) = \text{Right}\left(L^t(k_0, \ldots, k_{n-1}),\ b_t\right), \qquad t = 0, 1, 2, \ldots$$

with the left sides being of type K^d and right sizes of type $K \cup b$, and using only one b_i each. We have "double-decker equations" with degrees $(d, e, f) = (d, 1, 1)$.

The LFSR Synthesis. Let $S = \binom{n}{d}$. Let k' be a random key. From Section 4, we recall that the time to compute all the $L^i(k'), i = 0 \ldots 2S - 1$ is at most $\mathcal{O}(Sn^2)$, and in many practical cases even $\mathcal{O}(Sn)$.

Since the time to compute $Left(s)$ is $\mathcal{O}(n)$ computing $2S$ outputs $Left(L^t(k'))$ for $t = 0 \ldots 2S - 1$ will take time $\mathcal{O}(Sn)$. Then as in Section 5 we compute the LFSR connection polynomial α which takes about $\mathcal{O}(S \log S)$ operations.

Using the Pre-computed Information α. Again, as in Section 5, with a pre-computation in time at most $\mathcal{O}(Sn^2 + S \log S)$ we get an equation in which a linear combination of $S = \binom{n}{d}$ successive keystream bits b_i is equal to a linear combination of the k_i:

$$\forall t \sum_{i=0}^{S-1} \alpha_{t+i} b_{t+i} = \text{ResultingLinearCombination}(k_0, \ldots, k_{n-1}).$$

By using this equation n times for n successive windows of S keystream bits, we get a linear system that will give the key k. The complexity of the whole fast pre-computation attack is at most about $\mathcal{O}(Sn^2)$. In practice, it be we frequently

Table 3. Fast algebraic attack when f has degree d

Consecutive Data	$\binom{n}{d}$
Memory	$\mathcal{O}(n^{d+1})$
Pre-computation	$\mathcal{O}(n^{d+2})$
Complexity	$\mathcal{O}(n^{d+1})$

at most $\mathcal{O}(Sn)$ and at any rate this time can be thought as essentially linear in S, since $n \ll S$. We get the following attack:

Later, in Table 5, we will compare this attack to the attack from the next section, and other previously known general attacks.

Important Remark. This is a very simple linear attack on a stream cipher. Such equations of size $S = \binom{n}{d}$ do always exist. Let LC of the linear complexity of a cipher. We always have $LC \le \binom{n}{d}$ and when $S < LC$, the existence of such an equation of size S can be excluded. Conversely, if the linear complexity LC is less than expected $(LC < \binom{n}{d})$, the attack described here will still work with a complexity essentially linear in LC. This attack shows that, ciphers in which the Boolean functions f is of low degree and can be computed in constant time, can be broken in time essentially linear in the linear complexity LC of the cipher. This is, to the best of our knowledge, has never happened before.

7.2 Fast General Attack on Stream Ciphers Using a Subset of LFSR Bits

This section improves the general attack from [11]. We consider a stream cipher with n state bits, in which the keystream bit is derived by a Boolean function f using only a small subset of k state bits: $\{x_1, x_2, \ldots, x_k\} \subset \{s_0, s_1, \ldots, s_{n-1}\}$. We assume that k is a small constant and n is the security parameter. For example in LILI-128 $k = 10, n = 89$.

As in [11] we are looking for polynomials $g \ne 0$, such that fg is of low degree, and for this we check for linear dependencies in the set of polynomials $C = A \cup B$ defined as follows. A contains all possible monomials up to some maximum degree d (this part will later compose fg).

$$A = \{1, x_1, x_2, \ldots, x_1 x_2, \ldots\}.$$

Then we consider all multiples of f, multiplied by monomials of the degree up to e (this degree corresponds to the degree of g):

$$B = \{f(x), f(x) \cdot x_1, f(x) \cdot x_2, \ldots, f(x) \cdot x_1 x_2, \ldots\}.$$

Let $C = A \cup B$. All elements of A, B and C, can be seen as multivariate polynomials in the x_i: for this we need to substitute f with its expression in the x_i. A set of multivariate polynomials with k variables cannot have a dimension greater than 2^k. If there are more than 2^k elements in our set, linear dependencies will exist. Such linear combinations allow to find a function g of degree $\leq e$ such that $f \cdot g$ is of degree $\leq d$. More precisely, we will prove the following theorem that generalizes the Theorem given in [11]:

Theorem 7.2.1 (Tradeoff between the Degrees of fg and g).
Let f be any Boolean function $f : GF(2)^k \to GF(2)$. For any pair of integers (d, e) such that $d + e \geq k$ there is a Boolean function $g \neq 0$ of degree at most e such that: $f(x) \cdot g(x)$ is of degree at most d.

Proof: We have

$$|A| = \sum_{i=0}^{d} \binom{k}{i} \quad \text{and} \quad |B| = \sum_{i=0}^{e} \binom{k}{i}.$$

$$|C] = \sum_{i=0}^{d} \binom{k}{i} + \sum_{i=0}^{e} \binom{k}{i} = \sum_{i=0}^{d} \binom{k}{i} + \sum_{i=0}^{e} \binom{k}{k-i} > \sum_{i=0}^{k} \binom{k}{i} = 2^k.$$

Then, since the rank of $C = A \cup B$ cannot exceed 2^k, and $|C| > 2^k$, some linear dependencies must exist. Moreover, $g \neq 0$ because there are no linear dependencies in A, and therefore linear dependencies must combine either only the elements of B, or both A and B. This ends the proof. □

From this we see that for any pair of integers (d, e) such that $d + e \geq k$ and for any stream cipher with linear feedback, for which the non-linear filter uses k variables, it is possible to generate "double-decker equations" with degrees $(d, e, 1)$ in the n keystream bits.

With this assumption, computing Left(), and Right(), that by construction have at most 2^{k-1} monomials, will be very fast (using a table) and given 2^k bits of memory. this will be small, because we assumed that k is a small constant, for example for LILI-128 $k = 10$, $2^k = 1024$ is very small compared to the memory requirements of other parts of the attack.

Then the whole LFSR synthesis will take a time of at most $\mathcal{O}(Sn^2 + S \log S)$. By inspection we verify that our fast pre-computation attack gives roughly about:

Table 4. Fast algebraic attack when f has k inputs

$\forall(d, e)$ s.t. $d + e \geq k$	
Consecutive Data	$\binom{n}{d} + \binom{n}{e}$
Memory	$\mathcal{O}(2^k + \binom{n}{d}\binom{n}{e})$
Pre-computation	$\binom{n}{d}^{1+o(1)}$
Complexity	$\mathcal{O}(\binom{n}{d}\binom{n}{e}) + \binom{n}{e}^{\omega}$

Under the condition $d + e \geq k$, we may assume $d + e = k$, and we see that very roughly: $\mathcal{O}(\binom{n}{d}\binom{n}{e}) \approx n^d/d! \cdot n^e/e! \approx n^k \cdot \binom{k}{d}$. The complexity of this attack will never be essentially lower than n^k. However, to choose a bigger e will substantially decrease the amount of keystream needed for the attack. Many tradeoffs are possible, but two of them seem particularly interesting:

- It is easy to see that the maximum e that can be used in this attack without increasing the complexity, is such that $\binom{n}{e}^{\omega} \approx \mathcal{O}(\binom{n}{d}\binom{n}{e}) \approx n^k$. Thus we have $e \approx k(1/\omega)$ and then d is about $k(1 - 1/\omega)$. The amount of consecutive keystream required is then about $\binom{n}{d} = \binom{n}{k(1-1/\omega)}$.
- We may also try to achieve the smallest possible amount of keystream, which is achieved when $e = d \approx k/2$, with a slower attack. It can be seen that in this attack the fast pre-computation will not help because the final step will be slower than the pre-computation step. Moreover, sine $d = e$ the pre-computation step is not necessary at all to obtain equations of degree e, and this attack finally boils down to the general attack given in [11].

7.3   Summary – Fast General Attacks on Stream Ciphers Using Boolean Functions

In the following table we give a simplified analysis of the complexity of the attacks from Section 7.1 and 7.2 compared to previously known attacks.

Table 5. General attacks on LFSR-based stream ciphers using a Boolean function f

State bits	n	n	n	n
# inputs of f	k	k	k	k
d	k	$\lceil k/2 \rceil$	$d = deg(f) \leq k$	$k(1 - 1/\omega)$
e		$\lfloor k/2 \rfloor$	1	$k(1/\omega)$
Attack	[4,16]	[11]	Section 7.1	Section 7.2
Data	$\binom{n}{k}$	$\binom{n}{\lceil k/2 \rceil}$	$\binom{n}{d}$	$\binom{n}{k(1-1/\omega)}$
Consecutive	no	no	yes	yes
Memory	$\binom{n}{k}^2$	$\binom{n}{\lceil k/2 \rceil}^2$	$\mathcal{O}(n^{d+1})$	$\mathcal{O}(n^{k+1})$
Pre-computation			$\mathcal{O}(n^{d+2})$	$\mathcal{O}(n^{k+2})$
Attack Cxty	$\mathcal{O}(n^{k\omega})$	$\mathcal{O}(n^{k\omega/2})$	$\mathcal{O}(n^{d+1})$	$\mathcal{O}(n^{k+1})$

8   Conclusion

In this paper we have studied algebraic attacks on stream ciphers with linear feedback, using overdefined systems of algebraic multivariate equations over $GF(2)$. This gives many interesting attacks on both stateless combiners based on Boolean functions and also on combiners with memory. Using equations of a very special form, and by a novel application of known asymptotically fast results on LFSR reconstruction, we are able to propose the best attack known so far on three important stream ciphers. All these ciphers were believed quite secure up till very recently, now they can be broken quickly on a PC.

We also present two general algebraic attacks, for all regularly clocked ciphers with linear feedback (e.g. using LFSRs) filtered by a Boolean function. Both may be faster than all previously known attacks. In particular we show a surprising result to the effect that some ciphers may be broken in time essentially linear in their linear complexity.

There are good arguments to say that these attacks will work exactly as predicted, because one may generate as many equations as one wants, and if some are linearly dependent, one may generate more new equations. The computer simulations done in the extended version of [11] suggest that the number of equations that are linearly dependent in our attacks will be negligible.

Our attacks lead to, more or less the same design criterion for stream ciphers as proposed in [11]: the non-existence of multivariate relations of reasonable size and low degree linking the key bits and the output bits. This criterion turns out to be almost identical to the security criterion defined in Section 2 of [12] for multivariate trapdoor functions, and also to the requirements advocated in [13] for S-boxes of block ciphers. However, in this paper we show that some equations may be found and used in a time linear (or even sub-linear) in the size of the equations. Therefore such equations should be taken very seriously in the design of stream ciphers, and this even for systems of equations of sizes bigger than e.g. 2^{80}. Then, the existence of such big systems of equations cannot be a priori excluded by computer simulations, that would be too slow.

Therefore we do not recommend using stream ciphers with linear feedback with fairly simple clocking control. However, all algebraic attacks we are aware of, will not work, when the connections polynomials of the LFSRs are secret, or when the clocking is very complex and uses the full entropy of the key, and/or when the output sequence is decimated in a very complex way (as in shrinking generators).

Acknowledgments. The pre-computation attack on stream ciphers has been designed following an initial idea suggested by Philip Hawkes. I also thank Frederik Armknecht, Philip Hawkes, Willi Meier, Greg Rose, David Wagner, and above all, the anonymous referees of Crypto, for their very helpful comments.

References

1. Frederik Armknecht: *A Linearization Attack on the Bluetooth Key Stream Generator,* Available on http://eprint.iacr.org/2002/191/. 13 December 2002
2. Frederik Armknecht, Matthias Krause: *Algebraic Atacks on Combiners with Memory,* in these proceedings, Crypto 2003, LNCS 2729, Springer, 2003.
3. Ross Anderson: *Searching for the Optimum Correlation Attack,* FSE'94, LNCS 1008, pp 137–143, Springer. 1994.
4. Steve Babbage: *Cryptanalysis of LILI-128,* Nessie project internal report, https://www.cosic.esat.kuleuven.ac.be/nessie/reports/, 22 January 2001.
5. R. E. Blahut: *Theory and Practice of Error Control Codes,* Addison-Wesley, 1983.
6. Bluetooth CIG, Specification of the Bluetooth system, Version 1.1, February 22 2001, available from www.bluetooth.com.

7. R. P. Brent, F. G. Gustavson, D. Y. Y. Yun: *Fast solution of Toeplitz systems of equations and computation of Padé approximants.* J. Algorithms, 1:259–295, 1980.
8. Don Coppersmith, Shmuel Winograd: *Matrix multiplication via arithmetic progressions,* J. Symbolic Computation (1990), 9, pp. 251–280.
9. Paul Camion, Claude Carlet, Pascale Charpin and Nicolas Sendrier, *On Correlation-immune Functions,* Crypto'91, LNCS 576, pp. 86–100, Springer, 1991.
10. Nicolas Courtois: *Higher Order Correlation Attacks, XL algorithm and Cryptanalysis of Toyocrypt,* ICISC 2002, LNCS 2587, Springer. An updated version (2002) is available at http://eprint.iacr.org/2002/087/.
11. Nicolas Courtois and Willi Meier: *Algebraic Attacks on Stream Ciphers with Linear Feedback,* Eurocrypt 2003, Warsaw, Poland, LNCS 2656, pp. 345–359, Springer. An extended version is available at http://www.minrank.org/toyolili.pdf
12. Nicolas Courtois: *The security of Hidden Field Equations (HFE),* Cryptographers' Track Rsa Conference 2001, San Francisco 8-12 April 2001, LNCS 2020, Springer, pp. 266–281, April 2001.
13. Nicolas Courtois and Josef Pieprzyk, *Cryptanalysis of Block Ciphers with Overdefined Systems of Equations,* Asiacrypt 2002, LNCS 2501, Springer, a preprint with a different version of the attack is available at http://eprint.iacr.org/2002/044/.
14. Jean-Louis Dornstetter: *On the Equivalence Between Berlekamp's and Euclid's Algorithms.* IEEE Trans. on Information Theory. IT-33(3): 428–431. May 1987.
15. Eric Filiol: *Decimation Attack of Stream Ciphers,* Indocrypt 2000, LNCS 1977, pp. 31–42, 2000. Available on eprint.iacr.org/2000/040.
16. Jovan Dj. Golic: *On the Security of Nonlinear Filter Generators,* FSE'96, LNCS 1039, pp. 173–188, Springer, 1996.
17. Jovan Dj. Golic, Vittorio Bagini, Guglielmo Morgari: *Linear Cryptanalysis of Bluetooth Stream Cipher,* Eurocrypt 2002, LNCS 2332, pp. 238–255, Springer, 2002.
18. Bernhard Löhlein: *Attacks based on Conditional Correlations against the Nonlinear Filter Generator,* Available at http://eprint.iacr.org/2003/020/,
19. J. L. Massey: *Shift-register synthesis and BCH decoding,* IEEE Trans. Information Theory, IT-15 (1969), 122–127.
20. Willi Meier and Othmar Staffelbach: *Fast correlation attacks on certain stream ciphers,* Journal of Cryptology, 1(3):159–176, 1989.
21. Alfred J. Menezes, Paul C. van Oorschot, Scott A. Vanstone: *Handbook of Applied Cryptography,* Chapter 6, CRC Press.
22. M. Mihaljevic, H. Imai: *Cryptanalysis of Toyocrypt-HS1 stream cipher,* IEICE Transactions on Fundamentals, vol. E85-A, pp. 66–73, Jan. 2002. Available at http://www.csl.sony.co.jp/ATL/papers/IEICEjan02.pdf.
23. Jacques Patarin: *Cryptanalysis of the Matsumoto and Imai Public Key Scheme of Eurocrypt'88,* Crypto'95, Springer, LNCS 963, pp. 248–261, 1995.
24. Rainer A. Rueppel: *Analysis and Design of Stream Ciphers,* Springer, 1986.
25. Adi Shamir, Jacques Patarin, Nicolas Courtois, Alexander Klimov, *Efficient Algorithms for solving Overdefined Systems of Multivariate Polynomial Equations,* Eurocrypt'2000, LNCS 1807, pp. 392–407, Springer, 2000.
26. L. Simpson, E. Dawson, J. Golic and W. Millan: *LILI Keystream Generator,* SAC'2000, LNCS 2012, pp. 248–261, Springer, 2000. See www.isrc.qut.edu.au/lili/.
27. Markku-Juhani Olavi Saarinen: *A Time-Memory Tradeoff Attack Against LILI-128,* FSE 2002, LNCS 2365, pp. 231–236, Springer, 2002, available at http://eprint.iacr.org/2001/077/.

Cryptanalysis of SAFER++[*]

Alex Biryukov[1][**], Christophe De Cannière[1][* * *], and Gustaf Dellkrantz[1,2]

[1] Katholieke Universiteit Leuven, Dept. ESAT/SCD-COSIC,
Kasteelpark Arenberg 10,
B–3001 Heverlee, Belgium
{alex.biryukov, christophe.decanniere}@esat.kuleuven.ac.be
[2] Royal Institute of Technology,
Stockholm, Sweden
d98-gde@nada.kth.se

Abstract. This paper presents several multiset and boomerang attacks on SAFER++ up to 5.5 out of its 7 rounds. These are the best known attacks for this cipher and significantly improve the previously known results. The attacks in the paper are practical up to 4 rounds. The methods developed to attack SAFER++ can be applied to other substitution-permutation networks with incomplete diffusion.

1 Introduction

The 128-bit block cipher SAFER++ [9] is a 7-round substitution-permutation network (SPN), with a 128-bit key (the 256-bit key version[1] has 10 rounds). SAFER++ was submitted to the European pre-standardization project NESSIE [14] and was among the primitives selected for the second phase of this project.

SAFER [7] was introduced by Massey in 1993, and was intensively analyzed since then [4,6,8,11,13]. This resulted in a series of tweaks which lead to several ciphers in the family: SAFER-K (the original cipher), SAFER-SK (key schedule tweak), SAFER+ (key schedule and mixing transform tweak, increased number of rounds, AES candidate), SAFER++ (faster mixing tweak, key schedule tweak, fewer rounds due to more complex mixing). All these ciphers have common S-boxes derived from exponentiation and discrete logarithm functions. They share the Pseudo-Hadamard-like mixing transforms (PHT), although these are constructed in different ways in the different versions. The ciphers in the family also share the idea of performing key-mixing with two non-commutative operations.

The inventors claim that SAFER++ offers "further substantial improvement over SAFER+" [9]. The main feature is a new 4-point PHT transform in place

---

[*] The work described in this paper has been supported in part by the Commission of the European Communities through the IST Programme under Contract IST-1999-12324 and by the Concerted Research Action (GOA) Mefisto-666.

[**] F.W.O. Researcher, sponsored by the Fund for Scientific Research – Flanders.

[* * *] F.W.O. Research Assistant, sponsored by the Fund for Scientific Research – Flanders

[1] A legacy 64-bit block version was also proposed by the designers but is not studied in this paper.

D. Boneh (Ed.): CRYPTO 2003, LNCS 2729, pp. 195–211, 2003.
© International Association for Cryptologic Research 2003

of the 2-point PHT transform that was used previously in the SAFER family. The authors claim that "all 5-round characteristics have probabilities that are significantly smaller than 2^{-128}" and that SAFER++ is secure against differential cryptanalysis [1] after 5 rounds and against linear cryptanalysis [10] after 2.5 rounds.

The best previous attack on SAFER++ is linear cryptanalysis [12], which can break 3 rounds of SAFER++ (with 128-bit keys) with 2^{81} known plaintexts and 2^{101} steps for a fraction 2^{-13} of keys. For 256-bit keys the attack can break the 3.5-round cipher with 2^{81} known plaintexts and 2^{176} steps for a fraction 2^{-13} of keys.

In this paper we study only the 128-bit key version of SAFER++, since we would like to make our attacks as practical as possible. We design several very efficient multiset attacks on SAFER++ following the methodology of the structural attack on SASAS [2] and inspired by the collision attacks on RIJNDAEL [3]. These multiset attacks can break up to 4.5 rounds of SAFER++ with 2^{48} chosen plaintexts and 2^{94} steps, which is much faster than exhaustive search. Attacking 3 rounds is practical and was tested with an actual implementation running in milliseconds on a PC.

In the second half of the paper we show how to apply a cryptanalytic technique called the boomerang attack [15] to SAFER++. We start from ideas which are applicable to arbitrary SPNs with incomplete diffusion (such as RIJNDAEL, SAFER++ or SERPENT) and then extend our results using special properties of the SAFER S-boxes. The attacks thus obtained are more efficient then those we found via the multiset techniques, are practical up to 4 rounds and were confirmed experimentally on a mini-version of the cipher.

The average data complexity of the 5 round attack is 2^{78} chosen plaintexts/adaptive chosen ciphertexts with the same time complexity, most of which is spent encrypting the data. The attack completely recovers the 128-bit secret key of the cipher and can be extended to 5.5 rounds by guessing 30 bits of the secret key. See Table 1 for a summary of results presented in this paper and their comparison with the best previous attack.

This paper is organized as follows: Section 2 provides a short description of SAFER++ and Section 3 shows some interesting properties of the components. In Sections 4 and 5 we design our multiset attacks on SAFER++. Section 6 describes our application of boomerang techniques to SAFER++ reduced to 5 rounds and shows how to use the middle-round S-box trick to obtain even better results. Finally, Section 7 concludes the paper.

2   Description of Safer++

This section contains a short description of SAFER++. For more details, see [9]. In this paper, eXclusive OR (XOR) will be denoted by $\oplus$, addition modulo 256 by $\boxplus$ and subtraction modulo 256 by $\boxminus$. The notion of difference used is subtraction modulo 256. Throughout this paper we will number bytes and S-boxes from left to right, starting from 0.

Table 1. Comparison of our results with the best previous attack on SAFER++.

Attack	Key size	Rounds	Data[a]	Type[b]	Workload[c]	Memory[a]
Our Multiset attack	128	3 of 7	2^{16}	CC	2^{16}	2^4
Our Multiset attack	128	4 of 7	2^{48}	CP	2^{70}	2^{48}
Our Multiset attack	128	4.5 of 7	2^{48}	CP	2^{94}	2^{48}
Our Boomerang attack	128	4 of 7	2^{41}	CP/ACC	2^{41}	2^{40}
Our Boomerang attack	128	5 of 7	2^{78}	CP/ACC	2^{78}	2^{48}
Our Boomerang attack	128	5.5 of 7	2^{108}	CP/ACC	2^{108}	2^{48}
Linear attack[d] [12]	128	3 of 7	2^{81}	KP	2^{101}	2^{81}

[a] Expressed in number of blocks.
[b] KP – Known Plaintext, CP – Chosen Plaintext, ACC – Adaptive Chosen Ciphertext.
[c] Expressed in equivalent number of encryptions.
[d] Works for one in 2^{13} keys.

SAFER++ is an iterated product cipher in which every round consists of an upper key layer, a nonlinear layer, a lower key layer and a linear transformation. Fig. 1 shows the structure of one SAFER++ round. After the final round there is an output transformation that is similar to the upper key layer. The upper and lower key layers together with the nonlinear layer make up the *keyed nonlinear layer*, denoted by S. The linear layer is denoted by A.

2.1 The Keyed Nonlinear Layer

The upper key layer combines a 16 byte subkey with the 16 byte block. Bytes 0, 3, 4, 7, 8, 11, 12 and 15 of the subkey are XORed to the corresponding bytes of the block and bytes 1, 2, 5, 6, 9, 10, 13 and 14 are combined using addition modulo 256.

The nonlinear layer is based on two 8-to-8-bit functions, X and L defined as

$$X(a) = (45^a \bmod 257) \bmod 256,$$
$$L(a) = \log_{45}(a) \bmod 257,$$

with the special case that $L(0) = 128$, making X and L mutually inverse. In the nonlinear layer, bytes 0, 3, 4, 7, 8, 11, 12 and 15 are sent through the function X, and L is applied to bytes 1, 2, 5, 6, 9, 10, 13 and 14.

The lower key layer applies a 16 byte subkey to the 16 byte block using addition modulo 256 for bytes 0, 3, 4, 7, 8, 11, 12 and 15 and XOR for bytes 1, 2, 5, 6, 9, 10, 13 and 14.

2.2 The Linear Layer

The linear transformation of SAFER++ is built from a 4-point Pseudo Hadamard Transform (4-PHT) and a coordinate permutation. The 4-PHT can be implemented with six modular additions.

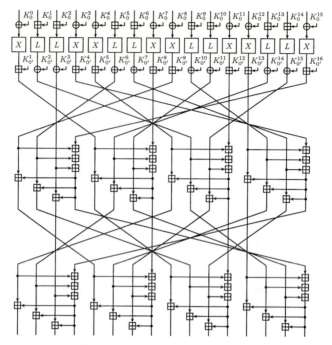

Fig. 1. One round of SAFER++.

The linear layer first reorders the input bytes and then applies the 4-PHT to groups of four bytes. The output of the linear layer is obtained after iterating this operation twice.

The linear layer and its inverse can be represented by the matrices A and A^{-1}. Since the linear layer consists of two iterations of one linear function the matrix A can be written as the square of a matrix $\sqrt{A}$. The matrices A and A^{-1} are shown in Appendix A.

2.3   The Key Schedule

The key schedule expands the 128 or 256-bit master key into the required number of subkeys. It consists of two similar parts differing only in the way the master key is used to fill the registers. The first part generates the subkeys for the upper key layer and the output transform and the second part generates subkeys for the lower key layer.

It can be noted that the key schedule provides no interaction between bytes of the key and furthermore, there is a big overlap between the key bytes used in different rounds. Therefore, we will not number the bytes of the subkeys according to the order in the subkeys, but according to which master key byte they depend on.

3   Properties of the Components

In this section we show some interesting properties of the components of SAFER++ which will be used later in our analysis.

3.1   Diffusion in the Linear Layer

In [8], the designers show that the choice of the components used in the linear layer provides "optimum transform diffusion" without sacrificing efficiency. In order to measure this diffusion, the authors compute the minimal number of output bytes that are affected by a change in a single input byte. In the case of SAFER++, for example, the linear layer guarantees that a single byte difference at the input of the layer will cause at least ten output bytes to be different.

While the "optimum transform diffusion" defined in this way is certainly a desirable property, it potentially allows some low-weight differentials that might still be useful for an attacker. For example, if two input bytes are changed simultaneously in SAFER++, the number of affected output bytes after the linear layer can be reduced to only three. The adversary might also consider to attack the layer in decryption direction, in which case single byte differences are only guaranteed to propagate to at least five bytes. Neither of these cases is captured by the diffusion criterion used in [8].

3.2   Symmetry of the Linear Layer

Due to the symmetry of the 4-PHT and the coordinate permutation used, there is a four byte symmetry in the linear layer. If the input difference to the linear layer is of the form

$$(a, b, c, d, a, b, c, d, a, b, c, d, a, b, c, d)$$

for any 8-bit values a, b, c, and d the output difference will be of the form

$$(x, y, z, t, x, y, z, t, x, y, z, t, x, y, z, t)$$

The nonlinear layer is symmetric in the same way and were it not for the subkeys, the property would hold for the whole SAFER++ cipher, with an arbitrary number of rounds.

A special illustration of this property are the two eigenvectors of the linear transformation corresponding to the eigenvalue 1:

$$(0, 1, -1, 0, 0, 1, -1, 0, 0, 1, -1, 0, 0, 1, -1, 0)$$
$$(1, 0, -1, 0, 1, 0, -1, 0, 1, 0, -1, 0, 1, 0, -1, 0)$$

These vectors and all linear combinations of them are fixed points of the linear transform.

3.3   Properties of the S-Boxes

The S-boxes of SAFER++ are constructed using exponentiations and logarithms as described previously. This provides us with some interesting mathematical properties. The following expressions hold for X and L:

$$X(a) + X(a \boxplus 128) = (45^a \bmod 257) + (45^{a+128} \bmod 257) = 257$$
$$\equiv \quad 1 \quad \bmod 256 \,,$$
$$L(a) - L(1 \boxminus a) \equiv 128 \quad \bmod 256 \,.$$

This is a very useful property that will be exploited in the boomerang attacks in Section 6. Though we discovered this property independently, it was known to Knudsen [6].

4   Multiset Attack on 3 Rounds Using Collisions

Multiset attacks are attacks where the adversary studies the propagation of multisets through the cipher, instead of actual values or differences as in other attacks. By multisets we mean sets where values can appear more than once. The multiplicity of each value is one important characteristic of a multiset.

In this and the following section we present two multiset attacks on 3 and 4 rounds of SAFER++. Both attacks exploit the asymmetry of the linear layer, but in different ways. The first attack considers the cipher in decryption direction and relies on the fact that the weaker diffusion in backward direction allows a stronger distinguisher. The second approach, described in Section 5, is built on a distinguisher in forward direction which, though weaker than the previous one, can more easily be extended with rounds at the top and the bottom. All multiset attacks presented in this paper are independent of the choice of the S-boxes.

4.1   A Two-Round Distinguisher in Decryption Direction

As mentioned earlier, changes propagate relatively slowly through the linear layer in decryption direction. Most byte positions at the input of a decryption round affect only 6 bytes at the output, and conversely, most byte positions at the output are affected by only 6 bytes at the input. After one more round, this property is destroyed, i.e., a change in any input byte can induce changes in any output byte after two rounds (complete diffusion on byte level). The number of different paths through which the changes propagate is still small, however, and this property allows us to build an efficient distinguisher.

The structure of a 2-round distinguisher is shown in Fig. 2. First, a multiset of 2^{16} texts is constructed, constant in all bytes except in byte 4 and 6, in which it takes all possible values. After one decryption round, these texts are still constant in 8 positions, due to the weak diffusion properties of the linear layer. We now focus on byte 13 at the top of the distinguisher. This byte is affected by 6 bytes of the preceding round, 5 of which are constant. This implies that

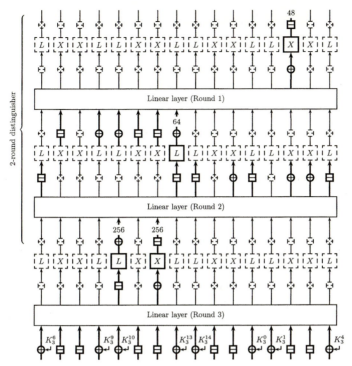

Fig. 2. A 3-round attack.

the changes in the two bytes at the input essentially propagate through a single path.

As a closer look reveals, this path has some additional properties: in Round 2, the two input bytes are first summed and then multiplied by 4 before they reach S-box 7. The byte at this position is then again multiplied by 4 in Round 1. The effect of these multiplications (modulo 256) is that the two most significant bits are lost, reducing the number of different values that can be observed at the top to 64. This is extremely unlikely to happen if the cipher were a random permutation, in which case all 256 possible values would almost certainly be observed. Moreover, due to the interaction with the special S-boxes of SAFER++, the number of different values is exactly 48.

4.2   Adding a Round at the Bottom

The distinguisher above is very strong, but it is hard to use it as such when a round is added at the bottom. Causing the multiset described above at the input of the distinguisher (see Fig. 2) would require half of the key bytes (those that are XORed) to be guessed in the final key addition layer.

In this paper we use a better approach and consider small multisets of 2^4 texts that are constant except in the two most significant bits of the fifth and the seventh byte. In order to cause such multisets from the bottom, we only need to guess the second most significant bit of the 8 key bytes that are XORed in the final key addition layer and generate sets of 2^4 ciphertexts of the form

$$(x \cdot A \boxplus y \cdot B \boxplus C) \oplus K \qquad (1)$$

with $(x, y) \in \{0, 64, 128, 192\}^2$, C any fixed 128-bit word, and

$$A = (4, 2, 2, 2, 1, 1, 2, 1, 1, 1, 1, 1, 2, 1, 1),$$
$$B = (1, 1, 1, 1, 1, 2, 1, 1, 2, 2, 4, 2, 2, 1, 1, 1),$$
$$K = (K_3^6, 0, 0, K_3^9, K_3^{10}, 0, 0, K_3^{13}, K_3^{14}, 0, 0, K_3^0, K_3^1, 0, 0, K_3^4).$$

Note that only the second most significant bits of the K_3^i are relevant (because of carries) and that the effect of the other bits can be absorbed in the constant C.

Now the question arises whether we can still distinguish the set of 16 values obtained in byte 13 of the plaintexts from a random set. An interesting characteristic that can be measured is the number of collisions in the sets, should they appear at all. If this number depends in a significant way on whether the key bits were correctly guessed or not, then this would allow us to recover the key. This question is easily answered by deriving a rough estimation of the number of collisions.

First, we deduce the expected number of collisions μ_0 in case the guess was correct. Assuming that all 48 possible values are equally likely, we obtain

$$\mu_0 = \binom{16}{2} \cdot \frac{1}{48} \approx 2.50 \quad \text{and} \quad \sigma_0 = \sqrt{\binom{16}{2} \cdot \frac{1}{48} \left(1 - \frac{1}{48}\right)} \approx 1.56,$$

with σ_0 the expected deviation. In order to estimate the number of collisions given a wrong guess, one could be tempted to assume that the sets at the output of the distinguisher are random. This is not the case however.[2] One can easily see, for example, that whether K_3^6 is correctly guessed or not matters only for half of the (x, y) pairs, i.e., when the second most significant bit of $4 \cdot x \boxplus y$ is set. In all these (and only these) cases an incorrect carry bit will appear in the most significant bit. If we absorb this wrong bit in a new constant C', we obtain two subsets of 8 texts, both of which still satisfy (1), but with two different fixed values C and C'. Hence we find

$$\mu_1 = 2 \cdot \binom{8}{2} \cdot \frac{1}{48} + 8^2 \cdot \frac{1}{256} \qquad \approx 1.42,$$

$$\sigma_1 = \sqrt{2 \cdot \binom{8}{2} \cdot \frac{1}{48} \left(1 - \frac{1}{48}\right) + 8^2 \cdot \frac{1}{256} \left(1 - \frac{1}{256}\right)} \qquad \approx 1.19.$$

The value of μ_1 is indeed considerably higher than what we would expect for a random set (about 0.47 collisions). Exactly the same situation occurs for 17

---

[2] Note that this property is useful if one does not care about recovering the key, but just needs a 3-round distinguisher.

other combinations of wrong key bits, and similar, but less pronounced effects can be expected for other guesses.

The result above can now be used to predict the complexity and success probability of our 3-round attack. The attack consists in running through all 2^8 possible partial key guesses and accumulating the total number of collisions observed after decrypting α sets of 16 texts. The maximum number of collisions is then assumed to correspond to the correct key. Taking this into consideration, we obtain the estimations below:

$$\text{Time and data complexity} \approx 2^8 \cdot 16 \cdot \alpha \,,$$

$$\text{Success probability} \approx 1 - 17 \cdot \Phi \left(\sqrt{\alpha} \cdot \frac{\mu_0 - \mu_1}{\sqrt{\sigma_0^2 + \sigma_1^2}} \right) \,.$$

Evaluating these expressions for $\alpha = 16$, we find a complexity of 2^{16} and a corresponding success probability of 77%. Similarly, for $\alpha = 32$ we get a complexity of 2^{17} and an expected probability as high as 98%. In order to verify these results, we performed a series of simulations and found slightly lower success probabilities: 70% and 89% for $\alpha = 16$ and $\alpha = 32$ respectively. This difference is due to the fact that our estimation only considers wrong guesses that have a high probability of producing many collisions.

5   Multiset Attack on 4 Rounds Using Structural Approach

The chosen ciphertext attack presented in the previous section is particularly efficient on 3 rounds, but extending it to 4 rounds turns out to be rather difficult. Either we could try to add a round at the bottom, which would require us to cause small bit changes after crossing a decryption round, or we could add a round at the top and try to recover the value of a particular byte from the output of this round. Both cases, however, are hindered by the asymmetry of the linear layer and this motivates the search for a distinguisher in forward direction.

Before describing the distinguisher we introduce some convenient notations to represent different types of multisets:

C – **Constant multiset.** A set containing a single value, repeated multiple times.

P – **Permutation multiset.** A set containing all possible values once, in an arbitrary order.

E – **Even multiset.** A set containing different values, each occurring an even number of times.

B – **Balanced multiset.** A set containing elements with arbitrary values, but such that their sum (modulo 256) is zero. We will write "B_0", if this property only holds for the least significant bit. Note that multisets satisfying the conditions C, P or E satisfy condition B_0 as well.

5.1   A Two-Round Distinguisher in Encryption Direction

The input to the 2-round distinguisher consists of a permutation multiset (P) at position 12, and constant multisets (C) elsewhere. This pattern is preserved after the first layer of S-boxes. It then crosses the linear layer, where it is transformed into a series of P and E multisets. Again, these properties remain unchanged after applying the second layer of S-boxes. Finally, we note that all multisets at this point satisfy condition B_0, and this property is preserved both by the second linear layer and the final key addition. The sixteen B_0-multisets at the output provide us with a 16-bit condition and allows the two round cipher to be distinguished from a random permutation.

5.2   Adding a Round at the Top

In order to add a round at the top, we must be capable of keeping all bytes constant after one encryption round, except for byte 12. This would have been very hard in decryption direction, but is fortunately relatively easy in this case. Due to the fact that the diffusion is incomplete in backward direction, we only need to cross the nonlinear layer in 6 active bytes. Moreover, we do not care about constants added to byte 12. This implies that 6 key bytes in the upper key layer (K_0^2, K_0^3, K_0^5, K_0^8, K_0^{12}, and K_0^{15}) and 2 XORed key bytes in the lower key layer ($K_{0'}^3$ and $K_{0'}^6$) need to be guessed. As K_0^3 and $K_{0'}^3$ are easily derived from each other, guessing 7 bytes will suffice.

In order to distinguish the wrong guesses from the correct one, 4 different multisets need to be examined for each key (yielding a 8-byte condition). Since only six bytes are active at the top, all multisets can be constructed from a pool of 2^{48} plaintexts. The time complexity of this 3-round attack is about $4 \cdot 2^{7 \cdot 8} = 2^{58}$ steps.

5.3   Adding a Round at the Bottom

The complexity derived above is a bit disappointing compared to the results in the previous section, but the good news is that adding an extra round at the bottom comes almost for free. This time, we will test a single byte at the output of the distinguisher (byte 8) and exploit the fact that this byte is only affected by 6 bytes at the output of the fourth round. The additional obstacles are 7 key bytes which need to be guessed. However, due to the key schedule and the special choice for the position of the active byte at the input of the distinguisher (byte 12), 6 of these key bytes can directly be determined from the keys guessed at the top. As a consequence, the total number of key bytes to guess is only increased by 1. Taking into account that each multiset provides a 1-bit condition in this case, we obtain a time complexity of $8 \cdot 8 \cdot 2^{8 \cdot 8} = 2^{70}$ steps.

The 4-round attack described above can easily be extended by another half round at the bottom (i.e., an S-box layer followed by a key addition). This would in principle require 6 more key bytes to be guessed, but again, 3 of them depend on previous guesses. Accordingly, the complexity increases to about $11 \cdot 8 \cdot 2^{11 \cdot 8} \approx 2^{94.5}$ steps.

6   Boomerang Attacks on Safer++

In this section we describe an attack on SAFER++ reduced to 5 rounds and then extend it to an attack on 5.5 rounds. The attack is a boomerang attack [15] using a combination of truncated differentials [5] and conventional differentials [1].

Contrary to conventional differentials which require full knowledge of the input difference and predict the full output difference, truncated (or wildcard) differentials restrict only parts of the input difference and may predict only parts of the output difference. It is thus natural to consider truncated differentials for ciphers which use operations on small sub-blocks (for example, bytes or nibbles).

Two natural ways of placing partial restrictions on differences are: to fix the difference in certain sub-blocks to constant values (a popular choice is zero) while allowing arbitrary differences in the other sub-blocks; or to place constraints on differences without restricting them to specific values. For example, one can consider differences of the form $(x, 0, -x, 2x, 0, -y, y, 4y)$, where the values x, y are not restricted and thus give *degrees of freedom*.

Truncated differentials have several distinct properties which make them different from conventional differentials. First of all, the probability of truncated differentials is usually not the same in forward and in backward direction. Secondly, due to the "wildcard" nature of truncated differentials plaintexts for truncated differentials may be efficiently packed into pools in which a large fraction of pairs satisfies the truncated difference. Such a *pool effect* often significantly reduces the data requirements for detecting truncated differentials. The data requirements of the truncated differential attack may be smaller than the inverse of the differential probability (which is usually a good measure for the data complexity of a conventional differential attack), and thus truncated differential attacks may still work in cases where the probability is lower than 2^{-n} where n is the block size. Finally, due to the large number of pairs in the pools and due to the partial prediction of the output difference for the *right pair*, filtration of wrong pairs (those pairs that have the predicted partial output difference, but do not follow the partial difference propagation expected by the truncated differential) becomes a crucial issue for truncated differential attacks.

The idea of the boomerang attack is to find good conventional (or truncated) differentials that cover half of the cipher but can not necessarily be concatenated into a single differential covering the whole cipher. The attack starts with a pair of plaintexts P and P' with a difference Δ which goes to difference Δ^* through the upper half of the cipher. The attacker obtains the corresponding ciphertexts C and C', applies the difference ∇ to obtain ciphertexts $D = C + \nabla$ and $D' = C' + \nabla$ and decrypts them to plaintexts Q and Q'. The choice of ∇ is such that the difference propagates to the difference ∇^* in the decryption direction through the lower half of the cipher. For the *right quartet* of texts, difference Δ^* is created in the middle of the cipher between partial decryptions of D and D' which propagates to the difference Δ in the plaintexts Q and Q'. This can be detected by the attacker.

Moreover, working with quartets (pairs of pairs) provides boomerang attacks with additional filtration power. If one partially guesses the keys of the top round one has two pairs of the quartet to check whether the uncovered partial differences follow the propagation pattern, specified by the differential. This effectively doubles the attacker's filtration power.

6.1   A Boomerang Distinguisher for 4.5 Rounds

In this part we build a distinguisher for 4.5 rounds of SAFER++. As mentioned before, the affine layer is denoted by A and the keyed nonlinear layer by S. Using this notation, the distinguisher will have the structure $[ASA - S - ASASA]$ (see Fig. 4, in which layers of rectangular boxes correspond to S-layers). The top part $[ASA-]$ will be covered top-down and the bottom part $[-ASASA]$ in the bottom-up direction. The part $[-S-]$ in the middle will come for free using the middle S-box trick.

Pairs of plaintexts (P, P'), that have difference $(0, x, 0, 0, x, x, 0, 0, 0, 0, 0, -4x, 0, 0, x, -x)$ are used from the top. This difference propagates to a single active S-box after one linear layer and is then diffused to up to 16 active bytes by the next linear layer.

We obtain the two ciphertexts C, C', modify them into $D = C \boxplus \nabla$, $D' = C' \boxplus \nabla$ and decrypt them to two new plaintexts Q and Q'. The difference ∇ is $(0, 0, 80_x, 0, 80_x, 80_x, 0, 0, 0, 0, 0, 0, 80_x, 0, 0, 0)$ which causes difference $(0, 0, 0, 0, 0, x, -x, 0, 0, 0, 0, 0, 0, 0, 0, 0)$ with approximate probability 2^{-7} after one decryption round. The probability is 2^{-7} since the input difference 80_x can only cause odd output differences through X. The two active bytes then cause bytes $3, 9, 11$ and 14 to be active after the next linear layer.

In a standard boomerang attack, we would now hope that the pairs (C, D) and (C', D') both have the same difference in the middle round, which would cause the difference of the pair (D, D') in the middle round to be the same as that for (P, P'). For the right quartet we would then have the same difference going back in the pair (D, D') as going forward in (P, P'), thus making (Q, Q') active in the same bytes as (P, P'). This can be detected by the attacker and used to recover parts of the key.

This version of the attack is reasonably efficient against four rounds of SAFER++ and can be applied to any substitution-permutation network with incomplete diffusion. However, the probability of the event used in the middle, where the parts of the boomerang meet, depends heavily on the number of active bytes and is often low. For the SAFER-family of ciphers we can do better by using special properties of the S-boxes.

6.2   The Middle Round S-Box Trick

Using the relation $X(a) \boxplus X(a \boxplus 80_x) = 1$, which holds with probability 1 for the exponentiation-based S-boxes (see Section 3.3), we can make the boomerang travel for free through the middle S-box layer.

Consider the middle S-box layer where the top and bottom part of the boomerang meet. Suppose that the difference coming from above and entering an X-box, is 80_x, i.e., a pair of values a and $a \boxplus 80_x$. After the X-box the values will always be of the form b and $1 \boxminus b$. Assume that on the two other faces of the boomerang we also have difference 80_x coming from below (see Fig. 3). Then the fourth face must have values $b \boxplus 80_x$ and $1 \boxminus b \boxplus 80_x$ on the lower side of the X-box. However, these values again sum to 1, since $80_x \boxplus 80_x = 0$ (actually, any pair of values that sum to 0 would work instead of 80_x here). As a consequence, we will observe values of the form c and $c \boxplus 80_x$ at the input of the X-box on the fourth face. See Fig. 3 for an illustration of this effect. The same

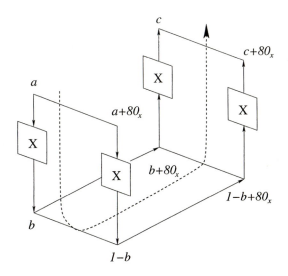

Fig. 3. The free pass of the boomerang through the middle SAFER++ S-boxes.

reasoning holds for the L-box since it is the inverse of X and the same differences are coming from the top and the bottom.

This shows that if we manage to produce texts with difference 80_x coming from both the top and the bottom in the boomerang, the boomerang travels through the middle S-box layer for free due to the special properties of the S-boxes.

6.3   Breaking 5 and 5.5 Rounds

We can break 5 rounds of SAFER++ with the distinguisher described in the previous sections. The truncated differentials used are shown in Fig. 4. From the top we use a difference with six active bytes that propagates to one active byte after one round with probability 2^{-40}. This one-byte difference then causes a difference of 80_x in bytes 0, 1, 2, 3, 8, 9, 11, 13, 14 and 15 after an additional round with probability 2^{-8}. All other bytes have zero-difference. This completes the upper part of the boomerang.

From the bottom we start with changes in the most significant bits of four bytes, which cause two active bytes after one decryption round. This then propagates to four active bytes with probability 2^{-7} and further to difference 80_x in all bytes except bytes 2, 4, 9 and 12 with probability $2^{-30.4}$. The total probability of the lower part of the boomerang is thus $\left(2^{-7} \cdot 2^{-30.4}\right)^2$, since we need the differential to hold in two pairs.

The top part of the boomerang in the decryption direction propagates with probability 1, due to the middle S-box trick. The total probability of the boomerang is thus $2^{-40} \cdot 2^{-8} \cdot \left(2^{-7} \cdot 2^{-30.4}\right)^2 = 2^{-122.8}$

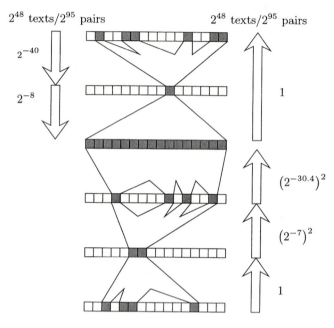

Fig. 4. The boomerang quartet for SAFER++ reduced to 5.5 rounds.

The procedure for attacking 5 rounds is as follows:

1. Prepare a pool of 2^{48} plaintexts P_i, $i = 0, \ldots, 2^{48} - 1$ that have all possible values in bytes 1, 4, 5, 11, 14 and 15 and are constant in the other bytes. Encrypt the pool to get a pool of 2^{48} ciphertexts C_i.
2. Create a new pool of ciphertexts D_i from the pool C_i by changing the most significant bits in bytes $2, 4, 5$ and 12. Decrypt the pool D_i to obtain a pool Q_i of 2^{48} plaintexts.
3. Sort the pool Q_i on the bytes that correspond to the constant bytes in the pool P_i and pick the pairs Q_j, Q_k that have zero difference in those ten bytes.
4. For each of the possibly good quartets P_j, P_k, Q_j, Q_k, guess the 3 key bytes K_0^4, K_0^{11} and K_0^{15}, do a partial encryption and check that the difference in the 3 bytes after the first nonlinear layer is of the form $(x, -4x, -x)$ with x odd, both in the P-pair and in the Q-pair. Note that we do not need to guess

the key bytes added at the bottom of this layer, as they do not influence the subtractive difference here. Keep the quartets that have the right difference, together with the key bytes that they suggest.

5. Guess the key byte K_0^{14}, do a partial encryption and check that the difference after XORing the key $K_{0'}^{15}$ (which we know from the previous step) at the bottom of the nonlinear layer is consistent with the difference found in Step 4. If no quartets survived, go to Step 1.

6. Guess the key bytes K_0^1 and $K_{0'}^2$, do a partial encryption and check that the difference after the first keyed nonlinear layer is the right one. Repeat this for the key bytes K_0^5 and $K_{0'}^6$.

7. Keep collecting suggestions for the 8 key bytes until one appears twice. For this suggestion, do an exhaustive search for the 8 remaining key bytes using trial encryption. If no consistent key is found, go to Step 1.

We now analyze the complexity of this attack. From the pool of 2^{48} plaintexts created in Step 1 we get approximately 2^{95} pairs. Since the probability of the boomerang is $2^{-122.8}$, the probability is approximately $2^{-27.8}$ that a pool contains a boomerang.

After Step 3 we expect to have about 2^{15} wrong quartets left since we have an 80-bit condition on 2^{95} pairs. Step 4 reduces the number of wrong quartets to 2^5, because after guessing 3 bytes of the key, we obtain a 17-bit restriction on each side of the boomerang, resulting in a 34-bit condition. Similarly, after Step 5, only about 2^{-3} quartets remain per pool. On the average, Step 6 will suggest 1 value for the 4 guessed key bytes per remaining quartet.

In order for the right key to be suggested twice, we need two boomerangs in Step 7. This will occur after having analyzed about 2^{29} pools on average. During this process, 2^{26} wrong keys will be suggested as well, but the chance that one of these 64-bit values is suggested twice is small (the probability is about $2^{51}/2^{64} = 2^{-13}$). This implies that the exhaustive search in Step 7 only needs to be performed for a single suggestion on average.

Since each of the 2^{29} pools required by the attack consist of 2^{48} chosen plaintexts and 2^{48} adaptively chosen ciphertexts, we obtain a data complexity of 2^{78}. Just collecting these texts will be the most time-consuming part of the attack, and the time complexity of the attack is therefore expected to be 2^{78}. These figures reflect the average case, but the complexities will exceed 2^{79} in only 5% of the cases.

The attack on 5 rounds can be extended to 5.5 rounds by guessing 30 bits of the key in positions $2, 4, 5$ and 12 at the end of the added half round. This increases the data and time complexities to 2^{108}. Note that 2 key bits have been saved by using the special properties of the L-box.

7 Conclusions

In this paper we have applied novel multiset attack techniques to round-reduced SAFER++ inspired by the recent structural analysis of the SASAS scheme and partial-function collision techniques. These multiset attacks are very efficient

up to 4.5 rounds and practical up to 3 rounds. This significantly improves the
previously known results.

In the second half of the paper we applied boomerang attacks to SAFER++
which allow for even more efficient attacks because they exploit special properties of the exponentiation and logarithmic S-boxes and their interaction with the
PHT-mixing layer. We presented an attack on 5.5 out of 7 rounds of SAFER++
requiring 2^{108} data blocks and steps of analysis. A 4-round variant of this
boomerang attack is practical and was tested on a 64-bit mini-version of the
cipher. See Table 1 for a summary of results presented in this paper and their
comparison with previously known attacks.

Finally, note that the methods developed in the second half of this paper can
be applied to arbitrary SPNs with incomplete diffusion, with the exception of the
middle round trick which exploits special properties of the SAFER++ S-boxes.

References

[1] E. Biham and A. Shamir, *Differential Cryptanalysis of the Data Encryption Standard*. Springer-Verlag, 1993.
[2] A. Biryukov and A. Shamir, "Structural cryptanalysis of SASAS," in *Advances in
Cryptology – EUROCRYPT 2001* (B. Pfitzmann, ed.), vol. 2045 of *Lecture Notes
in Computer Science*, pp. 394–405, Springer-Verlag, 2001.
[3] H. Gilbert and M. Minier, "A collision attack on seven rounds of Rijndael," in *Proceedings of the Third AES Candidate Conference*, pp. 230–241, National Institute
of Standards and Technology, Apr. 2000.
[4] J. Kelsey, B. Schneier, and D. Wagner, "Key-schedule cryptanalysis of 3-WAY,
IDEA, G-DES, RC4, SAFER, and Triple-DES," in *Advances in Cryptology –
CRYPTO'96* (N. Koblitz, ed.), vol. 1109 of *Lecture Notes in Computer Science*,
pp. 237–251, Springer-Verlag, 1996.
[5] L. R. Knudsen, "Truncated and higher order differentials," in *Fast Software Encryption, FSE'94* (B. Preneel, ed.), vol. 1008 of *Lecture Notes in Computer Science*, pp. 196–211, Springer-Verlag, 1995.
[6] L. R. Knudsen, "A detailed analysis of SAFER K," *Journal of Cryptology*, vol. 13,
no. 4, pp. 417–436, 2000.
[7] J. L. Massey, "SAFER K-64: A byte-oriented block-ciphering algorithm," in *Fast
Software Encryption, FSE'93* (R. J. Anderson, ed.), vol. 809 of *Lecture Notes in
Computer Science*, pp. 1–17, Springer-Verlag, 1994.
[8] J. L. Massey, "On the optimality of SAFER+ diffusion," in *Proceedings of the Second AES Candidate Conference*, National Institute of Standards and Technology,
Mar. 1999.
[9] J. L. Massey, G. H. Khachatrian, and M. K. Kuregian, "Nomination of SAFER++
as candidate algorithm for the New European Schemes for Signatures, Integrity,
and Encryption (NESSIE)." Primitive submitted to NESSIE by Cylink Corp.,
Sept. 2000.
[10] M. Matsui, "Linear cryptanalysis method for DES cipher," in *Advances in Cryptology – EUROCRYPT'93* (T. Helleseth, ed.), vol. 765 of *Lecture Notes in Computer
Science*, pp. 386–397, Springer-Verlag, 1993.
[11] S. Murphy, "An analysis of SAFER," *Journal of Cryptology*, vol. 11, no. 4,
pp. 235–251, 1998.

[12] J. Nakahara Jr, *Cryptanalysis and Design of Block Ciphers.* PhD thesis, Katholieke Universiteit Leuven, June 2003.

[13] J. Nakahara Jr, B. Preneel, and J. Vandewalle, "Linear cryptanalysis of reduced-round versions of the SAFER block cipher family," in *Fast Software Encryption, FSE 2000* (B. Schneier, ed.), vol. 1978 of *Lecture Notes in Computer Science*, pp. 244–261, Springer-Verlag, 2001.

[14] NESSIE Project – New European Schemes for Signatures, Integrity and Encryption. http://cryptonessie.org.

[15] D. Wagner, "The boomerang attack," in *Fast Software Encryption, FSE'99* (L. R. Knudsen, ed.), vol. 1636 of *Lecture Notes in Computer Science*, pp. 156–170, Springer-Verlag, 1999.

A   The Linear Layer

This appendix contains the matrices corresponding to the linear layer and its inverse.

$$A = \begin{pmatrix}
1 & 2 & 1 & 1 & 1 & 1 & 1 & 1 & 4 & 2 & 2 & 2 & 1 & 1 & 2 & 1 \\
2 & 1 & 1 & 1 & 1 & 1 & 2 & 1 & 1 & 1 & 1 & 1 & 2 & 4 & 2 & 2 \\
2 & 2 & 4 & 2 & 2 & 1 & 1 & 1 & 1 & 2 & 1 & 1 & 1 & 1 & 1 & 1 \\
1 & 1 & 1 & 1 & 1 & 2 & 1 & 1 & 1 & 1 & 2 & 1 & 2 & 1 & 1 & 1 \\
4 & 2 & 2 & 2 & 1 & 1 & 2 & 1 & 1 & 1 & 1 & 1 & 1 & 2 & 1 & 1 \\
1 & 1 & 2 & 1 & 2 & 1 & 1 & 1 & 2 & 4 & 2 & 2 & 1 & 1 & 1 & 1 \\
1 & 1 & 1 & 1 & 1 & 2 & 1 & 1 & 2 & 2 & 4 & 2 & 2 & 1 & 1 & 1 \\
1 & 2 & 1 & 1 & 1 & 1 & 1 & 1 & 2 & 1 & 1 & 1 & 1 & 1 & 2 & 1 \\
1 & 1 & 2 & 1 & 4 & 2 & 2 & 2 & 1 & 2 & 1 & 1 & 1 & 1 & 1 & 1 \\
1 & 1 & 1 & 1 & 2 & 4 & 2 & 2 & 1 & 1 & 2 & 1 & 2 & 1 & 1 & 1 \\
1 & 2 & 1 & 1 & 1 & 1 & 1 & 1 & 2 & 1 & 1 & 1 & 2 & 2 & 4 & 2 \\
2 & 1 & 1 & 1 & 1 & 1 & 2 & 1 & 1 & 1 & 1 & 1 & 1 & 2 & 1 & 1 \\
1 & 1 & 1 & 1 & 1 & 2 & 1 & 1 & 1 & 2 & 1 & 4 & 2 & 2 & 2 & 2 \\
2 & 4 & 2 & 2 & 1 & 1 & 1 & 1 & 2 & 1 & 1 & 1 & 1 & 1 & 2 & 1 \\
2 & 1 & 1 & 1 & 2 & 2 & 4 & 2 & 1 & 1 & 1 & 1 & 1 & 2 & 1 & 1 \\
1 & 1 & 2 & 1 & 2 & 1 & 1 & 1 & 1 & 2 & 1 & 1 & 1 & 1 & 1 & 1
\end{pmatrix}$$

$$A^{-1} = \begin{pmatrix}
0 & 0 & 0 & -4 & 1 & 0 & 1 & 0 & 0 & 1 & 0 & -1 & 1 & 0 & 0 & 0 \\
0 & 0 & 0 & -4 & 0 & 0 & 1 & -1 & 0 & 1 & 0 & 0 & 1 & 1 & 0 & 0 \\
0 & 0 & 1 & -4 & 0 & 0 & 1 & 0 & 0 & 1 & 0 & 0 & 1 & 0 & 0 & -1 \\
0 & 0 & -1 & 16 & -1 & 0 & -4 & 1 & 0 & -4 & 0 & 1 & -4 & -1 & 0 & 1 \\
1 & 0 & 0 & 0 & 0 & 0 & 0 & -4 & 1 & 0 & 1 & 0 & 0 & 1 & 0 & -1 \\
1 & 0 & 0 & -1 & 0 & 0 & 0 & -4 & 0 & 1 & 1 & 0 & 0 & 1 & 0 & 0 \\
1 & 0 & 0 & 0 & 0 & 0 & 0 & -4 & 0 & 0 & 1 & -1 & 0 & 1 & 1 & 0 \\
-4 & 0 & 0 & 1 & 0 & 0 & 0 & 16 & -1 & -1 & -4 & 1 & 0 & -4 & -1 & 1 \\
1 & 1 & 0 & 0 & 1 & 0 & 0 & -1 & 0 & 0 & 0 & -4 & 0 & 0 & 1 & 0 \\
0 & 1 & 0 & 0 & 1 & 1 & 0 & 0 & 0 & 0 & 0 & -4 & 0 & 0 & 1 & -1 \\
0 & 1 & 0 & -1 & 1 & 0 & 1 & 0 & 0 & 0 & 0 & -4 & 0 & 0 & 1 & 0 \\
-1 & -4 & 0 & 1 & -4 & -1 & -1 & 1 & 0 & 0 & 0 & 16 & 0 & 0 & -4 & 1 \\
0 & 0 & 1 & -1 & 0 & 1 & 0 & 0 & 1 & 0 & 0 & 0 & 1 & 0 & 0 & -4 \\
0 & 1 & 1 & 0 & 0 & 1 & 0 & 0 & 1 & 0 & 0 & -1 & 0 & 0 & 0 & -4 \\
0 & 0 & 1 & 0 & 0 & 1 & 0 & -1 & 1 & 0 & 1 & 0 & 0 & 0 & 0 & -4 \\
0 & -1 & -4 & 1 & 0 & -4 & 0 & 1 & -4 & 0 & -1 & 1 & -1 & 0 & 0 & 16
\end{pmatrix}$$

A Polynomial Time Algorithm for the Braid Diffie-Hellman Conjugacy Problem

Jung Hee Cheon[1] and Byungheup Jun[2]

[1] School of Mathematical Sciences, Seoul National University, Republic of Korea
jhcheon@math.snu.ac.kr
[2] Korea Institute for Advanced Study, Republic of Korea
bhjun@kias.re.kr

Abstract. We propose the first polynomial time algorithm for the braid Diffie-Hellman conjugacy problem (DHCP) on which the braid key exchange scheme and the braid encryption scheme are based [9]. We show the proposed method solves the DHCP for the image of braids under the Lawrence-Krammer representation and the solutions play the equivalent role of the original key for the DHCP of braids. Given a braid index n and a canonical length ℓ, the complexity is about $O(n^{14.4}\ell^{3.2})$ or $O(n^{4\tau+2\epsilon}\ell^{2\epsilon})$ bit operations for $\tau = \log_2 7 \approx 2.8$ and $\epsilon > \log_2 3 \approx 1.57$.

Keywords: Braid group, Non-abelian group, Conjugacy Problem

1 Introduction

In 2000, a key agreement and an encryption scheme based on braid groups were proposed by Ko *et. al* [9]. The schemes are analogous to the Diffie-Hellman key agreement scheme and the ElGamal encryption scheme on abelian groups. Their basic mathematical problem is the *Conjugacy Problem* (CP) on braids: For a braid group B_n, we are asked to find a braid a from $u, b \in B_n$ satisfying $b = aua^{-1} \in B_n$. The security is based on the *Diffie-Hellman Conjugacy Problem* (DHCP) to find $baua^{-1}b^{-1} \in B_n$ for given $u, aua^{-1}, bub^{-1} \in B_n$ for a and b in two commuting subgroups of B_n respectively. There are only brute-force attack and super-summit set attack as the analysis. Both yields a complexity of exponential time [9,4]. Recently, several heuristic algorithms were proposed using Burau representation. Though they may be implemented in quite efficient way, they do not solve the whole problem (their methods do not work for some parameters), so no theoretical bounds have been written yet [6,14].

One may approach the CP using a representation in another group whose structure we know better. As mathematicians have developed linear algebra for more than hundred years, linear algebraic groups are possible candidates. There are two candidates as linear representations of braid groups: Burau and Lawrence-Krammer representations. Burau representation was used in *loc. cit.* to make a quite reasonable records. Unfortunately, it is known to be unfaithful, they cannot bound the complexity of the scheme as we expected.

D. Boneh (Ed.): CRYPTO 2003, LNCS 2729, pp. 212–225, 2003.
© International Association for Cryptologic Research 2003

Lawrence-Krammer representation is now chosen to analyze the PKC. It has been proved faithful for arbitrary index of Braids, several times in independent ways by several authors. In general it increases the rank of the representations, so it is complicated to describe. Nevertheless, it is known, but not written clearly, one can easily recover the original braid from its matrix of the representation [13]. Under this assumption, we describe an algorithm to solve the CP.

1. Find the images of u and $v = aua^{-1}$ in $GL_{n(n-1)/2}$ ($\mathbb{Z}[t^{\pm 1}, q^{\pm 1}]$) via the Lawrence-Krammer representation $\mathcal{K} : B_n \to GL_{n(n-1)/2}(\mathbb{Z}[t^{\pm 1}, q^{\pm 1}])$.
2. Solve the CP for $\mathcal{K}(u)$ and $\mathcal{K}(v) = \mathcal{K}(a)\mathcal{K}(u)\mathcal{K}(a)^{-1}$ in $GL_{n(n-1)/2}(\mathbb{Z}[t^{\pm 1}, q^{\pm 1}])$.
3. Recover the braid a in B_n from the matrix obtained above.

The above algorithm contains a couple of difficulties. Firstly, direct applications of Gaussian elimination should deal with coefficients as large as 2^{2^n}. Secondly, a solution of the CP in the matrix group might not be in the image of the representation. It is not easy to choose a matrix in the solution space which lies inside the image of the representation.

To avoid these difficulties, we take the DHCP into our consideration, instead of the CP. The algorithm is modified, roughly as follows:

1. Assume $a \in LB_n$, $b \in RB_n$, and $u \in B_n$ where LB_n and RB_n are two commuting subgroups of B_n.
2. Find the images of u, $v = aua^{-1}$, and $w = bub^{-1}$ in $GL_{n(n-1)/2}$ ($\mathbb{Z}[t^{\pm 1}, q^{\pm 1}]$) via the Lawrence-Krammer representation $\mathcal{K} :$ $B_n \to GL_{n(n-1)/2}(\mathbb{Z}[t^{\pm 1}, q^{\pm 1}])$.
3. By estimating the entries of $\mathcal{K}(awa^{-1})$, take a prime p and irreducible polynomials $f(t)$ over $\mathbb{Z}/p$ and $g(q)$ over $\mathbb{Z}[t]/(p, f(t))$ satisfying

$$\mathcal{K}(awa^{-1}) = t^{-d}N^{-1}\{t^d N\mathcal{K}(awa^{-1}) \mod (p, f(t), g(q))\}$$

for some positive integer d and N.
4. Solve the simultaneous equations $\mathcal{K}(v)A = A\mathcal{K}(u)$ and $\mathcal{K}(\sigma_i)A = A\mathcal{K}(\sigma_i)$ with $n/2 < i \le n$ over a residue class field $k = \mathbb{Z}[t, q]/(p, f(t), g(q))$, where σ_i with $n/2 < i \le n$ generates RB_n.
5. This solution may not be $\mathcal{K}(a)$, but it plays an equivalent role of the key for the DHCP in braid groups. That is, any solution A of the above system of equations satisfies $A\mathcal{K}(w)A^{-1} = \mathcal{K}(b)\mathcal{K}(v)\mathcal{K}(b)^{-1} = \mathcal{K}(awa^{-1})$ since $\mathcal{K}(b)A = A\mathcal{K}(b)$ for $b \in RB_n$. The inverse of A can be computed in a similar way to the above method.
6. Recover the braid awa^{-1} in B_n by inverting the representation.

To reduce the complexity of this algorithm, we use $1/2$ instead of q (it is also faithful), reduce the bound of Krammer matrices, and remove several trivial variables and equations in the simultaneous equations. When ℓ is the Charney length of a, b, and u in B_n, the complexity of this algorithm analyzed in this article reaches about $2^{-3}n^{4\tau+2\epsilon}\ell^{2\epsilon}$ where $\tau = \log_2 7 \approx 2.8$ and $\epsilon > \log_2 3 \approx 1.57$.

This is not a feasible complexity for the parameters recommended in [9,4]. For example, for $n = 90$ and $\ell = 12$ as in [4] it is about 2^{103} bit operations. But even for $n = 10^5$ and $\ell = 10^4$, the complexity is just 2^{278}. Hence the braid encryption scheme can not be used in the future in this style. We would suggest that the protocol should be revised to use the full difficulty of CP to overcome the attack. In the near future, there may be modifications of this kind of attacks, since the chosen bounds of coefficients of the Krammer matrices are rather rough whereas an image of an Artin generator is almost sparse matrix with small coefficients. We also remark that the proposed algorithm does not give an answer to the CP. Thus the CP is still hard and unsolved.

The rest of the paper is composed as follows: In Section 2, we briefly review braid groups and braid cryptography. In Section 3, we introduce the Lawrence-Krammer representation and develop its properties. Also inverting algorithm is given in more concrete way with the complexity. In Section 4, we introduce an equivalent key which plays an equivalent role as the original braid and analyze the cryptosystem using this. Section 5 gives the conclusion of this paper.

2   An Overview of Braid Group Cryptography

2.1   Braid Groups

A *braid* is obtained by laying down a number of parallel strands and intertwining them so that they run in the same direction. The number of strands is called the *braid index*. The set B_n of isotopy classes of braids of index n is naturally equipped with a group structure, called the *n-braid group*, where the product of two braids x and y is nothing more than laying down the two braids in a row and then matching the end of x to the beginning of y.

Any braid can be decomposed as a product of simple braids. One type of simple braids is the *Artin generator* σ_i that have a single crossing between i-th and $(i+1)$-th strand. B_n is presented with the Artin generators $\sigma_1, \ldots, \sigma_{n-1}$ and relations $\sigma_i \sigma_j = \sigma_j \sigma_i$ for $|i-j| > 1$ and $\sigma_i \sigma_j \sigma_i = \sigma_j \sigma_i \sigma_j$ for $|i-j| = 1$. When a braid a is expressed as a product of Artin generators, the minimum number of terms in the product is called the *word length* of a.

We have still other presentations. Let S_n be the symmetric group of an n-element set $I_n = \{1, 2, \ldots, n\}$. Let Ref $= \{(i,j)|1 \leq i < j \leq n\}$ be the set of reflections (that interchange two elements and fix the other elements of I_n) in S_n and S the subset $\{(i, i+1)|1 \leq i < n\}$ of Ref. We define $\ell(x)$ the *length of a permutation* x in S_n as

$$\ell(x) := \min\{k|x = x_1 \cdots x_k \text{ for } x_i \in S\}.$$

B_n admits another presentation with generators $\{rx|x \in S_n\}$ and relations $r(xy) = (rx)(ry)$ if $\ell(xy) = \ell(x) + \ell(y)$. In this presentation, the longest permutation w_0 with $w_0(i) = n+1-i$ yields a braid Δ, which is called the *fundamental braid* or the *half-twist* depending on authors. Let B_n^+ denote the submonoid of B_n generated by S_n. A braid in B_n^+ is said to be positive. A braid x is written uniquely, $x = \Delta^k x'$ where x' is in $B_n^+ - \Delta B_n^+$. This is called the normal form of x.

There is a partial order on B_n^+: $x \leq y \Leftrightarrow y \in xB_n^+$. The ordering is inherited to S_n (We identify a permutation σ with the corresponding braid $r\sigma$ in B_n^+.). We denote rS_n by Ω for simplicity reason. For a braid $x \in B_n^+$, the greatest element of the set $\{y \in \Omega | y \leq x\}$ is called the *left most factor* of x and denoted by $\mathrm{LF}(x)$. A sequence of braids $(x_1, \ldots, x_k)$ in $\Omega - \{1\}$ is called the *greedy form* of x if $x_1 \cdots x_k = x$, $\mathrm{LF}(x_i x_{i+1}) = x_i$ for all i. The above k in the greedy form is called the *Charney length* of x. This length function is easily extended to general braids using Thurston normal form, but we don't need it so general for our purpose and we will omit the general definition.

2.2 Braid Cryptography

Let G be a non-abelian group and $u, a, b, c \in G$. In order to perform the Diffie-Hellman key agreement on G we need to choose a, b in G satisfying $ab = ba$ in the DHCP. Hence we introduce two commuting subgroups $G_1, G_2 \subset G$ satisfying $ab = ba$ for any $a \in G_1$ and $b \in G_2$. More precisely, the problems the braid cryptography are based on are as follows:

- Input: A non-abelian group G, two commuting subgroups $G_1, G_2 \subset G$
- Conjugacy Problem (CP): Given (u, aua^{-1}) with $u, a \in G$, compute a. (Note that if we denote aua^{-1} by u^a, it looks like the DLP.)
- Diffie-Hellman Conjugacy Problem (DHCP): Given (u, aua^{-1}, bub^{-1}) with $u \in G$, $a \in G_1$ and $b \in G_2$, compute $baua^{-1}b^{-1}$.
- Decisional Diffie-Hellman Conjugacy Problem (DDHCP): Given $(u, aua^{-1}, bub^{-1}, cuc^{-1})$ with $u, c \in G$, $a \in G_1$ and $b \in G_2$, decide whether $abub^{-1}a^{-1} = cuc^{-1}$.

In braids, we can easily take two commuting subgroups G_1 and G_2 of B_n (For simplicity, we only consider a braid group with an even braid index. But it is easy to extend this to an odd braid index.). For example, $G_1 = LB_n$ (resp. $G_2 = RB_n$) is the subgroup of B_n consisting of braids made by braiding left $n/2$ strands(resp. right $n/2$ strands) among n strands. Thus LB_n is generated by $\sigma_1, \ldots, \sigma_{n/2-1}$ and RB_n is generated by $\sigma_{n/2+1}, \ldots, \sigma_{n-1}$. Then we have the commutative property that for any $a \in G_1$ and $b \in G_2$, $ab = ba$.

[**Key agreement.**] This is the braid group version of the Diffie-Hellman key agreement.

1. **Initial setup:** (a) Choose system parameters n and ℓ from positive integers. (b) Select a sufficiently complicated positive braid $u \in B_n$ with ℓ canonical factors.
2. **Key agreement:** Perform the following steps each time a shared key is required.
   (a) A chooses a random secret positive braid $a \in LB_n$ with ℓ canonical

factors and sends $v_1 = aua^{-1}$ to B.

(b) B chooses a random secret braid $b \in RB_n$ with ℓ canonical factors and sends $v_2 = bub^{-1}$ to A.

(c) A receives v_2 and computes the shared key $K = av_2a^{-1}$.

(d) B receives v_1 and computes the shared key $K = bv_1b^{-1}$.

Since $a \in LB_n$ and $b \in RB_n$, $ab = ba$. It follows

$$av_2a^{-1} = a(bub^{-1})a^{-1} = b(aua^{-1})b^{-1} = bv_1b^{-1}.$$

Thus Alice and Bob obtain the same braid.

[**Public-key cryptosystem.**] Let $H\colon B_n \to \{0,1\}^k$ be a cryptographically secure hash function from the braid group to the message space.

1. **Initial setup:** (a) Choose system parameters n and ℓ from positive integers. (b) Select a sufficiently complicated positive braid $u \in B_n$ with ℓ canonical factors.
2. **Key generation:**
   (a) Choose a sufficiently complicated positive braid $u \in B_n$ with ℓ canonical factors.
   (b) Choose a positive braid $a \in LB_n$ with ℓ canonical factors.
   (c) Public key is (u, v), where $v = aua^{-1}$; Private key is a.
3. **Encryption:** Given a message $m \in \{0,1\}^k$ and the public key (u, v),
   (a) Choose a positive braid $b \in RB_n$ with ℓ canonical factors.
   (b) Ciphertext is (c, d), where $c = bub^{-1}$ and $d = H(bvb^{-1}) \oplus m$.
4. **Decryption:** Given a ciphertext (c, d) and private key a, compute $m = H(aca^{-1}) \oplus d$.

Since a and b commute, $aca^{-1} = abub^{-1}a^{-1} = baua^{-1}b^{-1} = bvb^{-1}$. So $H(aca^{-1}) \oplus d = H(bvb^{-1}) \oplus H(bvb^{-1}) \oplus m = m$ and the decryption recovers the original braid m.

We may take a non-positive braid for a system braid or secret braids. But since the problem in that case is reduced to the positive braid cases, positive braids are enough for the random braids in this cryptosystem.

3   The Lawrence-Krammer Representation

3.1   Definitions and Properties

Most definitions and facts in this section are taken from two papers [10] [11] of Krammer. Let us recall the Lawrence-Krammer representation of braid groups. This is a representation of B_n in $GL_m(\mathbb{Z}[t^{\pm 1}, q^{\pm 1}]) = Aut(V_0)$, where $m = n(n-1)/2$ and V_0 is the free module of rank m over $\mathbb{Z}[t^{\pm 1}, q^{\pm 1}]$. We shall denote the representation by $\mathcal{K}$. With respect to $\{x_{ij}\}_{1 \le i < j \le n}$ the free basis of V_0 the image of each Artin generator under $\mathcal{K}$ is written as

$$
\mathcal{K}(\sigma_k)(x_{ij}) = \begin{cases}
tq^2 x_{k,k+1}, & i = k, \quad j = k+1; \\
(1-q)x_{i,k} + qx_{i,k+1}, & j = k, \quad i < k; \\
x_{ik} + tq^{k-i+1}(q-1)x_{k,k+1}, & j = k+1, \quad i < k; \\
tq(q-1)x_{k,k+1} + qx_{k+1,j}, & i = k, \quad k+1 < j; \\
x_{kj} + (1-q)x_{k+1,j}, & i = k+1, \quad k+1 < j; \\
x_{ij}, & i < j < k \quad \text{or} \quad k+1 < i < j; \\
x_{ij} + tq^{k-i}(q-1)^2 x_{k,k+1}, & i < k < k+1 < j.
\end{cases}
\tag{1}
$$

The matrix $\mathcal{K}(\sigma_k)$ with respect to the basis x_{ij} will be called by the Krammer matrix of a braid σ_k.

To estimate the complexity of the algorithm proposed here, we need to estimate bounds for the entries of a Krammer matrix.

Two useful results in [11] follow below:

Fact 1 *[11]* $\mathcal{K}(\Delta)(x_{n+1-j,n+1-i}) = tq^{i+j-1}x_{ij}$ *for* $1 \le i < j \le n$.

Fact 2 *[11] Let* $x \in B_n$. *Consider the Laurent series of* $\mathcal{K}(x)$ *with respect to* t,

$$
\mathcal{K}(x) = \sum_{i=k}^{\ell} A_i(q)t^i, \quad A_i \in M_m(\mathbb{Z}[q^{\pm 1}]), \quad A_k \ne 0, A_\ell \ne 0.
\tag{2}
$$

Then $\ell_\Omega(x) = \max(\ell - k, -k, \ell)$.

If we consider a different generator $Q = \{s(i,j)|$ the permutation braid of the reflection $(i,j) \in S_n\}$, we can define another length function ℓ_Q with respect to Q. This length is the canonical length in the band generator presentation. Remark that $\ell_Q(x)$ is bounded by $(n-1)$-times of the canonical length in the Artin presentation, because a band generator is written with upto $(n-1)$ Artin generators.

Define the anti-automorphism of B_n, written $x \mapsto \bar{x}$, by giving $[ij] \mapsto [n+1-i, n+1-j]$. This preserves B_n^+ as well as the canonical length. Then the dual representation is defined as $\mathcal{K}^* : B_n \to GL(V_0)$ by $\mathcal{K}^*(x) = \mathcal{K}(\bar{x})^T$, where T denotes the transpose. Consider another basis $\{v_{ij}|1 \le i < j \le n\}$ of V_0. It is related to $\{x_{ij}\}$ by

$$
v_{ij} = x_{ij} + (1-q) \sum_{i<k<j} x_{kj}, \quad x_{ij} = v_{ij} + (q-1) \sum_{i<k<j} q^{k-1-i} v_{kj}.
\tag{3}
$$

Fact 3 *[10] Let* $x \in B_n$. *Consider the Laurent series of* $\mathcal{K}^*(x)$ *with respect to* q,

$$
\mathcal{K}^*(x) = \sum_{i=k}^{\ell} A_i(t)q^i, \quad A_i \in M_m(\mathbb{Z}[t^{\pm 1}]), \quad A_k \ne 0, A_\ell \ne 0.
\tag{4}
$$

Then $\ell_Q(x) = \frac{1}{2}\max(\ell - k, -k, \ell)$.

From the above three facts, we get the following theorem.

Theorem 1. *Let x be a braid with the canonical form $\Delta^k x_1 x_2 \cdots x_\ell$ where x_i is a permutation braid which is not the fundamental braid. Let δ be the minimal number of Artin generators in x. Then we have the following bounds for the coefficients of $\mathcal{K}(x)$:*

(a) *The degree in t is bounded below by k and above by $k + \ell$.*
(b) *The degree in q is bounded below by $2(n-1)\min(0, k) + (n-2)$ and above by $2(n-1)\max(k+\ell, k) + (n-2)$.*
(c) *The coefficients of each entry inside the Krammer matrix are bounded by 2^δ when we consider the entries as a polynomial in t, q, and $1 - q$.*

Proof. (a) It is clear from Fact 2.

(b) Since $\bar{x}$ has the same canonical length with x and the band canonical length is bounded by $(n-1)$ times the Artin canonical length, the degree in q is bounded below by $2(n-1)\min(0, k)$ and above by $2(n-1)\max(k+\ell, k)$ in the $\{v_{ij}\}$ basis. While taking the basis change from $\{v_{ij}\}$ to $\{x_{ij}\}$, we have at most $(n-2)$ increase in the degree of q. Hence we get (b).

(c) If we consider entries of a Krammer matrix as a polynomial in t, q, and $1-q$, every entry of any Artin generator is a monomial with coefficients in $\{0, \pm 1\}$. For any Artin generator σ, each column of $\mathcal{K}(\sigma)$ has at most two nonzero terms (See the equation (1)). Hence a multiplication by a Krammer matrix of an Artin generator results in the increase of the coefficients by at most 2 times for each entries. Note that it happens when the same monomial occurs twice at an entry in the result matrix. Hence the coefficients of entries is bounded by 2^δ for the number of Artin generator in x.

For any positive integer n, the Krammer representation is faithful even if q is a real number with $0 < q < 1$ [11]. Also the inverting algorithm does not change even if q is replaced by a real number with $0 < q < 1$. From now on, we will consider the modified Krammer representation $\mathcal{K}'(x) = \mathcal{K}(x)_{q=1/2}$. In that case, q is equal to $1 - q$.

Corollary 1. *Let x be a braid with the canonical form $\Delta^k x_1 x_2 \cdots x_\ell$ where x_i is a permutation braid which is not the fundamental braid. Let δ be the number of Artin generators in x. Then we have the following bounds for $\mathcal{K}'(x)$:*

(a) *The degree in t is bounded below by k and above by $\max(k+\ell, k)$.*
(b) *The coefficients of each entry inside $\mathcal{K}'(x)$ is given by a ratio of two integers. The absolute values of numerators and denominators are bounded by $2^{\delta - 2(n-1)k}$ and $2^{2(n-1)\max(k+\ell, k)}$, respectively.*

3.2   Inverting the Lawrence-Krammer Representation

Here we develop a way to recover a braid from its image matrix under the Lawrence-Krammer representation. As mentioned earlier, the faithfulness of the Lawrence-Krammer representation of B_n in a linear group has been proven in several ways by different authors. Moreover, it has been known to be so easy that it takes a polynomial time of low degree in braid length and the index but we haven't found any reference with an explicit complexity available at hand.

The proof of faithfulness was due to Krammer [11], which enables us to construct an algorithmic way to recover the original braid from a matrix of the representation.

From Fact 1 we can easily obtain the matrix of Δ as tA, for a matrix A whose entries are from $\mathbb{Z}[q^{\pm 1}]$. Together with Fact 2, it suffices to recover the original braid x' of the matrix $(tA)^{-d_0}\mathcal{K}(x)$. Note that x' lies in $B_n^+ - \Delta B_n^+$, which corresponds to the nontrivial part in the normal form of x. x' has obviously smaller Charney length than x.

Suppose now x is a positive braid. Let us take $\{v_{ij}\}$ as the basis of V_0. The Lawrence-Krammer representation $\mathcal{K}$ yields a natural action of the monoid B_n^+ over V_0. Let A be the subset of Ref, $\{(i,j) \in \text{Ref} | (x(1,\dots,1))_{(i,j),t=0} \neq 0\}$. This A corresponds to a permutation y in S_n which corresponds to the braid ry in Ω. It makes the left most factor of x, so one has $x = yx'$. Applying the same steps to $\mathcal{K}(x')$ recursively, we obtain the greedy form of x after all, as it decreases the Charney length.

In this way, given $\mathcal{K}(x) = \sum_{i=d_t}^{\ell} A_i(q)t^i$, we can recover $x \in B_n$ in polynomial time. We shall describe the algorithm roughly as follows:

Algorithm 1 *Invert the Lawrence-Krammer representation.*
> **Input:** *A matrix $\mathcal{K}(x) \in GL_m(t^{\pm 1}, q^{\pm 1})$ where $m = n(n-1)/2$*
> **Output:** *A braid $x \in B_n$.*

1. *Compute $\mathcal{K}(x') = \mathcal{K}(\Delta)^{-d_t}\mathcal{K}(x)$*
2. *Perform the basis change from $(v_{ij})_{ij}$ to $(x_{ij})_{ij}$.*
3. *For $k = 1$ to ℓ do*
   2.1 Take a nonzero element $y \in D_\phi$ and compute

$$A = \{c_{ij} | \mathcal{K}(x')y \text{ has a nonzero coefficient at the } ij \text{ coordinate}\}$$

   (For the definition of the set D_ϕ one can refer to [11].)
   2.2 Compute the maximal element $\tau_k \in S_n$ such that $L(z) \subset A$ as follows.
   *   – Find all reflections $\sigma_i = (i, i+1)$ with $1 \le i < n$ such that $L(\sigma_i) \subset A$*
   *   – Given $\tau\sigma_i$ and $\tau\sigma_j$, a greater element is selected from $\tau\sigma_i\sigma_j$ or $\tau\sigma_i\sigma_j\sigma_i$*
   *   – Find the maximal element $\tau_k \in S_n$ by the recursive use of the above method*
   2.3 Compute the positive braid x_k corresponding to τ_k
   2.4 Replace $\mathcal{K}(x')$ by $\mathcal{K}(x_k)^{-1}\mathcal{K}(x')$
4. *Output $x = \Delta^{d_t}x_1x_2\cdots x_k$*

The complexity of this algorithm is about d_t power of a matrix and $n^2\ell$ multiplication of permutations, which is dominated by d_t power of a matrix.

Theorem 2. *Given $\mathcal{K}(x) = \sum_{i=d_t}^{\ell} A_i(q)t^i$, we can recover $x \in B_n$ in $O(2m^3 \log d_t)$ multiplications of entries.*

Note that it works even when a (nonzero) constant multiple of $\mathcal{K}'(x)$ is given since we only check whether the coefficient is zero in each stage. Hence we may deal with integer coefficients instead of rational coefficients.

4   Cryptanalysis of Braid Cryptosystems

4.1   An Equivalent Key

The security of the key exchange scheme and the encryption scheme in braids are based on the DHCP. The DHCP asks to find $baua^{-1}b^{-1}$ from $u, v = aua^{-1}, w = bub^{-1}$ given two commuting subgroups LB_n and RB_n of B_n, $a \in LB_n$, $b \in RB_n$ and $u \in B_n$. In this section, firstly, we will show that we don't need the original key a but a "fake" key A to solve the DHCP. The DHCP on a linear group is equivalent to a system of linear equations, whose solutions acts as the fake key. Note that it breaks the encryption scheme and key agreement scheme, but does not solve the original conjugacy problem to the bottom. The conjugacy problem in a general non-commutative group is, nevertheless, still difficult.

Without solving the problem in B_n, we try to solve it in $\mathrm{GL}_m(\mathbb{Z}[t^{\pm 1}, q^{\pm 1}])$ for $q = 1/2$ and $m = n(n-1)/2$ via the modified Lawrence-Krammer representation. Denote by A, B, U, V, and W the image of a, b, u, v, and w under this representation $\mathcal{K}'$, respectively. We will compute a matrix A from $\mathrm{GL}_m(\mathbb{Z}[t])$ satisfying the following equations:

$$UA = AV \tag{5}$$

$$A\mathcal{K}'(\sigma_i) = \mathcal{K}'(\sigma_i)A, \quad n/2 < i < n. \tag{6}$$

The solutions in $\mathbb{Z}[t]^{m^2}$ make a nontrivial module $\mathcal{N}$ over $\mathbb{Z}[t]$, since we have already a nontrivial solution $\mathcal{K}(a)$. As the set of invertible matrices in $\mathcal{N}$ is dense under Zariski topology, we can take an invertible matrix over $\mathbb{Q}(t)$ from $\mathcal{N}$ with overwhelming probability. Let A' be an invertible matrix solution. Using A', one can compute $\mathcal{K}'(baua^{-1}b^{-1})$ in the matrix ring as follows:

$$A'WA'^{-1} = A'BUB^{-1}A'^{-1} = BA'UA'^{-1}B^{-1} = BVB^{-1} = \mathcal{K}'(baua^{-1}b^{-1}). \tag{7}$$

That is, the matrix A' plays the same role that the key a does. Thus we call such A' a *pseudo-key*.

4.2   A System of Linear Equations

We are able to change the above into an overdetermined system of linear equations of A. That is, we obtain the system of equations of the following form:

$$T_0 N = \begin{bmatrix} K \\ L_{n/2+1} \\ \vdots \\ L_{n-1} \end{bmatrix} X = 0, \tag{8}$$

where X is the column vector $[a_{11}, \dots, a_{1m}; a_{21}, \dots, a_{2m}; \dots; a_{m1} \dots, a_{mm}]^t$ made from $A = [a_{ij}]$ and K, L_i's are the $m^2 \times m^2$ matrix of the linear relations in Equation (5) and (6), respectively.

The system (8) has m^2 variables and $(n/2)m^2$ equations. However, by precise analysis of Krammer matrices, we can reduce the number of variables and equations as follows:

Theorem 3. *Equation (8) has at most $\frac{1}{7}n^4$ nontrivial variables and $\frac{1}{4}n^4$ nontrivial equations.*

Proof. Define V_k to be a subspace of V_0 generated by $\{x_{ij}|(i,j) \notin I_k\}$ where $I_k = \{(i,j)|1 \le i < j < k \quad \text{or} \quad k+1 < i < j \le n\}$. From Equation (1), we see that the Krammer matrix $\mathcal{K}(\sigma_k)$ transforms V_k to itself and acts as the identity on the basis element x_{ij} when $(i,j) \in I_k$. Thus it can be written as $\begin{bmatrix} M_k & 0 \\ 0 & I \end{bmatrix}$ by reordering of the basis, where M_k is a square matrix of size $k(n-k)+n$ $(= \binom{n}{2} - \binom{k-1}{2} - \binom{n-k-1}{2})$.

Since $\cap_{1 \le k < n/2} I_k = \{(i,j)|n/2 \le i < j \le n\}$, a Krammer matrix of any left-braid $a \in LB_n$ can be written as $\begin{bmatrix} M & 0 \\ 0 & I \end{bmatrix}$ where M is a square matrix of size $\frac{1}{8}(3n^2 - 2n - 8)$ $(= \binom{n}{2} - \binom{n/2+1}{2})$. Therefore only $\frac{1}{8^2}(3n^2 - 2n - 8)^2$ entries of A in Equation (5) are unknown.

This property of A reduces the number of equation in Equation (5) to $\binom{n}{2}^2 - \binom{n/2}{2}^2 \approx \frac{15}{64}n^4$. Also each equation in Equation (6) has only $k(n-k)+n$ nontrivial equations, whose sum for $n/2 \le k < n$ is about $\frac{1}{12}n^3$. Hence the total number of non-trivial equations are at most $\frac{1}{4}n^4$.

4.3   Estimate the Diffie-Hellman Key

Theorem 4. *Let $u \in B_n$, $a \in LB_n$, and $b \in RB_n$ with ℓ canonical factors. Then $abub^{-1}a^{-1}$ can be written as a product of at most ℓ number of Δ^{-1} and at most 3ℓ number of canonical factors. Further each entry inside $\mathcal{K}'(abub^{-1}a^{-1})$ is a Laurent polynomial of t*

$$\sum_{d=-\ell}^{4\ell} \frac{a_i}{b_i} t^d \quad \text{with } |a_i| \le 2^{\delta+2n\ell} \text{ and } |b_i| \le 2^{8n\ell},$$

where δ is the number of Artin generators in $abub^{-1}a^{-1}$ bounded by $2\ell n(n-1)$.

Proof. Denote by $\text{len}(x)$ the Charney length of x. Observe that $\text{len}(xy) \le \text{len}(x) + \text{len}(y)$ for $x,y \in B_n$ and $\text{len}(ab) \le \max(\text{len}(a), \text{len}(b))$ for $a \in LB_n$ and $b \in RB_n$. Also the inverse of x for $x \in B_n$ with r canonical factors is written as a product of at most r number of Δ^{-1} and at most r number of canonical factors. Since ab consists of at most ℓ canonical factors, we get the first assertion. The second assertion follows from Theorem 1.

Since u, v, and σ_k are positive braids, the entries of corresponding Krammer matrices are polynomial with rational coefficients. By multiplying the appropriate scalars to the both sides of Equations (5) and (6), we can consider

$U, V, \mathcal{K}'(\sigma_i)$, and even A as matrices whose entries are polynomials with integer coefficients.

Let p be a prime with $p > 2^{\delta + 10n\ell + 1}$ and $f(t)$ an irreducible polynomial of degree 5ℓ over $\mathbb{Z}/p$. Since each entry of $\mathcal{K}(abub^{-1}a^{-1})$ is a polynomial of degree 5ℓ and with coefficient $< p$, we know that

$$\mathcal{K}'(baua^{-1}b^{-1}) = t^{-\ell}2^{-8n\ell}\{t^{\ell}2^{8n\ell}\mathcal{K}'(baua^{-1}b^{-1}) \mod (p, f(t))\} \quad (9)$$

if we take a representative of a residue class for coefficients from the interval $(-p/2, p/2)$. Therefore we are enough to compute $A \mod (p, f(t))$ in Equation (5) and (6). From the famous Bertrand's postulate below, it is guaranteed that $p < 2^{\delta + 10n\ell + 2}$.

Fact 4 (Bertrand's postulate) *[7] There exists a prime between n and $2n$.*

4.4   Algorithm and Complexity

The proposed algorithm to solve the braid Diffie-Hellman problem is described roughly as follows:

Algorithm 2 *Find an equivalent key using Gaussian Elimination.*
  * **Input:** $u \in B_n$, $a \in LB_n$, $b \in RB_n$, $m = n(n-1)/2$, a prime p, and an irreducible polynomial $f(t)$ of the degree d satisfying Equation (9).
  * **Output:** $\mathcal{K}'(baua^{-1}b^{-1})$.

1. *Compute the images of u and $v = aua^{-1}$ in $GL_m(k)$ via $\mathcal{K}'$, where k is the residue field $k = \mathbb{Z}[t]/(p, f(t))$.*
2. *Induce a system $\frac{1}{4}n^4$ linear equations in $\frac{1}{7}n^4$ variables from the simultaneous equations $\mathcal{K}'(v)A = A\mathcal{K}'(u)$ and $\mathcal{K}'(\sigma_i)A = A\mathcal{K}'(\sigma_i)$ for $n/2 < i \leq n$ over k*
3. *Apply Gaussian elimination for the system in order to compute A. We may multiply an appropriate integer to the both side of each equation to get integer coefficients.*
4. *If A is nonsingular, compute A^{-1}. Otherwise, go back to the above step and take another solution.*
5. *Compute $\mathcal{K}'(w)$ for $w = bub^{-1}$ and output $A\mathcal{K}'(w)A^{-1} = \mathcal{K}'(baua^{-1}b^{-1})$*
6. *Use Algorithm 1 to compute $baua^{-1}b^{-1}$.*

To evaluate the complexity of Gaussian elimination step, we need the following two facts:

Fact 5 *[17, p.15] The Gaussian elimination of an $m \times m$ matrix takes $\frac{1}{3}m^{\tau}$ for $\tau = \log_2 7$, which can be reduced to 2.376 theoretically.*

Fact 6 *[16] One multiplication or one inversion in a finite field with cardinality p^d takes $O(\log^{\epsilon} p^d)$ bit operations where $\epsilon > \log_2 3$.*

Using the above facts, we can estimate the complexity of our algorithm as follows:

Theorem 5. *Assume LB_n and RB_n are two commuting subgroups of the n-braid group B_n. Given $u \in B_n, a^{-1}ua, b^{-1}ub$ for $a \in LB_n$ and $b \in RB_n$, $b^{-1}a^{-1}uab$ can be computed in about $2^{-5}n^{4\tau}(\ell\delta)^\epsilon$ (or $2^{-3}n^{4\tau+2\epsilon}\ell^{2\epsilon}$) bit operations where δ is the maximum word length of $abub^{-1}a^{-1}$ bounded by $2\ell n^2$.*

Proof. First, evaluate the complexity of Step 3. Since $p < 2^{\delta+10n\ell+2}$ and $d < 5\ell$, it is

$$\frac{1}{3}\left(\frac{1}{4}n^4\right)^\tau \log^\epsilon p^d \le 2^{-5}n^{4\tau}\ell^\epsilon(\delta + 10n\ell + 2)^\epsilon \approx 2^{-5}n^{4\tau}(\ell\delta)^\epsilon. \tag{10}$$

The inverse of A can be computed in $O(n^3 \log^\epsilon(p^d))$. From Theorem 2, we know that recovering the braid awa^{-1} takes $O(2m^3 \log \ell)$ multiplications in k, which is about $O(n^6(\ell\delta)^\epsilon)$. The remainder takes very little. Hence the complexity of this algorithm is dominated by that of Gaussian elimination.

If we take $\tau = 2.8$ and $\epsilon = 1.6$, the complexity is $O(n^{14.4}\ell^{3.2})$. Theoretically, we can take $\tau = 2.376$ so that the complexity is $O(n^{9.5}\ell^{3.2})$.

In Table 1, we compare the attack complexity of braid encryption scheme, where n is the braid index and ℓ is the canonical length of a, b and u. The column [9] shows the complexity of the brute force attack with complexity $(\frac{n}{2}!)^\ell$ (the first three numbers were cited from [9] and the remainder was computed by $2^{n\ell}$ roughly since it is enough for this large number.) and the column [4] shows the super-summit attack with complexity $(n/2)^\ell$. The complexity of the proposed algorithm is evaluated by $2^{-3}n^{4\tau+2\epsilon}\ell^{2\epsilon}$ for $\tau = \log_2 7$ and $\epsilon \approx \log_2 3$. The column for ECC means the key size of elliptic curve cryptography with corresponding complexity (which was estimated roughly by square-root attacks such as Pollard ρ).

Note that the super-summit attack [4] is efficient for small n, but the proposed attack is efficient for large n since it has a *polynomial* complexity. The table shows that it is very hard to increase the complexity of braid encryption scheme, for example, in order to obtain similar complexity to 556 bit elliptic curve cryptography, the braid index should be about 10^5 (huge!!). Also in this case one cipher text must be about $10^9 \approx 2^{30}$ bits.

5   Conclusion

In this paper we proposed a polynomial time algorithm to solve the DHCP in braid groups. Though the complexity is too large to break the encryption scheme with the proposed parameters in [9] in real time, the braid encryption scheme is considered to be insecure since increasing the key size increase the attack complexity only a little. For example, to get the same complexity with 556 bit elliptic curve cryptography, the braid index should be about 10^5, which is impossible since one cipher text must be more than 10^9 bits.

We expect that the complexity can be reduced by more precise analysis on the Lawrence-Krammer representation. Further this analysis can be applied even to the generalized scheme based the decomposition problem [4] with the similar

Table 1. The performance of the attack algorithm

n	ℓ	[9]	[4]	Proposed Alg.	Key size of corr. ECC
50	5	2^{251}	2^{13}	2^{86}	172
70	7	2^{665}	2^{35}	$2^{95.4}$	191
90	12	2^{1863}	2^{66}	2^{103}	206
200	30	2^{6000}	2^{199}	2^{125}	250
1000	100	2^{10^5}	2^{900}	2^{162}	324
10000	1000	2^{10^7}	2^{12330}	2^{221}	442
100000	10000	2^{10^9}	$2^{1566666}$	2^{278}	556

complexity since only the number of variables in the system of equations is doubled in the generalized version.

Since this cryptanalysis is based on the faithfulness of the Krammer representation, losing the group structure would be a possible way to avoid this kind of attacks. Currently, the key agreement scheme in [2] or the first key agreement scheme in [1] resists against this attack since it loses the group structure through the extractor map, so we cannot directly apply the same steps to obtain a pseudo-key [9].

Acknowledgements. The authors would like to thank Ki Hyoung Ko, Sang Jin Lee, Eonkyung Lee, Jae Choon Cha, and Sang Geun Hahn for initial discussions on this problem, and the anonymous referees for their helpful comments. This work was supported by KISA (*R&D* 2002-S-073) and IRIS.

References

1. I. Anshel, M. Anshel, B. Fisher, and D. Goldfeld, *New Key Agreement Protocols in Braid Group Cryptography*, Proc. of CT-RSA 2001, Lecture Notes in Computer Science, Vol. 2020, Springer-Verlag, pp 13–27, 2001.
2. I. Anshel, M. Anshel, and D. Goldfeld, *An Algebraic Method for Public-Key Cryptography*, Math. Res. Lett., Vol. 6, No. 3-4, pp. 287-291, 1999.
3. J. Birman, K. Ko and S. Lee, *A New Approach to the Word and Conjugacy Problem in the Braid Groups*, Advances in Mathematics, Vol. 139, pp. 322–353, 1998.
4. J. Cha, K. Koh, S. Lee, J. Han, and J. Cheon, *An Efficient Implementations of Braid Groups*, Proc. of Asiacrypt 2001, Lecture Notes in Computer Science, Vol. 2248, Springer-Verlag, pp. 144–156, 2001.
5. R. Gennaro and D. Micciancio, *Cryptanalysis of a Pseudorandom Generator Based on Braid Groups*, Proc. of Eurocrypt 2002, Lecture Notes in Computer Science, Vol. 2332, Springer-Verlag, pp. 1–13, 2002.
6. D. Hofheinz and R. Steinwandt, *A Practical Attack on Some Braid Group Based Cryptography Primitives*, Proc. of PKC 2003, Lecture Notes in Computer Science, Vol. 2567, Springer-Verlag, pp. 187–198, 2003.

7. G. H. Hardy, E. M. Wright, *An introduction to the Theory of Numbers*, Oxford Univ. Press, 1978

8. K. Koh *et. al New Signature Scheme Using Conjugacy Problem*, Preprint, 2002.

9. K. Ko, S. Lee, J. Cheon, J. Han, J. Kang, C. Park, *New Pulic-key Cryptosystem using Braid Groups*, Proc. of Crypto 2000, Lecture Notes in Computer Science, Vol. 1880, Springer-Verlag, pp. 166–183, 2000

10. D. Krammer, *The Braid group B_4 is Linear*, Inventiones Mathematics, Vol. 142, pp. 451–486, 2002.

11. D. Krammer, *Braid groups are Linear*, Annals of Mathematics, Vol. 155, pp. 131–156, 2002.

12. S. Lee, *The Trapdoor Oneway Functions in Braid Groups*, Workshop on Algbraic Methods in Cryptography, Slides are available in http://knot.kaist.ac.kr/ sjlee.

13. S. Lee and E. Lee, *Potential Weaknesses of the Commutator Key Agreement Protocol Based on Braid Groups*, Proc. of Eurocrypt 2002, Lecture Notes in Computer Science, Vol. 2332, Springer-Verlag, pp. 14–28, 2002.

14. E. Lee and J. Park, *Cryptanalysis of the Public-key Encryption based on Braid Groups*, Proc. of Eurocrypt 2003, Lecture Notes in Computer Science, Vol. 2656, Springer-Verlag, pp. 477–490, 2003.

15. E. Lee, S. J. Lee and S. G. Hahn, *Pseudorandomness from Braid Groups*, Proc. of Crypto 2001, Lecture Notes in Computer Science, Vol. 2139, Springer-Verlag, pp. 486–502, 2001.

16. A. Menezes, P. Oorschot, and S. Vanston, *Handbook of Applied Cryptography*, CRC Press, 1997.

17. G. Strang, *Linear Algebra and its Applications*, Harcourt, 1986.

The Impact of Decryption Failures on the Security of NTRU Encryption

Nick Howgrave-Graham[1], Phong Q. Nguyen[2], David Pointcheval[2],
John Proos[3], Joseph H. Silverman[1], Ari Singer[1], and William Whyte[1]

[1] NTRU Cryptosystems, 5 Burlington Woods, Burlington, MA 02144
{nhowgravegraham,jhs,asinger,wwhyte}@ntru.com
[2] CNRS/ENS–DI, 45 rue d'Ulm, 75005 Paris, France
{phong.nguyen,david.pointcheval}@ens.fr
[3] University of Waterloo, 200 University Ave. West, Waterloo, Canada N2L 3G1
japroos@math.uwaterloo.ca

Abstract. NTRUEncrypt is unusual among public-key cryptosystems in that, with standard parameters, validly generated ciphertexts can fail to decrypt. This affects the provable security properties of a cryptosystem, as it limits the ability to build a simulator in the random oracle model without knowledge of the private key. We demonstrate attacks which use decryption failures to recover the private key. Such attacks work for all standard parameter sets, and one of them applies to any padding. The appropriate countermeasure is to change the parameter sets and possibly the decryption process so that decryption failures are vanishingly unlikely, and to adopt a padding scheme that prevents an attacker from directly controlling any part of the input to the encryption primitive. We outline one such candidate padding scheme.

1 Introduction

An unusual property of the NTRU public-key cryptosystem is the presence of *decryption failures*: with standard parameters, validly generated ciphertexts may fail to decrypt. In this paper, we show the importance of decryption failures with respect to the security of the NTRU public-key cryptosystem. We believe this fact has been much overlooked in past research on NTRU.

First, we notice that decryption failures cannot be ignored, as they happen much more frequently than one would have expected. If one strictly follows the recommendations of the EESS standard [3], decryption failures happen as often as every 2^{12} messages with $N = 139$, and every 2^{25} messages with $N = 251$. It turns out that the probability is somewhat lower (around 2^{-40}) with NTRU products, as the key generation implemented in NTRU products surprisingly differs from the one recommended in [3]. In any case, decryption failures happen sufficiently often that one cannot dismiss them, even in NTRU products.

One may *a priori* think that the only drawback with decryption failures is that the receiver will not be able to decrypt. However, decryption failures have a big impact on the confidence we can have in the security level, because

D. Boneh (Ed.): CRYPTO 2003, LNCS 2729, pp. 226–246, 2003.

they limit the ability to build a simulator in the random oracle model without knowledge of the private key. This limitation is independent of the padding used. This implies that all the security proofs known (see [14,21]) for various NTRU paddings may not be valid after all, because decryption failures have been ignored in such proofs. This also means that existing generic padding schemes (such as REACT [23]) may not apply to NTRU as, to our knowledge, no existing padding scheme takes into account the possibility of decryption failures, perhaps because the only competitive cryptosystem that experiences decryption failures is NTRUEncrypt.

From a security point of view, the situation is even worse. It turns out that decryption failures not only influence the validity of a security proof, they leak information on the private key. We demonstrate this fact by presenting new efficient chosen-ciphertext attacks on NTRUEncrypt (with or without padding) that recover the private key. These chosen-ciphertext attacks are very different from chosen-ciphertext attacks [17,15] formerly known against NTRU: they only use valid ciphertexts, while the attacks of [17,15] use fake ciphertexts and can therefore be easily thwarted. Moreover, these chosen-ciphertext attacks do not use the full power of chosen-ciphertext attacks, but reaction attacks only [9]: Here, the attacker selects various messages, encrypts them, and checks whether the receiver is able to correctly decrypt the ciphertexts: eventually, the attacker has gathered sufficiently many decryption failures to be able to recover the private key. The only way to avoid such chosen-ciphertext attacks is to make sure that it is computationally infeasible to find decryption failures. This requires different parameter sets and certain implementation changes in NTRUEncrypt, which we hint in the final section of this paper.

The rest of the paper is organized as follows. In Sections 2 and 3, we recall NTRUEncrypt and the padding used in EESS [4]. In Section 4, we explain decryption failures, and their impact on security proofs for NTRU paddings. In Section 5, we present several efficient attacks based on decryption failures: some are tailored to certain paddings, while the most powerful one applies to any padding. In Appendix A, we give additional information on NTRUEncrypt, pointing out the difference between the specification of the EESS standard and the implementation in NTRU products. In Appendix B, we describe an alternative attack based on decryption failures that work against certain NTRU paddings.

2   The NTRU Encryption Scheme

2.1   Definitions and Notation

NTRUEncrypt operations take place in the quotient ring of polynomials $\mathcal{P} = \mathbb{Z}[X]/(X^N - 1)$. In this ring, addition of two polynomials is defined as pairwise addition of the coefficients of the same degree, and multiplication is defined as convolution multiplication. The convolution product $h = f * g$ of two polynomials f and g is given by taking the coefficient of X^k to equal

$h_k = \sum_{i+j\equiv k \bmod N} f_i \cdot g_j$. Several different measures of the size of a polynomial turn out to be useful. We define the *norm* of a polynomial f in the usual way, as the square root of the sum of the squares of its coefficients. We define the *width* of a polynomial f as the difference between its largest coefficient and its smallest coefficient.

The fundamental parameter in NTRUENCRYPT is N, the ring dimension. The parameter N is taken to be prime to prevent attacks due to Gentry [6], and sufficiently large to prevent lattice attacks. We also use two other parameters, p and q, which are relatively prime. Standard practice is to take q to be the power of 2 between $N/2$ and N, and p to be either the integer 3 or the polynomial $2 + X$. We thus denote p as a polynomial p in the following, and we focus on the case $p = 2 + X$ as $p = 3$ is no longer recommended in NTRU standards [3,4].

2.2 Overview

The basic NTRUENCRYPT key generation, encryption, and decryption primitives are as follows.

Key Generation: Requires a source of (pseudo-)random bits, and subspaces $\mathcal{L}_f, \mathcal{L}_g \subseteq \mathcal{P}$ from which the polynomials f and g are drawn. These subspaces have the property that all polynomials in them have small width – for example, they are now commonly taken to be the space of *all binary polynomials* with d_f, d_g 1s respectively.
- INPUT: Values N, p, q.
  - Randomly choose $f \in \mathcal{L}_f$ and $g \in \mathcal{L}_g$ in such a way that both f and g are invertible mod q;
  - Set $h = p * g * f^{-1} \bmod q$.
- PUBLIC OUTPUT: The input parameters and h.
- PRIVATE OUTPUT: The public output, f, and $f_p \equiv f^{-1} \bmod p$.

Encryption: Requires a source of (pseudo-)random bits, subspaces $\mathcal{L}_r$ and $\mathcal{L}_m$ from which the polynomials r and m are to be drawn, and an invertible means $\mathcal{M}$ of converting a binary message m to a message representative $m \in \mathcal{L}_m$. The subspaces $\mathcal{L}_r, \mathcal{L}_m$ also have the property that all polynomials in them have low width.
- INPUT: A message m and the public key h.
  - Convert m to the message representative $m \in \mathcal{L}_m$: $m = \mathcal{M}(m)$;
  - Generate random $r \in \mathcal{L}_r$.
- OUTPUT: The ciphertext $e = h * r + m \bmod q$.

Decryption:
- INPUT: The ciphertext e, the private key f, f_p and the parameters p and q.
  - Calculate $a = f * e \bmod q$. Here, "mod q" denotes reduction of a into the range $[A, A + q - 1]$, where the "centering value" A is calculated by a method to be discussed later;
  - Calculate $m = f_p * a \bmod p$, where the reduction is to the range $[0, 1]$.
- OUTPUT: The plaintext m, which is m converted from a polynomial in $\mathcal{L}_m$ to a message.

2.3   Decryption Failures

In calculating $a = f * e \bmod q$, one actually calculates

$$a = f * e = f * (r * h + m) = p * r * g + f * m \bmod q. \tag{1}$$

The polynomials p, r, g, m, and f are chosen to be small in $\mathcal{P}$, and so the polynomial $p * r * g + f * m$ will, with "very high probability", have width less than q. If this is the case, it is possible to reduce a into a range $[A, A + q - 1]$ such that the $\bmod q$ equality in the Equation (1) is an exact equality in $\mathbb{Z}$. If this equality is exact, the second convolution gives

$$f_p * a = p * (f_p * r * g) + f_p * f * m = 0 + 1 * m = m \bmod p, \text{ recovering m.}$$

Decryption only works if the equality $\bmod q$ in Equation (1) is also an equality in $\mathbb{Z}$. This condition will not hold if A has been incorrectly chosen so that some coefficients of $p * r * g + f * m$ lie outside the centering range, or if $p * r * g + f * m$ happens to have a width greater than q (so that there is no $\bmod q$ range that makes the Equation (1) an exact equality). In this case, the recovered m will differ from the encrypted m by some multiple of $q \bmod p$. These events are the *decryption failures* at the core of this paper.

Before NTRUEncrypt can be used, the subspaces $\mathcal{L}_f, \mathcal{L}_g, \mathcal{L}_r$ must be specified. See appendix A for details of the precise form of the polynomials f, g, r used in the standard [3] and the (slightly) different one deployed in NTRU Cryptosystems' products.

2.4   The NTRU Assumption

Among all the assumptions introduced in [21], the most important one is the one-wayness of the NTRU primitive, namely that the following problem is asymptotically hard to solve:

Definition 1 (The NTRU Inversion Problem). *For a given security parameter k, which specifies N, p, q and the spaces $\mathcal{L}_f$, $\mathcal{L}_g$, $\mathcal{L}_m$, and $\mathcal{L}_r$, as well as a random public key h and $e = h * r + m$, where $m \in \mathcal{L}_m$ and $r \in \mathcal{L}_r$, find m. We denote by $\mathsf{Succ}^{ow}_{ntru}(\mathcal{A})$ the success probability of any adversary $\mathcal{A}$.*

$$\mathsf{Succ}^{ow}_{ntru}(\mathcal{A}) = \Pr\left[\begin{array}{l} (h, \star) \leftarrow \mathcal{K}(1^k), m \in \mathsf{Messages}, r \in \mathsf{Random}, \\ e = h * r + m \bmod q : \mathcal{A}(e, h) = m \end{array} \right].$$

3   The NTRU Paddings

For clarity reasons, in the description of the paddings, we consider the encryption scheme $\Pi = (\mathcal{K}, \mathcal{E}, \mathcal{D})$, which is an improvement of the plain NTRU cryptosystem that includes the two public encodings $\mathcal{M}$ and $\mathcal{R}$:

$$\left\{ \begin{array}{ll} \mathcal{K}(1^k) = (\mathsf{pk} = h, \mathsf{sk} = (f, f_p)), & \text{with,} \\ \mathcal{E}_{pk}(m; r) = \mathcal{M}(m) + \mathcal{R}(r) * h \bmod q, & \mathcal{M} : \mathsf{Messages} = \{0,1\}^{mLen} \rightarrow \mathcal{L}_m \\ \mathcal{D}_{sk}(e) = \mathcal{M}^{-1}((e * f \bmod q)_A * f_p) & \mathcal{R} : \mathsf{Random} = \{0,1\}^{rLen} \rightarrow \mathcal{L}_r \end{array} \right.$$

Because of the encodings, without any assumption, recovering the bit-string m is as hard as recovering the polynomial $\mathsf{m} = \mathcal{M}(m)$. However, recovering ℓ bits of m does not necessarily provide ℓ bits of the polynomial $\mathsf{m} = \mathcal{M}(m)$, which is the reason why stronger assumptions were introduced in [21].

3.1   Description

Following the publication of [17], several padding schemes were proposed in [14, 13] to protect NTRUENCRYPT against adaptive chosen-ciphertext attacks. The resulting schemes were studied in [21], but under the assumption that the probability of decryption failures was negligible. We briefly review one of these schemes, known as SVES-1 and standardized in [3].

Let m be the original plaintext represented by a k_1-bit string. For each encryption, one generates a random string b, whose bit-length k_2 is between 40 and 80 [14]. However, $k_1 + k_2 \leq \mathsf{mLen}$. Let $\|$ denote bit-string concatenation.

To encrypt, first split each of m and b into equal size pieces $m = \overline{m} \| \underline{m}$ and $b = \overline{r} \| \underline{r}$. Then, use two hash functions F and G that map $\{0,1\}^{k_1/2+k_2/2}$ into itself, to compute:

$$m_1 = (\overline{m} \| \overline{r}) \oplus F(\underline{m} \| \underline{r}) \text{ and } m_2 = (\underline{m} \| \underline{r}) \oplus G(m_1).$$

A third hash function, $H : \{0,1\}^{\mathsf{mLen}} \to \{0,1\}^{\mathsf{rLen}}$, is applied to yield the ciphertext:

$$\mathcal{E}^3_{\mathsf{pk}}(m; b) = \mathcal{E}_{\mathsf{pk}}(m_1 \| m_2; H(m \| b)).$$

The decryption algorithm consists of recovering m, b from the plain NTRU decryption, and re-encrypting to check whether one obtains the given ciphertext. This is equivalent to extracting, from m and e, the alleged r, and check whether it is equal to $\mathcal{R}(H(m \| b))$. We denote by Π^3 the corresponding encryption scheme.

3.2   Chosen-Ciphertext Attacks

First of all, one may note that because of the random polynomial r that is generated from $H(m \| b)$, nobody can generate a valid ciphertext without knowing both the plaintext m and the random b, except with negligible probability, for well-chosen conversion $\mathcal{R}$ (which is not necessarily injective, according to [3,4]). Indeed, for a given ciphertext e, at most one r is acceptable. Without having asked $H(m \| b)$, the probability for $\mathcal{R}(H(m \| b))$ to be equal to r is less than ε_r:

$$\varepsilon_r = \max_{\mathsf{r}} \{ \Pr_{x \in \{0,1\}^{\mathsf{rLen}}} [\mathsf{r} = \mathcal{R}(x)] \}.$$

However, to lift any security level from the CPA scenario to the CCA-one, one needs a good simulation of the decryption oracle: since any valid ciphertext has been correctly constructed, one looks at the list of the query-answers to the H-oracle, and can re-encrypt each possible plaintext to check which one is the good one. This perfectly simulates the decryption of validly generated ciphertexts, unless a *decryption failure* occurs (Fail event). In the latter case, the output decrypted plaintext is not the encrypted one, whereas the output simulated plaintext is: the probability of bad simulation is finally less than $\varepsilon_r + p(m, r, h)$, where

$$p(m, r, h) = \Pr_{f,g}[\mathcal{D}_{f,f_p}(\mathcal{E}_h(m, r)) \text{ Fail } \mid h = p * g * f^{-1} \pmod{q}].$$

The security analyses in [21] were performed assuming that $p(m, r, h)$ is negligible (and even 0). But we shall see later that this is unfortunately not the case. We thus refine the security analyses and even show that the parameters have to be chosen differently.

4   Decryption Failures and Provable Security

4.1   Wrap Failures and Gap Failures

When decrypting, the recipient must place the coefficients of $a = f * e \bmod q$ into the correct range $[A, A + q - 1]$. To calculate A, we use the fact that convolution multiplication respects the homomorphism $(a * b)(1) = a(1) \cdot b(1)$, where $a(1)$ is the sum of the coefficients of the polynomial a. The decrypter knows $r(1)$ and $h(1)$, and so he can calculate $I = m(1) = e(1) - r(1) \cdot h(1) \bmod q$. Assuming $m(1)$ lies in the range $[N/2 - q/2, N/2 + q/2]$, we can calculate the average value of a coefficient of $p * r * g + f * m$ and set

$$A = \left\lfloor \frac{p(1) \cdot r(1) \cdot g(1) + f(1) \cdot I}{N} \right\rceil - \frac{q}{2}.$$

The assumption about the form of m is a reasonable one, because for randomly chosen m with $N = 251$ there is a chance of less than 2^{-56} that $m(1)$ will be less than $N/2 - q/2$ or greater than $N/2 + q/2$. In any case, the value of $m(1)$ is known to the encrypter, so they do not learn anything from a decryption failure based on m being too thick or too thin.

Having obtained A, we reduce a into the range $[A, A + q - 1]$. If the actual values of any of the coefficients of $p * r * g + f * m$ lie outside this range, decryption will produce the wrong message and this will be picked up in the re-encryption stage of decryption. However, if $p * r * g + f * m$ has a width less than q, there is still an A for which decryption is possible. This case, where the initial choice of A does not work but there is a choice of A which could work, is referred to as a "wrap failure".

Wrap failures are more common than "gap failures", where the width of $p * r * g + f * m$ is strictly greater than q. The standard [3] therefore recommends that, on the occurrence of a decryption failure, the decrypter adjusts the decryption range by setting A' successively equal to $A \pm 1, \pm 2, \ldots$, placing the coefficients of a into the new range $[A', A' + q - 1]$, and performing the mod p reduction. This is to be carried out until A' differs from A by some set T, the "wrapping tolerance"; if decryption has not succeeded at that point, the decryption function outputs the invalid symbol $\perp$.

This method increases the chance that a ciphertext will eventually decrypt; however, an attacker with access to timing information can tell when this re-centering has occurred. For standard $N = 251$ parameters and NTRUENCRYPT implemented as in NTRU products, a wrap failure on random m occurs once every 2^{21} messages, while a gap failure occurs about once every 2^{43} messages. The centering method above will therefore leak information at least once every million or so decryptions, and possibly more often if the attacker can carry out some precomputation as in [19]. For NTRUENCRYPT as implemented following the EESS standard, the number of messages is much lower: for $N = 251$, a gap failure occurs once every 2^{25} message.

4.2   Provable Security

In order to deal with any padding, one needs more precise probability informations than just $p(m, r, h)$:

$$p(m, h) = \Pr_r[p(m, r, h)]$$
$$p(h) = \Pr_m[p(m, h)] = \Pr_{m,r}[p(m, r, h)]$$

$$p_0 = E_h[\max_{m,r}\{p(m, r, h)\}];$$
$$p_1 = E_h[\max_m\{p(m, h)\}];$$
$$p_2 = E_h[p(h)].$$

Note that all of these probabilities are averages over the whole space of h. Implicit in these definitions is the assumption that, even if some keys (f, g) are more likely than others to experience decryption failures, an adversary cannot tell from the public key h which private key is more failure-prone.

Clearly, one cannot ensure that $p(m, r, h)$ is small, so p_0 is likely to be non-negligible. However, in several paddings, $r = \mathcal{R}(H(m \parallel b))$, where H is a random oracle, therefore the probability of bad simulation involves at least p_1, or even p_2. As discussed above, with recommended parameters, p_2 can be as small as 2^{-43}. Even this is not negligible, but better parameters may hopefully make these values negligible. However, there is a gap between the existence of such a pair (m, r) that makes a *decryption failure*, and the feasibility, for an adversary, to find/build some:

$$\mathrm{Succ}^{\mathcal{A}}_{\mathrm{fail}}(h) = \Pr_{f,g}[\mathcal{D}_{f,f_p}(\mathcal{E}_h(m, r)) \text{ Fail} \,|\, h = p * g * f_p \bmod q, (m, r) \leftarrow \mathcal{A}(h)].$$

As above, one needs to study the probabilities over some classes of adversaries, or when the adversary does not have the entire control over m or r:

$$\tilde{p}_0(t) = \max\{E_h[\mathrm{Succ}^{\mathcal{A}}_{\mathrm{fail}}(h)], |\mathcal{A}| \leq t\}$$
$$\tilde{p}_1(t, Q) = idem \text{ where } (m, y) \leftarrow \mathcal{A}(h), r = G(m, y)$$
$$\tilde{p}_2(t, Q) = idem \text{ where } (x, y) \leftarrow \mathcal{A}(h), m = F(x, y), r = G(x, y).$$

In the above bounds, for $\tilde{p}_0(t)$, we consider any adversary whose running time is bounded by t. For $\tilde{p}_1(t, Q)$ and $\tilde{p}_2(t, Q)$, F and G are furthermore assumed to be random oracles, to which the adversary can ask up to Q queries. Clearly, for any t and any Q,

$$\tilde{p}_0(t) \leq p_0 \quad \tilde{p}_1(t, Q) \leq Q \times p_1 \quad \tilde{p}_2(t, Q) \leq Q \times p_2.$$

We now reconsider the SVES-1 padding scheme, keeping these probabilities in mind. We note that the adversary controls m, by deriving m and b from the required m_1 and m_2. However, r is out of the adversary's control. Therefore after q_D queries to the decryption oracle and q_H queries to the random oracle H, the probability of a decryption failure is less than $q_D \times \tilde{p}_1(t, q_H)$, where t is the running time of the adversary. Denoting by $T_{\mathcal{E}}$ the time for one encryption, one gets the following improvement for a CCA attacker over a CPA one:

$$\mathsf{Adv}_{\Pi^3}^{\mathrm{ind-cca}}(t) \leq \mathsf{Adv}_{\Pi^3}^{\mathrm{ind-cpa}}(t + q_H T_{\mathcal{E}}) + 2q_D \times (\varepsilon_r + \tilde{p}_1(t, q_H)).$$

4.3 Improved Paddings

In [21], new paddings have been suggested, with better provable security (based on the NTRU inversion problem only). But decryption failures have been ignored again.

The OAEP-based Scheme: The first suggestion was similar to the SVES-1 padding, also using two more hash functions

$$F : \{0, 1\}^{k_1} \rightarrow \{0, 1\}^{k_2} \text{ and } G : \{0, 1\}^{k_2} \rightarrow \{0, 1\}^{k_1}.$$

One first computes $s = m \oplus G(b)$ and $t = b \oplus F(s)$. The ciphertext consists of $\mathcal{E}_{\mathrm{pk}}(s \parallel t; H(m \parallel b))$. Of course, the decryption checks the validity of r, relatively to $H(m \parallel b)$. The OAEP construction provides semantic security, while the H function strengthens it to chosen-ciphertext security (as already explained).

Here, the adversary can choose s, t directly, then reverse the OAEP construction to obtain m, b. However, they cannot control r. Therefore

$$\mathsf{Adv}_{\mathrm{oaep}'}^{\mathrm{ind-cca}}(t) \leq 2\mathsf{Succ}_{\mathrm{ntru}}^{\mathrm{ow}}(t + QT_{\mathcal{E}}) + 2q_D \times (\varepsilon_r + \tilde{p}_1(t, q_H)) + \frac{4q_H}{2^{k_1}} + \frac{2q_G}{2^{k_2}}.$$

where $Q = q_F q_G + q_H$. But this makes a quadratic reduction, as for any OAEP-based cryptosystem [2,5]. The particular above construction admits a better reduction, but under a stronger computational assumption.

SVES-2: The successor to SVES-1 proposed for standardization is a minor variant of the above [4], designed to handle variable length messages. In SVES-2, one uses two hash functions F and G, and form $M_1 = b \parallel \mathrm{len}(m) \parallel m_1$, $M_2 = m_2 \parallel 000\ldots$. In SVES-2, the message length is restricted to be an integer number of bytes, and the length is encoded in a single byte. The final $N \bmod 8$ bits of M_2 will always be zeroes (we use N_8 to denote $N \bmod 8$). We form $s = M_2 \oplus F(M_1)$, $t = M_1 \oplus G(s)$, and calculate the ciphertext as $\mathcal{E}_{\mathrm{pk}}(t \parallel s; H(m \parallel b))$. On decryption, the usual checks are performed, and in addition the decrypter checks that the length is valid and that M_2 consists of 0s from the end of the message onwards.

In this case, an attacker who chooses s, t and reverses the OAEP construction must get the correct length of m and the correct 0 bits at the end. However, an attacker can choose s, then select t such that $t \oplus G(s)$ has the correct form to be M_1, that is, such that $\mathrm{len}(m)$ is its maximum value. The reverse OAEP operation on s, t will then yield a valid M_1, M_2 if the last N_8 bits of $s \oplus F(M_1)$ are 0. Therefore an attacker can control all the bits of s, t with probability 2^{8+N_8}, or all but 8 of the bits of s, t with probability 2^{N_8}. Therefore

$$\mathsf{Adv}_{\mathsf{SVES-2}}^{\mathsf{ind-cca}}(t) \leq 2\mathsf{Succ}_{\mathsf{ntru}}^{\mathsf{ow}}(t + QT_\mathcal{E}) + 2q_D \times \left(\varepsilon_r + \frac{\tilde{p}_1(t, q_H)}{2^{N_8}} \right) + \frac{4q_H}{2^{k_1}} + \frac{2q_G}{2^{k_2}}.$$

where $Q = q_F q_G + q_H$.

NTRU-REACT: Thanks to the OW-PCA–security level of the NTRU primitive (granted the pseudo-inverse of h), one can directly use the REACT construction [23], in which the decryption algorithm of course checks the validity of c, but also the validity of r. The semantic security is clear, since the adversary has no advantage without having asked some crucial queries to the hash functions. With chosen-ciphertext attacks, the adversary cannot produce a valid ciphertext without having built correctly the authentication tag, except with probability $1/2^{k_2}$. Therefore, the simulation of the decryption oracle is perfect, unless a decryption failure occurs: the adversary knows b and thus m, but makes a decryption failure, that is not detected by the simulation. Since the adversary has the control over both m and r, the security against chosen-ciphertext attacks is not very high:

$$\mathsf{Adv}_{\mathsf{react}}^{\mathsf{ind-cca}}(t) \leq 2\mathsf{Succ}_{\mathsf{ntru}}^{\mathsf{ow}}(t + (q_G + q_H)T_\mathcal{E}) + 2q_D \times \left(\frac{1}{2^{k_2}} + \tilde{p}_0(t) \right).$$

In the *Improved NTRU-REACT*, the adversary completely loses control over r, which improves the security level, but not enough, since $\tilde{p}_0(t)$ is replaced by $\tilde{p}_1(t)$ only.

4.4   Comments

It is clear that decryption failures on valid ciphertexts mean that NTRUEN-CRYPT with the parameter sets given cannot have provable security to the level claimed. In the presence of decryption failures, it is impossible to correctly simulate the decryption oracle: the simulator will output a valid decryption on certain ciphertexts which the genuine decryption engine will fail to decrypt. Without knowledge of the private key, it is impossible to build a simulator; and if the simulator requires knowledge of the private key, it is impossible to have provable security. In the next section we will see how to use this information to recover the private key with considerably less effort than a standard lattice attack would take.

5   Some Attacks Based on Decryption Failures

In this section we present some attacks against NTRUENCRYPT as implemented in NTRU Cryptosystems' products for the $N = 251$ parameter set. See [4] for details of this parameter set, and Appendix A for information about the precise structure of f, g and r used. We stress that the attacks work even better on the EESS standard [3,4], because decryption failures arise much more frequently there. Although no paddings can prevent decryption failures, it turns out that some paddings are more prone to attacks based on decryption failures: this is because the attacker has more or less flexibility on the choice of the actual message and random nonce actually given as input to the encryption primitive, depending on the padding used.

5.1   Review: The Reversal of a Polynomial

Before outlining the attacks, we review the notion of the *reversal* $\bar{c}(X) \equiv c(X^{-1})$ of a polynomial c. If we represent c as the array $c = [c_0, c_1, c_2, \ldots, c_{N-1}]$ then its reversal is $\bar{c} = [c_0, c_{N-1}, c_{N-2}, \ldots, c_1]$: this is a ring automorphism. We denote by $\hat{c}$ the product of c and $\bar{c}$. This product has the property that $\hat{c}_i = c \cdot (X^i * c)$, or in other words that the successive terms of $\hat{c}$ are obtained by taking the dot product of c with successive rotations of itself. The significance of this is that we know that $\hat{c}_0 = \sum_i c_i^2 = \|c\|^2$, while the other terms of $\hat{c}$ will be $\mathcal{O}(\|c\|)$ in size.

   We therefore know that $f * \bar{f}$ has one term of size d_f, and the others are of size about $\sqrt{d_f}$. The polynomial $f * \bar{f}$ is therefore of great width compared to a product of two arbitrary polynomials of the same norm as f. We therefore assume that whenever $p * r * g + f * m$ is of great width, it means that r is correlated significantly with $\bar{g}$ and m is correlated significantly with $\bar{f}$.

5.2   A General Attack

We derive a powerful chosen-ciphertext attack that works independently of the padding used. Indeed, assume that an attacker is able to collect many triplets (m, r, e) such that e is an encryption of m with random nonce r, which cannot be correctly decrypted. If the probability of decryption failure is sufficiently high, an attacker could obtain such triplets by mounting a weak chosen-ciphertext attack, independently of the padding, by simply selecting random messages, encrypting them until decryption failures occur (which can be checked thanks to the decryption oracle). Interestingly, such an attack only uses valid ciphertexts.

   For such a triplet (m, r, e), we know that $p * r * g + f * m$ is of great width, and the previous section suggests that there is an integer i such that r somehow looks like $X^i * \bar{g}$ and m somehow looks like $X^i * \bar{f}$. Unfortunately, we do not know the value of i, otherwise it would be trivial to recover $\bar{f}$ by simply taking the average of $X^{-i} * m$. However, we can get rid of the unknown i by using the reversal: if m and r look like respectively $X^i * \bar{f}$ and $X^i * \bar{g}$, then $\hat{m}$ and $\hat{r}$ must look like respectively $\hat{f}$ and $\hat{g}$. Once $\hat{f}$ and $\hat{g}$ are derived by averaging methods, f and g themselves may be recovered in polynomial time using an algorithm due to Gentry and Szydlo [7]. Strictly speaking, to apply [7], one also needs to determine the ideal spanned by f which can be derived from $\hat{f} = f * \bar{f}$ and $f * \bar{g}$ (which is itself obtained by multiplying from $\bar{H}$ and $\hat{f} = f * \bar{f}$).

To check the validity of this attack, we would need to find a lot of decryption failures, which is relatively time-consuming, depending on the parameters. Instead, we checked that the attack worked, by experimenting with a weaker attack based on the following oracle $\mathcal{O}_B$: When the oracle $\mathcal{O}_B$ is queried on a valid ciphertext e, it indicates whether or not the width of $p * r * g + f * m$ is greater than B. Thus, the oracle $\mathcal{O}_q$ simply detects gap failures. By using values of B much smaller than q, we are able to verify the behavior of our attack in a reasonable time, with the following algorithm:

1. Set $u, v = 0$.
2. Generate a large number of valid ciphertexts $e = r * h + m \bmod q$. For each ciphertext e:
   a) Call $\mathcal{O}_B(e)$.
   b) If $\mathcal{O}_B(e)$ shows the width of a as being greater than B, set $u = u + \hat{m}$; set $v = v + \hat{r}$.
3. Divide u and v by the number of valid ciphertexts used.

Over a long enough transcript, u and v should converge to $\hat{f}$ and $\hat{g}$. We investigated this for f binary and for $f = 1 + p * F$, to see how many messages with width greater than B were necessary to recover $\hat{f}$ exactly. The results are shown in tables 1. Experimentally, we find that we approximate $\hat{f}$ best by $\hat{f}_i = (\langle \hat{m} \rangle_i - 61.24747)/0.007882857$.

Table 1. Messages to recover $\hat{f}$ for various values of B using $\mathcal{O}_B$.

B	Messages
18	100,000

binary f

B	Messages	Norm (guess - $\hat{f}$)
36	500,000	15.5
36	1,400,000	7.38

$f = 1 + p * F$

When the distance from the guess to $\hat{f}$ is about $\sqrt{N}/2 \approx 7.9$, we can essentially recover $\hat{f}$ by rounding. We can conclude that from a real decryption oracle ($k_2 \leq 128$) no more than a million decryption failures, and perhaps considerably fewer, are necessary to recover $\hat{f}$ and $\hat{g}$. This validates our general chosen-ciphertext attack which applies to all paddings, and shows that the security of NTRU encryption in the EESS standard [3,4] clearly falls far short of the hoped-for level of 2^{80}.

We note that one could imagine an even more powerful attack, where the attacker would simply average $\hat{e}$ for those e that cause wraps. For a sufficiently long transcript, this will converge to $A + B\hat{f} + C\hat{g}$. We have not investigated this idea in full – it will undoubtedly involve longer convergence times than the other attacks outlined above – but it is interesting that a successful attack may be mounted even by an entirely passive attacker.

5.3   A Specific Attack Based on Controlling m

The previous attack works against any padding and already emphasizes the importance of decryption failures on the security of NTRU encryption. Here, we describe a slightly more efficient chosen-ciphertext attack tailored to the SVES-1 padding scheme, based on the fact that an attacker essentially controls m directly (see above). This attack shows that certain paddings are weaker than others with respect to attacks based on decryption failures. We denote by r_m the value of r obtained from the m and b obtained from a given m in valid encryption. The strategy will be to try and cause wrap failures. We introduce the notation

$$\mathcal{B}_i = \{\text{binary polynomials with } i \text{ 1s and } N - i \text{ zeroes}\},$$

and denote by Flat(c) the operation of taking c and setting all terms that are 1 or more to be exactly equal to 1. Experimentally, the average width of the various polynomials is:

$$\langle \text{Width}(p * r * g) \rangle \approx 41; \quad \langle \text{Width}(f * m) \rangle \approx 47; \quad \langle \text{Width}(p * r * g + f * m) \rangle \approx 62.$$

If the attacker can increase the width of $p * r * g + f * m$ to 128, he will cause a gap failure; alternatively, if he can add about 33 to the largest term in $p * r * g + f * m$ while leaving the others essentially unchanged, this will cause a wrap failure. The following attack exploits this observation

Step 1: first cleaning of random strings. The attacker picks a random $F_{15} \in \mathcal{B}_{15}$, $D \in \mathcal{B}_5$, and forms

$$m = \text{Flat}((1 + X) * D * F_{15}).$$

On decryption (with $p = 2 + X$), a will include a term approximately equal to

$$(1 + X) * (2 + X) * f * D * F_{15} = (2 + 3X + X^2) * f * D * F_{15}.$$

If the 1s in F_{15} match a set of 15 1s in $\bar{f}$, then we know that at least one term in $p * F * m$ will be 45 or more because of the 3 in $(1 + X) * (2 + X)$. With high odds, this will mean that $p * F * m$ has greater than average width, and so greater than average chance of causing a decryption failure. This lets the attacker attempt to identify substrings of $\bar{F}$:

Attack: Step 1. The attacker picks a random $F_{15} \in \mathcal{B}_{15}$, $D \in \mathcal{B}_5$ and forms $m = \text{Flat}((1+X) * D * F_{15})$. For all rotations of m, he submits $e = r_m * h + m \bmod q$ to the decryption oracle. If any rotation of m causes a wrap, he stores F_{15}; otherwise, he discards it.

Obviously, there will be a large number of false positives in this step. An m might cause a decryption failure purely through luck; alternatively, an F_{15} which has an overlap of 14 rather than 15 with $\bar{F}$ will have a good chance of causing a wrap failure. We cannot distinguish between the cases immediately, so the strategy is to take an initial set of random F_{15}s, and use the decryption oracle to "clean" them so that the resulting set has a greater proportion of strings of high overlap with f.

Table 2 shows the effect of the first cleaning on a set of 2^{30} random F_{15}s. The wrap probabilities were determined experimentally, using our knowledge of f, by generating strings of a specific overlap and testing to see if they caused wraps. The final column is given by $2^{30} \cdot \text{Pr[overlap]} \cdot \text{Pr[wrap]} \cdot N$, as we try all the rotations of each of the 2^{30} ms. The total number of queries to the decryption oracle is $N \cdot 2^{30} \approx 2^{38}$.

Table 2. Effects of first cleaning on a set of 2^{30} F_{15}

Overlap	Pr[Overlap] (Theoretical)	Pr[Overlap]) (Experimental)	Pr[Wrap]	No. Left from 2^{30} F_{15}s
15	$2^{-20.7}$	–	$2^{-14.4}$	10
14	$2^{-15.135}$	$2^{-14.7}$	$2^{-15.5}$	200
13	$2^{-10.698}$	$2^{-10.70}$	$2^{-16.6}$	2,000
12	$2^{-6.981}$	$2^{-7.116}$	$2^{-17.5}$	20,000
11	$2^{-3.857}$	$2^{-4.030}$	$2^{-18.5}$	80,000
10	$2^{-1.423}$	$2^{-1.693}$	2^{-22}	40,000
9	–	$2^{-0.91}$	2^{-23}	25,000
8	–	$2^{-3.49}$	2^{-24}	2,000
7	–	$2^{-14.5}$	–	–

From table 2 we see that even the first cleaning is very effective: from a random set of size 2^{30} where most F_{15}s have an overlap of 9 with f, we have created a set of size about 2^{17} where the most commonly occurring overlap is 11. We now want to further improve the quality of our set of F_{15}s.

Step 2: second cleaning. The attacker now queries each surviving F_{15} by using different values of D to create ms, and observing whether or not these cause wraps.

Attack: Step 2. For each F_{15} that survived the first cleaning step, the attacker picks several $D \in \mathcal{B}_5$. For each D, he forms $m = \text{Flat}((1 + X) * D * F_{15})$. For all rotations of m, he submits $e = r_m * h + m \mod q$ to the decryption oracle. If any rotation of m causes a wrap, he stores F_{15}; otherwise, he discards it.

Table 3 shows the results of performing this step, choosing 2^8 Ds for each F_{15}. After this cleaning, there are almost no F_{15}s left with an overlap of less than 11. The total work in this stage is $2^8 \cdot 2^{17} \cdot N \approx 2^{33}$, and there are about 2^{12} F_{15}s left.

Now that the attacker has a relatively good set of F_{15}, he can try and assemble them to recover F.

Step 3: find correct relative rotations. Here the challenge is to find the correct rotations of the F_{15} relative to each other. One possibility would be to

Table 3. Effects of second cleaning on the set of F_{15}

Overlap	No. Tested	No. Left
15	10	3
14	200	40
13	2,000	120
12	20,000	800
11	80,000	1500
10	40,000	2
9	25,000	1
8	2,000	0

pick two F_{15}s and test the ms obtained by all rotations of the two against each other. However, it appears that we get better results by picking sets of three F_{15}s and trying all their relative rotations.

Attack: Step 3. Let F_1, F_2, F_3 be any three of the F_{15} that survived the second cleaning step. For each $0 \leq i, j < N$, set

$$F_{\sim45} = \mathsf{Flat}(F_1 + X^i F_2 + X^j F_3)$$

(Note that $F_{\sim45}$ will typically have slightly fewer than 45 1s). Set $m = \mathsf{Flat}((1 + X) * F_{\sim45})$. For all rotations of m, submit $e = r_m * h + m \bmod q$ to the decryption oracle. Store the number of wraps caused by m. Once all i, j pairs have been exhausted, pick another three F_1, F_2, F_3 and repeat. Continue until all the F_{15} have been used.

Figure 1 shows the wrap probability for m obtained as specified above. If F_1, F_2, F_3 are rotated correctly relative to each other, the overlap with $\bar{F}$ will typically be 33 or more, leading to a significantly greater chance of a wrap. Note that, because of the $(2 + X)$ in f, if $F_1 + X^i F_2 + X^j F_3$ is the correct relative alignment of F_1, F_2, F_3, a large number of wraps will also be caused by $F_1 + X^{i \pm 1} F_2 + X^{j \pm 1} F_3$. This helps us to weed out freak events: rather than simply taking the relative rotation of F_1, F_2, F_3 that gives the highest number of wraps, we look for the set of three consecutive rotations that give the highest total number of wraps and pick the rotation in the middle.

This step takes about $N^3 \cdot 2^{12} \approx 2^{36}$ work, and at the end of it we have about 2^{10} strings of length about 45, which will in general have an overlap of 33 or more with $\bar{F}$. The remaining task is to rotate these strings correctly relative to each other and recover $\bar{F}$ from them, but this is relatively trivial.

Step 4: from 45s to $\bar{F}$. We here use the fact that the $F_{\sim45}$ will be better correlated with each other when rotated correctly than in other rotations.

Attack: Step 4. Sort the $F_{\sim45}$ from the previous step in order by the number of wraps they caused. Set u equal to the $F_{\sim45}$ at the top of the list. For each other $F_{\sim45}$, find the X^i that maximizes the overlap of $X^i \cdot F$ and the reference F. Set

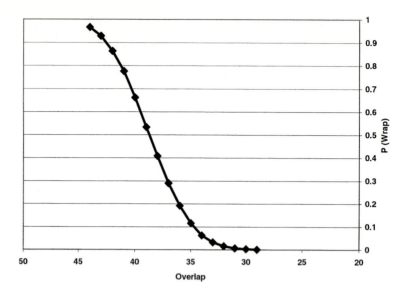

Fig. 1. Wrap probability on a single test for strings of length 45

$u = u + X^i \cdot F$. When all the Fs have been added, take the top d_f entries of u and set them equal to 1. Set the other entries to 0. This recovers F_{rev}.

Since there are 72 1s in $\bar{F}$ and 179 0s, and since the $F_{\sim 45}$ have typically 33 correct 1s and 12 incorrect 1s, we expect the entries in u corresponding to 1s in f_{rev} to have an average value of $33/72 \approx 0.46$, and the entries in u corresponding to 0s in f_{rev} to have an average value of $12/179 \approx 0.06$. This makes it easy to distinguish between the two. We have not implemented this part of the attack, but we do not anticipate any problems in its execution.

SVES-1 attack: Summary. We have presented an attack on the SVES-1 scheme that allows an attacker with access to decryption timing information to recover the private key in about 2^{40} queries to a decryption oracle with $N = 251$. This is a level of security that clearly falls far short of the hoped-for level of 2^{80}.

6   Countermeasures

NTRUENCRYPT as specified in [3] clearly falls short of the desired security levels, since it only involves the probability $\tilde{p}_1$. With the given parameters, even $\tilde{p}_2$ is likely to be non-negligible. One should thus recommend at least the following two countermeasures.

6.1   Changing the Parameters

The parameters, and perhaps the form of f, g, and r, should be altered so that decryption failures occur no more often than the claimed security level of the parameter set, so that the probability $\tilde{p}_2$, or even p_2, is indeed negligible. (For example, for $N = 251$, an attacker should be required to carry out 2^{80} work to find a single gap failure). Unfortunately, no efficient method is known to provably compute such a probability, though the paper [27] provides calculations under some simplifying assumptions.

6.2   Changing the Padding

A padding scheme with the appropriate provable security properties should be adopted. We have presented both theoretical and experimental reasons for preferring an NTRUENCRYPT padding scheme in which an attacker can control neither m nor r. Theoretically, only this padding scheme allows us to use p_2, the smallest of the expected decryption failure probabilities. Experimentally, we have demonstrated an attack which uses direct control of m to recover the private key faster than the attack which does not use control of m.

We therefore suggest the following padding scheme, which we call NAEP, as one that might be suitable for NTRUENCRYPT. The construction uses the hash functions

$$G : \{0,1\}^{\mathsf{mLen}} \to \{0,1\}^{\mathsf{rLen}} \text{ and } H : \mathcal{P} \to \{0,1\}^{\mathsf{mLen}}.$$

As before, m is the plaintext of length k_1 bits, and b is a random string, unique for each message, of length $k_2 = \mathsf{mLen} - k_1$ bits. One computes $\mathsf{r} = \mathcal{R}(G(m \,\|\, b))$ and $\mathsf{R} = \mathsf{r}*\mathsf{h} \bmod q$. Then the ciphertext consists of $\mathcal{E}_{\mathsf{pk}}((m \,\|\, b) \oplus H(\mathsf{R}); G(m \,\|\, b))$. Of course, the decryption checks the validity of r, relatively to $G(m \,\|\, b)$.

We do not make any claim on the provable security of this scheme. An analysis of the properties of a variant of this scheme, with a specific instantiation of H, appears in [16] and claims a security result which depends on $\tilde{p}_2$ only (and of course the intractability of the basic NTRU primitive.)

Acknowledgments. We would like to thank Jeff Hoffstein and Jill Pipher for fruitful discussions and contributions.

References

1. M. Bellare, A. Desai, D. Pointcheval, and P. Rogaway. Relations among Notions of Security for Public-Key Encryption Schemes. In *Crypto '98*, LNCS 1462, pages 26–45. Springer-Verlag, Berlin, 1998.
2. M. Bellare and P. Rogaway. Optimal Asymmetric Encryption – How to Encrypt with RSA. In *Eurocrypt '94*, LNCS 950, pages 92–111. Springer-Verlag, Berlin, 1995.
3. EESS: Consortium for Efficient Embedded Security. Efficient Embedded Security Standards #1: Implementation Aspects of NTRU and NSS. Draft Version 3.0 available at http://www.ceesstandards.org, July 2001.
4. EESS: Consortium for Efficient Embedded Security. Efficient Embedded Security Standards #1: Implementation Aspects of NTRUEncrypt and NTRUSign. Version 1.0 available at http://www.ceesstandards.org, November 2002.
5. E. Fujisaki, T. Okamoto, D. Pointcheval, and J. Stern. RSA–OAEP is Secure under the RSA Assumption. In *Crypto '01*, LNCS 2139, pages 260–274. Springer-Verlag, Berlin, 2001.

6. C. Gentry. Key Recovery and Message Attacks on NTRU-Composite. In *Eurocrypt '01*, LNCS 2045, pages 182–194. Springer-Verlag, Berlin, 2001.

7. C. Gentry and M. Szydlo. Cryptanalysis of the Revised NTRU Signature Scheme. In *Eurocrypt '02*, LNCS 2332, pages 299–320 Springer-Verlag, Berlin, 2002.

8. S. Goldwasser and S. Micali. Probabilistic Encryption. *Journal of Computer and System Sciences*, 28:270–299, 1984.

9. C. Hall, I. Goldberg, and B. Schneier. Reaction Attacks Against Several Public-Key Cryptosystems. In *Proc. of ICICS '99*, LNCS, pages 2–12. Springer-Verlag, 1999.

10. J. Hoffstein and J. Pipher and J. H. Silverman. NTRU: A Ring-Based Public Key Cryptosystem. In *Proc of ANTS 3*, LNCS 1423, pages 267–288. Springer-Verlag, 1998.

11. J. Hoffstein and J. H. Silverman. Random Small Hamming Weight Products With Applications To Cryptography. *Discrete Applied Mathematics*. To appear, available at [22].

12. J. Hoffstein and J. H. Silverman. Invertibility in Truncated Polynomial Rings. Technical report, NTRU Cryptosystems, October 1998. Report #009, version 1, available at [22].

13. J. Hoffstein and J. H. Silverman. Optimizations for NTRU. In *Public-key Cryptography and Computational Number Theory*. DeGruyter, 2000. To appear, available at [22].

14. J. Hoffstein and J. H. Silverman. Protecting NTRU against Chosen Ciphertext and Reaction Attacks. Technical report, NTRU Cryptosystems, June 2000. Report #16, version 1, available at [22].

15. J. Hong, J. W. Han, D. Kwon and D. Han. Chosen-Ciphertext Attacks on Optimized NTRU. Cryptology ePrint Archive: Report 2002/188.

16. N. Howgrave-Graham, J. H. Silverman, A. Singer and W. Whyte. NAEP: Provable Security in the Presence of Decryption Failures. Cryptology ePrint archive, http://eprint.iacr.org.

17. E. Jaulmes and A. Joux. A Chosen Ciphertext Attack on NTRU. In *Crypto '00*, LNCS 1880, pages 20–35. Springer-Verlag, Berlin, 2000.

18. A. May and J.H. Silverman. Dimension Reduction Methods for Convolution Modular Lattices. In *Proc. of CaCL 2001*, LNCS 2146, pages 110–125. Springer-Verlag, 2001.

19. T. Meskanen and A. Renvall. Wrap Error Attack Against NTRUEncrypt. To appear in *Proc. of WCC '03*.

20. M. Naor and M. Yung. Public-Key Cryptosystems Provably Secure against Chosen Ciphertext Attacks. In *Proc. of the 22nd STOC*, pages 427–437. ACM Press, New York, 1990.

21. P. Q. Nguyen and D. Pointcheval. Analysis and Improvements of NTRU Encryption Paddings. In *Crypto '02*, LNCS 2442, pages 210–225. Springer-Verlag, Berlin, 2002.

22. NTRU Cryptosystems. Technical reports. Available at http://www.ntru.com, 2002.

23. T. Okamoto and D. Pointcheval. REACT: Rapid Enhanced-security Asymmetric Cryptosystem Transform. In *CT – RSA '01*, LNCS 2020, pages 159–175. Springer-Verlag, Berlin, 2001.

24. J. Proos. Imperfect Decryption and an Attack on the NTRU Encryption Scheme. Cryptology ePrint Archive: Report 2003/002.

25. C. Rackoff and D. R. Simon. Non-Interactive Zero-Knowledge Proof of Knowledge and Chosen Ciphertext Attack. In *Crypto '91*, LNCS 576, pages 433–444. Springer-Verlag, Berlin, 1992.
26. J. H. Silverman. Estimated breaking times for NTRU lattices. Technical report, NTRU Cryptosystems, March 1999. Report #012, version 1, available at [22].
27. J. H. Silverman and W. Whyte, Estimating Decryption Failure Probabilities for NTRUEncrypt . Technical report, NTRU Cryptosystems, May 2003. Report #018, version 1, available at [22].

A   Standardized and Deployed Versions of NTRUEncrypt

A.1   Format of Objects

NTRUENCRYPT as standardized in [3] uses special forms for f, g and r, and specifies a padding method which is claimed to give provable security. We review these details here.

Private key: f. The private key f has two special features. First, f has the form $1 + p * F$, where F is a binary polynomial. This means that $f = 1$ mod p, and therefore that $f_p = 1$, eliminating the need for the second convolution on decryption [13]. Second, the binary polynomial F is of the form $f_1 * f_2 + f_3$, where f_1, f_2 and f_3 are chosen so that:

- f_1, f_2, f_3 are binary and have $d_{f_1}, d_{f_2}, d_{f_3}$ 1s respectively;
- $f_1 * f_2$ is binary;
- The 1s in f_3 are chosen such that they are not adjacent to any of the 1s in $f_1 * f_2$.

It should be pointed out that only the first of these restrictions is documented in [3]: the description above is of the form of private keys used in NTRU CRYPTOSYSTEMS' software and has the effect of decreasing the occurrence of decryption failures (but not to the point of making decryption failures sufficiently unlikely to avoid any security problems). For more details on the use of products of low Hamming weight polynomials in NTRU and other cryptosystems, see [13, 11].

Private key: g. The private key g is chosen to be binary, to have d_g 1s, and to have no consecutive 1s. Note that the last of these restrictions is not documented in [3], but is used in NTRU CRYPTOSYSTEMS' software.

Message: m. The message representative m is a binary polynomial of degree N. An algorithm for converting from octet strings to binary polynomials can be found in [3].

Blinding value: r. The blinding value r is chosen to be of the form $r_1 * r_2 + r_3$. Here, r_1, r_2, r_3 are generated by setting them to 0, then selecting $d_{r_1}, d_{r_2}, d_{r_3}$ indices between 0 and $N - 1$ and adding one to the coefficient at each index.

The difference between this and taking r_1, r_2, r_3 to be binary is that the indices used to generate them can repeat: for example, r_1 could consist of $d_{r_1} - 2$ 1s and one 2. A recent paper [19] uses this specific fact to recover g. The results presented here are more general and work for r of any form in the presence of decryption failures.

A.2   Encryption Schemes

There have been several encryption schemes associated with NTRU, but only two have been standardized in EESS #1. The first, SVES-1, proceeds as follows. It takes the number of random bits, d_b, as a parameter, and hash functions F, G, H.

Encryption: To encrypt a message m of length $|m| = N - d_b$ bits:

1. Generate the string b consisting of d_b random bits.
2. Set s equal to the first $|m|/2$ bits of m concatenated with the first $d_b/2$ bits of b. Set t equal to the last $|m|/2$ bits of m concatenated with the last $d_b/2$ bits of b.
3. Set $t' = t \oplus F(s)$. Set $s' = s \oplus G(t')$. Set $\mathsf{m}' = s'||t'$. Set $\mathsf{r} = H(m, b)$.
4. Output the ciphertext $\mathsf{e} = \mathsf{r} * \mathsf{h} + \mathsf{m}' \bmod q$.

Decryption: To decrypt the ciphertext e:

1. Recover m' from e using standard NTRUENCRYPT decryption.
2. Recover m, b from m' by reversing the masking defined above.
3. Set $\mathsf{r} = H(m, b)$ and calculate $\mathsf{e}' = \mathsf{r} * \mathsf{h} + \mathsf{m}' \bmod q$. If the result is the same as the received e, output m. Otherwise, output "invalid ciphertext".

SVES-1 was shown in [21] to have inadequate provable IND-CPA properties, due to the decision to split b into two parts. EESS #1 therefore also specifies SVES-2, which is similar to SVES-1 with the exception that all of b is included in the first hash function call. There are other minor differences between the two schemes – SVES-2 is designed to allow variable-length messages more gracefully, for example – but these are more engineering than cryptographic decisions.

One interesting fact to note is that in both SVES-1 and SVES-2 the message is randomized by the mask generation functions, but an adversary is free to choose the value of m' directly and then reverse the masking operation to find the m and b that would have given that m'.

B   Another Chosen-Ciphertext Attack

Here we present a brief overview of a second chosen-ciphertext attack against NTRUEncrypt. The attack is based on decryption failures; however, unlike the other attack presented in this paper, does not rely on the secret polynomial f being of the form $1 + (2 + X)F$. In fact, the new attack is not specific to the case of $\mathsf{p} = 2 + X$ and can also be applied against the originally proposed version of NTRU [10] which had $\mathsf{p} \in \mathbb{Z}$.

The attack assumes that the attacker can detect wrap errors and that the r values used during encryption must be selected at random. For the basic version of the attack we also assume that a message polynomial m can be encrypted with many random r (as is the case for the proposed NTRU-REACT padding schemes). We will discuss below the effect on the attack if each m yields a unique r_m. The basic version of the attack consists of repeating the following three steps until the secret key is revealed.

Step 1: finding a decryption failure. The goal of step 1 is to determine a valid pair m, r which lead to a decryption failure. The most straight forward approach is to simply select random m and r until the ciphertext they generate causes a decryption failure.

Instead of a random search, it is also possible to perform a systematic search for the required decryption failure. Given an m and r an attacker can determine exactly the set, $I_{m,r}$, of (f, g) pairs for which m, r will cause a decryption failure. Determining if m, r causes a decryption failure reveals whether or not (f, g) is in $I_{m,r}$. So instead of simply selecting m and r at random an attacker could perform some precomputation and obtain a list of m, r pairs for which $\bigcup I_{m,r}$ is larger than it would have been in a random search.

Step 2: search for more r's. For the majority of m, r pairs which lead to decryption failures, m * f will have one coefficient, i, which is both abnormally far from its expected value and further from the expected value than any other coefficient. We shall refer to the difference in the distances of the two coefficients of m * f furthest from their expected value as the *gap* of m * f. The true goal of step 1 is actually to find an m such that m * f has both a coefficient which is far from its expected value and a large gap.

Attack: Step 2. By repeatedly picking random r′ and determining if m, r′ leads to a decryption failure, the attacker can determine a list $r_0, r_1, \ldots, r_k$ of r values which cause decryption failures when used with m.

Suppose that step 1 found an m with the desired properties. The range $[A, A+ q-1]$ used during decryption is centered at the expected value of the coefficients of p * r′ * g + f * m. Thus, since the ith coefficient of m * f is abnormally far from its expected value, the rate at which the m, r′ cause decryption failures will be much higher than for random m, r. Furthermore, the expected value of every coefficient of p * r * g is $p(1)g(1)r(1)/N$. Thus when an m, r′ pair causes a decryption failure its most likely cause is the ith coefficient of p * r′ * g + f * m. The strength of this bias towards the i coefficient the will depend on the gap of m * f. This bias will cause a correlation between the $r_0, r_1, \ldots, r_k$ found in step 2 and $\bar{g}$.

Step 3: recovering the secret key. If k is sufficiently large then the value of $\bar{g}$ (and thus g) can be determined directly from the polynomials $r_0, r_1, \ldots, r_k$. However, it is possible to find the secret key with fewer r_j than would allow the direct recovery of $\bar{g}$. This is accomplished by using the r_j to determine some of coefficients of g and then using this partial knowledge of g in combination with the known lattice attacks on NTRU as in [18].

If the gap of m $*$ f is small then the bias towards the coefficient i may not be large enough to allow the recovery of the secret key. If this is the situation then the attack simply returns to step 1. Note that even if an iteration does not reveal the entire secret key some information may still have been determined.

Attack summary and variations. Two important questions arise regarding this attack. First, how much work is involved in one iteration of the steps? Second, how many iterations through the steps will be required? The number of iterations required depends on the maximum work allowed to be done in step 2 of an iteration. The more effort put into finding r_j's in step 2 the more likely step 3 is to succeed. Details on the running time of the attack against the p $= 3$ parameters suggested in [26] can be found in [24]. Below we include some details on the running time against the $N = 251$ parameter set of NTRUEncrypt as standardized in [4].

If, as with the SVES-1 padding scheme, each polynomial m only has one valid r then the basic attack described above can not be used. The problem arises in step 2, where if m is held constant then the r' used will also be constant. To overcome this problem the attacker can, instead of keeping m fixed, use the cyclic shifts of both m and m with minor changes applied to it. Care must be taken to record the shift amounts with the r_j found so that the shifts can be undone in step 3.

Table 4. $N = 251$ attack details

	m, r Pairs Checked		Decryption Failures	
	Avg Number	Std Dev	Avg Number	Std Dev
Total Attack	1991909.11	1706591.03	170.65	130.56
Successful Step 2	842589.58	767601.34	118.74	43.77

B.1   Implementation Results

The attack was implemented against 100 instances of the $N = 251$ parameter set of NTRUEncrypt as standardized in [4] taking f $= 1 + pF$, where F was a binary polynomial. Our implementation of the attack put a bound of 3 million on the number of r checked in step 2, checked to see if the secret key could be recover after every 25 decryption failures in step 2, and aborted iterations in step 2 if the rate at which the r_j were found was below a given threshold. The implementation assumed that the secret key could be recovered when the dimension of the lattice which would need to be reduced was less than one hundred. Of the 100 instances of the attacks the number of instances which found the secret key after $1, 2, 3, 4, 5, 6, 7$ and 8 iterations were $48, 23, 17, 4, 4, 2, 1$ and 1 respectively. Table 4 shows the average number of m, r pairs tested and decryption failures required over the 100 instances of the attack and during the step 2's of the successful iterations.

Universally Composable Efficient Multiparty Computation from Threshold Homomorphic Encryption

Ivan Damgård and Jesper Buus Nielsen

BRICS* Department of Computer Science
University of Aarhus
Ny Munkegade
DK-8000 Arhus C, Denmark
{ivan,buus}@brics.dk

Abstract. We present a new general multiparty computation protocol for the cryptographic scenario which is universally composable — in particular, it is secure against an active and *adaptive* adversary, corrupting any minority of the parties. The protocol is as efficient as the best known statically secure solutions, in particular the number of bits broadcast (which dominates the complexity) is $\Omega(nk|C|)$, where n is the number of parties, k is a security parameter, and $|C|$ is the size of a circuit doing the desired computation. Unlike previous adaptively secure protocols for the cryptographic model, our protocol does not use non-committing encryption, instead it is based on homomorphic threshold encryption, in particular the Paillier cryptosystem.

1 Introduction

The problem of multiparty computation (MPC) dates back to the papers by Yao [14] and Goldreich et al. [11]. What was proved there was basically that a collection of n parties can efficiently compute the value of an n-input function, s.t. everyone learns the correct result, but no other new information. More precisely, these protocols can be proved secure against a polynomial time bounded adversary who can *corrupt* a set of less than $n/2$ parties initially, and then make them behave as he likes, we say that the adversary is *active*. Even so, the adversary should not be able to prevent the correct result from being computed and should learn nothing more than the result and the inputs of corrupted parties. Because the set of corrupted parties is fixed from the start, such an adversary is called *static* or non-adaptive.

Proving security of the protocol from [11] requires a complexity assumption, such as existence of trapdoor one-way permutations. This is because the model of communication considered there is s.t. the adversary may see every message sent

---

* Basic Research in Computer Science,
  Centre of the Danish National Research Foundation.

D. Boneh (Ed.): CRYPTO 2003, LNCS 2729, pp. 247–264, 2003.

between parties, this is sometimes known as the *cryptographic model*. Later, *unconditionally* secure MPC protocols were proposed by Ben-Or et al. and Chaum et al. [2,6], in the model where *private* channels are assumed between every pair of parties. These protocols are secure, even against an *adaptive* adversary who may decide dynamically during the protocol who to corrupt.

Over the years, several protocols have been proposed which, under specific computational assumptions, improve the efficiency of general statically secure MPC, see for instance [10]. In particular, Cramer, Damgård and Nielsen [7] proposed a general MPC protocol that is secure against a static adversary corrupting any minority of the parties. The protocol assumes that keys for a threshold homomorphic cryptosystem have been set up, and has communication (broadcast) complexity $\Omega(nk|C|)$, where n is the number of parties, k is a security parameter, and $|C|$ is the size of a circuit computing the desired function. This is the so far the most efficient protocol known for the cryptographic model, tolerating a dishonest minority (which is the best possible if we want to guarantee termination). The homomorphic threshold cryptosystem can be the one of Paillier[13], or can be built from the quadratic residuosity assumption.

It is possible to get adaptive security, also in the cryptographic model: we can start from an adaptively secure protocol for the private channels model ([2, 6]), and then, instead of assuming perfect channels, we implement them using public-key encryption. While this will in general reduce adaptive security to static, using *non-committing encryption*, adaptive security carries over to the cryptographic model [4]. In [5], it is shown that non-committing encryption plus some additional techniques can provide MPC that is *universally composable*, an even stronger notion of security defined by Canetti [3]. All these protocols are, however, much less efficient than the best statically secure ones. An alternative approach that also gives adaptive security is to assume that parties can be trusted to securely erase certain critical data [1].

Our Results. In this paper, we present a new general MPC protocol for the cryptographic model, based on Paillier encryption. The protocol is universally composable, in particular, it is secure against an active *adaptive* adversary corrupting any minority of the parties. Up to a constant factor, it is as efficient as the (statically secure) protocol from [7]. It is therefore the first general MPC solution for the cryptographic model where going to adaptive security does not cause a major loss in efficiency (or costs an extra assumption, such as secure erasures). It is also the first that does not use non-committing encryption, which may be of separate interest from a technical point of view. (The protocol in [5] is not for the private channels model, but it uses non-committing encryption for building adaptively secure oblivious transfer.) Instead of non-committing encryption we use some ideas from the universally composable commitments of [9], and combine this with the protocol from [7]. Thus we work on the same principle that parties supply encrypted input, and the protocol produces encryptions of the outputs, that the parties can then cooperate to decrypt. To make this secure, we introduce a new technique for randomizing encryptions before they are decrypted. This means that the adversary has no control over the encryptions that are decrypted, and this turns out to be essential for our proof of security to go through.

We prove our protocols secure in the framework for universally composable (UC) security by Canetti[3]. This framework allows to define the security of general reactive tasks, rather than just evaluation of functions. This allows us to prove that our protocol does not just provide secure function evaluation, but is in fact equivalent to what we call an *arithmetic black box* (ABB). An ABB can be thought of as a secure general-purpose computer. Every party can in private specify inputs to the ABB, and any majority of parties can ask it to perform any feasible computational task and make the result (and only the result) public. Moreover the ABB can be invoked several times and keeps an internal state between invocations. This point of view allows for easier and more modular proofs, and also makes it easier to use our protocols as tools in other constructions.

2   An Informal Description

In this section, we give a completely informal introduction to some main ideas. All the concepts introduced here will be treated more formally later in the paper. We will assume that from the start, the following scenario has been established: we have a semantically secure threshold public-key system given, i.e., there is a public encryption key pk known by all parties, while the matching private decryption key has been shared among the parties, s.t. each party holds a share of it.

We will use a threshold version of the Paillier cryptosystem, so the message space is $\mathbf{Z}_N$ for some RSA modulus N. For a plaintext $a \in \mathbf{Z}_N$, we let $\overline{a}$ denote an encryption of a. We have certain homomorphic properties: from encryptions $\overline{a}, \overline{b}$, anyone can easily compute (deterministically) an encryption of $a + b$, which we denote $\overline{a} \boxplus \overline{b}$. We also require that from an encryption $\overline{a}$ and a constant $\alpha \in \mathbf{Z}_N$, it is easy to compute a random encryption of αa, which we denote $\alpha \boxdot \overline{a}$. This immediately gives us an algorithm $\boxminus$ for subtracting.

We can then sketch how a computation was performed securely in the (statically secure) protocol from [7]: we assume the desired computation is specified as a circuit doing additions and multiplications in $\mathbf{Z}_N$ — This allows us to simulate a Boolean circuit in a straightforward way using $0/1$ values in R. The protocol then starts by having each party publish encryptions of his input values and give zero-knowledge proofs that he knows these values. Then any operation involving addition or multiplication by constants can be performed with no interaction: if all parties know encryptions $\overline{a}, \overline{b}$ of input values to an addition gate, all parties can immediately compute an encryption of the output $\overline{a + b}$. This leaves only the problem of multiplications: Given encryptions $\overline{a}, \overline{b}$ (where it may be the case that no parties knows a nor b), compute securely an encryption of $c = ab$. This can be done by the following protocol (which is a slightly optimized version of the protocol from [7]):

1. Each party P_i chooses random $d_i \in \mathbf{Z}_N$ and broadcasts encryptions $\overline{d_i}$ and $\overline{d_i b}$.
2. All parties prove in zero-knowledge that they know their respective values of d_i, and that $\overline{d_i b}$ encrypts the correct value. Let S be the subset of the parties succeeding with both proofs.
3. All parties can now compute $\overline{a} \boxplus (\boxplus_{i \in S} \overline{d_i})$. This ciphertext is decrypted using threshold decryption, so all parties learn $a + \sum_{i \in S} d_i$.
4. All parties set $\overline{c} = (a + \sum_{i \in S} d_i) \boxdot \overline{b} \boxminus (\boxplus_{i \in S} \overline{d_i b})$.

At the final stage we know encryptions of the output values, which we just decrypt using threshold decryption. Intuitively this is secure if the encryption is secure because, other than the outputs, only random values or values that should be known to the adversary are ever decrypted. This intuition was proved for a static adversary in [7]. But in the UC framework, where the adversary is adaptive, it is well known that there are several additional problems:

Loosely speaking, a proof of security requires building a simulator that simulates in an indistinguishable way the adversary's view of a real attack, while having access only to data the adversary *is supposed to know at any given time*. This means that for instance in the input stage the simulator needs to show the adversary encryptions that are claimed to contain the inputs of honest parties. At this time the simulator does not know these inputs, so it must encrypt some arbitrary values. This is fine for the time being, but if one of the honest parties are later corrupted, the simulator learns the real inputs of this party and must reveal them to the adversary along with a simulation of all internal data of the party. The simulator is now stuck, since the real inputs most likely are not consistent with the arbitrary values encrypted earlier.

We handle this problem using a combination of two tricks: first, we include in the public key an encryption $K = \overline{1}$. Then, we redefine the encryption method, and fix the rule that to encrypt a value a, one computes $a \boxdot K$. Under normal circumstances, this will be an encryption of a. The point, however, is that in the simulation, the simulator can decide what K should be, and will set $K = \overline{0}$. Then all encryptions used in the simulation will contain 0, in fact — using the algebraic properties of the encryption scheme — the simulator can compute C as an encryption of 0, store the random coins used for this and later make it seem as if C was computed as $C = a \boxdot K$ for any a it desires.

However, the simulator must also be able to find the input values the adversary supplies, immediately as the encryptions are made public. This is not possible if it only sees encryptions of 0. We therefore redefine the way inputs are supplied: for each input value x of party P_i, P_i uses the UC commitment scheme of [9] to make a commitment $\mathtt{commit}(x)$ to x, and also sends an encryption $C = x \boxdot K$. Finally he proves in zero-knowledge that $\mathtt{commit}(x)$ contains the same value as C. This can be done efficiently because [9] is also based on Paillier encryption, and the UC property of the commitments precisely allows the simulator to extract inputs of corrupted parties and fake it on behalf of honest parties.

A final problem we face is that the simulator will not be able to "cheat" in the threshold decryption protocol. The key setup for this protocol fixes the shares of the private key even in the simulation, so a ciphertext can only be decrypted to the value it actually contains. Of course, when decrypting the outputs, the correct results should be produced both in simulation and real life, and so we have a problem since we just said that all ciphertexts in the simulation really contain 0. We solve this by randomizing all ciphertexts before they are decrypted: we include another fixed encryption $R = \overline{0}$ in the public key. Then, given ciphertext C, the parties cooperate to create an encryption $r \boxdot R$, where r is a (secret) randomly chosen value that depends on input from all parties. Then we compute $C \boxplus (r \boxdot R)$ and decrypt this ciphertext. Under normal circumstances, it will of course contain the same plaintext as C. But in the simulation, the simulator will set $R = \overline{1}$, and "cheat" in the process where r is chosen, s.t. $r = a$, where a is the value the simulator wants the decryption to return. This works, since in the simulation any C actually encrypts 0.

3   Preliminaries

Notation. Throughout the paper k will denote the security parameter. For a probabilistic polynomial time (PPT) algorithm A we will use $a \leftarrow A(x; r)$ to denote running A on input x and uniformly random bits $r \in \{0, 1\}^{p(|x|)}$ producing output a; Here $p(\cdot)$ is a polynomial upper bounding the running time of A.

Non-erasure Σ-protocols. Consider a binary relation $R \subseteq \{0, 1\}^* \times \{0, 1\}^*$ where $(x, w) \in R$ can be checked in PPT. By $L(R)$ we denote the set $\{x \in \{0, 1\}^* | \exists w \in \{0, 1\}^* ((x, w) \in R)\}$.

A non-erasure Σ-protocol for relation R is six PPT algorithms $(A, l, Z, B, \text{hvs}, \text{rbs}, \text{xtr})$ (and an integer l) specifying a three move, public randomness, honest verifier zero-knowledge protocol with special soundness. The prover has input $(x, w) \in R$ and the verifier has input x. The prover first computes a message $a \leftarrow A(x, w; r_A)$ and sends a to V. Then V returns a l-bit challenge e. The prover then computes a response to the challenge $z \leftarrow Z(x, w, r_a, e)$ and sends z to the verifier. The verifier then computes $b \leftarrow B(x, a, e, z)$, where $b \in \{0, 1\}$ indicates whether to believe that the prover knows a valid witness w or not. The algorithm hvs is called the **honest verifier simulator** and takes as input $x \in L(R)$, $e \in \{0, 1\}^l$ and a uniformly random bit-string r_{hvs} and produces as output (a, z) which is supposed to be distributed as the (a, z) produced by a honest prover with instance x receiving challenge e — this is defined formally below. The algorithm rbs is called the **random bits simulator**. It takes as input $(x, w) \in R$, a challenge $e \in \{0, 1\}^l$ and bits r_{hvs}, which we think of as the random bits used by hvs in a run $(a, z) \leftarrow \text{hvs}(x, e; r_{\text{hvs}})$, and it produces as output a bit-string r_A s.t. $a = A(x, w; r_A)$ and $z = Z(x, w, r_A, e)$. I.e. if (a, z) is the messages simulated using hvs given just $x \in L(R)$, then if w s.t. $(x, w) \in R$ later becomes known it is possible to construct random bits r_A s.t. it looks like (a, z) was generated as $a \leftarrow A(x, w; r_A)$ and $z \leftarrow Z(x, w, r_A, e)$. Finally xtr is a knowledge extractor, which given two correct conversations with the same first message can compute a witness. We now formalize these requirements along with completeness.

Completeness: For all $(x, w) \in R$, r_A and $e \in \{0,1\}^l$ we have that $B(x, A(x, w; r_a), e, Z(x, w, r_a, e)) = 1$.

Special non-erasure honest verifier zero-knowledge: The following two random variables are identically distributed for all $(x, w) \in R$ and $e \in \{0,1\}^l$:

$$\texttt{EXEC}(x, w, e) = [a \leftarrow A(x, w; r_A); z \leftarrow Z(x, w, r_A, e) : (x, w, a, r_A, e, z)]$$
$$\texttt{SIM}(x, w, e) = [(a, z) \leftarrow \texttt{hvs}(x, e; r_{\texttt{hvs}}); r_A \leftarrow \texttt{rbs}(x, w, e, r_{\texttt{hvs}}) :$$
$$(x, w, a, r_A, e, z)]$$

Special soundness: For all $x \in S$ and (a, e, z) and (a, e', z') where $e \neq e'$, $B(x, a, e, z) = 1$ and $B(x, a, e', z') = 1$, we have that $(x, \texttt{xtr}(x, a, e, z, e', z')) \in R$.

4   Non-erasure Concurrent Zero-Knowledge Proofs of Knowledge

In [8] a concurrent zero-knowledge proofs of knowledge is presented based on any Σ-protocol. The protocol assumes that the prover P and verifier V agree on a key K for a trapdoor commitment scheme. The prover has input $(x, w) \in R$ and the verifier knows x. The protocol proceeds as follows.

1. The prover generates $a \leftarrow A(x, w; r_A)$, commits $c \leftarrow \texttt{commit}_K(a; r_c)$ and sends c to the verifier.
2. The verifier sends uniformly random $e \in \{0,1\}^l$ to the prover.
3. The prover computes $z \leftarrow Z(x, w, r_A, e)$ and sends (a, r_c, z) to the verifier.
4. The verifier outputs 1 iff $c = \texttt{commit}_K(a, r_c)$ and $B(x, a, e, z) = 1$.

To simulate the protocol without w one assumes that the simulator knows the trapdoor t of K. The simulator can then let c be a fake commitment and when it receives e compute $(a, z) \leftarrow \texttt{hvs}(x, e; r_{\texttt{hvs}})$, use t to compute r_c s.t. $c = \texttt{commit}_K(a; r_c)$ and send (a, r_c, z). As observed in [9], if the Σ-protocol is non-erasure the above protocol is adaptively secure. Assume namely that P is corrupted and the simulator learns a witness. Then compute $r_A \leftarrow \texttt{rbs}(x, w, e, r_{\texttt{hvs}})$ and by special non-erasure honest verifier zero-knowledge the simulation is perfect. In the following we will call this to patch the state of the proof of knowledge. In [8] a knowledge extractor $\texttt{xtr}$ is given working as follows. Assume K is a random commitment key so that $\texttt{commit}_K$ is computational binding and assume a corrupt prover succeeds in giving an acceptable proof with probability p say. Then by rewinding and giving new random challenges e' until a proof is accepted again we get, e.w.n.p., values a, e, z, a, e', z' s.t. $B(x, a, e, z) = B(x, a, e', z') = 1$ and $e \neq e'$ and can compute a witness w for x. The expected number of rechallenges needed is $\frac{1}{p}$ so if we run a proof and extract if a correct proof is given the expected number of rechallenges used in the extraction is $p\frac{1}{p}$ where p is the probability that the proof is accepted in the first place. This generalizes to several proofs run concurrently. Each time a proof is accepted invoke $\texttt{xtr}$ and get a witness. If the context in which the proofs are run is PPT, then the running time, and the number of proofs, is bounded by a polynomial P, and so each invocation of xtr has expected running time less than P which with a total of at most P invocations, by the linearity of expectation, gives an expected polynomial running time of at most P^2.

Note that for the zero-knowledge simulator it is enough that K is a trapdoor commitment key, whereas for knowledge soundness it was needed that the key was random to ensure that commit$_K$ is computational binding. We can therefore let V pick the key K and send it to P. All we need to ensure is that is the simulator can get its hands on the trapdoor of K. We will use the protocol and simulator exactly this way later.

5   Universally Composable Security

In this section we give a sketch of the notion of universally composable security of synchronous protocols for the authenticated link model. Except for some minor technical differences it is the synchronous model described in [3].

A protocol π consists of n parties $P_1, \dots, P_n$, all PPT interactive Turing machines (ITMs). The execution of a protocol takes place in the presence of an environment $\mathcal{Z}$, also a PPT ITM, which supplies inputs to and receives outputs from the parties. $\mathcal{Z}$ also models the adversary of the protocol, and so schedules the activation of the parties and corrupts parties. In each round r each party P_i sends a message $m_{i,j,r}$ to each party P_j; The message $m_{i,i,r}$ is the state of P_i after round r, and $m_{i,i,0} = (k, r_i)$ will be the security parameter and the random bits used by P_i. We model open channels by showing the messages $\{m_{i,j,r}\}_{j\in[n]\setminus\{i\}}$ to $\mathcal{Z}$, where $[n] = \{1, \dots, n\}$. In each round $\mathcal{Z}$ inputs a value $x_{i,r}$ to P_i and receives and output $y_{i,r}$ from P_i. We write the r'th activation of P_i as $(\{m_{i,j,r}\}_{j\in[n]}, y_{i,r}) = P_i(\{m_{j,i,r-1}\}_{j\in[n]}, x_{i,r})$. We model that the parties cannot reliably erase their state by giving r_i to $\mathcal{Z}$ when P_i is corrupted. In the following C will denote the set of corrupted parties and $H = [n] \setminus C$. In detail the real-life execution proceeds as follows.

Init: The input is k, random bits $r = (r_1, \dots, r_n) \in (\{0,1\}^*)^n$ and an auxiliary input $z \in \{0,1\}^*$. Set $r = 0$ and $C = \emptyset$. For $i, j \in [n]$ let $m_{i,j,0} = \epsilon$ for $i \neq j$ and $m_{i,i,0} = (k, r_i)$. Then input k and z to $\mathcal{Z}$ and activate $\mathcal{Z}$.

Environment activation: When activated $\mathcal{Z}$ outputs one of the following commands: (activate $i, x_{i,r}, \{m_{j,i,r-1}\}_{j\in C}$) for $i \in H$; (corrupt i) for $i \in H$; (end round); or (guess b) for $b \in \{0,1\}$. Commands are handled as described below and the environment is then activated again. We require that all honest parties are activated between two (end round) commands. When a (guess b) command is given the execution stops.

Party activation: $\{m_{j,i,r-1}\}_{j\in H}$ were defined in the previous round; Add these to $\{m_{j,i,r-1}\}_{j\in C}$ from the environment and compute $(\{m_{i,j,r}\}_{j\in[n]}, y_{i,r}) = P_i(\{m_{j,i,r-1}\}_{j\in[n]}, x_{i,r})$. Then give $\{m_{i,j,r}\}_{j\in[n]\setminus\{i\}}$ to $\mathcal{Z}$.

Corrupt: Give r_i to $\mathcal{Z}$. Set $C = C \cup \{i\}$.

End round: Give the value $\{y_{i,r}\}_{i\in H}$ defined in **Party activation** to $\mathcal{Z}$ and set $r = r + 1$.

The result of the execution is the bit b output by $\mathcal{Z}$. We denote this bit by $\text{REAL}_{\pi,\mathcal{Z}}(k,r,z)$. This defines a Boolean distribution ensemble $\text{REAL}_{\pi,\mathcal{Z}} = \{\text{REAL}_{\pi,\mathcal{Z}}(k,z)\}_{k\in\mathbb{N}, z\in\{0,1\}^*}$, where we take r to be uniformly random.

To define the security of a protocol an ideal functionality $\mathcal{F}$ is specified. The ideal functionality is a PPT ITM with n input tapes and n output tapes which we think of as being connected to n parties. The input-output behavior of the ideal functionality defines the desired input-output behavior of the protocol. To be able to specify protocols which are allowed to leak certain information $\mathcal{F}$ has a special output tape (SOT) on which it writes this information. The ideal functionality also has the special input tape (SIT). Each time $\mathcal{F}$ is given an input for P_i it writes some value on the SOT modeling the part of the input which is not required to be kept secret. When $\mathcal{F}$ receives the input (activate v) a round is over, in response to which it writes a value on the output tape for each party. The value v models the inputs from the corrupted parties. When a party P_i is corrupted the ideal functionality receives the input (corrupt i) on the SIT.

We then say that a protocol π realizes an ideal functionality $\mathcal{F}$ if there exists an interface, also called simulator, $\mathcal{S}$ which given access to $\mathcal{F}$ can simulate a run of π with the same input-output behavior. In doing this $\mathcal{S}$ is given the inputs of the corrupted parties, and the information leaked on the SOT of $\mathcal{F}$, and can specify the inputs of corrupted parties. In detail the ideal process, proceeds as follows.

Init: The input is k, random bits $r = (r_\mathcal{F}, r_\mathcal{S}) \in (\{0,1\}^*)^2$ and an auxiliary input $z \in \{0,1\}^*$. Set $r = 0$ and $C = \emptyset$. Provide $\mathcal{S}$ with $r_\mathcal{S}$, provide $\mathcal{F}$ with $r_\mathcal{F}$ and give k and z to $\mathcal{Z}$ and activate $\mathcal{Z}$.

Environment activation: $\mathcal{Z}$ is defined exactly as in the real-word, but now commands are handled by $\mathcal{S}$, as described below.

Party activation: Give $\{m_{j,i,r-1}\}_{i\in C}$ to $\mathcal{S}$ and give $x_{i,r}$ to $\mathcal{F}$ on the input tape for P_i and run $\mathcal{F}$ to get some value $v_\mathcal{F}$ on the SOT. This value is given to $\mathcal{S}$ which is then required to compute some value $\{m_{i,j,r}\}_{j\in[n]\setminus\{i\}}$ which is given to $\mathcal{Z}$.

Corrupt: When $\mathcal{Z}$ corrupts P_i, $\mathcal{S}$ is given the values $x_{i,0}, y_{i,0}, x_{i,1}, \ldots$ exchanged between $\mathcal{Z}$ and $\mathcal{F}$ for P_i. Furthermore (corrupt i) is input on the SIT of $\mathcal{F}$ and $\mathcal{F}$ writes a value on the SOT; This value is given to $\mathcal{S}$. Then $\mathcal{S}$ outputs some value r_i which is given to $\mathcal{Z}$. Set $C = C \cup \{i\}$.

End round: $\mathcal{S}$ is activated and produces a value v. Then (activate v) is input to $\mathcal{F}$ which produces output $\{y_{i,r}\}_{i\in[n]}$. Then $\{y_{i,r}\}_{i\in C}$ is given to $\mathcal{S}$ and $\{y_{i,r}\}_{i\in H}$ is given to $\mathcal{Z}$. Set $r = r+1$.

The result of the ideal-process is the bit b output by $\mathcal{Z}$. We denote this bit by $\text{IDEAL}_{\mathcal{F},\mathcal{S},\mathcal{Z}}(k,r,z)$. This defines a Boolean distribution ensemble $\text{IDEAL}_{\mathcal{F},\mathcal{S},\mathcal{Z}} = \{\text{IDEAL}_{\mathcal{F},\mathcal{S},\mathcal{Z}}(k,z)\}_{k\in\mathbb{N}, z\in\{0,1\}^*}$.

Definition 1. *We say that π t-realizes $\mathcal{F}$ if there exists an interface $\mathcal{S}$ s.t. for all environments $\mathcal{Z}$ corrupting at most t parties it holds that $\text{IDEAL}_{\mathcal{F},\mathcal{S},\mathcal{Z}} \overset{c}{\approx} \text{REAL}_{\pi,\mathcal{Z}}$.*

We then define the hybrid models. The $\mathcal{G}$-hybrid model is basically the real-life model where the parties in addition to the communication lines also have access to an ideal functionality $\mathcal{G}$. In each round all parties give an input to $\mathcal{G}$ and receive the output from $\mathcal{G}$ in the following round. The inputs to the SIT of $\mathcal{G}$ are given by $\mathcal{Z}$ and $\mathcal{Z}$ receives the outputs on the SOT. When defining security of a protocol in a hybrid model the interface $\mathcal{S}$ must in addition to the communication and internal state of parties also simulate the values exchanged between $\mathcal{G}$ and $\mathcal{Z}$. We call this a hybrid interface.

Definition 2. *We say that π t-realizes $\mathcal{F}$ in the $\mathcal{G}$-hybrid model if there exists a hybrid interface $\mathcal{S}$ s.t. for all environments $\mathcal{Z}$ corrupting at most t parties it holds that $IDEAL_{\mathcal{F},\mathcal{S},\mathcal{Z}} \overset{c}{\approx} HYB^{\mathcal{G}}_{\pi,\mathcal{Z}}$.*

A universal composability theorem can be proven for this framework. I.e. if a protocol realizes a given functionality it can be replaced for the functionality in all contexts. The above description easily generalizes to the case of several ideal functionalities in a hybrid model. In particular we will in the protocols following assume that the parties have access to an ideal functionality $\mathcal{F}_{BA}$ for doing Byzantine Agreement. It expects a bit b_i as input from all parties. If all honest parties agree on a value v, then $\mathcal{F}_{BA}$ outputs v to all parties. Otherwise the environment is allowed to determine the output through the SIT. We also assume a broadcast channel, which can easily be model with an ideal functionality.

6   The Paillier Cryptosystem and Some Tools

In this section we describe the Paillier cryptosystem [13]. The public key is a k-bit RSA modulus $pk = N = pq$, where p an q are chosen s.t. $p = 2p' + 1, q = 2q' + 1$ for primes p', q', and both p and q have $k/2$ bits. The plaintext space is $\mathbf{Z}_N$ and the ciphertext space is $\mathbf{Z}^*_{N^2}$. To encrypt $a \in \mathbf{Z}_N$, one chooses $r \in \mathbf{Z}^*_N$ uniformly at random and computes the ciphertext as $E_{pk}(a; r) = g^a r^N \bmod N^2$, where the element $g = N + 1$ has order N in $\mathbf{Z}^*_{N^2}$. The encryption function is homomorphic in the following sense $E_{pk}(a_1; r_1)E_{pk}(a_2; r_2) \bmod N^2 = E_{pk}(a_1 a_2 \bmod N; r_1 r_2 \bmod N)$ and $E_{pk}(a; r)^b = E_{pk}(ab \bmod N; r^b \bmod N)$. The private key is e.g. $sk = \phi(N)(\phi(N)^{-1} \bmod N)$, and it is straightforward to verify that $((N + 1)^a r^N)^{sk} \bmod N^2 = Na + 1$ from which $a \bmod N$ can be computed. Given a one can then compute r^N and $r = (r^N)^{N^{-1} \bmod \phi(N)} \bmod N^2$. Under an appropriate complexity assumption — the DCRA — this system is semantic secure. The DCRA states that random elements in $\mathbf{Z}^*_{N^2}$ are computationally indistinguishable from random elements of form $r^N \bmod N^2$.

For any ciphertext $K = E_{pk}(a; r_K)$ we can consider the function $E_{pk,K}(m; r) = K^m r^N \bmod N^2 = E_{pk}(am; s^m r \bmod N)$. If r is uniformly random in $\mathbf{Z}^*_N$, then $s^m r \bmod N$ is uniformly random in $\mathbf{Z}^*_N$ and so $E_{pk,K}(m; r)$ is a uniformly random encryption of $am \bmod N$. Notice that if $a \in \mathbf{Z}^*_N$, then from K and $c = E_{pk,K}(m; r)$ and sk we can efficiently compute m. Therefore $E_{pk,K}(m; r)$ is again a semantically secure encryption function. If on the other hand $a = 0$, then $E_{pk,K}(m;r)$ is a uniformly random encryption of 0. So, $E_{pk,K}(m;r)$ can be seen as a perfect hiding commitment scheme $\mathtt{commit}_{pk,K}(m;r) = E_{pk,K}(m;r)$. It is trivial to see that $\mathtt{commit}_{pk,K}(m;r)$ is computationally binding. If $K \in E_{pk}(1)$, then $\mathtt{commit}_{pk,K}(m;r)$ would be perfect binding, so no algorithm can find two different openings. An algorithm finding two different openings when $K = E_{pk}(0)$ would therefore distinguish encryptions of 1 from encryptions K of 0.

Notice furthermore that if $K = E_{pk}(0;r_K)$ and r_K is known, then given arbitrary $m, m' \in \mathbf{Z}_N$ and $r \in \mathbf{Z}_N^*$ we can compute $r' = s^{m-m'}r \bmod N$, so that $E_{pk,K}(m;r) = E_{pk,K}(m';r')$. All in all we have argued that if $K = E_{pk}(0;r_K)$ is a random encryption, then $\mathtt{commit}_K(m;r)$ is a perfect hiding computationally binding trapdoor commitment scheme with trapdoor t_K. When considering values $K \in \mathbf{Z}_{N^2}^*$ as keys we will call keys of the form $K = E_{pk}(0;t_K)$ trapdoor keys and we will call keys of the form $K = E_{pk}(a;s)$ for $a \in \mathbf{Z}_n^*$ encryption keys. We now describe a commitment scheme from [9].

A UC Commitment Scheme. First we have to introduce a so-called double-trapdoor commitment scheme. Consider a trapdoor commitment scheme with key $K = (K_1, K_2)$, where K_1 and K_2 are both encryptions of 0. We commit as $\mathtt{commit}_{K_1,K_2}(m) = (E_{pk,K_1}(m_1), E_{pk,K_2}(m_2))$, where m_1 and m_2 are uniformly random values for which $m = m_1 + m_2 \bmod N$. To make a fake commitment just commit to random values m_1 and m_2, call the commitment (c_1, c_2). To trapdoor open such a commitment it is enough to know the trapdoor of one of K_1 and K_2. To open $c = (c_1, c_2)$ to m we can open c_1 to $m - m_2$ or we can open c_2 to $m - m_1$. We note for later use that the distribution of the random bits is independent of which trapdoor is used. Note that the domain of $\mathtt{commit}_K$ can be extended by committing block-wise. Here is the protocol from [9].

Setup: We assume that commitment sender S and receiver R agree on two independently chosen Paillier public-keys N and N' and a commitment key $K = (K_1, K_2) = (E_{N'}(0;t_1), E_{N'}(0,t_1))$. If R is honest at the beginning of the protocol we furthermore assume that t_1 and t_2 are uniformly random.

Commitment: S commits to $s \in \mathbf{Z}_N$ as follows.

1. S generate uniformly random $L_S \in \mathbf{Z}_{N^2}^*$, commits $c_L = \mathtt{commit}_{N',K}(L_S; r_L)$ and sends c_L to R.
2. R generates uniformly random $L_R \in \mathbf{Z}_{N^2}$ and sends L_R to R.
3. S computes $L = L_S L_R \bmod N^2$, commits to s under L as $c_s = \mathtt{commit}_L(s; r_s)$ for uniformly random $s \in \mathbf{Z}_N^*$ and sends c_s to R along with (L_S, r_L), and outputs (L, c_s, r_s).
4. R checks that $c_L = \mathtt{commit}_{N',K}(L_S; r_L)$ and if so computes $L = L_S L_R \bmod N^2$ and outputs (L, c_s).

In [9] a simulator $\mathcal{S}_{\mathtt{commit}}$ for the UC framework is given which has the property that given any so-called hitting commitment (L, c_s) it can simulate a run of the protocol which results in R outputting (L, c_s). All the simulator needs to do this is the trapdoor(s) of K when R is corrupted and an opening (s, r_s) of (L, c_s) when S is corrupted. The simulator has some properties given in Theorem 1.

Theorem 1. *Consider a simulation with hitting commitment (L, c_s) and output commitment (L', c_s') from R.*

1. *If L is uniformly random from $\mathbf{Z}_{N^2}^*$ and c_s is a uniformly random commitment to s, then the simulation is distributed* exactly *as the real-life protocol on input s.*
2. *If S is honest after the simulation, then $(L', c_s') = (L, c_s)$.*
3. *If N' is a modulus for which $\mathsf{commit}_{N',K}$ is computationally binding, then after a simulation where S was corrupted when c_L was sent and R was honest by the end of the simulation, L' is an encryption key, e.w.n.p.*

By simulating a trapdoor commitment under (pk, K) we mean the following. Generate $L = E_N(0; r_L)$ for uniformly random r_0 and compute $c_s = E_{N,L}(0; r_s)$ for uniformly random r_s. Then run $\mathcal{S}_{\mathrm{commit}}$ with hitting commitment (L, c_s). We write the resulting commitment as $(L, c_s, ?)$. If at some point any $s \in \mathbf{Z}_N$ is given, then use r_0 and r_s to compute r_s' s.t. $c_s = E_{pk,N}(s; r_s')$ and give (s, r_s') to $\mathcal{S}_{\mathrm{commit}}$ to patch the state of the simulation. We call this patching $(L, c_s, ?)$ to s.

A Non-Erasure Σ-Protocol. In [9] a non-erasure Σ-protocol for identical plaintext is given for the Paillier cryptosystem. The instance is (K_1, c_1, K_2, c_2) and a witness is (s, r_1, r_2) s.t. $c_1 = E_{pk,K_1}(s; r_1)$ and $c_2 = E_{pk,K_2}(s; r_2)$.

Escape-Secure Threshold Decryption. In this section we introduce the adaptively secure threshold decryption protocol from [12]. We first make a definition which will help us state the results precisely. We will introduce a weaker notion of UC security, which we call escape-security. An escape-simulator $\mathcal{S}$ is defined as a simulator in the UC framework augmented with a special escape state. An escape-simulation $\mathrm{IDEAL}_{\mathcal{F},\mathcal{S},\mathcal{Z}}(k, z)$ proceeds exactly as in the UC framework except that if $\mathcal{S}$ enters the escape state, then the simulation terminates with the output $\bot$ — we say that the simulation was escaped. For environment $\mathcal{Z}$ and values $k \in \mathbf{N}$ and $z \in \{0, 1\}^*$ of the security parameter we define the simulation probability by $s_{\mathcal{Z}}(k, z) = 1 - \Pr[\mathrm{IDEAL}_{\mathcal{F},\mathcal{S},\mathcal{Z}}(k, z) = \bot]$. When $s_{\mathcal{Z}}(k, z)$ is non-zero we define the conditional probabilities $c_{\mathcal{Z}}(k, z, b) = \frac{\Pr[\mathrm{IDEAL}_{\mathcal{F},\mathcal{S},\mathcal{Z}}(k,z)=b]}{s_{\mathcal{Z}}(k,z)}$. When $s_{\mathcal{Z}}(k, z)$ is zero we define $c_{\mathcal{Z}}(k, z, b) = \frac{1}{2}$ for $b \in \{0, 1\}$. Then $(c_{\mathcal{Z}}(k, z, 0), c_{\mathcal{Z}}(k, z, 1))$ defines a Boolean distribution ensemble which we denote by $[\mathrm{IDEAL}_{\mathcal{F},\mathcal{S},\mathcal{Z}}|\neg\bot]$.

Definition 3. *We say that π t-escape-securely realizes $\mathcal{F}$ if there exists an escape-simulator $\mathcal{S}$ s.t. $\mathrm{REAL}_{\pi,\mathcal{Z}} \overset{c}{\approx} [\mathrm{IDEAL}_{\mathcal{F},\mathcal{S},\mathcal{Z}}|\neg\bot]$ for all environments $\mathcal{Z}$ corrupting at most t parties.*

We now proceed to present the result from [12] in the above framework. Let (G, E, D) be a public-key cryptosystem and consider the following ideal functionality $\mathcal{F}_{(G,E,D)}$ for threshold decryption.

Init: In the first round it generates $(pk, sk) \leftarrow \mathcal{G}(k)$ and outputs pk to all parties and to the adversary.

Decryption: After a public key pk has been output we allow the parties to input a ciphertext C for decryption in any round and we also allow that more than one ciphertext is input in each round. Each ciphertext is handled as follows: If in some round r all honest parties input C, then send $(C, m = D_{sk}(C))$ to the adversary[1] and in round $r + 6$ return m to all parties.

In our terminology here, what is proved in [12] is:

Theorem 2. *There exists a protocol π_{pal} escape-securely realizing $\mathcal{F}_{pal}$ with simulation probability $\frac{1}{2}$.*

In [12] the key distribution is handled using a trusted party or a general MPC. Here we model this by assuming the key is generated by an ideal functionality $\mathcal{F}_{pal,key-gen}$; It generates a random threshold key $(N, pv, sk_1, \dots, sk_n)$ where N is a random Paillier public key, $sk_1, \dots, sk_n$ are the private key shares of the parties and pv is public values used in the protocol for a.o.t. checking decryption shares. In [12] a decryption protocol $\pi_{pal,dec}$ using this key is then described.

Their simulator consists of two parts. A simulator $\mathcal{S}_{pal,key-gen}$ which given the public key pk from the ideal functionality simulates a key $(N, pv, sk_1, \dots, sk_n) = \mathcal{S}_{pal,key-gen}(pk)$ with the property that (N, pv) is computationally indistinguishable from a real key. Furthermore a simulator $\mathcal{S}_{pal,dec}$ for the decryption protocol is given, which again is computationally indistinguishable from the real protocol as long as the simulation is not escaped. We need a last property of the protocol. Given a random Paillier key (pk, sk) it is possible to generate a threshold key $(pk, pv, sk_1, \dots, sk_n)$ with the same distribution as $\mathcal{F}_{pal,key-gen}$. We write $(pk, pv, sk_1, \dots, sk_n) = \texttt{SingleToThesh}(pk, sk)$.

7   An Arithmetic Black Box

The ABB is an ideal functionality $\mathcal{F}_{\text{ABB}}$. Initially $\mathcal{F}_{\text{ABB}}$ outputs a uniformly random key $pk = N$ to all parties, defining the ring $\mathbf{Z}_N$ that $\mathcal{F}_{\text{ABB}}$ does arithmetic over. In each activation it expects a command from all honest parties and carries out the command if all honest parties agree. Typically agreement means that the parties gave the same command, e.g. $(x \leftarrow x_1 + x_2)$, but this does not always make sense. If e.g. party P_i is to load a secret value s into variable x, using the command $(P_i : x \leftarrow s)$, of course the other parties cannot acknowledge by giving the same command. In this case they input $(P_i : x \leftarrow ?)$; The intended meaning is that P_i is allowed to define x using a value unknown to the other parties. If two honest parties ever disagree $\mathcal{F}_{\text{ABB}}$ goes corrupted in the following sense: It outputs its entire current state and all future inputs on the SOT. Besides this it lets the environment determine all future outputs through the SIT. The functionality $\mathcal{F}_{\text{ABB}}$ holds a list of defined variable names x and stores a value $\texttt{val}(x) \in \mathbf{Z}_N$ for each. For all commands we require that variables on the right

---

[1] We return these values to the adversary to specify that it is not part of the functionality to keep C or m secret

hand side of $\leftarrow$ are defined and that x is defined when a (output x) command is given. In the command $(P_i : x \leftarrow x_1 \cdot x_2)$ we require that x_1 was defined by a $(P_i : x_1 \leftarrow?)$ command. A violation of these requirements will make $\mathcal{F}_{\text{ABB}}$ go corrupted. Each time a variable x is defined $\mathcal{F}_{\text{ABB}}$ outputs (defined x) to all parties. Except for the s in $(P_i : x \leftarrow s)$ all input values are output on the SOT.

Init: Let c be the number of rounds used for setting up the keys in the real-life protocol. When the init command is given the ABB simply runs for c rounds ignoring all inputs. Then it generates a random Paillier key (N, sk) and outputs pk to all parties and outputs (N, sk) to the adversary.

Load: On $(P_i : x \leftarrow?)$ in round r, if P_i inputs $(P_i : x \leftarrow s)$ for $s \in \mathbf{Z}_N$, then set $\text{val}(x) = s$ in round $r + 8$. If in a round before round $r + 8$ P_i is corrupted and the adversary inputs (change s') on the SIT, then $\text{val}(x) = s'$, and if the adversary inputs fail, then $\text{val}(x)$ is not defined.

Linear combination: On $(x \leftarrow a_0 + \sum_{j=1}^{l} a_j x_j)$ for $a_j \in \mathbf{Z}_N$ define $\text{val}(x) = a_0 + \sum_{j=1}^{l} a_j \text{val}(x_j) \bmod N$ in the same round.

Private multiplication: On $(P_i : x \leftarrow x_1 \cdot x_2)$ in round r, if P_i also inputs $(P_i : x \leftarrow x_1 \cdot x_2)$, (this is an extra requirement as P_i might be corrupted), then in round $r + 4$ define $\text{val}(x) = \text{val}(x_1)\text{val}(x_2) \bmod N$. If in a round before round $r + 5$ the adversary inputs fail on the SIT and P_i is corrupted, then $\text{val}(x)$ is not defined.

Output: On (output x) in round r output (output $x = \text{val}(x)$) on the SIT and output (output $x = \text{val}(x)$) to all parties in round $r + 12$.

The functionality has no general multiplication command, but running in the $\mathcal{F}_{\text{ABB}}$-hybrid model, parties can do a multiplication of any two variable using the multiplication protocol from the introduction. We now describe a protocol π_{ABB} realizing $\mathcal{F}_{\text{ABB}}$. A variable x with $\text{val}(x) = s$ is represented by an encryption $\text{enc}(x) = E_{pk,K}(s)$ on which the parties agree.

Init: The protocol runs in the hybrid model with access to two copies of the functionality $\mathcal{F}_{\text{pal},key-gen}$ described in Section 6. We call the public key returned by the first copy $pk = N$, and we assume that the first copy also returns two random encryptions $K = E_{pk}(1)$ and $R = E_{pk}(0)$. We call the public key returned by the second copy $pk' = N'$, and we assume that the second copy also return $3n$ random encryptions $K_{i,l} = E_{pk'}(0)$ for $i \in [n], l \in \{0, 1, 2\}$. The key $K_{j,0}$ is for for running the proof of knowledge protocol from Section 4 with the relation in Section 6 with P_j as verifier. We will denote this by P_i proves to P_j. The double-trapdoor key $K_j = (N', K_{j,1}, K_{j,2})$ will be used for running the universally composable commitment scheme in Section 6 with P_j as receiver to generate commitments under N. When P_i is the committer with some value s we will denote this by P_i commits to s to P_j. In the first round the parties call the ideal functionalities and wait until all keys are returned.

Load: Below we use M as the key under which encryptions are made. When the parties carry out the load commands given from the environment they always use $M = K$. However the implementation of **Output** also uses the load command as an internal sub-routine, with $M = R$.

1. P_i computes $S = E_{pk,M}(s; r_S)$ and broadcasts S.
2. For $j \neq i$ party P_i commits to s to P_j. Denote the commitment by $(L_j, c_{s,j}, r_{s,j})$.
3. For $j \neq i$ party P_i proves to P_j with instance $x = (pk, M, S, L_j, c_{s,j})$ and witness $(s, r_S, s, r_{s,j})$.
4. For $j \neq i$ party P_j inputs 1 to $\mathcal{F}_{BA}$ iff the zero-knowledge proof given by P_i to P_j was accepted.
5. Wait until $\mathcal{F}_{BA}$ outputs a bit b. If $b = 1$ the parties set $\mathtt{enc}(x) = S$.

Linear combination: Set $\mathtt{enc}(x) = g^{a_0} \prod_{i=1}^{l} \mathtt{enc}(x_j)^{a_j} \bmod N^2$.

Private multiplication: We assume that P_i knows (s, r_s) s.t. $\mathtt{enc}(x_1) = E_{pk,K}(s; r_s)$.

1. P_i computes $X = E_{pk,\mathtt{enc}(x_2)}(s; r_X)$ and broadcasts X.
2. For $j \neq i$ party P_i proves to P_j with instance $x = (pk, K, \mathtt{enc}(x_1), \mathtt{enc}(x_2), X)$ and witness (s, r_S, s, r_X), and as above the parties run a BA to determine whether to accept the proof.
3. If the result of the BA is 1 the parties set $\mathtt{enc}(x) = X$.

Output:
1. P_i generates random $r_i \in \mathbf{Z}_N$ and loads it into variable x_i using $M = K$. Let $\{C_j\}_{j \neq i}$ denote the commitments used in Step 2 of the load.
2. Parti P_i loads r_i into variable y_i using $M = R$, but reuses the commitments $\{C_j\}_{j \neq i}$ from the previous load. If a party P_i input 0 to the BA in the previous load for P_j, then input 0 again in this load.
3. Let I be a size $t + 1$ subset of the set of indices i for which y_i is now defined. Let $\{\lambda_i^I\}_{i \in I}$ be degree t Lagrange coefficients interpolating from $\{f(i)\}_{i \in I}$ to $f(0)$ over $\mathbf{Z}_N$. Compute $S = \prod_{i \in I} \mathtt{enc}(y_i)^{\lambda_i^I} \bmod N^2$ and $T = \mathtt{enc}(x)S \bmod N^2$. We assume some fixed way of picking the subset I so that all parties agree on T.
4. The parties run $\pi_{\mathtt{pal},dec}$ from Section 6 on T and take as their output the value v returned by $\pi_{\mathtt{pal},dec}$.

Theorem 3. $\pi_{ABB} \lceil (n-1)/2 \rceil$-securely realizes $\mathcal{F}_{ABB}$.

Due to space limitations we can only sketch the proof of Theorem 3, a full proof will appear in the Ph.D. dissertation of the second author. In the proof we construct an interface S simulating π_{ABB} in the ideal process with access to $\mathcal{F}_{ABB}$. The simulator runs a copy of the protocol π_{ABB} and keeps it consistent with the inputs and outputs of $\mathcal{F}_{ABB}$ in ideal process (in which S is run). In simulating π_{ABB}, S must provide inputs to $\mathcal{F}_{ABB}$ on behalf of the corrupted parties. For all commands except **Load** and **Private Multiplication** only the inputs from honest parties are consider by $\mathcal{F}_{ABB}$. However for a load $(P_i : x \leftarrow ?)$ and a private multiplication $(P_i : x \leftarrow x_1 \cdot x_2)$ the input from S matters. In the first case S

must provide an input $(P_i : x \leftarrow s)$ for x to be defined and in the second case S must provide the input $(P_i : x \leftarrow x_1 \cdot x_2)$ for x to be defined. Whether the simulator supplies these inputs, and the value of s, depends on Z: If Z lets P_i participate in the simulation of π_{ABB} in a way which results in the honest parties defining $\text{enc}(x)$ to hold some encryption, then S will provide the input to $\mathcal{F}_{\text{ABB}}$ and in the case of a **Load** it will determine an appropriate value of s. The initialization is simulated as follows::

Init: For the first copy it receives the key $(pk = N, sk)$ from $\mathcal{F}_{\text{ABB}}$ and computes $(N, pv, sk_1, \ldots, sk_n) = \texttt{SingleToThesh}(pk, sk).(1)$ Then it computes $K = E_{pk}(0; r_K)$ and computes $R = E_{pk}(1; r_R)$ and outputs (pk, pv, sk_i, R, K) to party P_i. For the second copy it generates N' and all the keys $K_{j,l} = E_{N'}(0; r_{j,l})$ itself and distributes these as in the protocol. When running the simulator for the zero-knowledge proofs with P_j as verifier the key $K_{j,0}$ is given to the simulator, and if P_j is corrupted $t_{j,0}$ is given to the simulator. Notice that we do not use sk' so the commitment schemes $E_{pk', K_{j,0}}$ will be computationally binding as required. When running the simulator for a commitment from P_i to P_j the key $(K_{j,1}, K_{j,2})$ is given to the simulator, and if P_j is corrupted $t_{j,1}$ and $t_{j,2}$ are given to the simulator. Notice that except for K and R this initialization simulates perfectly the protocol.

The simulator maintains the invariant that a variable x is defined in the simulated π_{ABB} iff it is defined in $\mathcal{F}_{\text{ABB}}$. Furthermore the simulator maintains that for all defined variables x it knows $\text{ran}(x)$ s.t. $\text{enc}(x) = E_{pk}(0; \text{ran}(x))$. These values are used for trapdoor openings under K, e.g. in the simulation of the load command:

Load: We only describe how to simulate for $M = K$ as for all loads with key R we will actually know the value to load, so the 'simulation' can be done by running the protocol. On $(P_i : x \leftarrow ?)$, if P_i is honest we proceed as follows:

1. P_i computes $S = E_{pk}(0; r_0)$ and broadcasts S. If P_i is corrupted after this step, then we learn $(P_i : x \leftarrow s)$ from the ideal process. Then using r_K and r_0 compute r_S s.t. $S = E_{pk,K}(s; r_S)$ and uses r_S as the internal state of P_i.
2. For $j \neq i$ party P_i simulates a trapdoor commitment to P_j giving commitment $(L_j, c_{s,j}, ?)$. If P_i is corrupted in or after this step, then first patch as as above, and then patch $(L_j, c_{s,j}, ?)$ to s.
3. For each party P_j where $j \neq i$ party P_i run the simulator from Section 4 with instance $x = (pk, K, S, (L_j, c_{s,j}))$. If P_i is corrupted after this Step, then first patch as above. Then $w = (s, r_S, r_{s,j})$ is a witness to x. Give this witness to the simulator from Section 4 to patch the state of the proof of knowledge.
4. The parties execute the BA following the protocol. If the result is 1 the parties set $\text{enc}(x) = S$ and $\text{ran}(x) = r_0$.

For corrupted P_i the simulator inputs $(P_i : x \leftarrow 0)$ to $\mathcal{F}_{\text{ABB}}$ on behalf of P_i and then lets the honest parties follow the protocol. If the BA fails the simulator inputs `fail` on the SIT of $\mathcal{F}_{\text{ABB}}$. If the broadcast value $\text{enc}(x) = S$ is accepted, then the simulator must at the end of the load input (`change s`) on behalf of P_i on the SIT of $\mathcal{F}_{\text{ABB}}$. Since the BA output 1, at least $t+1$ parties input 1, so at least one party P_j which is still honest input 1. Let $(L_j, c_{s,j})$ be the commitment received by P_j. Since the proof received was accepted, P_i can, e.w.n.p., open $(L_j, c_{s,j})$ and S to the same value, s say — i.e. if we could rewind we could extract such a value. By Theorem 1 we can assume that if S was sent by the adversary, then L_j is a binding key, so (L_j, s_j) can only be opened to one value. Therefore we can using sk decrypt(2) $(L_j, c_{s,j})$ to obtain the value s to which P_i can open S. The intuition here is that if K had been an encryption of 1, then this value s, to which P_i can open S, would indeed be its plaintext value, so we have extracted the 'plaintext' of S.

If P_i is honest at the onset of the load but is corrupted before the third message of the commitment protocol is sent, i.e. in round $r + 2$, then the environment might send a commitment $(L_j, c_{s,j})$ to a value s' different from the s which has already been input to $\mathcal{F}_{\text{ABB}}$, and we have no guarantee that L_j is an encryption key. However, we have no need to decrypt either: Since the simulator sent S it patched the state of P_i to be consistent with $S = \text{enc}_{pk,K}(s; r_S)$. So, if K had been a binding key, then the state of P_i would be consistent with $\mathcal{F}_{\text{ABB}}$.

In all cases, if $\text{enc}(c)$ is accepted the simulator decrypts(3) $\text{enc}(x)$ to learn random bits $\text{ran}(x)$ s.t. $\text{enc}(x) = E_{pk}(0; \text{ran}(x))$.

A private multiplication is simulated similarly, letting $X = E_{pk}(0; r_0)$ and if s becomes known, computing r_X s.t. $X = E_{pk,\text{enc}(x_2)}(s; r_X)$ — in doing this we use that we know $\text{ran}(x_2)$ s.t. $\text{enc}(x_2) = E_{pk}(0; \text{ran}(x_2))$. Linear combination is simulated by letting $\text{enc}(x) = g^{a_0} \prod_{i=1}^{l} \text{enc}(x_j)^{a_j}$ and $\text{ran}(x) = a_0 \prod_{i=1}^{l} \text{ran}(x_j)^{a_j} \bmod N$. An output is simulated as follows:

Output: On input (`output` $x = v$) from $\mathcal{F}_{\text{ABB}}$ we know that all honest parties got the input (`output` x) and will output v in 12 rounds. We cannot use the simulator from Section 6 to simulate a decryption to v, as the inconsistent party might get corrupted. Instead we cheating in the randomization as to make T an encryption of v, and then decrypt honestly.

1. For honest P_i run the load command by committing to uniformly random elements r_i' under trapdoor keys.

2. Let J be the indices j of corrupted parties for which the load into x_j succeeded. If $|J| < t$, then add the indices of some uniformly random honest parties until $|J| = t$ and for those parties, let $r_i = r_i'$. Pick a random degree $t + 1$ polynomial $f(j)$ s.t. $f(j) = r_j$ for $j \in J$ and $f(0) = v$. Then for the honest parties set $r_i = f(i)$ and patch the load into x_i to be consistent with $s = r_i$. Then load r_i into y_i under R by running the load protocol honestly. This is possible as the previous load is now consistent with r_i.

3. This step is simulated by following the protocol. Since R is an encryption of 1, unless a corrupted party P_j was able to load different values into x_j and y_j, we will have that S is an encryption of $f(0) = v$ and so is then T.(4)
4. Run the decryption protocol as in the protocol.

We then prove $\text{IDEAL}_{\mathcal{F}_{\text{ABB}},\mathcal{S},\mathcal{Z}} \overset{c}{\approx} \text{REAL}_{\pi_{\text{ABB}},\mathcal{Z}}$ using a hybrids argument. Define $\mathbf{H_0'}$ by taking $\mathcal{F}_{\text{ABB}}$, $\mathcal{S}$ and $\mathcal{Z}$ and running $\text{IDEAL}_{\mathcal{F}_{\text{ABB}},\mathcal{S},\mathcal{Z}}$ with the modification that $\mathcal{S}_{gen,\text{pal}}(pk)$ is run at (1) to produce the key and $\mathcal{S}_{\text{pal},dec}(T, D_{sk}(T))$ is run at (4) instead of $\pi_{\text{pal},gen}$. Let $H_0 = [H_0'|\neg\bot]$. Doing this we got rid of the use of sk at (1) but introduced one at (4). By the results in Section 6 we will have that $\text{IDEAL}_{\mathcal{F}_{\text{ABB}},\mathcal{S},\mathcal{Z}} \overset{c}{\approx} H_0$. Define $\mathbf{H_1'}$ as H_0' but at $(2, 3)$ use xtr from Section 4 to extract the plaintext (and for (3), the random bits.) Let $H_1 = [H_1'|\neg\bot]$. By the results in Section 4 we will have $H_0 \overset{c}{\approx} H_1$. Define $\mathbf{H_2'}$ as H_1' but at (4) run $\mathcal{S}_{\text{pal},dec}(T, v)$ instead of $\mathcal{S}_{\text{pal},dec}(T, D_{sk}(T))$. Let $H_2 = [H_2'|\neg\bot]$. By using Sections 4 it can be proved that $v = D_{sk}(T)$, e.w.n.p. So, $H_2 \overset{c}{\approx} H_1$. Notice that H_2 can be produced without using sk, so now we can use the semantic security of E_{pk}. Define $\mathbf{H_3'}$ as H_2', but use $R = E_{pk}(0)$. It follows via semantic security that $H_3 = [H_3'|\neg\bot] \overset{c}{\approx} H_2$. Define $\mathbf{H_4'}$ as H_3' except that we use correct inputs to all honest parties and pick the r_i values uniformly at random for all honest parties. It can be seen that $H_4 = [H_4'|\neg\bot] = H_3$, as $K, R \in E_{pk}(0)$, so no information is leaked about the inputs and the environment sees at most t of the r_i values and so cannot distinguish random values from random values consistent with a degree $t+1$ polynomial. Define $\mathbf{H_5'}$ as H_4' but with $K = E_{pk}(1)$ and use semantic security to prove $H_5 = [H_5'|\neg\bot] \overset{c}{\approx} H_4$. Because $K = E(1)$ we cannot use the simulator from 6 for simulating the commitments anymore. However this not not needed either as from H_4' all honest parties use correct inputs, so at the same time we start running all commitments honestly. Define $\mathbf{H_6'}$ as H_5' but starting to use sk at (2) again and as for $H_0 \overset{c}{\approx} H_1$ get that $H_6 = [H_6'|\neg\bot] \overset{c}{\approx} H_5$. Define $\mathbf{H_7'}$ as H_6' but at (4) run $\mathcal{S}_{\text{pal},dec}$ on $(T, D_{sk}(T))$ instead. Proving $H_7 = [H_7'|\neg\bot] \overset{c}{\approx} H_6'$ basically involves proving the protocol correct, which is done using the results from Sections 4. The only difference between H_7 and $\text{REAL}_{\pi_{\text{ABB}},\mathcal{Z}}$ is between the use of $\mathcal{S}_{\text{pal},key-gen}$ and $\mathcal{S}_{\text{pal},dec}$ instead of $\mathcal{F}_{\text{pal},key-gen}$ and $\pi_{\text{pal},dec}$, and $H_7 \overset{c}{\approx} \text{REAL}_{\pi_{\text{ABB}},\mathcal{F}}$ follows as $\text{IDEAL}_{\mathcal{F}_{\text{ABB}},\mathcal{S},\mathcal{Z}} \overset{c}{\approx} H_0$.

References

1. D. Beaver and S. Haber. Cryptographic protocols provably secure against dynamic adversaries. In R. A. Rueppel, editor, *Advances in Cryptology – EuroCrypt '92*, pp. 307–323, Berlin, 1992. Springer-Verlag. LNCS Vol. 658.
2. M. Ben-Or, S. Goldwasser, and A. Wigderson. Completeness theorems for non-cryptographic fault-tolerant distributed computation (extended abstract). In *20th STOC*, pp. 1–10, Chicago, Illinois, May 1988.

3. R. Canetti. Universally composable security: A new paradigm for cryptographic protocols. In *42th FOCS*. IEEE, 2001.
4. R. Canetti, U. Feige, O. Goldreich, and M. Naor. Adaptively secure multi-party computation. In *28th STOC*, pp. 639–648, Philadelphia, Pennsylvania, May 1996.
5. R. Canetti, Y. Lindell, R. Ostrovsky, and A. Sahai. Universally composable two-party and multi-party secure computation. In *34th STOC*, pp. 494–503, Montreal, Quebec, Canada, 2002.
6. D. Chaum, C. Crépeau, and I. Damgård. Multiparty unconditionally secure protocols (extended abstract). In *20th STOC*, pp. 11–19, Chicago, Illinois, May 1988.
7. R. Cramer, I. Damgaard, and J. B. Nielsen. Multiparty computation from threshold homomorphic encryption. In *Advances in Cryptology - EuroCrypt 2001*, pp. 280–300, Berlin, 2001. Springer-Verlag. LNCS Vol. 2045.
8. I. Damgård. Efficient concurrent zero-knowledge in the auxiliary string model. In B. Preneel, editor, *Advances in Cryptology - EuroCrypt 2000*, pp. 418–430, Berlin, 2000. Springer-Verlag. LNCS Vol. 1807.
9. I. Damgård and J. B. Nielsen. Perfect hiding and perfect binding universally composable commitment schemes with constant expansion factor. In M. Yung, editor, *Advances in Cryptology - Crypto 2002*, pp. 581–596, Berlin, 2002. Springer-Verlag. LNCS Vol. 2442.
10. R. Gennaro, M. Rabin, and T. Rabin. Simplified VSS and fast-track multi-party computations with applications to threshold cryptography. In *PODC'98*, 1998.
11. O. Goldreich, S. Micali, and A. Wigderson. How to play any mental game or a completeness theorem for protocols with honest majority. In *19th STOC*, pp. 218–229, New York City, May 1987.
12. A. Lysyanskaya and C. Peikert. Adaptive security in the threshold setting: From cryptosystems to signature schemes. In C. Boyd, editor, *Advances in Cryptology - ASIACRYPT 2001*, pp. 331–350, Berlin, 2001. Springer. LNCS Vol. 2248.
13. P. Paillier. Public-key cryptosystems based on composite degree residue classes. In J. Stern, editor, *Advances in Cryptology - EuroCrypt '99*, pp. 223–238, Berlin, 1999. Springer-Verlag. LNCS Vol. 1592.
14. A. C. Yao. Protocols for secure computations (extended abstract). In *23rd FOCS*. IEEE. 1982. pp. 160–164, Chicago, Illinois, 3–5 Nov. 1982.

Universal Composition with Joint State

Ran Canetti and Tal Rabin

IBM T.J. Watson Research Center. {`canetti`,`talr`}`@watson.ibm.com`.

Abstract. Cryptographic systems often involve running multiple concurrent instances of some protocol, where the instances have some amount of joint state and randomness. (Examples include systems where multiple protocol instances use the same public-key infrastructure, or the same common reference string.) Rather than attempting to analyze the entire system as a single unit, we would like to be able to analyze each such protocol instance as stand-alone, and then use a general composition theorem to deduce the security of the entire system. However, no known composition theorem applies in this setting, since they all assume that the composed protocol instances have disjoint internal states, and that the internal random choices in the various executions are independent.

We propose a new composition operation that can handle the case where different components have some amount of joint state and randomness, and demonstrate sufficient conditions for when the new operation preserves security. The new operation, which is called *universal composition with joint state* (and is based on the recently proposed universal composition operation), turns out to be very useful in a number of quite different scenarios such as those mentioned above.

Keywords: Cryptographic protocols, protocol composition, security analysis

1 Introduction

Cryptographic systems often involve multiple concurrent instances of a large variety of different protocols, where different instances are being run by different sets of mutually suspicious parties; furthermore, they are often deployed in ever-changing and unpredictable environments that may involve additional unknown protocols. Directly analyzing the security of such a system as a single unit is often prohibitively complex. Instead, we would like to be able to de-compose a complex cryptographic system to simpler components, prove the security of each component in isolation, and then deduce the security of the re-composed system.

This "de-compositional approach" is indeed very attractive. However, its soundness hinges on our ability to deduce the security of the entire system from the security of the components. Here *secure composition theorems* come in handy. Roughly speaking, such theorems assert that if a protocol is secure when considered in isolation, then it remains secure even when multiple instances of this protocol are run in the same system, or (in some cases) even when the protocol is used as a component of an arbitrary system.

D. Boneh (Ed.): CRYPTO 2003, LNCS 2729, pp. 265–281, 2003.

A number of *composition operations* (i.e., ways to put together protocols in order to get a composite protocol) have been studied in the context of preservation of security. These include *sequential, parallel,* and *concurrent* composition, when the composed instances are run either by the same sets of parties, or by different sets of parties (e.g., [GK96,Bea91,DDN00,GO94,DNS98,Gol02]). They also include the more general operations of *modular* and *universal* composition, where protocols call other protocols as subroutines [MR91,Can00,DM00,PSW00, Can01]. However, all known composition theorems have the following limitation. They all assume, at least as far as the honest parties are concerned, that the local state of each one of the composed protocol instances is disjoint from the local states of all the other protocol instances run by the party. Furthermore, for each protocol instance, the honest parties are required to use a "fresh" random input that is independent of all the other random inputs. Thus, none of the known composition theorems is applicable when trying to de-compose a system into simpler components, while allowing the components to have some amount of joint state.

In contrast, many cryptographic systems consist of multiple concurrent instances of some (relatively simple) protocol, where all the instances have some limited amount of joint state and joint randomness. Prevalent examples include key-exchange and secure communication protocols, where multiple protocol instances use the same instance of a public-key signature or encryption scheme. Another set of examples include protocols in the common reference string model, where multiple instances use the same short reference string. Indeed, when attempting to analyze such systems, there was so far no alternative but to directly analyze the entire multi-instance system as a single unit.

We formulate a new composition operation for cryptographic protocols, that is applicable even in the case where multiple protocol instances have some amount of joint state. We also demonstrate sufficient conditions for when this operation preserves security. Our new operation, Universal Composition with joint state (JUC, pronounced "juicy"), is formulated within the Universal Composition (UC) framework [Can01], and extends its powers. The new operation drastically simplifies the analysis of systems where multiple instances have joint state, by allowing us to apply the de-composition methodology described above even to such systems. In fact, recent works which originally analyzed their security examining the entire protocol as a single unit have updated and simplified their proofs by utilizing the techniques presented in this paper [CK02,CLOS02].

1.1   The New Composition Theorem

We provide a very informal overview of the new composition theorem and its usage. Our system consists of a "high-level protocol", π, that uses multiple instances of a sub-protocol ρ, where the various instances of ρ have some joint state.

To be able to use the JUC theorem, we need to have in hand a protocol, $\hat{\rho}$, where a single instance of $\hat{\rho}$ has essentially the same functionality as multiple independent copies of ρ. We then proceed as follows. We first analyze the overall protocol π under the assumption that the copies of ρ are independent. (This can be done using known composition theorems, such as the UC theorem.) We then replace all instances of ρ within π with a single instance of $\hat{\rho}$. The JUC Theorem essentially states that protocol π behaves the same regardless of whether it is using multiple independent copies of ρ, or alternatively a single copy of $\hat{\rho}$.

The JUC theorem proves to be instrumental in de-composing complex systems. Using the terminology of the previous paragraph, it allows us to decompose our system into a "π part" plus a "$\hat{\rho}$ part," analyze each part in isolation, and then deduce the security of the re-composed system. The important thing to notice is that the "π part" treats all the copies of ρ as if they were independent copies without any joint state. This proves to be very useful in cases where the "π part" by itself consists of multiple copies of some other, simpler protocol ϕ, where different copies of ϕ call different copies of ρ. We can now analyze each copy of ϕ as stand-alone, then compose all copies of ϕ into a single protocol π (say, using the UC Theorem), and then use the JUC Theorem to compose π with all the copies of ρ as described above, in spite of the fact that the copies of ρ (and consequently the copies of ϕ) have joint state (see Figure 1).

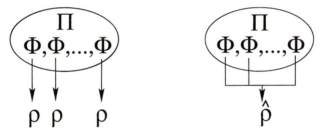

Fig. 1. Universal Composition with Joint State: Protocol π is analyzed under the assumption that the copies of ρ are independent (left). This is in spite of the fact that all copies of ρ are replaced by a single copy of $\hat{\rho}$ (right).

As said above, in order for the JUC Theorem to hold, protocol $\hat{\rho}$ must exhibit the same behavior as multiple "independent" instances of ρ. (A formalization of this intuitive requirement appears within.) Clearly, the protocol $\hat{\rho}$ that simply runs multiple independent copies of ρ would guarantee this "independence". However, such a protocol $\hat{\rho}$ would not be very interesting. The power of the JUC theorem is in the cases where protocol $\hat{\rho}$ is more efficient, and in particular makes meaningful use of joint state between copies of ρ. The rest of the introduction is dedicated to providing examples for the use of the JUC theorem.

1.2   Application to Protocols Using Digital Signatures

We exemplify the use of the JUC Theorem for de-composing systems where multiple protocol instances use the same instance of a signature scheme. Specifically, we concentrate on the prevalent example of key-exchange and secure channel protocols authenticated via digital signatures. Here, multiple parties run multiple instance of the key-exchange protocol in order to exchange multiple keys, while using the same public-key infrastructure which is the joint state. The exchanged keys are then used to establish secure communication channels. Previous analytical works of key-exchange and secure channel protocols treats the entire multi-party, multi-execution system as a single unit (see, e.g., [BR93, Sho99,CK01]).

Here we show how, using the JUC Theorem, one can de-compose the system to individual sessions, analyze each session independently of all others, and deduce the security of the entire multi-session system. We proceed as follows. Using the terminology of the previous subsection, the key-exchange protocol (for exchanging a single key) is denoted ϕ, and the multi-instance composition of ϕ is denoted π. We are given a protocol ρ which satisfies the security requirements from a digital signature scheme. The protocol π uses multiple calls to ρ, where different calls are made by different copies of ϕ within ρ. (We assume that each of these independent calls is assigned a unique identifier i. This assumption is central in our solution. See discussion about the identifiers in Section 2).

In order to be able to apply the JUC Theorem, we show how to construct a protocol $\hat{\rho}$ that behaves, within a single instance, like multiple independent copies of ρ. In fact, this can be done with essentially the complexity of a single instance of ρ. Protocol $\hat{\rho}$ runs a single copy of ρ. When the ith instance of ϕ invokes its instance of ρ to generate a signature on message m, protocol $\hat{\rho}$ uses its single instance of ρ to sign the message (i, m). Similarly, whenever asked to verify whether s is a signature of instance i on a message m, $\hat{\rho}$ uses its single instance of ρ to verify whether s is a signature on (i, m). We show that if ρ is a secure signature protocol then $\hat{\rho}$ satisfies the conditions required by the JUC Theorem (i.e., $\hat{\rho}$ behaves essentially like multiple independent copies of a signature scheme). This allows us to deduce the security of the entire, multi-session key exchange protocol even though we only analyzed the security of the single session protocol ϕ and the signature protocol $\hat{\rho}$.

The same methodology applies also to the treatment of secure session protocols. That is, it is possible to analyze the security of a single session protocol as stand alone, and then use the JUC theorem to deduce the security of the composite multi-session system — in spite of the fact that all sessions use the same instance of the signature scheme. Indeed, the updated version of [CK02] on key-exchange and secure session protocols has modified its presentation and analysis to utilize the JUC Theorem as proven here.

1.3 Application to Protocols in the Common Reference String Model

A similar phenomenon happens in the case of protocols in the common reference string (CRS) model, where all parties have access to a reference string taken from a predefined distribution. We often have protocols where multiple instances use the same short reference string. These instances may be run either by the same set of parties or by different sets. So far, the only known way to analyze such multi-instance systems was to directly analyze them as a single unit. Using universal composition with joint state, we show how one can de-compose a multi-instance system to individual instances, analyze each instance in isolation, and then deduce security of the entire multi-instance system — in spite of the fact that all the protocol instances use the same short reference string. We demonstrate several alternative ways to go about this de-composition.

The first and most general way to de-compose multi-instance systems in the CRS model proceeds as follows. We first recall that the CRS model can be captured as the model that provide the parties with access to an "idealized protocol" ρ that returns the same string to all parties, where the string is chosen from a predefined distribution. Let ϕ be a protocol that uses the CRS (i.e., ϕ invokes protocol ρ) and let π be some protocol that runs multiple instances of protocol ϕ. In order to prove the security of π, we proceed in four steps: (a) We prove the security of a single instance of ϕ using a single instance of ρ, in a stand-alone setting where no other protocol executions exist. (b) Using known composition theorems (e.g., the universal composition theorem of [Can01]), we deduce that the multi-instance protocol π is also secure. Here, however, protocol π uses multiple independent instances of ρ, which corresponds to having multiple independent copies of the reference string. (c) We construct a protocol, $\hat{\rho}$, that mimics the behavior of multiple independent copies of ρ, using only a single copy of ρ. In other words, $\hat{\rho}$ generates multiple "independent" copies of the reference string, given only a single copy of the string. (d) Using the JUC Theorem, we deduce that the entire composed protocol (consisting of the multi-instance protocol π, where all calls to all copies of ρ are replaced with calls to a single instance of protocol $\hat{\rho}$) is secure. We stress that here all the copies of ϕ within π use the same copy of protocol ρ, namely only a single instance of the reference string.

In order to complete the de-composition process, we need to come up with a protocol $\hat{\rho}$ that realizes multiple instances of ρ, given only a single instance of ρ. Our construction of protocol $\hat{\rho}$ is essentially the three-message coin-tossing-into-the-well protocol of Blum [Blu82], where the commitments are taken to be universally composable commitments e.g., those of [CF01,CLOS02,DN02].

However, while the above de-composition method is quite general, it is not completely satisfactory because of the need to run the additional interactive protocol $\hat{\rho}$. In particular, in the composed protocol each copy of ϕ is interactive — even if the original construction of ϕ is non-interactive. We would like to be able to carry out the de-composition paradigm without paying the price in communication rounds.

At a first glance, it may appear that the way to avoid adding rounds is to come up with better constructions of protocol $\hat{\rho}$, which would be non-interactive. However, we show that this is not possible. That is, we show that *any* protocol which realizes multiple "independent" copies of the CRS, given only a single copy of the CRS, must be interactive. Essentially, each party must send at least one message per each new instance of the reference string. Furthermore, this message must be essentially at least as long as the generated string.

Given the impossibility of a non-interactive solution for the general problem of generating multiple CRSs given a single short CRS, we turn to other, less general ways to de-compose multi-instance systems in the CRS model. Specifically, we describe how our methodology can be applied to Zero-Knowledge (ZK) protocols and commitment protocols in the CRS model. Let us first sketch how this works for ZK protocols. Here we let protocol ρ be a single-instance ZK protocol. That is, protocol ρ carries out a single ZK proof. Assume we have a protocol π that consists of multiple copies of some protocol ϕ, where each instance of ϕ uses (perhaps multiple) copies of ρ. We can now use the JUC theorem to replace all instances of ρ with a single instance of protocol $\hat{\rho}$ that realizes multiple ZK proofs within a single instance. Luckily, such protocols $\hat{\rho}$ exist, and use only a single short instance of the CRS for all instances of the ZK proof [CF01,DCO$^+$01]. In particular, the protocol of [DCO$^+$01] is non-interactive.

In the case of commitment protocols we follow the same steps, with the exception that protocol ρ is a commitment protocol for a single commitment-decommitment process. The "composite protocol" $\hat{\rho}$ now provides the functionality of multiple commitments and decommitments, while using only a single short instance of the CRS. Such protocols exist, e.g., those of [CF01,CLOS02, DN02].

We remark that our formalization and results for the CRS model play a central role in the updated proofs for the general construction in [CLOS02]. Earlier versions of the paper analyzed these constructions directly as multi-instance protocols and were considerably more cumbersome.

Organization. Section 2 reviews the notion of UC security and the UC theorem of [Can01]. Section 3 presents and proves the JUC Theorem. For lack of space, the paper contains only the application to protocols in the CRS model in Section 4. The application to protocols that use signature schemes appears in [CR03].

2   Review of the Universal Composition Theorem

We provide a brief review of the universally composable security framework [Can01]. The framework allows for defining the security properties of cryptographic tasks so that the security of protocols is maintained under a general composition operation with an unbounded number of instances of arbitrary protocols running concurrently in the system. This composition operation is called universal composition. Similarly, definitions of security in this framework are called universally composable (UC).

As in other general definitions (e.g., [GL90,MR91,Bea91,Can00,PSW00]), the security requirements of a given task (i.e., the functionality expected from a protocol that carries out the task) are captured via a set of instructions for a "trusted party" that obtains the inputs of the participants and provides them with the desired outputs (in one or more iterations). Informally, a protocol securely carries out a given task if running the protocol with a realistic adversary amounts to "emulating" an ideal process where the parties hand their inputs to a trusted party with the appropriate functionality and obtain their outputs from it, without any other interaction. We call the algorithm run by the trusted party an ideal functionality.

In order to allow proving the universal composition theorem, the notion of emulation in this framework is considerably stronger than previous ones. Traditionally, the model of computation includes the parties running the protocol and an adversary, $\mathcal{A}$, that controls the communication channels and potentially corrupts parties. "Emulating an ideal process" means that for any adversary $\mathcal{A}$ there should exist an "ideal process adversary" (or, simulator) $\mathcal{S}$ that causes the *outputs* of the parties in the ideal process to have similar distribution to the outputs of the parties in an execution of the protocol. In the UC framework the requirement on $\mathcal{S}$ is more stringent. Specifically, an additional entity, called the environment $\mathcal{Z}$, is introduced. The environment generates the inputs to all parties, reads all outputs, and in addition interacts with the adversary in an arbitrary way throughout the computation. A protocol is said to securely realize a given ideal functionality $\mathcal{F}$ if for any "real-life" adversary $\mathcal{A}$ that interacts with the protocol and the environment there exists an "ideal-process adversary" $\mathcal{S}$, such that *no environment* $\mathcal{Z}$ can tell whether it is interacting with $\mathcal{A}$ and parties running the protocol, or with $\mathcal{S}$ and parties that interact with $\mathcal{F}$ in the ideal process. In a sense, here $\mathcal{Z}$ serves as an "interactive distinguisher" between a run of the protocol and the ideal process with access to $\mathcal{F}$.

The following *universal composition theorem* is proven in [Can01]. Consider a protocol π that operates in the $\mathcal{F}$-hybrid model, where parties can communicate as usual, and in addition have ideal access to an unbounded number of *copies* of an ideal functionality $\mathcal{F}$. Let ρ be a protocol that securely realizes $\mathcal{F}$ as sketched above, and let π^ρ be identical to π with the exception that the interaction with *each copy* of $\mathcal{F}$ is replaced with an interaction with a *separate instance* of ρ. Then, π and π^ρ have essentially the same input/output behavior. In particular, if π securely realizes some ideal functionality $\mathcal{I}$ in the $\mathcal{F}$-hybrid model then π^ρ securely realizes $\mathcal{I}$ in the standard model (i.e. without ideal functionality).

On the Session Identifiers (SID's). Let us highlight one detail regarding the hybrid model that will become important in subsequent sections. Each copy of the ideal functionality $\mathcal{F}$ is assumed to have a unique identifier, called the session identifier (SID) of that copy. Each message sent to $\mathcal{F}$ in the hybrid model should contain a SID, and is then forwarded to the corresponding copy of $\mathcal{F}$. (If there is no such copy then a new one is invoked and given this SID.) Similarly, each message from a copy of $\mathcal{F}$ to a party contains the SID of that copy. The SIDs are determined by the protocol running in the hybrid model. Notice that this formalization allows each copy of $\mathcal{F}$, and each instance of the protocol that later replaces this copy, to know its own SID. It also guarantees that no two copies of $\mathcal{F}$ can ever have the same SID. This is essential for the composition theorem to hold. (See discussion in the Introduction.) Let $\mathcal{F}(sid)$ denote the copy of functionality $\mathcal{F}$ with SID $=(sid)$.

We stress that the model does not specified how the parties learn, or "agree" on the SID. While we cannot always assume that a set of uncoordinated parties can agree on an SID, there are many interesting and important settings where this assumption is reasonable. As an example consider a network in which two-party protocols are being executed by a set of uncoordinated parties. In this case a unique SID can be easily chosen by the two parties A and B themselves, by choosing locally unique strings r_A and r_B, respectively, which they exchange. The session id is defined to be $A \circ B \circ r_A \circ r_B$. Here an honest party is guaranteed to have a unique SID, even if the other party is cheating. Another example is within a closed network of multiple users executing multiple protocols, where the protocols have some ordering and thus can be allotted unique SIDs.

Nonetheless, it should be stressed that in some cases the requirement for unique SIDs is imperative for providing security. For instance, in [LLR02] it is shown that Byzantine Agreement can not be reached in some specific settings unless unique identifiers are provided.

3   Universal Composition with Joint State (JUC)

Recall that the universal composition operation requires replacing each copy of $\mathcal{F}$ with a different invocation of protocol ρ, where all the invocations of ρ in π^ρ must have disjoint states and independent random inputs. However, in our setting the copies of ρ have joint state. In fact, we wish to replace all the invocations of $\mathcal{F}$ with a *single* instance of some "joint protocol" $\hat{\rho}$. In order to do that, we essentially require that $\hat{\rho}$ has the same functionality as that of multiple independent copies of ρ. To formalize this requirement we define the multi-session extension of an ideal functionality. But first we define the composition operation.

The Composition Operation. The new composition operation, called universal composition with joint state (JUC), takes two protocols as arguments: a protocol π in the $\mathcal{F}$-hybrid model and a protocol $\hat{\rho}$ realizing the multi-session functionality. The result is a composed protocol, denoted $\pi^{[\hat{\rho}]}$, and described as follows. Essentially, universal composition with joint state is identical to universal composition, with two exception: First, each party P_i invokes only a single copy of $\hat{\rho}$ and replaces all calls to copies of $\mathcal{F}$ with activations of (the single copy of) $\hat{\rho}$. Second, now each activation of $\hat{\rho}$ includes *two* ids, the SID of $\hat{\rho}$ is set to some fixed, predefined value sid_0. The second id, SSID, is set to the original id of the invocation of $\mathcal{F}$. More specifically, protocol $\pi^{[\hat{\rho}]}$ behaves like π with the following changes.

1. *Modifications to the communication between $\pi^{[\hat{\rho}]}$ and $\hat{\rho}$ relative to the communication between π and $\mathcal{F}$:*
   a) When activated for the first time within party P_i, $\pi^{[\hat{\rho}]}$ initiates a copy of protocol $\hat{\rho}$ with SID= sid_0.
   b) Whenever π instructs party P_i to send a message (sid, v) to $\mathcal{F}(sid)$, protocol $\pi^{[\hat{\rho}]}$ instructs P_i to call $\hat{\rho}$ with input value (sid_0, sid, v).
   c) When (the single copy of) $\hat{\rho}$ generates an output value (sid_0, sid, v) within $\pi^{[\hat{\rho}]}$, then $\pi^{[\hat{\rho}]}$ proceeds just as π proceeds upon receiving output message (sid, v) from $\mathcal{F}(sid)$.
2. *Operations required for communications between two parties in the system:*

a) Whenever protocol $\hat{\rho}$ wishes to send a message m, generated by the computation relating to (sid_0, sid), to some party P_j, then P_i writes the message (sid_0, sid, m) on its outgoing communication tape.

b) Upon delivery of a message (sid_0, sid, m) from P_j, party P_i activates $\hat{\rho}$ with incoming message (sid_0, sid, m).

The Multi-session Extension of an Ideal Functionality. We formalize the security requirements from the "joint protocol" $\hat{\rho}$. Intuitively, the requirement is that it should have essentially the same functionality as multiple independent invocations of ρ. More formally, we define the following ideal functionality, $\hat{\mathcal{F}}$, which we want $\hat{\rho}$ to realize. Let $\mathcal{F}$ be an ideal functionality. (Intuitively, $\mathcal{F}$ is the functionality realized by a single instance of ρ.) According to the UC formalization, $\mathcal{F}$ expects each incoming message to contain a special field consisting of its session identifier (SID). All messages received by $\mathcal{F}$ are expected to have the same value of the SID. (Messages that have different session identifier than that of the first message are ignored.) Similarly, all outgoing messages generated by $\mathcal{F}$ carry the same SID.

The multi-session extension of $\mathcal{F}$, denoted $\hat{\mathcal{F}}$, is defined as follows. $\hat{\mathcal{F}}$ expects each incoming message to contain *two* special fields. The first is the usual session identifier field as in any ideal functionality. The second field is called the sub-session identifier (SSID) field. Upon receiving a message $(sid, ssid, v)$ (where sid is the SID, $ssid$ is the SSID, and v is an arbitrary value or list of values), $\hat{\mathcal{F}}$ first checks if there is a running copy of $\mathcal{F}$ whose (single) session identifier is $ssid$. If so, then $\hat{\mathcal{F}}$ activates that copy of $\mathcal{F}$ with incoming message $(ssid, v)$, and follows the instructions of this copy. Otherwise, a new copy of $\mathcal{F}$ is invoked (within $\hat{\mathcal{F}}$) and immediately activated with input $(ssid, v)$. From now on, this copy is associated with sub-session identifier $ssid$. (That is, $ssid$ is the session identifier of this copy of $\mathcal{F}$.) Whenever a copy of $\mathcal{F}$ sends a message $(ssid, v)$ to some party P_i, $\hat{\mathcal{F}}$ sends $(sid, ssid, v)$ to P_i, and sends $ssid$ to the adversary. Sending $ssid$ to the adversary implies that $\hat{\mathcal{F}}$ does not hide which copy of $\mathcal{F}$ is being activated within $\hat{\mathcal{F}}$.

It is stressed that $\hat{\mathcal{F}}$ is not explicitly used by π. It only serves as a criterion for the security of $\hat{\rho}$. Furthermore, while $\hat{\mathcal{F}}$ consists of several copies of $\mathcal{F}$ with disjoint states, there is no requirement that the protocol $\hat{\rho}$ that realizes $\hat{\mathcal{F}}$ would have such structure. Indeed, $\hat{\rho}$ may use some joint state for realizing all the copies of $\mathcal{F}$ within $\hat{\mathcal{F}}$. Clearly, the case where $\hat{\rho}$ uses some joint state is the case of interest for this work, as the other instance falls under the regular universal composition.

Theorem 1 (Universal Composition with Joint State). *Let $\mathcal{F}$ be an ideal functionality. Let π be a protocol in the $\mathcal{F}$-hybrid model, and let $\hat{\rho}$ be a protocol that securely realizes $\hat{\mathcal{F}}$, the multi-session extension of $\mathcal{F}$, in the real-life model. Then the composed protocol $\pi^{[\hat{\rho}]}$ in the real-life model emulates protocol π in the $\mathcal{F}$-hybrid model. That is, for any adversary $\mathcal{A}$ that interacts with parties running $\pi^{[\hat{\rho}]}$ in the real-life model, there exists an adversary $\mathcal{A}'$ that interacts with parties running π in the $\mathcal{F}$-hybrid model, such that no environment machine $\mathcal{Z}$ can tell whether it is interacting with $\pi^{[\hat{\rho}]}$ and $\mathcal{A}$, or alternatively with π and $\mathcal{A}'$.*

Proof. Let $\mathcal{F}, \pi, \hat{\rho}$ be as in the theorem statement. We show that $\pi^{[\hat{\rho}]}$ in the real-life model emulates protocol π in the $\mathcal{F}$-hybrid model. This is done in two steps: first we define a protocol $\tilde{\pi}$ and show that $\pi^{[\hat{\rho}]}$ in the real-life model emulates protocol $\tilde{\pi}$ in the $\hat{\mathcal{F}}$-hybrid model. Next we show that protocol $\tilde{\pi}$ in the $\hat{\mathcal{F}}$-hybrid model emulates protocol π in the $\mathcal{F}$-hybrid model.

Protocol $\tilde{\pi}$ is a slight variation of protocol π, operating in the $\hat{\mathcal{F}}$-hybrid model. Specifically, $\tilde{\pi}$ is identical to π, with the following exceptions. (Note that $\tilde{\pi}$ invokes only a single copy of $\hat{\mathcal{F}}$ throughout the computation.)

1. Whenever π instructs P_i to send a message (sid, v) to some copy $\mathcal{F}(sid)$ of $\mathcal{F}$, $\tilde{\pi}$ instructs P_i to send (sid_0, sid, v) to $\hat{\mathcal{F}}$.
2. Whenever some party P_i, running $\tilde{\pi}$, receives a message (sid_0, sid, v) from $\hat{\mathcal{F}}$, it follows the instructions of π upon receipt of the message (sid, v) from $\mathcal{F}(sid)$.

Recall that $\tilde{\pi}^{\hat{\rho}}$ is the protocol obtained by applying the universal composition operation of [Can01] to protocols $\tilde{\pi}$ and $\hat{\rho}$. Since protocol $\hat{\rho}$ securely realizes $\hat{\mathcal{F}}$, it follows from the universal composition theorem that protocol $\tilde{\pi}^{\hat{\rho}}$ in the real-life model emulates protocol $\tilde{\pi}$ in the $\hat{\mathcal{F}}$-hybrid model. Furthermore, it is easy to see that protocol $\tilde{\pi}^{\hat{\rho}}$ is identical to protocol $\pi^{[\hat{\rho}]}$. (These are two different descriptions of exactly the same protocol.) We thus have that protocol $\pi^{[\hat{\rho}]}$ in the real-life model emulates protocol $\tilde{\pi}$ in the $\hat{\mathcal{F}}$-hybrid model.

It remains to show that protocol $\tilde{\pi}$ in the $\hat{\mathcal{F}}$-hybrid model emulates protocol π in the $\mathcal{F}$-hybrid model. This is done as follows. Let $\hat{\mathcal{A}}$ be an adversary that interacts with parties running $\tilde{\pi}$ in the $\hat{\mathcal{F}}$-hybrid model. We construct an adversary $\mathcal{A}$ such that no environment will be able to tell whether it is interacting with $\mathcal{A}$ and π in the $\mathcal{F}$-hybrid model or with $\hat{\mathcal{A}}$ and $\tilde{\pi}$ in the $\hat{\mathcal{F}}$-hybrid model. Adversary $\mathcal{A}$ follows the instructions of $\hat{\mathcal{A}}$, with the following exceptions:

1. Whenever $\mathcal{A}$ is notified that some copy $\mathcal{F}(sid)$ of $\mathcal{F}$ has sent a message with identifier id to some party P_i, $\mathcal{A}$ records the pair (sid, id) and notifies $\hat{\mathcal{A}}$ that $\hat{\mathcal{F}}_{(sid_0)}$ has sent a message with identifier id to P_i. Note that $\mathcal{A}$ does not see the actual content of the message, but is only aware of the fact that a message has been sent.
2. Whenever $\hat{\mathcal{A}}$ delivers a message with identifier id from $\hat{\mathcal{F}}$ to P_i (in the $\hat{\mathcal{F}}$-hybrid model), $\mathcal{A}$ looks up the pair (sid, id) and delivers the message with identifier id from $\mathcal{F}(sid)$ to P_i.
3. When $\hat{\mathcal{A}}$ corrupts a party P_i (running $\tilde{\pi}$ in the $\hat{\mathcal{F}}$-hybrid model), $\mathcal{A}$ corrupts P_i (running π in the $\mathcal{F}$-hybrid model) and obtains the internal state of P_i for protocol π. It then 'translates' the internal state of P_i for protocol π to be consistent with an internal state of P_i for protocol $\tilde{\pi}$, and hand this information to $\hat{\mathcal{A}}$. (The translation is straightforward: calls to multiple copies of $\mathcal{F}$ are translated into calls to a single copy of $\hat{\mathcal{F}}$, with a fixed session identifier sid_0, and the corresponding SSIDs.)

It is straightforward to verify that the view of $\mathcal{Z}$ in an interaction with $\mathcal{A}$ and π in the $\mathcal{F}$-hybrid model is distributed identically to the view of $\mathcal{Z}$ in an interaction with $\hat{\mathcal{A}}$ and $\tilde{\pi}$ in the $\hat{\mathcal{F}}$-hybrid model.

We remark that if $\hat{\rho}$ operates in the $\mathcal{G}$-hybrid model for some ideal functionality $\mathcal{G}$, rather than in the real-life model, then $\pi^{[\rho]}$ also operated in the $\mathcal{G}$-hybrid model. In this case, we have that $\pi^{[\rho]}$ in the $\mathcal{G}$-hybrid model emulates protocol π in the $\mathcal{F}$-hybrid model.

4   Application to Protocols in the CRS Model

This section exemplifies the use of the JUC theorem for protocols in the common reference string model. There are two ways to approach this issue, the first is to take a single CRS and "stretch" it into multiple independent CRSs, which will later be utilized by protocols which require independent CRS's. The second is to examine a specific multi-session functionality which employs a CRS and to directly generate the multi-session functionality from the single CRS, and prove its security directly. While the first approach is more general, the second approach will result in more efficient protocols for specific functionalities.

Though the introduction starts with the first method, here we start with a specific example of commitments in the CRS model. This order of presentation is motivated by the fact that our protocol for stretching the CRS will need to employ a multi-session commitment functionality.

4.1   Commitment in the CRS Model

Let us first recall the ideal commitment functionality, $\mathcal{F}_{com}$, as defined in [CF01] (see Figure 2). Each copy of $\mathcal{F}_{com}$ handles the process of a single commitment followed by its single decommitment.

Functionality $\mathcal{F}_{com}$

$\mathcal{F}_{com}$ proceeds as follows, running with parties $P_1, ..., P_n$ and an adversary $\mathcal{S}$.

1. Upon receiving a value (Commit, sid, P_i, P_j, x) from P_i, record the value x and send the message (Receipt, sid, P_i, P_j) to P_j and $\mathcal{S}$. Ignore any subsequent Commit messages.
2. Upon receiving a value (Open, sid, P_i, P_j) from P_i, proceed as follows: If some value x was previously recoded, then send the message (Open, sid, P_i, P_j, x) to P_j and $\mathcal{S}$ and halt. Otherwise halt.

Fig. 2. The Ideal Commitment functionality

The universal composition theorem allows us to write a protocol π in the $\mathcal{F}_{com}$-hybrid model, and then replace each copy of $\mathcal{F}_{com}$ with an instance of a protocol ρ that securely realizes $\mathcal{F}_{com}$. However, we only know how to realize $\mathcal{F}_{com}$ in the common reference string model, or in other words in the $\mathcal{F}_{crs}$-hybrid model, where $\mathcal{F}_{crs}$ is the ideal functionality that provides all parties with a common string taken from some pre-specified distribution. Consequently, if the universal composition theorem is used as is, then each instance of ρ in the composed protocol π^ρ would use a different copy of $\mathcal{F}_{crs}$ (i.e., an independent copy of the reference string).

Instead, we are looking for a protocol where all copies of the commitment protocol use the same short reference string. In [CF01], this is solved as follows. First, they define an additional ideal functionality, $\mathcal{F}_{mcom}$, where a single copy handles multiple commitments and decommitments by different parties and to different messages. Next, they construct a protocol, UCC_{ReUse}, that securely realizes $\mathcal{F}_{mcom}$ and uses a single copy of $\mathcal{F}_{crs}$ for all the commitments. Thus, if the high-level protocol π is written in the $\mathcal{F}_{mcom}$-hybrid rather than the $\mathcal{F}_{com}$-hybrid model, then the composed protocol, $\pi^{UCC_{ReUse}}$, uses a single copy of the reference string for multiple commitments.

However, having the high-level protocol π use $\mathcal{F}_{mcom}$ rather than $\mathcal{F}_{com}$ comes at a price in the analysis of the protocol. In order to guarantee that a single copy of $\mathcal{F}_{crs}$ is used throughout the computation, π has to use only a single copy of $\mathcal{F}_{mcom}$ (since each copy of $\mathcal{F}_{mcom}$ uses a different copy of $\mathcal{F}_{crs}$). This puts a considerable restriction on the analysis of π, as the security needs to be proven for π as a single unit. This holds even if π consists of multiple instances of some simpler protocol ρ. Thus, much of the advantage of using the UC theorem is lost.

These restrictions can be avoided using universal composition with joint state. We first observe that the functionality $\mathcal{F}_{mcom}$ is nothing but a reformulation of $\hat{\mathcal{F}}_{com}$. The JUC Theorem thus says that if π is a protocol in the $\mathcal{F}_{com}$-hybrid model (and uses as many copies of $\mathcal{F}_{com}$ as it wishes) then the composed protocol $\pi^{[UCC_{ReUse}]}$ runs in the $\mathcal{F}_{crs}$-hybrid model, emulates π, *and uses only a single copy of $\mathcal{F}_{crs}$.* In other words, we can allow protocol π (and all the higher level protocols that may use π as a subroutine) to operate in an idealized model where commitments are completely independent of each other, and then use the JUC Theorem to implement all the commitments using a single short common string.

4.2   Protocols for Stretching the Common Reference String

This section investigates the possibility of realizing $\hat{\mathcal{F}}_{crs}$ in the $\mathcal{F}_{crs}$-hybrid model. More specifically, we present a protocol ρ that securely realizes $\hat{\mathcal{F}}_{crs}$ in the $\mathcal{F}_{crs}$-hybrid model, and uses only a single copy of $\mathcal{F}_{crs}$. Protocol ρ allows the parties to generate multiple, computationally independent copies of the random string, using a single copy of the string. Then we can design protocols in the $\mathcal{F}_{crs}$-hybrid model, where each protocol instance can assume that no other instances have access to the reference string it is using, and then replace all copies of $\mathcal{F}_{crs}$ with a single copy of the reference string.

For simplicity we consider only the case where the reference string is taken from the uniform distribution over $\{0,1\}^t$ for some t. Also, we restrict attention to protocols where only *two* parties need to have access to the reference string. It seems that this special case captures much of the essence of the problem.

We first show a simple protocol that securely realizes $\hat{\mathcal{F}}_{crs}$ (for two parties and with uniform distribution) using a single copy of $\mathcal{F}_{crs}$. The protocol requires interaction between the two parties in order to generate each new copy of the reference string. We then demonstrate that *any* protocol that realizes $\hat{\mathcal{F}}_{crs}$ must involve sending at least one message by each of the participants. Furthermore, a new message must be sent for obtaining essentially any new copy of the reference string.

Let us first formulate functionality $\mathcal{F}_{crs}$ for the restricted case described above (Fig. 3). The functionality is parameterized by t, the length of the reference string. Note that $\mathcal{F}_{crs}$ does not limit the identities of the parties that may obtain the common random value r. The number (and identities) or parties that actually obtain the string is determined by the protocol that realizes $\mathcal{F}_{crs}$.

Functionality $\mathcal{F}_{crs}^t$

$\mathcal{F}_{crs}$ proceeds as follows, running with parties $P_1, ..., P_n$, and an adversary $\mathcal{S}$, parameterized by an integer t.

1. When receiving (CRS,sid, P_i, P_j) from P_i, choose a value $r \xleftarrow{\text{R}} \{0,1\}^t$, send (CRS,$sid, r$) to P_i, and send (sid, P_i, P_j, r) to $\mathcal{S}$. Next, when receiving (CRS,sid, P_i, P_j) from P_j (and only P_j), send (CRS,sid, r) to P_j and $\mathcal{S}$, and halt.

Fig. 3. The Common Random String functionality

Realizing $\hat{\mathcal{F}}_{crs}$. We show how to realize $\hat{\mathcal{F}}_{crs}$, using a single copy of $\mathcal{F}_{crs}$, for the case where only two parties obtain each copy of the reference string. The protocol is essentially the coin-tossing-into-the-well protocol (ct) of Blum [Blu82], which employs a commitment scheme. We use a universally composable commitment that utilize the single common random string. We first present the protocol in the $\mathcal{F}_{com}$-hybrid model, and then use the JUC Theorem to compose this protocol with protocol UCC$_{\text{ReUse}}$ of [CF01] (as discussed in Section 4.1), and obtain the desired protocol. This means that the distribution needed for our protocol is the distribution needed for UCC$_{\text{ReUse}}$. The protocol in the $\mathcal{F}_{com}$-hybrid model is denoted ct^t. Whenever parties P_i and P_j are invoked to generate a new copy of the reference string (say, with SID sid), they proceed as follows.

Message 1: When activated with input CRS,sid, P_i, P_j), P_i chooses a random string $r_i \xleftarrow{\text{R}} \{0,1\}^t$ and commits to r_i for P_j, using a new copy of $\mathcal{F}_{com}$ with SID sid. (That is, P_i sends (Commit, sid, P_i, P_j, r_i) to $\mathcal{F}_{com}$).

Message 2: When activated with input CRS,sid, P_i, P_j), P_j waits to receive (Receipt, sid, P_i, P_j) from $\mathcal{F}_{com}$. Next, P_j chooses $r_j \xleftarrow{\text{R}} \{0,1\}^t$ and sends to P_i.

Message 3: Upon receiving r_j, P_i decommits to r_i. (That is, P_i sends (Open, sid, P_i, P_j) to $\mathcal{F}_{com}$).

Output: Both parties output the string $r_i \oplus r_j$.

Claim. Protocol ct^t securely realizes $\hat{\mathcal{F}}_{crs}^t$ in the $\mathcal{F}_{com}$-hybrid model.

Proof. Let $\mathcal{A}$ be an adversary that interacts with protocol ct^t in the $\mathcal{F}_{com}$-hybrid model. We construct an adversary $\mathcal{S}$ so that no environment can tell whether it is interacting with $\mathcal{S}$ in the ideal process for $\hat{\mathcal{F}}_{crs}^t$ or with $\mathcal{A}$ and ct in the $\mathcal{F}_{com}$-hybrid model. Adversary $\mathcal{S}$ runs a simulated copy of $\mathcal{A}$, and proceeds as follows. (Without loss of generality we assume that $\mathcal{A}$ does not run the protocol between two corrupted parties.)

1. Throughout, whenever $\mathcal{S}$ receives an input value from $\mathcal{Z}$, it copies this value to $\mathcal{A}$'s input tape. Whenever $\mathcal{A}$ writes a value on its output tape, $\mathcal{S}$ copies this value to its own output tape (to be read by $\mathcal{Z}$).

2. **Corrupted initiator:** If the simulated $\mathcal{A}$ generates a message (Commit, sid, P_i, P_j, r_i) from a corrupted party P_i to $\mathcal{F}_{com}$, then $\mathcal{S}$ records r_i, and sends a message (CRS, $0, sid, P_i, P_j$) from P_i to $\hat{\mathcal{F}}_{crs}$ in the ideal process. (That is, the SID of this message is 0, and the SSID is sid.) Upon receiving a value r from $\hat{\mathcal{F}}_{crs}$, $\mathcal{S}$ waits to receive a message (CRS,(sid, P_j) from $\hat{\mathcal{F}}_{crs}$. This message will notify $\mathcal{S}$ that P_j was activated with input (CRS, $0, sid, r$). Upon receiving this message, $\mathcal{S}$ sets $r_j = r \oplus r_i$ and activates $\mathcal{A}$ to receive the message r_j from P_j. Next, when $\mathcal{A}$ generates the message (Open, sid, P_i, P_j) from P_i to $\mathcal{F}_{com}$, $\mathcal{S}$ delivers the message that $\hat{\mathcal{F}}_{crs}$ sent to P_j. (This message contains the value r.)

3. **Corrupted responder:** If $\mathcal{S}$ receives the message $(0, sid, P_i, P_j, r)$ from $\hat{\mathcal{F}}_{crs}$ in the ideal process where P_j is corrupted, then $\mathcal{S}$ activates the simulated $\mathcal{A}$ to receive message (Receipt, sid, P_i, P_j) from $\mathcal{F}_{com}$. When $\mathcal{A}$ generates a message r_j from P_j, $\mathcal{S}$ delivers the message from $\hat{\mathcal{F}}_{crs}$ to P_i in the ideal process, and activates $\mathcal{A}$ to receive the message (Open, sid, P_i, P_j, r_i) from $\mathcal{F}_{com}$, where $r_i = r \oplus r_j$.

4. **Both parties uncorrupted:** If $\mathcal{S}$ receives the message $(0, sid, P_i, P_j, r)$ from $\hat{\mathcal{F}}_{crs}$ in the ideal process where both P_i and P_j are uncorrupted, then it simulates for $\mathcal{A}$ the information it sees when two uncorrupted parties run the protocol and obtain a common string r. This information consists of a notice from $\mathcal{F}_{com}$ that P_i committed to a value to P_j, the random value r_j sent by P_j, and the opening of the commitment to r_i such that $r_i \oplus r_j = r$.

5. **Party corruption:** If at any point the simulated $\mathcal{A}$ corrupts a party P_i then $\mathcal{S}$ corrupts P_i in the ideal process, and provides $\mathcal{A}$ with the simulated internal state of P_i. (It is easy to verify that this state is always implied by the information already known to $\mathcal{S}$ at the time of corruption.)

Let $\mathcal{Z}$ be an environment machine. It is straightforward to verify that the view of $\mathcal{Z}$ in the ideal process for $\hat{\mathcal{F}}_{crs}$ with $\mathcal{S}$ is distributed identically to its view of an execution of ct in the $\mathcal{F}_{com}$-hybrid model. Note that this holds even if $\mathcal{Z}$ is computationally unbounded.

Using the JUC Theorem, we have that protocol $ct^{[UCC_{ReUse}]}$ securely realizes $\hat{\mathcal{F}}_{crs}$ and uses only a single copy of $\mathcal{F}_{crs}$.

Limits on Protocols for Realizing $\hat{\mathcal{F}}_{crs}$. We show that any protocol that realizes $\hat{\mathcal{F}}_{crs}$ in the $\mathcal{F}_{crs}$-hybrid model, and uses only a few copies of $\mathcal{F}_{crs}$, must involve interaction. More specifically, we consider protocols that realize $\mathcal{F}_{crs}^s$ using a single copy of $\mathcal{F}_{crs}^t$, where $s > t$. (This is a considerable restriction of the original problem: Clearly, any protocol that securely realizes $\hat{\mathcal{F}}_{crs}$ realizes also $\mathcal{F}_{crs}^s$ for any s that is polynomial in t.) We show that any such protocol must require that each of the parties send at least $s - t$ bits of information. Translated back to the task of realizing $\hat{\mathcal{F}}_{crs}^t$ using a single copy of $\mathcal{F}_{crs}^t$, this bound implies that each party must send at least t bits in order to generate each new copy of the string. (Note that the bound holds also for protocols generating a string among more than two parties.) That is:

Claim. Let π be a protocol that securely realizes $\mathcal{F}_{crs}^s$ using a single copy of $\mathcal{F}_{crs}^t$, and assume that π requires one of the parties to send no more than u bits of information. Then $s \leq t + u$.

Proof. Let π, s, t, u be as in the premise of the claim. It follows that for any adversary $\mathcal{A}$ there exists a simulator $\mathcal{S}$ such that no environment $\mathcal{Z}$ can distinguish between an interaction with π and $\mathcal{A}$ in the $\mathcal{F}_{crs}^t$-hybrid model and an interaction with $\mathcal{S}$ in the ideal process for $\mathcal{F}_{crs}^s$. Let P_i be the party that sends at most u bits.

Consider the following adversary $\mathcal{A}$ and environment $\mathcal{Z}$. $\mathcal{Z}$ instructs $\mathcal{A}$ to corrupt all parties except for P_i, uniformly chooses random inputs $\alpha_1, ..., \alpha_n$ for the corrupted parties, and instructs $\mathcal{A}$ to have each corrupted party P_j run π with random input α_j. $\mathcal{A}$ follows the instructions of $\mathcal{Z}$ and reports the gathered information to $\mathcal{Z}$. This information consists of the t-bit value r_t obtained from $\mathcal{F}_{crs}^t$, and the u-bit concatenation, m, of all messages received from P_i. Next, $\mathcal{Z}$ obtains the s-bit output of P_i, r_s, picks a corrupted party P_j, and outputs 1 iff P_j outputs r_s after running π on random input α_j, and having received r_t from $\mathcal{F}_{crs}^t$ and messages m from P_i.

Clearly, if $\mathcal{Z}$ interacts with parties running π in the $\mathcal{F}_{crs}^t$-hybrid model, then it always outputs 1. On the other hand, let $\mathcal{S}$ be an ideal-process adversary. We claim that if $\mathcal{Z}$ interacts with $\mathcal{S}$ in the ideal process for $\mathcal{F}_{crs}^s$ then $\mathcal{Z}$ outputs 1 with probability at most 2^{t+u-s}. To see this, fix a value of α, and recall that $r_s \stackrel{R}{\leftarrow} \{0,1\}^s$ is chosen by $\mathcal{F}_{crs}^s$, and that the output of P_j is uniquely determined by r_t, α, and m. Then, the probability over the choice of r_s that there *exist* a u-bit value m and a t-bit value r_t such that the output of P_j is r_s is at most 2^{t+u-s}. Note that the claim holds even if $\mathcal{Z}$ and $\mathcal{A}$ are restricted to polynomial time and even if P_i and $\mathcal{S}$ are unbounded.

4.3   Zero-Knowledge in the CRS Model

We describe the use of the JUC theorem to Zero-Knowledge protocols in the CRS model. The use is very similar to the case of commitment. Let us first recall the zero-knowledge functionality, $\mathcal{F}_{zk}$, as defined in [Can01], (see Fig. 4).

Functionality $\mathcal{F}_{zk}^R$

$\mathcal{F}_{zk}$ proceeds as follows, running with parties $P_1, ..., P_n$ and an adversary $\mathcal{S}$, given a binary relation R.

1. Upon receipt of a value $(\mathtt{prover}, sid, P_i, P_j, x, w)$ from some party P_i, Send $(sid, P_i, x, R(x, w))$ to P_j and $\mathcal{S}$, and halt.

Fig. 4. The Zero-Knowledge functionality, $\mathcal{F}_{zk}$

As in the case of $\mathcal{F}_{com}$, we only know how to realize functionality $\mathcal{F}_{zk}$ in the CRS model (i.e., in the $\mathcal{F}_{crs}$-hybrid model). Also here, straightforward composition of a protocol π in the $\mathcal{F}_{zk}$-hybrid model with a protocol ρ that securely realizes $\mathcal{F}_{zk}$ in the $\mathcal{F}_{crs}$-hybrid model would result in a composed protocol π^ρ that is highly wasteful of the reference string. We solve the problem by using universal composition with joint state. That is, given a protocol $\hat{\rho}$ that securely realizes $\hat{\mathcal{F}}_{zk}$, the multi-session extension of $\mathcal{F}_{zk}$, we conclude that the composed protocol $\pi^{[\hat{\rho}]}$ runs in the $\mathcal{F}_{crs}$-hybrid model and emulates π. Furthermore, if $\hat{\rho}$ uses only few copies of $\mathcal{F}_{crs}$ then so does $\pi^{[\hat{\rho}]}$.

We complete the discussion by pointing to two protocols that securely realize $\hat{\mathcal{F}}_{zk}$ in the $\mathcal{F}_{crs}$-hybrid model, and use only a single copy of $\mathcal{F}_{crs}$. First, recall protocol hc in [CF01] that securely realizes $\mathcal{F}_{zk}$ in the $\mathcal{F}_{com}$-hybrid model. (We remark that the formalization of $\mathcal{F}_{zk}$ in [CF01] is slightly different than the one here. Nonetheless, it is easy to see that the two formalizations are equivalent.) We claim that this protocol in effect realizes also $\hat{\mathcal{F}}_{zk}$. (Simply run the protocol separately for each interaction.) Thus, using the JUC Theorem, we obtain that the composed protocol, $hc^{[\mathsf{UCC}_{\mathsf{ReUse}}]}$, securely realizes $\hat{\mathcal{F}}_{zk}$ in the $\mathcal{F}_{crs}$-hybrid model. That is, we have:

Claim. Protocol $hc^{[\mathsf{UCC}_{\mathsf{ReUse}}]}$ securely realizes $\hat{\mathcal{F}}_{zk}$ in the $\mathcal{F}_{crs}$-hybrid model. Furthermore, it uses only a single copy of $\mathcal{F}_{crs}$.

Next, we note that the simulation-sound non-interactive zero-knowledge proof of knowledge protocol of De-Santis et al. [DCO⁺01] can be written as a protocol that securely realizes $\hat{\mathcal{F}}_{zk}$ in the $\mathcal{F}_{crs}$-hybrid model, with respect to non-adaptive adversaries:

Claim. The protocol of [DCO⁺01] securely realizes $\hat{\mathcal{F}}_{zk}$ in the $\mathcal{F}_{crs}$-hybrid model, with respect to non-adaptive adversaries. Furthermore, it uses only a single copy of $\mathcal{F}_{crs}$.

References

[Bea91]   D. Beaver. Secure Multiparty Protocols and Zero-Knowledge Proof Systems Tolerating a Faulty Minority. *Journal of Cryptology*, 4:75–122, 1991.

[BGW88]   M. Ben-Or, S. Goldwasser, and A. Wigderson. Completeness Theorems for Noncryptographic Fault-Tolerant Distributed Computations. In *Proc. 20th STOC*, pages 1–10. ACM, 1988.

[Blu82]   M. Blum.   Coin Flipping by Telephone . In *IEEE Spring COMPCOM*, pages 133–137, 1982.

[BR93]   M. Bellare and P. Rogaway. Entity Authentication and Key Distribution. In *Crypto '93*, pages 232–249, 1993. LNCS No. 773.

[Can00]   R. Canetti. Security and Composition of Multiparty Cryptographic Protocols. *Journal of Cryptology*, 13(1):143–202, 2000.

[Can01]   R. Canetti. Universally Composable Security: A New Paradigm for Cryptographic Protocols. In *Proc. 42st FOCS*, pages 136–145. IEEE, 2001. http://eprint.iacr.org/2000/067.

[CF01]   R. Canetti and M. Fischlin. Universally Composable Commitments. In *Crypto '01*, pages 19–40. LNCS No. 2139.

[CK01]   R. Canetti and H. Krawczyk. Analysis of Key-Exchange Protocols and Their Use for Building Secure Channels . In *Eurocrypt '01*, pages 453–474, 2001. LNCS No. 2045.

[CK02]   R. Canetti and H. Krawczyk. Universally Composable Key Exchange and Secure Channels . In *Eurocrypt '02*, pages 337–351, 2002. LNCS No. 2332.

[CLOS02]   R. Canetti, Y. Lindell, R. Ostrovsky, and A. Sahai.   Universally Composable Two-Party and Multi-Party Secure Computation. In *Proc. 34th STOC*, pages 494–503.

[CR03]   R. Canetti and T. Rabin. Universal Composition with Joint State. Available online, http://eprint.iacr.org.

[DCO⁺01]   A. De Santis, G. Di Crescenzo, R. Ostrovsky, G. Persiano, and A. Sahai. Robust Non-interactive Zero-Knowledge. In *Crypto '01* LNCS No. 2139.

[DDN00]   D. Dolev, C. Dwork, and M. Naor. Non-malleable Cryptography. *SIAM J. Comput.*, 30(2):391–437, 2000.

[DM00]   Y. Dodis and S. Micali. Parallel Reducibility for Information-Theoretically Secure Computation. In *Crypto '00*, pages 74–92, 2000. LNCS No. 1880.

[DN02]   I. Damgard and J. Nielsen.   Universally Composable Commitment Schemes with Constant Expansion Factor. In *Crypto '02* LNCS No. 2442.

[DNS98]   C. Dwork, M. Naor, and A. Sahai. Concurrent zero-knowledge. In *Proc. 30th STOC*, pages 409–418.

[GK96]   O. Goldreich and H. Krawczyk. On the composition of zero-knowledge proof systems. *SIAM. J. Computing*, 25(1):169–192, 1996.

[GL90]   S. Goldwasser and L. Levin. Fair computation of general functions in presence of immoral majority. In *Crypto '90*, 1990. LNCS No. 537.

[GMW87]   O. Goldreich, S. Micali, and A. Wigderson. How to Play Any Mental Game. In *Proc. 19th STOC*, pages 218–229. ACM, 1987.

[GO94]   O. Goldreich and Y. Oren. Definitions and Properties of Zero-Knowledge Proof Systems. *Journal of Cryptology*, 7(1):1–32, 1994. Preliminary version by Y. Oren in FOCS87.

[Gol02]   O. Goldreich.   Concurrent Zero-Knowledge With Timing Revisited. In *Proc. 34th STOC*.

[LLR02]   Y. Lindell, A. Lysyanskya, and T. Rabin.   On the Composition of Authenticated Byzantine Agreement. In *Proc. 34th STOC*, pages 514–523.

[MR91]   S. Micali and P. Rogaway. Secure Computation. In *Crypto '91*, pages 392–404, 1991. Manuscript available.

[PSW00]   B. Pfitzmann, M. Schunter, and M. Waidner. Secure Reactive Systems. IBM Research Report RZ 3206 (#93252), IBM Research, Zurich, May 2000.

[Sho99]   V. Shoup. On Formal Models for Secure Key Exchange. Available at: http://www.shoup.org, 1999.

Statistical Zero-Knowledge Proofs with Efficient Provers: Lattice Problems and More

Daniele Micciancio[1]* and Salil P. Vadhan[2]**

[1] University of California, San Diego, La Jolla CA 92093, USA,
daniele@cs.ucsd.edu
[2] Harvard University, Cambridge MA 02138, USA,
salil@eecs.harvard.edu

Abstract. We construct several new statistical zero-knowledge proofs with *efficient provers*, i.e. ones where the prover strategy runs in probabilistic polynomial time given an **NP** witness for the input string.

Our first proof systems are for approximate versions of the SHORTEST VECTOR PROBLEM (SVP) and CLOSEST VECTOR PROBLEM (CVP), where the witness is simply a short vector in the lattice or a lattice vector close to the target, respectively. Our proof systems are in fact proofs of knowledge, and as a result, we immediately obtain efficient lattice-based identification schemes which can be implemented with arbitrary families of lattices in which the approximate SVP or CVP are hard.

We then turn to the general question of whether *all* problems in **SZK∩NP** admit statistical zero-knowledge proofs with efficient provers. Towards this end, we give a statistical zero-knowledge proof system with an efficient prover for a natural restriction of STATISTICAL DIFFERENCE, a complete problem for **SZK**. We also suggest a plausible approach to resolving the general question in the positive.

1 Introduction

Zero-knowledge proof systems, introduced in [1], have proven to be a powerful tool for constructing cryptographic protocols. They have also turned out to be a rich object of study from the perspective of complexity theory. In this paper, we focus on *statistical* zero knowledge (**SZK**), which is the form of zero knowledge that provides the strongest security guarantees and whose complexity-theoretic study has been most active in recent years. One significant gap between much of the recent theoretical study and the cryptographic applicability of **SZK** involves the *prover's efficiency*, i.e. whether the prover can be implemented in polynomial time (given some auxiliary information). This property is clearly essential for a zero-knowledge proof to be used in cryptographic protocols, but many of the theoretical results ignore this issue. Prover efficiency for **SZK** has been considered in the past, leading to the result of Bellare and Petrank [2] that any **SZK**

---

* Supported in part by NSF Career Award CCR-0093029.
** Supported in part by NSF Grant CCR-0205423 and a Sloan Research Fellowship.

D. Boneh (Ed.): CRYPTO 2003, LNCS 2729, pp. 282–298, 2003.

proof system admits a prover that runs in probabilistic polynomial time given an **NP** oracle. However, this notion of efficiency is insufficient for cryptography, as the **NP** oracle cannot be realized efficiently. In cryptographic applications, one would like the prover to run in probabilistic polynomial time given only the input string x (drawn from some **NP** language L) and an **NP**-witness w (the "secret key") that $x \in L$. We call a proof system with this property a proof system with an *efficient prover*. (These were called *prover-practical* proof systems in [3].) A number of the classic perfect and statistical zero-knowledge proof systems [1,4] have efficient provers, but not all problems in **SZK** ∩ **NP** are known to have such proof systems. Indeed, it remains an intriguing open problem to characterize the class of problems which have statistical zero-knowledge proofs with efficient provers and extend known results about statistical zero knowledge to this class.

In this paper, we construct statistical zero-knowledge proofs with efficient provers for several problems previously not known to have such proofs. We first do this for approximate versions of the CLOSEST VECTOR PROBLEM (CVP) and SHORTEST VECTOR PROBLEM (SVP) in lattices. These proof systems immediately yield efficient identification schemes based on the hardness of these problems. An interesting property of our schemes is that they allow us to use arbitrary lattices (where CVP and SVP are hard), which gives potential advantages both from the efficiency and security points of view; for example, there is no need to embed a "trapdoor basis" in the lattice. Then we construct a statistical zero-knowledge proof with an efficient prover for a natural restriction of STATISTICAL DIFFERENCE, which is known to be a complete problem for **SZK**. We view the latter result as progress towards characterizing the class of problems having statistical zero-knowledge proofs with efficient provers.

1.1   Statistical Zero Knowledge

Zero-knowledge proof systems are protocols by which a computationally unbounded *prover* can convince a probabilistic polynomial-time *verifier*, of an assertion, i.e. that some string x is a YES instance of some decision problem. The zero-knowledge property requires that the verifier "learns nothing" from this interaction other than the fact that the assertion being proven is true. In a *statistical* zero-knowledge *proof* system, the security for both parties is very strong. Specifically, it holds even with respect to *computationally unbounded* cheating provers or verifiers. Note that even though the security holds for computationally unbounded parties, the *prescribed* verifier strategy is always required to be polynomial time. We will discuss the prover's efficiency later. The class of problems possessing statistical zero-knowledge proofs is denoted **SZK**.

In addition to its cryptographic significance, **SZK** has turned out to be quite interesting from a complexity-theoretic perspective. On the one hand it is known to contain important computational problems, such as GRAPH NONISOMORPHISM [4] and QUADRATIC RESIDUOSITY [1]. On the other hand, it is contained in the class **AM** ∩ **co-AM** [5,6] and hence is unlikely to contain **NP**-hard problems. More recently, it was discovered that **SZK** is closed under complement [7] and has natural complete problems [8,9]. Moreover, a number of useful transformations of statistical zero-knowledge proof systems have been given, for example showing that every proof system which is statistical zero knowledge for the *honest verifier* can be transformed into one which is statistical zero knowledge even for cheating verifiers [7,10].

The above theoretical investigations focus on the traditional definition of **SZK**, whereby no computational restriction is placed on the prover strategy, and many manipulations used in the study of **SZK** do not preserve the prover's efficiency; indeed, this is inherent in the techniques used (namely, black-box transformations) [11]. Nevertheless, we consider it an important research direction to overcome this barrier and extend the study of **SZK** to protocols with efficient provers. In particular, can we characterize the subclass of **SZK** possessing statistical zero-knowledge proofs with efficient provers? Since the efficient prover property only makes sense for problems in **NP** (actually **MA**) and **SZK** is not known to be contained in **NP**,[1] so we do not hope to show that all of **SZK** has efficient provers. But do all problems in **SZK ∩ NP** have statistical zero-knowledge proofs with efficient provers?

1.2   Lattice Problems

A *lattice* is a subset of $\mathbb{R}^n$ consisting of all integer linear combinations of a set of linearly independent vectors. Two basic computational problems involving lattices are the SHORTEST VECTOR PROBLEM, finding the shortest nonzero vector in the lattice, and the CLOSEST VECTOR PROBLEM, finding the lattice vector closest to a given target vector. These problems have received a great deal of attention recently in both the cryptography and complexity theory literature. On the complexity side, approximate versions of both of these problems have been shown to be **NP**-hard [15,16,17,18], and variants of the approximate SHORTEST VECTOR PROBLEM have been shown to be related by a worst-case/average-case connection [19]. On the cryptography side, a number of cryptographic primitives have been proposed which implicitly or explicitly rely on the hardness of these problems. These include the one-way functions of [19,20], the collision-resistant hash functions of [21,22], the public-key encryption schemes of [23,24,25].

In [26], Goldreich and Goldwasser exhibited statistical zero-knowledge proofs for approximate versions of the *complements* of SHORTEST VECTOR PROBLEM and CLOSEST VECTOR PROBLEM.[2] That is, they gave protocols for proving that a lattice has no short vector (resp., has no vector close to the target vector). The Goldreich–Goldwasser proof systems do not have efficient provers. Indeed, the problems they consider are not known to be in **NP** and their main motivation was to prove that they are in **AM** (and, being also in **co-NP**, are thus unlikely to be **NP**-hard under standard types of reductions). However, since **SZK** is closed under complement [7], it follows from their result that the corresponding approximate versions of the SHORTEST VECTOR PROBLEM and the CLOSEST VECTOR

---

[1] Actually, there is some recent evidence that **AM** may equal **NP** [12,13,14] which would imply that **SZK ⊆ NP ∩ co-NP**.

[2] In fact, their proof systems are *perfect* zero knowledge (against an honest verifier).

PROBLEM themselves (rather than their complements) are also in **SZK**. Since these problems are in **NP**, we can hope to construct statistical zero-knowledge proofs with efficient provers for them. However, the **SZK** proofs obtained by applying the general result of [7] (or even later simplifications [8,9,27]) do not guarantee efficient provers, and in addition would be extremely cumbersome and impractical.

1.3   Our Results

We first construct statistical zero-knowledge proof systems with efficient provers for approximate versions of the SHORTEST VECTOR PROBLEM and CLOSEST VECTOR PROBLEM. The approximation factor for our proof system can be as small as in the Goldreich–Goldwasser proof systems, namely $\Theta(\sqrt{n/\log n})$ where n is the rank of the lattice. The prover strategy can be implemented in polynomial time given only a short lattice vector (resp., lattice vector close to the target vector). The proof systems are actually proofs of knowledge, and hence immediately give rise to identification schemes [28] provided one can efficiently generate lattices in which either of these problems is hard together with the corresponding witnesses. We remark that in order to efficiently prove that a target point is close to the lattice (or that the lattice contains short vectors) it is not necessary to know a short (trapdoor) basis, i.e., a basis consisting entirely of short vectors. On the security side, embedding a trapdoor basis has often been regarded as a weak point for many lattice and subset-sum based cryptosystems. Our identification schemes can be instantiated with any lattice, offering the highest degree of security. For example, one can use lattices derived from the random classes of [19] or [22]. This results in provably secure lattice-based identification (ID) schemes with an average-case/worst-case connection.[3,4] On the efficiency side, complete freedom in the choice of the lattice enables the use of lattices with special structure (e.g., the cyclic lattices of [20], or the convolutional modular lattices of NTRU [25]), or share the same lattice among different users, in order to get smaller key size or faster identification procedures. (See Section 5.)

We then return to the general question of efficient provers for **SZK**. We generalize techniques of Itoh, Ohta, and Shizuya [29] to show that a natural restriction of STATISTICAL DIFFERENCE has a statistical zero-knowledge proof

---

[3] In order to use these lattices in our construction one needs a procedure to generate a lattice together with a short vector, but this can be achieved as explained in [19] by slightly perturbing the lattice distribution.

[4] The results of [19,22] immediately give one way functions from worst case hardness assumptions, which, in turn, imply the existence of secure ID schemes. However, these generic constructions are pretty inefficient. Our constructions build ID schemes directly from the underlying lattice problems (i.e. without going through one-way functions), resulting in substantially more efficient ID schemes.

with an efficient (polynomial time) prover.[5] In the STATISTICAL DIFFERENCE problem, one is given two (suitably represented) probability distributions, and the question is to determine if they are relatively close (say, within statistical distance at most $1/2$) or are far apart (say, at statistical distance at least $1 - \epsilon$). This is a complete problem for **SZK** for any $0 < \epsilon < 1/\sqrt{2}$ [8]. STATISTICAL DIFFERENCE is not known to be in **NP**, so we cannot give a proof system with efficient provers for it. We consider the restriction of STATISTICAL DIFFERENCE obtained setting $\epsilon = 0$: determine if two distributions are within statistical distance $1/2$ or are completely disjoint. We observe that this problem is in **NP**, and show that it admits a statistical zero-knowledge proof system with efficient provers. Thus we view this as a step towards finding proof systems with efficient provers for all problems in **SZK** $\cap$ **NP**. In addition, the techniques we use (namely [29]) are not "black box," so this approach is not subject to the limitations in [11].

1.4   Related Work

The first zero-knowledge proof systems, namely those for QUADRATIC RESIDUOSITY and QUADRATIC NONRESIDUOSITY [1], and GRAPH ISOMORPHISM [4] had efficient provers and achieved perfect zero-knowledge. Subsequently, **SZK** proof systems with efficient provers have been found for a number of other number-theoretic problems (e.g., [3,30]), all random self-reducible problems [31] and monotone formulae over random self-reducible problems [32]).

Other notions of prover efficiency (mostly interesting from the perspective of computational complexity) have been considered before. Building upon previous work, Bellare and Petrank [2] show that for any **SZK** proof system, it is possible to implement the prover strategy in probabilistic polynomial time given an **NP** oracle. Notice that given an **NP** oracle for SATISFIABILITY, one can efficiently find **NP**-witnesses for arbitrary **NP** problems, by the self-reducibility of **NP**-complete problems (such as SATISFIABILITY). So, the provers considered in [2], are considerably more powerful than ours, and allow one to prove arbitrary **SZK** languages, even those outside **NP**.

A more restrictive notion of prover efficiency is considered in [33], where the prover is given oracle access to a decision oracle for the same language L underlying the proof system.[6] For example, the (honest-verifier) perfect zero-knowledge proof system for GRAPH NONISOMORPHISM [4] satisfies this notion of prover efficiency. The results of [33] are negative: there are **NP** languages for

---

[5] The prover in this proof system runs in polynomial time, but is not as practical as those for the lattice problems. In particular, our results about statistical difference should be regarded as a plausibility result aimed at characterizing the complexity class of statistical zero-knowledge proof systems with efficient provers, rather than a concrete proposal of a proof system to be used in cryptographic applications.

[6] When L is an **NP**-complete problem, then these provers are as powerful as those of [2]. However, **SZK** is not likely to contain any **NP**-complete problem. So, for an arbitrary language L in **SZK**, it is not clear how to efficiently prove membership in L given oracle access to a decision procedure for L.

which finding an **NP** witness for $x \in L$, or even proving membership $x \in L$ interactively (whether or not in zero-knowledge), cannot be efficiently reduced to deciding membership in L. This notion of proof system, called *competitive* in [33], is incomparable with ours. On the one hand, our provers are given an input string x together with an **NP**-witness for $x \in L$, and it is not clear how to efficiently compute such a witness given only a decision oracle for L when L is not **NP**-complete or self-reducible. On the other hand, the provers of [33] can make queries "$y \in L$?" to the oracle for arbitrary strings y (possibly different from the input string x), while our prover is only given a witness for the input string x.

In any case, the notions of prover efficiency considered by [2,33] and related papers, seem mostly interesting from a computational complexity perspective, and do not match the requirements of cryptographic applications. A crucial difference is that the notion we study here makes sense only for problems in **NP**, while the results of [2,33] apply to languages outside **NP** as well.

Organization. The rest of the paper is organized as follows. In Section 2 we give some basic definitions about statistical difference and the lattice problems studied in this paper. In Section 3 we present and analyze the proof system for CVP. The proof system for SVP is sketched in Section 4. Section 5 discusses our lattice based identification schemes. Finally, in Section 6 we study STATISTICAL DIFFERENCE, and the problem of designing **SZK** proofs with efficient provers for all problems in **SZK** ∩ **NP**. Because of space constraints, most proofs are not presented here, and can be found in the full version of the paper.

2 Preliminaries

In this section we recall some basic definitions and techniques that will be used in the rest of the paper. For more details the reader is referred to the books [34, 35] or the papers in the references.

2.1 Statistical Difference

The statistical distance between two discrete random variables X and Y over a (countable) set A is the quantity $\Delta(X, Y) = \frac{1}{2} \sum_{a \in A} |\Pr\{X = a\} - \Pr\{Y = a\}|$.

STATISTICAL DIFFERENCE is a collection of problems (parameterized by two real numbers $0 \le \alpha < \beta \le 1$) of the form: given two succinctly specified probability distributions, decide whether they are statistically close or statistically far apart. The probability distributions are specified by circuits which sample from them. That is, we are given a circuit $X : \{0,1\}^m \to \{0,1\}^n$ which we interpret as specifying the probability distribution $X(U_m)$ on $\{0,1\}^n$, where U_m is the uniform probability distribution over $\{0,1\}^m$. More formally, for $0 \le \alpha < \beta \le 1$, we define the following promise problem.

Definition 1 (STATISTICAL DIFFERENCE). *Instances of promise problem* $\text{SD}^{\alpha,\beta}$ *are pairs* (X, Y) *where* X *and* Y *are probability distributions.* (X, Y) *is a* YES *instance if* $\Delta(X, Y) \le \alpha$, *and a* NO *instance if* $\Delta(X, Y) \ge \beta$. *(We have* *defined these problems as the complements of those defined in [8], because this formulation is more convenient for our purposes.)*

In [8] it is shown that $\mathrm{SD}^{\alpha,\beta}$ is complete for **SZK** for all $0 < \beta/2 < \alpha < \beta^2 < 1$. In particular $\mathrm{SD}^{1/3,2/3}$ is **SZK**-complete, and $\mathrm{SD}^{1/2,1-\epsilon}$ is **SZK**-complete for all $0 < \epsilon < 1/\sqrt{2}$.

2.2   Lattice Problems and Technical Tools

Let $\mathbb{R}^m$ be the m-dimensional Euclidean space. A *lattice* in $\mathbb{R}^m$ is the set of all integral combinations of n linearly independent vectors $\mathbf{b}_1, \ldots, \mathbf{b}_n$ in $\mathbb{R}^m$ ($m \geq n$). The integers n and m are called the *rank* and *dimension* of the lattice, respectively. Using matrix notation, if $\mathbf{B} = [\mathbf{b}_1, \ldots, \mathbf{b}_n]$, the lattice generated by basis matrix $\mathbf{B}$ is $\mathcal{L}(\mathbf{B}) = \{\mathbf{Bx}: \mathbf{x} \in \mathbb{Z}^n\}$, where $\mathbf{Bx}$ is the usual matrix-vector multiplication. For computational purposes, $\mathbf{B}$ and $\mathbf{y}$ are usually restricted to have integer (or, equivalently, rational) entries. In this paper, we will occasionally use real vectors in order to simplify the exposition. However, the use of real numbers is not essential, and integer or rational approximations can always be substituted for real vectors whenever they occur. Moreover, we often assume that the lattice if full rank, i.e., $n = m$, as any lattice can be transformed into a full-rank *real* lattice.

Approximate versions of the SHORTEST VECTOR PROBLEM and CLOSEST VECTOR PROBLEM described in the introduction are captured by the promise problems GAPSVP$_\gamma$ and GAPCVP$_\gamma$ defined as follows.

Definition 2. *Instances of promise problem* GAPSVP$_\gamma$ *are pairs* $(\mathbf{B}, t)$ *where* $\mathbf{B} \in \mathbb{Z}^{m \times n}$ *is a lattice basis and* $t \in \mathbb{Q}$ *a rational number.* $(\mathbf{B}, t)$ *is a* YES *instance if* $\|\mathbf{Bx}\| \leq t$ *for some* $\mathbf{x} \in \mathbb{Z}^n \setminus \{\mathbf{0}\}$. $(\mathbf{B}, t)$ *is a* NO *instance if* $\|\mathbf{Bx}\| > \gamma t$ *for all* $\mathbf{x} \in \mathbb{Z}^n \setminus \{\mathbf{0}\}$.

Definition 3. *Instances of promise problem* GAPCVP$_\gamma$ *are triples* $(\mathbf{B}, \mathbf{y}, t)$ *where* $\mathbf{B} \in \mathbb{Z}^{m \times n}$ *is a lattice basis,* $\mathbf{y} \in \mathbb{Z}^m$ *is a vector and* $t \in \mathbb{Q}$ *is a rational number.* $(\mathbf{B}, \mathbf{y}, t)$ *is a* YES *instance if* $\|\mathbf{Bx} - \mathbf{y}\| \leq t$ *for some* $\mathbf{x} \in \mathbb{Z}^n$. $(\mathbf{B}, \mathbf{y}, t)$ *is a* NO *instance if* $\|\mathbf{Bx} - \mathbf{y}\| > \gamma t$ *for all* $\mathbf{x} \in \mathbb{Z}^n$.

In our proof systems for lattice problems we make extensive use of a modular reduction technique proposed in [36] to emulate the effect of selecting a point uniformly at random from a lattice. Any lattice $\mathcal{L}(\mathbf{B})$ defines a natural equivalence relation on $\mathrm{span}(\mathbf{B}) = \sum_i \mathbf{b}_i \cdot \mathbb{R}$, where two points $\mathbf{x}, \mathbf{y} \in \mathrm{span}(\mathbf{B})$ are equivalent if $\mathbf{x} - \mathbf{y} \in \mathcal{L}(\mathbf{B})$. For any lattice basis $\mathbf{B}$ define the half open parallelepiped $\mathcal{P}(\mathbf{B}) = \{\mathbf{Bx}: 0 \leq x_i < 1\}$. It is easy to see that for any point $\mathbf{x} \in \mathrm{span}(\mathbf{B})$, there exists a unique point $\mathbf{y} \in \mathcal{P}(\mathbf{B})$ such that $\mathbf{x}$ is equivalent to $\mathbf{y}$ modulo the lattice. This unique representative for the equivalence class of $\mathbf{x}$ is denoted $\mathbf{x} \bmod \mathbf{B}$. Intuitively, $\mathbf{x} \bmod \mathbf{B}$ is the displacement of $\mathbf{x}$ within the fundamental parallelepiped containing $\mathbf{x}$. Notice that if we fix a (small) perturbation vector $\mathbf{r}$, we add it to a lattice point $\mathbf{Bv}$ and reduce the result modulo $\mathbf{B}$, we get a vector $(\mathbf{Bv} + \mathbf{r}) \bmod \mathbf{B} = \mathbf{r} \bmod \mathbf{B}$ that does not depend on the lattice point $\mathbf{Bv}$ from which we started. In other words, if we start from the origin, and simply compute $\mathbf{r} \bmod \mathbf{B}$, we obtain exactly the same distribution.

3   The Closest Vector Problem

In this section we describe a statistical zero-knowledge proof system (in fact, a proof of knowledge) with efficient provers for approximating the closest vector problem.

Consider an instance $(\mathbf{B}, \mathbf{y}, t)$ of GAPCVP$_\gamma$. Look at a small ball around $\mathbf{y}$ and a small ball around a lattice point $\mathbf{Bw}$ closest to $\mathbf{y}$. If $\mathbf{y}$ and $\mathbf{Bw}$ are close to each other, the relative volume of the intersection of the two balls is quite large. So, if we pick a few random points from both balls, with high probability at least one of them will be in the intersection. The proof system works as follows: the prover picks random points from the two balls, reduces them modulo $\mathbf{B}$, and sends the reduced points to the verifier. Reducing the points modulo $\mathbf{B}$ has the nice effect that the resulting distribution can be efficiently sampled even without knowing the lattice point $\mathbf{Bw}$ closest to $\mathbf{y}$. (In fact, using two balls centered around $\mathbf{y}$ and the origin $\mathbf{0}$, results in exactly the same distribution after the reduction modulo $\mathbf{B}$. This is a crucial property to achieve zero-knowledge.) Let's say that the total number of points picked by the prover is even. Then, the verifier challenges the prover asking him to show that either (1) there is an even number of points from each ball; or (2) there is an odd number of points from each ball. If the prover can answer both challenges, then some point must belong to the intersection of the two balls, proving that the two balls intersect, and therefore their centers cannot be too far apart. Intuitively, the proof system is zero knowledge because all that the verifier sees is a set of random points from an efficiently samplable distribution.

Note that the proof system sketched above achieves neither perfect completeness nor perfect zero knowledge, but rather has a small (but negligible) completeness error and is *statistical* zero knowledge. The reason is that there is a nonzero probability that all the randomly chosen points will lie outside the intersection of the two balls, and in this case the prover will only be able to answer one of the two challenges. And intuitively, the verifier learns something in case the prover cannot answer, namely that none of the chosen points is in the intersection. Below, we achieve perfect completeness by having the prover modify the points chosen to ensure that at least one is in the intersection (if needed). However, this does not yield perfect zero knowledge, because now the points sent are no longer uniform in the two balls, but have a slightly skewed distribution that may be hard to sample exactly in polynomial time.

We now give the formal description of the proof system $(P_{\text{CVP}}, V_{\text{CVP}})$. In the description below k is a parameter to be determined that depends on the value of γ. In fact, the proof system is valid for any value of γ and k, and the choice of these parameters only affects the zero-knowledge property.

The Verifier. On input $(\mathbf{B}, \mathbf{y}, t)$, the verifier V_{CVP} proceeds as follows.

1. Receive k points $\mathbf{m}_1, \ldots, \mathbf{m}_k \in \mathbb{R}^n$ from the prover
2. Send a uniformly chosen random bit $q \in \{0, 1\}$ to the prover
3. Receive k bits $c_1, \ldots, c_k$ and k lattice points $\mathbf{Bv}_1, \ldots, \mathbf{Bv}_k$ and check that they satisfy $\sum_i c_i = q \pmod 2$ and $\|\mathbf{m}_i - (\mathbf{Bv}_i + c_i \mathbf{y})\| \leq \gamma t/2$ for all i.

The following lemma shows that the protocol defined by the verifier is sound, both as an interactive proof system and even as a proof of knowledge.

Lemma 4 (soundness). *If* $(\mathbf{B}, \mathbf{y}, t)$ *is a* NO *instance of* GAPCVP_γ, *then the verifier* V_{CVP} *rejects with probability at least* $1/2$ *when interacting with any prover strategy* P^*. *Moreover, there is a probabilistic algorithm* K *(the* knowledge extractor*) such that if a prover* P^* *makes* V_{CVP} *accept with probability* $1/2 + \epsilon$ *on some instance* $(\mathbf{B}, \mathbf{y}, t)$, *then* $K^{P^*}(\mathbf{B}, \mathbf{y}, t)$ *outputs a vector* $\mathbf{w} \in \mathbb{Z}^n$ *satisfying* $\|\mathbf{B}\mathbf{w} - \mathbf{y}\| \leq \gamma t$ *in expected time* $\mathrm{poly}(n)/\epsilon$.

The Prover. Now that we know that the above proof system is sound, we show that if $(\mathbf{B}, \mathbf{y}, t)$ is a YES instance, then it is always possible to make the verifier accept. Suppose $(\mathbf{B}, \mathbf{y}, t)$ is a YES instance of GAPCVP_γ, i.e., there exists an integer vector $\mathbf{w} \in \mathbb{Z}^n$ such that $\|\mathbf{y} - \mathbf{B}\mathbf{w}\| \leq t$. We describe a probabilistic polynomial time prover P_{CVP} that, given the witness $\mathbf{w}$ (or, equivalently, $\mathbf{u} = \mathbf{y} - \mathbf{B}\mathbf{w}$) as auxiliary input, makes the verifier accept with probability 1. The prover P_{CVP}, on input $(\mathbf{B}, \mathbf{y}, t)$ and $\mathbf{u} = \mathbf{y} - \mathbf{B}\mathbf{w}$, proceeds as follows:

1. Choose $c_1, \ldots, c_k \in \{0,1\}$ independently and uniformly at random. Also choose error vectors $\mathbf{r}_1, \ldots, \mathbf{r}_k \in \mathcal{B}(0, \gamma t/2)$ independently and uniformly at random. Then, check if there exists an index i^* such that $\|\mathbf{r}_{i^*} + (2c_{i^*} - 1)\mathbf{u}\| \leq \gamma t/2$. If not, set $i^* = 1$ and redefine $c_{i^*} = 0$ and $\mathbf{r}_{i^*} = \mathbf{u}/2$, so that $\|\mathbf{r}_{i^*} + (2c_{i^*} - 1)\mathbf{u}\| \leq \gamma t/2$ is certainly satisfied. Finally, compute points $\mathbf{m}_i = c_i \mathbf{y} + \mathbf{r}_i \bmod \mathbf{B}$ for all $i = 1, \ldots, k$ and send them to the verifier.
2. Wait for the verifier to reply with a challenge bit $q \in \{0,1\}$.
3. If $q = \oplus_i c_i$, then the prover completes the proof sending bits c_i and lattice vectors $\mathbf{B}\mathbf{v}_i = \mathbf{m}_i - (\mathbf{r}_i + c_i \mathbf{y})$ (for $i = 1, \ldots, k$) to the verifier. If $q \neq \oplus_i c_i$, then the prover sends the same messages to the verifier, but with c_{i^*} and $\mathbf{B}\mathbf{v}_{i^*}$ replaced by $1 - c_{i^*}$ and $\mathbf{B}\mathbf{v}_{i^*} + (2c_{i^*} - 1)(\mathbf{y} - \mathbf{u})$.

It is clear that P_{CVP} can be implemented in polynomial time. The reader can easily verify that if the honest verifier V_{CVP} interacts with prover P_{CVP}, then it always accepts.

The Simulator. We prove the zero knowledge property by exhibiting a probabilistic polynomial-time simulator that outputs the transcript of a conversation between a (simulated) prover and a given cheating verifier V^* with a probability distribution that (for appropriate values of γ, k) is statistically close to that between V^* and the real prover P_{CVP}.

The simulator S_{CVP}, on input $(\mathbf{B}, \mathbf{y}, t)$, and given black-box access to a (possibly cheating) verifier V^*, proceeds as follows:

1. Pick random $c_1, \ldots, c_k \in \{0,1\}$ and $\mathbf{r}_1, \ldots, \mathbf{r}_k \in \mathcal{B}(\mathbf{0}, \gamma t/2)$, and compute $\mathbf{m}_i = c_i \mathbf{y} + \mathbf{r}_j \bmod \mathbf{B}$ for all $i = 1, \ldots, k$.
2. Pass $\mathbf{m}_1, \ldots, \mathbf{m}_k$ to V^*, who replies with a query $q \in \{0,1\}$.[7]

---

[7] We can assume, without loss of generality, that the verifier always output a single bit answer. Any other message can be interpreted in some standard way.

3. If $q = \oplus c_i$, then output the transcript $(\{\mathbf{m}_i\}_{i=1}^{k}, q, \{(c_i, \mathbf{Bv}_i)\}_{i=1}^{k})$, where $\mathbf{Bv}_i = \mathbf{m}_i - (\mathbf{r}_i + c_i \mathbf{y})$. If $q \neq \oplus c_i$, then output `fail`.

Theorem 5. *If* $(\mathbf{B}, \mathbf{y}, t)$ *is a* YES *instance of* GAPCVP$_\gamma$, *then the statistical difference between the output of the simulator* S_{CVP} *(conditioned on the event that* S_{CVP} *does not fail), and the interaction between* V^* *and the real prover* P_{CVP}, *is at most* $2(1 - \beta(2/\gamma))^k$, *where* $\beta(\epsilon)$ *is the relative volume of the intersection of two unit spheres whose centers are at distance* ϵ.

Using the bound $\beta(\epsilon) \geq \max\left(\frac{3}{\exp(\epsilon^2 n/2)}, 1 - \epsilon\sqrt{n}\right)$ on the relative volume of the intersection of two spheres,[8] we immediately get the following corollary.

Corollary 6. $(P_{\mathrm{CVP}}, V_{\mathrm{CVP}})$ *is a statistical zero-knowledge proof system with perfect completeness and soundness error* $1/2$, *provided one of the following conditions holds true:*

- $\gamma = \Omega(\sqrt{n/\log n})$ *and* $k = \mathrm{poly}(n)$ *is a sufficiently large polynomial, or*
- $\gamma = \Omega(\sqrt{n})$ *and* $k = \omega(\log n)$ *is any superlogarithmic function of* n, *or*
- $\gamma = n^{0.5 + \Omega(1)}$ *and* $k = \omega(1)$ *is any superconstant function of* n.

Negligible Error. As is, the proof system has constant soundness error $(1/2)$, but it is often important to have negligible soundness error $(1/n^{\omega(1)})$. There are several approaches to reducing the soundness error, with different advantages:

(1) Repeat the proof system $\ell(n) = \omega(\log n)$ times in parallel. This unfortunately does not preserve the zero knowledge property, but does yield a constant-round statistically *witness-indistinguishable* proof of knowledge with negligible soundness error. (Witness indistinguishability means that for any two witness $\mathbf{w}$ and $\mathbf{w}'$, the verifier's view when the prover uses $\mathbf{w}$ is statistically close to its view when the prover uses $\mathbf{w}'$. See [34].)

(2) Repeat the proof system $\ell(n) = \Theta(\log n)$ times in parallel and then repeat the resulting protocol $\omega(1)$ times sequentially. This does preserve zero knowledge, yielding an $\omega(1)$-round statistical zero-knowledge proof of knowledge.

(3) In both of the approaches above, the ℓ-fold parallel repetition can be combined with the k-fold repetition already present in the original protocol to obtain more efficient protocols. Consider a modification of the original protocol $(P_{\mathrm{CVP}}, V_{\mathrm{CVP}})$, where in addition to sending k vectors in the first step, the prover also sends a random $k \times \ell$ matrix $\mathbf{M}$ over $\mathrm{GF}(2) = \{0, 1\}$. The verifier's challenge is then a random *vector* $\mathbf{q} \in \{0, 1\}^\ell$, and the condition $\oplus_i c_i = q$ is replaced with $\mathbf{Mc} = \mathbf{q}$. The advantage of this protocol is that it achieves both simulation and soundness error $2^{-\Omega(k)}$ with a protocol that involves only $O(k)$ n-dimensional vectors rather than $O(k^2)$ as achieved by independent repetitions of the original protocol.

---

[8] See [26] for a prove of the first inequality. The second one can be proved using similar techniques.

4   The Shortest Vector Problem

In this section we describe a statistical zero knowledge proof system $(P_{\text{SVP}}, V_{\text{SVP}})$ for GapSVP$_\gamma$. The reasons we are interested in the SHORTEST VECTOR PROBLEM are both theoretical (being SVP a different problem from CVP, it is interesting to know if it admits **SZK** proofs with efficient prover), and practical, as proofs of knowledge for SVP can be used in conjunction with the lattices of [19] to yield identification schemes with worst-case/average-case security guarantees. (See Section 5.) Intuitively, our proof system for GapSVP can be thought as a combination of the reduction from GapSVP$_\gamma$ to GapCVP$_\gamma$ of Goldreich, Micciancio, Safra and Seifert [37], followed by the invocation of the proof system for GapCVP described in the previous section. Things are not as simple because the reduction of [37] is not a Karp reduction, and in order to solve a shortest vector problem instance, it requires the solution of (polynomially) many closest vector problems. So, we combine all the GapCVP instances together using the Goldreich-Levin hardcore predicate [38]. This is just the intuition behind the proof system that we are going to describe. In fact, our proof system requires neither the explicit construction of many GapCVP instances, nor the complicated analysis of the Goldreich-Levin predicate. So, below we briefly describe the proof system without reference to those general tools. For a detailed description see the full version of this paper.

The basic idea is the same as the proof system for the closest vector problem, but this time instead of selecting points close to the origin or close to the target vector $\mathbf{y}$, we consider balls centered around all lattice points of the form $\mathbf{Bc}$, where $\mathbf{c} \in \{0,1\}^n$, and reduce the points modulo $2\mathbf{B}$. The prover starts the interaction by sending points $\mathbf{m}_i$ close to randomly chosen centers $\mathbf{Bc}_i$. For each such point, the prover also sends a binary vector $\mathbf{s}_i$. If the lattice does not contain short vectors, then balls centered around different $\mathbf{Bc}$ are disjoint (even after reduction modulo $2\mathbf{B}$), and the first message sent by the prover uniquely determines a bit $\sum_i \langle \mathbf{s}_i, \mathbf{c}_i \rangle \bmod 2$. Then the verifier asks the prover to show that $\sum_i \langle \mathbf{s}_i, \mathbf{c}_i \rangle \bmod 2 = q$, where q is a random bit chosen by the verifier. If the prover can answer both questions, then there must be a message $\mathbf{m}_i$ that is close to two different centers (modulo $2\mathbf{B}$), proving that the lattice contains short vectors.

5   Identification Schemes

An *identification scheme* is a protocol by which one party, Alice, can repeatedly prove her identity to other parties in such a way that these parties cannot later impersonate Alice. Following the now-standard paradigm of [28], ID schemes are immediately obtained from zero-knowledge proofs of knowledge. It should be remarked that the computational problems underlying our identification schemes are not likely to be **NP**-hard (cf. [26]). The same is true for most computational problems used in cryptography (e.g., factoring), so, in some sense, ours is as good a hardness assumption as any. However, factoring is a much more widely studied assumption than lattice problems, so our identification schemes should be used with caution. The discussion below concentrates on efficiency issues.

The proofs of knowledge from Section 3, give rise to $\omega(1)$-round ID schemes, because witness-indistinguishability is not enough to guarantee the security. However, we can obtain a 3-round identification scheme as follows. First, we consider a new problem OR-GAPCVP$_\gamma$ whose instances are *pairs* (x_1, x_2) of GAPCVP$_\gamma$ instances, and whose YES instances are those for which at least one of the x_i's is a YES instance of GAPCVP$_\gamma$. Using a technique from [32], we can convert our proof system into one for OR-GAPCVP$_\gamma$. Parallel repetition yields a constant-round statistically witness-indistinguishable proof of knowledge with negligible soundness error. For such 'OR' problems, witness indistinguishability implies "witness hiding," which suffices for the identification scheme [39] (cf., [34]). Details will be given in the full version of the paper.

We stress that, unlike all known cryptosystems based on lattice problems [24,23], these identification schemes only require the generation of lattices in which the approximate CLOSEST VECTOR PROBLEM (resp. SHORTEST VECTOR PROBLEM) is hard together with a close vector (resp. short vector). In particular, we do not need to generate an additional "short" basis, nor do we need "unique short vectors" or "hidden hyperplanes". In particular, this opens up more possibilities for using lattices with potential advantages both in terms of efficiency and security. As an example, for identification schemes based on SVP one can use the random class of lattices of [19,22], which, for appropriate choice of the parameters, results in identification schemes that are at least as hard to break (on the average) as the worst case instance of approximating GAPSVP in the worst case within factor $\tilde{O}(n^4)$, or approximating other lattice problems (shortest linearly independent vectors or covering radius) within factor $\tilde{O}(n^3)$. Alternatively, one can use lattices with special structure like the cyclic and quasi-cyclic lattices of [20], or the convolutional modular lattices of [25] (but possibly with different, more secure, values of the parameters, since we do not need to embed a decryption trapdoor), in which the basis has a more compact representation (almost linear in the security parameter, rather than the standard matrix representation, whose quadratic size has been a practical barrier for the use of lattice cryptosystems.) Another very interesting possibility for identification schemes based on our CVP proof system is to use lattices where CVP *with preprocessing* (CVPP) is hard. This is a variant of the standard CVP problem introduced in [40] and studied in [41, 42], where finding close lattice vectors is hard even if the lattice is fixed, and the only input is the target vector. This allows to use the same lattice $\mathbf{B}$ for all users, and hardwire the description of the lattice $\mathbf{B}$ in the key generation, identification and verification algorithms. When a new user wants to generate a key, he chooses a random short error vector $\mathbf{r}$ (the secret key) and computes $\mathbf{y} = \mathbf{r} \bmod \mathbf{B}$ as its public key. The security of the scheme relies on the fact that approximating CVP in the lattice generated by $\mathbf{B}$ (for appropriately constructed, but fixed, $\mathbf{B}$) is hard. The advantage is that both the secret and public keys are just a single vector which takes storage proportional to dimension of the lattice n (the security parameter),[9] rather than a matrix (representing the lattice) which in general takes storage at least proportional to n^2. There are still big gaps between our understanding of CVPP and its cryptographic applicability: the strongest inapproximability results known to date [42] only show that CVPP is hard to approximate within factors smaller than 3, while our system requires inapproximability within $\sqrt{n}$. More importantly, all known lower bounds [40,41,42] only establish the hardness in the worst-case (**NP**-hardness), while for cryptographic applications one needs average-case hardness. Still, the possibility that further developments about the complexity of lattice problems might lead to *practical* and *provably secure* identification schemes with worst-case/average-case guarantees is very appealing. In this perspective, establishing a worst-case/average-case connection for CVPP along the lines of [19,22] would be very interesting.

6   Statistical Difference

In this section, our focus will be the problem $SD^{\alpha,\beta}$ for various values of $0 \leq \alpha < \beta \leq 1$. Consider the **SZK**-complete problem $SD^{1/2,1-\epsilon}$, for $1/\sqrt{2} > \epsilon > 0$. Since we do not know if **SZK** $\subseteq$ **NP**, we do not hope to give a proof system with efficient provers for this language. Instead we consider the limit problem $SD^{1/2,1}$ obtained setting $\epsilon = 0$, i.e. deciding whether two distributions are statistically close or have disjoint supports. Unfortunately, this problem is not known to be complete for **SZK**. Note that $SD^{1/2,1}$ is in **NP**, since coin tosses r_X, r_Y for which the circuits produce identical samples (i.e. $X(r_X) = Y(r_Y)$) are a witness that (X, Y) is a YES instance. We will prove that $SD^{1/2,1}$ has a statistical zero-knowledge proof system with an efficient prover.

We now state a useful lemma that allows us to make the statistical difference exponentially small in YES instances of $SD^{1/2,1}$.

Lemma 7 (XOR Lemma [8]). *Given probability distributions X_0, X_1 and a parameter k, define probability distributions $Y_c = (x_1, \ldots, x_k)$ (for $c \in \{0,1\}$) obtained by uniformly choosing $(b_1, \ldots, b_k) \leftarrow \{0,1\}^k$ such that $b_1 \oplus \cdots \oplus b_k = c$, and then sampling each $x_i \leftarrow X_{b_i}$ independently. Then $\Delta(Y_0, Y_1) = \Delta(X_0, X_1)^k$.*

Thus, given an instance (X_0, X_1) of $SD^{1/2,1}$, this lemma shows how to construct circuits for a new pair of distributions (Y_0, Y_1) whose statistical difference is exponentially small if (X_0, X_1) is a YES instance, and whose supports are disjoint if (X_0, X_1) is a NO instance. We can use this to obtain simple statistical zero knowledge proof system for $SD^{1/2,1}$, mimicking the well-known proof systems for QUADRATIC RESIDUOSITY [1] and GRAPH ISOMORPHISM [4]: **(1)** First, the prover sends the verifier $y \leftarrow Y_0$, **(2)** and the verifier replies sending $b \leftarrow \{0,1\}$ to the prover; **(3)** then the prover sends $r \leftarrow \{s : Y_b(s) = y\}$ to

---

[9] This is obvious for the secret short vector **r**. The public vector **y** can be much bigger because it contains large integer entries. Fortunately, as shown in [36], it is possible to select the basis **B** in an optimally secure way that results also in reduced vectors **y** with small bit-size.

the verifier, (4) and the verifier accepts if $Y_b(r) = y$. It can be verified that the above proof system has soundness error $1/2$, completeness error $1/2^{k+1}$, and is statistical zero knowledge with simulator deviation $1/2^{k+1}$ (cf., [27]). However, even though $\mathrm{SD}^{1/2,1} \in \mathbf{NP}$, it does not appear that the prover strategy can be implemented in polynomial time given a witness. (If the verifier selects $b = 0$, the prover can respond with the coin tosses it used to generate y, but if the verifier selects $b = 1$, the prover must be able to find collisions between the circuits Y_0 and Y_1, which may be infeasible.)

To obtain efficient provers for $\mathrm{SD}^{1/2,1}$ itself, we use the ideas of Itoh, Ohta, and Shizuya [29]. The key concept is that of problem dependent commitment. This is a commitment scheme where the sender and receiver get as auxiliary input an instance x of a promise problem Π. The operations performed by the protocol depend on the value of x, and the protocol has different security properties depending on whether x is a YES or a NO instance of Π. Typically, the protocol is required to be secret when $x \in \Pi_{\mathrm{YES}}$ and unambiguous when $x \in \Pi_{\mathrm{NO}}$, or vice-versa. As usual, a problem dependent commitment is *statistically* secure if the secrecy and unambiguity properties hold in a statistical sense.

Itoh, Ohta, and Shizuya [29] considered only noninteractive problem-dependent commitment schemes in which both security properties are perfect. (Notice that any noninteractive statistically unambiguous commitment is necessarily perfectly unambiguous.) They proved that if a problem Π has a noninteractive problem-dependent commitment scheme which is perfectly secret on YES instances and perfectly unambiguous on NO instances, then Π has a perfect zero-knowledge proof system with an efficient prover. We observe that this result can be generalized as follows

Theorem 8 (generalizing [29]). *Suppose a promise problem Π is in $\mathbf{NP}$, with $\mathbf{NP}$ relation R, and that Π has a problem-dependent commitment scheme which is statistically secret on YES instances and statistically unambiguous on NO instances. Then Π has a statistical zero-knowledge proof system with an efficient prover (using any R-witness).*

We apply the theorem to $\mathrm{SD}^{1/2,1}$, by defining a problem dependent commitment for this problem as follows. On input $(b, (X_0, X_1), 1^k)$, the sender commits to b by sending the receiver $y \leftarrow Y_b$, where Y_b is obtained by applying the XOR Lemma (Lemma 7) to (X_0, X_1) with parameter k. In the reveal phase, the sender reveals b and the coin tosses used to generate y. The receiver checks that $Y_b(r) = y$. The reader can easily check that this commitment scheme is statistically secret on YES instances and perfectly unambiguous on NO instances. Using Theorem 8, we get the following result.

Theorem 9. $\mathrm{SD}^{1/2,1}$ *has a statistical zero-knowledge proof system with an efficient prover.*

6.1   Efficient Provers for All of SZK?

As discussed in the introduction, part of our motivation in this work is the general question of whether every problem in $\mathbf{SZK} \cap \mathbf{NP}$ has a statistical zero-knowledge proof system with an efficient prover. The following theorem suggests three possible approaches to solve this problem.

Theorem 10. *If any of the following conditions hold, every problem in* **SZK** $\cap$ **NP** *has a statistical zero-knowledge proof with an efficient prover:*

1. $SD^{1/3,2/3}$ *has a statistically secure problem-dependent commitment scheme.*

2. $SD^{1/3,2/3}$ *reduces to* $SD^{1/2,1}$ *via a randomized Karp reduction with* one-sided error *(even constant error).*

3. *Any* **NP** *problem that reduces to* $SD^{1/3,2/3}$, *also reduces to* $SD^{1/2,1}$.

The first approach is proved using the closure of **SZK** under complementation, and using the fact that if a promise problem Π reduces (via a randomized Karp reduction with one-sided *negligible* error probability) to a promise problem Γ, and Γ has a problem-dependent commitment scheme, then Π also has a problem-dependent commitment scheme with the same security properties. The second approach is just a way to prove the first condition, using the fact that one-sided error in Karp reductions to $SD^{1/2,1}$ can be made negligible. The last approach, essentially asks to prove that $SD^{1/2,1}$ is complete for **SZK** $\cap$ **NP**.

The proof systems described in this section differ in one important way from previous ones. All previous proof systems for variants of STATISTICAL DIFFERENCE e.g. [8,9,43], used the input circuits as "black boxes." That is, the use of the circuits by the verifier and prover consisted solely of evaluating the circuits on various inputs, and never referred to the internal structure of the circuits. It is not difficult to show, using constructions like [11], that no protocol of this form can be a statistical zero-knowledge proof with an efficient prover for even $SD^{0,1}$ (if one-way functions exist). The proof system of Theorem 9 is not black box, however, and does make use of the internal workings of the circuits (due to the techniques of [29], which in turn use [4]). This suggests that this approach does indeed have potential to resolve questions that may have previously seemed intractable.

We conclude this section by showing yet another relationship between problem-dependent commitment schemes and **SZK**. If we remove the assumption that problem Π is in **NP** from the hypothesis of Theorem 8, we can still conclude that Π has an **SZK** proof system, although not necessarily one with efficient prover.

Proposition 11. *If a problem Π has a statistically secure problem-dependent commitment scheme, then $\Pi \in$* **SZK**.

Acknowledgments. We are grateful to Oded Goldreich and Shafi Goldwasser for suggesting the problems on lattices. The second author thanks Danny Gutfreund, Moni Naor, and Avi Wigderson for discussions about **SZK** that influenced this work. We also thank Mihir Bellare for pointers to related work, and the anonymous reviewers and Minh Nguyen for helpful comments and corrections.

References

1. Goldwasser, S., Micali, S., Rackoff, C.: The knowledge complexity of interactive proof systems. SIAM J. Comput. **18** (1989) 186–208
2. Bellare, M., Petrank, E.: Making zero-knowledge provers efficient. In: 24th STOC. (1992) 711–722
3. Boyar, J., Friedl, K., Lund, C.: Practical Zero-Knowledge Proofs: Giving Hints and Using Deficiencies. J. Cryptology **4** (1991) 185–206
4. Goldreich, O., Micali, S., Wigderson, A.: Proofs that yield nothing but their validity or all languages in NP have zero-knowledge proof systems. J. ACM **38** (1991) 691–729
5. Fortnow, L.: The complexity of perfect zero-knowledge. In: Advances in Computing Research. Volume 5. JAC Press (1989) 327–343
6. Aiello, W., Håstad, J.: Statistical zero-knowledge languages can be recognized in two rounds. J. Comput. System Sci. **42** (1991) 327–345
7. Okamoto, T.: On relationships between statistical zero-knowledge proofs. J. Comput. System Sci. **60** (2000) 47–108
8. Sahai, A., Vadhan, S.: A complete problem for statistical zero knowledge. J. ACM **50** (2003) 196–249
9. Goldreich, O., Vadhan, S.: Comparing entropies in statistical zero-knowledge with applications to the structure of SZK. In: 14th CCC. (1999) 54–73
10. Goldreich, O., Sahai, A., Vadhan, S.: Honest verifier statistical zero-knowledge equals general statistical zero-knowledge. In: 30th STOC. (1998) 399–408
11. Vadhan, S.P.: On transformations of interactive proofs that preserve the prover's complexity. In: 32nd STOC. (2000) 200–207
12. Arvind, V., Köbler, J.: On pseudorandomness and resource-bounded measure. Theoret. Comput. Sci. **255** (2001) 205–221
13. Klivans, A.R., van Melkebeek, D.: Graph nonisomorphism has subexponential size proofs unless the polynomial-time hierarchy collapses. SIAM J. Comput. **31** (2002) 1501–1526
14. Miltersen, P.B., Vinodchandran, N.V.: Derandomizing Arthur-Merlin games using hitting sets. In: 40th FOCS. (1999) 71–80
15. Arora, S., Babai, L., Stern, J., Sweedyk, Z.: The hardness of approximate optima in lattices, codes, and systems of linear equations. J. Comput. System Sci. **54** (1997) 317–331
16. Ajtai, M.: The shortest vector problem in L_2 is NP-hard for randomized reductions (extended abstract). In: 30th STOC. (1998) 10–19
17. Micciancio, D.: The shortest vector problem is NP-hard to approximate to within some constant. SIAM J. Comput. **30** (2001) 2008–2035
18. Dinur, I., Kindler, G., Raz, R., Safra, S.: An improved lower bound for approximating CVP. Combinatorica (To appear) Preliminary version in FOCS '98.
19. Ajtai, M.: Generating hard instances of lattice problems (extended abstract). In: 28th STOC. (1996) 99–108
20. Micciancio, D.: Generalized compact knapsaks, cyclic lattices, and efficient one-way functions from worst-case complexity assumptions (extended abstract). In: 43rd FOCS. (2002) 356–365
21. Goldreich, O., Goldwasser, S., Halevi, S.: Collision-free hashing from lattice problems. Technical Report TR96-056, ECCC (1996)
22. Micciancio, D.: Improved cryptographic hash functions with worst-case/average-case connection (extended abstract). In: 34th STOC. (2002) 609–618

23. Goldreich, O., Goldwasser, S., Halevi, S.: Public-key cryptosystems from lattice reduction problems. In: CRYPTO '97. Volume 1294 of Springer LNCS. (1997) 112–131

24. Ajtai, M., Dwork, C.: A public-key cryptosystem with worst-case/average-case equivalence. In: 29th STOC. (1997) 284–293

25. Hoffstein, J., Pipher, J., Silverman, J.H.: NTRU: a ring based public key cryptosystem. In: Algorithmic number theory (ANTS III). Volume 1423 of Springer LNCS. (1998) 267–288

26. Goldreich, O., Goldwasser, S.: On the limits of nonapproximability of lattice problems. J. Comput. System Sci. **60** (2000) 540–563

27. Vadhan, S.P.: A Study of Statistical Zero-Knowledge Proofs. PhD thesis, MIT (1999)

28. Feige, U., Fiat, A., Shamir, A.: Zero-knowledge proofs of identity. J. Cryptology **1** (1988) 77–94

29. Itoh, T., Ohta, Y., Shizuya, H.: A language-dependent cryptographic primitive. J. Cryptology **10** (1997) 37–49

30. Gennaro, R., Micciancio, D., Rabin, T.: An efficient non-interactive statistical zero-knowledge proof system for quasi-safe prime products. In: 5th ACM CCS. (1998) 67–72

31. Tompa, M., Woll, H.: Random self-reducibility and zero knowledge interactive proofs of possession of information. In: 28th FOCS. (1987) 472–482

32. De Santis, A., Di Crescenzo, G., Persiano, G., Yung, M.: On monotone formula closure of SZK. In: 35th FOCS. (1994) 454–465

33. Bellare, M., Goldwasser, S.: The complexity of decision versus search. SIAM J. Comput. **23** (1994) 97–119

34. Goldreich, O.: Foundations of Cryptography: Basic Tools. Cambridge U. Press (2001)

35. Micciancio, D., Goldwasser, S.: Complexity of lattice problems: a cryptographic perspective. Volume 671 of Engineering and Computer Science. Kluwer (2002)

36. Micciancio, D.: Improving lattice based cryptosystems using the Hermite normal form. In: Cryptography and Lattices Conference. Volume 2146 of Springer LNCS. (2001) 126–145

37. Goldreich, O., Micciancio, D., Safra, S., Seifert, J.P.: Approximating shortest lattice vectors is not harder than approximating closest lattice vectors. Inf. Proc. Lett. **71** (1999) 55–61

38. Goldreich, O., Levin, L.: A hard predicate for all one-way functions. In: 21st STOC. (1989)

39. Feige, U., Shamir, A.: Witness indistinguishable and witness hiding protocols. In: 22nd STOC. (1990) 416–426

40. Micciancio, D.: The hardness of the closest vector problem with preprocessing. IEEE Trans. Inform. Theory **47** (2001) 1212–1215

41. Feige, U., Micciancio, D.: The inapproximability of lattice and coding problems with preprocessing. J. Comput. System Sci. (To appear) Preliminary version in CCC 2002.

42. Regev, O.: Improved Inapproximability of Lattice and Coding Problems with Preprocessing. In: 18th CCC. (2003)

43. Goldreich, O., Sahai, A., Vadhan, S.: Can statistical zero-knowledge be made non-interactive?, or On the relationship of SZK and NISZK. In: CRYPTO '99. Volume 1666 of Springer LNCS. (1999)

Derandomization in Cryptography

Boaz Barak[1,*], Shien Jin Ong[2,**], and Salil P. Vadhan[3,***]

[1] Weizmann Institute of Science, Rehovot, Israel, `boaz@wisdom.weizmann.ac.il`.
[2] Massachusetts Institute of Technology, Cambridge, MA, `shienjin@mit.edu`.
[3] Harvard University, Cambridge, MA, `salil@eecs.harvard.edu`.

Abstract. We give two applications of Nisan–Wigderson-type ("non-cryptographic") pseudorandom generators in cryptography. Specifically, assuming the existence of an appropriate NW-type generator, we construct:

1. A one-message witness-indistinguishable proof system for every language in **NP**, based on any trapdoor permutation. This proof system does not assume a shared random string or any setup assumption, so it is actually an "**NP** proof system."
2. A noninteractive bit commitment scheme based on any one-way function.

The specific NW-type generator we need is a hitting set generator fooling *nondeterministic circuits*. It is known how to construct such a generator if $\mathbf{E} = \mathbf{DTIME}(2^{O(n)})$ has a function of nondeterministic circuit complexity $2^{\Omega(n)}$ (Miltersen and Vinodchandran, FOCS '99). Our witness-indistinguishable proofs are obtained by using the NW-type generator to derandomize the ZAPs of Dwork and Naor (FOCS '00). To our knowledge, this is the first construction of an **NP** proof system achieving a secrecy property.

Our commitment scheme is obtained by derandomizing the interactive commitment scheme of Naor (J. Cryptology, 1991). Previous constructions of noninteractive commitment schemes were only known under incomparable assumptions.

1 Introduction

The computational theory of pseudorandomness has been one of the most fertile grounds for the interplay between cryptography and computational complexity. This interplay began when Blum, Micali, and Yao (BMY) [1,2], motivated by applications in cryptography, placed the study of pseudorandom generators on firm complexity-theoretic foundations. They gave the first satisfactory definition of pseudorandom generators along with constructions meeting that definition. Their notion quickly acquired a central position in cryptography, but it turned

* Supported by Clore Foundation Fellowship and Israeli Higher Education Committee Fellowship.
** Supported by MIT Eloranta Fellowship and MIT Reed UROP Fund.
*** Supported by NSF grants CCR-0205423 and CCR-0133096, and a Sloan Research Fellowship.

D. Boneh (Ed.): CRYPTO 2003, LNCS 2729, pp. 299–315, 2003.

out that the utility of pseudorandom generators was not limited to cryptographic applications. In particular, Yao [2] showed that they could also be used for *derandomization* — efficiently converting randomized algorithms into deterministic algorithms. Pseudorandom generators and their generalization, pseudorandom functions [3], also found a variety of other applications in complexity theory and the theory of computation (e.g., [4,5]).

Focusing on derandomization, Nisan and Wigderson (NW) [6] proposed a weakening of the BMY definition of pseudorandom generators which still suffices for derandomization. The benefit was that such NW-type pseudorandom generators could be constructed under weaker assumptions than the BMY ones (circuit lower bounds for exponential time, rather than the existence of one-way functions). Thus, a long body of work developed around the task of constructing increasingly efficient NW-type pseudorandom generators under progressively weaker assumptions. One of the highlights of this line of work is the construction of Impagliazzo and Wigderson [7] implying that $\mathbf{P} = \mathbf{BPP}$ under the plausible assumption that $\mathbf{E} = \mathbf{DTIME}(2^{O(n)})$ has a problem of circuit complexity $2^{\Omega(n)}$. More recently, the work on NW-type pseudorandom generators has also been found to be intimately related to randomness extractors [8], and has been used to prove complexity-theoretic results which appear unrelated to derandomization [9].

While allowing remarkable derandomization results such as the Impagliazzo–Wigderson result mentioned above, NW-type pseudorandom generators have not previously found applications in cryptography (for reasons mentioned below). In this work, we show that a stronger form of NW-type pseudorandom generators, namely ones fooling *nondeterministic circuits* [10,11,12,13], do have cryptographic applications. Using such pseudorandom generators (which can be constructed under plausible complexity assumptions), we:

1. Construct witness-indistinguishable "NP proofs" (i.e. one-message[1] proof systems, with no shared random string or other setup assumptions) for every language in NP, assuming the existence of trapdoor permutations.
2. Construct *noninteractive* bit commitment schemes from any one-way function.

Thus, each of these results requires two assumptions — the circuit complexity assumption for the NW-type pseudorandom generator (roughly, that $\mathbf{E}$ has a function of nondeterministic circuit complexity $2^{\Omega(n)}$) and a "cryptographic" assumption (one-way functions or trapdoor permutations).

Result 1 is the first construction of witness-indistinguishable NP proofs under any assumption whatsoever, and refutes the intuition that interaction is necessary to achieve secrecy in proof systems. It is obtained by derandomizing the ZAP construction of Dwork and Naor [14].

Result 2 is not the first construction of noninteractive commitment schemes, but is based on assumptions that appear incomparable to previous ones (which

---

[1] We use "messages" rather than "rounds", as the latter is sometimes used to refer to a pair of messages.

were based on the existence of one-to-one one-way functions). We obtain this result by derandomizing the Naor's interactive bit commitment scheme [15].

These two examples suggest that NW-type pseudorandom generators (and possibly other "non-cryptographic" tools from the derandomization literature) are actually relevant to the foundations of cryptography, and it seems likely that other applications will be found in the future.

NW-type Generators fooling Nondeterministic Circuits. The most important difference between BMY-type and NW-type pseudorandom generators is that BMY-type pseudorandom generators are required to fool even circuits with greater running time than the generator, whereas NW-type pseudorandom generators are allowed greater running time than the adversarial circuit. Typically, a BMY-type pseudorandom generator must run in some fixed polynomial time (say n^c), and fool all polynomial-time circuits (even those running in time, say, n^{2c}). In contrast, an NW-type pseudorandom generator may run in time $n^{O(c)}$ (e.g. n^{3c}) in order to fool circuits running in time n^c. BMY-type pseudorandom generators are well-suited for cryptographic applications, where the generator is typically run by the legitimate parties and the circuit corresponds to the adversary (who is always allowed greater running time). In contrast, NW-type pseudorandom generators seem non-cryptographic in nature. Nevertheless we are able to use them in cryptographic applications. The key observation is that, in the protocols we consider, (some of) the randomness is used to obtain a string that satisfies some fixed property *which does not depend on the adversary (or its running time)*. Hence, if this property can be verified in polynomial time, we can obtain the string using an NW-type pseudorandom generator of fixed polynomial running time. We then eliminate the randomness entirely by enumerating over all possible seeds. This is feasible because NW-type generators can have logarithmic seed length. Also, we show that in our specific applications, this enumeration does not compromise the protocol's security.

In the protocols we consider, the properties in question do not seem to be verifiable in polynomial time. However, they are verifiable in *nondeterministic* polynomial time. So we need to use a pseudorandom generator that fools non-deterministic circuits. Fortunately, it is possible for an *NW-type* pseudorandom generator to fool nondeterministic circuits, as realized by Arvind and Köbler [10] and Klivans and van Melkebeek [11].[2] Indeed, a sequence of works have constructed such pseudorandom generators under progressively weaker complexity assumptions [10,11,12,13]. Our results make use of the Miltersen–Vinodchandran construction [12] (which gives only a "hitting set generator" rather than a pseudorandom generator, but this suffices for our applications).

Witness Indistinguishable **NP** *Proofs.* In order to make zero-knowledge proofs possible, the seminal paper of Goldwasser, Micali, and Rackoff [17] augmented

---

[2] It is impossible for a BMY-type pseudorandom generator to fool nondeterministic circuits, as such a circuit can recognize outputs of the pseudorandom generator by guessing the corresponding seed and evaluating the generator to check. Some attempts to bypass this difficulty can be found in [16].

the classical notion of an **NP** proof with two new ingredients — interaction and randomization. Both were viewed as necessary for the existence of zero-knowledge proofs, and indeed it was proven by Goldreich and Oren [18] that without either, zero-knowledge proofs exist only for trivial languages (those in **BPP**). The role of interaction was somewhat reduced by the introduction of "noninteractive" zero-knowledge proofs [19,20], but those require a shared random string selected by a trusted third party, which can be viewed as providing a limited form of interaction. Given the aforementioned impossibility results [18], reducing the interaction further seems unlikely. Indeed, a truly noninteractive proof system, in which the prover sends a single proof string to the verifier, seems to be inherently incompatible with the intuitive notion of "zero knowledge": from such a proof, the verifier gains the ability to prove the same statement to others.

Despite this, we show that for a natural weakening of zero knowledge, namely *witness indistinguishability* [21], the interaction *can* be completely removed (under plausible complexity assumptions). Recall that a witness-indistinguishable proof system for a language $L \in$ **NP** is an interactive proof system for L that leaks no knowledge about which witness is being used by the prover (as opposed to leaking no knowledge at all, as in zero-knowledge proofs) [21]. Witness indistinguishability suffices for a number of the applications of zero knowledge [21], and also is a very useful intermediate step in the construction of zero-knowledge proofs [22].

Several prior results show that witness-indistinguishable proofs do not require the same degree of interaction as zero-knowledge proofs. Feige and Shamir [21] constructed 3-message witness-indistinguishable proofs for **NP** (assuming the existence of one-way functions), whereas the existence of 3-message zero-knowledge proofs is a long-standing open problem. More recently, the ZAPs of Dwork and Naor [14] achieve witness indistinguishability with just 2 messages (assuming trapdoor permutations), whereas this is known to be impossible for zero knowledge [18]. Dwork and Naor also showed that the interaction could be further reduced to one message at the price of *nonuniformity* (i.e. if the protocol can use some nonuniform advice of polynomial length); they interpret this as evidence that "*proving* a lower bound of two [messages] is unlikely."

We construct 1-message witness-indistinguishable proofs for **NP** in the "plain model", with no use of a shared random string or nonuniformity. Our proof system is obtained by derandomizing the Dwork–Naor ZAPs via an NW-type generator against nondeterministic circuits. Since our verifier is deterministic, we actually obtain a standard **NP** proof system with the witness indistinguishability property. More precisely, for any language $L \in$ **NP** with associated **NP**-relation R, we construct a new **NP**-relation R' for L. The relation R' has the property that one can efficiently transform any witness with respect to R into a distribution on witnesses with respect to R', such that the distributions corresponding to different witnesses are computationally indistinguishable.

Converting **AM** proof systems to **NP** proof systems was actually one of the original applications of NW-type generators versus nondeterministic circuits [10,

11]. The novelty in our result comes from observing that this conversion preserves the witness indistinguishability property.

The randomness requirements of *zero-knowledge* proofs have been examined in previous works. Goldreich and Oren [18] showed that only languages in **BPP** have zero-knowledge proofs in which either the prover or verifier is deterministic. Thus De Santis, Di Crescenzo, and Persiano [23,24,25] have focused on reducing the number of random bits. Specifically, under standard "cryptographic" assumptions, they constructed noninteractive zero-knowledge proofs with a shared random string of length $O(n^\epsilon + \log(1/s))$ and 2-message witness-indistinguishable proofs (actually, ZAPs) in which the verifier uses only $O(n^\epsilon + \log(1/s))$ random bits, where $\epsilon > 0$ is any constant and s is the soundness error. They posed the existence of 1-message witness-indistinguishable proofs for **NP** as an open problem. One of their main observations in [25] is that combinatorial methods for randomness-efficient error reduction, such as pairwise independence and expander walks, preserve witness indistinguishability. As mentioned above, we make crucial use of an analogous observation about NW-type generators.

Noninteractive Bit Commitment Schemes. Bit commitment schemes are one of the most basic primitives in cryptography, used pervasively in the construction of zero-knowledge proofs [26] and other cryptographic protocols. Here we focus on perfectly (or statistically) binding and computationally hiding bit commitment schemes. As usual, *noninteractive* bit commitment schemes, in which the commitment phase consists of a single message from the sender to the receiver, are preferred over interactive schemes. There is a simple construction of noninteractive bit commitment schemes from any *one-to-one* one-way function [27,2, 28]. From general one-way functions, the only known construction of bit commitment schemes, namely Naor's protocol [15] (with the pseudorandom generator construction of [29]), requires interaction.

We show how to use an NW-type pseudorandom generator against nondeterministic circuits to remove the interaction in Naor's protocol, yielding noninteractive bit commitment schemes under assumptions that appear incomparable to the existence of one-to-one one-way functions. In particular, ours is a "raw hardness" assumption, not requiring hard functions with any semantic structure such as being one-to-one.

From a different perspective, our result shows that "non-cryptographic" assumptions (nondeterministic circuit lower bounds for **E**) can reduce the gap between one-way functions and one-to-one one-way functions. In particular, a noninteractive bit commitment scheme gives rise to a "partially one-to-one one-way function": a polynomial-time computable function $f(x,y)$ such that x is uniquely determined by $f(x,y)$ and x is hard to compute from $f(x,y)$ (for random x, y). It would be interesting to see if this can be pushed further to actually construct one-to-one one-way functions from general one-way functions under a non-cryptographic assumption.

Perspective. The assumption required for the NW-type generators we use is a strong one, but it seems to be plausible (see Section 2.4). Perhaps its most significant feature is that it is very different than the assumptions typically used in cryptography (e.g. it is a worst-case assumption); nevertheless, our results show it has implications in cryptography. In our first result, we use it to demonstrate the plausibility of nontrivial 1-message witness-indistinguishable proofs, which will hopefully lead to efficient constructions for specific problems based on specific assumptions. As for our second result, the plausibility of noninteractive commitment schemes was already established more convincingly based on one-to-one one-way functions [27]. What we find interesting instead is that a "non-cryptographic" assumption can imply new relationships between basic cryptographic primitives, and in particular reduce the gap between one-way functions and one-to-one one-way functions.

2   Preliminaries

2.1   Pseudorandom Generators

A *pseudorandom generator* (PRG) is a deterministic algorithm $G: \{0,1\}^\ell \to \{0,1\}^m$, with $\ell < m$. Pseudorandom generators are used to convert a short random string into a longer string that looks random to any efficient observer.

Definition 1 (Pseudorandom generator). *We say that $G: \{0,1\}^\ell \to \{0,1\}^m$ is a (s,ϵ)-pseudorandom generator against circuits if for all circuits $C: \{0,1\}^m \to \{0,1\}$ of size at most s, it holds that $|\Pr[C(G(U_\ell)) = 1] - \Pr[C(U_m) = 1]| < \epsilon$, where U_k denotes the uniform distribution over $\{0,1\}^k$.*

BMY-type vs. NW-type Generators. As mentioned above, there are two main types of pseudorandom generators: Blum-Micali-Yao (BMY) [1,2] type and Nisan-Wigderson (NW) [6] type generator. Both can be defined for a wide range of parameters, but here we focus on the "classic" settings which we need. A BMY-type generator is the standard kind of pseudorandom generator used in cryptography.

Definition 2 (BMY-type generators). *A function $G = \bigcup_m G_m : \{0,1\}^\ell \to \{0,1\}^m$ is a BMY-type pseudorandom generator with seed length $\ell = \ell(m)$, if G is computable in time $\mathrm{poly}(\ell)$, and for every constant c, G_m is a $(m^c, 1/m^c)$-pseudorandom generator for all sufficiently large m.*

Note that a BMY-type generator is required to have running time that is a fixed polynomial, but must fool circuits whose running time is an arbitrary polynomial. Håstad, Impagliazzo, Levin, and Luby [29] proved that BMY-type pseudorandom generators with seed length $\ell(m) = m^\delta$ (for every $\delta > 0$) exist if and only if one-way functions exist.

NW-type generators differ from BMY-type generators most significantly in the fact that the generator has greater running time than the circuits it fools.

Definition 3 (NW-type generators). *A function $G = \bigcup_m G_m : \{0,1\}^\ell \to \{0,1\}^m$ is an NW-type pseudorandom generator with seed length $\ell = \ell(m)$, if G is computable in time $2^{O(\ell)}$ and G_m is a $(m^2, 1/m^2)$-pseudorandom generator for all m.[3]*

---

[3] One can replace m^2 in this definition with any *fixed* polynomial in m.

We will be interested "high end" NW-type generators, which have seed length $\ell(m) = O(\log m)$, and thus have running time which is a fixed polynomial in m.[4] Impagliazzo and Wigderson [7] proved that such a generator exists if $\mathbf{E} = \mathbf{DTIME}(2^{O(n)})$ has a function of circuit complexity $2^{\Omega(n)}$. Note that when the seed length is $\ell = O(\log m)$, all 2^ℓ seeds can be enumerated in time poly(m), and hence the generator can be used for complete derandomization. In particular, such a generator implies $\mathbf{BPP} = \mathbf{P}$.

2.2   Nondeterministic Computations and the Class AM

A significant advantage of NW-type generators that we will use is that they can fool *nondeterministic* circuits, because even if such a circuit can guess the seed, it does not have enough time to evaluate the generator on it.

Definition 4. *A* nondeterministic *Boolean circuit $C(x, y)$ is a circuit that takes x as its primary input and y as a witness. For each $x \in \{0,1\}^*$, we define $C(x) = 1$ if there exist a witness y such that $C(x, y) = 1$.*

A co-nondeterministic *Boolean circuit $C(x, y)$ is a circuit that takes x as its primary input and y as a witness. For each $x \in \{0,1\}^*$, we define $C(x) = 0$ if there exist a witness y such that $C(x, y) = 0$.*

Denote $S_{\mathrm{N}}(f)$ to be the minimal sized nondeterministic circuit computing f.

Nondeterministic and co-nondeterministic algorithms can be defined in a similar fashion, with the nonuniform circuit C being replaced by a *uniform algorithm*. Naturally, we measure the running time of a nondeterministic algorithm $A(x, y)$ in terms of the first input x.

The class $\mathbf{AM}$ [30] has two equivalent formulations. The first is as the class of languages with constant-message interactive proofs (see [31] for this definition). The second is as the class of languages decidable by polynomial-time *probabilistic* nondeterministic algorithms. Formally, a probabilistic nondeterministic algorithm $A(x, r, y)$ takes a random input r in addition to its regular input x and nondeterministic input y. We say A computes a function f if (a) when $f(x) = 1$, $\Pr_r[\exists y A(x, r, y) = 1] = 1$ and (b) when $f(x) = 0$, $\Pr_r[\exists y A(x, r, y) = 1] \leq \frac{1}{2}$. Then $\mathbf{AM}$ is the class of languages decidable by such algorithms $A(x, r, y)$ running in time poly$(|x|)$. The equivalence of the two definitions of $\mathbf{AM}$ is due to [30,32,33]. More generally, $\mathbf{AMTIME}(t(n))$ denotes the class of languages decidable by probabilistic nondeterministic algorithms running in time $t(n)$, and $\mathbf{i.o.{-}AMTIME}(t(n))$ is the class of languages decidable by probabilistic nondeterministic algorithms running in time $t(n)$ for *infinitely many input lengths*.

2.3   Hitting Set Generators

A *hitting set generator* (HSG) is a deterministic algorithm $H(1^m, 1^s)$ that outputs a *set* of strings of length m. We say H is *efficient* if its running time is polynomial (in m and s). Hitting set generators are weaker notions of pseudorandom generators.

---

[4] The running time of the generator is still greater than the size of the circuits it fools.

Definition 5 (Hitting set generators). *We say that H is an ϵ-hitting set generator against circuits, if for every circuit $C\colon \{0,1\}^m \to \{0,1\}$ of size at most s, the following holds: If $\Pr[C(U_m) = 1] > \epsilon$, then there exists $y \in H(1^m, 1^s)$ such that $C(y) = 1$.*

One can define analogously hitting set generators against *nondeterministic* and *co-nondeterministic* circuits, and also hitting set generators against nondeterministic and co-nondeterministic *uniform* algorithms. Hitting set generators against co-nondeterministic uniform algorithms will be used only in Section 4.

Note that a pseudorandom generator $G\colon \{0,1\}^\ell \to \{0,1\}^m$ fooling circuits of size s gives rise to a hitting set generator, by taking the set of outputs of G over all seeds. The hitting set generator will be efficient if G is computable in time $\mathrm{poly}(s,m)$ and has logarithmic seed length $\ell = O(\log m + \log s)$. In this sense hitting set generators are weaker than pseudorandom generators. Indeed, hitting set generators can be directly used to derandomize algorithms with one-sided error (i.e. **RP** algorithms), whereas pseudorandom generators can be used to derandomize circuits with two-sided error (**BPP** algorithms). Also note that we allow the hitting set generators to run in greater time than circuits it fools, so they correspond to NW-type generators. Since the error in **AM** proof systems can be made one-sided [33], the existence of an efficient $\frac{1}{2}$-HSG against co-nondeterministic circuits implies that **AM** = **NP**.

The first constructions of efficient HSG (in fact pseudorandom generators) against co-nondeterministic circuits was given by Arvind and Köbler [10]. Their construction was based on the assumption that there are languages in **E** that are hard on average for nondeterministic circuits of size $2^{\Omega(n)}$. Klivans and van Melkebeek [11] gave a construction based on a *worst-case* hardness assumption. Their assumption was the existence of languages in **E** with $2^{\Omega(n)}$ worst-case SAT-oracle circuit complexity, that is circuits with SAT-oracle gates. Miltersen and Vinodchandran [12] managed to relax the hardness condition to nondeterministic circuits (yet only obtained a hitting set generator rather than a pseudorandom generator). We state their main result.

Theorem 6 ([12]). [5] *If there exist a function $f \in \mathbf{E}$ such that $S_{\mathrm{N}}(f) = 2^{\Omega(n)}$, then there exist an efficient $\frac{1}{2}$-HSG against co-nondeterministic circuits.*

Shaltiel and Umans [13] subsequently extended Theorem 5 in two ways: First, they obtained a pseudorandom generator rather than a hitting set generator. Second, they obtained analogous results for quantitatively weaker assumption (e.g., when the $S_{\mathrm{N}}(f)$ is only superpolynomial rather than exponential) yielding correspondingly less efficient generators. However, we will not need these extensions in our paper.

Uniform Hitting Set Generators. Gutfreund, Shaltiel and Ta-Shma [34] extended Theorem 5 to give a hitting set generator against co-nondeterministic *uniform*

---

[5] [12] presented a $(1 - \delta)$-HSG for $\delta = 2^{m^\gamma}/2^m$, but it can be converted into a $\frac{1}{2}$-HSG using dispersers as done implicitly in their paper.

algorithms from *uniform* hardness assumptions. They used the same hitting set generator as Miltersen and Vinodchandran, but proceeded with a better analysis.

Theorem 7 (implicit in [34]). *If* $\mathbf{E} \not\subseteq$ *i.o.*$-\mathbf{AMTIME}(2^{\delta n})$ *for some* $\delta > 0$, *then an efficient* $\frac{1}{2}$-*HSG against co-nondeterministic uniform algorithms exists.*

Since nonuniformity can simulate randomness, the existence of a function $f \in \mathbf{E}$ such that $S_N(f) = 2^{\Omega(n)}$ (assumption of Theorem 5) implies that $\mathbf{E} \not\subseteq$ i.o.$-\mathbf{AMTIME}(2^{\delta n})$ for some $\delta > 0$ (assumption of Theorem 7).

2.4   Discussions

Are the Assumptions Reasonable? Our two results rely on the existence of hitting set generators as constructed in Theorems 5 and 7, which in turn make assumptions about $\mathbf{E}$ containing functions of high nondeterministic complexity. In our opinion, these assumptions are plausible. The two most common reasons to believe a hardness assumption are empirical evidence and philosophical (or structural) considerations. The widely held $\mathbf{P} \neq \mathbf{NP}$ assumption is supported by both. Empirically, much effort has been invested to finding efficient algorithms for $\mathbf{NP}$ problems. Philosophically, it seems unlikely that proofs should always be as easy to find as they are to verify. Other hardness assumptions, such as the hardness of factoring, are supported mainly by empirical evidence. Some, like $\mathbf{E} \not\subseteq \mathbf{NP}$ (equivalently, $\mathbf{EXP} \neq \mathbf{NP}$), are supported mainly by philosophical considerations: it seems unlikely that it should *always* be possible to prove the correctness of exponentially long computations with polynomial-sized proofs. The assumptions of Theorems 5 and 7 are natural strengthenings of this assumption, where we extend $\mathbf{NP}$ both by letting the running time grow from polynomial to subexponential and by allowing nonuniformity or randomization.

How do we find the function f? Once we accept the existence of *some* function $f \in \mathbf{E}$ such that $S_N(f) = 2^{\Omega(n)}$, can we find a *specific* function f satisfying that condition? The answer is yes. It is not hard to show that if there exists a function f satisfying the condition of Theorem 5, then *every* function that is $\mathbf{E}$-complete via linear-time reductions also satisfies that condition. In particular, we can take the bounded halting function $\mathrm{BH}(\cdot)$ defined as follows: $\mathrm{BH}(M, x, t) = 1$ if the Turing machine M outputs 1 on input x after at most t steps (where t is given in binary), and $\mathrm{BH}(M, x, t) = 0$ otherwise.

3   Witness Indistinguishable NP Proofs

In this section we use efficient hitting set generators against co-nondeterministic circuits to derandomize the *ZAP* construction of Dwork and Naor [14] and obtain a *noninteractive witness indistinguishable* (WI) proof system for any language in $\mathbf{NP}$. We call this an "$\mathbf{NP}$ proof system" because it consists of a single message from the prover to the verifier, as is the case in the trivial $\mathbf{NP}$ proof of simply sending the witness to the verifier.

As in the trivial **NP** proof system, our verifier algorithm will be deterministic. However, our prover algorithm will be *probabilistic*. We stress that our proof system is in the *plain model*, without assumptions of a shared random string or nonuniformity. As far as we know, this is the first noninteractive proof system for **NP** in the plain model that satisfies a secrecy property.

3.1   Definitions

Witness Relation. Let $W \subseteq \{0,1\}^* \times \{0,1\}^*$ be a relation. We define $W(x) = \{w \mid (x,w) \in W\}$. We define $L(W) = \{x \mid \exists w \text{ s.t. } (x,w) \in W\}$. If $w \in W(x)$ then we say that w is a *witness* for x. Recall that the class **NP** is the class of languages L such that $L = L(W)$ for a relation W that is decidable in time polynomial in the first input. If $L = L(W)$ is an **NP** language then we say that W is a *witness relation* corresponding to L.

Efficient Provers. Recall the definitions of interactive proofs. Let L be an **NP** language with witness relation W. We say that an interactive proof for L has an *efficient prover* if the prover strategy from the completeness condition can be implemented by an efficient algorithm that when proving that $x \in L$, gets $w \in W(x)$ as an auxiliary input. In this paper we will only be interested in interactive proofs for **NP** that have efficient provers.

NP *Proof Systems.* An **NP** *proof system* is an interactive proof system that is degenerate, in the sense that it consists of only a single message from the prover to the verifier, and that it has a deterministic verifier, and satisfies both perfect completeness and perfect soundness. Because the verifier is deterministic, an **NP** proof system for a language L induces a witness relation W corresponding to L by setting $W(x)$ to contain all the prover messages accepted by the verifier.

Witness Indistinguishability. We recall the notion of witness indistinguishability (WI), as defined by Feige and Shamir [21].

Definition 8 (witness indistinguishability, [21]). *Let L be an **NP** language with witness relation W_L. Let (P, V) be a proof system for L where P is an efficient (probabilistic polynomial-time) prover that gets a witness as auxiliary input.*

We say that (P, V) is witness indistinguishable (WI) if for every nonuniform polynomial-time verifier V^* and every $x \in L$, and for any $w, w' \in W_L(x)$, the view of V^* when interacting with $P(x, w)$ is computationally indistinguishable from its view when interacting with $P(x, w')$.

Feige and Shamir also proved that WI is closed under concurrent composition [21].

ZAPs. A *ZAP* [14] is a two-round public-coin interactive proof system that is witness indistinguishable. Dwork and Naor proved the following theorem.

Theorem 9 ([14]). *If trapdoor permutations[6] exist, then every language in* **NP** *has a ZAP.*

We note that the construction of ZAPs by [14] is actually based on the possibly weaker assumption that NIZK (noninteractive zero-knowledge in the shared random string model) systems exist for every language in **NP**. Thus, our construction can also be based on this possibly weaker assumption.

3.2 Our Result

The main theorem of this section follows.

Theorem 10. *Assume that there exists an efficient $\frac{1}{2}$-HSG against co-nondeterministic circuits and that trapdoor permutations exist. Then every language in* **NP** *has a witness-indistinguishable* **NP** *proof system.*

3.3 Proof of Theorem 10

We prove Theorem 10 by converting the ZAPs for languages in **NP** into WI **NP** proofs. Let L be an **NP** language with witness relation W_L, and let (P, V) be the ZAP for L. We denote the first message in a ZAP (the verifier's random coins sent to the prover) by r and denote the second message (sent by the prover to the verifier) by π. We let $\ell(n)$ denote the length of the verifier's first message in a proof for statements of length n. Let $x \in \{0,1\}^n \setminus L$. We say that $r \in \{0,1\}^{\ell(n)}$ is *sound* with respect to x if there does not exist a prover message π such that the transcript (x, r, π) is accepting. The statistical soundness of the ZAP scheme implies that for every $x \in \{0,1\}^n \setminus L$, the probability that $r \leftarrow \{0,1\}^{\ell(n)}$ is sound with respect to x is very high, and in particular it is larger than $\frac{1}{2}$.

Our construction is based on the following observation. Let $q(n)$ be a polynomial that bounds the running time of the honest ZAP verifier in a proof of statements of length n. For every $x \in \{0,1\}^n \setminus L$, there exists a *co-nondeterministic* circuit C_x of size less than $p(n) < q(n)^2$ that outputs 1 if and only if a string r is sound with respect to x. We stress that the time to verify the soundness of a string r only depends on the running time of the honest verifier (in our case it is $p(n)$).

On input r, the circuit C_x will output 1 if there does not exist a prover message π such that the transcript (x, r, π) is accepting, and 0 otherwise. Note that $\Pr[C_x(U_{\ell(n)}) = 1] > \frac{1}{2}$. Since H is a $\frac{1}{2}$-HSG against co-nondeterministic circuits, we have that for every $x \in \{0,1\}^n \setminus L$, there exists $r \in H(1^{\ell(n)}, 1^{p(n)})$ such that $C_x(r) = 1$. In other words, for every $x \in \{0,1\}^n \setminus L$, there exists a string $r \in H(1^{\ell(n)}, 1^{p(n)})$ such that r is sound with respect to x.

Our construction is as follows.

---

[6] We refer the reader to [31][Sec. 2.4.4] for the definition of trapdoor permutations. Actually, the definition we use is what is called by Goldreich an *enhanced* trapdoor permutation collection. See discussion on [35]. Such a collection is known to exist based on either the RSA or factoring hardness assumptions [36,37].

Protocol 11 (One-message WI NP proof for $L \in$ NP) *On common input $x \in \{0,1\}^n$ and auxiliary input w for the prover, such that $(x, w) \in W_L$, do the following.*

Prover's message
1. *Compute $(r_1, \ldots, r_m) \stackrel{\text{def}}{=} H(1^{\ell(n)}, 1^{p(n)})$.*
2. *Using the auxiliary input (witness) w and the ZAP prover algorithm, compute for every $i \in [1, m]$, a string π_i that is the prover's response to the verifier's message r_i in a ZAP proof for x.*
3. *Send to verifier $(\pi_1, \ldots, \pi_m)$.*

Verifier's Test
1. *Compute $(r_1, \ldots, r_m) \stackrel{\text{def}}{=} H(1^{\ell(n)}, 1^{p(n)})$.*
2. *Given prover's message $(\pi_1, \ldots, \pi_m)$, run the ZAP verifier on the transcript (x, r_i, π_i), for every $i \in [1, m]$.*
3. *Accept if the ZAP verifier accepts all these transcripts.*

Note that Protocol 11 is indeed a one-message system with a deterministic verifier, that satisfies the perfect completeness property. Thus, to prove Theorem 10, we need to prove that it satisfies both the perfect soundness and the witness indistinguishability property.

Lemma 12. *Protocol 11 is a perfectly sound proof system for L.*

Proof. Let $x \notin L$, with $|x| = n$. Since H is a HSG, there exists an $r_i \in H(1^{\ell(n)}, 1^{p(n)})$ that is sound with respect to x. This means that no prover's message π_i will make the ZAP verifier accept the transcript (x, r_i, π_i). Therefore, no string $\pi = (\pi_1, \ldots, \pi_m)$ will make the verifier of Protocol 11 accept. □

Lemma 13. *Protocol 11 is a witness indistinguishable (WI) proof system for L.*

Proof. This follows from the fact that witness indistinguishability is preserved under parallel composition. □

4   Noninteractive Bit Commitment

Bit commitment schemes are basic primitives in cryptography. Informally, a bit commitment scheme is a protocol that consists of two interacting parties, the sender and the receiver. The first step of the protocol involves the sender giving the receiver a commitment to a secret bit b. In the next step, the sender decommits the bit b by revealing a secret key. The commitment alone (without the secret key) must not reveal any information about b. This is called the *hiding* property. In addition, we require that the commitment to b be *binding*, that is the sender should not be able to decommit to a different bit $\bar{b}$. Note that given a bit-commitment scheme, a string-commitment scheme can be obtained by independently committing to the individual bits of the string (cf., [31]).

In an *interactive bit commitment scheme*, the sender and the receiver are allowed to interact during the commitment and decommitment steps. The formal definition of an interactive bit commitment scheme can be found in [31]. Often, however, *noninteractive* bit commitment schemes are preferred or even crucial. For these, a simpler definition can be given.

Definition 14 (noninteractive bit commitment). *A noninteractive bit commitment scheme is a polynomial-time algorithm S which takes a bit $b \in \{0,1\}$ and a random key $K \in \{0,1\}^{\mathrm{poly}(k)}$, where k is the security parameter, and outputs a commitment $C = S(b; K)$. The algorithm S must satisfy the following two conditions:*

1. *(Binding) There do not exist keys K, K' such that $S(0; K) = S(1; K')$.*
2. *(Hiding) The commitments to 0 and 1 are computationally indistinguishable. This means that the probability distributions $\{S(0; K)\}_{K \in \{0,1\}^{\mathrm{poly}(k)}}$ and $\{S(1; K)\}_{K \in \{0,1\}^{\mathrm{poly}(k)}}$ are computationally indistinguishable.*

There is a well known construction by Blum [27] of a noninteractive bit commitment scheme based on any *one-to-one* one-way function (using the function's hard-core predicate [2,28]). Naor [15] gave a construction of an *interactive* bit commitment scheme based on any one-way function (using pseudorandom generators [29]).

4.1   Our Result

The main result of this section is the following theorem.

Theorem 15. *Assume that there exists an efficient $\frac{1}{2}$-HSG against co-nondeterministic* uniform *algorithms and that one-way functions exist. Then there exists a noninteractive bit commitment scheme.*

The first condition is true if $\mathbf{E} \not\subseteq \mathrm{i.o.-}\mathbf{AMTIME}(2^{\Omega(n)})$ (by Theorem 7). We stress that the assumption of efficient $\frac{1}{2}$-HSG against co-nondeterministic *uniform* algorithms is sufficient, even if one wants to obtain a commitment scheme that is secure against *nonuniform* polynomial-sized circuits. However, to get such schemes it will be necessary to assume that the one-way function is secure against *nonuniform* polynomial-sized circuits.

Our result is incomparable to the previous results on bit commitment schemes. Our assumption is stronger than Naor's [15] (which only requires one-way functions), but we obtain a noninteractive commitment rather than an interactive one. Our assumption seems incomparable to assuming the existence of one-to-one one-way functions.

"Raw" Hardness vs. Hardness with Structure. Note that unlike the assumption of existence of *one-to-one* one-way functions, we do not assume in Theorem 15 that there exist a hard function with a particular structure. Rather, we only assume that there exists functions with "raw hardness" (i.e., a one-way function and a function in $\mathbf{E}$ with high $\mathbf{AM}$-complexity).

Even if one is told that one-to-one one-way functions exist, it is necessary to know a *particular* one-to-one one-way function to instantiate Blum's noninteractive commitment scheme. In contrast, we can construct a single noninteractive commitment scheme that is secure as long as there exists a one-way-function and a function $f \in \mathbf{E} \setminus \mathbf{i.o.} - \mathbf{AMTIME}(2^{\delta n})$. This is because we can instantiate our scheme with a universal one-way-function[7] and a function that is $\mathbf{E}$-complete via linear-time reductions such as the function $\mathrm{BH}(\cdot)$ (see discussion in Section 2.4).

4.2   Proof of Theorem 15

Our construction is based on derandomizing Naor's [15] *interactive* bit commitment scheme using a hitting set generator.

Let $G\colon \{0,1\}^k \to \{0,1\}^{3k}$ be BMY-type pseudorandom generator computable in time k^d for some constant d. Such a generator can be constructed based on any one-way function [29]. Naor [15] gave the following protocol for an interactive bit commitment scheme, based on the existence of such a generator.

Protocol 16 (interactive bit commitment scheme [15])
Input to receiver R: 1^k, where k is the security parameter.
Input to sender S: 1^k and a bit $b \in \{0,1\}$.

Commitment stage:

> **Receiver's step** *Select a random $r \leftarrow \{0,1\}^{3k}$ and sends r to S.*
> **Sender's step** *Select a random $s \leftarrow \{0,1\}^k$. If $b = 0$, send $\alpha = G(s)$ to R.*
> *Else, if $b = 1$, send $\alpha = G(s) \oplus r$ to R.*
Decommitment stage: *S reveals s and b. R accepts if $b = 0$ and $\alpha = G(s)$,*
> *or $b = 1$ and $\alpha = G(s) \oplus r$.*

Observe that when the sender commits to 0, the sender's message α is distributed according to $G(U_k)$. When the sender commits to 1, α is distributed according to $G(U_k) \oplus r$. For every $r \in \{0,1\}^{3k}$, the distributions $G(U_k)$ and $G(U_k) \oplus r$ are computationally indistinguishable. This implies that Protocol 16 is hiding. Define a string $r \in \{0,1\}^{3k}$ to be *good* for G if for all $s, s' \in \{0,1\}^k$, we have $G(s) \neq G(s') \oplus r$. Naor [15] showed that the probability that a random r in $\{0,1\}^k$ will be good is very high (e.g., at least $1 - 2^{-k}$). If the receiver selected a good r in the first step of the commitment stage of Protocol 16, then there do not exist $s, s' \in \{0,1\}^k$ such that $G(s) = G(s') \oplus r$, so no commitment α can be opened as both a 0 and 1. Since the probability of selecting a good r is high, Protocol 16 is binding.

Our Noninteractive Bit Commitment Scheme. Observe that the only interaction involved in Protocol 16 is in the receiver sending a random $r \in \{0,1\}^{3k}$ to the sender. However, one can see that the receiver does not have to send a random string, and it is enough to send a *good* string. This is because a good string r will make the distributions $G(U_k)$ and $G(U_k) \oplus r$ disjoint. As

---

[7] A construction of such a function appears in [38] (cf., [31][Sec. 2.4.1]).   It uses the observation that if there exists a one-way-function, then there exists a one-way function that is computable in time n^2.

we show in the proof of Lemma 18, testing whether r is good can be done by a polynomial-time *co-nondeterministic* uniform algorithm. Since the fraction of good r's is large, an efficient HSG against co-nondeterministic algorithms H can be used to select a candidate list of r's such that at least one element $r \in H$ is good. Thus, our protocol will be obtained by running the sender of Naor's protocol on each r in the hitting set. The resulting protocol follows.

Protocol 17 (noninteractive bit commitment scheme)
Input to receiver R: 1^k, where k is the security parameter.
Input to sender S: 1^k and a bit $b \in \{0,1\}$.

Commitment stage:

1. *Compute $r_1, \ldots, r_{p(k)} \stackrel{\text{def}}{=} H(1^{3k}, 1^{3k^d})$.*
2. *Choose $s_1, \ldots, s_{p(k)}$ at random from $\{0,1\}^k$.*
3. *If $b = 0$, send $\alpha = \langle G(s_1), \ldots, G(s_{p(k)}) \rangle$.*
   If $b = 1$, send $\alpha = \langle G(s_1) \oplus r_1, \ldots, G(s_{p(k)}) \oplus r_{p(k)} \rangle$.

Decommitment stage: *S reveals b and $\langle s_1, \ldots, s_{p(k)} \rangle$. R accepts if either of the following holds:*

1. *The bit $b = 0$ and $\alpha = \langle G(s_1), \ldots, G(s_{p(k)}) \rangle$.*
   or
2. *The bit $b = 1$ and $\alpha = \langle G(s_1) \oplus r_1, \ldots, G(s_{p(k)}) \oplus r_{p(k)} \rangle$.*

To show that Protocol 17 constitutes a bit commitment scheme (and hence proving Theorem 15), we first observe that the protocol has the hiding property. This means that the distributions $\langle G(U_k^1), G(U_k^2), \ldots, G(U_k^{p(k)}) \rangle$ and $\langle G(U_k^1) \oplus r_1, G(U_k^2) \oplus r_2, \ldots, G(U_k^{p(k)}) \oplus r_{p(k)} \rangle$ are computationally indistinguishable. This fact can be proved using a standard hybrid argument. The next lemma establishes the binding property.

Lemma 18. *Protocol 17 has the binding property.*

Proof. Define the co-nondeterministic algorithm A such that $A(r) = 1$ if $\forall s, s'$ $G(s) \oplus G(s') \neq r$. Note that $A(r) = 1$ if and only if r is good. Therefore $\Pr[A(U_{3k}) = 1] \geq 1 - 2^{-k} > 1/2$. In addition, the running time of A (on inputs of length k) is bounded by $3k^d$. Hence, there exists an $r_i \in H(1^{3k}, 1^{3k^d})$ such that $\forall s, s'$ $G(s) \oplus G(s') \neq r_i$. Therefore, there do not exist $s_1, \ldots, s_{p(k)}$ and $s_1', \ldots, s_{p(k)}'$ such that $\langle G(s_1), \ldots, G(s_{p(k)}) \rangle = \langle G(s_1') \oplus r_1, \ldots, G(s_{p(k)}') \oplus r_{p(k)} \rangle$. In other words, no commitment α can be opened as both a 0 and 1. Thus, Protocol 17 is perfectly binding. $\square$

4.3 Partially One-to-One One-Way Functions

Another interpretation of our result is as closing the gap between one-to-one and general one-way functions under a non-cryptographic assumption. We say that a function $f : \{0,1\}^* \times \{0,1\}^* \to \{0,1\}^*$ is a *partially one-to-one one-way function* if the value of x is uniquely determined from $f(x, y)$, yet no probabilistic polynomial-time algorithm can recover x from $f(x, y)$ (for random $x, y \xleftarrow{R} \{0, 1\}^k$) except with negligible probability (in k). It can be shown that partially one-to-one one-way functions exist if and only if noninteractive commitment schemes exist. Thus, a restatement of Theorem 15 is the following.

Corollary 19. *Assume that there exists an efficient $\frac{1}{2}$-HSG against co-nondeterministic uniform algorithms. Then one-way functions imply partially one-to-one one-way functions.*

An intriguing question is whether it can be shown that under a similar non-cryptographic assumption, one-way functions imply truly one-to-one one-way functions (rather than just partially one-to-one ones).

Acknowledgments. We thank the anonymous CRYPTO reviewers for helpful comments.

References

1. Blum, M., Micali, S.: How to generate cryptographically strong sequences of pseudo-random bits. SIAM J. Comput. **13** (1984) 850–864
2. Yao, A.C.: Theory and applications of trapdoor functions. In: Proc. 23rd FOCS, IEEE (1982) 80–91
3. Goldreich, O., Goldwasser, S., Micali, S.: How to construct random functions. JACM **33** (1986) 792–807
4. Razborov, A.A., Rudich, S.: Natural proofs. JCSS **55** (1997) 24–35
5. Valiant, L.G.: A theory of the learnable. Commun. ACM **27** (1984) 1134–1142
6. Nisan, N., Wigderson, A.: Hardness vs. randomness. JCSS **49** (1994) 149–167
7. Impagliazzo, R., Wigderson, A.: $P = BPP$ if E requires exponential circuits: Derandomizing the XOR lemma. In: Proc. 29th STOC, ACM (1997) 220–229
8. Trevisan, L.: Extractors and pseudorandom generators. JACM **48** (2001) 860–879
9. Impagliazzo, R., Kabanets, V., Wigderson, A.: In search of an easy witness: Exponential time vs. probabilistic polynomial time. In: Proc. 16th Conf. on Comp. Complexity, IEEE (2001) 2–12
10. Arvind, V., Köbler, J.: On pseudorandomness and resource-bounded measure. Theoret. Comput. Sci. **255** (2001) 205–221
11. Klivans, A.R., van Melkebeek, D.: Graph nonisomorphism has subexponential size proofs unless the polynomial-time hierarchy collapses. SIAM J. Comput. **31** (2002) 1501–1526
12. Miltersen, P.B., Vinodchandran, N.V.: Derandomizing Arthur-Merlin games using hitting sets. In: Proc. 40th FOCS, IEEE (1999) 71–80
13. Shaltiel, R., Umans, C.: Simple extractors for all min-entropies and a new pseudo-random generator. In: Proc. 42nd FOCS, IEEE (2001) 648–657
14. Dwork, C., Naor, M.: Zaps and their applications. In: Proc. 41st FOCS. (2000) 283–293
15. Naor, M.: Bit commitment using pseudorandomness. J. Cryptology **4** (1991) 151–158
16. Rudich, S.: Super-bits, demi-bits, and $N\widetilde{P}$/qpoly-natural proofs. In: Proc. 1st RANDOM, Springer (1997) 85–93

17. Goldwasser, S., Micali, S., Rackoff, C.: The knowledge complexity of interactive proof systems. SIAM J. Comput. **18** (1989) 186–208
18. Goldreich, O., Oren, Y.: Definitions and properties of zero-knowledge proof systems. J. Cryptology **7** (1994) 1–32
19. Blum, M., Feldman, P., Micali, S.: Non-interactive zero-knowledge and its applications (extended abstract). In: Proc. 20th STOC, ACM (1988) 103–112
20. Blum, M., De Santis, A., Micali, S., Persiano, G.: Noninteractive zero-knowledge. SIAM J. Comput. **20** (1991) 1084–1118
21. Feige, U., Shamir, A.: Zero knowledge proofs of knowledge in two rounds. In: Proc. 9th CRYPTO, Springer (1989) 526–545
22. Feige, U., Lapidot, D., Shamir, A.: Multiple non-interactive zero knowledge proofs under general assumptions. SIAM J. Comput. **29** (1999) 1–28
23. De Santis, A., Di Crescenzo, G., Persiano, G.: Randomness-efficient non-interactive zero-knowledge (extended abstract). In: Proc. 24th ICALP, Springer (1997) 716–726
24. De Santis, A., Di Crescenzo, G., Persiano, G.: Non-interactive zero-knowledge: A low-randomness characterization of NP. In: Proc. 26th ICALP, Springer (1999) 271–280
25. De Santis, A., Di Crescenzo, G., Persiano, G.: Randomness-optimal characterization of two NP proof systems. In: Proc. 6th RANDOM, Springer (2002) 179–193
26. Goldreich, O., Micali, S., Wigderson, A.: Proofs that yield nothing but their validity or all languages in NP have zero-knowledge proof systems. JACM **38** (1991) 691–729
27. Blum, M.: Coin flipping by phone. In: 24th IEEE Computer Conference (CompCon). (1982) 133–137
28. Goldreich, O., Levin, L.A.: A hard-core predicate for all one-way functions. In: Proc. 21st STOC, ACM (1989) 25–32
29. Hastad, J., Impagliazzo, R., Levin, L.A., Luby, M.: A pseudorandom generator from any one-way function. SIAM J. Comput. **28** (1999) 1364–1396
30. Babai, L., Moran, S.: Arthur-Merlin games: A randomized proof system and a hierarchy of complexity classes. JCSS **36** (1988) 254–276
31. Goldreich, O.: Foundations of cryptography. Cambridge University Press, Cambridge (2001)
32. Goldwasser, S., Sipser, M.: Private coins versus public coins in interactive proof systems. Advances in Computing Research **5** (1989) 73–90
33. Furer, Goldreich, Mansour, Sipser, Zachos: On completeness and soundness in interactive proof systems. Advances in Computing Research **5** (1989) 429–442
34. Gutreund, D., Shaltiel, R., Ta-Shma, A.: Uniform hardness vs. randomness tradeoffs for Arthur-Merlin games. In: Proc. 18th Conf. on Comp. Complexity, IEEE (2003)
35. Goldreich, O.: Foundations of cryptography : Corrections and additions for volume 1. Available from http://www.wisdom.weizmann.ac.il/~oded/foc-vol1.html\#err (2001)
36. Rivest, R.L., Shamir, A., Adleman, L.: A method for obtaining digital signatures and public key cryptosystems. Commun. ACM **21** (1978) 120–126
37. Rabin, M.: Digitalized signatures and public-key functions as intractable as factorization. Technical Report MIT/LCS/TR-212, Laboratory for Computer Science, Massachusetts Institute of Technology (1979)
38. Levin, L.: One-way functions and pseudorandom generators. Combinatorica **7** (1987) 357–363

On Deniability in the Common Reference String and Random Oracle Model

Rafael Pass

Department of Numerical Analysis and Computer Science
Royal Institute of Technology, Stockholm, Sweden
`rafael@nada.kth.se`

Abstract. We revisit the definitions of zero-knowledge in the Common Reference String (CRS) model and the Random Oracle (RO) model. We argue that even though these definitions syntactically mimic the standard zero-knowledge definition, they loose some of its spirit. In particular, we show that there exist a specific natural security property that is not captured by these definitions. This is the property of *deniability*. We formally define the notion of *deniable zero-knowledge* in these models and investigate the possibility of achieving it. Our results are different for the two models:

- Concerning the CRS model, we rule out the possibility of achieving deniable zero-knowledge protocols in "natural" settings where such protocols cannot already be achieved in plain model.
- In the RO model, on the other hand, we construct an efficient 2-round deniable zero-knowledge argument of knowledge, that preserves both the zero-knowledge property and the proof of knowledge property under concurrent executions (concurrent zero-knowledge and concurrent proof-of knowledge).

1   Introduction

Zero-knowledge proofs, i.e., interactive proofs that yield no other knowledge than the validity of the assertion proved, were introduced by Goldwasser, Micali and Rackoff [26] in 1982. Intuitively, the verifier of a zero-knowledge proof should not be able to do anything it could not have done before the interaction. Knowledge, thus, in this context means the ability to perform a task. The intuition is captured through a simulation definition: We say that a protocol is zero-knowledge if there exists a simulator (that does not have access to a prover) that can simulate a malicious verifier's output after interaction with a prover. The existence of such a simulator implies that if an adversary succeeds in a task after having communicated with a prover, the adversary could just as well have reached the same results without a prover by first running the simulator. This feature has made zero-knowledge a very powerful and useful tool for proving the security of cryptographic protocols.

For some applications, such as signature schemes [18] [39], voting systems, non-interactive zero-knowledge [5] [25], concurrent zero-knowledge [14], [9] etc.,

D. Boneh (Ed.): CRYPTO 2003, LNCS 2729, pp. 316–337, 2003.

it however seems hard, or is even impossible, to achieve efficient and secure schemes in the standard model. Stronger models, such as the Common Reference String (CRS) model [5], where a random string is accessible to the players, or the Random Oracle (RO) model [2], where a random function is accessible through oracle calls to the players, were therefore introduced to handle even those applications. Recently the CRS model has been extensively used in interactive settings to prove universal composability (e.g. [6] [7] [10]).

We note that an important part of the intuition behind zero-knowledge is lost in these two models in a multi-party scenario, if the CRS string or the random oracle may be reused. An easy way of seeing this is simply by noting that non-interactive zero-knowledge proofs are possible in both these model. A player having received a non-interactive proof of an assertion, it could not have proved before the interaction, can definitely do something new: it can simply send the same proof to someone else. This fact may seem a bit counter-intuitive since the intuition tells us that the simulation paradigm should take care of this. We note, however, that the simulator is much "stronger" in these models than in the plain model. As it is, the simulator is allowed to *choose* the CRS string, or random oracle, and this fact jeopardizes the zero-knowledge intuition. In fact the zero-knowledge property in these model only guarantees that the verifier will not be able to do anything *without* referring to the CRS string or the random oracle, it could not have done before. In the non-interactive setting, this problem has lead to the definition of *non-malleable non-interactive zero-knowledge* [37], and very recently *robust non-interactive zero-knowledge* [13]. In this paper we examine the problem in the more general interactive setting.

Deniable Zero-knowledge. In many interactive protocols (e.g. undeniable signatures [11], or deniable authentification [14]) it is essential that the transcript of the interaction does not yield any evidence of the interaction. We say that such protocols are *deniable*. We use the standard simulation paradigm to formalize this notion:

Definition 1. *[Informal meta-definition] A protocol is* deniable *if it is zero-knowledge and the zero-knowledge simulator can be run by the verifier.*[1]

The standard definition of zero-knowledge in the plain model certainly satisfies deniability, however this is *no longer* the case with the definitions of zero-knowledge in the CRS/RO models. This stems from the fact that in the real world the public information in the model, i.e., the CRS string or the random oracle, is fixed once and for all at start-up. When proving security, however, the simulator in these models is allowed to choose this public information in anyway it pleases as long as it "looks" ok. Thus, even though there exists a simulator for a protocol, there is no guarantee that a player can actually simulate a transcript using a certain predefined public information. Non-interactive proofs of a statement x are trivially proofs of an interaction with a party that can prove the

---

[1] Strictly speaking, the simulator is an algorithm and can therefore always be run by the verifier. What we mean here is that the output of the verifier when running this simulator algorithm should be "correctly" distributed.

assertion of the statement x, or else the soundness condition of the proof would be broken.

Indeed, the idea behind the simulation paradigm, and the reason for its widespread applicability, is that a verifier should be able to run the simulator by himself instead of interacting with a prover. The standard definitions of zero-knowledge in the CRS and RO models have not retained this spirit (since the simulator in these model is allowed to choose the public information, which evidently the verifier is not allowed to do), but only syntactically mimic the original zero-knowledge definition.

In the following we give formal definitions of deniable zero-knowledge in the CRS (see section 3) and RO (see section 4) models and investigate the possibility of achieving protocols satisfying the definitions.

When Does Deniability Matter. For some settings zero-knowledge and deniability is the goal (e.g. deniable authentification [14]). In such settings the standard definitions of zero-knowledge in the CRS/RO models clearly are not sufficient, since they do not guarantee deniability.

The issue of deniability also arises when a zero-knowledge protocol is used as a sub-protocol in a larger context where the CRS string or random oracle may be reused. In such a scenario it is no longer clear what security properties are guaranteed by the standard definitions of zero-knowledge in the CRS/RO models. More technically, general protocol composition becomes problematic since the simulator cannot be run when a specific CRS string or random oracle already has been selected.

Nevertheless, we mention that when "plugging-in" zero-knowledge protocols in the CRS/RO models into certain specific protocols, the standard definitions (that do not guarantee deniability) can in some cases be sufficient. For example in the construction of encryption schemes secure against chosen-ciphertext attacks [34], zero-knowledge protocols that do not satisfy deniability have been successfully used as sub-protocols.[2] (Looking ahead, the notion "unreplayability" introduced in section 1.1 is another example where zero-knowledge definitions that do not satisfy deniability can be sufficient).

Implications on the Framework for Universal Composability. A framework for universal composability (UC) was introduced by Canetti in [6]. The idea behind the framework is to put forward security definitions such that the security of a stand-alone component implies the security of a larger system where the component is plugged in, if the outer system is proven secure when having access to an "ideal" component. The UC framework allows for a modular design of cryptographic protocols, which facilitates the design of secure solutions, e.g. [7] [10].

---

[2] We mention that in the more complicated case of encryption schemes secure against *adaptive* chosen-cipher text attacks, the standard definition of zero-knowledge in the CRS model is not sufficient, but needs to be strengthened to guarantee *simulation-soundness*. [37]

The ideal zero-knowledge functionality was first defined in [6] and has later been used in several subsequent works. Due to the impossibility of implementing the ideal zero-knowledge functionailty in the plain model [6], the functionality was implemented in the CRS model [7] [13]. We note that the implementation of [13] is non-interactive, i.e., only a single message is send. Their protocol is, thus, not deniable and therefore constitutes an evidence that the ideal zero-knowledge functionality *does not* capture the concerns for deniability in the framework.

The example given shows the non-triviality of the task of defining ideal functionalities in the UC framework. At a first glance it seemed like the definition given of the ideal zero-knowledge functionality would satisfy deniability. Closer inspection of the framework shows, however, that the concern for transferability/deniability is not taken into account in the framework *when introducing public objects*, such as the CRS string. This can be seen as follows: The UC framework only guarantees security if a CRS string is not reused. A transferability/deniability attack, however, relies on the fact that an honest-party reuses a CRS that has been used in a different execution. In other words, such attacks are not ruled-out by the composition theorem of [6], since they involve honest-parties deviating from their prescribed protocols by reusing a CRS string.

A serious concern is born out of this discussion: Since the zero-knowledge proof functionality is both relatively simple and quite well understood, it should be easy to define an ideal functionality that satisfies the real spirit behind the concept. In particular, the ideal zero-knowledge functionality should be deniable. Given our understanding of the concept of zero-knowledge, the definition of the ideal zero-knowledge functionality given in [6] also seems to be the right one. However, as shown, this definition does not satisfy our expectations in the UC framework. We conclude that, in order to capture the spirit behind natural definitions of ideal functionalities, the introduction of public objects in the UC framework needs to be adapted. See section 3.3 for additional discussions.

1.1   Results Concerning the CRS Model

There could have been hope that the CRS model might be used to implement deniable zero-knowledge protocols in settings where the plain model is not sufficient. We show that in natural settings, where the usage of the CRS model seems meaningful, the demand for deniability makes the CRS model collapse down to the plain model:

- We show that known black-box impossibility result concerning zero-knowledge in the plain model also hold in the CRS model, with respect to deniable zero-knowledge. That is, we show the impossibility of non-trivial deniable black-box zero-knowledge arguments in the CRS model with either of the following extra requirements:
  - straight-line simulatatable (i.e., non-rewinding)
  - non-interactive
  - constant-round strict polynomial-time simulatable
  - constant-round public-coin

- • constant-round concurrent zero-knowledge
- • 3-round
- – We show an efficient transformation from deniable zero-knowledge protocols in CRS model to zero-knowledge protocols in the plain model using small overhead. This result thus rules out the possibility of constructing deniable zero-knowledge protocols in the CRS model that are much more efficient than protocols in the plain model.

Achieving a Weaker Form of Deniability. Although our results rule out the possibility of "interesting" deniable zero-knowledge protocols in many natural settings, we show that a limited form of deniability can be achieved in the CRS model by restricting the communication to a certain class of pre-specified protocols where the CRS string may be reused. Very loosely speaking, we say that a class of protocols is closed under *unreplayability* if an adversary cannot prove anything using a protocol in the class, after having interacted with a prover using a protocol in the class, that it could not have done before interacting with the prover. We show that a natural class of protocols is closed under unreplayability in the CRS model : If C is a class of interactive proofs (or arguments) *of knowledge*, with negligible soundness error, that are zero-knowledge in the CRS model, then C is closed under unreplayabilty. This result shows that restricting the communication to only arguments of knowledge that are zero-knowledge, eliminates the concern for deniability in the CRS model. We postpone these results to the full version of the paper.

1.2 Results Concerning the RO Model

While the results in the CRS model were mostly negative in nature, the situation in the RO model is rather different. Indeed we are able to construct "interesting" deniable zero-knowledge protocols.

More precisely, we show that 2 rounds are necessary and sufficient to construct deniable black-box zero-knowledge arguments for $\mathcal{NP}$ in the RO model. In fact, we construct an efficient 2-round deniable zero-knowledge argument for $\mathcal{NP}$ in the RO model that is both straight-line simulatable and witness extractable. This implies that both simulation of polynomially many concurrent executions (concurrent zero-knowledge) and simultaneous extraction of polynomially many witnesses under concurrent executions (concurrent proof of knowledge) can be performed. It was previously unknown how to simultaneously extract witnesses from polynomially many proofs in the RO model (let alone the question of deniability).

1.3 Other Models

We mention briefly that there are other models that are stronger than the plain model, such as the timing model of [14], or the on-line/off-line model of [35], that do not suffer from problems with deniability. We also note that in a public-key model, methods similar to those of designated verifiers [30] can be used to successfully implement non-trivial zero-knowledge protocols that are deniable. Indeed, the method of designated verifier shows how to convert zero-knowledge protocols that are not deniable into zero-knowledge protocols in a stronger model (namely the public-key model) that satisfy deniability.

1.4   A Computational Separation between the RO and CRS Model

An interesting, and (as far as we know) until now, open question has been to investigate if a plausibility result in the RO model implies a plausibility result in the CRS model. An information theoretical separation between the models follows from the difference in entropy of the random oracle and CRS string. However, the computational case, which is the relevant one when considering cryptographic applications, seems more complicated.

The existence of the powerful tool of pseudo-random functions [21] has shown that in some applications an object with low-entropy (the seed to the pseudo-random function) can be used to "simulate" the behavior of a high-entropy object (namely a random function). It, thus, might seem conceivable that methods of "stretching" randomness could be used to transform protocols in the RO model to protocols in the CRS model that achieve the same task.

A natural candidate to perform such a transformation would be to substitute the random oracle with a (hash) function chosen from a class of function according to the CRS string [2]. However, it was shown by Canetti, Goldreich and Halevi [8] that there exist schemes for which every transformations of this type results in an insecure schemes.

The question of the existence of other (more complicated) transformation has, nevertheless, remained open. A side-effect of our results settles this question by showing a computational separation between the RO model and the CRS model.[3]

In fact, by combining our negative results for the CRS model and the positive results in the RO model, we obtain applications (like for example 2-round deniable black-box zero-knowledge arguments) that can be achieved in the RO model but cannot be achieved in the CRS model.

1.5   Techniques

Although this paper is mostly conceptual in nature, we believe that some of the techniques used in the proofs might be of independent interest.

Tools for Constructing Protocols in the RO Model. In order to construct our 2-round deniable zero-knowledge argument in the RO model we define and construct efficient straight-line extractable (i.e., the extraction can be performed without rewinding) commitments and straight-line witness extractable

---

[3] We note that this is done without resorting to "heavy" machinery like for example the PCP theorem that is needed in [8].

arguments. We mention that the straight-line extraction feature implies two strong properties that were (as far as we know) previously unattained in the RO model:

- **Simultaneous extraction of polynomially many witnesses.** Previous methods to extract witnesses [39] relied on rewinding and could therefore not be used to extract witnesses under concurrent executions.
- **Tight security reductions for non-interactive proofs of knowledge.** Standard extraction techniques for non-interactive proofs of knowledge in the RO model [39] result in "loose" security reductions (see [27] for a discussion).[4] Using straight-line extraction, on the other hand, we obtain a linear and optimal security reduction.

We mention that this technique can be used also for standard zero-knowledge proofs in RO model that do not satisfy the stronger requirement of deniability.

Proofs of Protocol Security without the Simulation Paradigm. In the proof of Lemma 3 (in section 3.1) we show that a parallelized version of Blum's coin-tossing protocol [4] can be used to generate a pseudo-random string. The interesting part of the proof is that we show this *without resorting* to the standard simulation based definition of secure computation [24]. Previously, the only known constant-round coin-tossing protocol for generating a "random" string (and not a bit) is the protocol of Lindell [31] which relies on zero-knowledge proofs and is therefore not practical. (The protocol of Lindell is, however, simulatable). More details can be found in the full version.

1.6   Preliminaries

Due to lack of space in this abstract, we assume familiarity with the following notions: Zero-knowledge in the RO model (see [2]), Zero-knowledge in the CRS model, Witness relations, Commitment schemes, Hard instance ensembles, Witness Indistinguishability (WI), Witness Hiding (WH), Proofs of knowledge (see [19] for definitions), Special soundness (see [12]), Concurrent zero-knowledge (see [20] for a survey). Formal definitions are given in the full paper.

2   ZK in the CRS/RO Model Implies WH and WI

In this section, we show two lemmas concerning the witness hiding (WH) and witness indistinguishable (WI) properties of standard (not deniable) zero-knowledge proofs, or arguments, in the CRS and RO models. Due to lack of space the proofs are omitted and can be found in the full version of the paper.

---

[4] Roughly, in order to break the underlying assumption the "cheating prover" has to be run $O(q)$ times, where q is the running time of the cheating prover, thus resulting in a total running time of $O(q^2)$.

Lemma 1. *Suppose that Π is a zero-knowledge proof (argument), in the CRS/RO model, for the language L. Then, for all witness relations R_L for L, Π is witness hiding in the CRS/RO model.*

Remark 1. The lemma was proven for the plain model in [16].

Lemma 2. *Let the language $L \in \mathcal{NP}$, R_L be a witness relation for L, and Π be a zero-knowledge proof (argument) in the CRS/RO model for L with efficient prover for R_L. Then Π is witness indistinguishable for R_L in the CRS/RO model.*

Remark 2. The lemma was proven for the plain model in [16], and for non-interactive proofs in the CRS model in [15].

We note that in the case of WH, the proof of the lemma is a straight-forward adaptation of the proof in the plain model [16], but concerning WI such a simple adaptation can no longer be done, as was pointed out already for the non-interactive setting using a CRS model in [15]. The problem stems from the fact that WI in the CRS/RO model considers what happens when the prover uses different witnesses, but the *same* CRS string/random oracle.

Thus, although the lemmas show positive results concerning the security of protocols satisfying the standard definition of zero-knowledge in these models, the non-triviality of the adaptation needed in the case of WI, by itself, shows that special care has to be taken in models where the simulator is allowed to choose the public information.

Nevertheless, the essence of Lemma 1 and 2 is that in settings where only WH or WI is required as a security requirement, the standard definitions of zero-knowledge in the CRS or RO model are sufficient. Looking ahead, we will use the WH and WI properties of zero-knowledge proofs in the RO model in the construction of a deniable zero-knowledge protocol in the RO model.

3   On Deniable Zero-Knowledge Proofs in the CRS Model

To be able to obtain deniable zero-knowledge in the CRS model, we restrict the power of the simulator in the definition of zero-knowledge in the CRS model. The key to the problem seems to be the fact that the simulator in the CRS model *chooses* the CRS string. In fact, if the simulator was able to perform a simulation without choosing the CRS string, we would be sure that the verifier had not learnt anything, except the assertion of the statement being proved, even with respect to the CRS string. This leads us to a new zero-knowledge definition.

Definition 2. *We say that an interactive proof (P, V) for the language $L \in \mathcal{NP}$, with the witness relation R_L, is deniable zero-knowledge in the CRS model if for every PPT machine V^* there exists an expected polynomial time probabilistic simulator S such that the following two ensembles are computationally indistinguishable (when the distinguishing gap is a function in $|x|$)*

- $\{(r, \langle P(y_x), V^*(z)\rangle(x,r))\}_{z\in\{0,1\}^*, x\in L}$ *for arbitrary* $y_x \in R_L(x)$
- $\{(r, S(x,z,r)\}_{z\in\{0,1\}^*, x\in L}$

where r is a random variable uniformly distributed in $\{0,1\}^{poly(|x|)}$.
That is, for every probabilistic algorithm D running in time polynomial in the length of its first input, every polynomial p, all sufficiently long $x \in L$, *all* $y_x \in R_L(x)$ *and all auxiliary inputs* $z \in \{0,1\}^*$ *it holds that*

$$|Pr[D(x,z,r,\langle P(y_x), V^*(z)\rangle(x,r))) = 1]$$
$$-Pr[D(x,z,r,S(x,z,r)) = 1]| < \frac{1}{p(|x|)}$$

where r is a random variable uniformly distributed in $\{0,1\}^{poly(|x|)}$.

3.1   On the Impossibility of More Efficient Deniable ZK Protocols

We show that if there exist an interactive deniable zero-knowledge proof (or argument) with negligible soundness error for a language L in the CRS model then there exists an interactive zero-knowledge proof (or argument) with negligible soundness error for L in the plain model using essentially the same communication complexity. In fact, we show a general transformation that only uses an overhead of twice the length of the CRS string plus the length of a statistically binding commitment to a string of the same length as the CRS string.

The construction. Suppose that protocol Π is an interactive deniable zero-knowledge proof (or argument) with negligible soundness error, for the language L, in the CRS model. Suppose further that a CRS string of length $p(n)$, where $p(n)$ is a polynomial, is used for proving membership of instances in L of size n, using Π. Now consider the protocol Π' in the plain model (without a CRS string), for proving membership of instances in L of length n, constructed by simply adding a coin-tossing phase to the protocol Π:

Protocol Π'
Phase one:
P $\rightarrow$ V: Commits, using a statistically binding commitment scheme, that is non-uniformly computationally hiding, to a random string of length $p(n)$.
V $\rightarrow$ P: Sends a random string of length $p(n)$.
P $\rightarrow$ V: Opens up the commitment.
Phase two:
P $\leftrightarrow$ V: Both parties thereafter use the XOR of the strings as a CRS string and execute the protocol Π.

In the full version of the paper we show the following lemma:

Lemma 3. *If Π is an interactive deniable zero-knowledge proof (or argument) with negligible soundness error, for the language L, in the CRS model, then the protocol Π', resulting from the above transformation, is an interactive zero-knowledge proof (or argument) with negligible soundness error for the language L.*

Remark 3. The existence of statistically binding commitment schemes that are non-uniformly computationally hiding is implied by the existence of non-uniform one-way functions by combining the results of [29] and [33].

Remark 4. We note that we do not show that the coin-tossing protocol in phase one is simulatable. Indeed, for our construction to work we simply have to show that the output of the coin-tossing is pseudo-random.

This result, thus, rules out the possibility of finding deniable zero-knowledge protocols that can be implemented much more efficiently in the CRS model than in the plain model.

3.2 On the Impossibility of "Non-trivial" Deniable ZK Protocols

In this section we show that known black-box impossibility result concerning zero-knowledge in the plain model also hold in the CRS model with respect to deniable zero-knowledge. That is we show that for known settings where it seems interesting to resort to the CRS model the demand for deniability makes the CRS model collapse down to the plain model. (The proofs that are left out are given in the full version).

Theorem 1. *If Π is a straight-line black-box simulatable deniable zero-knowledge proof (or argument), in the CRS model, for the language L with negligible soundness error, then $L \in \mathcal{BPP}$.*

We continue with two impossibility results that follow from Lemma 3:

Theorem 2. *Assume the existence of statistically binding commitment schemes that are non-uniformly computationally hiding. If Π is a constant-round strict polynomial-time black-box simulatable deniable zero-knowledge proof (or argument) with negligible soundness in the CRS model for the language L, then $L \in \mathcal{BPP}$.*

Theorem 3. *Assume the existence of statistically binding commitment schemes that are non-uniformly computationally hiding. If Π is a constant-round black-box simulatable public-coin deniable zero-knowledge proof (or argument) with negligible soundness in the CRS model for the language L, then $L \in \mathcal{BPP}$.*

Proof. It is clear from the construction that the transformation in section 3.1 preserves the public-coin property of the protocol. Now, since Goldreich and Krawczyk [22] have shown the impossibility of non-trivial constant-round black-box public-coin zero-knowledge arguments, $L \in \mathcal{BPP}$. $\square$

As a sanity check to the definition we also note the impossibility of non-trivial non-interactive zero-knowledge arguments,

Theorem 4. *If Π is a non-interactive deniable zero-knowledge argument, in the CRS model, for the language L with negligible soundness error, then $L \in \mathcal{BPP}$.*

Proof. Follows directly from Theorem 1 since non-interactive arguments need to be black-box straight-line simulatable. □

Indeed, non-interactive proofs are the most obvious violation of deniability, in the CRS model, since they can be passed on.

Goldreich-Krawczyk Reductions. In 1990, Goldreich and Krawczyk [22] showed that if a language L has an interactive zero-knowledge argument with negligible soundness, using less than 4 rounds, with a blackbox simulator, then $L \in \mathcal{BPP}$. The method of Goldreich-Krawczyk has later been used to show black-box impossibility results in the case of constant-round concurrent zero-knowledge [9], and very recently in the case of strict polynomial time simulatable zero-knowledge [1]. On a high-level, the Goldreich-Krawczyk method is a constructive reduction from a machine deciding the language L to a simulator of the zero-knowledge argument. That is, the existence of a simulator implies the existence of a machine deciding the language, which in turn implies that the language is in $\mathcal{BPP}$.

Indeed, since the reduction is black-box and constructive, the same reduction can be used for protocols that are deniable zero-knowledge in the CRS model. The machine deciding the language, would simply first choose a random string and thereafter run the original deciding machine using the random string as a CRS string. Careful examination of the proofs of [22] and [9] thus gives:

Theorem 5. *If Π is a 3-round black-box simulatable deniable zero-knowledge proof (or argument) in the CRS model, for the language L, with negligible soundness error, then $L \in \mathcal{BPP}$.*

Theorem 6. *If Π is a constant-round black-box simulatable deniable concurrent zero-knowledge argument in the CRS model, for the language L, with negligible soundness error, then $L \in \mathcal{BPP}$.*

3.3   Conclusions and Directions for Future Research

We have shown that for currently known settings, the CRS model cannot be used to implement deniable black-box zero-knowledge protocols for languages in $\mathcal{NP}$, that cannot already be implemented in the plain model. In the full version of the paper we, nevertheless, show that a limited form of deniability (called *unreplayability*) can be achieved by restricting the communication of honest-parties to a certain class of protocols (see section 1.1).

Concerning the UC framework [6], we have shown that the ideal zero-knowledge functionality is not deniable. Thus, in order to be able to model universally composable deniable zero-knowledge, either a new definition has to be given or the incorporation of public objects in the framework modified.

A possible approach would be to only allow composition with ideal functionalities that physically cannot be reused, thus ruling out the use of the ideal CRS functionality and other functionalities that model public information. However, since the plain model is too weak to construct even universally composable commitment [7], some extra set-up assumptions need to be incorporated into the security definitions, in such a way that the simulator can be run by the parties themselves. For example, if incorporating a CRS string in the framework, the simulator should be able to carry out the simulation for all but a negligible fraction of CRS strings, in analogy with Definition 2. We note, however, that Theorem 1, which states the impossibility of straight-line simulatable deniable zero-knowledge in the CRS model, yields the impossibility of universally composable zero-knowledge in this setting, since a protocol implementing the ideal zero-knowledge functionality must have a straight-line simulator.[5] On the other hand, if incorporating a public-key infrastructure in the framework, methods similar to those of designated verifier [30] could possibly be used to achieve universally composable deniable zero-knowledge.

An altogether different approach was taken in [32] [36] where it is shown how to realize the ideal zero-knowledge functionality *without* resorting to set-up assumptions (such as a CRS string), by trading universal composability for the weaker notion of concurrent composability.

Open Problems. An interesting open problem is to find a type of deniable zero-knowledge protocol that can be achieved in the CRS but not in the plain model. Since most of our results only apply in the black-box setting, a direction would be to investigate the non-black-box setting.

4   On Deniable Zero-Knowledge Proofs in the RO Model

As in the CRS model, in order to obtain interactive proofs and arguments, with random oracles, that capture the spirit of zero-knowledge, we need to resort to a weaker simulation model, where the simulator no longer is allowed to choose the random oracle, but should be able to perform the simulation for all but a negligible fraction of random oracles. Such a simulator can therefore be run by a verifier, assuring that the intuitive interpretation of zero-knowledge holds, i.e., that the verifier cannot do anything except to assert the validity of the statement proved, that it could not have done before the interaction with a prover.

Definition 3. *We say that an interactive proof (P, V) for the language $L \in \mathcal{NP}$, with witness relation R_L, is* deniable zero-knowledge in the RO model *if for every*

---

[5] For those familiar with the UC framework, this is due to the fact that the environment cannot be rewound. Now, supposing a real-life adversary that simply forwards messages between the environment and the simulator, shows that the simulator needs to be straight-line. More details in [6].

PPT verifier V^* there exists an expected polynomial time probabilistic simulator S such that the following two ensembles are computationally indistinguishable (when the distinguishing gap is a function in $|x|$):

- $\{(RO, \langle P^{RO}(y_x), V^{*RO}(z)\rangle(x))\}_{z \in \{0,1\}^*, x \in L}$ *for arbitrary* $y_x \in R_L(x)$
- $\{RO, S^{RO}(z, x))\}_{z \in \{0,1\}^*, x \in L}$

where $RO : \{0,1\}^{poly(|x|)} \to \{0,1\}^{poly(|x|)}$ is a uniformly distributed random variable.

That is, for every probabilistic algorithm D running in time polynomial in the length of its first input, every polynomial p, all sufficiently long $x \in L$, all $y_x \in R_L(x)$ and all auxiliary inputs $z \in \{0,1\}^*$ it holds that

$$|Pr[D^{RO}(x, z, \langle P^{RO}(y_x), V^{*RO}(z)\rangle(x))) = 1]$$
$$-Pr[D^{RO}(x, z, S^{RO}(x, z)) = 1]| < \frac{1}{p(|x|)}$$

where $RO : \{0,1\}^{poly(|x|)} \to \{0,1\}^{poly(|x|)}$ is a uniformly distributed random variable.

We note that when proving security according to the standard zero-knowledge definition in the RO model, the simulator has two advantages over a plain model simulator, namely,

- The simulator can see what values parties query the oracle on.
- The simulator can answer these queries in whatever way it chooses as long as the answers "look" ok.

The definition of deniable zero-knowledge in the RO model restricts the power of the simulator and *only* allows it to see on what value the parties query the oracle (thus out of the two advantages only the first remains). This is due to the fact that in the definition of deniable zero-knowledge in the RO model, the distinguisher is given access to the random oracle and can thus verify if the simulator has answered the oracle queries in accordance to the pre-specified oracle. We, however, use this first property in a novel fashion, and show that it alone is an extremely powerful tool. Looking ahead, we use the random oracle to construct commitment schemes where the simulator, gaining access to all oracle calls, will be able to extract the committed values, without rewinding the committer.

As a sanity check to the definition we start by noting: (proof is given in the full version)

Theorem 7. *If Π is a one-round deniable zero-knowledge argument, in the RO model, for the language $L \in \mathcal{NP}$ with negligible soundness error, then $L \in \mathcal{BPP}$.*

On the positive side we show that 2 rounds are necessary to construct efficient and "robust" deniable zero-knowledge protocols for $\mathcal{NP}$. In fact we construct a protocol that is both concurrent zero-knowledge and concurrent proof of knowledge through a transformation from any special-sound honest-verifier zero-knowledge (HVZK) public-coin argument. We here briefly outline the construction.

Outline of the Construction of 2-round Deniable ZK Arguments. On a very high level the protocol follows the paradigm of Feige-Shamir [17]. The verifier start by sending a "challenge" and a witness hiding proof of knowledge of the answer to the challenge, to the prover. The prover thereafter shows using a WI argument that either it has a witness for the statement it wishes to prove or that it has the answer to the challenge.

The difficulty in constructing such a protocol relies in the fact that each of these steps must be implemented in a single message.[6]

The main technical ingredient that allows us to achieve this goal is the introduction of straight-line extractable commitments in the RO model (see section 4.1). On a high level, these are commitments where the value committed to can be extracted by a simulator without the use of rewinding techniques. We construct such commitment schemes by letting the committer use the random oracle to commit. It follows from the random properties of the oracle that the committer, in order to succeed in opening a commitment must have applied the oracle on it, which means that by simply observing all the queries the adversary makes, the committed values can be extracted without rewinding.

Having established this powerful tool, in section 4.2 we construct a one-round straight-line witness extractable zero-knowledge arguments for Graph-3-Coloring in the RO model, by implementing the commitment scheme in the GMW protocol [23] with straight-line extractable commitments and thereafter applying the Fiat-Shamir transformation [18] [2] to "collapse" it down to a one-round zero-knowledge argument in the RO model (see Lemma 6). Straight-line witness extraction here means that a witness to the statement proved can be extracted without rewinding the prover. Lemma 1 and 2 can now be applied to show that the one-round protocol, which is zero-knowledge in the RO model, is both WH and WI in the RO model.

In order to achieve an efficient protocol, in Lemma 7, we show how to construct a WH and WI one-round straight-line witness extractable argument from any special-sound HVZK public-coin argument. Essentially this is done by transforming the special-sound HVZK argument into a cut-and-choose argument and thereafter applying the same transformation as was done for Graph-3-Coloring.

In section 4.3 we finally put everything together to achieve the 2-round deniable zero-knowledge argument (see Theorem 8). We here rely on the efficient OR transformation of [12] to implement the second message of the protocol.

We mention that some technical problems related to the malleability of the commitments arise in the security proof. Nevertheless, since we have access to

---

[6] Technically, it is actually sufficient that the first step is implemented with a single message. The second step could conceivable be implemented using 2 rounds (see [35]). Nevertheless, our construction implements both steps using one-round solutions.

a random oracle these problems can be resolved in a rather straightforward manner.

4.1   Straight-Line Extractable Commitments

We construct efficient commitment schemes with strong properties, without allowing the simulator to choose the random oracle. We start by defining the notion of straight-line extractable commitments schemes in the RO model. For simplicity we only state the definition for non-interactive commitment schemes.

Definition 4. *Let a PPT committer C commit to a string using a non-interactive commitment scheme, sending c to the receiver, where $|c| = poly(n)$. We say that the non-interactive commitment scheme is straight-line extractable in the RO model if there exists a PPT extractor machine E such that for all c, if C succeeds in decommitting to x with non-negligible probability, then $E(c, l) = x$ with overwhelming probability, where l is a list of all the random oracle queries and answers performed by C during and before the commit phase.*

Remark 5. We note that the extractor E is not given access to the random oracle, but instead receives both the queries and the answers to those queries.

When having access to a random oracle it is easy to construct efficient commitment schemes that are straight-line extractable. Let l be a super-logarithmic polynomially bounded function, i.e., $\omega(log(n)) \leq l(n) \leq poly(n)$, and $RO : \{0,1\}^{2n} \to \{0,1\}^{l(n)}$ be a random oracle. Consider the following commitment scheme:

SLCom
Commit phase (A sends a commitment, to $x \in \{0,1\}^n$, to B)
A randomly picks $r \in \{0,1\}^n$
$A \to B : c = RO(x, r)$

Reveal phase
$A \to B : x, r$
B checks that $c = RO(x, r)$

Lemma 4. *SLCom is a straight-line extractable non-interactive commitment scheme in the RO model.*

Proof. The hiding and binding properties are proven in the full version.
Straight-line extraction follows: The extractor simply goes through the list $\{(x_i, r_i), c_i\}_{i=1..poly(n)}$ and checks if there is an i such that $c_i = c$. If so it returns x_i, and otherwise nothing. Since the committer, in order to succeed in opening the commitment with probability that is non-negligible, must have used the random oracle on the value it committed to, the extractor always succeeds if the committer succeeds with probability that is non-negligible. (A cheating

committer that has not used the random oracle on the value committed to has a probability of $\frac{T(n)}{2^{l(n)}}$, where $T(n)$ is the number of oracle calls during the decommit phase, of decommitting.) □

In fact, SLCom can be used either as a statistically binding or statistically hiding commitment scheme depending on the parameter l: (proof is given in the full version)

Lemma 5. *If $l(n) = 4n$ then SLCom is a statistically binding non-interactive commitment scheme in the RO model. If $l(n) = n/8$ then SLCom is a statistically hiding non-interactive commitment scheme in the RO model.*

Extractable Commitments with Oracle Restrained to a Prefix. To be able to construct multiple commitments that are non-malleable with respect to each other we generalize the notion of straight-line extractability. We say that a commitment scheme in the RO model is *straight-line extractable with oracle restrained to the prefix s* if the commitment scheme is straight-line extractable and there exists an extractor that succeeds in extracting witnesses using only oracle queries that begin with the prefix s. We will in the following let different parties use different prefixes allowing for individual extraction of the committed values.

We note that SLCom can be changed in a straight-forward manner to become straight-line extractable with oracle restrained to the prefix s, by simply concatenating the string s to the oracle queries, i.e., $RO(s, x, r)$ becomes a commitment to the string x, where $RO : \{0, 1\}^{2n+|s|} \to \{0, 1\}^{l(n)}$.

4.2   Straight-Line Witness Extractable Proofs

All previously known proofs of knowledge in the RO model (e.g. [39]) relied on rewinding and could therefore not be applied to simultaneously extract polynomially many witnesses. We introduce a stronger notion of proofs of knowledge, namely proofs where witnesses can be extracted without rewinding the prover. More formally,

Definition 5. *We say that an interactive proof with negligible soundness (P, V) for the language $L \in \mathcal{NP}$, with the witness relation R_L, is straight-line witness extractable in the RO model if for every PPT machine P^* there exists a PPT witness extractor machine E such that for all $x \in L$, all $y, r \in \{0, 1\}^*$, if $P^*_{x,y,r}$ convinces the honest verifier with non-negligible probability, on common input x, then $E(\text{view}_V[(P^*_{x,y,r}, V(x))], l) \in R_L(x)$ with overwhelming probability, where $P^*_{x,y,r}$ denotes the machine P^* with common input fixed to x, auxiliary input fixed to y and random tape fixed to r, $\text{view}_V[(P^*_{x,y,r}, V(x))]$ is V's view including its random tape, when interacting with $P^*_{x,y,r}$, and l is a list of all oracle queries and answers posed by $P^*_{x,y,r}$ and V.*

We show two constructions to achieve efficient straight-line witness extractable arguments in the RO model. First, we show how the GMW [23] protocol for proving the existence of a 3 coloring to a graph directly can be turned into a straight-line witness extractable, WH and WI, one-round argument in the RO model, by applying the Fiat-Shamir transformation [18] to "collapse" it down to one round, and using straight-line extractable commitments. Secondly we show how to transform any three round special-sound HVZK public-coin argument into a straight-line witness extractable, WH and WI, one-round argument. The second construction is of interest as it allows us to construct efficient protocols without going through Cook's transformation.

An Argument System for Graph-3-Coloring. We start off with the three round protocol of GMW (Goldreich, Micali, Widgerson) [23]:

Protocol Π (GMW's Graph 3-coloring proof):
Common input: a directed graph $G = (V_G, E_G)$, with $n = |V_G|$
Auxiliary input to the prover: a 3-coloring of G, $c_0, c_1, .., c_n \in \{1, 2, 3\}$.

P uniformly chooses a permutation π over 1,2,3.
P $\to$ V: Commits to $\pi(c_0), \pi(c_1), .., \pi(c_n)$ using any statistically binding commitment scheme.
V $\to$ P: Uniformly selects an edge $(i, j) \in E$.
P $\to$ V: Reveals c_i, c_j.
V checks that c_i and c_j are different colors.

As is shown in [2] the protocol can be collapsed down to a one-round zero-knowledge argument, Π', in the RO model by running $t = 2n * |E_G|$ parallel versions of the protocol and applying the random oracle to all the t first messages, to "simulate" the honest verifier. This transformation is called the Fiat-Shamir transformation [18].

Protocol Π':
P $\to$ V: $a' = a'_1, a'_2, .., a'_t$, $c' = c'_1, c'_2, .., c'_t$.
V checks that for all $1 \leq i \leq t$, $(a'_i, RO(a')_i, c'_i)$ is an accepting execution of the protocol Π, where $RO(a')_i$ signifies the i'th part of the random oracle's reply, such that each part has the appropriate size of the verifier's challenge in protocol Π.

Since Π' is zero-knowledge in the RO model it is, by Lemma 1 and 2, also WH and WI. Now, if the commitment scheme chosen has the property of being straight-line extractable, the resulting protocol is straight-line witness extractable. (proof is given in the full version)

Lemma 6. *If the protocol Π' is instantiated with a straight-line extractable commitment scheme the resulting protocol is straight-line witness extractable, witness hiding and witness indistinguishable in the RO model.*

A Transformation from HVZK Protocols. Suppose $\Pi = (a, b, c)$ is a three round special-sound HVZK public-coin argument for the language $L \in \mathcal{NP}$.

In order to achieve a one-round witness extractable, WH and WI argument for L we transform the protocol Π into a cut-and-choose protocol Π' and thereafter use the same transformation as was done in the case of the proof of Graph-3-Coloring. Consider the following protocol:

Protocol Π':
P → V: a, two different random numbers $b_0, b_1 \in B$, commitments to c_0, and c_1 where c_i is the answer to the query b_i with a as first message in the protocol Π
V → P: chooses q randomly from $\{0, 1\}$
P → V: Decommits to c_q
V checks that (a, b_q, c_q) is a consistent execution of the protocol Π

Now, let Π'' be the protocol obtained after applying the Fiat-Shamir transformation on Π', i.e., running $2n$ versions of the protocol in parallel, and simulating the verifier's challenge by applying the random oracle to the first message:

Protocol Π'':
P → V: $a' = a_1', a_2', .., a_t'$, $c' = c_1', c_2', .., c_t'$.
V checks that for all $1 \leq i \leq 2n$, $(a_i', RO(a')_i, c_i')$ is an accepting execution of the protocol Π', where $RO(a')_i$ signifies the i'th bit of the random oracle's reply.

In the full version of the paper we show,

Lemma 7. *If the protocol Π'' is instantiated with a straight-line extractable commitment scheme the resulting protocol is a straight-line witness extractable, witness hiding and witness indistinguishable argument for L in the RO model.*

Witness Extraction by an Oracle Restrained to a Prefix. As with the commitments schemes, the above mentioned protocols can easily be turned into arguments that are witness extractable by an oracle restrained to a certain prefix, by using commitment schemes that are straight-line witness extractable by oracle restrained to the prefix.

4.3   Deniable Concurrent Zero-Knowledge Proofs of Knowledge

In this section we use the witness extractable, WH and WI, one-round arguments in a way similar to the Feige-Shamir construction [17] to construct a 2-round straight-line simulatable deniable zero-knowledge argument of knowledge for $\mathcal{NP}$ in the RO model. Since the protocol is straight-line simulatable it is also deniable concurrent zero-knowledge:

Theorem 8. *Assuming the existence of polynomially computable one-way functions, there exists a two round deniable black-box concurrent zero-knowledge argument for languages in $\mathcal{NP}$ in the RO model. Furthermore the argument is both straight-line witness extractable, and straight-line simulatable.*

Proof. Let $f : \{0,1\}^n \rightarrow \{0,1\}^{poly(n)}$ be a one-way function, and let Π' be a special-sound HVZK public-coin argument for proving the knowledge of a pre-image to f. Such argument systems exists for every one-way function, by reducing the one-way function to an instance of the graph hamiltonicity problem, using Cook's theorem, and thereafter using Blum's protocol [3]. We emphasize, however, that if a specific one-way function is used, the HVZK argument can be tailored for the function to get an efficient implementation. Examples of such protocols are the Guillou-Quisquater scheme [28] for the RSA function, and the Schnorr scheme [38] for the discrete logarithm.

Let the witness relation $R_{L'}$, where $(x,y) \in R_{L'}$ if $f(x) = y$, characterize the language L'.

Let $RO : \{0,1\}^{poly(n)} \rightarrow \{0,1\}^{poly(n)}$ be a random oracle, and the language $L \in \mathcal{NP}$. Consider the following protocol for proving that $x \in L$:

Protocol SLZK
V chooses a random number $r \in \{0,1\}^n$.
V $\rightarrow$ P: $c = f(r)$, a one-round WH, straight-line witness extractable, by oracle restrained to prefix "0", argument of the statement "$\exists r'$ s.t $c = f(r')$" for the witness relation $R_{L'}$.
P $\rightarrow$ V: a one-round WI, straight-line witness extractable, by oracle restrained to prefix "1", argument of the statement "$\exists r'$ s.t $c = f(r') \vee x \in L$" for the witness relation $R_{L \vee L'}(c,x) = \{(r',w)|r' \in R_{L'}(c) \vee w \in R_L(x)\}$.

To implement the first message, we use the transformation described in section 4.2 to turn Π', i.e., the special-sound HVZK zero-knowledge argument for L', into the needed one-round argument for L'.
The second message is implemented as follows: Assuming that we have a special-sound HVZK public-coin argument for L, we can use the efficient OR transformation in [12] to yield a special-sound HVZK public-coin argument for $L \vee L'$ and the witness relation $R_{L \vee L'}$.[7] We can thereafter apply the transformation in section 4.2.

Completeness of the protocol is clear. In order to prove soundness, we start by noting that the prover sends an argument that is straight-line witness extractable by oracle restrained to prefix "1". But since the honest verifier has not used the oracle with prefix "1", a witness can be extracted using only the prover's oracle queries. If a malicious prover succeeds in convincing the honest verifier, he must thus have either an r' s.t $c = f(r')$ or a witness for $x \in L$. We will show that the prover needs to have a witness for x: Let the probability ensemble U be uniform on $\{0,1\}^n$, and let $X = f(U)$ be a probability ensemble for the language L'. Then since f is a one-way function, X is a hard instance ensemble. Now, if the prover, after having received the verifier's first message was able to a find a witness to a randomly chosen instance in the hard-instance

---

[7] The resulting argument uses less communication than the argument for L plus the argument for L'.

ensemble X, this would violate the witness hiding property of the verifier's message. The claim that the prover must have a witness for x follows. The protocol is thus straight-line witness extractable for the statement $x \in L$. Soundness follows automatically.

Straight-line zero-knowledge: The simulator simply extracts r from the verifier's first message and then uses it as a "fake" witness to send its proof. Since the prover's message is a WI argument, the simulator's output is indistinguishable from the honest prover's. □

Remark 6. We note that since the protocol is straight-line witness extractable it is also witness extractable under concurrent executions, i.e., witnesses to all concurrent executions can be simultaneously extracted. Indeed, this feature is of great importance in, for example, authentification schemes. We note that it was previously unknown how to simultaneously extract witnesses from polynomially many proofs in the RO model.

Remark 7. Even though we have access to a random oracle we need to rely on the existence of one-way functions since our protocol uses the one-way function in a non-blackbox way. In fact, we either apply Cook's transformation on the function, or use specially tailored protocols for specific one-way functions.

A Note on the Efficiency of SLZK. Although the protocol SLZK is constructed through an efficient transformation from any special-sound HVZK argument, the transformation turns the HVZK protocol into a cut-and-choose protocol inducing a blow up in communication complexity of n. In the full version of the paper we also show the existence of a more efficient protocol considering communication complexity, which in particular is not cut-and-choose. The protocol that is also a proof and not an argument, as SLZK, however uses four rounds instead of the optimal two.

4.4   Conclusions and Directions for Future Research

We have shown the facility of constructing efficient and powerful protocols that are deniable zero-knowledge in the RO model.

Open problems. The most urgent open problem is to find a more efficient construction of one-round witness extractable arguments that do not rely on cut-and-choose techniques. Secondly, our 2-round protocol relies on the existence of one-way functions, while our 4-round protocol (given in the full version) does not. We wonder if it is possible to construct 2-round straight-line simulatable deniable zero-knowledge protocols without any further assumptions than the random oracle.

Acknowledgments. First, I wish to thank Johan Håstad for his invaluable help and comments. I am also very grateful to Ran Canetti for helpful discussions. Thanks also to Shafi Goldwasser, Tal Rabin, Alon Rosen, Victor Shoup and the anonymous referees for helpful comments.

References

1. B. Barak and Y. Lindell. Strict Polynomial-Time in Simulation and Extraction. In *34th STOC*, pages 484–493, 2002.
2. M. Bellare and P. Rogaway. Random Oracles are Practical: A Paradigm for Designing Efficient Protocols. In *1st ACM Conf. on Computer and Communications Security*, pages 62–73, 1993.
3. M. Blum. How to prove a Theorem So No One Else Can Claim It. *Proc. of the International Congress of Mathematicians,* Berekeley, California, USA, pages 1444–1451, 1986.
4. M. Blum. Coin Flipping by Telephone. In *Crypto81*, ECE Report 82-04, ECE Dept., UCSB, pages 11–15, 1982
5. M. Blum, P. Feldman and S. Micali. Non-Interactive Zero-Knowledge and Its Applications. In *20th STOC*, pages 103–112, 1988
6. R. Canetti. Universally Composable Security: A New Paradigm for Cryptographic Protocols. In *34th STOC*, pages 494–503, 2002.
7. R. Canetti and M. Fischlin. Universally Composable Commitments. In *Crypto2001*, Springer LNCS 2139, pages 19–40, 2001.
8. R. Canetti, O. Goldreich and S. Halevi. The Random Oracle Methodology, Revisited. In *30th STOC*, pages 209–218, 1998
9. R. Canetti, J. Kilian, E. Petrank and A. Rosen. Black-Box Concurrent Zero-Knowledge Requires (almost) Logarithmically Many Rounds. *SIAM Jour. on Computing*, Vol. 32(1), pages 1–47, 2002.
10. R. Canetti, Y. Lindell, R. Ostrovsky and A. Sahai. Universally Composable Two-Party and Multy-Party Computation. In *34th STOC*, pages 494–503,2002.
11. D. Chaum and H. van Antwerpen. Undeniable Signatures. In *Crypto89*, Springer LNCS 435, pages. 212–216, 1989.
12. R. Cramer, I. Damgård and B. Schoenmakers. Proofs of Partial Knowledge and Simplified Design of Witness Hiding Protocols. In *Crypto94*, Springer LNCS 839, pages. 174–187, 1994
13. A. De Santis, G. Di Crescenzo, R. Ostrovsky, G. Persiano and A. Sahai. Robust Non-interactive Zero Knowledge. In *Crypto2001*, Springer LNCS 2139, pages 566–598, 2001.
14. C. Dwork, M. Naor and A. Sahai. Concurrent Zero-Knowledge. In *30th STOC*, pages 409–418, 1998.
15. U. Feige, D. Lapidot and A. Shamir. Multiple Noninteractive Zero Knowledge Proofs under General Assumptions. *Siam Jour. on Computing 1999*, Vol. 29(1), pages 1–28.
16. U. Feige and A. Shamir. Witness Indistinguishability and Witness Hiding Protocols. In *22nd STOC*, pages 416–426, 1990.
17. U. Feige and A. Shamir. Zero Knowledge Proofs of Knowledge in Two Rounds. In *Crypto89*, Springer LNCS 435, pages. 526–544, 1989.
18. A. Fiat and A. Shamir. How to Prove Yourself: Practical Solutions to Identification and Signature Problems. In *Crypto86*, Springer LNCS 263, pages 181–187, 1987

19. O. Goldreich. *Foundations of Cryptography – Basic Tools*. Cambridge University Press, 2001.
20. O. Goldreich. Zero-knowledge twenty years after their invention. Weizmann Institute, 2002.
21. O. Goldreich, S. Goldwasser and S. Micali. How to Construct Random Functions. *JACM*, Vol. 33(4), pages 210–217, 1986.
22. O. Goldreich and H. Krawczyk. On the Composition of Zero-Knowledge Proof Systems. *SIAM Jour. on Computing*, Vol. 25(1), pages 169–192, 1996.
23. O. Goldreich, S. Micali and A. Wigderson. Proofs that Yield Nothing But Their Validity or All Languages in NP Have Zero-Knowledge Proof Systems. *JACM*, Vol. 38(1), pp. 691–729, 1991.
24. O. Goldreich, S. Micali and A. Wigderson. How to Play any Mental Game – A Completeness Theorem for Protocols with Honest Majority. In *19th STOC*, pages 218–229, 1987.
25. O. Goldreich and Y. Oren. Definitions and Properties of Zero-Knowledge Proof Systems. *Jour. of Cryptology*, Vol. 7, No. 1, pages 1–32, 1994.
26. S. Goldwasser, S. Micali and C. Rackoff. The Knowledge Complexity of Interactive Proof Systems. *SIAM Jour. on Computing*, Vol. 18(1), pp. 186–208, 1989.
27. E. Goh and S. Jarecki. A Signature Scheme as Secure as the Diffie-Hellman Problem. In *EuroCrypt2003*, Springer LNCS 2656, pages 401–415, 2003.
28. L.C. Guillou and J. Quisquater. A Practical Zero-Knowledge Protocol Fitted to Security Microprocessor Minimizing Both Trasmission and Memory. In *EuroCrypt88*, Springer LNCS 330, pages 123–128, 1988.
29. J. Håstad, R. Impagliazzo, L.A. Levin and M. Luby. Construction of Pseudorandom Generator from any One-Way Function. *SIAM Jour. on Computing*, Vol. 28 (4), pages 1364–1396, 1999.
30. M. Jakobsson, K. Sako and R. Impagliazzo. Designated Verifier Proofs and Their Applications. In *EuroCrypt96*, Springer LNCS 1070, pages 143–154.
31. Y. Lindell. Parallel Coin-Tossing and Constant-Round Secure Two-Party Computation. In *Crypto2001*, Springer LNCS 2139, pages 171–189, 2001.
32. Y. Lindell. Bounded-Concurrent Secure Two-Party Computation Without Setup Assumptions. To appear in *34th STOC*, 2003.
33. M. Naor. Bit Commitment using Pseudorandomness. *Jour. of Cryptology*, Vol. 4, pages 151–158, 1991.
34. M. Naor and M. Yung. Universal One-Way Hash Functions and their Cryptographic Applications. In *21st STOC*, pages 33–43, 1989.
35. R. Pass. Simulation in Quasi-polynomial Time and its Application to Protocol Composition. In *EuroCrypt2003*, Springer LNCS 2656, pages 160–176, 2003.
36. R. Pass and A. Rosen. Bounded-Concurrent Two-Party Computation in Constant Number of Rounds. Submitted.
37. A. Sahai. Non-Malleable Non-Interactive Zero Knowledge and Adaptive Chosen-Ciphertext Security. In *40th FOCS*, pages 543–553, 1999.
38. C.P. Schnorr. Efficient Identification and Signatures for Smart Cards. In *Crypto89*, Springer LNCS 435, pages 235–251, 1989.
39. J. Stern and D. Pointcheval. Security Arguments for Digital Signatures and Blind Signatures. *Jour. of Cryptology*, Vol. 13, No. 3, pages 361–396, 2000.

Primality Proving via One Round in ECPP and One Iteration in AKS

Qi Cheng*

School of Computer Science
The University of Oklahoma
Norman, OK 73019, USA
qcheng@cs.ou.edu

Abstract. On August 2002, Agrawal, Kayal and Saxena announced the first deterministic and polynomial time primality testing algorithm. For an input n, the AKS algorithm runs in heuristic time $\tilde{O}(\log^6 n)$. Verification takes roughly the same amount of time. On the other hand, the Elliptic Curve Primality Proving algorithm (ECPP), runs in random heuristic time $\tilde{O}(\log^6 n)$ ($\tilde{O}(\log^5 n)$ if the fast multiplication is used), and generates certificates which can be easily verified. More recently, Berrizbeitia gave a variant of the AKS algorithm, in which some primes cost much less time to prove than a general prime does. Building on these celebrated results, this paper explores the possibility of designing a more efficient algorithm. A random primality proving algorithm with heuristic time complexity $\tilde{O}(\log^4 n)$ is presented. It generates a certificate of primality which is $O(\log n)$ bits long and can be verified in deterministic time $\tilde{O}(\log^4 n)$. The reduction in time complexity is achieved by first generalizing Berrizbeitia's algorithm to one which has higher density of easily-proved primes. For a general prime, one round of ECPP is deployed to reduce its primality proof to the proof of a random easily-proved prime.

1 Introduction

Testing whether a number is prime or not is one of the fundamental problems in computational number theory. It has wide applications in computer science, especially in cryptography. After tremendous efforts invested by researchers in about two hundred years, it was finally proved by Agrawal, Kayal and Saxena [3] that the set of primes is in the complexity class **P**. For a given integer n, the AKS algorithm runs in time no longer than $\tilde{O}(\log^{12} n)$, while the best deterministic algorithm before it has subexponential complexity [2]. Under a conjecture concerning the density of the Sophie-Germain primes, The AKS algorithm should give out answer in time $\tilde{O}(\log^6 n)$.

Notation: In this paper, we use "ln" for logarithm base e and "log" for logarithm base 2. We write $r^\alpha||n$, if $r^\alpha|n$ but $r^{\alpha+1} \nmid n$. By $\tilde{O}(f(n))$, we mean $O(f(n)\mathrm{polylog}(f(n)))$.

---

* This research is partially supported by NSF Career Award CCR-0237845

D. Boneh (Ed.): CRYPTO 2003, LNCS 2729, pp. 338–348, 2003.

The AKS algorithm is based on the derandomization of a polynomial identity testing. It involves many iterations of polynomial modular exponentiation. To test the primality of a integer n, the algorithm first searches for a suitable r, which is provably $O(\log^6 n)$, or heuristically $O(\log^2 n)$. Then the algorithm will check for l from 1 to $L = \lceil 2\sqrt{r} \log n \rceil$, whether

$$(x + l)^n = x^n + l \quad (\text{mod } n, x^r - 1). \tag{1}$$

The algorithm declares that n is a prime if all the checks pass. The computing of $(x + l)^n \pmod{n, x^r - 1}$ takes time $\tilde{O}(r \log^2 n)$ if we use the fast multiplication. The total time complexity is thus $\tilde{O}(rL \log^2 n)$.

While the AKS algorithm is a great accomplishment in the theory, the current version is very slow. Unless its time complexity can be dramatically improved, it cannot replace random primality testing algorithms with better efficiency. In most of applications in cryptography, an efficient random algorithm is sufficient, as long as the algorithm can generate a certificate of primality, which in deterministic time convinces a verifier who does not believe any number theory conjectures. A primality testing algorithm which generates a certificate of primality is sometimes called *primality proving algorithm*. Similarly a primality testing algorithm which generates a certificate of compositeness is sometimes called *compositeness proving algorithm*. Very efficient random compositeness proving algorithms have long been known. Curiously, primality proving algorithms [5, 1] lag far behind of compositeness proving algorithms in term of efficiency and simplicity.

Recently, Berrizbeitia [7] proposed a brilliant modification to the AKS original algorithm. He used the polynomial $x^{2^s} - a$ instead of $x^r - 1$ in equation (1), where $2^s \approx \log^2 n$. Among others, he was able to prove the following proposition:

Proposition 1. *Given an integer $n \equiv 1 \pmod 4$. Denote $s = \lceil 2 \log \log n \rceil$. Assume that $2^k || n - 1$ and $k \geq s$. If there exists an integer a, such that $\left(\frac{a}{n}\right) = -1$ and $a^{\frac{n-1}{2}} \equiv -1 \pmod n$, then*

$$(1 + x)^n \equiv 1 + x^n \quad (\text{mod } n, x^{2^s} - a)$$

iff n is a power of a prime.

Unlike the AKS algorithm, where each prime costs roughly the same, there are "easily-proved primes" in Berrizbeitia's algorithm, namely, the primes p where $p-1$ has a factor of a power of two larger than $\log^2 n$. For those primes, one iteration of polynomial modular exponentiation, which runs in time $\tilde{O}(\log^4 n)$, establishes the primality of p, provided that a suitable a exists. In fact, a can be found easily if n is indeed a prime and randomness is allowed in the algorithm. It serves as a prime certificate for n.

Definition 1. *In this paper, for a primality proving algorithm, we call a prime p easily-proved, if the algorithm runs in expected time $\tilde{O}(\log^4 p)$ on p.*

What is the density of the easily-proved primes in Berrizbeitia's algorithm? The number of primes of form $2^s x + 1$ less than b is about $\frac{\pi(b)}{\phi(2^s)}$. Hence heuristically for a random prime p, $p - 1$ has a factor $2^s \approx \log^2 p$ with probability $\frac{1}{2\log^2 p}$, in the other words, the easily-proved primes have density $\frac{1}{2\log^2 p}$ around p in his algorithm.

1.1   Increasing the Density of Easily-Proved Primes

We prove the following theorem in Section 5, which can be regarded as a generalization of Proposition 1.

Theorem 1. *(Main) Given a number n which is not a power of an integer. Suppose that there exists a prime r, $r^\alpha || n - 1 (\alpha \geq 1)$ and $r \geq \log^2 n$. In addition, there exists a number $1 < a < n$, such that $a^{r^\alpha} \equiv 1 \pmod{n}$, $\gcd(a^{r^{\alpha-1}} - 1, n) = 1$, and*

$$(1 + x)^n = 1 + x^n \pmod{n, x^r - a},$$

then n is a prime.

The number a can be found easily if n is a prime and randomness is allowed. It serves as a prime certificate for n. Base on this theorem, we propose a random algorithm which establishes the primality of p in time $\tilde{O}(\log^4 p)$ if $p - 1$ contains a prime factor between $\log^2 p$ and $C \log^2 p$ for some small constant C.

Definition 2. *We call a positive integer n C-good, if $n - 1$ has a prime factor p such that $\log^2 n \leq p \leq C \log^2 n$.*

What is the density of C-good primes? Clearly the density should be higher than the density of easily-proved primes in Berrizbeitia's algorithm. Let $m = \prod_{p \text{ prime}, b_1 \leq p \leq Cb_1} p$. First we count the number of integers between 1 and m which have a prime factor between b_1 and Cb_1. This is precisely the number of nontrivial zero-divisors in ring $\mathbf{Z}/m\mathbf{Z}$:

$$(m - 1) - m \prod_{p \text{ prime}, b_1 \leq p \leq Cb_1} (1 - \frac{1}{p}).$$

We will prove in Section 4 that this number is greater than $\frac{m}{\ln b_1}$ for $C = \mathbf{c}$ and b_1 sufficiently large, where $\mathbf{c}$ is an absolute constant to be determined later. We need fix a explicit value for C in the algorithm. Without loss of generosity, set $C = 2$. For simplicity, we call a number *good*, when it is 2-good. Since compared with $\log^2 n$, n is very big, we expect that

Conjecture 1. There exists an absolute constant λ, such that for any sufficiently large integer n,

$$\frac{Number\ of\ 2 - good\ primes\ between\ n - 2\sqrt{n} + 1\ and\ n + 2\sqrt{n} + 1}{Number\ of\ primes\ between\ n - 2\sqrt{n} + 1\ and\ n + 2\sqrt{n} + 1} > \frac{\lambda}{\ln(\log^2 n)}.$$

We are unable to prove this conjecture however, but we present in the paper some numerical evidences. We comment that questions about the prime distribution in a short interval are usually very hard.

1.2 Algorithm for the General Primes

For general primes, we apply the idea in the Elliptic Curve Primality Proving algorithm (ECPP). ECPP was proposed by Goldwasser, Kilian [8] and Atkin [4] and implemented by Atkin and Morain [5]. In practice, ECPP performs much better than the current version of AKS. It has been used to prove primality of numbers up to thousands of decimal digits [10].

In ECPP, if we want to prove that an integer n is a prime, we reduce the problem to the proof of primality of a smaller number (less than $n/2$). To achieve this, we try to find an elliptic curve with $\omega n'$ points over $\mathbf{Z}/n\mathbf{Z}$, where ω is completely factored and n' is a probable prime greater than $(\sqrt[4]{n} + 1)^2$. Once we have such a curve and a point on the curve with order n', the primality of n' implies the primality of n. Since point counting on elliptic curves is expensive, we usually use the elliptic curves with complex multiplications of small discriminants. Nonetheless, it is plausible to assume that the order of the curve has the desired form with the same probability as a random integer does. ECPP needs $O(\log n)$ rounds of reductions to eventually reduce the problem to a primality proof of a very small prime, say, less than 1000. As observed in [9], one round of reduction takes heuristic time $\tilde{O}(\log^5 n)$, or $\tilde{O}(\log^4 n)$ if we use the fast multiplication. To get the time complexity, it is assumed that the number of primes between $n - 2\sqrt{n} + 1$ and $n + 2\sqrt{n} + 1$ is greater than $\sqrt{n}/\log^2 n$, and the number of points on an elliptic curve with small discriminant complex multiplication behaves like a random number in the Hassa range. We refer the assumption as the ECPP heuristics. Rigorous proof of the time complexity seems out of reach, as it involves the study of the prime distribution in a short interval.

Our algorithm can be decomposed into two stages. In the first stage, for a general probable prime n, we will use one round of ECPP to reduce its proof of primality to a good probable prime n' near n. For convenience, we require that $n - 2\sqrt{n} + 1 \le n' \le n + 2\sqrt{n} + 1$ (See section 6 for implementation issues). Note that up to a constant factor, the time complexity of one round reduction in ECPP is equivalent to the time complexity of finding a curve with a prime order. In the set of primes between $n - 2\sqrt{n} + 1$ and $n + 2\sqrt{n} + 1$, the density of good primes is $\frac{\lambda}{\ln(\log^2 n)}$ by conjectures. Hence heuristically *the extra condition on n' (that n' should be good) will increase the time complexity merely by a factor of* $O(\log \log n)$. Therefore for all the primes, without significant increase of time complexity, we reduce its primality proving to the proof of a good prime. In the second stage, we find a primality certificate for n'. To do this, we search for a which satisfies the conditions in the main theorem, and compute the polynomial modular exponentiation. The total expected running time of the first and the second stages becomes $\tilde{O}(\log^4 n)$. However, because of the reasons given above, it seems difficult to obtain the rigorous time complexity. Put it altogether, we now have a general purpose prime proving algorithm, which has following properties:

1. it runs very fast ($\tilde{O}(\log^4 n)$) assuming reasonable conjectures.
2. For many primes, ECPP subroutine is not needed.

3. The certificate, which consists of the curve, a point on the curve with order n', n' and a, is very short. It consists of only $O(\log n)$ bits as opposed to $O(\log^2 n)$ bits in ECPP.

4. A verifier can be convinced in deterministic time $\tilde{O}(\log^4 n)$. In fact, the most time consuming part in the verification is the computation of one polynomial modular exponentiation.

This paper is organized as following: In Section 2, we review the propositions used by AKS and ECPP to prove primality. In Section 3, we describe our algorithm and present the time complexity analysis. In Section 4, we prove a theorem which can be regarded as an evidence for the density heuristics. The main theorem is proved in Section 5. We conclude this paper with some discussions on the implementation of the algorithm.

2   Proving Primality in AKS and ECPP

The ECPP algorithm depends on rounds of reductions of the proof of primality of a prime to the proof of primality of a smaller prime. The most remarkable feature of ECPP is that a verifier who does not believe any conjectures can be convinced in time $\tilde{O}(\log^3 n)$ if the fast multiplication is used. It is based on the following proposition [5].

Proposition 2. *Let N be an integer prime to 6, E be an elliptic curve over $\mathbf{Z}/N\mathbf{Z}$, together with a point P on E and two integers m and s with $s|m$. Denote the infinite point on E by O. For each prime divisor q of s, denote $(m/q)P$ by $(x_q : y_q : z_q)$. Assume that $mP = O$ and $gcd(z_q, N) = 1$ for all q. If $s > (\sqrt[4]{N} + 1)^2$, then N is a prime.*

The certificate for N in ECPP consists of the curve E, the point P, m, s and the certificate of primality of s. Usually the ECPP algorithm uses elliptic curves with complex multiplications of small discriminants. For implementation details, see [5].

The AKS algorithm proves a number is a prime through the following proposition.

Proposition 3. *Let n be a positive integer. Let q and r be prime numbers. Let S be a finite set of integers. Assume*

1. *that q divides $r - 1$;*
2. *that $n^{\frac{r-1}{q}} \not\equiv 0, 1 \pmod{r}$;*
3. *that $gcd(n, b - b') = 1$ for all the distinct $b, b' \in S$;*
4. *that $\binom{q + |S| - 1}{|S|} \geq n^{2\lfloor \sqrt{r} \rfloor}$;*
5. *that $(x + b)^n \equiv x^n + b \pmod{x^r - 1, n}$ for all $b \in S$.*

Then n is a power of a prime.

3   Description and Time Complexity Analysis of Our Algorithm

Now we are ready to sketch our algorithm.

   Input: a positive integer n

   Output: a certificate of primality of n, or "composite".

1. If n is a power of an integer, return "composite".
2. Run a random compositeness proving algorithm, for example, the Rabin-Miller testing [6, Page 282], on n. If a proof of compositeness is found, output the proof, return "composite" and exit;
3. If $n - 1$ contains a prime factor between $\log^2 n$ and $2 \log^2 n$, skip this step. Otherwise, call ECPP to find an elliptic curve over $\mathbf{Z}/n\mathbf{Z}$ with n' points, where n' is a probable prime and n' is 2-good. Set $n = n'$;
4. Let r be the prime factor of $n - 1$ satisfying $\log^2 n \le r \le 2 \log^2 n$;
5. Randomly select a number $1 < b < n$. If $b^{n-1} \ne 1 \pmod{n}$, exit;
6. Let $a = b^{\frac{n-1}{r^\alpha}} \pmod{n}$; If $a = 1$, or $a^{r^{\alpha-1}} = 1$, go back to step 5;
7. If $gcd(a^{r^{\alpha-1}} - 1, n) \ne 1$, exit;
8. If $(1 + x)^n \ne 1 + x^n \pmod{n, x^r - a}$, exit;
9. Use ECPP procedure to construct the curve and the point and compute the order. Output them with a. Return "prime".

On any input integer n, this algorithm will either output "composite" with a compositeness proof, or "prime" with a primality proof, or nothing. As we can see, it output nothing, only when the probable prime n in Step 4 is actually composite. The chance should be extremely small.

Now we analyze its time complexity. Testing whether a number n is good or not can be done in time $\tilde{O}(\log^3 n)$. The step 3 takes time $\tilde{O}(\log^4 n)$, if the ECPP heuristics is true, Conjecture 1 in the introduction section is true, and the fast multiplication algorithm is used. If n is indeed a prime, then the probability of going back in step 6 is at most $1/r$. The step 7 takes time at most $\tilde{O}(\log^2 n)$. The step 8 takes time $\tilde{O}(\log^4 n)$, since $r \le 2 \log^2 n$. Hence the heuristic expected running time of our algorithm is $\tilde{O}(\log^4 n)$. Obviously the verification algorithm takes deterministic time $\tilde{O}(\log^4 n)$.

4   Density of Good Numbers

What is the probability that a random number has a prime factor between b_1 and $b_2 = Cb_1$? Let $m = \prod_{p \text{ prime}, b_1 \le p \le b_2} p$. We first compute the density of integers between 1 and $m - 1$ which has a prime factor between b_1 and b_2. Those numbers are precisely the zero-divisors in $\mathbf{Z}/m\mathbf{Z}$. The number of non-zero-divisors between 1 and m is $\phi(m) = m \prod_{p \text{ prime}, b_1 \le p \le b_2} (1 - \frac{1}{p})$, where ϕ is the Euler phi-function. First we estimate the quantity:

$$\beta_{b_1, b_2} = \prod_{p \text{ prime}, b_1 \le i \le b_2} (1 - \frac{1}{p})$$

It is known [11] that $\prod_{p<x,p \text{ prime}}(1 - \frac{1}{p}) = \frac{e^{-\gamma}}{\ln x}(1 + O(\frac{1}{\ln x}))$, where γ is the Euler constant. There must exist two absolute constants c_1, c_2, such that

$$\frac{e^{-\gamma}}{\ln x}(1 + \frac{c_1}{\ln x}) \leq \prod_{p<x,p \text{ prime}} (1 - \frac{1}{p}) \leq \frac{e^{-\gamma}}{\ln x}(1 + \frac{c_2}{\ln x})$$

Set $C = \mathbf{c}$ where $\mathbf{c}$ represents $e^{c_2-c_1+2}$.

$$\prod_{p \text{ prime}, b_1 \leq p \leq b_2} (1 - \frac{1}{p}) = \frac{\prod_{p \text{ prime}, p \leq b_2}(1 - \frac{1}{p})}{\prod_{p \text{ prime}, p \leq b_1}(1 - \frac{1}{p})}$$

$$\leq \frac{\ln b_1}{\ln \mathbf{c}b_1} \frac{1 + \frac{c_2}{\ln \mathbf{c}b_1}}{1 + \frac{c_1}{\ln b_1}}$$

$$= \frac{\ln^3 b_1 + (\ln \mathbf{c} + c_2) \ln^2 b_1}{\ln^3 b_1 + (2\ln \mathbf{c} + c_1)\ln^2 b_1 + (\ln^2 \mathbf{c} + 2c_1 \ln \mathbf{c})\ln b_1 + c_1 \ln^2 \mathbf{c}}$$

Thus $1 - \beta_{b_1, b_2} \geq \frac{(\ln \mathbf{c} + c_1 - c_2) \ln^2 b_1 - (\ln^2 \mathbf{c} + 2c_2 \ln \mathbf{c})\ln b_1 - c_2 \ln^2 \mathbf{c}}{\ln^3 b_1 + (2 \ln \mathbf{c} + c_1)\ln^2 b_1 + (\ln^2 \mathbf{c} + 2c_1 \ln \mathbf{c})\ln b_1 + c_1 \ln^2 \mathbf{c}} > \frac{1}{\ln b_1}$, when b_1 is sufficiently large. It is expected that the density of good primes in the set of primes in a large interval should not be very far away from $\frac{1}{\ln b_1}$. See Table 1 for numerical data concerning the density of 2-good primes around 2^{500}. Notice that

$$\beta_{250000,500000} = 0.9472455$$

$$1 - \beta_{250000,500000} = 0.0527545$$

$$\frac{1}{\ln 250000} = 0.0804556$$

5   Proof of the Main Theorem

In this section we prove the main theorem. It is built on a series of lemmas. Some of them are straight-forward generalizations of the lemmas in Berrizbeitia's paper [7]. We include slightly different proofs of those lemmas, though, for completeness. Some of the proofs are brief, for details see [7].

Lemma 1. *Let r, p be primes, $r|p - 1$. If $a \in \mathbf{F}_p$ is not a r-th power of any element in $\mathbf{F}_p$, then $x^r - a$ is irreducible over $\mathbf{F}_p$.*

Proof. Let θ be one of the roots of $x^r - a = 0$. Certainly $[\mathbf{F}_p(\theta) : \mathbf{F}_p] > 1$. Let $\xi \in \mathbf{F}_p$ be one of the r-th primitive roots of unity.

$$x^r - a = x^r - \theta^r = \prod_{0 \leq i \leq r-1} (x - \xi^i \theta).$$

Let $[\mathbf{F}_p(\theta) : \mathbf{F}_p] = r'$. Then for all i, $[\mathbf{F}_p(\xi^i \theta) : \mathbf{F}_p] = r'$. Hence over $\mathbf{F}_p$, $x^r - a$ will be factored into polynomials of degree r' only. Since r is a prime, this is impossible, unless that $r' = r$.

Table 1. Number of 2-good primes around 2^{500}

From	To	Number of primes	Number of 2-good primes	Ratio
$2^{500} + 0$	$2^{500} + 200000$	576	35	6.07%
$2^{500} + 200000$	$2^{500} + 400000$	558	38	6.81%
$2^{500} + 400000$	$2^{500} + 600000$	539	30	5.56%
$2^{500} + 600000$	$2^{500} + 800000$	568	23	4.05%
$2^{500} + 800000$	$2^{500} + 1000000$	611	39	6.38%
$2^{500} + 1000000$	$2^{500} + 1200000$	566	26	4.59%
$2^{500} + 1200000$	$2^{500} + 1400000$	566	38	6.71%
$2^{500} + 1400000$	$2^{500} + 1600000$	526	27	5.13%
$2^{500} + 1600000$	$2^{500} + 1800000$	580	26	4.48%
$2^{500} + 1800000$	$2^{500} + 2000000$	563	20	3.55%
$2^{500} + 2000000$	$2^{500} + 2200000$	562	22	3.91%
$2^{500} + 2200000$	$2^{500} + 2400000$	561	21	3.74%
$2^{500} + 2400000$	$2^{500} + 2600000$	609	34	5.58%
$2^{500} + 2600000$	$2^{500} + 2800000$	601	28	4.66%
$2^{500} + 2800000$	$2^{500} + 3000000$	603	33	5.47%
$2^{500} + 3000000$	$2^{500} + 3200000$	579	37	6.39%
$2^{500} + 3200000$	$2^{500} + 3400000$	576	31	5.38%
$2^{500} + 3400000$	$2^{500} + 3600000$	604	35	5.79%
$2^{500} + 3600000$	$2^{500} + 3800000$	612	40	6.53%
$2^{500} + 3800000$	$2^{500} + 4000000$	588	29	4.93%
$2^{500} + 4000000$	$2^{500} + 4200000$	574	33	5.75%
$2^{500} + 4200000$	$2^{500} + 4400000$	609	27	4.43%
$2^{500} + 4400000$	$2^{500} + 4600000$	549	35	6.37%
$2^{500} + 4600000$	$2^{500} + 4800000$	561	30	5.34%
$2^{500} + 4800000$	$2^{500} + 5000000$	545	29	5.32%
$2^{500} + 5000000$	$2^{500} + 5200000$	590	20	3.39%
$2^{500} + 5200000$	$2^{500} + 5400000$	557	27	4.84%
$2^{500} + 5400000$	$2^{500} + 5600000$	591	28	4.73%
$2^{500} + 5600000$	$2^{500} + 5800000$	517	33	6.38%
$2^{500} + 5800000$	$2^{500} + 6000000$	566	18	3.18%
$2^{500} + 6000000$	$2^{500} + 6200000$	575	30	5.21%
$2^{500} + 6200000$	$2^{500} + 6400000$	573	26	4.53%
$2^{500} + 6400000$	$2^{500} + 6600000$	558	36	6.45%
$2^{500} + 6600000$	$2^{500} + 6800000$	574	32	5.57%
$2^{500} + 6800000$	$2^{500} + 7000000$	594	22	3.70%
$2^{500} + 7000000$	$2^{500} + 7200000$	596	31	5.20%
$2^{500} + 7200000$	$2^{500} + 7400000$	567	26	4.58%
$2^{500} + 7400000$	$2^{500} + 7600000$	619	28	4.52%
$2^{500} + 7600000$	$2^{500} + 7800000$	565	25	4.42%
$2^{500} + 7800000$	$2^{500} + 8000000$	561	25	4.45%
$2^{500} + 8000000$	$2^{500} + 8200000$	570	26	4.56%

Lemma 2. *Let $n > 2$ be an integer. Let r be a prime and $r^{\alpha}\|n - 1$. Suppose that there exists a integer $1 < a < n$ such that*

1. $a^{r^\alpha} \equiv 1 \pmod{n}$;
2. $gcd(a^{r^{\alpha-1}} - 1, n) = 1$;

Then there must exist a prime factor p of n, such that $r^\alpha || p - 1$ and a is not a r-th power of any element in $\mathbf{F}_p$.

Proof. For any prime factor q of n, $a^{r^\alpha} \equiv 1 \pmod{q}$ and $a^{r^{\alpha-1}} \not\equiv 1 \pmod{q}$, so $r^\alpha | q - 1$. If $r^{\alpha+1} | q - 1$ for all the prime factors, then $r^{\alpha+1} | n - 1$, contradiction. Hence there exists a prime factor p, such that $r^\alpha || p - 1$. Let g be a generator in $\mathbf{F}_p^*$. If $a = g^t$ in $\mathbf{F}_p$, then $p - 1 | t r^\alpha$, and $p - 1 \nmid t r^{\alpha-1}$. Hence $r \nmid t$.

In the following text, we assume that n is an integer, $n = p^l d$ where p is a prime and $gcd(p, d) = 1$. Assume r is a prime and $r | p - 1$. Let $x^r - a$ be an irreducible polynomial in $\mathbf{F}_p$. Let θ be one of the roots of $x^r - a$. For any element in the field $\mathbf{F}_p(\theta)$, we can find a unique polynomial $f \in \mathbf{F}_p[x]$ of degree less than r such that the element can be represented by $f(\theta)$. Define $\sigma_m : \mathbf{F}_p(\theta) \to \mathbf{F}_p(\theta)$ as $\sigma(f(\theta)) = f(\theta^m)$.

Lemma 3. *We have that* $a^m = a$ *in* $\mathbf{F}_p$ *iff* $\sigma_m \in Gal(\mathbf{F}_p(\theta)/\mathbf{F}_p)$.

Proof. ($\Leftarrow$): Since $\sigma_m \in Gal(\mathbf{F}_p(\theta)/\mathbf{F}_p)$, θ^m must be a root of $x^r - a$. Hence $a = (\theta^m)^r = a^m$ in $\mathbf{F}_p$.

($\Rightarrow$): For any two elements $a, b \in \mathbf{F}_p(\theta)$, we need to prove that $\sigma_m(a + b) = \sigma_m(a) + \sigma_m(b)$ and $\sigma_m(ab) = \sigma_m(a)\sigma_m(b)$. The first one is trivial from the definition of σ_m. Let $a = f_a(\theta)$ and $b = f_b(\theta)$ where $f_a(x), f_b(x) \in \mathbf{F}_p[x]$ has degree at most $r - 1$. If $deg(f_a(x)f_b(x)) \le r - 1$, it is easy to see that $\sigma_m(ab) = \sigma_m(a)\sigma_m(b)$. Now assume that $deg(f_a(x)f_b(x)) \ge r$. Then $f_a(x)f_b(x) = h(x) + (x^r - a)p(x)$ where $h(x), p(x) \in \mathbf{F}_p[x]$ and $deg(h(x)) < r$. Then $\sigma_m(ab) = \sigma_m(h(\theta)) = h(\theta^m) = h(\theta^m) + (a^m - a)p(\theta^m) = h(\theta^m) + (\theta^{mr} - a)p(\theta^m) = f_a(\theta^m)f_b(\theta^m) = \sigma_m(a)\sigma_m(b)$.

This shows that σ_m is a homomorphism. To complete the proof, we need to show that it is also one-to-one. This is obvious since θ^m is a root of $x^r - a = 0$.

Define $G_m = \{f(\theta) \in \mathbf{F}_p(\theta)^* | f(\theta^m) = f(\theta)^m\}$. It can be verified that G_m is a group when σ_m is in $Gal(\mathbf{F}_p(\theta)/\mathbf{F}_p)$.

Lemma 4. *Suppose* $\sigma_n \in Gal(\mathbf{F}_p(\theta)/\mathbf{F}_p)$. *Then for any* $i, j \ge 0$, $\sigma_{d^i p^j} \in Gal(\mathbf{F}_p(\theta)/\mathbf{F}_p)$ *and* $G_n \subseteq G_{d^i p^j}$.

Proof. Notice that the map $x \to x^{p^l}$ is a one-to-one map in $\mathbf{F}_p(\theta)$. The equation $a^n = a$ implies that $(a^d)^{p^l} = a$, hence $a^d = a$, and $a^{d^i p^j} = a$. We have $\sigma_{d^i p^j} \in Gal(\mathbf{F}_p(\theta)/\mathbf{F}_p)$.

Let $f(\theta) \in G_n$. Thus $f(\theta^n) = f(\theta)^n$, this implies $f(\theta^{p^l d}) = f(\theta)^{p^l d} = f(\theta^{p^l})^d$. So θ^{p^l} is a solution of $f(x^d) = f(x)^d$. Since it is one of the conjugates of θ, θ must be a solution as well. This proves that $f(\theta^d) = f(\theta)^d$. Similarly since θ^d is also one of the conjugates of θ, as $\sigma_d \in Gal(\mathbf{F}_p(\theta)/\mathbf{F}_p)$, we have $f(\theta^{d^2}) = f(\theta^d)^d = f(\theta)^{d^2}$. By reduction, $f(\theta^{d^i}) = f(\theta)^{d^i}$ for $k \ge 0$. Hence $f(\theta^{d^i p^j}) = f(\theta^{d^i})^{p^j} = f(\theta)^{d^i p^j}$. This implies that $f(\theta) \in G_{d^i p^j}$.

Lemma 5. *If* $\sigma_{m_1}, \sigma_{m_2} \in Gal(\mathbf{F}_p(\theta)/\mathbf{F}_p)$ *and* $\sigma_{m_1} = \sigma_{m_2}$, *then* $|G_{m_1} \cap G_{m_2}|$ *divides* $m_1 - m_2$.

This lemma is straight forward from the definition.

Lemma 6. *Let* $A = a^{r^{\alpha-1}}$. *If* $(1 + \theta) \in G_n$, *so is* $1 + A^i\theta$ *for any* $i = 1, 2, 3, \cdots, r-1$. *And* $|G_n| \geq 2^r$.

Proof. If $(1 + \theta) \in G_n$, this means that $(1 + \theta)^n = 1 + \theta^n$. It implies that $(1+\theta')^n = 1+\theta'^n$ for any conjugate θ' of θ. Since A is a primitive root of unity in $\mathbf{F}_p$, hence $A^i\theta$ are conjugates of θ. We have $(1+A^i\theta)^n = 1+(A^i\theta)^n = 1+(A^n)^i\theta^n$ and we know that $A^n = A$. This proves that $1+A^i\theta \in G_n$. The group G_n contains all the elements in the set

$$\{\prod_{i=0}^{r-1}(1 + A^i\theta)^{\epsilon_i} \mid \sum_{i=0}^{r-1} \epsilon_i < r\},$$

by simple counting we have $|G_n| \geq 2^r$.

Finally we are ready to give the proof of the main theorem (Theorem 1) of this paper.

Proof. Since $|Gal(\mathbf{F}_p(\theta)/\mathbf{F}_p)| = r$, hence there exist two different pairs (i_1, j_1) and (i_2, j_2) with $0 \leq i_1, j_1, i_2, j_2 \leq \lfloor\sqrt{r}\rfloor$, such that $\sigma_{d^{i_1}p^{j_1}} = \sigma_{d^{i_2}p^{j_2}}$. According to Lemma 4, $G_n \subseteq G_{d^{i_1}p^{j_1}}$, $G_n \subseteq G_{d^{i_2}p^{j_2}}$, this implies that $G_n \subseteq G_{d^{i_1}p^{j_1}} \cap G_{d^{i_2}p^{j_2}}$. Therefore $|G_n|$ divides $d^{i_1}p^{j_1} - d^{i_2}p^{j_2}$, but $d^{i_1}p^{j_1} - d^{i_2}p^{j_2} < n^{\lfloor\sqrt{r}\rfloor} \leq 2^{\sqrt{r}\log n} \leq 2^r$. hence $d^{i_1}p^{j_1} - d^{i_2}p^{j_2} = 0$, which in turn implies that n is a power of p.

6   Implementation and Conclusion

In this paper, we propose a random primality proving algorithm which runs in heuristic time $\tilde{O}(\log^4 n)$. It generates a certificate of primality of length $O(\log n)$ which can be verified in deterministic time $\tilde{O}(\log^4 n)$.

When it comes to implement the algorithm, space is a bigger issues than time. Assume that n has 1000 bit, which is in the range of practical interests. To compute $(1 + x)^n \pmod{n, x^r - a}$, we will have an intermediate polynomial of size 2^{30} bit, or 128M bytes. As a comparison, ECPP is not very demanding on space. In order to make the algorithm available on a desktop PC, space efficient exponentiation of $1+x$ is highly desirable. This is the case for the original version of the AKS algorithm as well.

For the sake of theoretical clarity, we use just one round of ECPP reduction in the algorithm. To implement the algorithm, it may be better to follow the ECPP algorithm and launch the iteration of AKS as soon as an intermediate prime becomes good. Again assuming that the intermediate primes are distributed randomly, the expected number of rounds will be $\log\log n$. It is a better strategy since the intermediate primes get smaller.

We can certainly incorporate small time-saving features suggested by various researchers on the original version of AKS. The details will be included in the full version of this paper.

Acknowledgments. We thank Professors Pedro Berrizbeitia and Carl Pomerance for very helpful discussions and comments.

References

1. L. M. Adleman and M.A. Huang. *Primality Testing and Abelian Varieties Over Finite Fields*. Lecture Notes in Mathematics. Springer-Verlag, 1992.
2. L. M. Adleman, C. Pomerance, and R. S. Rumely. On distinguishing prime numbers from composite numbers. *Annals of Mathematics*, 117:173–206, 1983.
3. M. Agrawal, N. Kayal, and N. Saxena. Primes is in P. http://www.cse.iitk.ac.in/news/primality.pdf, 2002.
4. A.O.L. Atkin. Lecture notes of a conference in Boulder (Colorado), 1986.
5. A.O.L. Atkin and F. Morain. Elliptic curves and primality proving. *Mathematics of Computation*, 61:29–67, 1993.
6. Eric Bach and Jeffrey Shallit. *Algorithmic Number theory*, volume I. The MIT Press, 1996.
7. Pedro Berrizbeitia. Sharpening "primes is in p" for a large family of numbers. http://lanl.arxiv.org/abs/math.NT/0211334, 2002.
8. S. Goldwasser and J. Kilian. Almost all primes can be quickly certified. In *Proc. 18th ACM Symp. on Theory of Computing*, pages 316–329, Berkeley, CA, 1986. ACM.
9. A. Lenstra and H. W. Lenstra Jr. *Handbook of Theoretical Computer Science A*, chapter Algorithms in Number Theory, pages 673–715. Elsevier and MIT Press, 1990.
10. F. Morain. Primality proving using elliptic curves: An update. In *Proceedings of ANTS III*, volume 1423 of *Lecture Notes in Computer Science*, 1998.
11. G. Tenenbaum. *Introduction to analytic and probabilistic number theory (English Translation)*. Cambridge University Press, 1995.

Torus-Based Cryptography

Karl Rubin[1][*] and Alice Silverberg[2]

[1] Department of Mathematics
Stanford University
Stanford CA, USA
rubin@math.stanford.edu

[2] Department of Mathematics
Ohio State University
Columbus, OH, USA
silver@math.ohio-state.edu

Abstract. We introduce the concept of torus-based cryptography, give a new public key system called CEILIDH, and compare it to other discrete log based systems including Lucas-based systems and XTR. Like those systems, we obtain small key sizes. While Lucas-based systems and XTR are essentially restricted to exponentiation, we are able to perform multiplication as well. We also disprove the open conjectures from [2], and give a new algebro-geometric interpretation of the approach in that paper and of LUC and XTR.

1 Introduction

This paper accomplishes several goals. We introduce a new concept, namely torus-based cryptography, and give a new torus-based public key cryptosystem that we call CEILIDH. We compare CEILIDH with other discrete log based systems, and show that it improves on Diffie-Hellman and Lucas-based systems and has some advantages over XTR. Moreover, we show how to use the mathematics underlying XTR and Lucas-based systems to interpret them in terms of algebraic tori. We also show that a certain conjecture about algebraic tori has as a consequence new torus-based cryptosystems that would generalize and improve on CEILIDH and XTR. Further, we disprove the open conjectures from [2], and thereby show that the approach to generalizing XTR that was suggested in [2] cannot succeed.

The Lucas-based systems, the cubic field system in [5], and XTR have the discrete log security of the field $\mathbb{F}_{p^n}$, for $n = 2$, 3, and 6, resp., while the data required to be transmitted consists of $\varphi(n)$ elements of $\mathbb{F}_p$. Since these systems have $n \log p$ bits of security when exchanging $\varphi(n) \log p$ bits of information, they are more efficient than Diffie-Hellman by a factor of $n/\varphi(n) = 2$, 3/2, and 3, respectively. See [10,15,16,20,21,1] for Lucas-based systems and LUC, and [3,7, 8] for XTR and related work.

---

[*] Rubin was partially supported by NSF grant DMS-0140378.

What makes discrete log based cryptosystems work is that they are based on the mathematics of algebraic groups. An algebraic group is both a group and an algebraic variety. The group structure allows you to multiply and exponentiate. The variety structure allows you to express all elements and operations in terms of polynomials, and therefore in a form that can be efficiently handled by a computer.

In classical Diffie-Hellman, the underlying algebraic group is $\mathbb{G}_m$, the multiplicative group. Algebraic tori (not to be confused with complex tori of elliptic curve fame) are generalizations of the multiplicative group. By definition, an algebraic torus is an algebraic variety that over some extension field is isomorphic to $(\mathbb{G}_m)^d$, namely, d copies of the multiplicative group. For the tori we consider, the group operation is just the usual multiplication in a (larger) finite field.

The cryptosystems based on algebraic tori introduced in this paper accomplish the same goal as Lucas-based systems, XTR, and [5] of attaining the full security of the field $\mathbb{F}_{p^n}$ while requiring the transmission of only $\varphi(n)$ elements of $\mathbb{F}_p$. However, they additionally take advantage of the fact that an algebraic torus is a multiplicative group. For every n one can define an algebraic torus T_n with the property that $T_n(\mathbb{F}_p)$ consists of the elements in $\mathbb{F}_{p^n}^{\times}$ whose norms are 1 down to every intermediate subfield. This torus T_n has dimension $\varphi(n)$. When the torus is "rational", then its elements can be compactly represented by $\varphi(n)$ elements of $\mathbb{F}_p$. Doing cryptography inside this subgroup of $\mathbb{F}_{p^n}^{\times}$ has the security of the Diffie-Hellman problem in $\mathbb{F}_{p^n}^{\times}$ (see Lemma 7 below), but only $\varphi(n)$ elements of $\mathbb{F}_p$ need to be transmitted.

The CEILIDH[1] public key system is Compact, Efficient, Improves on LUC, and Improves on Diffie-Hellman. It also has some advantages over XTR. The system is based on the 2-dimensional algebraic torus T_6. The CEILIDH system does discrete log cryptography in a subgroup of $\mathbb{F}_{p^6}^{\times}$ while representing the elements in $\mathbb{F}_p^2$, giving a savings comparable to that of XTR, and having exactly the same security proof. While XTR and the Lucas-based cryptosystems are essentially restricted to exponentiation, CEILIDH allows full use of multiplication, thereby enabling a wider range of applications. In particular, where XTR uses a hybrid ElGamal encryption scheme that exchanges a key and then does symmetric encryption with that shared key, CEILIDH can do an exact analogue of (non-hybrid) ElGamal, since it has group multiplication at its disposal. Because of this multiplication, any cryptographic application that can be done in an arbitrary group can be done in a torus-based cryptosystem such as CEILIDH.

We also show that XTR, rather than being based on the torus T_6, is based on a quotient of this torus by the symmetric group S_3. The reason that XTR does not have a straightforward multiplication is that this quotient variety is not a group. (We note, however, that XTR has additional features that permit efficient computations.)

We exhibit a similar, but easier, construction based on the 1-dimensional torus T_2, obtaining a system similar to LUC but with the advantage of being

---

[1] The Scots Gaelic word *ceilidh*, pronounced "kayley", means a traditional Scottish gathering. This paper is dedicated to the memory of a cat named Ceilidh.

able to efficiently perform the group operation (in fact, directly in $\mathbb{F}_p$). This system has the security of $\mathbb{F}_{p^2}$ while transmitting elements of the field $\mathbb{F}_p$ itself.

The next case where $n/\varphi(n)$ is "large" is when $n = 30$ (and $\varphi(n) = 8$). Here, the 8-dimensional torus T_{30} is not known to be rational, though this is believed to be the case. An explicit rational parametrization of T_{30} would give a compact representation of this group by 8 elements of $\mathbb{F}_p$, with the security of the field $\mathbb{F}_{p^{30}}$. It would also refute the statement made in the abstract to [2] that "it is unlikely that such a compact representation of elements can be achieved in extension fields of degree thirty."

Conjectures were made in [2] suggesting a way to generalize LUC and XTR to obtain the security of the field $\mathbb{F}_{p^{30}}$ while transmitting only 8 elements of $\mathbb{F}_p$. In addition to showing that a rational parametrization of the torus T_{30} would accomplish this, we also show that the method suggested in [2] for doing this cannot. The reason is that, reinterpreting the conjectures in [2] in the language of algebraic tori, they say that the coordinate ring of the quotient of T_{30} by a certain product of symmetric groups is generated by the first 8 of the symmetric functions on 30 elements. (This would generalize the fact that the coordinate ring of T_6/S_3 is generated by the trace, which is what enables the success of XTR.) In §2 we disprove the open conjectures from [2]. This confirms the idea in [2] that the approach in [2] is unlikely to work.

Section 2 gives counterexamples to the open questions in [2]. Section 3 gives background on algebraic tori, defines the tori T_n, shows that $T_n(\mathbb{F}_q)$ is the subgroup of $\mathbb{F}_{q^n}^{\times}$ of order $\Phi_n(q)$, and shows that the security of cryptosystems based on this group is the discrete log security of $\mathbb{F}_{q^n}^{\times}$. Section 4 discusses rational parametrizations and compact representations, while §5 gives explicit rational parametrizations of T_6 and T_2. In §6 we introduce torus-based cryptography, and give the CEILIDH system (based on the torus T_6), a system based on T_2, and conjectured systems based on T_n for all n (most interesting for $n = 30$ or 210). In §7 we reinterpret the Lucas-based cryptosystems, XTR, and the point of view in [2] in terms of algebraic tori, and compare these systems to our torus-based systems.

Note that [12] gives another example, this time in the context of elliptic curves rather than multiplicative groups of fields, where the Weil restriction of scalars is used to obtain $n \log(q)$ bits of security from $\varphi(n) \log(q)$ bit transmissions.

1.1   Notation

Let $\mathbb{F}_q$ denote the finite field with q elements, where q is a prime power. Write φ for the Euler φ-function. Write Φ_n for the n-th cyclotomic polynomial, and let $G_{q,n}$ be the subgroup of $\mathbb{F}_{q^n}^{\times}$ of order $\Phi_n(q)$. Let $\mathbb{A}^n$ denote n-dimensional affine space, i.e., the variety whose $\mathbb{F}_q$-points are $\mathbb{F}_q^n$ for every q.

2   Counterexamples to the Open Questions in [2]

Four conjectures are stated in [2]. The two "strong" conjectures are disproved there. Here we disprove the two remaining conjectures (Conjectures 1 and 3 of

[2], which are also called (d, e)-**BPV** and n-**BPV**). In fact, we do better. We give examples that show not only that the conjectures are false, but also that weaker forms of the conjectures (i.e., with less stringent conclusions) are also false.

Fix an integer $n > 1$, a prime power q, and a factorization $n = de$ with $e > 1$. Recall that $G_{q,n}$ is the subgroup of $\mathbb{F}_{q^n}^\times$ of order $\Phi_n(q)$, where Φ_n is the n-th cyclotomic polynomial. Let $S_{q,n}$ be the set of elements of $G_{q,n}$ not contained in any proper subfield of $\mathbb{F}_{q^n}$ containing $\mathbb{F}_q$. For $h \in G_{q,n}$, let $P_h^{(d)}$ be the characteristic polynomial of h over $\mathbb{F}_{q^d}$, and define functions $a_j : G_{q,n} \to \mathbb{F}_{q^d}$ by

$$P_h^{(d)}(X) = X^e + a_{e-1}(h)X^{e-1} + \cdots + a_1(h)X + a_0(h).$$

Then $a_0(h) = (-1)^e$, and if also n is even then

$$a_j(h) = (-1)^e(a_{e-j}(h))^{q^{n/2}} \tag{1}$$

for all $j \in \{1, \ldots, e-1\}$ (see for example Theorem 1 of [2]).

The following conjecture is a *consequence* of Conjecture (d, e)-**BPV** of [2].

Conjecture (p, d, e)-BPV′ ([2]) *Let* $u = \lceil \varphi(n)/d \rceil$. *There are polynomials* $Q_1, \ldots, Q_{e-u-1} \in \mathbb{Z}[x_1, \ldots, x_u]$ *such that for all* $h \in S_{p,n}$ *and* $j \in \{1, \ldots, e - u - 1\}$,

$$a_j(h) = Q_j(a_{e-u}(h), \ldots, a_{e-1}(h)).$$

We will prove below the following result.

Theorem 1 *Conjecture* (p, d, e)-**BPV′** *is false when* (p, d, e) *is any one of the triples* $(7, 1, 30), (7, 2, 15), (11, 1, 30), (11, 2, 15)$.

If $n > 1$ is fixed, then Conjecture n-**BPV** of [2] says that there exists a divisor d of both n and $\varphi(n)$ such that $(d, n/d)$-**BPV** holds. Since $\gcd(30, \varphi(30)) = 2$, when $n = 30$ we need only consider $d = 1$ and 2. Since $(d, n/d)$-**BPV** implies $(p, d, n/d)$-**BPV′** for every p, the following is an immediate consequence of Theorem 1.

Corollary 2 *Conjectures* $(1, 30)$-**BPV**, $(2, 15)$-**BPV**, *and* 30-**BPV** *of [2] are false. Thus, Conjectures 1 and 3 of [2] are both false.*

Remark 3 The case $n = 30$ is particularly relevant for cryptographic applications, because this is the smallest value of n for which $n/\varphi(n) > 3$. If Conjecture 30-**BPV** of [2] were true it would have had cryptographic applications.

Proof of Theorem 1. If Conjecture (p, d, e)-**BPV′** were true, then for every $h \in S_{p,n}$ the values $a_{e-u}(h), \ldots, a_{e-1}(h)$ would determine $a_j(h)$ for *every* j. We will disprove Conjecture (p, d, e)-**BPV′** by exhibiting two elements $h, h' \in S_{p,n}$ such that $a_j(h) = a_j(h')$ whenever $e - u \leq j \leq e - 1$ but $a_j(h) \neq a_j(h')$ for at least one $j < e - u$.

Let $n = 30$, and $p = 7$ or 11. Note that $\Phi_{30}(7) = 6568801$ (a prime) and $\Phi_{30}(11) = 31 \times 7537711$. Since $\Phi_{30}(p)$ is relatively prime to 30, by Lemma 1 of [2] we have $S_{p,30} = G_{p,30} - \{1\}$. We view the field $\mathbb{F}_{p^{30}}$ as $\mathbb{F}_p[x]/f(x)$ with an irreducible polynomial $f(x) \in \mathbb{F}_p[x]$, and we fix a generator g of $G_{p,n}$. Specifically, let $r = (p^{30} - 1)/\Phi_{30}(p)$ and let

$$f(x) = x^{30} + x^2 + x + 5, \qquad g = x^r, \qquad \text{if } p = 7,$$
$$f(x) = x^{30} + 2x^2 + 1, \qquad g = (x+1)^r, \qquad \text{if } p = 11.$$

Case 1: $n = 30$, $e = 30$, $d = 1$. Then $u = \lceil \varphi(n)/d \rceil = 8$. For $h \in S_{p,30} = G_{p,30} - \{1\}$ and $1 \le j \le 29$ we have $a_j(h) = a_{30-j}(h)$ by (1), so we need only consider $a_j(h)$ for $15 \le j \le 29$.

By constructing a table of g^i and their characteristic polynomials $P_{g^i}^{(d)}$ for $i = 1, 2, \ldots$, and checking for matching coefficients, we found the examples in Tables 1 and 2. The examples in Table 1 disprove Conjecture $(7, 1, 30)$-**BPV′** and the examples in Table 2 disprove Conjecture $(11, 1, 30)$-**BPV′**.

Table 1. Values of $a_j(h) \in \mathbb{F}_7$ for several $h \in G_{7,30}$

$h \setminus j$	15	16	17	18	19	20	21	22	23	24	25	26	27	28	29
g^{2754}	3	2	0	6	4	4	2	5	4	0	2	2	1	4	4
g^{6182}	5	4	4	5	5	3	1	5	4	0	2	2	1	4	4
g^{5374}	2	0	5	2	1	6	4	6	1	1	5	6	4	2	6
g^{23251}	4	2	0	2	3	6	4	6	1	1	5	6	4	2	6

Table 2. Values of $a_j(h) \in \mathbb{F}_{11}$ for several $h \in G_{11,30}$

$h \setminus j$	15	16	17	18	19	20	21	22	23	24	25	26	27	28	29
g^{7525}	10	2	9	7	7	5	6	9	2	1	8	10	4	1	10
g^{31624}	10	2	2	4	2	3	10	9	2	1	8	10	4	1	10
g^{46208}	9	9	6	10	6	10	10	8	1	3	2	7	4	6	5
g^{46907}	7	8	0	0	1	7	10	8	1	3	2	7	4	6	5

Case 2: $n = 30$, $e = 15$, $d = 2$. Then $u = \lceil \varphi(n)/d \rceil = 4$. For $h \in S_{p,30} = G_{p,30} - \{1\}$ and $1 \le j \le 14$ we have $a_j(h) = \overline{a_{15-j}}(h)$ by (1), where $\bar{a}$ denotes conjugation in $\mathbb{F}_{p^2}$. Thus we need only consider $a_j(h)$ for $8 \le j \le 14$. View $\mathbb{F}_{p^2}$ as $\mathbb{F}_p(i)$ where $i^2 = -1$. A computer search as above leads to the examples in Tables 3 and 4. The examples in Table 3 disprove Conjecture $(7, 2, 15)$-**BPV′** and the examples in Table 4 disprove Conjecture $(11, 2, 15)$-**BPV′**.

This concludes the proof of Theorem 1.

Table 3. Values of $a_j(h) \in \mathbb{F}_{49}$ for certain $h \in G_{7,30}$

$h \setminus j$	8	9	10	11	12	13	14
g^{173}	$4+4i$	$5+i$	$1+6i$	$4i$	$2+3i$	$6+3i$	$3+i$
g^{2669}	6	$6+3i$	$5+i$	$4i$	$2+3i$	$6+3i$	$3+i$
g^{764}	$6+6i$	5	5	0	0	6	2
g^{5348}	$6+i$	5	5	0	0	6	2

Table 4. Values of $a_j(h) \in \mathbb{F}_{121}$ for certain $h \in G_{11,30}$

$h \setminus j$	8	9	10	11	12	13	14
g^{9034}	$10+i$	$10i$	$3+3i$	$1+4i$	$8+9i$	$5+4i$	9
g^{18196}	$6+8i$	$9+10i$	$8+i$	$1+4i$	$8+9i$	$5+4i$	9

Remark 4 Using these examples and some algebraic geometry, we prove in Theorem 5.3 of [13] that Conjectures $(p, 1, 30)$-**BPV′** and $(p, 2, 15)$-**BPV′** are each false for almost every prime p.

Remark 5 For $d = 1$ and $e = 30$, the last two lines of Table 1 (resp., Table 2) show that even the larger collection of values $a_{18}(h)$, $a_{20}(h)$, $\ldots$, $a_{29}(h)$ (resp., $a_{21}(h)$, $\ldots$, $a_{29}(h)$) does not determine any of the other values when $p = 7$ (resp., $p = 11$). We also found that no 8 coefficients determine all the rest; we found 64 pairs of elements so that given any set of 8 coefficients, one of these 64 pairs match up on these coefficients but not everywhere. In fact, we computed additional examples that show that when $p = 7$, no ten coefficients determine all the rest. We also show that when $p = 7$ no set of eight coefficients determines even one additional coefficient.

Suppose now $d = 2$, $e = 15$, and $p = 7$. Then the last two lines of Table 3 show that even the larger collection of values $a_9(h)$, $\ldots$, $a_{14}(h)$ does not determine the remaining value $a_8(h) \in \mathbb{F}_{49}$. We have computed additional examples that show that *no* choice of four of the values $a_8(h), \ldots, a_{14}(h)$ determines the other three.

3 Algebraic Tori

Good references for algebraic tori are [11,17].

Definition 6 An *algebraic torus* T over $\mathbb{F}_q$ is an algebraic group defined over $\mathbb{F}_q$ that over some finite extension field is isomorphic to $(\mathbb{G}_m)^d$, where $\mathbb{G}_m$ is the multiplicative group and d is necessarily the dimension of T. If T is isomorphic to $(\mathbb{G}_m)^d$ over $\mathbb{F}_{q^n}$, then one says that $\mathbb{F}_{q^n}$ *splits* T.

Let $k = \mathbb{F}_q$ and $L = \mathbb{F}_{q^n}$. Writing $\mathrm{Res}_{L/k}$ for the Weil restriction of scalars from L to k (see §3.12 of [17] or §1.3 of [19] for the definition and properties), then

$\mathrm{Res}_{L/k}\mathbb{G}_m$ is a torus. The universal property of the Weil restriction of scalars gives an isomorphism:

$$(\mathrm{Res}_{L/k}\mathbb{G}_m)(k) \cong \mathbb{G}_m(L) = L^\times. \tag{2}$$

If $k \subset F \subset L$ then the universal property also gives a norm map:

$$\mathrm{Res}_{L/k}\mathbb{G}_m \xrightarrow{N_{L/F}} \mathrm{Res}_{F/k}\mathbb{G}_m$$

which makes the following diagram commute:

$$
\begin{array}{ccc}
(\mathrm{Res}_{L/k}\mathbb{G}_m)(k) & \xrightarrow{N_{L/F}} & (\mathrm{Res}_{F/k}\mathbb{G}_m)(k) \\
{\scriptstyle \cong}\Big\downarrow & & {\scriptstyle \cong}\Big\downarrow \\
L^\times & \xrightarrow{\quad N_{L/F} \quad} & F^\times
\end{array}
\tag{3}
$$

(recall that the norm of an element is the product of its conjugates).

Define the torus T_n to be the intersection of the kernels of the norm maps $N_{L/F}$, for all subfields $k \subset F \subsetneq L$.

$$T_n := \ker \left[\mathrm{Res}_{L/k}\mathbb{G}_m \xrightarrow{\oplus N_{L/F}} \bigoplus_{k \subseteq F \subsetneq L} \mathrm{Res}_{F/k}\mathbb{G}_m \right].$$

By (3), for k-points we have:

$$T_n(k) \cong \{\alpha \in L^\times : N_{L/F}(\alpha) = 1 \text{ whenever } k \subset F \subsetneq L\}. \tag{4}$$

The dimension of T_n is $\varphi(n)$ (see [17]).

The group $T_n(\mathbb{F}_q)$ is a subgroup of the multiplicative group $\mathbb{F}_{q^n}^\times$. Lemma 7 below identifies $T_n(\mathbb{F}_q)$ with the cyclic subgroup $G_{q,n} \subset \mathbb{F}_{q^n}^\times$ of order $\Phi_n(q)$, and shows that the security of discrete log-based cryptosystems on the group T_n is really that of the multiplicative group of $\mathbb{F}_{q^n}$ and not any smaller field.

Lemma 7 *(i)* $T_n(\mathbb{F}_q) \cong G_{q,n}$.
(ii) $\#T_n(\mathbb{F}_q) = \Phi_n(q)$.
(iii) If $h \in T_n(\mathbb{F}_q)$ is an element of prime order not dividing n, then h does not lie in a proper subfield of $\mathbb{F}_{q^n}/\mathbb{F}_q$.

Proof. The group $\mathbb{F}_{q^n}^\times$ is cyclic of order $q^n - 1$, and $\mathrm{Gal}(\mathbb{F}_{q^n}/\mathbb{F}_q)$ is generated by the Frobenius automorphism which sends $x \in \mathbb{F}_{q^n}^\times$ to x^q. Hence if t divides n, then $N_{\mathbb{F}_{q^n}/\mathbb{F}_{q^t}}(x) = x^{(q^n-1)/(q^t-1)}$. Thus by (4),

$$T_n(\mathbb{F}_q) \cong \{x \in \mathbb{F}_{q^n}^\times : x^c = 1\} \tag{5}$$

where $c = \gcd\{(q^n-1)/(q^t-1) : t \mid n \text{ and } t \neq n\}$. Since $q^t - 1 = \prod_{j\mid t} \Phi_j(q)$, we have that $\Phi_n(q)$ divides c. There are polynomials $a_t(u) \in \mathbb{Z}[u]$ such that

$$\sum_{t\mid n, t \neq n} a_t(u) \frac{u^n - 1}{u^t - 1} = \Phi_n(u)$$

(see for example Theorem 1 of [4] or Theorem 2 of [14][2]), and so c divides $\Phi_n(q)$ as well. Thus $c = \Phi_n(q)$, so $T_n(\mathbb{F}_q) \cong G_{q,n}$ by (5) and the definition of $G_{q,n}$. Part (ii) follows from (i). Part (iii) now follows from Lemma 1 of [2].

4   Rationality of Tori and Compact Representations

Definition 8 Suppose T is an algebraic torus over $\mathbb{F}_q$ of dimension d. Then T is *rational* if and only if there is a birational map $\rho : T \to \mathbb{A}^d$ defined over $\mathbb{F}_q$. In other words, T is rational if and only if, after embedding T in an affine space $\mathbb{A}^t$, there are Zariski open subsets $W \subset T$ and $U \subset \mathbb{A}^d$, and (rational) functions $\rho_1, \dots, \rho_d \in \mathbb{F}_q(x_1, \dots, x_t)$ and $\psi_1, \dots, \psi_t \in \mathbb{F}_q(y_1, \dots, y_d)$ such that $\rho = (\rho_1, \dots, \rho_d) : W \to U$ and $\psi = (\psi_1, \dots, \psi_t) : U \to W$ are inverse isomorphisms. Call such a map ρ a *rational parametrization* of T.

A rational parametrization of a torus T gives a *compact representation* of the group $T(\mathbb{F}_q)$, i.e., a way to represent every element of the subset $W(\mathbb{F}_q) \subset T(\mathbb{F}_q)$ by d coordinates in $\mathbb{F}_q$. In general this is "best possible" (in terms of the number of coordinates), since a rational variety of dimension d has approximately q^d points over $\mathbb{F}_q$, and thus cannot be represented by fewer than d elements of $\mathbb{F}_q$.

Letting $X = T - W$, then $\dim(X) \leq d - 1$, so $|X(\mathbb{F}_q)| = O(q^{d-1})$. Thus the fraction of elements in $T(\mathbb{F}_q)$ that are "missed" by a compact representation is $|X(\mathbb{F}_q)|/|T(\mathbb{F}_q)| = O(1/q)$. For cryptographically interesting values of q this will be very small, and in special cases (by describing X explicitly as in the examples below) we obtain an even better bound.

Conjecture 9 (Voskresenskii [17]) *The torus T_n is rational.*

The conjecture is true for n if n is a prime power (see Chapter 2 of [17]) or a product of two prime powers ([6]; see also §6.3 of [17]). In the next section we will exhibit explicit rational parametrizations when $n = 6$ and 2.

When n is divisible by more than two distinct primes the conjecture is still open. Note that [18] claims a proof of a result that would imply that for every n, T_n is rational over $\mathbb{F}_q$ for almost all q. However, there is a serious flaw in the proof. Even the case $n = 30$, which would have interesting cryptographic applications, is not settled.

5   Explicit Rational Parametrizations

5.1   Rational Parametrization of T_6

Next we obtain an explicit rational parametrization of the torus T_6, thereby giving a compact representation of $T_6(\mathbb{F}_q)$. More precisely, we will show that T_6 is birationally isomorphic to $\mathbb{A}^2$, and therefore every element of $T_6(\mathbb{F}_q)$ can be represented by two elements of $\mathbb{F}_q$.

---

[2] The authors thank D. Bernstein and H. Lenstra for pointing out references [4,14].

Fix $x \in \mathbb{F}_{q^2} - \mathbb{F}_q$, so $\mathbb{F}_{q^2} = \mathbb{F}_q(x)$, and choose an $\mathbb{F}_q$-basis $\{\alpha_1, \alpha_2, \alpha_3\}$ of $\mathbb{F}_{q^3}$. Then $\{\alpha_1, \alpha_2, \alpha_3, x\alpha_1, x\alpha_2, x\alpha_3\}$ is an $\mathbb{F}_q$-basis of $\mathbb{F}_{q^6}$. Let $\sigma \in \mathrm{Gal}(\mathbb{F}_{q^6}/\mathbb{F}_q)$ be the element of order 2. Define a (one-to-one) map $\psi_0 : \mathbb{A}^3(\mathbb{F}_q) \hookrightarrow \mathbb{F}_{q^6}^{\times}$ by

$$\psi_0(u_1, u_2, u_3) = \frac{\gamma + x}{\gamma + \sigma(x)}$$

where $\gamma = u_1\alpha_1 + u_2\alpha_2 + u_3\alpha_3$. Then $N_{\mathbb{F}_{q^6}/\mathbb{F}_{q^3}}(\psi_0(\mathbf{u})) = 1$ for every $\mathbf{u} = (u_1, u_2, u_3)$. Let $U = \{\mathbf{u} \in \mathbb{A}^3 : N_{\mathbb{F}_{q^6}/\mathbb{F}_{q^2}}(\psi_0(\mathbf{u})) = 1\}$. By (4), $\psi_0(\mathbf{u}) \in T_6(\mathbb{F}_q)$ if and only if $\mathbf{u} \in U$, so restricting ψ_0 to U gives a morphism $\psi_0 : U \to T_6$. It follows from Hilbert's Theorem 90 that every element of $T_6(\mathbb{F}_q) - \{1\}$ is in the image of ψ_0, so ψ_0 defines an isomorphism

$$\psi_0 : U \xrightarrow{\sim} T_6 - \{1\}.$$

We will next define a birational map from $\mathbb{A}^2$ to U. A calculation in Mathematica shows that U is a hypersurface in $\mathbb{A}^3$ defined by a quadratic equation in u_1, u_2, u_3. Fix a point $\mathbf{a} = (a_1, a_2, a_3) \in U(\mathbb{F}_q)$. By adjusting the basis $\{\alpha_1, \alpha_2, \alpha_3\}$ of $\mathbb{F}_{q^6}$ if necessary, we can assume without loss of generality that the tangent plane at $\mathbf{a}$ to the surface U is the plane $u_1 = a_1$. If $(v_1, v_2) \in \mathbb{F}_q \times \mathbb{F}_q$, then the intersection of U with the line $\mathbf{a} + t(1, v_1, v_2)$ consists of two points, namely $\mathbf{a}$ and a point of the form $\mathbf{a} + \frac{1}{f(v_1, v_2)}(1, v_1, v_2)$ where $f(v_1, v_2) \in \mathbb{F}_q[v_1, v_2]$ is an explicit polynomial that we computed in Mathematica. The map that takes (v_1, v_2) to this latter point is an isomorphism

$$g : \mathbb{A}^2 - V(f) \xrightarrow{\sim} U - \{\mathbf{a}\},$$

where $V(f)$ denotes the subvariety of $\mathbb{A}^2$ defined by $f(v_1, v_2) = 0$. Thus $\psi_0 \circ g$ defines an isomorphism

$$\psi : \mathbb{A}^2 - V(f) \xrightarrow{\sim} T_6 - \{1, \psi_0(\mathbf{a})\}.$$

For the inverse isomorphism, suppose that $\beta = \beta_1 + \beta_2 x \in T_6(\mathbb{F}_q) - \{1, \psi_0(\mathbf{a})\}$ with $\beta_1, \beta_2 \in \mathbb{F}_{q^3}$. One checks easily that $\beta_2 \neq 0$, and if $\gamma = (1 + \beta_1)/\beta_2$ then $\gamma/\sigma(\gamma) = \beta$. Write $(1 + \beta_1)/\beta_2 = u_1\alpha_1 + u_2\alpha_2 + u_3\alpha_3$ with $u_i \in \mathbb{F}_q$, and define

$$\rho(\beta) = \left(\frac{u_2 - a_2}{u_1 - a_1}, \frac{u_3 - a_3}{u_1 - a_1} \right).$$

It follows from the discussion above that $\rho : T_6(\mathbb{F}_q) - \{1, \psi_0(\mathbf{a})\} \xrightarrow{\sim} \mathbb{A}^2 - V(f)$ is the inverse of the isomorphism ψ. We obtain the following.

Theorem 10 *The above maps ρ and ψ induce inverse birational maps between T_6 and $\mathbb{A}^2$.*

To implement the CEILIDH system, one must choose a finite field $\mathbb{F}_q$ and compute the rational maps ρ and ψ explicitly. We do this in two families of examples. Note that in each family the coefficients of the rational maps ρ and ψ are independent of q. When $(n, q) = 1$, write ζ_n for a primitive n-th root of unity.

Example 11 Fix $q \equiv 2$ or $5 \pmod 9$. Let $x = \zeta_3$ and $y = \zeta_9 + \zeta_9^{-1}$. Then $\mathbb{F}_{q^6} = \mathbb{F}_q(\zeta_9)$, $\mathbb{F}_{q^2} = \mathbb{F}_q(x)$, and $\mathbb{F}_{q^3} = \mathbb{F}_q(y)$. The basis we take for $\mathbb{F}_{q^3}$ is $\{1, y, y^2 - 2\}$, and we take $\mathbf{a} = (0, 0, 0)$. Then $\psi_0(\mathbf{a}) = \zeta_3^2$, and a calculation gives $f(v_1, v_2) = 1 - v_1^2 - v_2^2 + v_1 v_2$. Thus

$$\psi(v_1, v_2) = \frac{1 + v_1 y + v_2(y^2 - 2) + f(v_1, v_2)x}{1 + v_1 y + v_2(y^2 - 2) + f(v_1, v_2)x^2}.$$

For $\beta = \beta_1 + \beta_2 x \in T_6(\mathbb{F}_q) - \{1, \zeta_3^2\}$, we have

$$\rho(\beta) = (u_2/u_1, u_3/u_1) \quad \text{where } (1 + \beta_1)/\beta_2 = u_1 + u_2 y + u_3(y^2 - 2).$$

Example 12 Fix $q \equiv 3$ or $5 \pmod 7$. Let $x = \sqrt{-7}$ and $y = \zeta_7 + \zeta_7^{-1}$. Then $\mathbb{F}_{q^6} = \mathbb{F}_q(\zeta_7)$, $\mathbb{F}_{q^2} = \mathbb{F}_q(x)$, and $\mathbb{F}_{q^3} = \mathbb{F}_q(y)$. The basis we take for $\mathbb{F}_{q^3}$ is $\{1, y, y^2 - 1\}$, and we take $\mathbf{a} = (1, 0, 2)$. A calculation gives $f(v_1, v_2) = (2v_1^2 + v_2^2 - v_1 v_2 + 2v_1 - 4v_2 - 3)/14$. Thus

$$\psi(v_1, v_2) = \frac{\gamma + f(v_1, v_2)x}{\gamma - f(v_1, v_2)x}$$

where $\gamma = f(v_1, v_2) + 1 + v_1 y + (2f(v_1, v_2) + v_2)(y^2 - 1)$. If $\beta = \beta_1 + \beta_2 x \in T_6(\mathbb{F}_q) - \{1, \psi_0(\mathbf{a})\}$, then

$$\rho(\beta) = \left(\frac{u_2}{u_1 - 1}, \frac{u_3 - 2}{u_1 - 1}\right) \quad \text{where } (1 + \beta_1)/\beta_2 = u_1 + u_2 y + u_3(y^2 - 1).$$

5.2   Rational Parametrization of T_2

We give an explicit birational isomorphism between T_2 and $\mathbb{P}^1$. For simplicity we assume that q is not a power of 2, and we write $\mathbb{F}_{q^2} = \mathbb{F}_q(\sqrt{d})$ for some non-square $d \in \mathbb{F}_q^\times$. Let σ be the non-trivial automorphism of $\mathbb{F}_{q^2}/\mathbb{F}_q$, so $\sigma(\sqrt{d}) = -\sqrt{d}$.

Define a map $\psi : \mathbb{A}^1(\mathbb{F}_q) \to T_2(\mathbb{F}_q)$ by

$$\psi(a) = \frac{a + \sqrt{d}}{a - \sqrt{d}} = \frac{a^2 + d}{a^2 - d} + \frac{2a}{a^2 - d}\sqrt{d}.$$

Conversely, suppose $\beta = \beta_1 + \beta_2\sqrt{d} \in T_2(\mathbb{F}_q)$, with $\beta \neq \pm 1$ (so $\beta_2 \neq 0$). Then

$$\beta = \frac{1 + \beta}{1 + \sigma(\beta)} = \psi\left(\frac{1 + \beta_1}{\beta_2}\right).$$

Thus if we let $\rho(\beta) = (1 + \beta_1)/\beta_2$, then ρ and ψ define inverse isomorphisms

$$T_2 - \{\pm 1\} \underset{\psi}{\overset{\rho}{\rightleftarrows}} \mathbb{A}^1 - \{0\}.$$

In fact, these maps extend naturally to give an isomorphism $T_2(\mathbb{F}_q) \xrightarrow{\sim} \mathbb{F}_q \cup \{\infty\}$ by sending 1 to ∞ and -1 to 0. A simple calculation shows that if $a, b \in \mathbb{F}_q$ and $a \neq -b$, then

$$\psi(a)\psi(b) = \psi\left(\frac{ab+d}{a+b}\right). \tag{6}$$

Therefore instead of doing cryptography in the subgroup T_2 of $\mathbb{F}_{q^2}$, we can do all operations (i.e., multiplications and exponentiations in T_2) directly in $\mathbb{F}_q$ itself, where now multiplication in T_2 has been translated into the map $(a,b) \mapsto \frac{ab+d}{a+b}$ from $\mathbb{F}_q \times \mathbb{F}_q$ to $\mathbb{F}_q$.

6   Torus-Based Cryptosystems

Next we introduce public key cryptosystems based on a torus T_n with a rational parametrization. The case $n = 6$ is the CEILIDH system. By Lemma 7(iii), $T_n(\mathbb{F}_q)$ has the same cryptographic security as $\mathbb{F}_{q^n}^\times$. However, thanks to the compact representation that allows us to represent an element of $T_n(\mathbb{F}_q)$ by $\varphi(n)$ elements of $\mathbb{F}_q$, the size of any data represented by a group element is decreased by a factor of $\varphi(n)/n$ compared to classical cryptosystems using $\mathbb{F}_{q^n}^\times$. This give an improvement of a factor of 3 (resp., 2) using CEILIDH (resp., T_2).

Any discrete log based cryptosystem for a general group can be done using a torus T_n with a rational parametrization. Below we describe torus-based versions of Diffie-Hellman key exchange, ElGamal encryption, and ElGamal signatures. Other examples where this can be done in a straightforward way include DSA and Nyberg-Rueppel signatures (see also §5 of [8]).

Note that it is easy to turn any torus-based cryptosystem into an RSA-like system whose security is based on the difficulty of factoring, analogous to the LUC system of [15]. Here, one views the torus T_n over a ring $\mathbb{Z}/N\mathbb{Z}$. However, as shown in [1], such RSA-based systems do not seem to have significant advantages over RSA.

Parameter selection: Choose a prime power q and an integer n such that the torus T_n over $\mathbb{F}_q$ has an explicit rational parametrization, $n \log(q) \approx 1024$ (to obtain 1024 bit security), and $\Phi_n(q)$ is divisible by a prime ℓ that has at least 160 bits. Let $m = \varphi(n)$, and fix a birational map $\rho : T_n(\mathbb{F}_q) \to \mathbb{F}_q^m$ and its inverse ψ. Choose $\alpha \in T_n$ of order ℓ (taking an arbitrary element of $\mathbb{F}_{q^n}^\times$ and raising it to the power $(q^n - 1)/\ell$ will usually work), and let $g = \rho(\alpha) \in \mathbb{F}_q^m$. Note that n is a small number (2, 6, ...). For the protocols below, the public data is n, q, ρ, ψ, ℓ, and either g or $\alpha = \psi(g)$.

Key agreement scheme (torus-based Diffie-Hellman):
1. Alice chooses a random integer a in the range $1 \leq a \leq \ell - 1$. She computes $P_A := \rho(\alpha^a) \in \mathbb{F}_q^m$ and sends it to Bob.
2. Bob chooses a random integer b in the range $1 \leq b \leq \ell - 1$. He computes $P_B := \rho(\alpha^b) \in \mathbb{F}_q^m$ and sends it to Alice.
3. Alice computes $\rho(\psi(P_B)^a) \in \mathbb{F}_q^m$.
4. Bob computes $\rho(\psi(P_A)^b) \in \mathbb{F}_q^m$.

Since $\psi \circ \rho$ is the identity, we have $\rho(\psi(P_B)^a) = \rho(\alpha^{ab}) = \rho(\psi(P_A)^b)$, and this is Alice's and Bob's shared secret.

Encryption scheme (torus-based ElGamal encryption):

1. **Key Generation:** Alice chooses a random integer a in the range $1 \le a \le \ell - 1$ as her private key. Her public key is $P_A := \rho(\alpha^a) \in \mathbb{F}_q^m$.
2. **Encryption:** Bob represents the message M as an element of the group generated by α, selects a random integer k in the range $1 \le k \le \ell - 1$, computes $\gamma = \rho(\alpha^k) \in \mathbb{F}_q^m$ and $\delta = \rho(M\psi(P_A)^k) \in \mathbb{F}_q^m$, and sends the ciphertext (γ, δ) to Alice.
3. **Decryption:** Alice computes $M = \psi(\delta)\psi(\gamma)^{-a}$.

Signature scheme (torus-based ElGamal signatures):

1. **Key Generation:** Alice chooses a random integer a in the range $1 \le a \le \ell - 1$ as her private key. Her public key is $P_A := \rho(\alpha^a) \in \mathbb{F}_q^m$. The system requires a public cryptographic hash function $H : \{0,1\}^* \to \mathbb{Z}/\ell\mathbb{Z}$.
2. **Signature Generation:** Alice selects a random integer k in the range $1 \le k \le \ell - 1$, computes $\gamma = \rho(\alpha^k) \in \mathbb{F}_q^m$ and $\delta = k^{-1}(H(M) - aH(\gamma)) \pmod{\ell}$. Alice's signature on the message M is the pair (γ, δ).
3. **Verification:** Bob accepts Alice's signature (γ, δ) on M if and only if

$$\psi(P_A)^{H(\gamma)}\psi(\gamma)^{\delta} = \alpha^{H(M)}$$

in T_n.

The torus-based encryption scheme is the generalized ElGamal protocol (see p. 297 of [9]) applied to T_n, and the torus-based signature scheme is the generalized ElGamal signature scheme (see p. 458 of [9]) for the group T_n, where the maps ρ and ψ are used to go back and forth between the group law on T_n and the compact representation in $\mathbb{F}_q^m$.

Diffie-Hellman and ElGamal fail when any of the computed quantities is 1 or a small power of the generator, and RSA fails when one obtains something that is not relatively prime to the modulus. Similarly, a torus-based cryptosystem fails when one tries to apply ρ or ψ to a point where the map is not defined. Since there are very few such points (none for T_2 and only two for T_6 in the examples in §5), the probability of this occurring is negligible, and can ignored (or such points can be checked for and discarded). Lemma 7 shows that torus-based cryptosystems have exactly the same security as that of a multiplicative group $\mathbb{F}_{q^n}^{\times}$, and an attack on a T_n-cryptosystem gives an attack on an $\mathbb{F}_{q^n}^{\times}$.

Note that the shared key sizes for key agreement, the public key and ciphertext sizes for encryption, and the public key sizes for the signature schemes are all $\varphi(n)/n$ as long as those for the corresponding classical schemes, for the same security. Further, torus-based signatures have $\varphi(n)\log(q) + \log(\ell)$ bits, while the corresponding classical ElGamal signature scheme with the same security using a subgroup of order ℓ has $n\log(q) + \log(\ell)$ bit signatures.

The CEILIDH key exchange, encryption, and signature schemes are the above protocols with $n = 6$ and with ρ and ψ as in §5.1. Note that $\Phi_6(q) = q^2 - q + 1$ and $m = 2$, and q and ℓ can be chosen as in XTR.

The T_2 key exchange, encryption, and signature schemes are the above protocols with $n = 2$ and with ρ and ψ as in §5.2. However, we obtain an extra savings in the T_2 case, since there is no need to go back and forth between T_2 and $\mathbb{F}_q$ using the functions ρ and ψ. Using (6), all the group computations can be done directly and simply in $\mathbb{F}_q$, rather than in the group $T_2(\mathbb{F}_q)$.

The T_n cryptosystem uses the above protocols, whenever we have an n for which the torus T_n has an explicit and efficiently computable rational parametrization ρ and inverse map ψ. Conjecture 9 states that for every n, the torus T_n is rational. This is most interesting in the case $n = 30 = 2 \cdot 3 \cdot 5$, where $n/\varphi(n) = 3\frac{3}{4}$, but might also be of interest when $n = 210 = 2 \cdot 3 \cdot 5 \cdot 7$, where $n/\varphi(n) = 4\frac{3}{8}$. An explicit rational parametrization of the 8-dimensional torus T_{30} (analogous to the maps ρ and ψ of the CEILIDH and T_2 systems) would allow us to represent elements of $T_{30}(\mathbb{F}_q)$ by 8 elements of $\mathbb{F}_q$.

7 Understanding LUC, XTR, and "Beyond" in Terms of Tori

The Lucas-based systems, the cubic field system in [5], and XTR have the security of $\mathbb{F}_{p^2}$, $\mathbb{F}_{p^3}$, and $\mathbb{F}_{p^6}$, respectively, while representing elements in $\mathbb{F}_p$, $\mathbb{F}_p^2$, and $\mathbb{F}_{p^2}$, respectively. However, unlike the above torus-based systems, they do not make full use of the field multiplication. Here, we give a conceptual framework that explains why. We interpret these schemes in terms of varieties that are quotients of tori, and compare these schemes to the torus-based schemes of §3.

Consider two cases: $n = 2$ (the LUC case) and $n = 6$ (the XTR case). (It is straightforward to do the cubic case of [5] similarly.) Let F be $\mathbb{F}_q$ in the LUC case and $\mathbb{F}_{q^2}$ in the XTR case. Let $t = [\mathbb{F}_{q^n} : F]$, so $t = 2$ for LUC and $t = 3$ for XTR. In LUC and XTR, instead of $g \in G_{q,n}$ one considers the trace

$$Tr(g) := Tr_{\mathbb{F}_{q^n}/F}(g) \in F,$$

where the trace is the sum of the conjugates. One can show that for $g \in G_{q,n}$, the trace $Tr(g)$ determines the entire characteristic polynomial of g over F. In other words, knowing the trace of g is equivalent to knowing its unordered set of conjugates (but not the conjugates themselves). Let

$$C_g = \{g^\tau : \tau \in \mathrm{Gal}(\mathbb{F}_{q^n}/F)\},$$

the set of Galois conjugates of g.

Given a set $C = \{c_1, \dots, c_t\} \subset \mathbb{F}_{q^n}$, let $C^{(j)} = \{c_1^j, \dots, c_t^j\}$. If $C = C_g$, then $C^{(j)} = C_{g^j}$. In place of exponentiation ($g \mapsto g^j$), the XTR and LUC systems compute $Tr(g^j)$ from $Tr(g)$. In the above interpretation, they compute C_{g^j} from C_g, without needing to distinguish between the elements of C_g.

On the other hand, given sets of conjugates $\{g_1, \dots, g_t\}$ and $\{h_1, \dots, h_t\}$, it is not possible (without additional information) to multiply them to produce a new set of conjugates, because we do not know if we are looking for $C_{g_1 h_1}$, or $C_{g_1h_2}$, for example, which will be different. Therefore, XTR and LUC do not have straightforward multiplication algorithms.

However, XTR includes a partial multiplication algorithm (see Algorithm 2.4.8 of [7]). Given $Tr(g)$, $Tr(g^{j-1})$, $Tr(g^j)$, $Tr(g^{j+1})$, and a and b, the algorithm outputs $Tr(g^{a+bj})$. Thus for an XTR-based system, any transmission of data that needs to be multiplied requires sending three times as much data, effectively negating the improvement of $3 = 6/\varphi(6)$ that comes from XTR's compact representation. An analogous situation holds true for the signature scheme LUCELG DS in [16].

The CEILIDH system, since its operations take place in the group $G_{q,6}$, can do both multiplication and exponentiation, while taking full advantage of the compact representation for transmitting data. In particular, XTR-ElGamal encryption is key exchange followed by symmetric encryption with the shared key, while CEILIDH has full-fledged ElGamal encryption and signature schemes.

In the torus-based systems above, the information being exchanged is (a compact representation of) an element of a torus T_n. Further, the computations that are performed are multiplications in this group. We will see below that for XTR, the information being exchanged corresponds to an element of the quotient of T_6 by a certain action of the symmetric group on three letters, S_3. Similarly for LUC, the elements being exchanged correspond to elements of T_2/S_2. The set of equivalence classes T_6/S_3 is not a group, because multiplication in T_6 does not preserve S_3-orbits. This explains why XTR does not have a straightforward way to multiply. However, exponentiation in T_6 *does* preserve S_3-orbits, and it induces a well-defined exponentiation in T_6/S_3, and therefore in the set of XTR traces (the set $XTR(q)$ defined below).

What XTR takes advantage of is the fact that the quotient variety T_6/S_3 is rational, and the trace map to the quadratic subfield gives an explicit rational parametrization. This rational parametrization embeds T_6/S_3 in $\mathbb{A}^2$, as shown in Theorem 13 below, and therefore gives a compact representation of T_6/S_3.

Let $k = \mathbb{F}_q$, $L = \mathbb{F}_{q^6}$, and $F = \mathbb{F}_{q^2}$. If G is a group and V is a variety, then G acts on $\oplus_{\gamma \in G} V$ by permuting the factors. We have

$$\text{Res}_{L/k}\mathbb{G}_m \xrightarrow{\sim} \bigoplus_{\gamma \in \text{Gal}(L/k)} \mathbb{G}_m \xrightarrow{\sim} \left(\bigoplus_{\gamma \in \text{Gal}(F/k)} \mathbb{G}_m \right)^3 \qquad (7)$$

where the first isomorphism is defined over L and preserves the action of the Galois group $\text{Gal}(L/k)$ on both sides. The symmetric group S_3 acts naturally on $(\oplus_{\gamma \in \text{Gal}(F/k)} \mathbb{G}_m)^3$. Pulling back this action via the above composition defines an action of S_3 on $\text{Res}_{L/k}\mathbb{G}_m$ that preserves the torus $T_6 \subset \text{Res}_{L/k}\mathbb{G}_m$. The quotient map $T_6 \to T_6/S_3$ induces a (non-surjective) map on k-points $T_6(k) \to (T_6/S_3)(k)$. Let

$$XTR(q) = \{Tr_{L/F}(\alpha) : \alpha \in T_6(k)\} \subset F,$$

the set of traces used in XTR.

Theorem 13 *The set* $\mathrm{XTR}(q)$ *can be naturally identified with the image of* $T_6(k)$ *in* $(T_6/S_3)(k)$. *More precisely, there is a birational embedding*

$$T_6/S_3 \hookrightarrow \mathrm{Res}_{F/k}\mathbb{A}^1 \cong \mathbb{A}^2$$

such that $\mathrm{XTR}(q)$ *is the image of the composition*

$$T_6(k) \longrightarrow (T_6/S_3)(k) \hookrightarrow (\mathrm{Res}_{F/k}\mathbb{A}^1)(k) \cong F.$$

Proof. Let $k = \mathbb{F}_q$, $L = \mathbb{F}_{q^6}$, and $F = \mathbb{F}_{q^2}$. We have a commutative diagram (see (7))

$$
\begin{array}{ccccccc}
T_6 & \hookrightarrow & \mathrm{Res}_{L/k}\mathbb{G}_m & \hookrightarrow & \mathrm{Res}_{L/k}\mathbb{A}^1 & \xrightarrow{\sim} & \left(\bigoplus_{\gamma\in\mathrm{Gal}(F/k)}\mathbb{A}^1\right)^3 \\
 & & & & \downarrow{\scriptstyle Tr_{L/F}} & & \downarrow \\
 & & & & \mathrm{Res}_{F/k}\mathbb{A}^1 & \xrightarrow{\sim} & \bigoplus_{\gamma\in\mathrm{Gal}(F/k)}\mathbb{A}^1
\end{array}
\qquad (8)
$$

where the top and bottom isomorphisms are defined over L and F, respectively, and the right vertical map is the "trace" map $(\alpha_1, \alpha_2, \alpha_3) \mapsto \alpha_1 + \alpha_2 + \alpha_3$.

The morphism $Tr_{L/F} : \mathrm{Res}_{L/k}\mathbb{A}^1 \to \mathrm{Res}_{F/k}\mathbb{A}^1$ of (8) factors through the quotient $(\mathrm{Res}_{L/k}\mathbb{A}^1)/S_3$, so by restriction it induces a morphism $Tr : T_6/S_3 \to \mathrm{Res}_{F/k}\mathbb{A}^1$. By definition $\mathrm{XTR}(q)$ is the image of the composition $T_6(k) \to (T_6/S_3)(k) \to (\mathrm{Res}_{F/k}\mathbb{A}^1)(k) \cong F$, and T_6 and $\mathrm{Res}_{F/k}\mathbb{A}^1$ are both 2-dimensional varieties, so to prove the theorem we need only show that $Tr : T_6/S_3 \to \mathrm{Res}_{F/k}\mathbb{A}^1$ is injective. Suppose $g \in T_6(\bar{k})$. Using (7) we can view $g = (g_1, g_2, g_3) \in (\bigoplus_{\gamma\in\mathrm{Gal}(F/k)}\bar{k}^\times)^3$. Let σ be the non-trivial element of $\mathrm{Gal}(F/k)$. Since $g \in T_6(\bar{k})$, we have $g_1 g_2 g_3 = N_{L/F}(g) = 1$ and $g_i g_i^\sigma = 1$ for $i = 1, 2, 3$ by the definition of T_6. Hence we also have

$$g_1 g_2 + g_1 g_3 + g_2 g_3 = 1/g_3 + 1/g_2 + 1/g_1 = g_3^\sigma + g_2^\sigma + g_1^\sigma = Tr(g)^\sigma.$$

Thus the trace of g determines all the symmetric functions of $\{g_1, g_2, g_3\}$. Hence if $h = (h_1, h_2, h_3) \in T_6(\bar{k})$ and $Tr(h) = Tr(g)$, then $\{h_1, h_2, h_3\} = \{g_1, g_2, g_3\}$, i.e., h and g are in the same orbit under the action of S_3. Thus Tr is injective.

Similarly for LUC, the trace map induces a birational embedding $T_2/S_2 \hookrightarrow \mathbb{A}^1$, the variety T_2/S_2 is not a group, and

$$\mathrm{LUC}(q) = \{Tr_{\mathbb{F}_{q^2}/\mathbb{F}_q}(\alpha) : \alpha \in T_2(k)\} \subset \mathbb{F}_q$$

is the image of $T_2(\mathbb{F}_q)$ under the trace map $T_2 \to T_2/S_2 \hookrightarrow \mathbb{A}^1$.

7.1   Beyond XTR

As in [2] and §2 above, let $n = de$. Assume that n is square-free, and let $k = \mathbb{F}_q$, $L = \mathbb{F}_{q^n}$, and $F = \mathbb{F}_{q^d}$. We have

$$T_n \subset \mathrm{Res}_{L/k}\mathbb{G}_m \xrightarrow{\sim} \bigoplus_{\gamma\in\mathrm{Gal}(L/k)}\mathbb{G}_m \xrightarrow{\sim} \left(\bigoplus_{\gamma\in\mathrm{Gal}(F_\ell/k)}\mathbb{G}_m\right)^\ell \xrightarrow{\sim} \left(\bigoplus_{\gamma\in\mathrm{Gal}(F/k)}\mathbb{G}_m\right)^e$$

where the first isomorphism is defined over L and preserves the action of the Galois group $\mathrm{Gal}(L/k)$ on both sides, ℓ is any prime divisor of n, and $F_\ell = \mathbb{F}_{q^{n/\ell}}$. The symmetric group S_e acts naturally on $(\oplus_{\gamma \in \mathrm{Gal}(F/k)} \mathbb{G}_m)^e$. Pulling back this action via the above composition defines an action of S_e on $\mathrm{Res}_{L/k}\mathbb{G}_m$. Note that this action does not necessarily preserve the torus T_n. Similarly, S_ℓ acts naturally on $(\oplus_{\gamma \in \mathrm{Gal}(F_\ell/k)} \mathbb{G}_m)^\ell$. Since $N_{L/F_\ell}(g) = 1$ for every $g \in T_n$, it follows that T_n is in fact fixed under the induced action of S_ℓ.

Definition 14 Let $B_{(d,e)}$ denote the image of T_n in $(\mathrm{Res}_{L/k}\mathbb{G}_m)/S_e$.

If the variety $B_{(d,e)}$ is rational, then one can do cryptography. For example, this was done for the cases $(d,e) = (6,1)$ and $(2,1)$ in this paper (CEILIDH and T_2, respectively), for $(1,2)$ in the LUC papers, and for $(2,3)$ in XTR. Note that $(1,1)$ gives the usual Diffie-Hellman. Our (conjectural when n is a product of more than two primes) T_n cryptosystems are the cases $(n,1)$, and [2] discusses the cases $(d,e) = (1,30)$ and $(2,15)$. The variety $B_{(d,e)}$ is not generally a group. However, when $e = 1$, then $B_{(d,e)} = T_n$ which is a group.

Theorem 3.7 of [13] shows that the variety $B_{(d,e)}$ is birationally isomorphic to the quotient of T_n by the action of $\prod_{\text{primes } \ell \mid e} S_\ell$. Thus, the conjectures in [2] can be interpreted in this language as asking about the rationality of the varieties $T_{30}/(S_3 \times S_5)$ and $T_{30}/(S_2 \times S_3 \times S_5)$, and asking in particular if the morphisms from $B_{(1,30)}$ (resp., $B_{(2,15)}$) to $\mathbb{A}^8$ induced by the first $8/d$ (for $d = 1$ or 2, respectively) symmetric functions for the field extension L/F define rational parametrizations. We saw in §2 that these symmetric functions do not generate the coordinate ring of $B_{(1,30)}$ (resp., $B_{(2,15)}$).

The definitions in §3 can be easily extended to apply to an arbitrary cyclic extension L/k, not necessarily of finite fields. In particular, for $k = \mathbb{Q}$ and L a cyclic degree 30 extension of $\mathbb{Q}$, consider the above morphisms from characteristic zero versions of $B_{(1,30)}$ and $B_{(2,15)}$ to $\mathbb{A}^8$. We show in [13] that these maps are not birational, and (by reducing mod p) that for all but finitely many primes p, Conjecture $(p,1,30)$-**BPV'** (resp., Conjecture $(p,2,15)$-**BPV'**) is false (see Remark 4 above).

References

1. D. Bleichenbacher, W. Bosma, A. K. Lenstra, *Some remarks on Lucas-based cryptosystems*, in Advances in cryptology — CRYPTO '95, Lect. Notes in Comp. Sci. **963**, Springer, Berlin, 1995, 386–396.
2. W. Bosma, J. Hutton, E. R. Verheul, *Looking beyond XTR*, in Advances in Cryptology — Asiacrypt 2002, Lect. Notes in Comp. Sci. **2501**, Springer, Berlin, 2002, 46–63.
3. A. E. Brouwer, R. Pellikaan, E. R. Verheul, *Doing more with fewer bits*, in Advances in Cryptology — Asiacrypt '99, Lect. Notes in Comp. Sci. **1716**, Springer, Berlin, 1999, 321–332.
4. N. G. de Bruijn, *On the factorization of cyclic groups*, Nederl. Akad. Wetensch. Proc. Ser. A **56** (= Indagationes Math. **15**) (1953), 370–377.

5. G. Gong, L. Harn, *Public-key cryptosystems based on cubic finite field extensions*, IEEE Trans. Inform. Theory **45** (1999), 2601–2605.

6. A. A. Klyachko, *On the rationality of tori with cyclic splitting field*, in Arithmetic and geometry of varieties, Kuybyshev Univ. Press, Kuybyshev, 1988, 73–78 (Russian).

7. A. K. Lenstra, E. R. Verheul, *The XTR public key system*, in Advances in Cryptology — CRYPTO 2000, Lect. Notes in Comp. Sci. **1880**, Springer, Berlin, 2000, 1–19.

8. A. K. Lenstra, E. R. Verheul, *An overview of the XTR public key system*, in Public-key cryptography and computational number theory (Warsaw, 2000), de Gruyter, Berlin, 2001, 151–180.

9. A. J. Menezes, P. C. van Oorschot, S. A. Vanstone, Handbook of applied cryptography, CRC Press, Boca Raton, FL, 1997.

10. W. B. Müller, W. Nöbauer, *Some remarks on public-key cryptosystems*, Studia Sci. Math. Hungar. **16** (1981), 71–76.

11. T. Ono, *Arithmetic of algebraic tori*, Ann. of Math. **74** (1961), 101–139.

12. K. Rubin, A. Silverberg, *Supersingular abelian varieties in cryptology*, in Advances in Cryptology — CRYPTO 2002, Lect. Notes in Comp. Sci. **2442**, Springer, Berlin, 2002, 336–353.

13. K. Rubin, A. Silverberg, *Algebraic tori in cryptography*, to appear in High Primes and Misdemeanours: lectures in honour of the 60th birthday of Hugh Cowie Williams, Fields Institute Communications Series, American Mathematical Society, Providence, RI.

14. I. J. Schoenberg, *A note on the cyclotomic polynomial*, Mathematika **11** (1964), 131–136.

15. P. J. Smith, M. J. J. Lennon, *LUC: A New Public Key System*, in Proceedings of the IFIP TC11 Ninth International Conference on Information Security IFIP/Sec '93, North-Holland, Amsterdam, 1993, 103–117.

16. P. Smith, C. Skinner, *A public-key cryptosystem and a digital signature system based on the Lucas function analogue to discrete logarithms*, in Advances in Cryptology — Asiacrypt 1994, Lect. Notes in Comp. Sci. **917**, Springer, Berlin, 1995, 357–364.

17. V. E. Voskresenskii, Algebraic groups and their birational invariants, Translations of Mathematical Monographs **179**, American Mathematical Society, Providence, RI, 1998.

18. V. E. Voskresenskii, *Stably rational algebraic tori*, Les XXèmes Journées Arithmétiques (Limoges, 1997), J. Théor. Nombres Bordeaux **11** (1999), 263–268.

19. A. Weil. Adeles and algebraic groups. Progress in Math. **23**, Birkhäuser, Boston (1982).

20. H. C. Williams, *A $p + 1$ method of factoring*, Math. Comp. **39** (1982), 225–234.

21. H. C. Williams, *Some public-key crypto-functions as intractable as factorization*, Cryptologia **9** (1985), 223–237.

Efficient Universal Padding Techniques for Multiplicative Trapdoor One-Way Permutation

Yuichi Komano[*1] and Kazuo Ohta[2]

[1] TOSHIBA,
Komukai Toshiba-cho 1, Saiwai-ku, Kawasaki-shi, Kanagawa, Japan
yuichi1.komano@toshiba.co.jp
[2] The University of Electro-Communications,
Chofugaoka 1-5-1, Chofu-shi, Tokyo, Japan
ota@ice.uec.ac.jp

Abstract. Coron et al. proposed the ES-based scheme PSS-ES which realizes an encryption scheme and a signature scheme with a unique padding technique and key pair. The security of PSS-ES as an encryption scheme is based on the *partial-domain one-wayness* of the encryption permutation. In this paper, we propose new ES schemes OAEP-ES, OAEP++-ES, and REACT-ES, and prove their security under the assumption of *only* the *one-wayness* of encryption permutation. OAEP-ES, OAEP++-ES, and REACT-ES suit practical implementation because they use the same padding technique for encryption and for signature, and their security proof guarantees that we can prepare one key pair to realize encryption and signature in the same way as PSS-ES. Since *one-wayness* is a weaker assumption than *partial-domain one-wayness*, the proposed schemes offer tighter security than PSS-ES. Hence, we conclude that OAEP-ES, OAEP++-ES, and REACT-ES are more effective than PSS-ES. REACT-ES is the most practical approach in terms of the tightness of security and communication efficiency.

1 Introduction

Since the invention of the RSA encryption scheme [11], there have been a lot of interest in standardization and investigations into public key cryptosystems, in particular those for encryption and signature schemes. The encryption scheme OAEP (*Optimal Asymmetric Encryption Padding*, [2]) and the signature scheme PSS (*Probabilistic Signature Scheme*, [3]) are considered to be practical because they offer the strongest security level: IND-CCA2 (*indistinguishability against adaptive chosen ciphertext attack*) and EUF-ACMA (*existentially unforgeable against adaptive chosen message attack*).

OAEP first pads and then encrypts the plaintext while PSS pads and then signs the message; for encryption (signature), the trapdoor one-way permutation is applied in the direct (inverse) direction. Coron et al. [4] proposed the ES

---

[*] Work Performed at Major in Mathematical Science, Graduate School of Engineering and Science, Waseda University.

D. Boneh (Ed.): CRYPTO 2003, LNCS 2729, pp. 366–382, 2003.

scheme (Encryption-Signature scheme[1]) PSS-ES, which is based on the message recovery signature scheme PSS-R [3], and proved its security. For encryption and signature, PSS-ES uses a shared padding scheme and key pair; the public key and the private key are chosen adequately for encryption and signing, respectively. Hence this scheme is useful in terms of implementation. The security proofs in [4], however, have some technical mistakes. Moreover, even if these mistakes are corrected, the fact that the security of PSS-ES as an encryption scheme is based on *partial-domain one-wayness* of the encryption permutation, decreases the reduction efficiency; it must use long keys to achieve adequate security.

This paper gives the exact security of PSS-ES by correcting the problems in [4]. Moreover, this paper introduces new ES schemes, OAEP-ES and REACT-ES, that are based on OAEP+ [12] and REACT [10], respectively[2]. The proposed schemes satisfy IND-CCA2&ACMA (*Indistinguishability against adaptive chosen ciphertext attack and adaptive chosen message attack*) as an encryption scheme and EUF-CCA2&ACMA (*Existentially unforgeable against adaptive chosen ciphertext attack and adaptive chosen message attack*) as a signature scheme under the assumption of *only* the *one-wayness* of the permutation, while PSS-ES relies upon the *partial-domain one-wayness* of the encryption permutation for its security as an encryption scheme.

The rest of this paper is organized as follows. Section 2 recalls the definitions of the ES scheme and its security notations. Section 3 proposes new ES schemes, OAEP-ES and REACT-ES, and gives their security. In section 4, we point out the problems of original security proof of PSS-ES, given by Coron et al. [4], and give its exact security. In sections 5 and 6, we compare reduction efficiency of proposed schemes with the one of PSS-ES following the estimation of Nakashima and Okamoto [9] and discuss the reason why our schemes are more practical than PSS-ES. Furthermore, Appendices A and B present the security proofs of REACT-ES.

As a result, OAEP-ES, OAEP++-ES, and REACT-ES can realize secure encryption-signature scheme (ES scheme) with a unique padding technique and key pair; their reduction efficiency are much better than those of PSS-ES. Due to the high reduction efficiency of its security proof and its improved communication efficiency, REACT-ES is the most practical approach.

2   Definitions

2.1   ES Scheme with Universal Padding Technique

We describe a model of the ES scheme[3] (*Encryption-Signature scheme*) and its security. Since the ES scheme realizes an encryption scheme and a signature

---

[1] The ES scheme differs from *signcryption* [14]; the ES scheme realizes both encryption and signature schemes with a common padding technique and key pair (*encrypt or sign*), while signcryption realizes *encrypt then sign* or *sign then encrypt* scheme.

[2] We can construct another ES scheme, OAEP++-ES, based on OAEP++ [7]. In this paper, we will omit the detail of OAEP++-ES.

scheme with a common padding technique and key pair, we introduce attack model CCA2&ACMA following [4], where adversary $\mathcal{A}$ (forger $\mathcal{F}$) can freely use both decryption oracle D and signing oracle Σ. We extend notions of security IND-CCA2 [1] and EUF-ACMA [6] to create IND-CCA2&ACMA (*Indistinguishability against adaptive chosen ciphertext attack and adaptive chosen message attack*) and EUF-CCA2&ACMA (*Existentially unforgeable against adaptive chosen ciphertext attack and adaptive chosen message attack*), respectively.

Definition 1 (ES scheme with a unique padding technique). *If μ is a padding technique, then the ES scheme $(\mathcal{K}, \mathcal{E}, \mathcal{D}, \mathcal{S}, \mathcal{V})$ with μ is defined as follows:*

— *Key generation algorithm $\mathcal{K}$ is probabilistic algorithm which, given security parameter k, outputs the pair of public and private keys, $\mathcal{K}(1^k) = (\mathsf{pk}, \mathsf{sk})$. We regard pk as f and sk as f^{-1}, hereafter.*
— *Encryption algorithm $\mathcal{E}$ takes plaintext x and public key pk, calculates $z = \mu(x, r)$ with some random integer r, and returns ciphertext[3] $y = f(z) = \mathcal{E}_{\mathsf{pk}}(x)$. This algorithm is probabilistic[4].*
— *Decryption algorithm $\mathcal{D}$ takes ciphertext y and private key sk, calculates $z = f^{-1}(y)$ and $\mu^{-1}(z) = x \| r$ (un-padding), and returns plaintext $x = \mathcal{D}_{\mathsf{sk}}(y)$ if y is a valid ciphertext. Otherwise $\mathcal{D}$ returns Reject. This algorithm is deterministic.*
— *Signing algorithm $\mathcal{S}$ takes message x and private key sk, calculates $z = \mu(x, r)$ with some random integer r, and returns signature[5] $\sigma = f^{-1}(z) = \mathcal{S}_{\mathsf{sk}}(x)$. This algorithm is probabilistic.*
— *Verification algorithm $\mathcal{V}$ takes signature σ and public key pk, calculates $z = f(\sigma)$ and $\mu^{-1}(z) = x \| r$ (un-padding), and returns message $x = \mathcal{V}_{\mathsf{pk}}(\sigma)$ if σ is a valid signature. Otherwise $\mathcal{V}$ returns Reject. This algorithm is deterministic.*

We denote the ES scheme for encryption and for signature by ES(E) and ES(S), respectively (*e.g.*, OAEP-ES(E) and OAEP-ES(S) mean the OAEP-ES using in an encryption and a signature, respectively).

Definition 2 (IND-CCA2&ACMA). *Let $\mathcal{A}$ be an adversary of the encryption scheme. The attack scenario is described as follows:*

1. *$\mathcal{A}$ receives public key pk with $\mathcal{K}(1^k) = (\mathsf{pk}, \mathsf{sk})$.*
2. *$\mathcal{A}$ submits decryption queries for ciphertext y of his choice to decryption oracle D and gets corresponding plaintext x. Moreover, $\mathcal{A}$ submits signing queries for message x' of his choice to signing oracle Σ and gets corresponding signature σ.*
3. *$\mathcal{A}$ generates two plaintexts x_0, x_1 of identical length, and sends them to encryption oracle E as a challenge.*

---

[3] The input of f may be a part of z, *i.e.*, we allow to regard $y = f(z_1) \| z_2$ ($\sigma = f^{-1}(z_1) \| z_2$) as the ciphertext (signature) for $z = z_1 \| z_2$.
[4] Since padding technique μ is probabilistic, encryption permutation f may be deterministic(*e.g.*, RSA).

4. E chooses $b \xleftarrow{R} \{0,1\}$ and returns $y^* = \mathcal{E}_{pk}(x_b)$ to $\mathcal{A}$ as a target ciphertext.
5. $\mathcal{A}$ continues to submit decryption queries for ciphertext y of his choice to D and gets corresponding plaintext x. Moreover, $\mathcal{A}$ continues to submit signing queries for message x' of his choice to Σ and gets corresponding signature σ. In this phase, the only restriction is that $\mathcal{A}$ cannot issue a query for y^* to D.
6. $\mathcal{A}$ guesses b in this attack and outputs $\hat{b}$.

The adversary's advantage is defined as $\mathsf{Adv}(\mathcal{A}) = |2\Pr[b = \hat{b}] - 1|$. We say that the encryption scheme is $(t, q_D, q_\Sigma, q_H, \epsilon)$-secure in the sense of IND-CCA2&ACMA if an arbitrary adversary[5], whose running time is bounded by t, cannot achieve an advantage more than ϵ after making at most q_D decryption queries, q_Σ signing queries, and q_H hash queries.

Definition 3 (EUF-CCA2&ACMA).   Let $\mathcal{F}$ be a forger of the signature scheme. The attack scenario is described as follows:

1. $\mathcal{F}$ receives public key pk with $\mathcal{K}(1^k) = (\mathsf{pk}, \mathsf{sk})$.
2. $\mathcal{F}$ submits signing queries for message x of his choice to signing oracle Σ and gets corresponding signature σ. Moreover, $\mathcal{F}$ submits decryption queries for ciphertext y' of his choice to decryption oracle D and gets corresponding plaintext x'.
3. $\mathcal{F}$ outputs forgery σ^* with $\mathcal{V}_{pk}(\sigma^*) = x^*$ for some x^* ($x^* \neq x$ for any signing query x).

The forger's success probability is defined as $\epsilon = \Pr[\mathcal{V}_{pk}(\sigma^*) = x^*]$. We say that the signature scheme is $(t, q_D, q_\Sigma, q_H, \epsilon)$-secure in the sense of EUF-CCA2&ACMA if an arbitrary forger[7], whose running time is bounded by t, cannot achieve a success probability more than ϵ after making at most q_D decryption queries, q_Σ signing queries, and q_H hash queries.

Note that the security proof of the ES scheme with a unique padding technique comes in two parts, first as an encryption scheme and then as a signature scheme.

2.2   Assumption of One-Way Permutation

We classify trapdoor one-way permutations according to the difficulty of inverting them as follows [5]:

Definition 4.   Let $f : \{0,1\}^{k_0} \times \{0,1\}^{k_1} \to \{0,1\}^{k_0} \times \{0,1\}^{k_1}$ be a permutation. We say that

---

[5] We restrict the adversary (forger) by upper bounding the running time and the number of decryption, signing, and hash queries. We denote that $\mathcal{A}$ ($\mathcal{F}$) breaks an encryption scheme (signature scheme) in $(t, q_D, q_\Sigma, q_H, \epsilon)$ if $\mathcal{A}$ can distinguish b ($\mathcal{F}$ can outputs a forgery) within the time bound t and the advantage (success probability) more than ϵ using, at most, q_D decryption, q_Σ signing, and q_H hash queries.

— f is (τ, ϵ)-one-way, if an arbitrary adversary whose running time is bounded by τ has success probability $\mathsf{Succ}^{\mathsf{ow}}(\mathcal{A})$ that does not exceed ϵ. Here, $\mathsf{Succ}^{\mathsf{ow}}(\mathcal{A}) = \Pr_{s,t}[\mathcal{A}(f(s,t)) = (s,t)]$.

— f is (τ, ϵ)-partial-domain one-way, if an arbitrary adversary whose running time is bounded by τ has success probability $\mathsf{Succ}^{\mathsf{pd-ow}}(\mathcal{A})$ that does not exceed ϵ. Here, $\mathsf{Succ}^{\mathsf{pd-ow}}(\mathcal{A}) = \Pr_{s,t}[\mathcal{A}(f(s,t)) = s]$.

Moreover, we define $\mathsf{Succ}^{\mathsf{ow}}(\tau) = \max_{\mathcal{A}} \mathsf{Succ}^{\mathsf{ow}}(\mathcal{A})$ and $\mathsf{Succ}^{\mathsf{pd-ow}}(\tau) = \max_{\mathcal{A}} \mathsf{Succ}^{\mathsf{pd-ow}}(\mathcal{A})$, for all $\mathcal{A}$, whose running time is bounded by τ.

By the above definition, we have $\mathsf{Succ}^{\mathsf{pd-ow}}(\tau) \geq \mathsf{Succ}^{\mathsf{ow}}(\tau)$ for any τ. This inequality means that *partial-domain one-wayness* is a stronger assumption than *one-wayness*.

Through this paper, we assume that permutation f is multiplicative[6]. The multiplicative property of the permutation is described below.

Definition 5. *If f is a function, we call it a multicative function if*

$$f(ab) = f(a)f(b)$$

for arbitrary a and b.

3 Proposal Schemes

Coron et al. [4] used PSS-R to construct PSS-ES which realizes both an encryption and a signature with a common padding technique and key pair. PSS-ES is suitable for implementation, however, its security as an encryption scheme relies on the *partial-domain one-wayness* of f. Since the *partial-domain one-wayness* is stronger assumption than the *one-wayness*, the reduction efficiency is not tight and it must use long keys to achieve adequate security.

We propose new ES schemes, OAEP-ES, OAEP++-ES, and REACT-ES, which overcome this problem, and describe their security results. Since the security proofs of OAEP-ES, OAEP++-ES are similar to that of REACT-ES, we give the proofs of REACT-ES in Appendix A and B.

3.1 Methodology

We will give new ES schemes based on several encryption schemes which have a padding technique; OAEP+, OAEP++, and REACT. The simplest method[7] of

---

[6] Though the security of ES(S) can be ensured without the multiplicative property of f (which is not used in the security proof of ES(E) at all) as in [4], the reduction is not tight. Our interest is the comparison among ES schemes discussed in Section 5 in the practical situation, where RSA scheme is adopted as (f, f^{-1}), which satisfies the multiplicative property.

[7] Coron et al. simply constructed an encryption scheme by replacing signing permutation f^{-1} of PSS-R with f and proposed PSS-ES which has the same padding technique as PSS-R.

constructing an ES scheme from encryption schemes seems by replacing encryption permutation f with its inverse f^{-1}.

Unfortunately, if we construct a new signature scheme from an encryption scheme by simple replacement of permutation of f with f^{-1}, its security is not ensured. For example, it is easy for a known-message attacker to generate an existential forgery under the *one-way* permutation with a special property in the similar way of Shoup's attack.

This is a formal explanation of this situation. In the security proof of a signature, in order to invert f on an input of integer η (*i.e.*, to calculate $f^{-1}(\eta)$), we embed η into some random oracle query about message x and random integer r (*e.g.*, consider the query $r||x$ to H' in OAEP+) and simulate another random oracle about r (*e.g.*, $G(r)$ in OAEP+). In this strategy, if the random oracle value about r (*e.g.*, $G(r)$) is already defined, we abort the simulation (fail to simulate). However, when the adversary can freely choose the query r, it implies that we fail to simulate this case with a high probability.

Therefore, there might be a possibility that we could generally construct a provably secure ES scheme from an encryption scheme as follows[8]: (i) we replace the r, which is an input for random oracle G, by a hash value of x and a new $r'(e.g., r = w = H'(x||r'))$, and (ii) we replace x with $x||r'$.

In this paper, we create ES schemes from OAEP+ and REACT, following this methodology.

3.2 OAEP-ES

A simple ES scheme can be created using OAEP+ [12], OAEP-ES. OAEP-ES relies for its security upon *only* the *one-wayness* of the permutation, so it is more practical than PSS-ES. OAEP-ES has, however, worse reduction efficiency than OAEP++-ES and REACT-ES as we will show. A description of OAEP-ES and its security results are as follows.

OAEP-ES with hash functions $G : \{0,1\}^{k_1} \rightarrow \{0,1\}^{n+k_0}$ and $H, H' : \{0,1\}^{n+k_0} \rightarrow \{0,1\}^{k_1}$, and the common padding scheme μ_1 (Figure 1) and key pair (f, f^{-1})[9], is executed as follows:

—*Encryption and Signing*: In order to encrypt or to sign x, we choose $r \xleftarrow{R} \{0,1\}^{k_0}$, set $w = H'(x||r) \in \{0,1\}^{k_1}$, and calculate $s = (x||r) \oplus G(w)$, $t = H(s) \oplus w$, and $\mu_1(x, r) = s||t$. We then return $y = f(\mu_1(x, r))$ as the ciphertext or $\sigma = f^{-1}(\mu_1(x, r))$ as the signature, respectively.

—*Decryption and Verification*: For ciphertext y or signature σ, we recover $s||t = f^{-1}(y)$ or $s||t = f(\sigma)$ ($|s| = n + k_0$, $|t| = k_1$), respectively. Next, we calculate $w = t \oplus H(s)$, divide $x||r = s \oplus G(w)$ ($|x| = n$, $|r| = k_0$), and check whether $w = H'(x||r)$. If the check passes, we return x; otherwise Reject.

The security results of OAEP-ES are as follows:

---

[8] Reference [8] gives a detailed explanation of this methodology.

[9] In the general model, we assume that $f : \{0,1\}^k \rightarrow \{0,1\}^k$ is a multiplicative permutation. If the implementation uses RSA permutation: $\mathbb{Z}_n \rightarrow \mathbb{Z}_n$, we put "0" in front of the padding data to make the domain k bit integer. In this case, the model

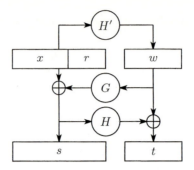

Fig. 1. Padding Techniques μ_1 for ES Schemes

Theorem 1 (Security result of OAEP-ES(E)).   *Let $\mathcal{A}$ be an adversary that breaks OAEP-ES in $(\tau, q_D, q_\Sigma, q_G, q_{H'}, q_H, \epsilon)$ in the sense of IND-CCA2&ACMA. Then:*

$$\begin{cases} \mathsf{Succ}^{\mathsf{ow}}(\tau') \geqq \epsilon - \frac{q_{H'} + q_\Sigma}{2^{k_0}} - \frac{(q_{H'} + q_\Sigma + 1)(q_G + q_{H'} + q_\Sigma) + q_D}{2^{k_1}} \\ \tau' \leqq \tau + \{(q_G + q_{H'} + q_\Sigma)(q_H + q_{H'} + q_\Sigma) + q_{H'} + q_\Sigma\}T_f \end{cases}$$

where T_f denotes the time complexity of f.

Theorem 2 (Security result of OAEP-ES(S)).   *Let $\mathcal{F}$ be a forger that breaks OAEP-ES in $(\tau, q_D, q_\Sigma, q_G, q_{H'}, q_H, \epsilon)$ in the sense of EUF-CCA2&ACMA. Then:*

$$\begin{cases} \mathsf{Succ}^{\mathsf{ow}}(\tau') \geqq \epsilon - \frac{q_{H'} q_\Sigma}{2^{k_0}} - \frac{(q_{H'} + q_\Sigma)(q_G + q_{H'} + q_\Sigma) + q_D + 1}{2^{k_1}} \\ \tau' \leqq \tau + (2q_{H'} + 2q_\Sigma + 1)T_f \end{cases}$$

where T_f denotes the time complexity of f.

3.3   REACT-ES

REACT was proposed by Okamoto and Pointcheval [10]. To use REACT for encryption, we first generate random integer r and encrypt plaintext x by a symmetric encryption scheme with the hash value of r as the key. Second, we encrypt r by an asymmetric encryption scheme and send it with ciphertext of x and a check code.

Therefore, in REACT, once we encrypt r with the asymmetric encryption scheme, we can send a long plaintext using the symmetric encryption scheme with high speed (which, so-called, is KEM (Key Encapsulation Mechanism, [13])); REACT is more practical in terms of communication efficiency than OAEP, OAEP+, and OAEP++. Moreover, Nakashima and Okamoto [9] showed that REACT has tighter security than OAEP or OAEP+.

---

and theorems will need to be adjusted. We adopt the same discussion for PSS-ES, OAEP++-ES, and REACT-ES.

REACT-ES with hash functions $G : \{0,1\}^{k_1} \rightarrow \{0,1\}^{k_3}$, $H' : \{0,1\}^{n+k_0} \rightarrow \{0,1\}^{k_1}$, and $H : \{0,1\}^{2(n+k_0+k_1)} \rightarrow \{0,1\}^{k_2}$ $(k = k_1)$, symmetric encryption scheme E_{key}^{sym}, where key length is k_3, public key f, and private key f^{-1}, is executed as follows (Figure 2):

—*Encryption and Signing*: In order to encrypt or to sign x, we choose $r \xleftarrow{R} \{0,1\}^{k_0}$, set $w = H'(x||r) \in \{0,1\}^{k_1}$, and calculate $c_2 = E_{G(w)}^{sym}(x||r)$. Next, we set $c_1 = f(w)$ for encryption or $c_1 = f^{-1}(w)$ for signing, and return $(c_1, c_2, c_3 = H(x||r, w, c_1, c_2))$ as the ciphertext or signature, respectively.

—*Decryption and Verification*: For ciphertext (c_1, c_2, c_3) or signature (c_1', c_2, c_3), we recover $w = f^{-1}(c_1)$ or $w = f(c_1')$, respectively. Next, we calculate $x||r$ from $E_{G(w)}^{sym}(c_2)$, and check whether both "$w = H'(x||r)$ and $c_3 = H(x||r, w, c_1, c_2)$" or both "$w = H'(x||r)$ and $c_3 = H(x||r, w, c_1', c_2)$", respectively. If the check passes, we return x as the plaintext or the message, respectively; otherwise Reject.

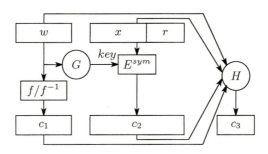

Fig. 2. REACT-ES

We use the following theorems to examine the security of REACT-ES. The proof are described in Appendix A and B, respectively.

Theorem 3 (Security result of REACT-ES(E)).   *Let the symmetric encryption scheme be (τ', ν)-secure[10], and let $\mathcal{A}$ be an adversary that breaks REACT-ES in $(\tau, q_D, q_\Sigma, q_G, q_{H'}, q_H, \epsilon)$ in the sense of IND-CCA2&ACMA. Then:*

$$\begin{cases} \mathsf{Succ}^{\mathsf{ow}}(\tau') \geqq \epsilon - \nu - \frac{q_{H'} + q_\Sigma}{2^{k_0}} - \frac{q_D}{2^{k_1}} \\ \tau' \leqq \tau + (q_G + q_H + 2q_{H'} + 2q_\Sigma)T_f \end{cases}$$

where T_f denotes the time complexity of f.

Theorem 4 (Security result of REACT-ES(S)).   *Let $\mathcal{F}$ be a forger that breaks REACT-ES in $(\tau, q_D, q_\Sigma, q_G, q_{H'}, q_H, \epsilon)$ in the sense of EUF-CCA2&ACMA. Then:*

$$\begin{cases} \mathsf{Succ}^{\mathsf{ow}}(\tau') \geqq \epsilon - \frac{q_{H'} q_\Sigma}{2^{k_0}} - \frac{q_D + 1}{2^{k_1}} \\ \tau' \leqq \tau + (2q_{H'} + 2q_\Sigma + 1)T_f \end{cases}$$

where T_f denotes the time complexity of f.

---

[10] See the definition of security model of symmetric encryption scheme, §2.2 of [10].

4   PSS-ES

4.1   Security of PSS-ES(E)

The security proof of PSS-ES(E) in [4] (Theorem 2 and Lemma 4) has two minor technical mistakes as follows: (i) the number of queries (about w) to G is not $q_{H'} + q_\Sigma$ (the last line in page 14 of [4]) but $q_G + q_{H'} + q_\Sigma$ because $G(w)$ may be defined by query w to G directly, (ii) this proof overlooks calculation time $(q_{H'} + q_\Sigma)T_f$ as part of the cost of querying the decryption oracle (line 10 in page 14 of [4], reading in Lemma 1's results into proof of Lemma 4). This consideration of these problems yields the following security result.

Theorem 5 (Security result of PSS-ES(E)).   *Let $\mathcal{A}$ be an adversary that breaks PSS-ES(E) in $(\tau, q_D, q_\Sigma, q_G, q_{H'}, \epsilon)$ in the sense of IND-CCA2&ACMA. Then:*

$$
\begin{cases}
\mathsf{Succ}^{\mathsf{pd-ow}}(\tau') \geq \frac{1}{q_G + q_{H'} + q_\Sigma}\left(\epsilon - \frac{q_{H'} + q_\Sigma}{2^{k_0}} - \frac{(q_{H'} + q_\Sigma)(q_G + q_{H'} + q_\Sigma) + q_D}{2^{k_1}}\right) \\
\tau' \leq \tau + 2(q_{H'} + q_\Sigma)T_f
\end{cases}
$$

where T_f denotes the time complexity of f.

4.2   Security of PSS-ES(S)

The proof of Theorem 3 in [4] has three minor technical mistakes as follows: (i) it misses the probability $\frac{q_\Sigma}{2^{k_0}}$ that appears because $\mathcal{I}$ cannot answer the signing query for the pair of message and random integers implanting η previously[11], (ii) the number of queries w to G is not $q_{H'} + q_\Sigma$ (line 18 in page 16 of [4]) but $q_G + q_{H'} + q_\Sigma$ because $G(w)$ may be defined by the query w to G directly, (iii) this proof overlooks the calculation time $(q_{H'} + q_\Sigma)T_f$ as part of the cost of querying the decryption oracle (line 9 in page 16 of [4], reading in Lemma 1's results into proof of Theorem 3). This consideration of these problems yields the following security result.

Theorem 6 (Security result of PSS-ES(S)).   *If $\mathcal{F}$ is a forger that breaks PSS-ES(S) in $(\tau, q_D, q_\Sigma, q_G, q_{H'}, \epsilon)$ in the sense of EUF-CCA2&ACMA, then:*

$$
\begin{cases}
\mathsf{Succ}^{\mathsf{ow}}(\tau') \geq \epsilon - \frac{q_{H'}q_\Sigma}{2^{k_0}} - \frac{(q_{H'} + q_\Sigma)(q_G + q_{H'} + q_\Sigma) + q_D + 1}{2^{k_1}} \\
\tau' \leq \tau + (2q_{H'} + 2q_\Sigma + 1)T_f
\end{cases}
$$

where T_f denotes the time complexity of f.

5   Reduction Efficiency

We evaluate the security of RSA-OAEP-ES and RSA-REACT-ES following the approach taken by Nakashima and Okamoto [9] and compare them to RSA-PSS-ES. For each scheme, we consider the usages of encryption and signature.

---

[11] In our results, since η is embedded $q_{H'}$ times, the corresponding probability is $\frac{q_{H'}q_\Sigma}{2^{k_0}}$.

Reference [9] uses the recommended key size in order to confirm that no adversary has the ability to break the 1024, 2048 bits factoring problem. In estimating the key size, we use Lemma 4 of [5] to modify the security statement of RSA-PSS-ES; that is, f's *partial-domain one-wayness* is replaced by *one-wayness* of RSA permutation paying the cost of running time and decreasing the success probability.

Throughout this evaluation, we assume that breaking the RSA problem is equivalent to solving the factoring problem, and that k_0 and k_1 are enough large so that factors that suppress the reduction efficiency can be ignored. The complexity of the factoring problem is measured by applying a number field sieve. Table 1 shows the recommended key size that achieves the same complexity as the 1024, 2048 bits factoring problem.

Table 1. Recommended key size

Scheme		1024bit	2048bit
PSS-ES	Encryption	6221	12452
	Signature	1363	2596
OAEP-ES	Encryption	5252	10838
	Signature	1363	2596
REACT-ES	Encryption	1363	2596
	Signature	1363	2596

As in Table 1, OAEP-ES has better reduction efficiency than PSS-ES because the security of PSS-ES(E) is based on *partial-domain one-wayness*. Therefore, compared to PSS-ES, OAEP-ES can decrease the key size by more than 950 bits for the 1024 bits factoring problem and by more than 1600 bits for 2048 bits factoring problem.

Moreover, as in Table 1, REACT-ES offers much better reduction efficiency than PSS-ES and OAEP-ES, and the key size does not increase comparing with the number of bits in the factoring problem[12]. This is because the running time of the permutation inverter of REACT-ES is of the order of $q_{H'}$ while that of OAEP-ES is of order of $q_G q_H$. This means that the key length of REACT-ES is shorter than that of OAEP-ES.

6   Discussion

REACT-ES is superior to OAEP-ES in terms of the running time of the permutation inverter, as shown by Theorem 3 (moreover, since PSS-ES owes its security to *partial-domain one-wayness*, its reduction efficiency is not good).

More precisely, when inverting the permutation for PSS-ES and OAEP-ES, the inverter should locate the preimage using the product of two hash functions'

---

[12] OAEP++-ES has the same reduction efficiency as REACT-ES, *i.e.*, the recommended key size for OAEP++-ES is the same as the one for REACT-ES.

input/output lists[13] (G-List and H-List). The inverter of REACT-ES, however, locates the preimage using the sum of two lists (H-List and G-List).Accordingly, the running time of the above theorem on REACT-ES is less than those on PSS-ES and OAEP-ES.

Therefore, as described in Section 5, the recommended key sizes that provide the same complexity as the 1024, 2048 bits factoring problem are, for OAEP++-ES and REACT-ES, much shorter than those of PSS-ES and OAEP-ES, and are about the same as the bit size of the factoring problem.

With regard to communication efficiency, the length of plaintext or message, in PSS-ES, OAEP-ES, and OAEP++-ES, is restricted to the key size. REACT-ES, however, allows us to encrypt (sign) arbitrary length of plaintext (message) by using symmetric encryption; it follows that REACT-ES is the most practical technique giving the high reduction efficiency of its security proof and its improved communication efficiency.

7   Conclusion

This paper first gave the general methodology to construct an ES scheme from an encryption scheme with a padding technique and proposed new ES schemes, OAEP-ES, OAEP++-ES, and REACT-ES, which use a unique padding technique and key pair to realize encryption and signature. It also proved that these two usages of proposed schemes satisfy IND-CCA2&ACMA and EUF-CCA2&ACMA, respectively. These schemes are suitable for implementation because they need only one padding technique and key pair.

Moreover, OAEP++-ES and REACT-ES offer much better reduction efficiency than PSS-ES and OAEP-ES. Using the evaluation of [9], the difficulty of breaking OAEP++-ES and REACT-ES is almost equal to that of the key size factoring problem. Hence, we conclude that OAEP++-ES and REACT-ES are more efficient than PSS-ES or OAEP-ES. Furthermore, from the view of the communication efficiency, REACT-ES allows us to encrypt (sign) a plaintext (message) arbitrary length through the use of symmetric encryption; we can conclude that REACT-ES is the most practical candidate due to the tightness of its security and its improved communication efficiency.

This paper also corrected the original mistakes made in proving the security of PSS-ES.

References

1. M. Bellare, A. Desai, D. Pointcheval, and P. Rogaway. Relations among notions of security for public-key encryption schemes. In H. Krawczyk, editor, *Advances in Cryptology — CRYPTO'98*, pages 26–45. Springer, 1998. Lecture Notes in Computer Science No. 1462.

---

[13] For PSS-ES, when replacing the *partial-domain one-wayness* to the *one-wayness* as in Lemma 4 of [5], we ought to run the adversary twice and get two input/output lists (two G-Lists).

2. M. Bellare and P. Rogaway. Optimal asymetric encryption — how to encrypt with RSA. In A.D. Santis, editor, *Advances in Cryptology — EUROCRYPT'94*, volume 950 of *Lecture Notes in Computer Science*, pages 92–111, Berlin, Heidelberg, New York, 1995. Springer-Verlag.

3. M. Bellare and P. Rogaway. The exact security of digital signatures –how to sign with RSA and Rabin. In U. Maurer, editor, *Advances in Cryptology — EUROCRYPT'96*, volume 1070 of *Lecture Notes in Computer Science*, pages 399–416, Berlin, Heidelberg, New York, 1996. Springer-Verlag.

4. J. S. Coron, M. Joye, D. Naccache, and P. Paillier. Universal padding schemes for RSA. In M. Yung, editor, *Advances in Cryptology — CRYPTO 2002*, volume 2422 of *Lecture Notes in Computer Science*, pages 226–241, Berlin, Heidelberg, New York, 2002. Springer-Verlag.

5. E. Fujisaki, T. Okamoto, D. Pointcheval, and J. Stern. RSA-OAEP is chosen-ciphertext secure under the RSA assumption. *Journal of Cryptology*, 2002.

6. S. Goldwasser, S. Micali, and R. Rivest. A digital signature scheme against adaptive chosen message attack. *Journal of Computing (Society for Industrial and Applied Mathematics)*, 17(2):281–308, 1988.

7. K. Kobara and H. Imai. OAEP++ : A very simple way to apply OAEP to deterministic OW-CPA primitives. 2002. Available at
   http://eprint.iacr.org/2002/130/.

8. Y. Komano and K. Ohta. OAEP-ES – Methodology of universal padding technique. *manuscript*, 2003.

9. T. Nakashima and T. Okamoto. Key size evaluation of provably secure RSA-based encryption schemes. *SCIS 2002, The 2002 Symposium on Cryptography and Information Security*, 2002.

10. T. Okamoto and D. Pointcheval. REACT: Rapid Enhanced-security Asymmetric Encryptosystem Tranceform. In D. Naccache, editor, *CT – RSA '2001*, volume 2020 of *Lecture Notes in Computer Science*, pages 159–175, Berlin, Heidelberg, New York, 2001. Springer-Verlag.

11. R. L. Rivest, A. Shamir, and L. Adleman. A method for obtaining digital signatures and public key cryptosystems. *Communications of the ACM*, 21(2):120–126, 1978.

12. V. Shoup. OAEP reconsidered. In J. Kilian, editor, *Advances in Cryptology — CRYPTO'2001*, volume 2139 of *Lecture Notes in Computer Science*, pages 239–259, Berlin, Heidelberg, New York, 2001. Springer-Verlag.

13. V. Shoup. A proposal for an ISO standard for public key encryption (version 2.1). In *manuscript*, 2001. http://shoup.net/papers/.

14. Y. Zheng. Degital signcryption or how to achieve cost(signature & encryption) $<<$ cost(signature) + cost(encryption). In *Advances in Cryptology — CRYPTO'97*, volume 1294 of *Lecture Notes in Computer Science*, pages 165–179, Berlin, Heidelberg, New York, 1997. Springer-Verlag.

A   Proof of Theorem 3

We follow the definition of symmetric encryption scheme and its security model from [10]. In the security proof of Theorem 3, assume that the symmetric encryption scheme is (τ', ν)-secure.

A.1   Construction of Inverter $\mathcal{I}$

We give the construction of inverter $\mathcal{I}$ that breaks the *one-wayness* of f about c^+, by using adversary $\mathcal{A}$ that breaks REACT-ES(E) in $(\tau, q_D, q_\Sigma, q_G, q_{H'}, q_H, \epsilon)$ in the sense of IND-CCA2&ACMA, as follows: we input public key f to $\mathcal{A}$, answer the queries that $\mathcal{A}$ asks to the random oracles, to the decryption oracle, and to the signing oracle in the following way, and receive challenge (x_0, x_1). We then choose $b \xleftarrow{R} \{0,1\}$, $r^+ \xleftarrow{R} \{0,1\}^{k_0}$, and $\mathsf{k} \xleftarrow{R} \{0,1\}^{k_3}$ and put $c_2^+ = E_{\mathsf{k}}^{sym}(x_b||r^+)$. Moreover, we answer the queries that $\mathcal{A}$ asks in the following way, and finally, receive $\hat{b}$ (or stop $\mathcal{A}$ after its running time τ is over).

In simulating random oracles G, H', and H, we construct input/output lists, G-List, H'-List, and H-List, respectively. In G-List, we preserve pair $(w, G(w))$ of query w and answer $G(w)$. In H'-List, we keep seven-tuple $(x||r, H'(x||r), z, c_1, c_2, c_3, c_3')$ of query $x||r$, answer $H'(x||r)$, guarantee z, c_2, c_3' for signing queries, and pledge c_1, c_2, c_3 for decryption queries. In H-List, we preserve sextuplet $(w, x||r, c_1, c_2, H(w, x||r, c_1, c_2))$ of query $w, x||r, c_1, c_2$ and answer $H(w, x||r, c_1, c_2)$.

Answering the random oracle queries to G, H', and H: For new query w to G, we choose a random integer from $\{0,1\}^{k_3}$, put it to $G(w)$, answer to $\mathcal{A}$, and add $(w, G(w))$ to G-List. If w has already been queried to G, we locate $(w, G(w)) \in$ G-List and answer $G(w)$.

For new query $(w, x||r, c_1, c_2)$ to H, we choose random integer c_3 from $\{0,1\}^{k_2}$, put it to $H(w, x||r, c_1, c_2)$, answer to $\mathcal{A}$, and add $(w, x||r, c_1, c_2, c_3)$ to H-List. Moreover, we simulate $G(w)$ in the above way[14]. If $(w, x||r, c_1, c_2)$ has already been queried to H, we locate $(w, x||r, c_1, c_2, c_3) \in$ H-List and answer c_3.

For new query $x||r$ to H', we get $z \xleftarrow{R} \{0,1\}^{k_1}$, set $f(z) = w$, and calculate $c_1 = f(w)$. Next, we simulate $G(w)$ in the same way described above, calculate $c_2 = E_{G(w)}^{sym}(x||r)$. Finally, we put $c_3 = H(w, x||r, c_1, c_2)$ and $c_3' = H(w, x||r, z, c_2)$ by simulating H in the same way described above, answer w as $H'(x||r)$ to $\mathcal{A}$, and add $(x||r, w, z, c_1, c_2, c_3, c_3')$ to H'-List. If $x||r$ has already been queried to H', we locate $(x||r, w, *, *, *, *, *) \in$ H'-List and answer w.

Answering the decryption queries to D: In order for decryption query $y = (c_1, c_2, c_3)$ to be valid ciphertext, $(x||r, *, *, c_1, c_2, c_3, *)$ must be contained in H'-List. In this case, we can answer with the corresponding plaintext x. Otherwise, we answer Reject since the probability of $H'(x||r) = w$ is negligible.

Answering the signing queries to Σ: For signing query x to Σ, we get $r \xleftarrow{R} \{0,1\}^{k_0}$ and check whether $(x||r, *, z, *, c_2, *, c_3')$ is in H'-List. If so, we answer $\sigma = (z, c_2, c_3')$ to $\mathcal{A}$ as a signature. Otherwise, we choose $z \xleftarrow{R} \{0,1\}^{k_1}$, set $f(z) = w$, and calculate $c_1 = f(w)$. Next, we simulate $G(w)$ in the same way described above, calculate $c_2 = E_{G(w)}^{sym}(x||r)$. Finally, we put $c_3 = H(w, x||r, c_1, c_2)$ and $c_3' = H(w, x||r, z, c_2)$ by simulating H in the same way described above, add $(x||r, w, z, c_1, c_2, c_3, c_3')$ to H'-List, and answer $\sigma = (z, c_2, c_3')$ as a signature to $\mathcal{A}$.

---

[14] We simulate G because we want to collect the information of input/output on w in G-List; this makes the estimation of the success probability of the permutation inversion easy.

A.2   Analysis

Let $y^+ = (c^+, c_2^+, c_3^+)$ be a target ciphertext that we answer to $\mathcal{A}$ deviating the protocol, and w^+, r^+, and x^+ be corresponding elements. In order to analyze the success probability of $\mathcal{I}$, we use following notations: AskG and AskH' are events for which $(w^+, *) \in$ G-List, and $(*||r^+, *, *, *, *, *, *) \in$ H'-List, respectively, and moreover, let EBad be an event[15] that $\text{AskH}' \wedge [H'(x_i||r^+) \neq w^+$ for $i = 0, 1]$, let DBad be an event that we fail to simulate in D, and let Bad = EBad $\vee$ DBad[16]. Our aim in setting these notations is to estimate the probability of AskG. At first, we divide this event as follows:

$$\Pr[\text{AskG}] = \Pr[\text{AskG} \wedge \text{Bad}] + \Pr[\text{AskG} \wedge \neg\text{Bad}]. \tag{1}$$

With regard to $\Pr[\text{AskG} \wedge \text{Bad}]$ in equation (1), from the definition of Bad, we have

$$\begin{aligned}
\Pr[\text{AskG} \wedge \text{Bad}] &= \Pr[\text{Bad}] - \Pr[\neg\text{AskG} \wedge \text{Bad}] \\
&\geq \Pr[\text{Bad}] - \Pr[\text{EBad}|\neg\text{AskG}] - \Pr[\text{DBad}|\neg\text{AskG}]. \tag{2}
\end{aligned}$$

We can estimate $\Pr[\text{EBad}|\neg\text{AskG}]$ in inequality (2) because, by the definition of EBad, we have $\Pr[\text{EBad}|\neg\text{AskG}] \leq \Pr[\text{AskH}'|\neg\text{AskG}]$. Here, $\Pr[\text{AskH}'|\neg\text{AskG}] \leq \frac{q_{H'} + q_{\Sigma}}{2^{k_0}}$, because if $\neg\text{AskG}$ occurs, $G(w^+)$ and r^+ are random integers for $\mathcal{A}$ and it is only by accident that $*||r^+$ is queried to H'.

Moreover, $\Pr[\text{DBad}|\neg\text{AskG}]$ in inequality (2) is less than $\frac{q_D}{2^{k_1}}$. Note that in answering to decryption query (c_1, c_2, c_3), we search H'-List for corresponding plaintext x, therefore we fail to simulate the decryption oracle if $\mathcal{A}$ does not query H' about $x||r$ and ciphertext (decryption query) y output by $\mathcal{A}$ is valid. However, if $\mathcal{A}$ does not query H' about $x||r$, $H'(x||r)$ is uniformly distributed in $\{0, 1\}^{k_1}$, and then, it is only by accident (with probability $\frac{1}{2^{k_1}}$) that $w = f^{-1}(c_1)$ equals $H'(x||r)$.

Hence, we can evaluate $\Pr[\text{AskG} \wedge \text{Bad}]$ in equation (1) by

$$\Pr[\text{AskG} \wedge \text{Bad}] \geq \Pr[\text{Bad}] - \frac{q_{H'} + q_{\Sigma}}{2^{k_0}} - \frac{q_D}{2^{k_1}}. \tag{3}$$

With regard to the second term of equation (1), it is meaningful to consider the advantage of $\mathcal{A}$ because of the condition $\neg\text{Bad}$. We can do this by evaluating $\Pr[\text{AskG} \wedge \neg\text{Bad}]$ as follows:

$$\begin{aligned}
\Pr[\text{AskG} \wedge \neg\text{Bad}] &\geq \Pr[\mathcal{A} = b \wedge \text{AskG} \wedge \neg\text{Bad}] \\
&= \Pr[\mathcal{A} = b \wedge \neg\text{Bad}] - \Pr[\mathcal{A} = b \wedge \neg\text{AskG} \wedge \neg\text{Bad}]. \tag{4}
\end{aligned}$$

In inequality (4), both

---

[15] In this event, $\mathcal{A}$ may notice that we answer y^+ as a target ciphertext deviating the protocol.

[16] Note that we never fail to simulate the answer to the signing query, described in section A.1, and do not need to consider event ΣBad.

$$\Pr[\mathcal{A} = b \wedge \neg \mathsf{Bad}] \geqq \Pr[\mathcal{A} = b] - \Pr[\mathsf{Bad}] = (\frac{\epsilon}{2} + \frac{1}{2}) - \Pr[\mathsf{Bad}]$$

and[17]

$$\Pr[\mathcal{A} = b \wedge \neg \mathsf{AskG} \wedge \neg \mathsf{Bad}] = \Pr[\mathcal{A} = b | \neg \mathsf{AskG} \wedge \neg \mathsf{Bad}] \Pr[\neg \mathsf{AskG} \wedge \neg \mathsf{Bad}]$$
$$= (\frac{\nu}{2} + \frac{1}{2})(1 - \Pr[\mathsf{Bad}] - \Pr[\mathsf{AskG} \wedge \neg \mathsf{Bad}])$$
$$\leqq \frac{1}{2}(1 - \Pr[\mathsf{Bad}] - \Pr[\mathsf{AskG} \wedge \neg \mathsf{Bad}]) + \frac{\nu}{2} \cdot 1.$$

hold[18]. Therefore, by substituting above two inequalities into (4),

$$\Pr[\mathsf{AskG} \wedge \neg \mathsf{Bad}] \geqq \frac{\epsilon - \nu - \Pr[\mathsf{Bad}] + \Pr[\mathsf{AskG} \wedge \neg \mathsf{Bad}]}{2}$$

holds and this inequality leads to

$$\Pr[\mathsf{AskG} \wedge \neg \mathsf{Bad}] \geqq \epsilon - \nu - \Pr[\mathsf{Bad}]. \qquad (5)$$

Hence, the considerations of equation (1) and inequalities (3) and (5) conclude the proof of Theorem 3.

The running time τ' of $\mathcal{I}$ is the sum of the following terms: (i) the running time τ of $\mathcal{A}$ because we run $\mathcal{A}$ once, (ii) in order to find corresponding pair from G-List to c^+, we compute f at most $q_G + q_{H'} + q_H + q_\Sigma$ times, i.e., $(q_G + q_{H'} + q_H + q_\Sigma)T_f$, (iii) in order to be able to simulate D and Σ, we calculate both $f(z)$ and $f(w)$ in simulation of H' and Σ[19] $q_{H'} + q_\Sigma$ times, i.e., $(q_{H'} + q_\Sigma)T_f$. Hence, $\tau' \leqq \tau + (q_G + q_H + 2q_{H'} + 2q_\Sigma)T_f$ holds.

---

[17] Note that in our simulation, if $\mathcal{A}$ notices the deviation (i.e., if event Bad occurs), it does not run for some pairs of random coins of $\mathcal{A}$ and $\mathcal{I}$. Therefore, $\Pr[\mathcal{A} = b]$ in this inequality is taken over the random coins of $\mathcal{A}$ and $\mathcal{I}$ in which $\mathcal{A}$ does not notice the deviation. Though the probabilistic space is restricted and smaller than the entire probabilistic space, the probability of the event that $\mathcal{A} = b$ is equal to the one taken over the entire probabilistic space, from the definition of the random oracle model; $\Pr[\mathcal{A} = b] = \frac{\epsilon}{2} + \frac{1}{2}$.

[18] Note that the probability that $\mathcal{A} = b$ holds under the condition of $\neg \mathsf{AskG}$ and $\neg \mathsf{Bad}$ is equal to the probability that $\mathcal{A}$ can distinguish b from x_0, x_1 and c_2^+, without secret key k; $\Pr[\mathcal{A} = b | \neg \mathsf{AskG} \wedge \neg \mathsf{Bad}] = \frac{\nu}{2} + \frac{1}{2}$. This is because from $\neg \mathsf{Bad}$, $\mathcal{A}$ cannot notice the deviation and performs the same way as in the real run. Moreover, from $\neg \mathsf{AskG}$, $\mathcal{A}$ cannot know k $= G(w^+)$.

[19] This seems to require the calculation of f $2(q_{H'} + q_\Sigma)$ times, but $q_{H'} + q_\Sigma$ calculations are sufficient. Indeed, when we add an element including w to G-List or H-List, we check whether $f(w) = c^+$ holds. This action plays the role of preparing for the simulation of D and is already counted in (ii). Therefore, we consider only the preparation for the signing oracle queries in (iii).

B   Proof of Theorem 4

B.1   Construction of Inverter $\mathcal{I}$

We give the construction of inverter $\mathcal{I}$ that breaks the *one-wayness* of f about η, by using forger $\mathcal{F}$ that breaks REACT-ES(S) in $(\tau, q_D, q_\Sigma, q_G, q_{H'}, q_H, \epsilon)$ in the sense of EUF-CCA2&ACMA as follows: we input public key f to $\mathcal{F}$, answer the queries that $\mathcal{F}$ asks to the random oracles, to the decryption oracle, and to the signing oracle in the same way in section A.1, except those to H' and Σ (described below). Finally, we receive forgery $\sigma^* = (c_1^*, c_2^*, c_3^*)$ (or stop $\mathcal{F}$ after its running time τ is over.)

In simulating random oracles G, H', and H, we construct input/output lists, G-List, H′-List, and H-List, respectively. G-List holds $(w, G(w))$, the pairing of query w and answer $G(w)$. H′-List holds $(b, x||r, H'(x||r), z, c_1, c_2, c_3.c_3')$, the bit $b = 0/1$, query $x||r$, answer $H'(x||r)$, guarantee z, c_2, c_3' for signing queries, and pledge c_1, c_2, c_3 for decryption queries. H-List holds $(w, x||r, c_1, c_2, H(w, x||r, c_1, c_2))$, the pairing of query $w, x||r, c_1, c_2$ and answer $H(w, x||r, c_1, c_2)$.

Answering the random oracle queries to H': For new query $x||r$ to H', we get $z \xleftarrow{R} \{0,1\}^{k_1}$, set $f(z)\eta = w$, and calculate $c_1 = f(w)$. Next, we simulate $G(w)$ in the same way as in section A.1, calculate $c_2 = E_{G(w)}^{sym}(x||r)$. Finally, we put $c_3 = H(w, x||r, c_1, c_2)$ and $c_3' = H(w, x||r, z, c_2)$ by simulating in the same way as in section A.1, answer w as $H'(x||r)$ to $\mathcal{F}$, and add $(1, x||r, w, z, c_1, c_2, c_3, c_3')$ to H′-List. If $x||r$ has already been queried to H', we locate $(*, x||r, w, *, *, *, *, *) \in$ H′-List and answer w.

Answering the signing queries to Σ: For signing query x to Σ, we get $r \xleftarrow{R} \{0,1\}^{k_0}$ and check whether $(0, x||r, *, z, *, c_2, *, c_3')$ is in H′-List. If so, we answer $\sigma = (z, c_2, c_3')$ to $\mathcal{F}$ as a signature. Moreover, if $(1, x||r, *, *, *, *, *, *)$ is in H′-List, we abort. Otherwise, we choose $z \xleftarrow{R} \{0,1\}^{k_1}$, put $f(z) = w$, and calculate $c_1 = f(w)$. Next, we simulate $G(w)$ in the same way as in section A.1, calculate $c_2 = E_{G(w)}^{sym}(x||r)$. Finally, we put $c_3 = H(w, x||r, c_1, c_2)$ and $c_3' = H(w, x||r, z, c_2)$ by simulating in the same way as in section A.1, add $(0, x||r, w, z, c_1, c_2, c_3, c_3')$ to H′-List, and answer $\sigma = (z, c_2, c_3')$ as a signature to $\mathcal{F}$.

B.2   Analysis

Let $\sigma^* = (c_1^*, c_2^*, c_3^*)$ be a forgery output by $\mathcal{F}$; w^*, r^*, and x^* are the corresponding elements. In order to analyze the success probability of $\mathcal{I}$, let DBad be the same event as in A.2, ΣBad an event that $\mathcal{I}$ fails to simulate in Σ, and Bad = DBad $\vee$ ΣBad. Moreover, let S be an event that $\mathcal{V}_{pk}(\sigma^*) = x^*$, and let AskH′ be one that $\mathcal{F}$ queries directly H' about $x^*||r^*$.

At first, we consider

$$1 = \Pr[\text{Bad}] + \Pr[\neg\text{Bad}]. \tag{6}$$

With regard to $\Pr[\text{Bad}] \leq \Pr[\text{DBad}] + \Pr[\Sigma\text{Bad}]$ in equation (6), we have

$$\Pr[\text{Bad}] \leqq \frac{q_{H'}q_\Sigma}{2^{k_0}} + \frac{q_D}{2^{k_1}}. \tag{7}$$

In fact, $\Pr[\mathsf{DBad}]$ is evaluated in the same way as in section A.2. On the other hand, $\Pr[\varSigma\mathsf{Bad}]$ is bounded by $q_\varSigma(\frac{q_{H'}}{2^{k_0}})$. Note that in simulating the answer signing query x, we first choose random integer r and z for the candidate of the signature, and simulate H' about $x\|r$. In this phase, $\varSigma\mathsf{Bad}$ occurs if $x\|r$ has already queried to H' by $\mathcal{F}$ directly, because we can not calculate $f^{-1}(\eta)$. For a signing query, the probability that $x\|r$ is queried to H' is bounded by $\frac{q_{H'}}{2^{k_0}}$ [20] because of randomness of r, and then, we can estimate $\Pr[\varSigma\mathsf{Bad}]$ by $q_\varSigma(\frac{q_{H'}}{2^{k_0}})$.

With regard to $\Pr[\neg\mathsf{Bad}]$ in equation (6), we divide event $\neg\mathsf{Bad}$ by S and have

$$\Pr[\neg\mathsf{Bad}] = \Pr[\mathsf{S}\wedge\neg\mathsf{Bad}] + \Pr[\neg\mathsf{S}\wedge\neg\mathsf{Bad}]. \tag{8}$$

In this equation (8),

$$\Pr[\neg\mathsf{S}\wedge\neg\mathsf{Bad}] \leqq \Pr[\neg\mathsf{S}|\neg\mathsf{Bad}] = 1 - \Pr[\mathsf{S}|\neg\mathsf{Bad}] = 1 - \epsilon \tag{9}$$

holds [21].

Next, we estimate $\Pr[\mathsf{S}\wedge\neg\mathsf{Bad}]$ in equation (8) by dividing event $\mathsf{S}\wedge\neg\mathsf{Bad}$ by $\mathsf{AskH'}$:

$$\Pr[\mathsf{S}\wedge\neg\mathsf{Bad}] = \Pr[\mathsf{S}\wedge\neg\mathsf{Bad}\wedge\mathsf{AskH'}] + \Pr[\mathsf{S}\wedge\neg\mathsf{Bad}\wedge\neg\mathsf{AskH'}].$$

In this equality, $\Pr[\mathsf{S}\wedge\neg\mathsf{Bad}\wedge\neg\mathsf{AskH'}]$ is bounded by $\frac{1}{2^{k_1}}$ because it is an incident that $H'(x^*\|r^*) = w^*$ if $(1, x^*\|r^*, *, *, *, *, *, *) \notin H'$-List holds. On the other hand, we have $\Pr[\mathsf{S}\wedge\neg\mathsf{Bad}\wedge\mathsf{AskH'}] \leqq \Pr[\mathsf{S}\wedge\mathsf{AskH'}] \leqq \mathsf{Succ}^{\mathsf{ow}}(\tau')$ because if both $(1, x^*\|r^*, *, z^*, *, *, *, *) \in H'$-List and S hold, then we can compute $\frac{c_1^*}{z^*} = \frac{f^{-1}(f(z^*)\eta)}{z^*} = f^{-1}(\eta)$ from the multiplicative property of f. Therefore, we have

$$\Pr[\mathsf{S}\wedge\neg\mathsf{Bad}] \leqq \mathsf{Succ}^{\mathsf{ow}}(\tau') + \frac{1}{2^{k_1}}. \tag{10}$$

By substituting inequalities (9) and (10) into equation (8), we have

$$\Pr[\neg\mathsf{Bad}] \leqq \mathsf{Succ}^{\mathsf{ow}}(\tau') + \frac{1}{2^{k_1}} + 1 - \epsilon. \tag{11}$$

Finally, if we substitute inequalities (7) and (11) into equation (6), we can conclude the proof of Theorem 4.

The running time τ' of $\mathcal{I}$ is the sum of the following terms: (i) the running time τ of $\mathcal{F}$ because we run $\mathcal{F}$ once, (ii) in the simulation of $\varSigma$, we have to prepare the answer for queries to D and $\varSigma$, i.e., $2q_\varSigma T_f$, (iii) in the simulation of H', we have to prepare the answer for queries to D and to implant η, i.e., $2q_{H'}T_f$, (iv) we have to find z^* corresponding to c_1^* by computing $f(c_1^*)$ once, i.e., T_f. Hence, $\tau' \leqq \tau + (2q_{H'} + 2q_\varSigma + 1)T_f$ holds.

---

[20] Note that for signing query x, if we choose random integer r such that $x\|r$ is queried to H' through the past signing query, we can reply this query by locating corresponding signature from H'-List. Therefore, we only consider the case that $x\|r$ has already queried to H' by $\mathcal{F}$ directly, in the estimation of $\varSigma\mathsf{Bad}$.

[21] Note that the success probability of $\mathcal{F}$ under the condition that $\mathcal{F}$ does not notice the simulation is equal to the one in real run and this leads $\Pr[\mathsf{S}|\neg\mathsf{Bad}] = \epsilon$.

Multipurpose Identity-Based Signcryption
A Swiss Army Knife for Identity-Based Cryptography

Xavier Boyen

IdentiCrypt, 420 Florence, Palo Alto, California — `crypto@boyen.org`

Abstract. Identity-Based (IB) cryptography is a rapidly emerging approach to public-key cryptography that does not require principals to pre-compute key pairs and obtain certificates for their public keys—instead, public keys can be arbitrary identifiers such as email addresses, while private keys are derived at any time by a trusted private key generator upon request by the designated principals. Despite the flurry of recent results on IB encryption and signature, some questions regarding the security and efficiency of practicing IB encryption (IBE) and signature (IBS) as a joint IB signature/encryption (IBSE) scheme with a common set of parameters and keys, remain unanswered.

We first propose a stringent security model for IBSE schemes. We require the usual strong security properties of: (for confidentiality) indistinguishability against adaptive chosen-ciphertext attacks, and (for non-repudiation) existential unforgeability against chosen-message insider attacks. In addition, to ensure as strong as possible ciphertext armoring, we also ask (for anonymity) that authorship not be transmitted in the clear, and (for unlinkability) that it remain unverifiable by anyone except (for authentication) by the legitimate recipient alone.

We then present an efficient IBSE construction, based on bilinear pairings, that satisfies all these security requirements, and yet is as compact as pairing-based IBE and IBS in isolation. Our scheme is secure, compact, fast and practical, offers detachable signatures, and supports multi-recipient encryption with signature sharing for maximum scalability.

1 Introduction

Recently, Boneh and Franklin [5] observed that bilinear pairings on elliptic curves could be used to make identity-based encryption possible and practical. Following this seminal insight, the last couple of years have seen a flurry of results on a number of aspects of what has now become the nascent field of Identity-Based (IB) cryptography.

1.1 Identity-Based Cryptography

The distinguishing characteristic of IB cryptography is the ability to use any string as a public key; the corresponding private key can only be derived by a trusted Private Key Generator (PKG), custodian of a master secret. For encryption purposes, this allows Alice to securely send Bob an encrypted message,

D. Boneh (Ed.): CRYPTO 2003, LNCS 2729, pp. 383–399, 2003.

using as public key any unambiguous name identifying Bob, such as Bob's email address, possibly before Bob even knows his own private key. For signature purposes, Alice may sign her communications using a private key that corresponds to an unambiguous name of hers, so that anybody can verify the authenticity of the signature simply from the name, without the need for a certificate. Revocation issues are handled by using short-lived time-dependent identities [5].

An inherent limitation of IB cryptography is the trust requirement that is placed on the PKG, as an untrustworthy PKG will have the power to forge Alice's signature, and decrypt Bob's past and future private communications. This can be partially alleviated by splitting the master secret among several PKGs under the jurisdiction of no single entity, as explained in [5]. The window of vulnerability can also be reduced by periodically changing the public parameters, and purging any master secret beyond a certain age, effectively limiting the interval during which IB cryptograms can be decrypted. Traditional public-key cryptography is not completely immune to the problem, either: in a public key infrastructure, the certification authority has the power to issue fake certificates and impersonate any user for signature purposes; it can similarly spoof encryption public key certificates in order to decrypt future ciphertexts addressed to targeted users, albeit not in a manner not amenable to easy detection.

The idea of IB cryptography first emerged in 1984 [24], although only an IB signature (IBS) scheme was then suggested, based on conventional algebraic methods in $\mathbb{Z}_n$. Other IBS and identification schemes were quick to follow [13,12]. However, it is only in 2001 that a practical IB encryption (IBE) mechanism was finally suggested [5], based on the much heavier machinery of bilinear pairings on elliptic curves, whose use in cryptography had slowly started to surface in the few years prior, e.g., for key exchange [18] and IBS [23]. Interestingly, a more conventional approach to IBE was proposed shortly thereafter [10], albeit not as efficient as the pairing-based IBE scheme of Boneh and Franklin.

1.2 Motivation and Contribution

Following the original publication of the BF-IBE scheme [5], a number of authors have proposed various new applications of pairing-based IB cryptography. These include various IB signature schemes [22,16,8], key agreement [25], a 2-level hierarchical IB encryption [17], and a general hierarchical IB encryption that can also be used to produce signatures [15]. More specifically focusing on joint authentication and encryption, we note a repudiable authenticated IBE [20], an authenticated key agreement scheme [9], and a couple of IB signcryption schemes that efficiently combine signature and encryption [21,19].

What the picture is currently missing is an algorithm that combines (existing or new) IBE and IBS in a practical and secure way. Indeed, it would be of great practical interest to be able to use the same IB infrastructure for signing and encrypting. A possibility is to combine some existing IBE and IBS using black-box composition techniques, such as [1]; this is however rather suboptimal.

A better approach would be to exploit the similarities between IBE and IBS, and elaborate a dual-purpose IB Encryption-Signature (IBSE) scheme based on a shared infrastructure, toward efficiency increases and security improvements. Doing so, we would have to ensure that no hidden weakness arises from the combination, which is always a risk if the same parameters and keys are used. The issues that arise from this approach are summarized as follows:

- Can IBE and IBS be practiced in conjunction, in a secure manner, sharing infrastructure, parameters, and keys, toward greater efficiency?
- What emerging security properties can be gained from such a combination?

Our contributions to answering these questions are twofold. We first specify a security model that a strong IBSE combination should satisfy. Our model specifies the IBSE version of the strongest notions of security usually considered in public-key cryptography. For confidentiality, we define a notion of ciphertext indistinguishability under adaptive chosen-ciphertext attacks. For non-repudiation, we define a notion of signature unforgeability under chosen-message attacks, in the stringent case of an 'insider' adversary, *i.e.*, with access to the decryption private key, as considered in [1]. We also specify the additional security features of ciphertext authentication, anonymity, and unlinkability, that, if less conventional, are highly desirable in practice: together, they convince the legitimate recipient of the ciphertext origin, and conceal it from anyone else.

We then propose a fast IBSE scheme satisfying our strong security requirements, which we prove in the random oracle model [3]. Our scheme uses the properties of bilinear pairings to achieve a two-layer sign-then-encrypt combination, featuring a detachable randomized signature, followed by anonymous deterministic encryption. The scheme is very efficient, more secure than what we call *monolithic* signcryption—in which a single operation is used for decryption and signature verification, as in the original signcryption model of [27]—and more compact than generic compositions of IBE and IBS. Our two-layer design is also readily adapted to provide multi-recipient encryption of the same message with a shared signature and a single bulk message encryption.

Performance-wise, our dual-purpose optimized IBSE scheme is as compact as most existing single-purpose IBE and IBS taken in isolation. It is also about as efficient as the monolithic IB signcryption schemes of [21] and [19], with the added flexibility and security benefits that separate anonymous decryption and signature verification layers can provide. A comparative summary of our scheme with competing approaches can be found in §6, Table 2.

1.3   Outline of the Paper

We start in §2 by laying out the abstract IBSE specifications. In §3, we formalize the various security properties sought from the cryptosystem. In §4, we review the principles of IB cryptography based on pairings. In §5, we describe an implementation of our scheme. In §6, we make detailed performance and security comparisons with the competition. In §7, we prove compliance of our implementation with the security model. In §8, we study a few extensions of practical significance. Finally, in §9, we draw some conclusions.

2   Specification of the Cryptosystem

An *Identity-Based Signature/Encryption* scheme, or IBSE, consists of a suite
of six algorithms: Setup, Extract, Sign, Encrypt, Decrypt, and Verify. In essence,
Setup generates random instances of the common public parameters and master
secret; Extract computes the private key corresponding to a given public identity
string; Sign produces a signature for a given message and private key; Encrypt
encrypts a signed plaintext for a given identity; Decrypt decrypts a ciphertext
using a given private key; Verify checks the validity of a given signature for a
given message and identity. Messages are arbitrary strings in $\{0,1\}^*$.

The functions that compose a generic IBSE are thus specified as follows.

Setup On input 1^n, produces a pair $\langle \sigma, \pi \rangle$ (where σ is a randomly generated
master secret and π the corresponding common public parameters, for the
security meta-parameter n).

Extract$_{\pi,\sigma}$ On input id, computes a private key pvk (corresponding to the identity
id under $\langle \sigma, \pi \rangle$).

Sign$_\pi$ On input $\langle \text{pvk}_A, \text{id}_A, m \rangle$, outputs a signature s (for pvk$_A$, under π), and
some ephemeral state data r.

Encrypt$_\pi$ On input $\langle \text{pvk}_A, \text{id}_B, m, s, r \rangle$, outputs an anonymous ciphertext c (containing the signed message $\langle m, s \rangle$, encrypted for the identity id$_B$ under π).

Decrypt$_\pi$ On input $\langle \text{pvk}_B, \hat{c} \rangle$, outputs a triple $\langle \hat{\text{id}}_A, \hat{m}, \hat{s} \rangle$ (containing the purported sender identity and signed message obtained by decrypting $\hat{c}$ by the
private key pvk$_B$ under π).

Verify$_\pi$ On input $\langle \hat{\text{id}}_A, \hat{m}, \hat{s} \rangle$, outputs $\top$ 'true' or $\bot$ 'false' (indicating whether $\hat{s}$
is a valid signature for the message $\hat{m}$ by the identity $\hat{\text{id}}_A$, under π).

Since we are concerned with sending messages that are simultaneously encrypted
and signed, we allow the encryption function to make use of the private key of the
sender. Accordingly, we assume that Encrypt is always used on an output from
Sign, so that we may view the Sign/Encrypt composition as a single 'signcryption'
function; we keep them separate to facilitate the treatment of multi-recipient
encryption with shared signature in §8.1. We also insist on the dichotomy Decrypt
vs. Verify, to permit the decryption of anonymous ciphertexts, and to decouple
signature verification from the data that is transmitted over the wire, neither of
which would be feasible had we used a monolithic 'unsigncryption' function.

It is required that these algorithms jointly satisfy the following consistency
constraints.

Definition 1. *For all master secret and common parameters* $\langle \sigma, \pi \rangle \leftarrow$
Setup$[1^n]$, *any identities* id$_A$ *and* id$_B$, *and matching private keys* pvk$_A$ $=$
Extract$_{\pi,\sigma}[\text{id}_A]$ *and* pvk$_B$ $=$ Extract$_{\pi,\sigma}[\text{id}_B]$, *we require for consistency that,*
$\forall m \in \{0,1\}^*$:

$$\left\{ \begin{array}{r} \langle s, r \rangle \leftarrow \text{Sign}_\pi[\text{pvk}_A, \text{id}_A, m] \\ c \leftarrow \text{Encrypt}_\pi[\text{pvk}_A, \text{id}_B, m, s, r] \\ \langle \hat{\text{id}}_A, \hat{m}, \hat{s} \rangle \leftarrow \text{Decrypt}_\pi[\text{pvk}_B, c] \end{array} \right\} \implies \left\{ \begin{array}{c} \hat{\text{id}}_A = \text{id}_A \\ \hat{m} = m \\ \text{Verify}_\pi[\text{id}_A, \hat{m}, \hat{s}] = \top \end{array} \right\}$$

In the sequel, we omit the subscripted parameters π and σ when understood
from context.

3   Formal Security Model

Due to the identity-based nature of our scheme, and the combined requirements on confidentiality and non-repudiation, the security requirements are multi-faceted and quite stringent. For example, for confidentiality purposes, one should assume that the adversary may obtain any private key other than that of the targeted recipient, and has an oracle that decrypts any valid ciphertext other than the challenge. For non-repudiation purposes, we assume that the forger has access to any private key other than that of the signer, and can query an oracle that signs and encrypts any message but the challenge. These assumptions essentially amount to the 'insider' model in the terminology of [1].

We also consider the notions of *ciphertext unlinkability* and *ciphertext authentication*, which allow the legitimate recipient to privately verify—but not prove to others—that the ciphertext addressed to him and the signed message it contains were indeed produced by the same entity. We note that these properties are not jointly achieved by other schemes that combine confidentiality and non-repudiation, such as the signcryption of [21] and [19]. We also ask for *ciphertext anonymity*, which simply means that no third party should be able to discover whom a ciphertext originates from or is addressed to, if the sender and recipient wish to keep that a secret.

All these properties are recapitulated as follows.

1. *message confidentiality* (§3.1): allows the communicating parties to preserve the secrecy of their exchange, if they choose to.
2. *signature non-repudiation* (§3.2): makes it universally verifiable that a message speaks in the name of the signer (regardless of the ciphertext used to convey it, if any). This implies message authentication and integrity.
3. *ciphertext unlinkability* (§3.3): allows the sender to disavow creating a ciphertext for any given recipient, even though he or she remains bound to the valid signed message it contains.
4. *ciphertext authentication* (§3.4): allows the legitimate recipient, alone, to be convinced that the ciphertext and the signed message it contains were crafted by the same entity. This implies ciphertext integrity.
5. *ciphertext anonymity* (§3.5): makes the ciphertext appear anonymous (hiding both the sender and the recipient identities) to anyone who does not possess the recipient decryption key.

For simplicity of the subsequent analysis, we disallow messages from being addressed to the same identity as authored them—a requirement that we call the *irreflexivity assumption*. Remark that if such a mode of operation is nonetheless desired, it can easily be achieved, either, (1) by endowing each person with an additional 'self' identity, under which they can encrypt messages signed under their regular identity, or, (2) by splitting each identity into a 'sender' identity and a 'recipient' identity, to be respectively used for signature and encryption purposes. This can be done, *e.g.*, by prepending an indicator bit to all identity strings; each individual would then be given two private keys by the PKG.

For clarity, and regardless of which of the above convention is chosen, if any, we use the subscripts 'A' for Alice the sender and 'B' for Bob the recipient.

3.1   Message Confidentiality

Message confidentiality against adaptive chosen-ciphertext attacks is defined in terms of the following game, played between a challenger and an adversary. We combine signature and encryption into a dual-purpose oracle, to allow Encrypt to access the ephemeral random state data r from Sign.

Start. The challenger runs the Setup procedure for a given value of the security parameter n, and provides the common public parameters π to the adversary, keeping the secret σ for itself.

Phase 1. The adversary makes a number of queries to the challenger, in an adaptive fashion (*i.e.*, one at a time, with knowledge of the previous replies). The following queries are allowed:

Signature/encryption queries in which the adversary submits a message and two distinct identities, and obtains a ciphertext containing the message signed in the name of the first identity and encrypted for the second identity;

Decryption queries in which the adversary submits a ciphertext and an identity, and obtains the identity of the sender, the decrypted message, and a valid signature, provided that (1) the decrypted identity of the sender differs from that of the specified recipient, and (2) the signature verification condition Verify $= \top$ is satisfied; otherwise, the oracle only indicates that the ciphertext is invalid for the specified recipient;

Private key extraction queries in which the adversary submits an identity, and obtains the corresponding private key;

Selection. At some point, the adversary returns two distinct messages m_0 and m_1 (assumed of equal length), a signer identity id_A, and a recipient identity id_B, on which it wishes to be challenged. The adversary must have made no private key extraction query on id_B.

Challenge. The challenger flips $b \in \{0,1\}$, computes $\mathrm{pvk}_A = \mathrm{Extract}[\mathrm{id}_A]$, $\langle s, r \rangle \leftarrow \mathrm{Sign}[\mathrm{pvk}_A, m_b]$, $c \leftarrow \mathrm{Encrypt}[\mathrm{pvk}_A, \mathrm{id}_B, m_b, s, r]$, and returns the ciphertext c as challenge to the adversary.

Phase 2. The adversary adaptively issues a number of additional encryption, decryption, and extraction queries, under the additional constraint that it not ask for the private key of id_B or the decryption of c under id_B.

Response. The adversary returns a guess $\hat{b} \in \{0,1\}$, and wins the game if $\hat{b} = b$.

It is emphasized that the adversary is allowed to know the private key pvk_A corresponding to the signing identity, which gives us *insider-security* for confidentiality [1]. On the one hand, this is necessary if confidentiality is to be preserved in case the sender's private key becomes compromised. On the other hand, this will come handy when we study a 'repudiable' IBSE variant in §8.2.

This game is very similar to the IND-ID-CCA attack in [5]; we call it an IND-IBSE-CCA attack.

Definition 2. *An identity-based joint encryption and signature (IBSE) scheme is said to be semantically secure against adaptive chosen-ciphertext insider* *attacks, or* IND-IBSE-CCA *secure, if no randomized polynomial-time adversary has a non-negligible advantage in the above game. In other words, any randomized polynomial-time IND-IBSE-CCA adversary* $\mathcal{A}$ *has an advantage* $\mathbf{Adv}_{\mathcal{A}}[n] = |\mathbf{P}[\hat{b} = b] - \frac{1}{2}|$ *that is* $o[1/\mathrm{poly}[n]]$ *for any polynomial* $\mathrm{poly}[n]$ *in the security parameter.*

Remark that we insist that the decryption oracle perform a validity check before returning a decryption result, even though Decrypt does not specify it. This requirement hardly weakens the model, and allows for stronger security results. We similarly ask that the oracles enforce the irreflexivity assumption, *e.g.*, by refusing to produce or decrypt non-compliant ciphertexts.

3.2   Signature Non-repudiation

Signature non-repudiation is formally defined in terms of the following game, played between a challenger and an adversary.

Start. The challenger runs the Setup procedure for a given value of the security parameter n, and provides the common public parameters π to the adversary, keeping the secret σ for itself.

Query. The adversary makes a number of queries to the challenger. The attack may be conducted adaptively, and allows the same queries as in the Confidentiality game of §3.1, namely: signature/encryption queries, decryption queries, and private key extraction queries.

Forgery. The adversary returns a recipient identity id_B and a ciphertext c.

Outcome. The adversary wins the game if the ciphertext c decrypts, under the private key of id_B, to a signed message $\langle \mathrm{id}_A, \hat{m}, \hat{s} \rangle$ that satisfies $\mathrm{id}_A \neq \mathrm{id}_B$ and Verify$[\mathrm{id}_A, \hat{m}, \hat{s}] = \top$, provided that (1) no private key extraction query was made on id_A, and (2) no signature/encryption query was made that involved $\hat{m}$, id_A, and some recipient $\mathrm{id}_{B'}$, and resulted in a ciphertext c' whose decryption under the private key of $\mathrm{id}_{B'}$ is the claimed forgery $\langle \mathrm{id}_A, \hat{m}, \hat{s} \rangle$.

Such a model is very similar to the usual notion of existential unforgeability against chosen-message attacks [11,26]; we call it an EUF-IBSE-CMA attack.

Definition 3. *An IBSE scheme is said to be existentially signature-unforgeable against chosen-message insider attacks, or* EUF-IBSE-CMA *secure, if no randomized polynomial-time adversary has a non-negligible advantage in the above game. In other words, any randomized polynomial-time EUF-IBSE-CMA adversary* $\mathcal{A}$ *has an advantage* $\mathbf{Adv}_{\mathcal{A}}[n] = \mathbf{P}[\mathrm{Verify}[\mathrm{id}_A, \hat{m}, \hat{s}] = \top]$ *that behaves as* $o[1/\mathrm{poly}[n]]$ *for any polynomial* $\mathrm{poly}[n]$.

In the above experiment, the adversary is allowed to obtain the private key pvk_B for the forged message recipient id_B, which corresponds to the stringent requirements of *insider-security* for authentication [1]. There is one important difference, however: in [1], non-repudiation applies to the ciphertext itself, which is the only sensible thing to do in the context of a signcryption model with a monolithic 'unsigncryption' function. Here, given our two-step Decrypt/Verify specification, we define non-repudiation with respect to the decrypted signature, which is more intuitive and does not preclude ciphertext unlinkability (see §3.3).

3.3   Ciphertext Unlinkability

Ciphertext unlinkability is the property that makes it possible for Alice to deny having sent a given ciphertext to Bob, even if the ciphertext decrypts (under Bob's private key) to a message bearing Alice's signature. In other words, the signature should only be a proof of authorship of the plaintext message, and not the ciphertext. (We shall make one exception to this requirement in §3.4, where we seek that the legitimate recipient be able privately authenticate the ciphertext, in order to be convinced that it is indeed addressed to him or her.)

Ciphertext unlinkability allows Alice, *e.g.*, as a news correspondent in a hostile area, to stand behind the content of her reporting, but conceal any detail regarding the particular channel, method, place, or time of communication, lest subsequent forensic investigations be damaging to her sources. When used in conjunction with the multi-recipient technique of §8.1, this property also allows her to deniably provide exact copies of her writings to additional recipients.

We do not present a formal experiment for this property. Suffice it to say that it is enough to ask that, given a plaintext message signed by Alice, Bob be able to create a valid ciphertext addressed to himself for that message, that is indistinguishable from a genuine ciphertext from Alice.

Definition 4. *An IBSE scheme is said to be ciphertext-unlinkable if there exists a polynomial-time algorithm that, given an identified signed message $\langle \text{id}_A, m, s \rangle$ such that* $\mathsf{Verify}[\text{id}_A, m, s] = \top$*, and a private key* $d_B = \mathsf{Extract}[\text{id}_B]$*, assembles a ciphertext c that is computationally indistinguishable from a genuine encryption of* $\langle m, s \rangle$ *by* id_A *for* id_B*.*

As mentioned earlier, ciphertext unlinkability is the reason why we considered the notion of signature unforgeability in §3.2, instead of the usual notion of ciphertext unforgeability as studied in the signcryption model of [1]. Indeed, if a ciphertext were unforgeable, surely it would be undeniably linkable to its author.

Note also that ciphertext unlinkability only makes sense in a two-layer signcryption model like ours, as opposed to the monolithic model of [27] used in [21, 19]. Indeed, if part of the ciphertext itself is needed to verify the authenticity of the plaintext, ciphertext indistinguishability is lost as soon as the recipient is compelled to prove authenticity to a third party.

3.4   Ciphertext Authentication

Ciphertext authentication is, in a sense, the complement to unlinkability. Authentication requires that the legitimate recipient be able to ascertain that the ciphertext did indeed come from the same person who signed the message it contains. (Naturally, he or she cannot prove this to anyone else, per the unlinkability property.)

We define ciphertext authentication in terms of the following game.

Start. The challenger runs the Setup procedure for a given value of the security parameter n, and provides the common public parameters π to the adversary, keeping the secret σ for itself.

Query. The adversary makes a number of queries to the challenger, as in the Confidentiality game of §3.1 and the Non-repudiation game of §3.2.

Forgery. The adversary returns a recipient identity id_B and and a ciphertext c.

Outcome. The adversary wins the game if c decrypts, under the private key of id_B, to a signed message $\langle \mathsf{id}_A, \hat{m}, \hat{s} \rangle$ such that $\mathsf{id}_A \neq \mathsf{id}_B$ and that satisfies $\mathsf{Verify}[\mathsf{id}_A, \hat{m}, \hat{s}] = \top$, provided that (1) no private key extraction query was made on either id_A or id_B, and (2) c did not result from a signature/encryption query with sender and recipient identities id_A and id_B.

We contrast the above experiment, which is a case of 'outsider' security for authentication on the whole ciphertext, with the scenario for signature non-repudiation, which required insider security on the signed plaintext only. We call the above experiment an AUTH-IBSE-CMA attack.

Definition 5. *An IBSE scheme is said to be existentially ciphertext-unforgeable against chosen-message outsider attacks, or* AUTH-IBSE-CMA *secure, if no randomized polynomial-time adversary has a non-negligible advantage in the above game. In other words, any randomized polynomial-time EUF-IBSE-CMA adversary $\mathcal{A}$ has an advantage $\mathbf{Adv}_{\mathcal{A}}[n] = \mathbf{P}[\mathsf{Verify}[\mathsf{id}_A, \hat{m}, \hat{s}] = \top]$ that behaves as $o[1/\mathrm{poly}[n]]$ for any polynomial $\mathrm{poly}[n]$.*

3.5 Ciphertext Anonymity

Finally, we require ciphertext anonymity, which is to say that the ciphertext must contain no information in the clear that identifies the author or recipient of the message (and yet be decipherable by the intended recipient without that information).

Ciphertext anonymity against adaptive chosen-ciphertext attacks is defined as follows.

Start. The challenger runs the Setup procedure for a given value of the security parameter n, and provides the common public parameters π to the adversary, keeping the secret σ for itself.

Phase 1. The adversary is allowed to make adaptive queries of the same types as in the Confidentiality game of §3.1, *i.e.:* signature/encryption queries, decryption queries, and private key extraction queries.

Selection. At some point, the adversary returns a message m, two sender identities id_{A_0} and id_{A_1}, and two recipient identities id_{B_0} and id_{B_1}, on which it wishes to be challenged. The adversary must have made no private key extraction query on either id_{B_0} or id_{B_1}.

Challenge. The challenger flips two random coins $b', b'' \in \{0, 1\}$, computes $\mathsf{pvk} = \mathsf{Extract}[\mathsf{id}_{A_{b'}}]$, $\langle s, r \rangle \leftarrow \mathsf{Sign}[\mathsf{pvk}, m]$, $c \leftarrow \mathsf{Encrypt}[\mathsf{pvk}, \mathsf{id}_{B_{b''}}, m, s, r]$, and gives the ciphertext c to the adversary.

Phase 2. The adversary adaptively issues a number of additional encryption, decryption, and extraction queries, under the additional constraint that it not ask for the private key of either id_{B_0} or id_{B_1}, or the decryption of c under id_{B_0} or id_{B_1}.

Response. The adversary returns two guesses $\hat{b}', \hat{b}'' \in \{0, 1\}$, and wins the game if $\langle \hat{b}', \hat{b}'' \rangle = \langle b', b'' \rangle$.

This game is the same as for confidentiality, except that the adversary is challenged on the identities instead of the message; it is an insider attack. We call it an ANON-IBSE-CCA attack.

Definition 6. *An IBSE is said to be ciphertext-anonymous against adaptive chosen-ciphertext insider attacks, or* ANON-IBSE-CCA *secure, if no randomized polynomial-time adversary has a non-negligible advantage in the above game. In other words, any randomized polynomial-time ANON-IBSE-CCA adversary $\mathcal{A}$ has an advantage $\mathbf{Adv}_\mathcal{A}[n] = |\mathbf{P}[\hat{b} = b] - \frac{1}{4}|$ that is $o[1/\mathrm{poly}[n]]$ for any polynomial $\mathrm{poly}[n]$ in the security parameter.*

We emphasize that anonymity only applies to the ciphertext, against non-recipients, and is thus consistent with both non-repudiation (§3.2) and authentication (§3.4). To illustrate the difference bewteen unlinkability and anonymity, note that the authenticated IBE scheme of [20] is unlinkable but not anonymous, since the sender identity must be known prior to decryption.

4   Review of IB Cryptography from Pairings

We now give a brief summary of the Boneh-Franklin algorithm for identity-based cryptography based on bilinear pairings on elliptic curves.

Let $\mathbb{G}_1$ and $\mathbb{G}_2$ be two cyclic groups of prime order p, writing the group action multiplicatively (in both cases using 1 to denote the neutral element).

Definition 7. *An (efficiently computable, non-degenerate) map $\mathbf{e} : \mathbb{G}_1 \times \mathbb{G}_1 \to \mathbb{G}_2$ is called a* bilinear pairing *if, for all $x, y \in \mathbb{G}_1$ and all $a, b \in \mathbb{Z}$, we have $\mathbf{e}[x^a, y^b] = \mathbf{e}[x, y]^{ab}$.*

Definition 8. *The (computational)* bilinear Diffie-Hellman problem *for a bilinear pairing as above is described as follows: given $g, g^a, g^b, g^c \in \mathbb{G}_1$, where g is a generator and $a, b, c \in \mathbb{F}_p^\star$ are chosen at random, compute $\mathbf{e}[g, g]^{abc}$. The advantage of an algorithm $\mathcal{B}$ at solving the BDH problem is defined as $\mathbf{Adv}_\mathcal{B}[\mathbf{e}] = \mathbf{P}[\mathcal{B}[g, g^a, g^b, g^c] = \mathbf{e}[g, g]^{abc}]$.*

Definition 9. *Let $\mathcal{G}$ be a polynomial-time randomized function that, on input 1^n, returns the description of a bilinear pairing $\mathbf{e} : \mathbb{G}_1 \times \mathbb{G}_1 \to \mathbb{G}_2$ between two groups $\mathbb{G}_1$ and $\mathbb{G}_2$ of prime order p. A BDH parameter generator $\mathcal{G}$ satisfies the* bilinear Diffie-Hellman assumption *if there is no randomized algorithm $\mathcal{B}$ that solves the BDH problem in time $\mathcal{O}[\mathrm{poly}[n]]$ with advantage $\Omega[1/\mathrm{poly}[n]]$. The probability space is that of the randomly generated parameters $\langle \mathbb{G}_1, \mathbb{G}_2, p, \mathbf{e} \rangle$, the BDH instances $\langle g, g^a, g^b, g^c \rangle$, and the randomized executions of $\mathcal{B}$.*

The Boneh-Franklin system provides a concrete realization of the above definitions. It is based on an elliptic-curve implementation of the BDH parameter generator $\mathcal{G}$, which we describe following [2,14] as recently generalized in [6].

Let $E/\mathbb{F}_q$ be an elliptic curve defined over some ground field $\mathbb{F}_q$ of prime characteristic χ. For any extension degree $r \geq 1$, let $E(\mathbb{F}_{q^r})$ be the group of points in $\{\langle x, y \rangle \in (\mathbb{F}_{q^r})^2\} \cup \{\infty\}$ that satisfy the curve equation over $\mathbb{F}_{q^r}$. Let $\nu = \#E(\mathbb{F}_q)$, the number of points on the curve including ∞. Let p be a prime $\neq \chi$ and $\nmid \chi - 1$, such that $p \mid \nu$ and $p^2 \nmid \nu$. Thus, there exists a subgroup $\mathbb{G}_1'$ of order p in $E(\mathbb{F}_q)$. Let κ be the *embedding degree* of $\mathbb{G}_1'$ in $E(\mathbb{F}_q)$, *i.e.*, the smallest integer ≥ 1 such that $p \mid q^\kappa - 1$, but $p \nmid q^r - 1$ for $1 \leq r \leq \kappa$. Under those conditions, there exist a subgroup $\mathbb{G}_1''$ of order p in $E(\mathbb{F}_{q^\kappa})$, and a subgroup $\mathbb{G}_2$ of order p in the multiplicative group $\mathbb{F}_{q^\kappa}^\star$. For appropriately chosen curves, one can then construct a non-degenerate bilinear map $\mathbf{e} : \mathbb{G}_1 \times \mathbb{G}_1 \to \mathbb{G}_2$ believed to satisfy the BDH assumption, where $\mathbb{G}_1$ is either $\mathbb{G}_1'$ or $\mathbb{G}_1''$.

Specifically, [7] show how to obtain a non-degenerate pairing $\bar{\mathbf{e}} : \mathbb{G}_1' \times \mathbb{G}_1'' \to \mathbb{G}_2$, based on the Tate or the Weil pairing, which can then be combined with a computable isomorphism $\psi : \mathbb{G}_1'' \to \mathbb{G}_1'$, called the *trace map*, to obtain a suitable bilinear map $\mathbf{e} : \mathbb{G}_1 \times \mathbb{G}_1 \to \mathbb{G}_2$ with $\mathbb{G}_1 = \mathbb{G}_1''$. Alternatively, selected curves afford efficiently computable isomorphisms $\phi : \mathbb{G}_1' \to \mathbb{G}_1''$, called *distortion maps*, which can be combined with $\bar{\mathbf{e}}$ to yield pairings of the form $\mathbf{e} : \mathbb{G}_1 \times \mathbb{G}_1 \to \mathbb{G}_2$ with $\mathbb{G}_1 = \mathbb{G}_1'$. The benefit of the latter construction is that the elements of $\mathbb{G}_1'$ have more compact representations than those of $\mathbb{G}_1''$.

It is desired that p and q^κ be large enough for the discrete logarithm to be intractable in generic groups of size p and in the multiplicative group $\mathbb{F}_{q^\kappa}^\star$. Most commonly, q is a large prime or power of 2 or 3, and $\log p \geq 160$, $\log q^\kappa \geq 1000$. We refer the reader to [4] for background information, and to [5] and [14] for details on the concrete implementation.

In the sequel, we treat the above notions as abstract mathematical objects satisfying the properties summarized in Definitions 7, 8, and 9.

Based on this setup, the Boneh-Franklin system defines four operations, the first two for setup and key extraction purposes by the PKG, the last two for encryption and decryption purposes. The two PKG functions are recalled below.

bfSetup On input a security parameter $n \in \mathbb{N}$: obtain $\langle \mathbb{G}_1, \mathbb{G}_2, p, \mathbf{e} \rangle \leftarrow \mathcal{G}[1^n]$ from the BDH parameter generator; pick two random elements $g \in \mathbb{G}_1^\star$ and $\sigma \in \mathbb{F}_p^\star$, set $g^\sigma = (g)^\sigma \in \mathbb{G}_1^\star$; and construct the hash function $H_0 : \{0,1\}^\star \to \mathbb{G}_1^\star$. Finally, output the common public parameters $\pi = \langle \mathbb{G}_1, \mathbb{G}_2, p, \mathbf{e}, g, g^\sigma, H_0 \rangle$ and the master secret $\sigma = \sigma$.

bfExtract On input id $\in \{0,1\}^\star$: hash the given identity into a public element $i_{\mathsf{id}} = H_0[\mathsf{id}] \in \mathbb{G}_1^\star$, and output $d_{\mathsf{id}} = (i_{\mathsf{id}})^\sigma \in \mathbb{G}_1^\star$ as the private key $\mathsf{pvk}_{\mathsf{id}}$.

5   Encryption-Signature Scheme

We now present an efficient realization of the abstract IBSE specifications of §2.

Table 1 details the six algorithms of our scheme. The Setup and Extract functions are essentially the same as in the original Boneh-Franklin system [5].

Table 1. The IBSE algorithms. The hash functions are modeled as random oracles. The output of H_4 is viewed as a stream that is truncated as dictated by context, $viz.$, $H_4[\text{key}] \oplus \text{data}$ perfoms a length-preserving "one-time pad" encryption or decryption.

Setup On input a security parameter $n \in \mathbb{N}$: establish the Boneh-Franklin parameters $\mathbb{G}_1, \mathbb{G}_2, p, \mathbf{e}, g, g^\sigma, \sigma$ as in bfSetup, and select five hash functions $H_0 : \{0,1\}^* \to \mathbb{G}_1^*$, $H_1 : \mathbb{G}_1^* \times \{0,1\}^* \to \mathbb{F}_p^*$, $H_2 : \mathbb{G}_2^* \to \{0,1\}^{\lceil \log p \rceil}$, $H_3 : \mathbb{G}_2^* \to \mathbb{F}_p^*$, $H_4 : \mathbb{G}_1 \to \{0,1\}^*$; then, output the common public parameters $\langle \mathbb{G}_1, \mathbb{G}_2, p, \mathbf{e}, g, g^\sigma, H_0, H_1, H_2, H_3, H_4 \rangle$ and the master secret σ.

Extract On input $\text{id} \in \{0,1\}^*$: proceed as in bfExtract.

Sign On input the private key d_A of some sender identity id_A, and a message m:
derive $i_A = H_0[\text{id}_A]$ (so $d_A = (i_A)^\sigma$),
pick a random $r \in \mathbb{F}_p^*$,
let $j = (i_A)^r \in \mathbb{G}_1^*$,
let $h = H_1[j, m] \in \mathbb{F}_p^*$,
let $v = (d_A)^{r+h} \in \mathbb{G}_1$,
then, output the signature $\langle j, v \rangle$; also forward $\langle m, r, \text{id}_A, i_A, d_A \rangle$ for further use by Encrypt.

Encrypt On input a recipient identity id_B, and $\langle j, v, m, r, \text{id}_A, i_A, d_A \rangle$ from Sign as above:
derive $i_B = H_0[\text{id}_B]$,
compute $u = \mathbf{e}[d_A, i_B] \in \mathbb{G}_2^*$,
let $k = H_3[u] \in \mathbb{F}_p^*$;
set $x = j^k \in \mathbb{G}_1^*$,
let $w = u^{k\,r} \in \mathbb{G}_2^*$,
set $y = H_2[w] \oplus v$,
set $z = H_4[v] \oplus \langle \text{id}_A, m \rangle$;
then, output the ciphertext $\langle x, y, z \rangle$.

Decrypt On input a private key d_B for id_B, and an anonymous ciphertext $\langle \hat{x}, \hat{y}, \hat{z} \rangle$:
derive $i_B = H_0[\text{id}_B]$,
compute $\hat{w} = \mathbf{e}[\hat{x}, d_B]$,
recover $\hat{v} = H_2[\hat{w}] \oplus \hat{y}$,
recover $\langle \hat{\text{id}}_A, \hat{m} \rangle = H_4[\hat{v}] \oplus \hat{z}$,
derive $\hat{i}_A = H_0[\hat{\text{id}}_A]$,
compute $\hat{u} = \mathbf{e}[\hat{i}_A, d_B]$,
let $\hat{k} = H_3[\hat{u}]$,
set $\hat{j} = \hat{x}^{\hat{k}^{-1}}$;
then, output the decrypted message $\hat{m}$, the signature $\langle \hat{j}, \hat{v} \rangle$, and the purported identity of the originator $\hat{\text{id}}_A$.

Verify. On input a signed message $\langle \hat{m}, \hat{j}, \hat{v} \rangle$ by purported sender identity id_A:
derive $\hat{i}_A = H_0[\hat{\text{id}}_A]$,
let $\hat{h} = H_1[\hat{j}, m]$,
check whether $\mathbf{e}[g, \hat{v}] \overset{?}{=} \mathbf{e}[g^\sigma, (\hat{i}_A)^{\hat{h}} \hat{j}]$;
then, output $\top$ if the equality holds, output $\bot$ otherwise.

Sign and Encrypt implement the IBS of [8], although other randomized signature schemes could be substituted for it. Encrypt and Decrypt are less conventional.

Intuitively, Sign implements a randomized IBS whose signatures comprise a commitment j to some random r chosen by the sender, and a closing v that depends on r and the message m. Encrypt superposes two layers of (expansionless) deterministic encryption. The inner layer encrypts j into x using a minimalist authenticated IBE built from zero-round pairing-based key agreement. The outer layer concurrently determines the value w that encrypts to the same x under a kind of anonymous IBE, derandomized to rely on the entropy already present in x. Then, w is hashed into a one-time pad to encrypt the second half of the signature v, which in turn seeds a one-time pad for the bulk encryption of m.

It is helpful to observe that the exponentiations $\star^r$ and $\star^k$ used in Sign for commitment and in Encrypt for authenticated encryption, as well as the key extraction $\star^\sigma$, and the bilinear pairing $\mathbf{e}[\star, i_B]$ that intervenes in the determination of w, all commute. The legitimate recipient derives its ability to decrypt x from the capacity to perform all of the above operations (either explicitly or implicitly)—but it can only do so in a specific order, different than the sender.

The results of §7 show the scheme is secure. We now prove its consistency.

Theorem 10. *The IBSE scheme of Table 1 is consistent.*

Proof. For decryption, if $\langle \hat{x}, \hat{y}, \hat{z} \rangle = \langle x, y, z \rangle$, it follows that $\hat{w} = \mathbf{e}[i_A{}^{rk}, i_B{}^\sigma] = \mathbf{e}[i_A{}^\sigma, i_B]^{rk} = w$ (in $\mathbb{G}_2^\star$), and thus $\hat{v} = v$ and $\langle \hat{\mathsf{id}}_A, \hat{m} \rangle = \langle \mathsf{id}_A, m \rangle$; we also have $\hat{u} = \mathbf{e}[\hat{i}_A, i_B]^\sigma = u$ (in $\mathbb{G}_2^\star$), hence $\hat{k} = k$ (in $\mathbb{F}_p^\star$), and thus $\hat{j} = (j^k)^{\hat{k}^{-1}} = j$ (in $\mathbb{G}_1^\star$). For verification, if $\langle \hat{m}, \hat{\mathsf{id}}_A, \hat{j}, \hat{v} \rangle = \langle m, \mathsf{id}_A, j, v \rangle$, we have $\mathbf{e}[g, \hat{v}] = \mathbf{e}[g, i_A]^{\sigma\,(r+h)} = \mathbf{e}[g^\sigma, (\hat{i}_A)^h\,(\hat{i}_A)^r] = \mathbf{e}[g^\sigma, (\hat{i}_A)^h\,\hat{j}]$ (in $\mathbb{G}_2$), as required. □

6   Competitive Performance

Table 2 gives a comparison between various IB encryption and signature schemes, in terms of size, performance, and security properties.

Our comparisons include most relevant pairing-based IB schemes for encryption, authenticated encryption, signature, and signcryption. We also include a suite of hybrid schemes, obtained by combining IBS [8] with either IBE [5] or AuthIBE [20]; each pair is composed in three different ways depending on the order of application of the primitives: encrypt-then-sign ($\mathcal{EtS}$), sign-then-encrypt ($\mathcal{StE}$), and commit-then-parallel-encrypt-and-sign ($\mathcal{CtS\&E}$), as per [1]. Roughly speaking, in $\mathcal{CtS\&E}$, the plaintext m is reversibly transformed into a redundant pair $\langle a, b \rangle$, where a is a commitment to m that reveals "no information" about m; then, a is signed and b encrypted using the given primitives, in parallel.

For fairness, the size comparison factors out the overhead of explicitly including the sender identity to the signed plaintext prior to encryption; our scheme does this to avoid sending the identity in the clear. Note that all authenticated communication schemes require the recipient to get hold of that information, but most simply assume that it is conveyed using a different channel.

Evidently, the proposed scheme offers an interesting solution to the problem of identity-based signed encryption: it offers an unmatched combination of security features that not only provide the usual confidentiality/non-repudiation requirements, but also guarantee authentication, anonymity, and unlinkability of the ciphertext. Our scheme achieves all this at a cost comparable to that of monolithic IB signcryption, and in a significantly tighter package than any generic combination of existing IB encryption and signature algorithms.

By comparison, the two listed signcryption schemes have comparable spatial and computational overheads but, by the very nature of monolithic signcryption, cannot offer ciphertext anonymity. As for the suite of generic compositions, they have a slight advantage in terms of cost, but incur a large size penalty, and require us to choose between ciphertext authentication and anonymity.

We also note that, in the original Boneh-Franklin setup, the IBSE ciphertexts and signed plaintexts are essentially as compact as that of IBE or IBS taken in

Table 2. Comparison between various IB encryption, signature, signcryption, and multipurpose schemes. Times are expressed as triples $\langle \#b, \#m, \#e \rangle$, where $\#b$ is the number of bilinear pairings, $\#m$ is the number of $\mathbb{G}_1$ exponentiations, and $\#e$ is the number of $\mathbb{G}_2$ or $\mathbb{F}_p$ exponentiations (simple group operations in $\mathbb{G}_1$ and multiplications and inversions in $\mathbb{F}_p$ or $\mathbb{G}_2$ are omitted). Sizes are reported as pairs $\langle \#p, \#q \rangle$, where $\#p$ is the number of $\mathbb{G}_1$ elements, and $\#q$ is the number of $\mathbb{F}_p$ or $\mathbb{G}_2$ elements, in excess of the original unsigned message size $\|m\|$ taken as baseline (treating the sender identity as part of m, if included); the 'cipher' size is the ciphertext overhead, $\|c\| - \|m\|$, while the 'plain' size is the signature overhead after decryption, or $\|\langle m, s \rangle\| - \|m\|$. Security is indicated as follows: message <u>C</u>onfidentiality, signature <u>N</u>on-repudiation, and ciphertext <u>A</u>uthentication, <u>U</u>nlinkability, and an<u>O</u>nymity; for non-IBSE schemes, an uppercase denotes an analogous security notion, a lowercase a weaker notion.

Scheme	Security: Conf, Nrep,Auth,Ulnk,anOn	Size:#el.$\mathbb{G}_1$,$\mathbb{G}_2$+$\mathbb{F}_p$ Cipher	Plain	Time:#pair.,exp.$\mathbb{G}_1$,$\mathbb{G}_2$+$\mathbb{F}_p$ Sign	Encrypt	Decrypt	Verify
IB Encryption [5]	C,–,–,U,O	1,1	—	—	1,0,0	1,1,0	—
IB Auth.Encr. [20]	C,–,A,U,–	0,2	—	—	1,0,0	1,0,0	—
IB Signature [8]	–,N,A,–,–	—	2,0	0,2,0	—	—	2,1,0
[a] IB Signature [22]	–,N,A,–,–	—	2,0	0,4,0	—	—	2(3),0,2
IB Sign. [16, #3]	–,N,A,–,–	—	1,1	1,2,1	—	—	2,0,1
IB Sign. [16, #4]	–,N,A,–,–	—	2,0	0,2,0	—	—	2,0,1
[b] IB SignCrypt. [21]	*,N,A,–,–	2,0	2,0	$\cdots$ 1,3,0 $\cdots$			$\cdots$ 4,0,1 $\cdots$
IB SignCrypt. [19]	C,N,A,–,–	1,1	1,1	$\cdots$ 2,2,2 $\cdots$			$\cdots$ 4,0,2 $\cdots$
[c] IB E–then–S	c,N,A,–,–	3,1	2,0	0,2,0	1,0,0	1,1,0	2,1,0
[c] IB S–then–E	C,n,–,U,O	3,1	2,0	0,2,0	1,0,0	1,1,0	2,1,0
[c] IB commit–E&S	C,N,A,U,–	3,1,+	2,0,+	0,2,0	1,0,0	1,1,0	2,1,0
[d] IB AE–then–S	c,N,A,U,–	2,2	2,0	0,2,0	1,0,0	1,0,0	2,1,0
[d] IB S–then–AE	C,n,A,U,–	2,2	2,0	0,2,0	1,0,0	1,0,0	2,1,0
[d] IB commit–AE&S	C,N,A,U,–	2,2,+	2,0,+	0,2,0	1,0,0	1,0,0	2,1,0
IBSE: this paper	C,N,A,U,O	2,0	2,0	0,2,0	1,0,2	2,1,0	2,1,0

[a] Signature verification in [22] requires 3 pairings, one of which may be precomputed.
[b] The signcryption scheme of [21] is not adaptive CCA-secure, see [19] for details.
[c] These are compositions of IBE [5] and IBS [8,16] using $\mathcal{EtS}, \mathcal{StE}, \mathcal{CtS\&E}$ from [1].
[d] These are compositions of AuthIBE [20] and IBS [8,16] using $\mathcal{EtS}, \mathcal{StE}, \mathcal{CtS\&E}$ [1]. $\mathcal{EtS}$ and $\mathcal{StE}$ respectively degrade the CCA indistinguishability and CMA unforgeability of its constituents in the insider model; the '+' are a reminder that the more secure $\mathcal{CtS\&E}$ incurs extra overhead due to the commitment redundancy. See [1] for details.

isolation; this is generally true when $p \approx q$, and when $\mathbb{G}_1 = \mathbb{G}_1'$ so that its points can be represented as elements of $\mathbb{F}_q$ using point compression [4]. However, the schemes of [5], [19], and especially [20] have smaller ciphertexts and signatures, as the case may be, in generalized setups where $p \ll q$, or $\mathbb{G}_1 = \mathbb{G}_1''$.

7 Security Analysis

We now state our security results for the scheme of §5 in the models of §3.

Theorem 11. *Let $\mathcal{A}$ be a polynomial-time IND-IBSE-CCA attacker that has advantage $\geq \epsilon$, and makes $\leq \mu_i$ queries to the random oracles H_i, $i = 0, 1, 2, 3, 4$. Then, there exists a polynomial-time algorithm $\mathcal{B}$ that solves the bilinear Diffie-Hellman problem with advantage $\geq \epsilon/(\mu_0 \, \mu_2)$.*

Theorem 12. *Let $\mathcal{A}$ be an EUF-IBSE-CCA attacker that makes $\leq \mu_i$ queries to the random oracles H_i, $i = 0, 1, 2, 3, 4$, and $\leq \mu_{\mathrm{se}}$ queries to the signature/encryption oracle. Assume that, within a time span $\leq \tau$, $\mathcal{A}$ produces a successful forgery with probability $\geq \epsilon = 10 \, (\mu_{\mathrm{se}} + 1) \, (\mu_{\mathrm{se}} + \mu_1)/2^n$, for a security parameter n. Then, there exists an algorithm $\mathcal{B}$ that solves the bilinear Diffie-Hellman problem in expected time $\leq 120686 \, \mu_0 \, \mu_1 \, \tau/\epsilon$.*

Theorem 13. *There exists a polynomial-time algorithm that, given an identifier id_A, a signed plaintext $\langle m, j, v \rangle$ from id_A, and a private key d_B, creates a ciphertext $\langle x, y, z \rangle$ that decrypts to $\langle m, j, v \rangle$ under d_B, with probability 1.*

Theorem 14. *Let $\mathcal{A}$ be a polynomial-time AUTH-IBSE-CMA attacker with advantage $\geq \epsilon$, that makes $\leq \mu_i$ queries to the random oracles H_i, $i = 0, 1, 2, 3, 4$. Then, there exists a polynomial-time algorithm $\mathcal{B}$ that solves the bilinear Diffie-Hellman problem with advantage $\geq 2 \, \epsilon/(\mu_0 \, (\mu_0 - 1) \, (\mu_1 \, \mu_2 + \mu_3))$.*

Theorem 15. *Let $\mathcal{A}$ be a polynomial-time ANON-IBSE-CCA attacker that has advantage $\geq \epsilon$, and makes $\leq \mu_i$ queries to the random oracles H_i, $i = 0, 1, 2, 3, 4$. Then, there exists a polynomial-time algorithm $\mathcal{B}$ that solves the bilinear Diffie-Hellman problem with advantage $\geq 3 \, \epsilon/(\mu_0 \, (\mu_0 - 1) \, (\mu_1 \, \mu_2 + 2 \, \mu_2 + \mu_3))$.*

8   Practical Extensions

We now mention a few straightforward generalizations of practical interest.

8.1   Encrypting for Multiple Recipients

Encrypting the same message m for a set of n recipients $\mathrm{id}_{B_1}, ..., \mathrm{id}_{B_n}$ is easily achieved as follows. The Sign operation is carried out once (which establishes the randomization parameter r), then the Encrypt operation is performed independently for each recipient, based on the output from Sign.

Since the message m and the randomization parameter r are invariant for all the Encrypt instances, it is easy to see that the z component of the ciphertext also remains the same. Thus, the multi-recipient composite ciphertext is easily assembled from one instance of $\langle x_i, y_i \rangle \in \mathbb{G}_1^\star \times \mathbb{G}_1^\star$ for each recipient B_i, plus a single instance of $z \in \{0, 1\}^*$ to be shared by all. Thus, a multi-recipient ciphertext is compactly encoded in the form $c = \langle \langle x_1, y_1 \rangle, ..., \langle x_n, y_n \rangle, z \rangle$.

The security models of §3 have to support two additional types of queries: *multi-recipient signature/encryption queries*, in which a given message, sender, and list of recipients, are turned into a multi-recipient ciphertext, and *multi-recipient decryption queries*, in which the individual elements of a multi-recipient ciphertext are decrypted, under a given identity, and a valid plaintext is returned, if there is any. The modified security analysis is deferred to the full paper.

8.2   Integrity without Non-repudiation

The scheme of §5 is trivially modified to provide message integrity without non-repudiation or authentication. To do this, the sender merely substitutes the public parameters $\langle g, g^\sigma \rangle$ for $\langle i_A, d_A \rangle$, wherever the sender's key pair is used in the Sign and Encrypt operations. The sender also tags the message as 'anonymous', instead of specifying an identity. Similarly, the Decrypt and Verify operations are performed substituting g^σ for $\hat{i}_A$ wherever it appears as a function argument.

This is valid since the key pair relation $d_A = (i_A)^\sigma$ is paralleled by $g^\sigma = (g)^\sigma$, but authentication is meaningless since the signing 'private' key g^σ is public.

9   Conclusion

In this paper, we have proposed a comprehensive security model for multi-purpose identity-based encryption-signature cryptosystems. Our security model defines five core properties that we believe precisely capture what a consumer of cryptography intuitively expects when he or she wishes to engage in "secure signed communication" with a remote party. It bears repeating that these do not only include the standard confidentiality and non-repudiation requirements, but also the much less commonly offered features of ciphertext authentication, ciphertext deniability or unlinkability, and true ciphertext anonymity with respect to third parties. We have given precise definitions for all these properties in the context of identity-based cryptography.

As second contribution, we have presented a new cryptographic scheme that precisely implements all facets of the aforementioned notion of "secure signed communication", in the certificate-free world of identity-based cryptography. Our scheme offers efficient security bounds in all the above respects; it is fast, compact, scalable, and practical—as we have illustrated through detailed comparisons with most or all mainstream identity-based cryptosystems to date.

Acknowledgements. The author would like to thank Dan Boneh, Jonathan Katz, and the anonymous referees of Crypto 2003 for many helpful suggestions and comments. Credit goes to Guido Appenzeller for suggesting the "Swiss Army Knife" moniker.

References

1. J.H. An, Y. Dodis, and T. Rabin. On the security of joint signature and encryption. In *Proc. Eurocrypt '02, LNCS 2332*, 2002.
2. P.S.L.M. Barreto, H.Y. Kim, B. Lynn, and M. Scott. Efficient algorithms for pairing-based cryptosystems. In *Proc. Crypto '02, LNCS 2442*, 2002.
3. M. Bellare and P. Rogaway. Random oracles are practical: A paradigm for designing efficient protocols. In *Proc. Conf. Computer and Communication Security*, 1993.
4. I. Blake, G. Seroussi, and N. Smart. *Elliptic Curves in Cryptography*. Cambridge University Press, 1999.

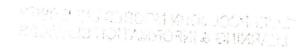

5. D. Boneh and M. Franklin. Identity-based encryption from the Weil pairing. In *Proc. Crypto '01, LNCS 2139*, pages 213–229, 2001. See [6] for the full version.
6. D. Boneh and M. Franklin. Identity based encryption from the weil pairing. Cryptology ePrint Archive, Report 2001/090, 2001. http://eprint.iacr.org/.
7. D. Boneh, B. Lynn, and H. Shacham. Short signatures from the Weil pairing. In *Proc. Asiacrypt '01, LNCS 2248*, pages 514–532, 2001.
8. J.C. Cha and J.H. Cheon. An identity-based signature from gap Diffie-Hellman groups. Cryptology ePrint Archive, Report 2002/018, 2002. http://eprint.iacr.org/.
9. L. Chen and C. Kudla. Identity based authenticated key agreement from pairings. Cryptology ePrint Archive, Report 2002/184, 2002. http://eprint.iacr.org/.
10. C. Cocks. An identity based encryption scheme based on quadratic residues. In *Proc. 8th IMA Int. Conf. Cryptography and Coding*, pages 26–28, 2001.
11. U. Feige, A. Fiat, and A. Shamir. A digital signature scheme secure against adaptive chosen-message attacks. *SIAM J. Computing*, 17(2):281–308, 1988.
12. U. Feige, A. Fiat, and A. Shamir. Zero-knowledge proofs of identity. *J. Cryptology*, 1:77–94, 1988.
13. A. Fiat and A. Shamir. How to prove yourself: Practical solutions to identification and signature problems. In *Proc. Crypto '86, LNCS 263*, pages 186–194, 1984.
14. S.D. Galbraith, K. Harrison, and D. Soldera. Implementing the Tate pairing. Technical Report HPL-2002-23, HP Laboratories Bristol, 2002.
15. C. Gentry and A. Silverberg. Hierarchical ID-based cryptography. Cryptology ePrint Archive, Report 2002/056, 2002. http://eprint.iacr.org/.
16. F. Hess. Exponent group signature schemes and efficient identity based signature schemes based on pairings. Cryptology ePrint Archive, Report 2002/012, 2002. http://eprint.iacr.org/.
17. J. Horwitz and B. Lynn. Toward hierarchical identity-based encryption. In *Proc. Eurocrypt '02, LNCS 2332*, pages 466–481, 2002.
18. A. Joux. A one round protocol for tripartite Diffie-Hellman. In *Proc. 4th Alg. Numb. Th. Symp., LNCS 1838*, pages 385–294, 2000.
19. B. Libert and J.-J. Quisquater. New identity based signcryption schemes based on pairings. Cryptology ePrint Archive, Report 2003/023, 2003. http://eprint.iacr.org/.
20. B. Lynn. Authenticated identity-based encryption. Cryptology ePrint Archive, Report 2002/072, 2002. http://eprint.iacr.org/.
21. J. Malone-Lee. Identity-based signcryption. Cryptology ePrint Archive, Report 2002/098, 2002. http://eprint.iacr.org/.
22. K.G. Paterson. ID-based signatures from pairings on elliptic curves. Cryptology ePrint Archive, Report 2002/004, 2002. http://eprint.iacr.org/.
23. R. Sakai, K. Ohgishi, and M. Kasahara. Cryptosystems based on pairings. In *Proc. SCIS '00*, pages 26–28, Okinawa, Japan, 2000.
24. A. Shamir. Identity-based cryptosystems and signature schemes. In *Proc. Crypto '84, LNCS 196*, pages 47–53, 1984.
25. N.P. Smart. An identity based authenticated key agreement protocol based on the Weil pairing. Cryptology ePrint Archive, Report 2001/111, 2001. http://eprint.iacr.org/.
26. D. Pointcheval J. Stern. Security arguments for digital signatures and blind signatures. *J. Cryptology*, 13:361–396, 2000.
27. Y. Zheng. Digital signcryption or how to achieve $cost\,(signature\,\&\,encryption) \ll cost\,(signature) + cost\,(encryption)$. In *Proc. Crypto '97, LNCS 1294*, 1997.

SIGMA: The 'SIGn-and-MAc' Approach to Authenticated Diffie-Hellman and Its Use in the IKE Protocols

Hugo Krawczyk

EE Department, Technion, Haifa, Israel, and IBM T.J. Watson Research Center.
`hugo@ee.technion.ac.il`

Abstract. We present the SIGMA family of key-exchange protocols and the "SIGn-and-MAc" approach to authenticated Diffie-Hellman underlying its design. The SIGMA protocols provide perfect forward secrecy via a Diffie-Hellman exchange authenticated with digital signatures, and are specifically designed to ensure sound cryptographic key exchange while providing a variety of features and trade-offs required in practical scenarios (such as optional identity protection and reduced number of protocol rounds). As a consequence, the SIGMA protocols are very well suited for use in actual applications and for standardized key exchange. In particular, SIGMA serves as the cryptographic basis for the signature-based modes of the standardized Internet Key Exchange (IKE) protocol (versions 1 and 2).

This paper describes the design rationale behind the SIGMA approach and protocols, and points out to many subtleties surrounding the design of secure key-exchange protocols in general, and identity-protecting protocols in particular. We motivate the design of SIGMA by comparing it to other protocols, most notable the STS protocol and its variants. In particular, it is shown how SIGMA solves some of the security shortcomings found in previous protocols.

1 Introduction

In this paper we describe the SIGMA family of key-exchange protocols, with emphasis on its design features and rationale. The SIGMA protocols introduce a general approach to building authenticated Diffie-Hellman protocols using a *careful* combination of digital signatures and a MAC (message authentication) function. We call this the *"SIGn-and-MAc" approach* which is also the reason for the SIGMA acronym.

SIGMA serves as the cryptographic basis for the Internet Key Exchange (IKE) protocol [11,16] standardized to provide key-exchange functionality to the IPsec suite of security protocols [17]. More precisely, SIGMA is the basis for the signature-based authenticated key exchange in IKE [11], which is the most commonly used mode of public-key authentication in IKE, and the basis for the only mode of public-key authentication in IKEv2 [16].

D. Boneh (Ed.): CRYPTO 2003, LNCS 2729, pp. 400–425, 2003.

This paper provides the first systematic description of the development and rationale of the SIGMA protocols. The presentation is intended to motivate the design choices in the protocol by comparing and contrasting it to alternative protocols, and by learning from the strong and weak aspects of previous protocols. It also explains how the different variants of the SIGMA protocol follow from a common design core. In particular, it explains the security basis on which the signature-based modes of IKE, and its current revision IKEv2, are based. The presentation is informal and emphasizes rationale and intuition rather than rigorous analysis. A formal analysis of the SIGMA protocol has been presented in [7] where it is shown that the basic SIGMA design and its variants are secure under a complexity-theoretic model of security. While this rigorous analysis is essential for gaining confidence in the security design of SIGMA, it does not provide an explicit understanding of the design process that led to these protocols, and the numerous subtleties surrounding this design. Providing such an understanding is a main goal of this paper which will hopefully be beneficial to cryptographers and security protocol designers (as well as for those engineering security solutions based on these protocols).

The basic guiding requirements behind the design of SIGMA are (a) to provide a secure key-exchange protocol based on the Diffie-Hellman exchange (for ensuring "perfect forward secrecy"), (b) use digital signatures as the means for public-key authentication of the protocol, and (c) provide the option to protect the identities of the protocol peers from being learned by an attacker in the network. These were three basic requirements put forth by the IPsec working group for its preferred key-exchange protocol. The natural candidate for satisfying these requirements is the well-known STS key-exchange protocol due to Diffie, van Oorschot and Wiener [8]. We show, however, that this protocol and some of its variants (including a variant adopted into Photuris [14], a predecessor of IKE as the key-exchange protocol for IPsec) suffer from security shortcomings that make them unsuited for some practical scenarios, in particular in the wide Internet setting for which the IPsec protocols are designed. Still, the design of SIGMA is strongly based on that of STS: both the strengths of the STS design principles (very well articulated in [8]) as well as the weaknesses of some of the STS protocol choices have strongly influenced the SIGMA design.

One point that is particularly important for understanding the design of SIGMA (and other key-exchange protocols) is the central role that the requirement for identity protection has in this design. As it turns out, the identity protection functionality conflicts with the essential requirement of peer authentication. The result is that both requirements (authentication and identity protection) can be satisfied simultaneously, at least to some extent, but their coexistence introduces significant subtleties both in the design of the protocol and its analysis. In order to highlight this issue we compare SIGMA to another authenticated Diffie-Hellman design, a variant of the ISO protocol [12], that has been shown to be secure [6] but which is not well-suited to support identity protection. As we will see SIGMA provides a satisfactory and flexible solution to this problem by supporting identity protection as an optional feature of the protocols, while keeping the number of communication rounds and cryptographic operations to a minimum. As a result SIGMA can suit the identity protection scenarios as well as those that do not require this functionality; in the later case there is no penalty (in terms of security, computation, communication, or general complexity) relative to protocols that do not offer identity protection at all. We thus believe that SIGMA is well suited as a "general purpose" authenticated Diffie-Hellman protocol that can serve a wide range of applications and security scenarios.

History of the SIGMA protocols. The SIGMA approach was introduced by the author in 1995 [19] to the IPsec working group as a possible replacement for the Photuris key-exchange protocol [14] developed at the time by that working group. Photuris used a variant of the STS protocol that we showed [19] to be flawed through the attack presented in Section 3.3. In particular, this demonstrated that the Photuris key exchange, when used with optional identity protection and RSA signatures (or any signature scheme allowing for message recovery), was open to the same attack that originally motivated the design of STS (see Section 3.1). Eventually, the Photuris protocol was replaced with the Internet Key Exchange (IKE) protocol which adopted SIGMA (unnamed at the time) into its two signature-based authentication modes: main mode (that provides identity protection) and aggressive mode (which does not support identity protection). The IKE protocol was standardized in 1999, and a revised version (IKEv2) is currently under way [16] (the latter also uses the SIGMA protocol as its cryptographic key exchange).

Related work. There is a vast amount of work that deals with the design and analysis of key-exchange (and authentication) protocols and which is relevant to the subject of this paper. Chapter 12 of [27] provides many pointers to such works, and additional papers can be found in the recent security and cryptography literature. There have been a few works that provided analysis and critique of the IKE protocol (e.g., [9,29]). Yet, these works mainly discuss issues related to functionality and complexity trade-offs rather than analyzing the core cryptographic design of the key exchange protocols. A formal analysis of the IKE protocols has been carried by Meadows [26] using automated analysis tools. In addition, as we have already mentioned, [7] provides a formal analysis of the SIGMA protocols (and its IKE variants) based on the complexity-theoretic approach to the analysis of key-exchange protocols initiated in [2]. A BAN-logic analysis of the STS protocols is presented in [31], and attacks on these protocols that enhance those reported in [19] are presented in [4] (we elaborate on these attacks in Section 3.3). Finally, we mention the SKEME protocols [20] which served as the basis for the cryptographic structure of IKE and its non-signature modes of authentication, but did not include a signature-based solution as in SIGMA.

Organization. In Section 2 we informally discuss security requirements for key-exchange protocols in general and for SIGMA in particular, and present specific requirements related to identity protection. Section 3 presents the STS protocol and its variants, and analyzes the strengths and weaknesses of these protocols. Section 4 discusses the ISO protocol as a further motivation for the design of SIGMA (in particular, this discussion serves to stress the role of identity protection in the design of SIGMA). Finally, Section 5 presents the SIGMA protocols together with their design rationale and security properties. In particular, Section 5.4 discusses the SIGMA variants used in the IKE protocols.

Note: In the full version of this paper [24] some additional information is provided as appendices. Specifically, Appendix A includes a simplified (and somewhat informal) definition of key-exchange security. Appendix B presents a "full fledge" instantiation of SIGMA which includes some of the elements omitted in the simplified presentation of Section 5 but which are crucial for a full secure implementation of the protocols. Appendix C discusses key-derivation issues and presents the specific key-derivation technique designed for, and used in, the IKE protocols. This technique is of independent interest since it applies to the derivation of keys in other key-exchange protocols; in particular, it includes a mechanism for "extracting randomness" from Diffie-Hellman keys.

2   Preliminaries: On the Security of Key-Exchange Protocols

Note: This section is important for understanding the design goals of SIGMA; yet, the impatient reader may skip it in a first reading (but see the notation paragraph at the end of the section).

In this paper we present an informal exposition of the design rationale behind the development of the SIGMA protocols. This exposition is intended to serve crypto protocol designers and security engineers to better understand the subtle design and analytical issues arising in the context of key-exchange (KE for short) protocols in general, and in the design of SIGMA in particular. This exposition, however, is not a replacement for a formal analysis of the protocol. A serious analysis work requires a formal mathematical treatment of the underlying security model and protocol goals. This essential piece of work for providing confidence in the security of the SIGMA protocols is presented in a companion paper [7]. The interested reader should consult that work for the formal foundations of security on which SIGMA is based. Yet, before going on to present the SIGMA protocols and some of its precursors we discuss informally some of the salient aspects of the analytical setting under which we study and judge KE protocols. This presentation will also provide a basis for the discussion of some of the techniques, strengths and weaknesses showing up in the protocols studied in later sections.

We start by noting that *there is no ultimate security model*. Security definitions may differ depending on the underlying mathematical methodology, the intended application setting, the consideration of different properties as more or less important, etc. The discussion below focuses on the core security properties of KE protocols as required in most common settings. These requirements stem from the the quintessential application of KE protocols, namely, the supply of shared keys to pairs of parties which later use these keys to secure (via integrity and secrecy protection) their pairwise communications. In addition, we deal with some more specific design goals of SIGMA motivated by requirements put forth by the IPsec working group: the use of the Diffie-Hellman exchange as the basic technique for providing "perfect forward secrecy", the use of digital signatures for authenticating the exchange, and the (possibly optional) provision of "identity protection".

2.1   Overview of the Security Model and Requirements

In spite of being a central (and "obvious") functionality in many cryptographic and security applications, the notion of a "secure key-exchange protocol" remains a very complex notion to formalize correctly. Here we state very informally some basic requirements from KE protocols that we will use as a basis for later discussion of security issues arising in the design of KE protocols. These requirements are in no way a replacement for a formal treatment carried in [7], but are consistent (at least at the intuitive level) with the notion of security in that work.

Authentication. Each party to a KE execution (referred to as a *session*) needs to be able to uniquely *verify the identity* of the peer with which the session key is exchanged.

Consistency. If two honest parties establish a common session key then both need to have a consistent view of who the peers to the session are. Namely, if a party A establishes a key K and believes the peer to the exchange to be B, then if B establishes the session key K then it needs to believe that the peer to the exchange is A; and vice-versa.

Secrecy. If a session is established between two honest peers then no third party should be able to learn any information about the resultant session key (in particular, no such third party, watching or interfering with the protocol run, should be able to distinguish the session key from a random key).

While the "authentication" and "secrecy" requirements are very natural and broadly accepted, the requirement of "consistency" is much trickier and many times overlooked. In Section 3.1 we exemplify this type of failure through an attack first discovered in [8]. This attack, to which we refer as an "identity misbinding attack", applies to many seemingly natural and intuitive protocols. Avoiding this form of attack and guaranteeing a consistent binding between a session key and the peers to the session is a central element in the design of SIGMA.

One important point to observe is that the above requirements are not absolute but exist only in relation to a well-defined attack model. The adversarial model from [7] assumes that each party holds a long-term private authentication key that is used to uniquely identify and authenticate this party (in the context of this paper we can concretely think of this long-term authentication key as being a secret digital signature key.) It also assumes the existence of a trusted certification authority, or any other trusted mechanism (manual distribution, web of trust, etc), for binding identities with public keys. Parties communicate over a public network controlled by a fully-active man-in-the-middle attacker which may intercept, delete, delay, modify or inject messages at will. This attacker also controls the scheduling of protocol sessions (a *session* is an execution instance of the protocol) which may run concurrently at the same or different parties.

In addition, the attacker may learn some of the secret information held by the parties to the protocol. Specifically, the attacker may learn the long-term secret information held by a party, in which case this party is considered as controlled by the attacker (and referred to as *corrupted*). There is no requirement about the security of sessions executed by a corrupted party (since the attacker may impersonate it at will), however, it is required that session keys produced (and erased from memory) before the party corruption happened will remain secure (i.e. no information on these keys should be learned by the attacker). This protection of past session keys in spite of the compromise of long-term secrets is known as **perfect forward secrecy (PFS)** and is a fundamental property of the protocols discussed here. The attacker may also learn session-specific information such as the value of a session key or some secret information contained in the internal state of a session (e.g., the exponent x of an ephemeral Diffie-Hellman exponential g^x used in that session). In this case, there is no requirement on the security of the compromised session but we do require that this leakage has no effect on other (uncompromised) sessions. This models resistance to a variety of attacks, including **known-key attacks** and **replay attacks** (see [27]), and emphasizes the need for **key independence** between different sessions.

The analysis of protocols under this model is carried on the basis of the generic properties required from the cryptographic primitives used in the protocol, rather than based on the properties of specific algorithms. This **algorithm independence** (or **generic security**) principle is important in case that specific crypto algorithms need to be replaced (for better security or improved performance), and to support different combinations of individually secure algorithms.

Discussion: sufficiency of the above security requirements. One important question is whether the above security requirements (and more precisely the formal security requirements from [7]), under which we judge the security of protocols in this work, are necessary and/or sufficient to guarantee "key-exchange security". Necessity is easy to show through natural examples in which the removal of any one of the above required properties results in explicit and clearly harmful attacks against the security of the exchanged key (either by compromising the secrecy of the key or by producing an inconsistent binding between the key and the identities of the holders of that key). Sufficiency, however, is harder to argue. We subscribe to the approach put forth in [6] (and followed by [7]) by which a minimal set of requirements for a KE protocol must ensure the security of the quintessential application of KE protocols, namely, the provision of "secure channels" (i.e., the sharing of a key between peers that subsequently use this key for protecting the secrecy and integrity of the information transmitted between them). It is shown in [7] that their definition (outlined here) is indeed sufficient (and actually minimalistic) for providing secure channels.

Also important to stress is that this definitional approach dispenses of some requirements that some authors (e.g., [25]) consider vital for a sound definition of security. One important example is the aliveness requirement, namely, if A completes a session with peer B then A has a proof that B was "alive" during the execution of the protocol (e.g., by obtaining B's unique authentication on some nonce freshly generated by A). This property is *not* guaranteed by our (or [7]) definition of security. Moreover, some natural key-transport protocols (e.g., the ENC protocol formally specified in [6]) are useful key-exchange protocols that guarantee secure channels yet do not provide a proof of aliveness. The only possible negative aspect of a KE protocol that lacks the aliveness guarantee is that a party may establish a session with a peer that did not establish the corresponding session (and possibly was not even operational at the time); this results in a form of "denial of service" for the former party but not a compromise of data transmitted and protected under the key. However, DoS attacks with similar effects are possible even if aliveness guarantees are provided, for example by the attacker preventing the arrival of the last protocol message to its destination.

A related (and stronger) property not guaranteed by our basic definition of security is peer awareness. Roughly speaking, a protocol provides peer awareness for A if when A completes a session with peer B, A has a guarantee that (not only is B alive but) B has initiated a corresponding session with peer A. Adding aliveness and peer awareness guarantees to a KE that lacks these properties is often very simple, yet it may come at a cost (e.g., it may add messages to the exchange or complicate other mechanisms such as identity protection). Therefore, it is best to leave these properties as optional rather than labeling as "insecure" any protocol that lacks them.[1]

All the protocols discussed in this paper provide aliveness proofs to both parties but only the ISO protocol and the 4-message SIGMA-I with added ACK (Section 5.2) provide peer awareness to both parties. In particular, the IKE protocols (Section 5.4) do not provide peer awareness to one of the peers. As said, this property can be added, when required, at the possible expense of extra messages or other costs.

2.2   Identity Protection

As discussed in Section 2.1, key-exchange protocols require strong mutual authentication and therefore they must be designed to communicate the identity of each participant in the protocol to its session peer. This implies that the identities must be transmitted as part of the protocol. Yet some applications require to prevent the disclosure of these identities over the network. This may be the case in settings where the identity (for the purpose of authentication) of a party is not directly derivable from the routing address that must appear in the clear in the protocol messages. A common example is the case of mobile

---

[1] We stress that in contrast to the key-exchange setting, the aliveness requirements, and sometimes peer awareness, is essential in "entity authentication" protocols whose sole purpose may be to determine the aliveness of a peer.

devices wishing to prevent an attacker from correlating their (changing) location with the logical identity of the device (or user). Note that such an application may not just need to hide these identities from passive observers in the network but may require to conceal the identity even from active attackers. In this case the sole encryption of the sender's identity is not sufficient and it is required that the peer to the session proves its own identity before the encrypted identity is transmitted. As it turns out the requirement to support identity protection adds new subtleties to the design of KE protocols; these subtleties arise from the conflicting nature of identity protection and authentication. In particular, *it is not possible* to design a protocol that will protect both peer identities from active attacks. This is easy to see by noting that the first peer to authenticate itself (i.e. to prove its identity to the other party) must disclose its identity to the other party before it can verify the identity of the latter. Therefore the identity of the first-authenticating peer cannot be protected against an active attacker. In other words, KE protocols may protect both identities from passive attacks and may, at best, protect the identity of one of the peers from disclosure against an active attacker.

This best-possible level of identity protection is indeed achievable by some KE protocols, and in particular is attained by the SIGMA protocols. The underlying design of SIGMA allows for a protocol variant where the initiator of the exchange is protected against active attacks and the responder's identity is protected against passive attacks (we refer to this variant as SIGMA-I), and it also allows for another variant where the responder's id is protected against active attacks and the initiator's against passive attacks only (SIGMA-R). Moreover, providing identity protection has been a main motivating force behind the design of SIGMA which resulted from the requirement put forth by the IPsec working group to support (at least optionally) identity protection in its KE protocol. The SIGMA protocols thus provide the best-possible protection against identity disclosure. The choice of SIGMA-I or SIGMA-R depends on which identity is considered as more sensitive and requires protection against active attacks. On the other hand, SIGMA also offers full KE security also in cases where identity protection is not needed. That is, the core security of the protocol does not depend on hiding identities but rather this protection of identities is a functionality added on top of the core protocol.

A related issue which is typical of settings where identity protection is a concern, but may also appear elsewhere, is that parties to the protocol may not know at the beginning of a session the specific identity of the peer but rather learn this identity as the protocol proceeds. (This may even be the case for the initiator of the session which may agree to establish the initiated session with one of a set of peers rather than with one predefined peer). This adds, in principle, more attack avenues against the protocol and also introduces some delicate formal and design issues (e.g., most existing formalisms of key-exchange protocols do assume that the peer identities are fixed and known from the start of the session). In [7] this more general and realistic setting is formalized under the name of the post-specified peer setting and the SIGMA protocols are shown to be secure in this model. See [7] for the technical details.

Finally we comment on one additional privacy aspect of KE protocols. In some scenarios parties may wish to keep their privacy protected not only against attackers in the network but also to avoid leaving a "provable trace" of their communications in the hands (or disks) of the peers with which they communicate. A protocol such as ISO (see Section 4) in which each party to the protocol signs the peer's identity is particularly susceptible to this privacy concern (since these signatures can serve to prove to a third party the fact that the communication took place). In the SIGMA protocols, however, this proof of communication is avoided to a large extent by not signing the peer's identity, thus providing a better solution to this problem.

Note: Some may consider the non-repudiation property of a protocol such as ISO (Section 4) as an advantage. However, we consider that non-repudiation using digital signatures does not belong to the KE protocol realm but as a functionality that needs to be dealt with carefully in specific applications, and with full awareness of the signer to the non-repudiation consequences.

2.3 Further Remarks and Notation

Denial of Service. Key-exchange protocols (including SIGMA) open opportunities for Denial-of-Service attacks since the responder to an exchange is usually required to generate state and/or perform costly computations before it has the opportunity to authenticate the peer to the exchange. This type of attacks cannot be prevented in a strong sense but can be mitigated by using some fast-to-verify measures. One such technique has been proposed by Phil Karn [14] via the use of "cookies" that the responder to a KE protocol uses to verify that the initiator of the exchange is being able to receive messages directed to the IP address from which the exchange was initiated (thus preventing some form of trivial DoS attacks in which the attacker uses forged origin addresses, and also improving the chances to trace back a DoS attack). This and other techniques are orthogonal to the cryptographic details of the KE protocol and then can be adopted into SIGMA. In particular, recent proposals for revising IKE [16,1] incorporate Karn's technique into SIGMA. Other forms of denial of service are possible (and actually unavoidable) such as an active attacker that prevents the completion of sessions or lets one party complete the session and the other not.

A word of caution. It is important to remark that all the protocols discussed in this paper are presented in their most basic form, showing only their cryptographic core. When used in practice it is essential to preserve this cryptographic core but also to take care of additional elements arising in actual settings. For example, if the protocol negotiates some security parameters or uses the protocol messages to send some additional information then the designers of such full-fledge protocol need to carefully expand the coverage of authentication also to these additional elements. We also (over) simplify the protocol presentation by omitting the explicit use of "session identifiers": such identifiers are needed for the run of a protocol in a multi-session setting where they are used to match incoming protocol messages with open KE sessions. Moreover, the binding of messages to specific session id's is required for core security reasons such as preventing interleaving attacks. Similarly, nonces may need to be included in the protocol to ensure freshness of messages (e.g. to prevent replay attacks). In our presentation, however, these elements are omitted by over-charging the Diffie-Hellman exponentials used in the protocols with the additional functionality of also serving as session ids and nonces. For the level of conceptual discussion in this paper, simplifying the presentation by reducing the number of elements in the protocol is useful (and also in line with the traditional presentation of protocols in the cryptographic literature, in particular with [8]). But when engineering a real-world protocol we recommend to clearly separate the functionality of different elements in the protocol. For illustration purposes, we present a version of a "full fledge" SIGMA protocol in [24].

Notation. All the protocols presented here use the Diffie-Hellman exchange. We use the traditional exponential notation g^x where g is a group generator. However, all the treatment here applies to any group in which the Diffie-Hellman problem is hard. (A bit more precisely, groups in which the so called "Decisional Diffie-Hellman Assumption (DDH)" holds, namely, the infeasibility to distinguish between quadruples of the form (g, g^x, g^y, g^{xy}) and quadruples (g, g^x, g^y, g^z) where x, y, z are random exponents.) We use the acronym DH to denote Diffie-Hellman, and use the noun "exponential" for elements such as g^x and the word "exponent" for x. In the description of our protocols the DH group and generator g are assumed to be fixed and known in advance to the parties or communicated at the onset of the protocol (in the later case, the DH parameters need to be included in the information authenticated by the protocol).

Throughout the paper we will also use the notation $\{\cdots\}_K$ to denote encryption of the information between the brackets under a symmetric encryption function using key K. Other cryptographic primitives used in the paper are a MAC (message authentication code) which is assumed to be unforgeable against chosen message attack by any adversary that is not provided the MAC key, and a digital signature scheme SIG assumed to be secure against chosen message attacks. By $\text{SIG}_A(msg)$ we denote the signature using A's private key on the message msg. The letters A and B denote the parties running a KE protocol, while Eve (or E) denotes the (active) attacker. We also use A, B, E to denote the identities used by these parties in the protocols.

3   The STS Protocols

Here we discuss the STS protocol (and some of its variants) which constitutes one of the most famous and influential protocols used to provide authenticated DH using digital signatures, and of particular appeal to scenarios where identity protection is a concern. The STS protocol, due to Diffie, van Oorschot and Wiener, is presented in [8] where a very instructive description of its design rationale is provided. In particular, this work is the first to observe some of the more intricate subtleties related to the authentication of protocols in general and of the DH exchange in particular. The STS protocol served as the starting point for the SIGMA protocols described in this paper. Both the strengths of the STS design principles as well as the weaknesses of some of the protocol choices have motivated the design of SIGMA. These aspects are important to be understood before presenting SIGMA. We analyze several variants of the protocol proposed in [8,27,14].

Remark. The attacks on the STS protocol and its variants presented here originate with the communications by the author to the IPsec working group in 1995 [19]. Since then some of these attacks were recalled elsewhere (e.g. [30]) and enhancements of the attack against the MAC variant have been provided in [4].

3.1   BADH and the Identity-Misbinding Attack: A Motivating Example

As the motivation for the STS protocol (and later for SIGMA too) we present a proposal for an "authenticated DH protocol" which intuitively provides an authenticated KE solution but is actually flawed. We denote this protocol by BADH ("badly authenticated DH").

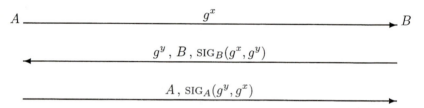

The output of the protocol is a session key K_s derived from the DH value g^{xy}. (Note: the identity of A may also be sent in the first message, this is immaterial to the discussion here.)

This protocol provides the most natural way to authenticate a DH exchange using digital signatures. Each party sends its DH exponential signed under its private signature key. The inclusion of the peer's exponential under one's signature is required to prove freshness of the signature for avoiding replay attacks (we will discuss more about this aspect in the context of SIGMA, in particular the possibility to replace the signature on the peer's exponential with the signature on a peer-generated nonce). One of the important contributions of [8] was to demonstrate that this protocol, even if seemingly natural and intuitively correct, does not satisfy the important *consistency requirement* discussed in Section 2.1. Indeed, [8] present the following attack against the BADH protocol. An active ("person-in-the-middle") attacker, which we denote by Eve (or E), lets the first two messages of the protocol to go unchanged between A and B, and then it replaces the third message from A to B with the following message from Eve to B:

$$E \xrightarrow{\qquad\qquad E,\ \text{SIG}_E(g^y,g^x)\qquad\qquad} B$$

The result of the protocol is that A records the exchange of the session key K_s with B, while B records the exchange of the *same* key K_s with Eve. In this case, any subsequent application message arriving to B and authenticated under the key K_s will be interpreted by B as coming from Eve (since from the point of view of B the key K_s represents Eve not A). Note that this attack does not result in a breach of secrecy of the key (since the attacker does not learn, nor influence, the key in any way) but it does result in a severe breach of authenticity since the two parties to the exchange will use the same key with different understandings of who the peer to the exchange is, thus breaking the consistency requirement. To illustrate the possible adverse effects of this attack we use the following example from [8]: imagine B being a bank and A a customer sending to B a monetary element, such as an electronic check or digital cash, encrypted and authenticated under K_s. From the point of view of B this is interpreted as coming from Eve (which we assume to also be a customer of B) and thus the money is considered to belong to Eve rather than to A (hopefully for Eve the money will go to her account!).

The essence of the attack is that Eve succeeds in convincing the peers to the DH exchange (those that chose the DH exponentials) that the exchange ended successfully yet the derived key is bound by each of the parties to a different peer. Thus the protocol fails to provide an authenticated binding between the key and the honest identities that generated the key. We will refer to this attack against the consistency requirement of KE protocols as an identity misbinding attack (or just "misbinding attack" for short).[2]

3.2   The Basic STS Protocol

Having discovered the misbinding attack on the "natural" authenticated DH protocol BADH, Diffie et al. [8] designed the STS protocol intended to solve this problem. The basic STS protocol is:

$$A \xrightarrow{\hspace{3cm} g^x \hspace{3cm}} B$$

$$\xleftarrow{\hspace{2cm} g^y \,,\, B \,,\, \{\, \mathrm{SIG}_B(g^x, g^y)\,\}_{K_s} \hspace{2cm}}$$

$$\xrightarrow{\hspace{2cm} A \,,\, \{\, \mathrm{SIG}_A(g^y, g^x)\,\}_{K_s} \hspace{2cm}}$$

where the notation $\{\cdots\}_K$ denotes encryption of the information between the brackets under a symmetric encryption function using key K. In the STS protocol the key used for encryption is the same as the one output as the session key produced by the exchange[3].

---

[2] This type of attack appears in the context of other authentication and KE protocols. It is sometimes referred to as the "unknown key share attack" [4,15].

[3] This is a weakness of the protocol since the use of the session key in the protocol leaks information on the key (e.g., the key is not anymore indistinguishable from

Is this protocol secure? In particular, is the introduction of the encryption of the signatures sufficient to thwart the identity misbinding attack? This at least has been the intention of STS. The idea was that by using encryption under the DH key the parties to the exchange "prove" knowledge of this key something which the attacker cannot do. Yet, no proof of security of the STS protocol exists (see more on this below). Even more significantly we show here that the misbinding attack applies to this protocol in any scenario where parties can register public keys without proving knowledge of the corresponding signature key. (We note that while such "proof of possession" is required by some CAs for issuing a certificate, this is not a universal requirement for public key certificates; in particular it is not satisfied in many "out-of-band distribution" scenarios, webs of trust, etc.) In this case Eve can register A's public key as its own and then simply replace A's identity (or certificate) in the third message of STS with her own. B verifies the incoming message and accepts it as coming from Eve. Thus, in this case the STS protocol fails to defend against the misbinding attack. Thus, for the STS to be secure one must assume that a secure external mechanism for proof of possession of signature keys is enforced. As we will see both the ISO protocol discussed in Section 4 and the SIGMA protocols presented here do not require such a mechanism. Moreover, even under the assumption of external "proof of possession" the above STS protocol has not been proven secure.

Note. In [31] an analysis of the STS protocol based on an extension of BAN logic [5] is presented. However, the modeling of the encryption function in that analysis is as a MAC function. Therefore this analysis holds for the MAC variant of STS presented in the next subsection. However, as we will see, for considering that protocol secure one needs to assume that the CA verifies that the registrant of a public key holds the corresponding private key (proof of possession) and, moreover, that "on-line registration" attacks as discussed below are not possible.

What is the reason for this protocol failure? The main reason is to assume that the combination of proof of possession of the session key together with the signature on the DH exponentials provide a sufficient binding between the identities of the (honest) peers participating in the exchange and the resultant key. However, as the above attack shows this is not true in general. Can this shortcoming be corrected? One first observation is that encryption is not the right cryptographic function to use for proving knowledge of a key. Being able to encrypt a certain quantity under a secret key is no proof of the knowledge of that key. Such a "proof of key possession" is not guaranteed by common modes of encryption such as CBC and is explicitly violated by any mode using XOR of a (pseudo) random pad with the plaintext (such as counter or feedback

---

random). In addition, this can lead to the use of the same key with two different algorithms (one inside the KE protocol, and another when using the exchanged session key in the application that triggered the key exchange), thus violating the basic cryptographic principle of key separation (see, e.g., [20]). These weaknesses are easily solved by deriving different, and computationally independent, keys from the DH value g^{xy}, one used internally in the protocol for encryption and the other as the session key output by the protocol.

modes, stream ciphers, etc.). To further illustrate this point consider a seemingly stronger variant of the protocol in which not only the signature is encrypted but also the identity (or full certificate) of the signer is encrypted too. In this case the above attack against STS is still viable if the encryption is of the XOR type discussed above. In this case, when A sends the message $\{A, \text{SIG}_A(g^y, g^x)\}_{K_s}$, Eve replaces A's identity (or certificate) by just XORing the value $A \oplus E$ in the identity location in the ciphertext. When decrypted by B this identity is read as E's and the signature verified also as E's. Thus we see that even identity encryption does not necessarily prevent the attack. As we will see in the next section replacing the encryption with a MAC function, which is better suited to prove possession of a key, is still insufficient to make the protocol secure.

3.3 Two STS Variants: MACed-Signature and Photuris

In [8] a variant of the basic STS protocol is suggested in which the encryption function in the protocol is replaced with a message authentication (MAC) function. Namely, in this STS variant each party in the protocol applies its signature on the DH exponentials plus it concatenates to it a MAC on the signature using the key K_s. For example, the last message from A to B in this protocol consists of the triple (A, b, c) where $b = \text{SIG}_A(g^y, g^x)$ and $c = \text{MAC}_{K_s}(b)$. In [8] this variant is not motivated as a security enhancement but as an alternative for situations –such as export control restrictions– in which the use of a strong encryption function is not viable. However, considering that a MAC function is more appropriate for "proving knowledge of a key" than an encryption function (as exemplified above) then one could expect that this variant would provide for a more secure protocol. This is actually incorrect too. The above attack on basic STS (where Eve records the public key of A under her name) can be carried exactly in the same way also in this MAC-based variant of the protocol. Same for the case where on top of the signature and identities (or even on top of the MAC) one applies an encryption function of the XOR type.

Moreover, if (as it is common in many application) A and B communicate their public key to each other as part of the KE protocol (i.e., the identities A and B sent in the protocol are their public-key certificates), then this MAC-ed signature variant is not secure even if the system does ensure that the registrant of a public key knows the corresponding private key! This has been shown by Blake-Wilson and Menezes [4] who present an ingenious *on-line registration attack* against the protocol. In this form of attack, the attacker Eve intercepts the last message from A to B and then registers a public key (for which she knows the private key) that satisfies $\text{SIG}_E(g^y, g^x) = \text{SIG}_A(g^y, g^x)$. Eve then replaces the certificate of A with her own in the intercepted message and forwards it to B (leaving the signature and mac strings unchanged from A's original message). Clearly, B will accept this as a valid message from Eve since both signature and mac will pass verification. In other words, Eve successfully mounted an identity-misbinding attack against the MACed-signature protocol. In [4] it is shown that this on-line registration attack can be performed against natural signature schemes. In particular, it is feasible against RSA signatures provided that the registrant of the public key can choose her own public RSA exponent.[4] While the full practicality of such an attack is debatable, it certainly suffices to show that one cannot prove this protocol to be secure on the basis of generic cryptographic functions, even under the assumption that the CA verifies possession of the private signature key. As a final note on this attack, we point out that this attack is possible even if the protocol is modified in such a way that each peer includes its own identity under the signature (something that can be done to avoid the need for "proof of possession" in the public-key registration stage).

From the above examples we learn that the failure to the misbinding attack is more essentially related to the *insufficiency* of binding the DH key with the signatures. Such a binding (e.g., via a MAC) provides a proof that *someone* knows the session key, but does not prove *who* this *someone* is. As we will see later, the essential binding here needs to be done between the signature and the recipient's identity (the ISO protocol) or between the DH key and the sender's identity (the SIGMA protocol).

We finish this section by showing the insecurity of another variant of the STS protocol described in [27] and used as the core cryptographic protocol in Photuris [14] (an early proposal for a KE protocol for IPsec). As the previous variants, this one is also illustrative of the subtleties of designing a good KE protocol. This variant dispenses of the use of encryption or MAC; instead it attempts at binding the DH key to the signatures by including the DH key g^{xy} under the signature:

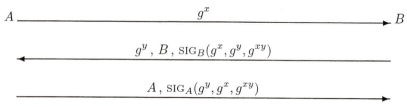

An obvious, immediate, complaint about this protocol is that the DH key g^{xy} is included under the signature, and therefore any signature that leaks information on the signed data (for example, any signature scheme that provides "message recovery") will leak information on g^{xy}. This problem is relatively easy to fix: derive two values from g^{xy} using a one-way pseudorandom transformation; use one value to place under the signature, and the other as the generated session key. A more subtle weakness of the protocol is that it allows, even with the above enhancement, for an an identity misbinding attack whenever the signature scheme allows for message recovery (e.g. RSA). In this case the attacker, Eve, proceeds as follows: it lets the protocol proceed normally between A and B for the first two messages, then it intercepts the last message from A to B and replaces it with the message

---

[4] In this case, Eve uses an RSA public modulus equal to the product of two primes p and q for which computing discrete logarithms is easy (e.g., all factors of $p-1$ and $q-1$ are small), and calculates the private exponent d for which $(hash(g^y, g^x))^d$ equals the signature sent by A.

$$E \xrightarrow{\qquad\qquad E,\; \mathrm{SIG}_E(g^y, g^x, g^{xy}) \qquad\qquad} B$$

But how can E sign the key g^{xy} (or a value derived from it) if it does not know g^{xy}? For concreteness assume that $\mathrm{SIG}_A(M) = RSA_A(hash(M))$, for some hash (and encoding) function $hash$. Since Eve knows A's public key it can invert A's signature to retrieve $hash(g^y, g^x, g^{xy})$ and then apply its own signature $RSA_E(hash(g^y, g^x, g^{xy}))$ as required to carry the above attack! (Note that this attack does not depend on any of the details of the public-key registration process; the attacker uses its legitimately generated and registered public key.)

Photuris included the above protocol as an "authentication only" solution, namely, one in which identities are not encrypted. It also offered optional identity protection by applying encryption on top of the above protocol. In the later case the above simple misbinding attack does not work. Yet, even in this case no proof of security for such a protocol is known. The above protocol (without encryption) is also suggested as an STS variant in [27] where it is proposed to explicitly hash the value g^{xy} before including it under the signature.

Remark: In this STS variant [27] the value g^{xy} under the signature is replaced with $h(g^{xy})$ where h is a hash function. This explicit hashing of g^{xy} seems to be intended to protect the value g^{xy} in case that the signature in use reveals its input. While this is not sufficient to defend against our identity misbinding attack, it is interesting to check whether revealing the value $h(g^{xy})$ may be of any use to an eavesdropper (note that in this case the attacker has the significantly simpler task of monitoring the protocol's messages rather than actively interfering with the protocol as required to carry the misbinding attack). Certainly, learning $h(g^{xy})$ is sufficient for distinguishing the key g^{xy} from random (even if the hash function acts as an ideal "random oracle"). But can the attacker obtain more than that? To illustrate the subtle ways in which security deficiencies may be exploited, consider the following practical scenario in which the function h is implemented by SHA-1 and the key derivation algorithm defines the session key to be $K_s = \mathrm{HMAC\text{-}SHA1}_{g^{xy}}(v)$, where v is a non-secret value. The reader can verify (using the definition of HMAC in [21]) that in this case the attacker does not need to find g^{xy} for deriving the session key K_s, but it suffices for her to simply know SHA-1(g^{xy}). Therefore if this later value is revealed by the signature then the security of the protocol is totally lost. Not only this example shows the care required in designing these protocols, but it also points to the the potential weaknesses arising from protocols whose security cannot be claimed in a generic (i.e. algorithm-independent) way.

4   The ISO Protocol

Here we recall the ISO KE protocol [12] which similarly to STS uses digital signatures to authenticate a DH exchange[5]. However, the ISO protocol resolves the problem of key-identity binding demonstrated by the misbinding attack on

---

[5] Strictly speaking, the protocol presented here is a simplification of the protocol in [12]. The latter includes two elements that are redundant and do not contribute

the BADH protocol (see Section 3.1) differently. The protocol simply adds the identity of the *intended recipient* of the signature to the signed information. Specifically, the protocol is:

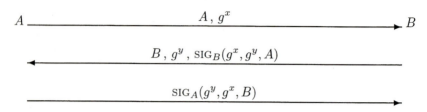

$$A \xrightarrow{\quad\quad\quad\quad\quad\quad\quad A,\, g^x \quad\quad\quad\quad\quad\quad\quad} B$$

$$\xleftarrow{\quad\quad\quad\quad B,\, g^y,\, \text{SIG}_B(g^x, g^y, A) \quad\quad\quad\quad}$$

$$\xrightarrow{\quad\quad\quad\quad\quad \text{SIG}_A(g^y, g^x, B) \quad\quad\quad\quad\quad}$$

It is not hard to see that the *specific* identity misbinding attack as described in Section 3.1 is avoided by the inclusion of the identities under the signatures. Yet having seen the many subtleties and protocol weaknesses related to the STS protocols in the previous section it is clear that resolving one specific attack is no guarantee of security. Yet the confidence in this protocol can be based on the analytical work of [6] where it is shown that this is a secure KE protocol (under the security model of that work). It is shown there that any feasible attack in that model against the security of the ISO protocol can be transformed into an efficient cryptanalytical procedure against the DH transform or against the digital signature function scheme in use.

The ISO protocol is simple and elegant. It uses a minimal number of messages and of cryptographic primitives. It allows for delaying computation of the DH key g^{xy} to the end of the interaction (since the key is not used inside the protocol itself) thus reducing the effect of computation on protocol latency. The protocol is also minimal in the sense that the removal of any of its elements would render the protocol insecure. In particular, as demonstrated by the BADH protocol, the inclusion of the recipient's identity under the signature is crucial for security. It is also interesting to observe that replacing the recipient's identity under the signature with the signer's identity results in an insecure protocol, open to the identity-misbinding attack exactly as in the case of BADH.

Therefore, it seems that we have no reason to look for other DH protocols authenticated with digital signatures. This is indeed true as long as "identity protection" is not a feature to be supported by the protocol. As explained below, in spite of all its other nice properties the ISO protocol does *not* satisfactorily accommodate the settings in which the identities of the participants in the protocol are to be concealed from attackers in the network (especially if such a protection is sought against active attacks).

The limitation of the ISO protocol in providing identity protection comes from the fact that in this protocol each party needs to know the identity of the peer before it can produce its own signature. This means that no party to the

---

significantly to the security of the protocol and are therefore omitted here. These elements are the inclusion of the signer's identity under the signature and an additional MAC value. In contrast to SIGMA, where the additional MAC is essential for security, the MAC in [12] serves only for explicit key confirmation (which adds little to the implicit key confirmation provided in the simplified variant discussed here).

protocol (neither A or B) can authenticate the other party before it reveals its own name to that party. This leaves *both* identities open to active attacks. If the only protection sought in the protocol is against passive eavesdroppers then the protocol can be built as a 4-message protocol as follows:

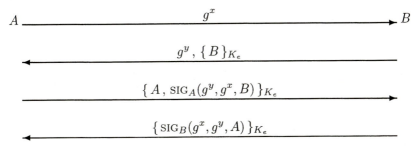

where K_e is an encryption key derived from the DH key g^{xy}. We note that with this addition of encryption the ISO protocol looses several of its good properties (in particular, the minimality discussed above and the ability to delay the computation of g^{xy} to the end of the protocol) while it only provides partial protection of identities since both identities are trivially susceptible to active attacks.

Another privacy (or lack of privacy) issue related to the ISO protocol which is worth noting is that by signing the peer's identity each party to the protocol leaves in the hands of the peer a signed (undeniable) trace that the communication took place (see the discussion at the end of Section 2.2).

The SIGMA protocol presented in the next section provides better, and more flexible, support for identity protection with same or less communication and computational cost, and with an equivalent proof of security.

Remark (*an identity-protection variant of the ISO protocol*):   We end this section by suggesting an adaptation of the ISO protocol to settings requiring identity protection (of one of the peers) to active attacks. We only sketch the idea behind this protocol. The idea is to run the regular ISO protocol but instead of A sending its real identity in the first message it sends an "alias" computed as $\hat{A} = hash(A, r)$ for a random r. Then B proceeds as in the basic protocol but includes the value $\hat{A}$ under its signature instead of A's identity; it also uses the key g^{xy} to encrypt its own identity and signature. In the third message A reveals its real identity 'A' and the value r used to compute $\hat{A}$. It also sends its signature (with B's identity signed as in the regular ISO protocol). This whole message is privacy-protected with encryption under K_e. The above protocol can be shown to be secure under certain assumptions on the hash function $hash$. Specifically, this function needs to satisfy some "commitment" properties similar to those presented in [22].

We omit further discussion of this protocol and proceed to present the SIGMA protocol that provides a satisfactory and flexible solution to the KE problem suitable also for settings with identity protection requirements, and with less requirements on the underlying cryptographic primitives than the above "alias-based" ISO variant.

5   The SIGMA Protocols

The weaknesses of the STS variants (which provide identity protection but not full security in general) and the unsuitability of the ISO protocol for settings where identity protection is a requirement motivated our search for a solution that would provide solid security for settings where identity protection is or is not a requirement. The result is the SIGMA protocols that we present here and whose design we explain based on the design lessons learned through the examples in previous sections (and many other in the literature). SIGMA takes from STS the property that each party can authenticate to the other without needing to know the peer's identity (recall that the lack of this property in the ISO protocol makes that protocol inappropriate to support identity protection). And it takes from ISO the careful binding between identities and keys, but it implements this binding in a very different way. More specifically, SIGMA decouples the authentication of the DH exponentials from the binding of key and identities. The former authentication task is performed using digital signatures while the latter is done by computing a MAC function keyed via g^{xy} (or more precisely, via a key derived from g^{xy}) and applied to the sender's identity. This "SIGn-and-MAc" approach is the essential technique behind the design of these protocols and the reason for the SIGMA acronym.

As pointed out in Section 2.3, we focus on the cryptographic core of the protocol leaving important system and implementation details out of the discussion. In particular, as we will also note below, in the following presentation we overcharge the DH exponentials with the added functionality of session id's and freshness nonces. (A "full fledge" SIGMA instantiation with a more careful treatment of these elements is presented in [24].)

5.1   The Basic SIGMA Protocol

The most basic form of SIGMA (without identity protection) is the following:

$$A \xrightarrow{\qquad\qquad\qquad\qquad g^x \qquad\qquad\qquad\qquad} B$$

$$\xleftarrow{\qquad g^y,\, B,\, \mathrm{SIG}_B(g^x, g^y),\, \mathrm{MAC}_{K_m}(B) \qquad}$$

$$\xrightarrow{\qquad A,\, \mathrm{SIG}_A(g^y, g^x),\, \mathrm{MAC}_{K_m}(A) \qquad}$$

The output of the protocol is a session key K_s derived from the DH value g^{xy} while the key K_m used as a MAC key in the protocol is also derived from this DH value. It is essential for the protocol security that the keys K_m and K_s be "computationally independent" (namely no information on K_s can be learned from K_m and vice-versa).[6] Note that this basic protocol does not provide identity protection. This will be added on top of the above protocol using encryption

---

[6] We discuss specific ways to derive these values from g^{xy} using pseudorandom functions in the full version of this paper [24].

(see following sections). The important point is that SIGMA's security is built in a modular way such that its core cryptographic security is guaranteed independently of the encryption of identities. Thus the same design serves for scenarios requiring identity protection but also for the many cases where such protection is not an issue (or is offered only as an option). We note that the identities A and B transmitted in messages 2 and 3 may be full public-key certificates; in this case the identities included under the MAC may be the certificates themselves or identities bound to these certificates.

The first basic element in the logic of the protocol is that the DH exponential chosen by each party is protected from modification (or choice) by the attacker via the signature that the party applies to its own exponential. We note that the inclusion of the peer's exponential under the signature is not mandatory and can be replaced with a nonce freshly chosen and communicated by the peer. Yet, either the peer's exponential (if chosen fresh and anew in each session) or a fresh nonce must be included under the signature; otherwise the following *replay attack* is possible. It would suffice for the attacker to learn the exponent x of a single ephemeral exponential g^x used by a party A in one session for the attacker to be able to impersonate A on a KE with any other party (simply by replaying the values g^x and $\mathrm{SIG}_A(g^x)$). Thus, in this case A's impersonation by the attacker is possible even without learning A's long-term signature key. This violates the security principle (see Section 2.1) by which the exposure of ephemeral secrets belonging to a specific session should not have adverse effects on the security of other sessions.

The second fundamental element in SIGMA's design is the MACing of the sender's identity under a key derived from the DH key. This can be seen as a "proof of possession" of the DH key but its actual functionality is to *bind* the session key to the identity of each of the protocol participants in a way to provide the "consistency" requirement of KE protocols. As discussed in Section 2.1, this is a fundamental requirement needed, in particular, to avoid attacks such as the identity misbinding attack from Section 3.1. Note that without this MACing the protocol "degenerates" into the BADH protocol from Section 3.1 which is susceptible to this attack. Therefore we can see that all the elements in the protocol are mandatory (up to replacement of the peer's exponential under the signature with a fresh nonce).

We note that the above SIGMA protocol, as well as all the following variants, satisfy all the security guarantees discussed in Section 2.1. In particular, they provide "perfect forward secrecy (PFS)" due to the use of the Diffie-Hellman exchange. This assumes that DH exponentials are chosen anew and independently for each session, that the exponents x, y used in a DH exponentials g^x, g^y are erased as soon as the computation of the key g^{xy} is completed, and that these exponents are not derivable from any other quantity stored in the party's computer (in particular, if x is generated pseudorandomly then the value of past exponents x should not be derivable from the present state of the PRG). We note that SIGMA can allow for re-use of DH exponentials by the same party across different sessions. However, in this case the forward secrecy property is lost (or at least confined to hold only after all sessions using the exponent x are completed and the exponent x erased). In case of re-use of DH exponents one must derive the keys used by the session (e.g. K_m, K_s) in a way that depends on some session-specific non-repeating quantity (a nonce or session-id). Also, as discussed before, in this case such a fresh nonce needs to be included under the peer's signature. There are other, more theoretical, issues concerning the re-use of DH exponents that are not treated here.

As we have stressed before, this informal outline of the design rationale for SIGMA does not constitute a proof of security for the protocol. The formal analysis in which we can base our confidence in the protocol appears in the companion analysis paper [7].

5.2   Protecting Identities: SIGMA-I

As said, SIGMA is designed to serve as a secure key-exchange protocol both in settings that do not require identity protection (in which case the above simple protocol suffices) or those where identity protection is a requirement. The main point behind SIGMA's design that allows for easy addition of identity protection is that the peer's identity is not needed for own authentication. In particular, one of the peers can delay communicating its own identity until it learns the peer's identity in an authenticated form. Specifically, to the basic SIGMA protocol we can add identity protection by simply encrypting identities and signatures using a key K_e derived from g^{xy} (K_e must be computationally independent from the authentication key K_m and the session key K_s):

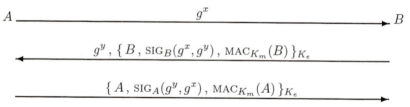

This protocol has the property that it protects the identity of the initiator from active attackers and the identity of the responder from passive attackers. Thus, the protocol is suitable for situations where concealing the identity of the initiator is considered of greater importance. A typical example is when the initiator is a mobile client connecting to a remote server. There may be little or no significance in concealing the server's identity but it may be of prime importance to conceal the identity of the mobile device or user. We stress that the encryption function (as applied in the third message) must be resistant to active attacks and therefore must combine some form of integrity. Combined secrecy-integrity transforms such as those from [13] can be used, or a conventional mode of encryption (e.g. CBC) can be used with a MAC function computed on top of the ciphertext [3, 23]. Due to the stronger protection of the identity of the Initiator of the protocol we denote this variant by SIGMA-I.

We remark that while this protocol has the minimal number of messages that any KE protocol resistant to replay attacks (and not based on trusted timestamps) can use, it is sometimes desirable to organize the protocol in full round-trips with each pair of message containing a "request message" and a "response message". If so desired, the above protocol can add a fourth message from B to A with a simple ACK authenticated under the authentication key K_m. This ACK message serves to A as a proof that B already established the key and communications protected under the exchanged key K_s can start. It also provides the flexibility for A to either wait for the ACK or start using the session key as soon as it sent the third protocol message. (Depending on B's policy this traffic may be accepted by B if the channel – or "security association" in the language of IKE – was already established by B, or discarded if not, or queued until the key establishment is completed.) Finally, it is worth noting that this ACK-augmented protocol provides the *peer awareness* property discussed in Section 2.1. (This is in contrast to the other variants of SIGMA presented here which do not enjoy this property.)

5.3   A Four Message Variant: SIGMA-R

As seen, SIGMA-I protects the initiator's identity against active attacks and the responder's against passive attacks. Here we present SIGMA-R which provides defense to the responder's identity against active attacks and to the initiator's only against passive attacks. We start by presenting a simplified version of SIGMA-R without encryption:

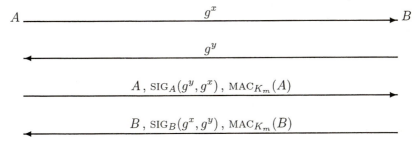

The logic of the protocol is similar to that of the basic SIGMA. The difference is that B delays the sending of its identity and authentication information to the fourth message after it verified A's identity and authentication in message 3. This "similarity" in the logic of the protocol does not mean that its security is implied by that of the 3-message variants. Indeed, the protocol as described above is open to a *reflection attack* that is not possible against the 3-message variant. Due to the full symmetry of the protocol an attacker can simply replay each of the messages sent by A back to A. If A is willing to accept a key exchange with itself then A would successfully complete the protocol.[7] Therefore, to prevent this attack the protocol needs to ensure some "sense of direction" in the authenticated

---

[7] The only damage of this attack seems to be that it forces A to use a key derived from the distribution g^{x^2} rather than g^{xy}. These distributions may be distinguishable depending on the DH groups.

information. This can be done by explicitly adding different "tags" under the MAC for each of the parties (e.g., A would send $\text{MAC}_{K_m}(\text{"0"}, A)$ while B would send $\text{MAC}_{K_m}(\text{"1"}, B)$), or by using different MAC keys in each direction (i.e., instead of deriving a single key K_m from g^{xy} one would derive two keys, K_m and K'_m, where the former is used by A to compute its MAC and the latter by B). Any of these measures are sufficient to prevent the reflection attack and make the protocol secure [7] (another defense is for A to check that the peer's DH exponential is different than her own.)

The full protocol SIGMA-R (with identity protection) is obtained by encrypting the last two messages in the above depicted protocol (and adding a reflection defense as discussed before). A "full fledge" illustration of protocol SIGMA-R is presented in [24].

Remark (*the inter-changeability property of SIGMA*). It is worth noting that the last two messages in the above protocol can be interchanged. Namely, B may proceed as described in SIGMA-R and wait for the reception of A's message (message 3 in the above picture) before sending his last message. But B may also decide to send his last message (signature and mac) immediately after, or together with, message 2 (which results in SIGMA-I). In this way, B may control if he is interested in protecting his own identity from active attacks or if he prefers to favor a faster exchange. The protocol may also allow for messages 3 and 4 to cross in which case the protocol is still secure but both identities may be open to active attacks.

5.4   Further Variants and the Use of SIGMA in IKE

As seen above the MAC of the sender's identity is essential for SIGMA's security. Here we present a variant of the protocol that differs from the above descriptions by the way the MAC value is placed in the protocol's messages. Specifically, the idea is to include the MAC value under the signature (i.e., as part of the signed information). The interest on this variant is that it saves in message length by avoiding explicit sending of the MAC value, and more significantly because it is the variant of SIGMA adopted into the IKE protocol (both IKE version 1 [11] and version 2 [16]).

The MAC moved under the signature may cover just the identity of the sender or the whole signed information. For example, in B's message the pair $(\text{SIG}_B(g^x, g^y), \text{MAC}_{K_m}(B))$ is replaced with either (i) $\text{SIG}_B(g^x, g^y, \text{MAC}_{K_m}(B))$ or (ii) $\text{SIG}_B(\text{MAC}_{K_m}(g^x, g^y, B))$. In this way the space for an extra MAC outside the signature is saved, and the verification of the MAC is merged with that of the signature. In either case, *as long as the* MAC *covers the identity of the signer* then the same security of the basic SIGMA protocol (as well as SIGMA-I and SIGMA-R) is preserved[8] [7]. Variant (ii) is used in the IKE protocol (version

---

[8] A technicality here is that moving the MAC inside is possible only for MAC functions whose verification is done by recomputation of the MAC value; this is the case for almost all practical MAC functions, in particular when the MAC is implemented via a pseudorandom function as in IKE [11,16].

1) [11] in two of its authentication modes: the signature-based exchange of IKE uses the basic 3-message SIGMA protocol (without identity encryption) as presented in Section 5.1 for its aggressive mode, and it uses the 4-message SIGMA-R in its main mode. (In the later case, the use of SIGMA-R in IKE is preceded by two extra messages for negotiating security parameters.) In IKE the MAC function is implemented via a pseudorandom function which is also used in the protocol for the purpose of key expansion and derivation.[9] IKE version 2 [16] uses variant (i) with SIGMA-R as its single key exchange method authenticated with public keys. In this protocol the peer's DH exponential is not signed; the essential freshness guarantee is provided by signing a nonce chosen by the peer (see Section 5.1).

The SIGMA-R protocol has also been adopted in the JFK protocol [1] which has been proposed in the context of the undergoing revision of the IKE protocol. We note that in both [16,1] protocol SIGMA-R is augmented with mechanisms that provide some defense against Denial-of-Service attacks as discussed in Section 2.3.

Acknowledgment. I wish to thank Ran Canetti for embarking with me on a long and challenging journey towards formalizing and proving the security of key-exchange protocols (in particular, the SIGMA protocols). Special thanks to Dan Harkins for being receptive to my design suggestions when specifying the signature modes of IKE, and to Charlie Kaufman (and the IPsec WG) for incorporating this design into IKEv2. Thanks to Sara Bitan and Paul Van Oorschot for many useful discussions, and to Pau-Chen Cheng for sharing much of his implementation and system experience with me. This research was partially funded by the Irwin and Bethea Green & Detroit Chapter Career Development Chair.

References

1. B. Aiello, S. Bellovin, M. Blaze, R. Canetti, J. Ioannidis, A. Keromytis, O. Reingold, "Efficient, DoS-Resistant Secure Key Exchange for Internet Protocols," *ACM Computers and Communications Security conference (CCS)*, 2002.
   http://www.research.att.com/~smb/papers/jfk-ccs.pdf
2. M. Bellare and P. Rogaway, "Entity authentication and key distribution", *Advances in Cryptology, - CRYPTO'93*, Lecture Notes in Computer Science Vol. 773, D. Stinson ed, Springer-Verlag, 1994, pp. 232–249.
3. S. M. Bellovin, "Problem Areas for the IP Security Protocols", *Proceedings of the Sixth Usenix Unix Security Symposium*, 1996.
4. S. Blake-Wilson and A. Menezes, "Unknown key-share attacks on the station-to-station (STS) protocol", *Proceedings of PKC '99*, Lecture Notes in Computer Science, 1560 (1999), 154–170.

---

[9] This use of a prf –as a MAC – under the signature has been a source of confusion among analysts of the IKE protocol; the prf was sometimes believed to have some other functionality related to the signature. It is important then to realize that its functionality under the signature is simply (and essentially!) that of a MAC covering the signer's identity.

5. M. Burrows, M. Abadi and R. Needham, "A logic for authentication," *ACM Trans. Computer Systems* Vol. 8 (Feb. 1990), pp. 18–36

6. Canetti, R., and Krawczyk, H., "Analysis of Key-Exchange Protocols and Their Use for Building Secure Channels", Eurocrypt'2001, LNCS Vol. 2045. Full version in: *Cryptology ePrint Archive* (http://eprint.iacr.org/), Report 2001/040.

7. Canetti, R., and Krawczyk, H., "Security Analysis of IKE's Signature-based Key-Exchange Protocol", *Crypto 2002*. LNCS Vol. 2442. Full version in: *Cryptology ePrint Archive* (http://eprint.iacr.org/), Report 2002/120.

8. W. Diffie, P. van Oorschot and M. Wiener, "Authentication and authenticated key exchanges", *Designs, Codes and Cryptography*, 2, 1992, pp. 107–125. Available at http://www.scs.carleton.ca/~paulv/papers/sts-final.ps.

9. N. Ferguson and B. Schneier, "A Cryptographic Evaluation of IPSec", http://www.counterpane.com/ipsec.html, 1999.

10. O. Goldreich, "Foundations of Cryptography: Basic Tools", Cambridge Press, 2001.

11. D. Harkins and D. Carrel, ed., "The Internet Key Exchange (IKE)", *RFC 2409*, Nov. 1998.

12. ISO/IEC IS 9798-3, "Entity authentication mechanisms — Part 3: Entity authentication using asymmetric techniques", 1993.

13. C. Jutla, "Encryption Modes with Almost Free Message Integrity", *Advances in Cryptology – EUROCRYPT 2001 Proceedings*, Lecture Notes in Computer Science, Vol. 2045, Springer-Verlag, B. Pfitzmann, ed, 2001.

14. Karn, P., and Simpson W.A., "The Photuris Session Key Management Protocol", draft-ietf-ipsec-photuris-03.txt, Sept. 1995.

15. B. Kaliski, "An unknown key-share attack on the MQV key agreement protocol", *ACM Transactions on Information and System Security (TISSEC)*. Vol. 4 No. 3, 2001, pp. 275–288.

16. C. Kaufman, "Internet Key Exchange (IKEv2) Protocol", draft-ietf-ipsec-ikev2-07.txt, April 2003 (to be published as an RFC).

17. S. Kent and R. Atkinson, "Security Architecture for the Internet Protocol", *Request for Comments 2401*, Nov. 1998.

18. S. Kent and R. Atkinson, "IP Encapsulating Security Payload (ESP)", *Request for Comments 2406*, Nov. 1998.

19. H. Krawczyk, Communication to IPsec WG, *IPsec mailing list archives*, April-October 1995. http://www.vpnc.org/ietf-ipsec/

20. H. Krawczyk, "SKEME: A Versatile Secure Key Exchange Mechanism for Internet,", *Proceedings of the 1996 Internet Society Symposium on Network and Distributed System Security*, Feb. 1996, pp. 114–127.
http://www.ee.technion.ac.il/~hugo/skeme-lncs.ps

21. Krawczyk, H., Bellare, M., and Canetti, R., "HMAC: Keyed-Hashing for Message Authentication", *RFC 2104*, February 1997.

22. Krawczyk, H., "Blinding of Credit Card Numbers in the SET Protocol", *Proceedings of Financial Cryptography'99*, LNCS Vol. 1648, 1999.

23. H. Krawczyk, "The order of encryption and authentication for protecting communications (Or: how secure is SSL?)", Crypto'2001, LNCS Vol. 2139. Full version in: *Cryptology ePrint Archive* (http://eprint.iacr.org/), Report 2001/045.

24. H. Krawczyk, "SIGMA: the 'SIGn-and-MAc' Approach to Authenticated Diffie-Hellman and its Use in the IKE Protocols", full version.
http://www.ee.technion.ac.il/~hugo/sigma.html

25. G. Lowe, "Some New Attacks upon Security Protocols", *9th IEEE Computer Security Foundations Workshop*, IEEE Press 1996, pp.162–169.

26. Meadows, C., "Analysis of the Internet Key Exchange Protocol Using the NRL Protocol Analyzer", *Proc. of the 1999 IEEE Symposium on Security and Privacy,* IEEE Computer Society Press, 1999.
27. A. Menezes, P. Van Oorschot and S. Vanstone, "Handbook of Applied Cryptography," CRC Press, 1996.
28. Orman, H., "The OAKLEY Key Determination Protocol", *Request for Comments 2412,* Nov. 1998.
29. Perlman, R., and Kaufman, C., "Analysis of the IPsec key exchange Standard", *WET-ICE Security Conference,* MIT, 2001.
30. V. Shoup, "On Formal Models for Secure Key Exchange", Theory of Cryptography Library, 1999. Available at:
    `http://philby.ucsd.edu/cryptolib/1999/99-12.html`.
31. P. van Oorschot, "Extending cryptographic logics of belief to key agreement protocols", *Proceedings, 1st ACM Conference on Computer and Communications Security,* Nov. 1993, Fairfax, Virginia, pp. 232–243,

On Memory-Bound Functions for Fighting Spam

Cynthia Dwork[1], Andrew Goldberg[1], and Moni Naor[2]*

[1] Microsoft Research, SVC
1065 L'Avenida
Mountain View, CA 94043
{dwork,goldberg}@microsoft.com
[2] Weizmann Institute of Science
Rehovot 76100, Israel
naor@wisdom.weizmann.ac.il

Abstract. In 1992, Dwork and Naor proposed that e-mail messages be accompanied by easy-to-check *proofs of computational effort* in order to discourage junk e-mail, now known as spam. They proposed specific CPU-bound functions for this purpose. Burrows suggested that, since memory access speeds vary across machines much less than do CPU speeds, *memory-bound* functions may behave more equitably than CPU-bound functions; this approach was first explored by Abadi, Burrows, Manasse, and Wobber [3].
We further investigate this intriguing proposal. Specifically, we

1. Provide a formal model of computation and a statement of the problem;
2. Provide an abstract function and prove an asymptotically tight amortized lower bound on the number of memory accesses required to compute an acceptable proof of effort; specifically, we prove that, on average, the sender of a message must perform many unrelated accesses to memory, while the receiver, in order to verify the work, has to perform significantly fewer accesses;
3. Propose a concrete instantiation of our abstract function, inspired by the RC4 stream cipher;
4. Describe techniques to permit the receiver to verify the computation with *no* memory accesses;
5. Give experimental results showing that our concrete memory-bound function is only about four times slower on a 233 MHz settop box than on a 3.06 GHz workstation, and that speedup of the function is limited even if an adversary knows the access sequence and uses optimal off-line cache replacement.

1 Introduction

Unsolicited commercial e-mail, or spam, is more than just an annoyance. At two to three *billion* daily spams worldwide, or close to 50% of all e-mail, spam incurs huge infrastructure costs, interferes with worker productivity, devalues the internet, and is ruining e-mail.

---

* Incumbent of the Judith Kleeman Professorial Chair. Research supported in part by a grant from the Israel Science Foundation. Part of this work was done while visiting Microsoft Research, SVC.

D. Boneh (Ed.): CRYPTO 2003, LNCS 2729, pp. 426–444, 2003.

This paper focuses on the computational approach to fighting spam, and, more generally, to combating denial of service attacks, initiated by Dwork and Naor [11] (also discussed by Back; see [18,9]). The basic idea is:

"If I don't know you and you want to send me a message, then you must prove that you spent, say, ten seconds of CPU time, just for me and just for this message."

The "proof of effort" is cryptographic in flavor; as explained below, it is a moderately hard to compute (but very easy to check) function of the message, the recipient's address, and a few other parameters. Dwork and Naor called such a function a *pricing function* because the proposal is fundamentally an economic one: machines that currently send hundreds of thousands of spam messages each day, could, at the 10-second price, send only eight thousand. To maintain the current 2-3 billion daily messages, the spammers would require 250,000–375,000 machines.

CPU-bound pricing functions suffer from a possible mismatch in processing speeds among different types of machines (desktops *vs.* servers), and in particular between old machines and the presumed new, top of the line, machines that could be used by a high-tech spam service. In order to remedy these disparities, Burrows proposed an alternative computational approach, first explored in [3], based on memory latency. His creative suggestion is to design a pricing function requiring a moderately large number of scattered memory accesses. Since memory latencies vary much less across machines than do clock speeds, memory-bound functions should prove more equitable than CPU-bound functions.

Our Contributions. In the current paper we explore Burrows' suggestion. After reviewing the computational approach (Section 2) and formalizing the problem (Section 3), we note that the known time/space tradeoffs for inverting one-way functions [21,14] (where space now refers to cache) constrain the functions proposed in [3] (Section 4). We propose an abstract function, using random oracles, and give a lower bound on the amortized complexity of computing an acceptable proof of effort (Section 5)[1]. We suggest a very efficient concrete implementation of the abstract function, inspired by the RC4 stream cipher (Section 6). We present experimental results showing that our concrete memory-bound function is only about four times slower on a 233 MHz settop box than on 3.06 GHz workstation (Section 7). Finally, we modify our concrete proposal to free the receiver from having to make memory accesses, with the goal of allowing small-memory devices to be protected by our computational anti-spam protocol. A more complete version of the paper is available at
www.wisdom.weizmann.ac.il/~naor/PAPERS/dgn.html.

2   Review of the Computational Approach

In order to send a message m, software operating on behalf of the sender computes a *proof of computational effort* $z = f(m, sender, receiver, date)$ for a moderately hard

---

[1] None of [11,18,9,3] obtains a lower bound.

to compute "pricing" function f. The message m is transmitted together with the other arguments to f and the resulting proof of effort z^2. Software operating on behalf of the receiver checks that the proof of effort has been properly computed; a missing proof can result in some user-prespecified action, such as placing the message in a special folder, marking it as spam, subjecting it to further filtering, and so on. Proof computation and verification should be performed automatically and in the background, so that the typical user e-mail experience is unchanged.

The function f is chosen so that (1) f is not amenable to amortization; in particular, computing $f(m, sender, Alice, date)$ does not help in computing $f(m, sender, Bob, date)$. This is key to fighting spam: the function must be recomputed for each recipient (and for any other change of parameters). (2) There is a "hardness" parameter to vary the cost of computing f, allowing it to grow as necessary to accommodate Moore's Law. (3) There is an important difference in the costs of computing f and of checking f: the cost of sending a message should grow much more quickly as a function of the hardness parameter than the cost of checking that a proof of effort is correct. This allows us to keep verification very cheap, ensuring that the ability to wage a denial of service attack against a receiver is not exacerbated by the spam-fighting tool. In addition, if verification is sufficiently cheap, then it can be carried out by the receiver's mail (SMTP) server.

Remark 1. With the right architecture, the computational approach permits *single-pass send-and-forget* e-mail: once the mail is sent the sender never need take any further action, once the mail is received the proof of effort can be checked locally; neither sender nor receiver ever need contact a third party. In other words, single-pass send-and-forget means that e-mail, the killer application of the Internet, is minimally disturbed.

Remark 2. We briefly remark on our use of the date as an argument to the pricing function. The receiver temporarily stores valid proofs of effort. The date is used to control the amount of storage needed. When a new proof of effort, together with its parameters, is received, one first checks the date: if the date is, say, over a week old, then the proof is rejected. Otherwise, the receiver checks the saved proofs of effort to see if the newly received proof is among them. If so, then the receiver rejects the message as a duplicate. Otherwise, the proof is checked for validity.

In [11], f is a forged signature in a careful weakening of the Fiat-Shamir signature scheme. Back's proposal, called *HashCash*, is based on finding hash collisions. It is currently used to control access to bulletin boards [18]; verification is particularly simple in this scheme.

3   Computational Model and Statement of the Problem

The focus on memory-bound functions requires specification of certain details of a computational model not common in the theory literature. For example, in real contemporary

---

[2] Having m as an argument to the function introduces some practical difficulties in real mail systems. One can instead use the following three arguments: receiver's e-mail address, date, and a nonce. However, the intuition is more clear if we include the message.

hardware there is (at least) two kinds of space: ordinary memory (the vast majority) and *cache* – a small amount of storage on the same integrated circuit chip as the central processing unit[3]. Cache can be accessed roughly 100 times more quickly than ordinary memory, so the computational model needs to differentiate accordingly. In addition, when a desired value is not in cache (a *cache miss*), and an access to memory is made, a small block of adjacent words (a *cache line*), is brought into the cache simultaneously. So in some sense values nearby the desired one are brought into cache "for free". Our model is an abstraction that reflects these considerations, among others.

When arguing the security of a cryptographic scheme one must specify two things: the power of the adversary and what it means for the adversary to have succeeded in breaking the scheme. In our case defining the adversary's power is tricky, since we have to consider many possible architectures. Nevertheless, for concreteness we assume the adversary is limited to a "standard architecture" as follows:

1. There is a large memory, partitioned into m blocks (also called cache lines) of b bits each;
2. The adversary's cache is small compared to the memory. The cache contains at most s (for "space") *words*; a cache line typically contains a small number (for example, 16) of words;
3. Although the memory is large compared to the cache, we assume that m is still only polynomial in the largest feasible cache size s;
4. Each word contains w bits (commonly, $w = 32$);
5. To access a location in the memory, if a copy is not already in the cache (a *cache miss*), the contents of the block containing that location must be brought into the cache – a *fetch*; since every cache miss results in a fetch, we use these terms interchangeably;
6. We charge one unit for each fetch of a memory block. Thus, if two adjacent blocks are brought into cache, we charge two units (there is no discount for proximity at the block level).
7. Computation on data *in the cache* is essentially *free*. By not (significantly) charging the adversary for this computation, we are increasing the power of the adversary; this strengthens the lower bound.

Thus, the challenge is to design a pricing function f as described in Section 2, together with algorithms for computing and checking f, in which the costs of the algorithms are measured in terms of memory fetches and the "real" time to compute f on currently available hardware is, say, about 10 seconds (in fact, f may be parameterized, and the parameters tuned to obtain a wide range of target computation times).

The adversary's goal is to maximize its production of (message, proof of computational effort) pairs while minimizing the number of cache misses incurred. The adversary is considered to have won if it has a strategy that produces many (message, proof) pairs with an *amortized* number of fetches (per message plus proof) which is substantially less than the expected number of fetches for a single computation obtained in the analysis of the algorithm. We do not care if the messages are sensical or not.

We remark that it may be possible to defeat a memory-bound function with specific parameters by building a special-purpose architecture, such as a processor with a huge,

---

[3] In fact, there are multiple levels of cache; Level 1 is on the chip.

fast, on-chip cache. However, since the computational approach to fighting spam is essentially an economic one, it is important to consider the cost of designing and building the new architecture. These issues are beyond the scope of this paper.

4   Simple Suggestions and Small-Space Cryptanalyses

In the full paper we show that pricing functions based on meet in the middle and subset sum can be computed with very few memory access, and hence do not solve our problem. In these proceedings we confine our attention to the proposal in [3], described next.

Easy-to-Compute Functions. These functions are essentially iterates of a single basic "random-looking" function g. They vary in their choice of basic function. The basic function has the property that a single function inversion is more expensive than a memory look-up.

Let n and ℓ be parameters and let $g : \{0,1\}^n \longrightarrow \{0,1\}^n$. Let g_0 be the identity function and for $i = 1 \ldots \ell$, let the function $g_i(x) = g(g_{i-1}(x)) \oplus i$.

Input. $y = g_\ell(x)$ *for some* $x \in \{0,1\}^n$ *and* α, *a hash of the values* $x, g_1(x), \ldots, g_\ell(x)$.
Output. $x' \in g_\ell^{-1}(y)$ *such that the string* $x', g_1(x'), \ldots, g_\ell(x')$ *hashes to* α.

The hope is that the best way to resolve the challenge is to build a table for g^{-1} and to work backwards from y, exploring the tree of pre-images[4]. Since forward computation of g is assumed to be quite easy, constructing the inverse table should require very little total time compared to the memory accesses needed to carry out the proof of effort.

The limitation of this approach is that, since g can be computed with no memory accesses, there is a time/space tradeoff for *inverting* g in which no memory accesses are performed (in our context, space refers to cache size, since we are interested in what can be done without going to memory) [21,14,26]. Those results imply that g can be inverted at the cost of two forward computations of g, *with no memory accesses*.

This suggests basing computational challenges on functions that are (in some sense) *hard in both directions*.

5   An Abstract Function and Lower Bound on Cache Misses

In this section we describe an "abstract" pricing function and prove a tight lower bound on the number of memory accesses that must be made in order to produce a message acceptable to the receiver, in the model defined in Section 3. The function is "abstract" in that it uses idealized hash functions, also known as random oracles. A concrete implementation is proposed in Section 6.

Meaning of the Model and the Abstraction: Our computational model implicitly constrains the adversary by constraining the architecture available to the adversary. Our use of random oracles for the lower bound argument similarly constrains the adversary, as there are some things it cannot compute without accessing the oracles. We see two

---

[4] The root of the tree is labelled with y. A vertex at distance $d \geq 0$ from the root having label $z \in Range(g_{\ell-d})$ has one child labeled with each $z' \in g^{-1}(z \oplus (\ell - d)) \in Range(g_{\ell-d-1})$.

advantages in such modelling: (i) It provides rationale to the design of algorithms such as those of Section 6, this is somewhat similar to what Luby and Rackoff [22] did for the application of Feistel Permutations in the design of DES; (ii) If there is an attack on the simplified instantiation of the algorithm of Section 6, then the model provides guidelines for modifications. Note that we assume that the arguments to the random oracle must be in cache in order to make the oracle call.

The inversion techniques of [21,14] do not apply to truly random functions, as these have large Kolmogorov complexity (no short representation). Accordingly, our function involves a large *fixed forever* table T of truly random w-bit integers[5]. The table should be approximately twice as large as the largest existing caches, and will dominate the space needs of our memory-bound function.

We want to force the legitimate sender of a message to take a random walk "through T," that is, to make a series of random accesses to T, each subsequent location determined, in part, by the contents of the current location.

Such a walk is called a *path*. The algorithm forces the sender to explore many different paths until a path with certain desired characteristics is found. We call this a *successful path*. Once a successful path has been identified, information enabling the receiver to check that a successful path has been found is sent along with the message. Verification requires work proportional to the path length, determined by a parameter ℓ. Each path exploration is called a *trial*. The expected number of trials to find a successful path is 2^e, where e (for "effort") is a parameter. The expected amount of work performed by the sender is proportional to 2^e times the path length.

5.1 Description of the Abstract Algorithm

The algorithm uses a modifiable array A, initialized for each trial, of size $|A|w > b$ bits (recall that b is the number of bits in a memory block, or cache line)[6].

Before we present the abstract algorithm, we introduce a few hash functions H_0, H_1, H_2, H_3, of varying domains and ranges, that we model as idealized random functions (random oracles). The function H_0 is only used during initialization of a path. It takes as input a message m, sender's name (or address) S, receiver's name (or address) R, and date d, together with a trial number k, and produces an array A. The function H_1 takes an array A as input and produces an index c into the table T. The function H_2 takes as input an array A and an element of T and produces a new array, which gets assigned to A. Finally, the function H_3 is applied to an array A to produce a string of $4w$ bits.

A word on notation: For arrays A and T, we denote by $|A|$ (respectively, $|T|$) the number of elements in the array. Since each element is a word of w bits, the numbers of *bits* in these arrays are $|A|w$ and $|T|w$, respectively.

The path in a generic trial is given by:

---

[5] "Fixed forever" means fixed until new machines have bigger caches, in which case the function must be updated.

[6] The intuition for requiring $|A|w > b$ is that, since A cannot fit into a single memory block, it is more expensive to fetch A into cache than it is to fetch an element of T into cache.

```
Initialization:
  A = H₀(m, R, S, d, k)
Main Loop: Walk for ℓ steps (ℓ is the path length):
  c ← H₁(A)
  A ← H₂(A, T[c])
Success occurs if:
  after ℓ steps the last e bits of H₃(A) are all zero.
```

Path exploration is repeated for $k = 1, 2, \ldots$ until success occurs. The information for identifying the successful path is simply all five parameters and the final $H_3(A)$ obtained during the successful trial[7].

Verification that the path is indeed successful is trivial: the verifier simply carries out the exploration of the one path and checks that success indeed occurs with the given parameters and that the reported hash value $H_3(A)$ is correct.

The connection to Algorithm MBound, described in Section 6, will be clear: we need only specify the four hash functions. To keep computation costs low in MBound, we will not invoke full-strength cryptographic functions in place of the random oracles, nor will we even modify all entries of array A at each step.

The size of A also needs consideration. If A is too small, say, a pointer into T, then the spammer can mount an attack in which many different paths (trials for either the same or different messages) can be explored at a low amortized cost, as we now informally describe. At any point, the spammer can have many different A's (that is, A's for different trials) in the cache. The spammer then fetches a memory block containing several elements of T, and advances along each path for which some element in the given memory block was needed. This allows exploitation of locality in T. Thus, intuitively, we should choose $|A|$ sufficiently large that it is infeasible to store many different A's in the cache.

5.2   Lower Bound on Cache Misses

We now prove a lower bound on the amortized number of block transfers that any adversary constrained as described in Section 3 must incur in order to find a successful path. Specifically, we show that the *amortized* complexity (measured in the number of memory fetches per message) of the abstract algorithm is asymptotically tight.

The computation on each message must follow a specific sequence of oracle calls in order to make progress. The adversary may make any oracle calls it likes; however, to make progress on a path it must make the specified calls. *By watching an execution unfold, we can observe when paths begin, and when they make progress.* Calls to the oracle that make progress (as determined by the history) are called *progress calls*.

Theorem 1. *Consider an arbitrarily long but finite execution of the adversary's program – we don't know what the program is, only that the adversary is constrained to use an architecture as described in Section 3. Under the following additional conditions, the amortized complexity of generating a proof of effort that will be accepted by a verifier is $\Omega(2^e \ell)$:*

---

[7] The value of $H_3(A)$ is added to prevent the spammer from simply guessing k, which has probability $1/2^e$ of success.

- $|T| \geq 2s$ (recall that the cache contains s words of w bits each)
- $|A|w \geq bs^{1/5}$ (recall that b is the block size, in bits).
- $\ell > 8|A|$
- The total amount of work by the spammer (measured in oracle calls) per successful path is no more than $2^{o(w)}2^e\ell$.
- ℓ is large enough so that the spammer cannot call the oracle 2^ℓ times.

Remark 3. First note that $|A|$ is taken to be much larger than b/w. We already noted that if $|A|$ is very small than a serious attack is possible. However, even if $|A|$ is roughly b/w, it is possible to attack the algorithm by storing many copies of T under various permutations. In this case the adversary can hope to concurrently be exploring about $\log s$ paths for which a single memory block contains the value in T needed by all $\log s$ paths. Hence, if (for some reason) it is important that $|A| \leq O(b/w)$ we can only get a lower bound of the form $\Omega(2^e\ell/\log s)$.

Proof. (of Theorem 1) We start with an easy lemma regarding the number of oracle calls needed to find a successful path.

Lemma 1. *The amortized number of calls to H_1 and H_2 per proof of effort that will be accepted by a verifier is $\Omega(2^e\ell)$.*

Lemma 2. *Let $b_1 \ldots b_m$ be independent unbiased random bits and let $k \leq m$. Suppose we have a system that, given a hint of length $B < k$ (which may be based on the value of $b_1 \ldots b_m$), produces a subset S of k indices and a guess of the values of $\{b_i \mid i \in S\}$. Then the probability that all k guesses are correct is at most $2^B/2^k$, where the probability is over the random variables and the coin flips of the hint generation and the guessing system.*

We now get to the main content of the lower bound and to the key lemma (Lemma 3): We break the execution into intervals in which, we argue, the adversary is, forced to learn a large number of elements of T. That is, there will be a large number of scattered elements of T which the adversary will need in order to make progress during the interval, and very little information about these elements is in the cache at the start of the interval.

We first motivate our definition of an interval. We want to think of each A as incompressible, since it is the output of a random function. However, if, say, this is the beginning of a path exploration, and $A = H_0(m, S, R, d, k)$, then it may require less space simply to list the arguments to H_0; since our model does not charge (much) for oracle calls, the adversary incurs no penalty for this. For this reason, we will focus on the values of A only in the second half of a path. Recall that A is modified at each step of the Main Loop; intuitively, since these modifications require many elements of T, these "mature" A's cannot be compressed. Our definition of an interval will allow us to focus on progress on paths with "mature" A's.

Let $n = s/|A|$; it is helpful to think of n as the number of A's that can simultaneously fit into cache (assuming they are incompressible). A progress call is *mature* if it is the jth progress call of the path, for $j > \ell/2$ (recall that ℓ is the length of a path). An *interval* is defined by fixing an arbitrary starting point in an execution of the adversary's algorithm (which may involve the simultaneous exploration of many paths), and running the execution until $8n$ mature progress calls (spread over any number of paths) have been made to oracle H_1.

Lemma 3. *The average number of memory accesses made during an interval is $\Omega(n)$, where the average is taken over the choice of T, the responses of the random oracles, and the random choices made by the adversary.*

It is an easy consequence of this lemma that the amortized number of memory accesses to find a successful path is $\Omega(2^e\ell)$. This is true since by Lemma 1, success requires an expected $\Omega(2^e\ell)$ mature progress calls to H_1, and the number of intervals is the total number of mature progress calls to H_1 during the execution, divided by $8n$, which is $\Omega(2^e\ell/n)$. (Note that we have made no attempt to optimize the constants involved.)

Proof. (of Lemma 3) Intuitively, the spammer's problem is that of *asymmetric communication complexity* between memory and the cache. Only the cache has access to the functions H_1 and H_2 (the arguments must be brought into cache in order to carry out the function calls). The goal of the (spammer's) cache is to perform *any* $8n$ mature progress calls. Since by definition the progress calls to H_1 are calls in which the arguments have not previously been given to H_1 in the current execution, we can assume the values of H_1's responses on these calls are uniform over $\{1, \ldots, |T|\}$ given all the information currently in the system (memory and cache contents and queries made so far). The cache must tell the memory which blocks are needed for the subsequent call to H_2. Let β be the average number of blocks sent by the main memory to the cache during an interval, and we assume for the sake of contradiction that $\beta = o(n)$ (the lemma asserts that $\beta = \Omega(n)$). We know that the cache sends the memory $\beta \log m$ bits to specify the block numbers (which is by assumption $o(n \log m)$ bits), and gets in return βb bits altogether from the memory. The key to the lemma is, intuitively, that the relatively few possibilities in requesting blocks by the cache imply that many different elements of T indicated by the indices returned by the $8n$ mature calls to H_1 have to be stored in the same set of blocks. We will argue that this implies that a larger than s part of T can be reconstructed from the cache contents alone, which is a contradiction given the randomness of T.

We now proceed more formally. Lemma 3 will follow from a sequence of claims. The first is that there are many entries of T for which many possible values are consistent with the cache contents at the beginning of the interval. That is, T is largely unexplored from the cache's point of view. The proof is based on Sauer's Lemma (see [6])

Claim 1. There exist $\gamma, \delta \geq 1/2$ such that: given the cache contents at the beginning of the interval, it is expected that there exists a subset of the entries of T, called T', of size at least $\delta|T|$ such that for each entry i in T' there is a set S_i of $2^{\gamma w}$ possible values for $T[i]$ and all the S_i's are mutually consistent with the cache contents.

From now on we assume that we have cache content consistent with a large number of possibilities for T' as in the claim and use this cache configuration to show that it is possible to extract many entries of T'.

Claim 2. If the number β of memory accesses is $o(n)$, then the number of different paths on which a mature progress call is made during an interval is at most $3n$.

It therefore follows that in a typical interval there are at least $8n - 3n = 5n$ pairs of consecutive mature progress calls to H_1 on a common path. Thus, for example, one path may experience $5n + 1$ mature progress calls, or each of n paths may experience at least 6 mature progress calls, or something in between. Each such pair of calls to H_1 is separated by a call to H_2 which requires the contents of the location of T specified by the first H_1 call in the pair. It is these interstitial calls to H_2 that are of interest to us: because their preceding calls to H_1 first occur during the interval, and H_1 is random, it cannot be known at the start of the interval which elements of T will be needed as arguments to these calls to H_2. Intuitively, the adversary *must* go to main memory to find an expected $(|T| - s)/|T| > 1/2$ of them.

Consider the set of $8n$-tuples over $\{1, \ldots, |T|\}$ as the set of possible answers H_1 returns on the mature progress calls in the interval; there are $|T|^{8n}$ such tuples. Fix all other random choices: the value of T, the previous calls to H_1 and H_2 and the random tape of the spammer). The spammer's behavior in an interval is now determined solely by this $8n$-tuple. If the spammer can defeat our algorithm, then, for some fixed $\epsilon > 0$, the spammer completes the interval retrieving at most 2β blocks with at least ϵ probability, over the choice of $8n$-tuple. Call these tuples the *good* ones. By Markov's inequality, for at least half of these good $8n$-tuples the spammer retrieves at least β blocks. We first claim that in most of those tuples the spammer goes frequently into H_2 with values $T[i]$ where $i \in T'$.

Claim 3. Let T' be any subset of the entries of T of size at least δn. Consider the set of good $8n$-tuples over $\{1, \ldots, |T|\}$ as the set of possible answers H_1. Then except for at most an exponential in n fraction of them the spammer must use an entry in T' for a call to at H_2 least n times during an interval.

Claim 4. Suppose that we have subset X of size x of entries in T. Then the probability over H_1 that a $8n$-tuple contains more than $n/2$ entries in X is at most $(2^8 x/|T|)^{n/2}$.

Claim 5. Suppose that we have a collection of good $8n$-tuples and we want to cover at least x values in T' using only a few members of the collection, say $2x/n$ (assume that the collection is at least that large). If this is impossible then there is a set $X \subset T'$ of size x such that every member of the collection has at least $n/2$ entries in X.

The idea for deriving the contradiction to the fact that only $\beta = o(n)$ blocks are brought from memory to cache is that there should be many good $8n$-tuples that share the same set of blocks (that is, by retrieving one set of blocks all elements appearing in many good $8n$-tuples can be reconstructed in the cache). In fact, since the memory size is m, a $1/\binom{m}{2\beta}$ fraction of them share the same set of blocks (the factor of 2 comes from the definition of a good $8n$-tuple). Consider such a collection and suppose that there are $2x/n$ tuples in this collection whose union covers x entries in T'. Then the "memory" can use these $2x/n$ tuples to transfer the value of x entries in T' by sending the $2\beta b$ bits describing the content of the common blocks and in addition for each tuple in the cover: (1) Specifying the $8n$-tuple: this takes $8n \log |T|$ bits; (2) Specifying which calls to H_2 in the execution have the correct parameters (there may be some "bogus" calls to

H_2 in which the wrong values for elements of T are used as parameters). If the interval contains z calls to H_2 then this takes $\log \binom{z}{n/2}$ bits which is $O(n \log z)$.

So altogether it suffices for $2\beta b + 16x \log |T| + 2x \log z$ bits to be sent from the memory to the cache. In return, the cache learns γw bits for each of x entries in T', or $x\gamma w$ bits altogether. To derive the contradiction, since w was taken to be much larger than $\log |T|$ and $2^{w/2}$ much larger than the amortized number of oracle calls per interval, $\log z$ is much smaller than w and we only have to worry about the $2\beta b$ term.

Assume that $\beta \leq 1/20n = s/20|A|$ and, for simplicity that m, the memory size is $|T|^2$ (recall that in our model m is polynomial in s, and in our theorem $|T| = \Theta(s)$). Set $x = 4\beta b/w$. Of all good tuples, pick the largest collection agreeing with a set of β blocks, i.e., consisting of at least a $1/\binom{T^2}{\beta}$ fraction of the good tuples. We now claim that this collection has $2x/n$ $8n$-tuples whose union is of size at least x (this will be sufficient for a contradiction).

Suppose that this is not the case and the $2x/n$ tuples covering x do not exist. Then as we have seen above in Claim 5 there is a set X of size x where each tuple in the collection has at least $n/2$ entries in X. But we know from Claim 4 that the fraction (among all tuples) of such a collection can be at most $(2^8 x/|T|)^{n/2}$. Taking into account ϵ (the faction of all tuples that are good) we must compare $((2^8 x/|T|)^{n/2})(1/\epsilon)$ to $1/\binom{T^2}{2\beta}$ and if the latter is larger we know that the collection is too large to be compressed into X. For simplicity take $\epsilon = 1$. Indeed

$$\frac{(2^8 x/|T|)^{n/2}}{\binom{T^2}{2\beta}} = \frac{(2^8 x)^{n/2}}{T^{n/2 - 4\beta}}$$

taking logs we get that we need to compare $\log 2^8 x$ and

$$(\log T)\frac{n - 4\beta}{n} = (\log T)\frac{s/|A| - 4s/20|A|}{s/|A|} = (\log T)\frac{4}{5}.$$

But since $x = 4\beta b/w = 4sb/(20|A|w)$ and $|A| \geq s^{1/5}b/w$ we get that $x \leq 1/5s^{4/5}$ and indeed $8 + \log x$ is smaller than $4/5 \log |T|$.

This concludes the proofs of Lemma 3 and Theorem 1

6   A Concrete Proposal

In this section we describe a concrete implementation of the abstract algorithm of Section 5, which we call Algorithm MBound. As in the abstract algorithm, our function involves a large *fixed forever* array T, now of 2^{22} truly random 32-bit integers[8]. In terms of the parameters of Section 5, we have $|T| = 2^{22}$ and $w = 32$. This array requires 16 MB and dominates the space needs of our memory-bound function, which requires less than 18 MB total space.[9] The algorithm requires in addition a fixed-forever truly random array A_0 containing 256 32-bit words. A_0 is used in the definition of H_0. Note that A_0 is incompressible.

---

[8] "Fixed forever" means fixed until new machines have bigger caches, in which case the function must be updated.

[9] To send mail, a machine must be able to handle a program of this size.

6.1   Description of MBound

Our proposal was inspired by the (alleged) RC4 pseudo-random generator (see, *e.g.*, the descriptions of RC4 in [16,23,24]).

Description of H_0. Recall that we have a fixed-forever array A_0 of 256 truly random 32-bit words. At the start of the kth trial, we compute $A = H_0(m, S, R, d, k)$ by first computing (using strong cryptography) a 256-word mask and then XORing A_0 together with the mask. Here is one way to define H_0:

1. Let $\alpha_k = h(m, S, R, d, k)$ ($|\alpha_k| = 128$), for a cryptographically strong hash function h such as, say, SHA-1.
2. Let $\eta(\alpha_k)$ be the 2^{13}-bit string obtained by *concatenating* the 2^7-bit α_k with itself 2^6 times[10]. Treating the array A as a 2^{13}-bit string (by concatenating its entries in row-major order), we let $A = A_0 \oplus \eta(\alpha_k)$. Note that, unlike in the case of RC4, our array A is *not* a permutation of elements $\{1, 2, \ldots, 256\}$, and its entries are 32 bits, rather than 8 bits.

We initialize c, the *current* location in T, to be the last 22 bits of A (when A is viewed as a bit string). In the sequel, whenever we say $A[i]$ we mean $A[i \bmod 2^8]$; similarly, by $T[c]$ we mean $T[c \bmod 2^{22}]$.

The path in a generic trial is given by:

```
Initialize Indices:
    i = 0;  j = 0
Walk for ℓ steps (ℓ is the path length):
    i = i + 1
    j = j + A[i]
    A[i] = A[i] + T[c]
    A[i] = RightCyclicShift(A[i], 11) (shift forces all 32 bits into
play)
    Swap(A[i], A[j])
    c = T[c] ⊕ A[A[i] + A[j]]
Success occurs if the last e bits of h(A) are all 0.
```

In the last line, the hash function h can again be SHA-1. It is applied to A, treated as a bit string.

The principal difference with the RC4 pseudo-random generator is in the use of T: bits from T are fed into MBound's pseudo-random generation procedure, both in the modification of A and in the updating of c.

In terms of the abstract function, we can tease our proposal apart to obtain, *roughly*:

Description of H_1 (updates c, leaves A unchanged). The function H_1 is essentially
$$i = i + 1$$
$$j = j + A[i]$$

---

[10] The reason we concatenate the string in order to generate $\eta(\alpha_k)$, rather than generate a cryptographically strong string of length 2^{13} is to save CPU cycles - this is an operation that is done many times and if each bit of $\eta(\alpha_k)$ is strong it could make the scheme CPU bound.

$v = A[i] + T[c]$ (v is a temporary variable)
$v = RightCyclicShift(v, 11)$
$c = T[c] \oplus A[A[j] + v]$

Description of H_2 (updates A).
   $A[i] = A[i] + T[c]$
   $A[i] = RightCyclicShift(A[i], 11)$
   $Swap(A[i], A[j])$

Description of H_3. The hash function $H_3(A)$ is simply some cryptographically strong hash function with 128 bits of output, such as SHA-1.

This all but completes the description of Algorithm MBound and its connection to our abstract function; it remains to choose the parameters.

6.2   Parameters for MBound

We can define the computational puzzle solved by the sender as follows.

Input. A message m, a sender's alias S, a receiver's alias R, a time t, the table T and the auxiliary table A_0.

Output. m, S, R, d, i and α such that $1 \leq i \leq 2e$ and the ith path (that is, the path with trial number $k = i$), is successful and α is the result of hashing the final value of A in the successful path.

If $i > 2^{2e}$, the receiver rejects the message (with overwhelming probability one of the first 2^{2e} trials should be successful).

To be specific in the following analysis, we make several assumptions. These assumptions are reasonable for current technology, and our analysis is sufficiently robust to tolerate substantial changes in many of these parameters. Let P be the desired expected time for computing the proof of effort and let τ be the memory latency. We assume that P is 10 seconds and τ is .2 microseconds. We also assume that the maximum size of the fast cache is 8 MB and that cache lines (memory blocks) are 64 bytes wide (so blocks contain $b = 512$ bits).

The output conditions ensure that for a random starting point, the probability of a successful output is $1/2^e$. The expected number of walks to be checked is 2^e. Therefore the expected value of P is

$$E[P] = 2^e \cdot \ell \cdot \tau.$$

The cost of verification by the receiver is essentially ℓ cache misses, by following the right path. (In Section 8 we discuss how to reduce or eliminate these cache misses.)

We have not yet set the values for e and ℓ. Choosing one of these parameters forces the value of the other one. Consider the choice of e: one possibility might be to make e very large, and the paths short, say, even of length 1. This would make verification extremely cheap. However, while the good sender will explore the paths sequentially, a cheating sender may try several paths in parallel, hoping to exploit locality by batching several accesses to T, one from each of these parallel explorations. In addition, A changes slowly, and to get to the point in which many "mature" values of A cannot be compressed requires that many entries of A have been modified. For our concrete proposal, therefore, we let $\ell = 2048$. Then $2^e = P/\ell\tau = 10/(2048 * 2 * 10^{-7}) \approx 24,414$.

Table 1. Computational Platforms, sorted by CPU speed.

name	class	model	processor	CPU clock	OS
P4-3060	workstation	DELL XW8000	Intel Pentium 4	3.06 Ghz	Linux
P4-2000	desktop	Compaq Evo W6000	Intel Pentium 4	2.0 Ghz	Windows XP
P3-1200	laptop	DELL Latitude C610	Intel Pentium 3M	1.2 Ghz	Windows XP
P3-1000	desktop	Compaq DeskPro EN	Intel Pentium 3	1.0 Mhz	Windows XP
Mac-1000	desktop	Power Mac G4	PowerPC G4	1000 Mhz	OSX
P3-933	desktop	DELL Dimension 4100	Intel Pentium 3	933 Mhz	Linux
SUN-900	server	SUN Ultra 60	UlraSPARC III+	900 Mhz	Solaris
SUN-450	server	SUN Ultra 60	UlraSPARC II	450 Mhz	Solaris
P2-266	laptop	Compaq Armada 7800	Intel Pentium 2	266 Mhz	Windows 98
S-233	settop	GCT-AllWell STB3036N	Nat. Semi. Geode GX1	233 Mhz	Linux

7   Experimental Results

In this section we describe several experiments aimed at establishing practicality of our approach and verifying it experimentally. First we compare our memory-bound function performance to that of the CPU-intensive HashCash function [18] on a variety of computer architectures. We confirm that the memory-bound function performance is significantly more platform-independent. We also measure the solution-to-verification time ratio of our function. Then we run simulations showing how the number of cache misses during the execution of our memory-bound function depends on the cache size and the cache replacement strategy. We observe that even if an adversary knows future accesses, this does not help much unless the cache size is close to the size of T. Finally, we study how the running time depends on the size of the big array T.

Table 2. Memory hierarchy.

machine	L2 cache	L2 line	memory
P4-3060	256 KB	128 bytes	4 GB
P4-2000	256 KB	128 bytes	512 MB
P3-1200	256 KB	64 bytes	512 MB
P3-1000	256 KB	64 bytes	512 MB
P3-933	256 KB	64 bytes	512 MB
Mac-1000	256 KB	64 bytes	512 MB
SUN-900	8 MB	64 bytes	8 Gb
SUN-450	8 MB	64 bytes	1 Gb
P2-266	512 KB	32 bytes	96 MB
S-233	16 KB	16 bytes	128 MB

7.1   Different Architectures

We conducted tests on a variety of platforms, summarized in Table 1. These platforms vary from the popular Pentium 3 and Pentium 4 systems and a Macintosh G4 to SUN servers with large caches. We even tested our codes on a settop box, which is an example of a low-power device. The P2-266 laptop is an example of a "legacy" machine and is representative of a low-end machine among those widely used for e-mail today (that is, in 2003). Table 2 gives sizes of the relevant components of the memory hierarchy, including L2 cache size, L2 cache line size, and memory size. With one exception, all machines have two levels of cache and memory. The exception is the Macintosh, which has a 2 MB off-chip L3 cache in addition to the 256 KB on-chip L2 cache.

Table 3. Program timings. Times are averages over 20 runs, measured in units of the smallest average. For HashCash, the smallest average is 4.44 sec.; for MBound, it is 9.15 sec.%.

machine	HashCash	MBound	
name	time	time	sol./ver.
P4-3060	1.00	1.01	2.32 E4
P4-2000	1.91	1.33	1.65 E4
P3-1200	2.21	1.00	2.55 E4
P3-1000	2.67	1.06	2.48 E4
Mac-1000	1.86	1.96	2.61 E4
P3-933	2.15	1.06	2.51 E4
SUN-900	1.82	2.24	2.50 E4
SUN-450	5.33	2.94	2.02 E4
P2-266	10.17	2.67	1.84 E4
S-233	43.20	4.62	1.50 E4

7.2   Memory- vs. CPU-Bound

The motivation behind memory-bound functions is that their performance is less dependent on processor speed than is the case for CPU-bound functions. Our first set of experiments compares an implementation of our memory-bound function, *MBound*, to our implementation of *HashCash* [18]. HachCash repeatedly appends a trial number to the message and hashes the resulting string, until the output ends in a certain number zero bits (22 in our experiments). For MBound, with its slower iteration time, we set the required number of zero bits to 15.

Table 3 gives running times for HashCash and MBound, normalized by the fastest machine time. Note that HashCash times are closely correlated with processor speed. Running times for MBound show less variation. The difference between the P2-266 laptop and the fastest machine used in our tests for HashCash is a factor of 10.17, while the difference for MBound is only a factor of 2.67. The HashCash vs. MBound gap is even larger for the S-233 settop box.[11]

Modern Pentium-based machines perform well in memory-bound computations. The Macintosh does not do so well; we believe that this is due to its poor handling of the translation lookahead buffer (TLB) misses. SUN servers do poorly in spite of their large

---

[11] Note that S-233 is a special-purpose device and code produced by the C compiler may be poorly optimized of the processor. This may be one of the reasons why this machine was so slow in our tests.

caches. This is due to their poor handling of TLB misses and the penalty for their ability of handle large memories.

8 Freeing the Receiver from Accessing T

Since the spam-protected receiver will sometimes act also as an e-mail sender, he will have access to the array T. However, we would like receiving mail not to have to involve accessing T at all. For example, one might wish to be able to receive mail on a cell phone. In this section we explore the possibility that the sender adds some information to its message that will permit the receiver to efficiently verify the proof of effort with no accesses to T. Of course, the *conceptually* simplest method for freeing R from accessing T is for the creator of T to sign all the elements of T (more precisely, the signature is on the pair $(c, T[c])$, to disallow permuting the table). However, this requires too much storage at the sender, even using the signature scheme yielding the shortest signatures [8].

Compressed RSA Signatures. Here we use properties of the RSA scheme previously exploited in the literature [13,12]. Let (N, e) be the public key of an RSA signature scheme chosen by the creator of T^{12}. Let F be a function mapping pairs $(c, T[c])$ into Z_N^*, that is, a mapping from $32 + 22 = 54$-bit strings into Z_N^*. In our analysis we will model F as a random oracle. For all $1 \leq c \leq |T|$ let $v_c = F(c, T[c])$ and let $w_c = v_c^{1/e} \bmod N$. Thus, v_c is a hash of the pair $(c, T[c])$ and w_c is a signature on the string v_c.

The sender's protocol contains, in addition to T, the public modulus N, the description of F, and the w_c's. The receiver's protocol uses only the description of F and the public key (N, e), together with a description of the sender's path exploration algorithm (minus the array T itself).

Let the sender's successful path be the sequence $c_1, c_2, \ldots c_\ell$ of locations in T. The proof of effort contains two parts:

1. $T[c_1], T[c_2], \ldots, T[c_\ell]$, (a total of about 4 KB), and
2. $w = \prod_{i=1}^{\ell} w_{c_i} \bmod N$ (about 1 KB).

Note that there is no need to include the indices $c_1, \ldots, c_\ell$ in the first part, as these are implicit from the algorithm. Similarly, there is no need to send the v_c's, since these are implicit from F and the $(c_i, T[c_i])$ pairs. Let $t_1, \ldots, t_\ell$ be the first part of the proof, and w the second part (each t_i is *supposed* to be $T[c_i]$, but the verifier cannot yet be certain this is the case). The proof is checked as follows.

1. Compute $v'_{c_1}, v'_{c_2}, \ldots v'_{c_\ell}$, where $v'_{c_i} = F(c_i, t_i)$.
2. Check whether $w^e = (\prod_{i=1}^{\ell} v_{c_i}) \bmod N$.

The security of the scheme rests on the fact that it is possible to translate a forged signature on

$$(c_1, T[c_1]), \ldots (c_\ell, T[c_\ell])$$

---

[12] The signing key d is a valuable secret!

into an inversion of the RSA function on a specific instance. This is summarized as follows:

Theorem 2. *If F is a random oracle, then any adversary attempting to produce a set of* claimed *values*

$$T[c_1], T[c_2], \ldots T[c_\ell]$$

that is false yet acceptable to the receiver can be translated into an adversary for breaking RSA with the same run time and probability of success (to preserve probability of success we need that e be a prime larger than ℓ).

Although transmission costs are low, the drawback of the compressed RSA scheme is again the additional storage requirements for the sender: each w_c is at least $1,000$ bits (note, however, that these extra values are not needed until after a successful path has been found). This extra storage requirement might discourage a user from embracing the scheme. We address this next.

Storage-Optimized Compressed RSA. We optimize storage with the following storage / communication / computation tradeoff: Think of T as an $a \times b$ matrix where $a \cdot b = |T|$; the amount of extra communication will be a elements of T. The amount of extra storage required by the sender will be b signatures.

At a high level, given a path using values $T[c_1], T[c_2], \ldots, T[c_\ell]$, values in the same row of T will be verified together as in the compressed RSA scheme. The communication costs will therefore be at most a elements, one per row of T. However, as we will see below, *there is no need to store the w_c's explicitly.* Instead, we can get away with storing a relatively small number of signatures (one per column), from which it will be possible to efficiently reconstruct the w_c values as needed.

Instead of a single exponent e, both sending and receiving programs will contain a (common) list of primes $e_1, e_2, \ldots e_a$. For $1 \le i \le a$, e_i is used for verifying elements of row i of the table. Although we don't need to store the w_c values explicitly, for elements v_c appearing in row i we define $w_c = v_c^{1/e_i} \bmod N$.

The compressed RSA scheme is applied to the entries in each row independently. It only remains to describe how the needed w_c values are constructed on the fly.

The b "signatures", one per column, used in the sending program are computed by the creator as follows. For each column $1 \le j \le b$, the value for column j is $u_j = \prod_{i=1}^{a} w_{c_{j_i}} \bmod N$. Here, c_{j_i} is the index of the element $T[i, j]$, when T is viewed as a matrix rather than as an array (that is, assuming row-major order, $c_{j_i} = (i-1)a + j - 1$). Thus, $v_{c_{j_i}} = T[(i-1)a + j - 1]$ and $w_{c_{j_i}} = (v_{c_{j_i}})^{1/e_i}$. As in Batch RSA [13], one can efficiently extract any $w_{c_{j_i}}$ from u_j using a few multiplications and exponentiations.

Set $a = 16$. The number of data bits in a column is $2^4 \cdot 2^5 = 2^9$. The number of "signature bits" is 2^{10} per column. Thus storage requirement just more than doubles, rather than increasing by a factor of 5-10, at the cost of sending 16 elements of Z_N^* (*i.e.*, 2 KB).

9   Concluding Remarks

We have continued the discussion, initiated in [3], of using memory-bound rather than CPU-bound pricing functions for computational spam fighting. We considered and analyzed several potential approaches. Using insights gained in the analyses, we proposed a different approach based on truly random, incompressible, functions, and obtained both a rigorous analysis and experimental results supporting our approach.

From a theoretical perspective, however, the work is not complete. First, we have the usual open question that arises whenever random oracles are employed: can a proof of security (in our case, a lower bound on the average number of cache misses in a path) be obtained without recourse to random oracles? Second, much more unusually, can we prove security without cryptographic assumptions? Note that we did not make cryptographic assumptions in our analysis.

One of the more interesting challenges suggested by this work is to apply results from complexity theory in order to be able to make rigorous statements about proposed schemes. One of the more promising directions in recent years is the work on lower bounds for branching program and the RAM model by Ajtai s [4,5] and Beame et al [7]. It is not clear how to directly apply such results.

At first blush egalitarianism seems like a wonderful property in a pricing function. However, on reflection it may not be so desirable. Since the approach is an economic one it may be counterproductive to design functions that can be computed just as quickly on extremely cheap processors as on supercomputers – after all, we are trying to force the spammers to expend resources, and it is the volume of mail sent by the spammers that should make their lives intolerable while the total computational effort expended by ordinary senders remains benign. So perhaps less egalitarian is better, and users with weak or slow machines, including PDAs and cell phones, could subscribe to a service that does the necessary computation on their behalf. In any case, small-memory machines cannot be supported, since the large caches are so very large, so in any real implementation of computational spam fighting some kind of computation service must be made available.

References

1. M. Abadi and M. Burrows, *(multiple) private communication(s)*
2. M. Abadi, *private communication.*
3. M. Abadi, M. Burrows, M. Manasse and T. Wobber, *Moderately Hard, Memory-Bound Functions*, Proceedings of the 10th Annual Network and Distributed System Security Symposium, February, 2003.
4. M. Ajtai, *Determinism versus Non-Determinism for Linear Time RAMs*, STOC 1999, pp. 632–641.
5. M. Ajtai, *A Non-linear Time Lower Bound for Boolean Branching Programs*, FOCS 1999: 60–70
6. N. Alon, and J. Spencer, *The Probabilistic Method*, Wiley & Sons, New-York, 1992
7. P. Beame, M. E. Saks, X. Sun, E. Vee, *Super-linear time-space tradeoff lower bounds for randomized computation,* FOCS 2000, pp. 169–179.
8. D. Boneh, B. Lynn and H. Shacham, *Short signatures from the Weil pairing*, ASIACRYPT 2001, pp. 514–532.

9. www.camram.org/mhonarc/spam/msg00166.html.
10. W. Diffie and M.E. Hellman, Exhaustive cryptanalysis of the NBS Data Encryption Standard, Computer 10 (1977), 74-84.
11. C. Dwork and M. Naor, Pricing via Processing, Or, Combatting Junk Mail, Advances in Cryptology – CRYPTO'92, Lecture Notes in Computer Science No. 740, Springer, 1993, pp. 139–147.
12. C. Dwork and M. Naor, *An Efficient Existentially Unforgeable Signature Scheme and Its Applications*, Journal of Cryptology 11(3): 187-208 (1998)
13. A. Fiat, *Batch RSA*, . Journal of Cryptology 10(2): 75-88 (1997).
14. A. Fiat and M. Naor, Rigorous Time/Space Tradeoffs for Inverting Functions, STOC'91, pp. 534–541
15. A. Fiat and A. Shamir, How to Prove Yourself, *Advances in Cryptoglogy – Proceedings of CRYPTO'84*, pp. 641–654.
16. Fluhrer, I. Mantin and Shamir, *Attacks on RC4 and WEP*,. Cryptobytes 2002.
17. A. J. Menezes, P. C. van Oorschot and S. A. Vanstone, *Handbook of Applied Cryptography*, CRC Press, 1996. Also available: http://www.cacr.math.uwaterloo.ca/hac/
18. A. Back, Hashcash - A Denial of Servic Counter-Measure, available at http://www.cypherspace.org/hashcash/hashcash.pdf.
19. M. Bellare, J. A. Garay and T. Rabin, *Fast Batch Verification for Modular Exponentiation and Digital Signatures*, EUROCRYPT 1998, pp. 236–250.
20. M. Bellare, J. A. Garay and T. Rabin, *Batch Verification with Applications to Cryptography and Checking*, LATIN 1998, pp. 170–191.
21. M. Hellman, A Cryptanalytic Time Memory Trade-Off, *IEEE Trans. Infor. Theory 26*, pp. 401–406, 1980
22. M. Luby and C. Rackoff, How to Construct Pseudorandom Permutations and Pseudorandom Functions, *SIAM J. Computing 17*(2), pp. 373–386, 1988
23. I. Mantin, *Analysis of the Stream Cipher RC4*, Master's Thesis, Weizmann Institute of Science, 2001. Available www.wisdom.weizmann.ac.il/~itsik/RC4/rc4.html
24. I. Mironov, (Not So) Random Shuffles of RC4, *Proc. of CRYPTO'02*, 2002
25. M. Naor and M. Yung, *Universal One-Way Hash Functions and their Cryptographic Applications*, STOC 1989, pp. 33–43
26. P. Oechslin, *Making a faster Cryptanalytic Time-Memory Trade-Off, these proceedings*.
27. P. C. van Oorschot and M. J. Wiener, *Parallel Collision Search with Cryptanalytic Applications*, Journal of Cryptology, vol. 12, no. 1, 1999, pp. 1–28.
28. R. Schroeppel and A. Shamir, $A \; T = O(2^{(n/2)})$, $S = O(2^{(n/4)})$ *Algorithm for Certain NP-Complete Problems*, SIAM J. Comput. 10(3): 456–464 (1981). 1979.

Lower and Upper Bounds on Obtaining History Independence

Niv Buchbinder and Erez Petrank[*]

Computer Science Department, Technion, Haifa, Israel,
{nivb,erez}@cs.technion.ac.il

Abstract. History independent data structures, presented by Micciancio, are data structures that possess a strong security property: even if an intruder manages to get a copy of the data structure, the memory layout of the structure yields no additional information on the data structure beyond its content. In particular, the history of operations applied on the structure is not visible in its memory layout. Naor and Teague proposed a stronger notion of history independence in which the intruder may break into the system several times without being noticed and still obtain no additional information from reading the memory layout of the data structure.

An open question posed by Naor and Teague is whether these two notions are equally hard to obtain. In this paper we provide a separation between the two requirements for comparison based algorithms. We show very strong lower bounds for obtaining the stronger notion of history independence for a large class of data structures, including, for example, the heap and the queue abstract data structures. We also provide complementary upper bounds showing that the heap abstract data structure may be made weakly history independent in the comparison based model without incurring any additional (asymptotic) cost on any of its operations. (A similar result is easy for the queue.) Thus, we obtain the first separation between the two notions of history independence. The gap we obtain is exponential: some operations may be executed in logarithmic time (or even in constant time) with the weaker definition, but require linear time with the stronger definition.

Keywords: History independent data-structures, Lower bounds, Privacy, The heap data-structure, The queue data-structure.

1 Introduction

1.1 History Independent Data Structures

Data structures tend to store unnecessary additional information as a side effect of their implementation. Though this information cannot be retrieved via the 'legitimate' interface of the data structure, it can sometimes be easily retrieved by inspecting the actual memory representation of the data structure. Consider,

---

[*] This research was supported by the E. AND J. BISHOP RESEARCH FUND.

D. Boneh (Ed.): CRYPTO 2003, LNCS 2729, pp. 445–462, 2003.

for example, a simple linked list used to store a wedding guest-list. Using the simple implementation, when a new invitee is added to the list, an appropriate record is appended at the end of the list. It can be then rather discomforting if the bride's "best friend" inspects the wedding list, just to discover that she was the last one to be added. History independent data structures, presented by Micciancio [8], are meant to solve such headaches exactly. In general, if privacy is an issue, then if some piece of information cannot be retrieved via the 'legitimate' interface of a system, then it should not be retrievable even when there is full access to the system. Informally, a data structure is called History independent if it yields no information about the sequence of operations that have been applied on it.

An abstract data structure is defined by a list of operations. Any operation returns a result and the specification defines the results of sequence of operations. We say that two sequences S_1, S_2 of operations yield the same content if for any suffix T, the results returned by T operations on the data structure created by S_1 and on the data structure created by S_2 are the same. For the heap data structure the content of the data structure is the set of values stored inside it.

We assume that in some point an adversary gains control over the data structure. The adversary then tries to retrieve some information about the sequence of operations applied on the data structure. The data structure is called History independent if the adversary cannot retrieve any more information about the sequence other than the information obtainable from the content itself.

Naor and Teague [10] strengthen this definition by allowing the adversary to gain control more than once without being noted. In this case, one must demand for any two sequences of operations and two lists of 'stop' points in which the adversary gain control of the data structure, if in all 'stop' points, the content of the data structure is the same (in both sequences), then the adversary cannot gain information about the sequence of operations applied on the data structure other than the information yielded by the content of the data structure in those 'stop' points. For more formal definition of History independent data structure see section 3.

An open question posed by Naor and Teague is whether the stronger notion is harder to obtain than the weaker notion. Namely, is there a data structure that has a weakly history independent implementation with some complexity of operations, yet any implementation of this data structure that provides strong history independence has a higher complexity.

1.2   The Heap

The heap is a fundamental data structure taught in basic computer science courses and employed by various algorithms, most notably, sorting. As an abstract structure, it implements four operations: build-heap, insert, remove-max and increase-key. The basic implementations require a worst case time of $O(n)$

for the build-heap operation (on n input values), and $O(\log n)$ for the other three operations[1]. The standard heap is sometimes called *binary heap*.

The heap is a useful data structure and is used in several important algorithms. It is the heart of the *Heap-Sort* algorithm suggested by Williams [12]. Other applications of heap use it as a *priority queue*. Most notable among them are some of the basic graph algorithms: Prim's algorithm for finding Minimum Spanning Tree [11] and Dijkstra's algorithm for finding Single-Source Shortest Paths [4].

1.3   This Work

In this paper we answer the open question of Naor and Teague in the affirmative for the comparison based computation model. We start by providing strong and general lower bounds for obtaining strong history independence. These lower bounds are strong in the sense that some operations are shown to require linear time. They are general in the sense that they apply to a large class of data structures, including, for example, the heap and the queue data structures. The strength of these lower bounds implies that strong data independence is either very expensive to obtain, or must be implemented with algorithms that are not comparison based.

To establish the complexity separation, we also provide an implementation of a weakly history independent heap. A weakly history independent queue is easy to construct and an adequate construction appears in [10]. Our result on the heap is interesting in its own sake and constitutes a second contribution of this paper. Our weakly history independent implementation of the heap requires no asymptotic penalty on the complexity of the operations of the heap. The worst case complexity of the build-heap operation is $O(n)$. The worst case complexity of the increase-key operation is $O(\log n)$. The expected time complexity of the operations insert and extract-max is $O(\log n)$, where expectation is taken over all possible random choices made by the implementation in a single operation. This construction turned out to be non-trivial and it requires an understanding of how uniformly chosen random heaps behave. To the best of our knowledge a similar study has not appeared before.

The construction of the heap and the simple implementation of the queue are within the comparison based model. Thus, we get a time complexity separation between the weak and the strong notions of history independent data structure. Our results for the heap and the queue appear in table 1. The lower bound for the queue is satisfied for either the insert-first or the remove-last operations. The upper bounds throughout this paper assume that operations on keys and pointers may be done in constant time. If we use a more prudent approach and consider the bit complexity of each comparison, our results are not substantially affected. The lower bound on the queue was posed as an open question by Naor and Teague.

---

[1] The more advanced Fibonacci heaps obtain better amortized complexity and seem difficult to be made History independent. We do not study Fibonacci heaps in this paper.

Table 1. Lower and upper bounds for the heap and the queue

Operation	Weak History Independence	Strong History Independence
heap:insert	$O(\log n)$	$\Omega(n)$
heap:increase-key	$O(\log n)$	$\Omega(n)$
heap:extract-max	$O(\log n)$	No lower bound
heap:build-heap	$O(n)$	$\Omega(n \log n)$
queue: max {insert-first, remove-last}	$O(1)$	$\Omega(n)$

1.4   Related Work

History independent data structures were first introduced by Micciancio [8] in the context of incremental cryptography. Micciancio has shown how to obtain an efficient History independent 2-3 tree. In [10] Naor and Teague have shown how to implement a History independent hash table. They have also shown how to obtain a history independent memory allocation. Naor and Teague note that all known implementations of strongly independent data structures are canonical. Namely, for each possible content there is only one possible memory layout. A proof that this must be the case has been shown recently by [5] (and independently proven by us). Andersson and Ottmann showed lower and upper bounds on the implementation of unique dictionaries [1]. However, they considered a data structure to be unique if for each content there is only one possible representing graph (with bounded degree) which is a weaker demand than canonical. Thus, they also obtained weaker lower bounds for the operations of a dictionary.

There is a large body of literature trying to make data structures *persistent*, i.e. to make it possible to reconstruct previous states of the data structure from the current one [6]. Our goal is exactly the opposite, that no information whatsoever can be deduced about the past.

There is considerable research on protecting memories. Oblivious RAM [9] makes the address pattern of a program independent on the actual sequence. it incurs a cost of $poly \log n$. However, it does not provide history independence since it assumes that the CPU stores some secret information; this is an inappropriate model for cases where the adversary gains complete control.

1.5   Organization

In section 2 we provide some notations to be used in the paper. In section 3 we review the definitions of History independent data structures. In section 4 we present the first lower bounds for strongly history independence data structures. As a corollary we state lower bounds on some operations of the heap and queue data structures. In section 5 we review basic operations of the heap. In Section 5.1 we present some basic properties of randomized heaps. In section 6 we show how to obtain a weak history independent implementation of the heap data structure with no asymptotic penalty on the complexity of the operations.

2   Preliminaries

Let us set the notation for discussing events and probability distributions. If S is a probability distribution then $x \in S$ denotes the operation of selecting an element at random according to S. When the same notation is used with a set S, it means that x is chosen uniformly at random among the elements of the set S. The notation $Pr\,[R_1; R_2; \ldots; R_k : E]$ refers to the probability of event E after the random processes $R_1, \ldots, R_k$ are performed in order. Similarly, $E\,[R_1; R_2; \ldots; R_k : v]$ denotes the expected value of v after the random processes $R_1, \ldots, R_k$ are performed in order.

3   History Independent Data Structures

In this section we present the definitions of History independent data structures. An implementation of a data structure maps the sequence of operations to a memory representation (i.e an assignment to the content of the memory). The goal of a history independent implementation is to make this assignment depend only on the content of the data structure and not on the path that led to this content. (See also a motivation discussion in section 1.1 above).

   An abstract data structure is defined by a list of operations. We say that two sequences S_1 and S_2 of operations on an abstract data structure yield the same content if for all suffixes T, the results returned by T when the prefix is S_1, are the same as the results returned when the prefix is S_2. For the heap data structure, its content is the set of values stored inside it.

Definition 1. *A data structure implementation is history independent if any two sequences S_1 and S_2 that yield the same content induce the same distribution on the memory representation.*

This definition [8] assumes that the data structure is compromised once. The idea is that, when compromised, it "looks the same" no matter which sequence led to the current content. After the structure is compromised, the user is expected to note the event (e.g., his laptop was stolen) and the structure must be re-randomized.

   A stronger definition is suggested by Naor and Teague [10] for the case that the data structure may be compromised several times without any action being taken after each compromise. Here, we demand that the memory layout looks the same at several points, denoted *stop points* no matter which sequences led to the contents at these points. Namely, if at ℓ stop points (break points) of sequence σ the content of the data structure is $C_1, C_2, \ldots, C_\ell$, then no matter which sequences led to these contents, the memory layout joint distribution at these points must depend only on the contents $C_1, C_2, \ldots, C_\ell$. The formalization follows.

Definition 2. *Let S_1 and S_2 be sequences of operations and let $P_1 = \{i_1^1, i_2^1, \ldots i_l^1\}$ and $P_2 = \{i_1^2, i_2^2, \ldots i_l^2\}$ be two list of points such that for all*

$b \in \{1, 2\}$ and $1 \leq j \leq l$ we have that $1 \leq i_j^b \leq |S_b|$ and the content of data structure following the i_j^1 prefix of S_1 and the i_j^2 prefix of S_2 are identical. A data structure implementation is strongly history independent if for any such sequences the distributions of the memory representations at the points of P_1 and the corresponding points of P_2 are identical.

It is not hard to check that the standard implementation of operations on heaps is not History independent even according to definition 1.

4   Lower Bounds for Strong History Independent Data Structures

In this section we provide lower bounds on strong history independent data structures in the comparison based model. Naor and Teague noted that all implementations of strong history independent data structure were canonical. In a canonical implementation, for each given content, there is only one possible memory layout. It turns out that this observation may be generalized. Namely, all implementations of (well-behaved) data structure that are strongly independent, are also canonical. This was recently proven in [5] (and independently by us). See section 4.1 below for more details. For completeness, we include the proof in [2].

We use the above equivalence to prove lower bounds for canonical data structures. In subsection 4.2 below, we provide lower bounds on the complexity of operations applied on a canonical data structures in the comparison based model. We may then conclude that these lower bounds hold for strongly history independent data structures in the comparison based model.

4.1   Strong History Independence Implies Canonical Representation

Canonical representation is implied by strongly history independent data structures for well-behaved data structures. We start by defining well-behaved data structures, via the *content graph* of the structure. Let C be some possible content of an abstract data-structure. For each abstract data-structure we define its *content graph* to be a graph with a vertex for each possible content C of the data structure. There is a directed edge from a content C_1 to a content C_2 if there is an operation OP with some parameters that can be applied on C_1 to yields the content C_2. Notice that this graph may contain an infinite number of nodes when the elements in the data-structure are not bounded. It is also possible that some vertices have an unbounded degree. We say that a content C is *reachable* there is a sequence of operations that may applied on the empty content and yield C. For our purposes only reachable nodes are interesting. In the sequel, when we refer to the content graph we mean the graph induced by all reachable nodes.

We say that an abstract data structure is *well-behaved* if its content graph is strongly connected. That is, for each two possible contents C_i, C_j, there exists a finite sequence of operations that when applied on C_i yields the content C_j. We may now phrase the equivalence between the strong history independent definition and canonical representations. This lemma appears in [5] and was proven independently by us. For completeness, we include the proof in the full version [2]

Lemma 1. *Any strongly history independent implementation of a well-behaved data-structure is canonical, i.e., there is only one possible memory representation for each possible content.*

4.2   Lower Bounds on Comparison Based Data Structure Implementation

We now proceed to lower bounds on implementations of canonical data structures. Our lower bounds are proven in the *comparison based* model. A *comparison based* algorithm may only compare keys and store them in memory. That is, the keys are treated by the algorithm as 'black boxes'. In particular, the algorithm may not look at the inner structure of the keys, or separate a key into its components. Other then that the algorithm may, of-course, save additional data such as pointers, counters etc. Most of the generic data-structure implementations are comparison based. An important data structure that is implemented in a non-comparison-based manner is hashing, in which the value of the key is run through the hash function to determine an index. Indeed, for hashing, strongly efficient history independent implementations (which are canonical) exist and the algorithms are not comparison based [10]. Recall that we call an implementation of data structure *canonical* if there is only one memory representation for each possible content.

We assume that a data structure may store a set of keys whose size is unbounded $k_1, k_2, \ldots, k_i, \ldots$. We also assume that there exists a total order on the keys. We start with a general lower bound that applies to many data structures (lemma 2 below). In particular, this lower bound applies to the heap. We will later prove a more specific lemma (see lemma 3 below) that is valid for the queue, and another specific lemma (lemma 4 below) for the operation build-heap of the heap.

In our first lemma, we consider data structures whose content is the set of keys stored in it. This means that the set of keys in the data structure completely determines its output on any sequence of (legitimate) operations applied on the data structure. Examples of such data structures are: a heap, a tree, a hash table, and many others. However, a queue does not satisfy this property since the output of operations on the queue data structure depends on the order in which the keys were inserted into the structure.

Lemma 2. *Let $k_1, k_2, \ldots$ be an infinite set of keys with a total order between them. Let D be an abstract data structure whose content is the set of keys stored inside it. Let I be any implementation of D that is comparison based and canonical Then the following operations on D*

- insert(D, v)
- extract(D, v)
- increase-key(D, v_1, v_2) *(i.e. change the value from v_1 to v_2)*

require time complexity

1. $\Omega(n)$ *in worst case,*
2. $\Omega(n)$ *amortized time.*

Remark 1. Property (ii) implies property (i). We separate them for clarity of the representation.

Proof. We start with the first part of the lemma (worst case lower bound) for the insert operation. For any $n \in \mathbb{N}$, let $k_1 < k_2 < \ldots < k_{n+1} < k_{n+2}$ be $n+2$ keys. Consider any sequence of insert operations inserting n of these keys to D. Since the implementation I is comparison based, and the content of the data structure is the set of keys stored inside it, the keys must be stored in the data structure. Since the implementation I is canonical, then for any such set of keys, the keys must be stored in D in the same addresses regardless of the order in which they were inserted into the data structure. Furthermore, since I is comparison based, then the address of each key does not depend on its value, but only on its order within the n keys in the data structure. Denote by d_1 the address used to store the smallest key, by d_2 the address used to store the second key, and so forth, with d_n being the memory address of the largest key (If there is more than one address used to store a key choose one arbitrarily). By a similar argument, any set of $n+1$ keys must be stored in the memory according to their order. Let these addresses be $d'_1, d'_2, \ldots d'_{n+1}$. Next, we ask how many of these addresses are different. Let Δ be the number of indices for which $d_i \neq d'_i$ for $1 \leq i \leq n$.

Now we present a challenge to the data structure which cannot be implemented efficiently by I. Consider the following sequences of operations applied on an empty data-structure: $S = \mathsf{insert}(k_2), \mathsf{insert}(k_3) \ldots \mathsf{insert}(k_{n+1})$. After this sequence of operations k_i must be located in location d_{i-1} in the memory. We claim that at this state either $\mathsf{insert}(k_{n+2})$ or $\mathsf{insert}(k_1)$ must move at least half of the keys from their current location to a different location. This must take at least $n/2 = \Omega(n)$ steps.

If $\Delta > n/2$ then we concentrate on $\mathsf{insert}(k_{n+2})$. This operation must put k_{n+2} in address d'_{n+1} and must move all keys k_i ($2 \leq i \leq n+1$) from location d_{i-1} to location d'_{i-1}. There are $\Delta \geq n/2$ locations satisfying $d_{i-1} \neq d'_{i-1}$ and we are done. Otherwise, if $\Delta \leq n/2$ then we focus on $\mathsf{insert}(k_1)$. This insert must locate k_1 in address d'_1 and move all keys k_i, $2 \leq i \leq k+1$ from location d_{i-1} to location d'_i. For any i satisfying $d_{i-1} = d'_{i-1}$, it holds that $d_{i-1} \neq d'_i$ (since d'_i must be different from d'_{i-1}). The number of such cases is $n - \Delta \geq n/2$. Thus, for more than $n/2$ of the keys we have that $d_i \neq d'_{i+1}$, thus the algorithm must move them, and we are done.

To show the second part of the lemma for insert, we extend this example to hold for an amortized analysis as well. We need to show that for any integer $\ell \in \mathbb{N}$, there exists a sequence of ℓ operations that require time complexity

$\Omega(n \cdot \ell)$. We will actually show a sequence of ℓ operations each requiring $\Omega(n)$ steps. We start with a data structure containing the keys $l+1, l+2, \ldots, l+n+1$. Now, we repeat the above trick ℓ times. Since there are at least ℓ keys smaller than the smallest key in the structure, the adversary can choose in each step between entering a key larger than all the others or smaller than all the keys in the data structure.

The proof for the extract operation is similar. We start with inserting $n + 1$ keys to the structure and then extract either the largest or the smallest, depending on Δ. Extracting the largest key cause a relocation of all keys for which $d'_i \neq d_i$. Extracting the smallest key moves all the keys for which $d_i = d'_i$. One of them must be larger than $n/2$. The second part of the lemma may be achieved by inserting $n + \ell$ keys to the data structure, and then run ℓ steps, each step extracting the smallest or largest value, whichever causes relocations to more than half the values.

Finally, we look at increase-key. Consider an increase-key operation that increases the smallest key to a value larger than all the keys in the structure. Since the implementation is canonical this operation should move the smallest key to the address d_n and shift all other keys from d_i to d_{i-1}. Thus, n relocations are due and a lower bound of n steps is obtained. To show the second part of the lemma for increase-key we may repeat the same operation ℓ times for any $\ell \in \mathbb{N}$.

We remark that the above lemma is tight (up to constant factors). We can implement a canonical data structure that keeps the keys in two arrays. The $n/2$ smaller keys are sorted bottom up at the first array and the other $n/2$ keys are sorted from top to bottom in the other array. Using this implementation, inserting or extracting a key will always move at most half of the keys.

We now move to showing a lower bound on a canonical implementation of the queue data structure. Note that lemma 2 does not hold for the queue data structure since its content is not only the set of values inside it. Recall that a queue has two operations: insert-first and remove-last.

Lemma 3. *In any comparison based canonical implementation of a queue either* insert-first *or* remove-last *work in* $\Omega(n)$ *worst time complexity. The amortized complexity of the two operations is also* $\Omega(n)$.

Proof. Let $k_1 < k_2 < \ldots < k_{n+1}$ be $n + 1$ keys. Consider the following two sequences of operations applied both on an empty queue: $S_1 = $ insert-first(k_1), insert-first$(k_2) \ldots$ insert-first(k_n) and $S_2 = $ insert-first(k_2), insert-first$(k_3) \ldots$ insert-first(k_{n+1}). Since the implementation is comparison based it must store the keys in the memory layout in order to be able to restore them. Also, since the implementation is comparison based, it cannot distinguish between the two sequences and as the implementation is also canonical the location of each key in the memory depends only on its order in the sequence. Thus, the address (possibly more than one address) of k_1 in the memory layout after running the first sequence must be the same as the address used to store k_2 in the second sequence. In general, the address used to store k_i in the first sequence is the same as the address used to store the key k_{i+1} in the second sequence. This means that after running sequence S_1, each of the keys $k_2, k_3, \ldots, k_n$ must reside in a different location than its location after running S_2.

Consider now two more operations applied after S_1: insert-first(k_{n+1}), remove-last (i.e., remove k_1). The content of the data structure after these two operations is the same as the content after running the sequence S_2. Thus, their memory representations must be the same. This means that $n-1$ keys (i.e $k_2, k_3, \ldots, k_n$) must have changed their positions. Thus, either insert or remove-last operation work in worst time complexity of $\Omega(n)$. This trick can be repeated l times showing a series of insert and remove-last such that each pair must move $\Omega(n)$ keys resulting in the lower bound on the amortized complexity.

Last, we prove a lower bound on the build-heap operation in a comparison based implementation of the heap.

Lemma 4. *For any comparison based canonical implementation of a heap the operation* build-heap *must perform* $\Omega(n \log n)$ *operations.*

Proof. Similarly to sorting, we can view the operation of build-heap in terms of a decision tree. Note that the input may contain any possible permutation on the values $k_1, \ldots, k_n$ but the output is unique: it is the canonical heap with $k_1, \ldots, k_n$. The algorithm may be modified to behave in the following manner: first, run all required comparisons between the keys (the comparisons can be done adaptively), and then, based on the information obtained, rearrange the input values to form the canonical heap. We show a lower bound on the number of comparisons. Each comparison of keys separates the possible inputs to two subsets: those that agree and those that disagree with the comparison made. By the end of the comparisons, each of the $n!$ possible inputs must be distinguishable from the other inputs. Otherwise, the algorithm will perform the same rearrangement on two different inputs, resulting in two different heaps. Thinking of the comparisons as a decision tree, we note that the tree must contain at least $n!$ leaves, each representing a set with a single possible input. This means that the height of the decision tree must be $\Omega(\log(n!)) = \Omega(n \log n)$ and we are done.

4.3   Translating the Lower Bounds to Strong History Independence

We can now translate the results of section 4.2 and state the following lemmas:

Lemma 5. *Let D be a well behaved data structure for which its content is the values stored inside it. Let I be any implementation of D which is comparison based and strongly history independent. Then the following operations on D*

- insert(D, v)
- extract(D, v)
- increase-key(D, v_1, v_2) *(i.e. change the value from v_1 to v_2)*

require time complexity

1. *$\Omega(n)$ in worst case,*
2. *$\Omega(n)$ amortized time.*

Proof. The lemma follows directly from lemma 2 and 1.

Corollary 1. *For any strongly history independent comparison based implementation of the heap data structure, the operations* insert *and* increase-key *work in* $\Omega(n)$ *amortized time complexity. The time complexity of the* build-heap *operation is* $\Omega(n \log n)$.

Proof. The lower bounds on insert and increase-key follow from lemma 5. This is true since the content of the heap data structure is the keys stored inside it and the heap abstract data structure is well behaved. The lower bound on the build-heap operation follows directly from lemma 4 and 1.

Lemma 6. *For any strong history independent comparison based implementation of the queue data structure the worst time complexity of either* insert-first *or* remove-last *is* $\Omega(n)$. *Their amortized complexity is* $\Omega(n)$.

Proof. The lemma follows directly from lemma 3 and 1.

5   The Heap

In this section we review the basics of the heap data structure and set up the notation to be used in the rest of this paper. A good way to view the heap, which we adopt for the rest of this paper, is as an almost full binary tree condensed to the left. Namely, for heaps of $2^\ell - 1$ elements (for some integer ℓ), the heap is a full tree, and for sizes that are not a power of two, the lowest level is not full, and all leaves are at the left side of the tree. Each node in the tree contains a value. The important property of the heap-tree is that for each node i in the tree, its children contain values that are smaller or equal to the value at the node i. This property ensures that the maximal value in the heap is always at the root. Trees of this structure that satisfy the above property are denoted *well-formed heaps*. We denote by $parent(i)$ the parent of a node i and by v_i the content of node i. In a tree that represents a heap, it holds that for each node except for the root:

$$v_{parent(i)} \geq v_i$$

We will assume that the heap contains distinct elements, $v_1, v_2, \ldots, v_n$. Previous work (see [10]) justified using distinct values by adding some total ordering to break ties. In general, the values in the heap are associated with some additional data and that additional data may be used to break ties. The nodes of the heap will be numbered by the integers $\{1, 2, \ldots, n\}$, where 1 is the root 2 is the left child of the root 3 is the right child of the root etc. In general the left child of node i is node $2i$, and the right child is node number $2i + 1$. We denote the number of nodes in the heap H by $size(H)$ and its height by $height(H)$.

We will denote the rightmost leaf in the lowest level *the last leaf.* The position next to the last leaf, where the next leaf would have been had there been another value, is called *the first vacant place*. These terms are depicted in figure 1. Given

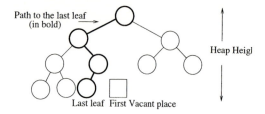

Fig. 1. The height of this heap is 4. The path to the last leaf is drawn in bold. In this example, the path to the first vacant place is the same except for the last edge.

a heap H and a node i in the heap, we use H^i to denote the sub-heap (or sub-tree) containing the node i and all its descendants. Furthermore, the sub-heap rooted by the left child of i is denoted H_L^i and the sub-heap rooted by the right child is denoted H_R^i The standard implementation of a heap is described in [3].

5.1   Uniform Heaps and Basic Machinery

In this section we investigate some properties of randomized heaps and present the basic machinery required for making heaps History independent. One of the properties we prove in this section is that the following distributions are equal on any given n distinct values $v_1, \ldots, v_n$.

Distribution Ω_1: Pick uniformly at random a heap among all possible heaps with values $v_1, \ldots, v_n$.

Distribution Ω_2: Pick uniformly at random a permutation on the values $v_1, \ldots, v_n$. Place the values in an (almost) full tree according to their order in the permutation. Invoke build-heap on the tree.

Note that the shape of a size n heap does not depend on the values contained in the heap. It is always the (almost) full tree with n vertices. The distributions above consider the placement of the n values in this tree.

In order to investigate the above distributions, we start by presenting a procedure that inverts the build-heap operation (see [3] and [2] for the definition of build-heap). Since build-heap is a many-to-one function, the inverse of a given heap is not unique. We would like to devise a randomized inverting procedure build-heap$^{-1}(H)$ that gets a heap H of size n as input and outputs a uniformly chosen inverse of H under the function build-heap. Such an inverse is a permutation π of the values $v_1, \ldots, v_n$ satisfying build-heap$(v_{\pi(1)}, \ldots, v_{\pi(n)}) = H$. It turns out that a good understanding of the procedure build-heap^{-1} is useful both for analyzing History independent heaps and also for the actual construction of its operations.

Recall that the procedure build-heap invokes the procedure heapify repeatedly in a bottom-up order on all vertices in the heap. The inverse procedure build-heap^{-1} invokes a randomized procedure heapify^{-1} on all vertices in the heap in a top-bottom order, i.e., from the roots to the leaves. We begin by defining the

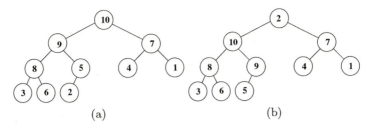

Fig. 2. An example of invoking heapify$^{-1}(H, 10)$. Node number 10 is the node that contains the value 2. In (b) we can see the output of invoking heapify^{-1} on the proper heap in (a). The value 2 is put at the root, the path from the root to the father of 2 is shifted down. Note that the two sub-trees in (b) are still well-formed heaps. Applying Heapify on (b) will cause the value 2 at the root to float down back to its position in the original H as in (a)

randomized procedure heapify^{-1}. This procedure is a major player in most of the constructions in this paper.

Recall that heapify gets a node and two well-formed heaps as sub-trees of this node and it returns a unified well-formed heap by floating the value of the node down always exchanging values with the larger child. The inverse procedure gets a proper heap H. It returns a tree such that at the root node there is a random value from the nodes in the heap and the two sub-trees of the root are well-formed sub-heaps. The output tree satisfies the property that if we run heapify on it, we get the heap H back. We make the random selection explicit and let the procedure heapify^{-1} get as input both the input heap H and also the random choice of an element to be placed at the root.

The operation of heapify^{-1} on input (H, i) is as follows. The value v_i of the node i in H is put in the root and the values in all the path from the root to node i are shifted down so as to fill the vacant node i and make room for the value v at the root. The resulting tree is returned as the output. Let us first check that the result is fine syntactically, i.e., that the two sub-trees of the root are well-formed heaps. We need to check that for any node, but the root, the values of its children are smaller or equal to its own value. For all vertices that are not on the shifted path this property is guaranteed by the fact that the tree was a heap before the shift. Next, looking at the last (lower) node in the path, the value that was shifted into node i is the value that was held in its parent. This value is at least as large as v and thus at least as large as the values at the children of node i. Finally, consider all other nodes on this path. One of their children is a vertex of the path, and was their child before the shift and cannot contain a larger value. The other child was a grandchild in the original heap and cannot contain a smaller value as well.

Claim. Let n be an integer and H be any heap of size n, then for any $1 \le i \le n$,

$$\text{heapify}\left(\text{heapify}^{-1}(H, i)\right) = H.$$

Proof. Proof omitted (the proof appears in [2]).

---

procedure build-heap^{-1}(H : **Heap**) : **Tree**
begin
1.   if ($size(H) = 1$) then return(H)
2.   Choose a node i uniformly at random among the nodes in the heap H.
3. $H \leftarrow$ heapify^{-1}(H, i)
4.   Return $TREE(root(H),$ build-heap^{-1}(H_L), build-heap^{-1}(H_R))
end

Fig. 3. The procedure build-heap^{-1}(H)

An example of invoking heapify^{-1}(H, i) is depicted in figure 2. The complexity of heapify^{-1}(H, i) is linear in the difference between the height of node i and the height of the input heap (or sub-heap), since this is the length of the shifted path. Namely, the complexity of heapify^{-1}(H, i) is $O(height(H) - height(i))$.

Using heapify^{-1}(H, i) we now describe the procedure build-heap^{-1}(H), a randomized algorithm for inverting the build-heap procedure. The output of the algorithm is a permutation of the heap values in the same (almost) full binary tree T underlying the given heap H. The procedure build-heap^{-1} is given in Figure 3. In this procedure we denote by $TREE(root, T_L, T_R)$ the tree obtained by using node "root" as the root and assigning the tree T_L as its left child and the tree T_R as its right child. The procedure build-heap^{-1} is recursive. It uses a pre-order traversal in which the root is visited first (and $heapify^{-1}$ is invoked) and then the left and right sub-heaps are inverted by applying build-heap^{-1} recursively.

Claim. For any heap H and for any random choices of the procedure build-heap^{-1},

$$\text{build-heap} \left(\text{build-heap}^{-1}(H) \right) = H$$

Proof Sketch: The claim follows from the fact that for any i and a heap H, heapify $\left(\text{heapify}^{-1}(H, i)\right) = H$, and from the fact that the traversal order is reversed. The heapify operations cancel one by one the heapify^{-1} operations performed on H in the reversed order and the same heap H is built back from the leaves to the root. $\square$

In what follows, it will sometimes be convenient to make an explicit notation of the randomness used by build-heap^{-1}. In each invocation of the (recursive) procedure, a node is chosen uniformly in the current sub-heap. The procedure build-heap^{-1} can be thought of as a traversal of the graph from top to bottom, level by level, visiting the nodes of each level one by one and for each traversed node i, the procedure chooses uniformly at random a node x_i in the sub-heap H^i and invokes heapify^{-1}(H^i, x_i). Thus, the random choices of this algorithm include a list of n choices $(x_1, \ldots, x_n)$ such that for each node i in the heap, $1 \le i \le n$, the chosen node x_i is in its sub-tree. The x_i's are independent of the actual values in the heap. They are randomized choices of locations in the heap. Note, for example, that for any leaf i it must hold that $x_i = i$ since there is only one node in the sub-heap H^i. The vector $(x_1, \ldots, x_n)$ is called *proper* if for all i, $i \leq i \leq n$, it holds that x_i is a node in the heap H^i. We will sometimes let the procedure build-heap$^{-1}(H)$ get its random choices explicitly in the input and use the notation build-heap$^{-1}(H, (x_1, \ldots, x_n))$.

We are now ready to prove some basic lemmas regarding random heaps with n distinct values. In the following lemmas we denote by $\Pi(n)$ the set of all permutations on the values $v_1, v_2 \ldots, v_n$.

Lemma 7. *Each permutation $\pi \in \Pi(n)$ has one and only one heap H and a proper vector $\boldsymbol{X}_n = (x_1, \ldots, x_n)$ such that $(v_{\pi(1)}, \ldots, v_{\pi(n)}) = $ build-heap$^{-1}(H, \boldsymbol{X}_n)$.*

Proof. Proof omitted (the proof appears in [2]).

Corollary 2. *If H is picked up uniformly among all possible heaps with the same content then $T = $ build-heap$^{-1}(H)$ is a uniform distribution over all $\pi \in \Pi(n)$.*

Proof. As shown, for any permutation π in *support*(H), i.e., a permutation that satisfies build-heap$(v_{\pi(1)}, \ldots, v_{\pi(n)}) = H$, there is a unique random vector $(x_1, \ldots, x_n)$, that creates the permutation. Each random (proper) vector has the same probability. Thus, π is chosen uniformly among all permutation in *support*(H). Since H is chosen uniformly among all heaps the corollary follows.

Lemma 8. *Let n be an integer and $v_1, \ldots, v_n$ be a set of n distinct values. Then, for heap H that contains the values $v_1, \ldots, v_n$ it holds that:*

$$Pr\left[\pi \in \Pi(n) : \text{build-heap}(v_{\pi(1)}, v_{\pi(2)}, \ldots, v_{\pi(n)}) = H\right] = p(H)$$

Where $p(H)$ is a function depending only on n (the size of H). Furthermore, $p(H) = N(H)/|\Pi(n)|$ where $N(H)$ can be defined recursively as follows:

$$N(H) = \begin{cases} 1 & \text{if } size(H) = 1 \\ size(H) \cdot N(H_L) \cdot N(H_R) & \text{otherwise} \end{cases}$$

Proof. For any H the probability that build-heap$(v_{\pi(1)}, v_{\pi(2)}, \ldots, v_{\pi(n)}) = H$ is the probability that the permutation π belongs to *support*(H). According to lemma 7 the size of *support*(H) is the same for any possible heap H. This follows from the fact that any random vector $(x_1, x_2, \ldots, x_n)$ result in different permutation in *support*(H) and each permutation in *support*(H) has a vector that yield it. The size of *support*(H) is exactly the number of possible random (proper) vectors. This number can be formulated recursively as $N(H)$ depending only on the size of the heap. The probability for each heap now follows.

Corollary 3. *The following distributions Ω_1 and Ω_2 are equal.*

Distribution Ω_1: *Pick uniformly at random a heap among all possible heaps with values $v_1, \ldots, v_n$.*

Distribution Ω_2: *Pick uniformly at random permutation $\pi \in \Pi(n)$ and invoke build-heap$(v_{\pi(1)}, v_{\pi(2)}, \ldots, v_{\pi(n)})$.*

Proof. As shown in lemma 8, distribution Ω_2 gives all heaps containing the values $v_1, \ldots, v_n$ the same probability. By definition, this is also the case in Ω_1.

6   Building and Maintaining History Independent Heap

In this section we present our main theorem and provide some overview of the proof. The full proof appears in [2].

Theorem 1. *There exists a History independent implementation of the heap data structure with the following time complexity. The worst case complexity of the* build-heap *operation is $O(n)$. The worst case complexity of the* increase-key *operation is $O(\log n)$. The expected time complexity of the operations* insert *and* extract-max *is $O(\log n)$, where expectation is taken over all possible random choices made by the implementation.*

Our goal is to provide an implementation of the operations build-heap, insert, extract-max, and increase-key that maintains history independence without incurring an extra cost on their (asymptotic) time complexity. We obtain history independence by preserving the uniformity of the heap. When we create a heap, we create a uniform heap among all heaps on the given values. Later, each operation on the heap assumes that the input heap is uniform and the operation maintains the property that the output heap is still uniform for the new content. Thus, whatever series of operation is used to create the heap with the current content, the output heap is a uniform heap with the given content. This means that the memory layout is history independent and the set of operations make the heap History independent. The rest of the proof describes the implementation for each of the 4 operations. In what follows, we try to highlight the main issues. The full implementation with proof of history independence and time complexity analysis is given in [2].

The operation build-heap. This is the simplest since time complexity $O(n)$ is allowed. Given n input values, we permute them uniformly at random and run the (non-oblivious) build-heap. Using the basic machinery from section 5.1, this can be shown to yield a uniformly chosen heap on the input values.

The operation increase-key. Here we extend the operation to allow both increasing and decreasing the given value. This extension is useful when implementing the other the operations. To increase a key the standard implementation turns out to work well, i.e., it maintains the uniformity of the distribution on the current heap. To decrease a key, it is enough to use heapify. We show that the scenario is appropriate for the procedure heapify and that uniformity is preserved.

6.1   The Operation Extract-Max

We start with a naive implementation of extract-max which we call extract-max-try-1. This implementation has complexity $O(n)$. Of-course, this is not an acceptable complexity for the extract-max operation but this first construction will be later modified to make the real History independent extract-Max. The procedure extract-max-try-1 goes as follows. We run build-heap^{-1} on the heap H

---

procedure extract-max-try-1(H: **Heap**) : **Heap**
begin
1.   Choose uniformly at random a proper randomization vector $(x_1, \ldots, x_{n+1})$
            for the procedure build-heap^{-1}.
2. $T = $ build-heap$^{-1}(H, (x_1, x_2, \ldots, x_{n+1}))$
3.   Let T' be the tree obtained by removing the last node with value v_i from T.
4.   H' = build-heap(T')
5.   if v_i is the maximum then return (H'). Otherwise:
6.   Modify the value at the root to v_i.
7. $H'' = $ heapify(H, 1) (i.e. apply heapify on the root)
8.   Return (H'')
end

Fig. 4. The procedure extract-max-try-1

to get a uniform permutation π on the $n + 1$ values. Next, we remove the value at the last leaf $v_{\pi(n+1)}$. After this step we get a uniformly chosen permutation of the n values excluding the one we have removed. Next, we run build-heap on the n values to get a uniformly chosen heap among the heaps without $v_{\pi(n+1)}$. If $v_{\pi(n+1)}$ is the maximal value then we are done. Otherwise, we continue by replacing the value at the root (the maximum) with the value $v_{\pi(n+1)}$ and running heapify on the resulting tree to "float" the value $v_{\pi(n+1)}$ down and get a well-formed heap. We will show that this process results in a uniformly chosen heap without the maximum value. The pseudo code of the naive extract-max-try-1 appears in figure 4.

The time consuming lines of the procedure extract-max-try-1 are lines 2,3, and 4 in which we un-build the heap, remove the last leaf and then build it back again. The next step is to note that many of the steps executed are redundant. In particular, build-heap^{-1} runs heapify^{-1} on each node of the heap from top to bottom. Later, after removing the last node, build-heap runs heapify on each node of the heap from bottom to top. These operations cancel each other except for the removed leaf that causes a distortion in the reversed operation. We first claim that it is enough to run heapify^{-1} and heapify on the nodes that belong to the path from the root to the last leaf. Proving this requires some care, but the outcome is a procedure with time complexity $O(\log^2 n)$, since heapify^{-1} and heapify both requires logarithmic time and we need to run them on $O(\log n)$ nodes.

By now, we have an implementation of extract-max requiring worst case time $O(\log^2 n)$. Our last step is to check exactly how the heap is modified by each of these heapify^{-1} and heapify invocations. It turns out that we may save more operations by running these only on vertices "that affect the identity of the last leaf". Identifying these vertices require some technical work, but it turns out that their number equals the height of the vertex that is moved to the last leaf by build-heap^{-1}. Analysis of the expected height of this vertex in our scenario gives a constant number. Thus, on average, we only need to perform heapify^{-1}

and heapify on a constant number of vertices and we are done with operation extract-Max. The details and proofs appear in [2].

6.2   The Operation Insert

Here, again, we start by providing a procedure insert-try-1 implementing the operation insert with complexity $O(n)$. Again, this allows a construction of a simple and useful implementation that will be improved later. We first choose the location i, $1 \leq i \leq n + 1$ to which we insert the new value a. (The choice $i = n + 1$ means no insertion.) We put the value a in the node i and remember the value v_i that was replaced at node i. This may yield a tree which is not a well-formed heap because the value a may not "fit" the node i. Hence, we apply increase-key-oblivious on the location i with the new value a. After the new value a is properly placed in the heap, we run build-heap^{-1}. We will show that this yields a uniform permutation of the values $(v_1, v_2, \ldots, v_{i-1}, a, v_{i+1}, \ldots, v_n)$. Now, we add the value v_i at the end of this ordering, getting a uniform permutation on the $n + 1$ values $v_1, v_2, \ldots, v_n, a$. Running build-heap on this order of the values yields a random heap on the $n + 1$ values.

Then, we scrutinize this simple procedure to identify all redundant steps. Indeed, we are able to obtain a modified procedure that runs in expected time $O(\log n)$ and outputs the same distribution as this simple procedure. The details and proofs appear in [2].

References

1. A. Andersson, T. Ottmann. Faster Uniquely Represented Dictionaries. Proc. 32nd IEEE Sympos. Foundations of Computer Science, pages 642–649, 1991
2. Niv Buchbinder, Erez Petrank. Lower and Upper Bounds on Obtaining History Independence http://www.cs.technion.ac.il/~erez/publications.html
3. Thomas H. Cormen, Charles E. Leiserson, and Ronald L. Rivest. Introduction to Algorithms. MIT Press and McGraw-Hill Book Company, 6th edition, 1992.
4. E.W.Dijkstra A note on two problems in connexion with graphs. Numerische Mathematik, 1:269–271, 1959
5. J.D. Hartline, E.S. Hong, A.E. Mohr, W.R. Pentney, and E.C. Rocke. Characterizing History independent Data Structures. ISAAC 2002 pp. 229–240, 2002.
6. J.R.Driscoll, N. Sarnak, D.D. Sleator, and R.E. Tarjan. Making data structures persistent. Journal of Computer and System Sciences, 38(1):86–124, 1989.
7. R.W. Floyd. Algorithm 245 (TREESORT). Communications of the ACM, 7, 1964.
8. D. Micciancio. Oblivious data structures: Applications to cryptography. In Proc. 29th ACM Symp. on Theory of computing, pages 456–464, 1997.
9. O. Goldreich and R. Ostrovsky. Software protection and simulation on oblivious rams. Journal of the ACM, 43(3):431–473, 1996.
10. M.Naor and V.Teague. Anti-persistence: History Independent Data Structures. Proc. 33rd ACM Symp. on Theory of Computing, 2001.
11. R.C.Prim. Shortest connection networks and some generalizations. Bell System Technical Journal, 36:1389–1401, 1957
12. J.W.J Williams Algorithm 232 (HEAPSORT). Communication of the ACM, 7:347–348, 1964

Private Circuits: Securing Hardware against Probing Attacks

Yuval Ishai[1]*, Amit Sahai[2], and David Wagner[3]

[1] Technion — Israel Institute of Technology, yuvali@cs.technion.ac.il
[2] Princeton University, sahai@cs.princeton.edu
[3] University of California, Berkeley, daw@cs.berkeley.edu

Abstract. Can you guarantee secrecy even if an adversary can eavesdrop on your brain? We consider the problem of protecting privacy in *circuits*, when faced with an adversary that can access a bounded number of wires in the circuit. This question is motivated by side channel attacks, which allow an adversary to gain partial access to the inner workings of hardware. Recent work has shown that side channel attacks pose a serious threat to cryptosystems implemented in embedded devices. In this paper, we develop theoretical foundations for security against side channels. In particular, we propose several efficient techniques for building *private circuits* resisting this type of attacks. We initiate a systematic study of the complexity of such private circuits, and in contrast to most prior work in this area provide a formal threat model and give proofs of security for our constructions.

Keywords: Cryptanalysis, side channel attacks, provable security, secure multiparty computation, circuit complexity.

1 Introduction

This paper concerns the following fascinating question: Is it possible to maintain secrecy even if an adversary can eavesdrop on your brain? A bit more precisely, can we guarantee privacy when one of the basic assumptions of cryptography breaks down, namely, when the adversary can gain access to the insides of the hardware that is making use of our secrets? We formalize this question in terms of protecting privacy in *circuits*, where an adversary can access a bounded number of wires in the circuit. We initiate the study of this problem and present several efficient techniques for achieving this new type of privacy. Before describing the model and our contribution in more detail, we motivate the problem by providing some necessary background.

1.1 Background

Our understanding of cryptography has made tremendous strides in the past three decades, fueled in large part by the success of analysis- and proof-driven design. Most such work has analyzed algorithms, not implementations: typically one thinks of a cryptosystem as a black box implementing some mathematical function and implicitly assumes the

---

* Work done in part while at Princeton University.

D. Boneh (Ed.): CRYPTO 2003, LNCS 2729, pp. 463–481, 2003.

implementation faithfully outputs what the function would (and nothing else). However, in practice implementations are not always a true black box: partial information about internal computations can be leaked (either directly or through *side-channels*), and this may put security at risk.

This difference between implementations and algorithms has led to successful attacks on many cryptographic implementations, even where the underlying algorithm was quite sound. For instance, the power consumed during an encryption operation or the time it takes for the operation to complete can leak information about intermediate values during the computation [25,26], and this has led to practical attacks on smartcards. Electromagnetic radiation [33,16,34], compromising emanations [36], crosstalk onto the power line [37,35], return signals obtained by illuminating electronic equipment [3, 35], magnetic fields [32], cache hit ratios [24,30], and even sounds given off by rotor machines [23] can similarly give the attacker a window of visibility on internal values calculated during the computation. Also of interest is the *probing attack*, where the attacker places a metal needle on a wire of interest and reads off the value carried along that wire during the smartcard's computation [2]. In general, side channel attacks have proven to be a significant threat to the security of embedded devices.

The failure of proof-driven cryptography to anticipate these risks comes from an implicit assumption in many[1] currently accepted definitions in theoretical cryptography, namely, the *secrecy assumption*. The secrecy assumption states that legitimate participants in a cryptographic computation can keep intermediate values and key material secret during a local computation. For instance, by modeling a chosen-plaintext attack on the encryption scheme E as an algorithm A^{E_k} with oracle access to E_k, we implicitly assume that the device implementing E_k outputs only $E_k(x)$ on input x, and does not leak anything else about the computation of $E_k(x)$. Thus the 'Standard Model' in theoretical cryptography often takes the secrecy assumption for granted, but as we have seen, there are a bevy of ways that the secrecy assumption can fail in real systems.

One possible reaction is to study implementation techniques that ensure the secrecy assumption will always hold. For instance, we can consider adding large capacitors to hide the power consumption, switch to dual-rail logic so that power consumption will be independent of the data, shield the device in a tamper-resistant Faraday cage to prevent information leakage through RF emanations, and so on. Many such hardware countermeasures have been proposed in the literature. However, a limitation of such approaches is that, generally speaking, each such countermeasure must be specially tailored for the set of side channels it is intended to defeat, and one can only plan a defense if one knows in advance what side channels an attacker might try to exploit. Consequently, if the designer cannot predict all possible ways in which information might leak, hardware countermeasures cannot be counted on to defend reliably against side channel attacks.

This leaves reason to be concerned that hardware countermeasures may not be enough on their own to guarantee security. If the attacker discovers a new class of side channel attacks not anticipated by the system designer, all bets are off. Given the wide variety of side channel attacks that have been discovered up till now, this seems like a significant

---

[1] This implicit assumption is definitely not universal. For instance, the field of secure multi-party computation asks for security even when some parties can be corrupted or observed.

risk: As a general rule of thumb, wherever three or four such vulnerabilities are known, it would be prudent to assume that there may be another, similar but unknown vulnerability lurking in the wings waiting to be discovered. In particular, it is hard to predict what other types of side channel pitfalls might be discovered in the future, and as a result, it is hard to gain confidence that any given implementation will be free of side channels. This is a "risk of the unknown", rather than a known risk[2], and risks of the unknown are the worst kind of risks to assume. Consequently, the secrecy assumption seems optimistic, and we submit that hardware countermeasures may not be the final answer.

A different possible response is to design algorithms that, when implemented, will be inherently robust against side channel attacks. For instance, Daemen and Rijmen proposed replacing each wire of a circuit by two wires, one carrying the original bit and the other its complement [15]; Messerges proposed "data masking", where each value is split into two shares using a 2-out-of-2 secret sharing scheme [27]; Goubin and Patarin suggested a "duplication" method based on similar methods [21]; and many other proposals can be found in the literature. However, none of those schemes have been proven secure, and unsurprisingly, some have since been broken [11,14]. This experience suggests that the field needs to be put on solid theoretical foundations. For obvious reasons, we would prefer a principled approach that has been proven secure over an ad-hoc countermeasure.

1.2 Our Contribution

In this paper, we take on this challenge. Working in the context of Boolean circuits, we show how to implement cryptosystems (or any algorithm) in a way that can tolerate the presence of a large class of side channel attacks without loss of security. In particular, we show how to transform any circuit implementing some cryptographic algorithm into another, larger circuit that implements the same functionality but that will remain secure even if the attacker can observe up to any t internal bits produced during the computation within one clock cycle.

As a result, our constructions provide a generic defense against probing attacks. They are generic in the sense that we defend against a large class of attacks. To defend against information leakage, we do not need to know how the information might leak; rather, we only need to predict how much information might leak or at what rate. Our constructions are also generic in the sense that they apply to any cryptosystem of interest: rather than trying to secure just, say, AES encryption, we show that any circuit whatsoever can be made robust against probing attacks.

Also, we emphasize that our constructions are provably secure. We develop a formal model of the adversary, propose definitions of security against probing attacks, and prove that our constructions meet these definitions. This puts the field on a principled theoretical footing and removes fears that our proposals might be broken by cryptanalysis.

OUR MODEL. Ideally, we would like to achieve security against an all-powerful attacker, *i.e.* one that can observe every internal value produced during the computation. However, this task is generally impossible to achieve, as follows from the impossibility of

---

[2] We thank Mark Miller for introducing us to this turn of phrase.

Table 1. A summary of our main results. Here n denotes the size of the original circuit and t the number of adversarial probes we wish to tolerate. All uses of $O()$ notation hide small constants. We use $\tilde{O}()$ to hide large constants, polylogarithmic factors, or polynomials in a security parameter.

Applies to	Privacy type	Size	Sec.	Comments
any circuit	perfect	$O(nt^2)$	§4	our basic scheme
PRG circuits	computational	$O(nt)$	§6	only applies to pseudorandom generators
any circuit	computational	$O(nt^2) + \tilde{O}(t^3)$	§6	derandomized version of basic scheme
any circuit	statistical	$\tilde{O}(nt)$	§5	
any circuit	statistical	$\tilde{O}((w + t)d)$	§5	layered circuit of width w and depth d

obfuscation [4]. Instead, we settle for achieving security against adversaries that are limited in their power to observe the computation. There are many ways we could consider limiting the adversary, but in this paper we choose a simple metric: a t-limited adversary is one that can observe at most t wires of the circuit within a certain time period (such as during one clock cycle).[3] We believe this is a reasonable restriction, as most side channels give the attacker only partial information about the computation. In particular, in probing attacks the cost of micro-probing equipment is directly related to the number of needles one can manipulate at one time—a station with five probes is considerably more expensive than one with only a single probe—and so an attacker is limited in the number of wires that can be observed at any one time. Consequently, the value t is a good measure of the cost of a probing attack. We refer the reader to Section 2 for a more detailed treatment of the model, in particular for the useful case of *stateful* circuits which carry state information from one invocation to the next.

Our model can be compared to that of Chari, et al., who took a first step by analyzing k-out-of-k secret sharing in a model where the attacker can obtain a noisy view of all circuit elements [11], with applications to security against power analysis. That work, however, did not provide security against probing attacks or other side channels where the attacker can view any t wires of his choosing, and our constructions are quite different from theirs. Also of relevance are works on exposure-resilient functions and all-or-nothing transforms (e.g. [9]), which attempt to efficiently secure *storage* (but not computation) against probing attacks, and work on oblivious RAM (cf. [20]) aimed at protecting *software* by hiding the access pattern of a (trusted) CPU.

MPC ON SILICON? There is an interesting relation between the problem we study and that of secure multi-party computation (MPC). In some sense, our contribution may be viewed as a novel application of MPC techniques to the design of secure hardware. We would like to stress, however, that our focus and goals are quite different from the traditional ones in the MPC literature, and that our main results are not derived from state-of-the-art results in this area. We refer the reader to Appendix A for a detailed discussion of the relation between our problem and the MPC problem.

---

[3] By default, we allow the adversary to adaptively move its t probes between time periods, but not *within* a time period. See Section 2 for more details.

OUR RESULTS. Our basic results are as follows. We show that any circuit with n gates can be transformed into a circuit of size $O(nt^2)$ that is perfectly secure against all probing attacks leaking up to t bits at a time (see Section 4). This general transformation increases circuit size by a factor of $O(t^2)$, but for some specific cryptosystems we can do better. For PRG's, we can find constructions that yield an $O(nt)$ transformed circuit size, rather than $O(nt^2)$ (Section 6). Finally, we present statistically private transformations which significantly improve the asymptotic efficiency of previous constructions, but whose concrete efficiency becomes better only when t is quite large. See Table 1 for a summary of the main results. Additional results, such as a trading circuit size for increased latency, will be included in the full version of this paper.

We do not know how practical our constructions will be. However, our results already show that the cost of security is not too high. Since many cryptosystems can be implemented quite efficiently in hardware (e.g., $n \approx 10^3$ or 10^4 gates), and since our use of big-$O()$ notation typically does not hide any large constants, it seems that security using our techniques is within the reach of modern systems. We leave a more thorough performance analysis to others.

2   Definitions

Circuits. We will examine probing attacks in the setting of Boolean circuits. A *deterministic circuit* C is a directed acyclic graph whose vertices are Boolean gates and whose edges are wires. We will assume without loss of generality that every gate has fan-in at most 2 and fan-out at most 3. A *randomized circuit* is a circuit augmented with random-bit gates. A *random-bit gate* is a gate with fan-in 0 that produces a random bit and sends it along its output wire; the bit is selected uniformly and independently of everything else afresh for each invocation of the circuit.

The size of a circuit (usually denoted by n) is defined as the number of gates and its depth is the length of the longest path from an input to an output. We will sometimes consider a width-w depth-d *layered* circuit, where the underlying graph is a depth-d layered graph with at most w wires connecting two adjacent layers.

A *stateful circuit* is a circuit augmented with memory cells. A *memory cell* is a stateful gate with fan-in 1: on any invocation of the circuit, it outputs the previous input to the gate, and stores the current input for the next invocation. Thus, memory cells act as delay elements. We extend the usual definition of a circuit by allowing stateful circuits to possibly contain cycles, so long as every cycle traverses at least one memory cell. When specifying a stateful circuit, we must also specify an initial state for the memory cells. When C denotes a circuit with memory cells and s_0 an initial state for the memory cells, we write $C[s_0]$ for the circuit C with memory cells initially filled with s_0. Stateful circuits can also have external input and output wires. For instance, in an AES circuit the internal memory cells contain the secret key, the input wires a plaintext, and the output wires produce the corresponding ciphertext.

We define two distinct notions of security, for stateless and stateful circuits. While we view the stateful model as more interesting from an application point of view, the stateless model is somewhat cleaner and solutions for this model are used as the basis for solutions for the stateful model.

Privacy for stateful circuits. Let T be an efficiently computable *randomized* algorithm mapping a (stateful) circuit C along with an initial state s_0 to a (stateful) circuit C' along with an initial state s_0'. We say that T is a *t-private stateful transformer* if it satisfies soundness and privacy, defined as follows:

SOUNDNESS. The input-output functionality of C initialized with s_0 is indistinguishable from that of C' initialized with s_0'. This should hold for any sequence of invocations on an arbitrary sequence of inputs. In other words, $C[s_0]$ and $C'[s_0']$ are indistinguishable to an *interactive* distinguisher.

PRIVACY. We require that C' be private against a *t-limited* interactive adversary. Specifically, the adversary is given access to C' initialized with s_0' as its internal state. Then, the adversary may invoke C' multiple times, adaptively choosing the inputs based on the observed outputs. Prior to each invocation, the adversary may fix an arbitrary set of t internal wires to which it will gain access in that invocation. We stress that while this choice may be adaptive between invocations, i.e., may depend on the outputs and on wire values observed in previous invocations, the adversary is assumed to be too slow to move its probes while the values propagate through the circuit.[4] To define privacy against such a t-limited adversary, we require the existence of a *simulator* which can simulate the adversary's view using only a black-box access to C', i.e., without having access to any internal wires.[5]

Note that randomization is vital for stateful transformers, for otherwise it is impossible to hide the initial state from the adversary. However, apart from the (trusted) randomized initialization, the circuit C' may be deterministic.

We distinguish between three types of transformers: *perfect, statistical,* and *computational,* corresponding to the quality of indistinguishability in the soundness requirement and the type of emulation provided by the simulator. For the latter two types of security, we assume that T is also given a *security parameter* k in terms of which the indistinguishability is defined and the complexity of T is measured.

Privacy for stateless circuits. In contrast to the stateful case, where inputs and outputs are considered public and it is only the internal state that is hidden, privacy for stateless circuits should keep both inputs and outputs hidden in every invocation. To make this possible, we allow the use of a randomized *input encoder* I and an *output decoder* O, a pair of circuits whose internal wires cannot be probed by the adversary. Both I and O should be independent of the circuit C being transformed, and will typically require a small number of gates to compute. Thus, they may be thought of as being implemented by expensive tamper-resistant hardware components. A private stateless transformer can now be defined similarly to the stateful case.

Let T be an efficiently computable *deterministic* function mapping a stateless circuit C to a stateless circuit C', and let I, O be as above. We say that (T, I, O) is a *t-private stateless transformer* if it satisfies soundness and privacy, defined as follows:

---

[4] Most of our constructions are in fact secure even against a fully adaptive adversary, that can also move its probes *within* an invocation, as long as the total number of probes in each invocation does not exceed t.

[5] In a case where C is randomized, the adversary's view should be simulated jointly with the circuit's outputs. This is necessary to capture information learned about the outputs.

SOUNDNESS. The input-output functionality of $O \circ C' \circ I$ (i.e., the iterated application of I, C', O in that order) is indistinguishable from that of C. Note that in the deterministic case this implies functional equivalence.

PRIVACY. We require that the view of any t-limited adversary, which attacks $O \circ C' \circ I$ by probing at most t wires in C', can be simulated from scratch, i.e. without access to any wire in the circuit. As in the stateful case, the identity of the probed wires has to be chosen in advance by the adversary.

3   Perfect Privacy for Stateless Circuits

In this section we present our first construction for protecting privacy in stateless circuits. In the next section we will show how to use this to achieve protection for the more useful model of stateful circuits, where the contents of memory are to be protected.

Similarly to interactive protocols for secure multi-party computation (e.g., [6,19]), our construction makes use of a simple secret-sharing scheme. The new twist in the circuit setting is that the atomic unit of information observable by the adversary is any *intermediate* computation rather than an entire party in the protocol setting. We achieve our result through a careful choice of intermediate computations, which allows us to obtain privacy without losing efficiency. The constants involved in the result we present here are quite small, and this construction may be of practical value. We now establish:

Theorem 1. *There exists a perfectly t-private stateless transformer (T, I, O) such that T maps any stateless circuit C of size n and depth d to a randomized stateless circuit of size $O(nt^2)$ and depth $O(d \log t)$.*

Proof. For simplicity, we focus on the case that C is deterministic. We start by describing the construction of the transformer (T, I, O). Let[6] $m = 2t$.

INPUT ENCODER I: Each binary input x is mapped to $m + 1$ binary values: First, m random binary values $r_1, \ldots, r_m$ are chosen using m random-bit gates. The encoding is then these m random values together with $r_{m+1} = x \oplus r_1 \oplus \cdots \oplus r_m$. The circuit I computes the encoding of each input bit independently in this way.

OUTPUT DECODER O: Corresponding to each output bit of C will be $m + 1$ bits $y_1, \ldots, y_{m+1}$ produced by $T(C)$. The associated output bit of C computed by O will be $y_1 \oplus \cdots \oplus y_{m+1}$.

CIRCUIT TRANSFORMER T: Assume without loss of generality that the circuit C consists of only NOT and AND gates. We will construct a transformed circuit C', maintaining the invariant that corresponding to each wire in C will be $m + 1$ wires in C' carrying an additive $m + 1$ out of $m + 1$ secret sharing of the value on that wire of C. The circuit C' is obtained by transforming the gates of C as follows.

For a NOT gate acting on a wire w, we merely take the $m + 1$ wires $w_1, \ldots, w_{m+1}$ associated with w in C', and put a NOT gate on w_1.

---

[6] Note that there is a way to slightly modify this construction which requires $m = t$ instead of $m = 2t$. See below.

Consider an AND gate in C with inputs a, b and output c. In C', we will have corresponding wires $a_1, \ldots, a_{m+1}$ and $b_1, \ldots, b_{m+1}$. Recall that $a = \sum_i a_i \bmod 2$ and $b = \sum_i b_i \bmod 2$. Thus $c = (a \text{ AND } b) = \sum_{i,j} a_i b_j \bmod 2$. The difficulty is in computing shares of c by grouping together elements from the summation so that t intermediate values do not reveal any information to the adversary. We now describe our technique for doing so: In the transformation of this gate, we first compute intermediate values $z_{i,j}$ for $i \neq j$. For each $1 \leq i < j \leq m + 1$, we introduce a random-bit gate producing a random bit $z_{i,j}$. Then we compute $z_{j,i} = (z_{i,j} \oplus a_i b_j) \oplus a_j b_i$. Note that individually each $z_{i,j}$ is distributed uniformly, but any pair $z_{i,j}$ and $z_{j,i}$ depend on a_i, a_j, b_i, and b_j. Now, we compute the output bits $c_1, \ldots, c_m$ in C' of this AND gate in C as

$$c_i = a_i b_i \oplus \bigoplus_{j \neq i} z_{i,j}.$$

In this way, each AND gate in C is expanded to a "gadget" of $O(m^2)$ gates in C', and the gadgets in C' are connected in the same way that the AND gates of C are connected. The resulting circuit, call it C', is the transformed version of C produced by T: i.e., we define $T(C) = C'$, with C' as above. This completes the description of T.

Clearly, this construction preserves the functionality of the original circuit. To prove t-privacy, we must show how to simulate the view of the t-limited adversary without knowing the input values for C. The simulation will proceed by running the adversary, and providing it with answers to its t queries. We will show that the distribution of answers our simulation provides is *identical* to the distribution the adversary would obtain in a real attack on C'.

The simplest description of the simulator is just this: Answer all adversary queries based on the evaluation of the circuit C' when fed uniform and independent bits as input. In order to prove that this simulation works, we give a different description of the simulator. We first describe the simulator for a circuit C consisting of a single AND gate, and then extend the proof and simulation to the general case.

SIMULATION FOR A SINGLE GATE. Let C' be the transformed circuit, consisting of a single gadget, with input wires $\{a_i\}$ and $\{b_i\}$, and outputs $\{c_i\}$. Recall that in a true evaluation of C', the a_i's and b_i's are additive secret shares with the property that any m shares from the a_i's are distributed as uniform independent random bits, and similarly for the b_i's. We will argue that a perfect simulation of the adversary's query responses is possible based on knowledge of m or fewer shares from the a_i's and the b_i's. Since such a collection of shares is distributed uniformly, this will establish our result.

Suppose an adversary corrupts wires $w_1, \ldots, w_t$ in C'. We will define a set $I \subset [m + 1]$ of indices such that the joint distribution of values assigned to the wires w_h (for any specific inputs a and b to the original circuit C) can be perfectly and efficiently simulated given the values of $a|_I := (a_i)_{i \in I}$ and $b|_I$. As mentioned above, the values $a|_I$ and $b|_I$, in turn, can be perfectly simulated by picking them uniformly and independently at random, as long as $|I| \leq m$. Hence, it suffices to describe a procedure for constructing the set I and simulating the values of the t corrupted wires w_h given $a|_I$ and $b|_I$. We describe such a procedure now.

1. Initially, I is empty and all w_h are unassigned.
2. For every wire w_h of the form $a_i, b_i, a_i b_i, z_{i,j}$ (for any $i \neq j$), or a sum of values of the above form (including c_i as a special case), add i to I. Note that this covers all wires in C' except for wires corresponding to $a_i b_j$ or $z_{i,j} \oplus a_i b_j$ for some $i \neq j$. For such wires, add both i and j to I.[7]
3. Now that the set I has been determined—and note that since there are at most t wires w_h, the cardinality of I can be at most $m = 2t$—we show how to complete a perfect simulation of the values on w_h using only the values $a|_I$ and $b|_I$. Assign values to the $z_{i,j}$ as follows:

   - If $i \notin I$ (regardless of j), then $z_{i,j}$ does not enter into the computation for any w_h. Thus, its value can be left unassigned.
   - If $i \in I$, but $j \notin I$, then $z_{i,j}$ is assigned a random independent value. Analysis: Note that if $i < j$ this is what would have happened in the real circuit C'. If $i > j$, however, we are making use of the fact that by construction, $z_{j,i}$ will never be used in the computation of any w_h. Hence we can treat $z_{i,j}$ as a uniformly random and independent value.
   - If both $i \in I$ and $j \in I$, then we have access to a_i, a_j, b_i, and b_j. Thus, we compute $z_{i,j}$ and $z_{j,i}$ exactly as they would have been computed in the actual circuit C'; i.e., one of them (say $z_{j,i}$) is assigned a random value and the other $z_{i,j}$ is assigned $z_{j,i} \oplus a_i b_j \oplus a_j b_i$.

4. For every wire w_h of the form $a_i, b_i, a_i b_i, z_{i,j}$ (for any $i \neq j$), or a sum of values of the above form (including c_i as a special case), we know that $i \in I$, and all the needed values of $z_{i,j}$ have already been assigned in a perfect simulation. Thus, w_h can be computed in a perfect simulation.
5. The only types of wires remaining are $w_h = a_i b_j$ or $w_h = z_{i,j} \oplus a_i b_j$. But by Step 2, both $i, j \in I$, and by Step 3, $z_{i,j}$ has been assigned, thus the value of w_h can be simulated perfectly.
6. Note that all c_i values for $i \in I$ can be simulated perfectly by the argument above. This completes the simulation and the argument of correctness.

SIMULATION FOR A GENERAL CIRCUIT. The simulation for a general transformed circuit C' proceeds very similarly to the above. First, examining each gadget g in C', we compute the set I. Note that since a total of t wires can be corrupted throughout the circuit C', the size of the set I will still be bounded by m. Next we perform the simulation as above, working our way from the inputs of C' to the outputs. Note that by the observation in Step 6 above, we maintain the invariant that for each gadget g, the shares of the inputs to g with indices belonging to I are perfectly simulated. Thus, inductively, the values of all corrupted wires in C' are simulated perfectly.

RE-RANDOMIZED OUTPUTS. We observe that as long as every output of C' has passed through one AND gadget (if this is not the case, we can artificially AND an output bit with itself), then for each original output bit, the encoded outputs are m-wise independent even

---

[7] We note that by changing the construction slightly, namely by computing $(a_i + r)b_j$ and rb_j where r is a fresh random value, we could have avoided increasing I by 2 indices rather than just 1 for any single wire observed by the adversary. This would have allowed us to choose $m = t$ rather than $m = 2t$ as we have chosen now.

given the entire encoding of the inputs. This can be used to prove that the construction is in fact secure against a stronger type of adversary who may observe at most t' wires in *each gadget*, where $t' = \Omega(t)$.[8]

IMPROVEMENT IN RANDOMNESS USE. We note, omitting the proof in this abstract, that in the above construction the same randomness (*i.e.*, the choices of $z_{i,j}$ for $i < j$) could be used in all the gadgets. This reduces the number of random bits to $O(t^2)$.

UNPROTECTED INPUTS AND OUTPUTS. We have described the construction above for protecting *all* inputs and outputs. It is easy to modify the construction so that certain inputs and outputs are unencoded, and may be observed by both the adversary and the simulator. This is useful in the stateful model, discussed next.

4   Perfect Privacy for Stateful Circuits

In this section we show how to achieve privacy in the stateful model, as defined in Section 2. This model is perhaps much more natural and realistic than the stateless model we considered in the previous section; however, as we show below, achieving privacy in this model is easy once privacy has been achieved in the stateless model.

Our goal is to transform a stateful circuit C into a t-private stateful circuit C' by using a privacy transformer for the stateless case. We now describe the construction. Recall that a stateless privacy transformer must encode the input in some way; we assume that the output is encoded using the same encoding. We also assume that the stateless transformer enjoys the *re-randomized outputs* property, namely that the output encoding for each original output bit is t-wise independent even given all encodings of input bits. Let us refer to the encoding of the stateless transformer as $E_t(x)$, where t is the privacy threshold of the stateless transformer, and x is the input being transformed. We represent each memory cell in C using the same representation. Relying on our stateless transformer as a building block, a stateful transformer $T = (T_C, T_s)$ can proceed as follows. The memory x of C is stored in C' in encoded form $E_{2t}(x)$.[9] C' will work by considering the transformed memory $E_{2t}(x)$ as an input to the original circuit C, which is transformed using the stateless $2t$-privacy transformation. We also modify C so that the next state of the memory is always an output. Then, these encoded outputs are fed back into memory for the next clock cycle. The regular inputs and outputs of C are unprotected, and need not be encoded. This completes the description of T_C.

A simulation argument proving the correctness of this transformer proceeds very similarly to the stateless case analyzed above. In fact, a sequence of invocations of a stateful circuit may be unwound into a larger stateless circuit with an equivalent functionality. Here, the initial state is viewed as a hidden input, and the final state as a hidden output. Thus, the security proof for the stateful case essentially reduces to that of the

---

[8] The ratio between t and t' depends on the maximal fan-out of C (which we fixed to 3 by default). This dependence can be eliminated by slightly modifying the construction.

[9] Note that the use of $2t$ as a threshold is critical, since the adversary could observe t bits of the inputs to the memory at the end of one clock cycle, and then another t bits of the outputs of the memory in the next clock cycle; in this way the adversary would observe $2t$ bits of the state of the memory.

stateless case. [10] In the "unwound" circuit, the adversary can corrupt up to t wires in each of the concatenated circuits Q produced by the stateless transformation. The simulation proof proceeds exactly as before; the additional corruptions do not obstruct the proof because of the re-randomization property: the outputs of one Q are t-wise independent conditioned on all the values of the inputs to Q; thus in order to provide a full joint simulation of the entire unwound circuit, we need only be able to recover a bounded set of inputs from each component Q. To summarize, we have shown:

Theorem 2. *There exists a perfectly t-private stateful circuit transformer which maps any stateful circuit C of size n and depth d to a randomized stateful circuit of size $O(nt^2)$ and depth $O(d \log t)$.*

5   Statistically Private Transformers

In this section we obtain statistically-private transformers which improve the previous constructions when the privacy threshold t is large. For the description and analysis of these transformers, it is convenient to rely on the following notion of *average-case security*.

Definition 1. *A circuit transformer $T = T(C, k)$ is said to be (statistically) p-private in the average case if $C' = T(C, k)$ is statistically private against an adversary which corrupts each wire in C' with independent probability p. That is, the joint distribution of the random set of corrupted wires and the values observed by the adversary can be simulated up to a $k^{-\omega(1)}$ statistical distance.*

We note that a p-adversary as above is roughly the same as an adversary that corrupts a uniformly random subset of $p|C'|$ wires in C'. Intuitively, average-case privacy in this sense should be easier to realize than the standard (worst-case) notion of privacy. Indeed, the circuit transformer from the previous section with k additive shares (i.e., $m = k$) is perfectly private with respect to any adversary corrupting $k/4$ wires in each gadget. It follows that the view of an adversary corrupting each wire with probability, say, $1/(10k)$ can be perfectly simulated except with negligible failure probability. Thus, we have:

Lemma 1. *There exists a circuit transformer $T(C, k)$ producing a circuit C' of size $O(k^2|C|)$, such that T is $\Omega(1/k)$-private in the average case.*

In contrast, achieving *worst-case* privacy against an adversary corrupting $\Omega(|C'|/k)$ of the wires in C' appears to be much harder; in particular, the constructions from the previous section are very far from achieving this when $|C'| \gg k$. The key idea underlying the asymptotic improvements in this section is the following *reduction* from worst-case privacy to average-case privacy.

We start with an efficient circuit transformer T guaranteeing p-privacy in the average case. We then transform its output $C' = T(C, k)$ into a larger circuit $\tilde{C}'$, which in a sense may be viewed as a "sparse" implementation of C'. The circuit $\tilde{C}'$ will carry out the same

---

[10] One technical difference between the two models is that the inputs and outputs in the stateful model are known to the adversary. However, these values are given to the simulator "for free" and can thus be easily incorporated into the simulation.

computation performed by C' in essentially the same way, but will effectively utilize only a small random subset of its wires; all remaining wires of $\tilde{C}'$ will be independent of the inputs and thus rendered useless to the adversary. We stress that the subset of useful wires in $\tilde{C}'$ will only be determined during the invocation of $\tilde{C}'$ and will therefore be independent of the set of corrupted wires. Hence, for an appropriate choice of parameters, the (worst-case) t-privacy of $\tilde{C}'$ will reduce to the average-case p-privacy of C'.

We will describe two distinct instantiations of the above approach. The first is somewhat simpler, but incurs an $\tilde{O}(t) \cdot k^{O(1)}$ multiplicative blowup to the circuit size (see Remark 1). When $t \gg k$, this already provides an asymptotic improvement over the previous solutions, which incur an $O(t^2)$ overhead. In the second construction, which is only sketched in this abstract, we manage to avoid the dependence on t by amortizing it over multiple gates.

Both instantiations make use of *sorting networks* as a building block. A sorting network is a layered circuit from ℓ integer-valued input wires to ℓ integer-valued output wires, which outputs its input sequence in a sorted order. The internal gates in a sorting network are of a very special type: each such gate, called a comparator, has two inputs and two outputs and returns its pair of inputs in a sorted order. The celebrated AKS network [1] achieves the optimal parameters of $O(\ell \log \ell)$ size and $O(\log \ell)$ depth. However, in terms of practical efficiency it is preferable to use simpler sorting networks, such as Batcher's [5], whose slightly inferior asymptotic complexity ($O(\ell \log^2 \ell)$ size and $O(\log^2 \ell)$ depth) hides much smaller constants.

A gate-by-gate approach. Our initial construction transforms the circuit $C' = T(C, k)$ to a circuit $\tilde{C}'$ as follows. With each wire i of C' there are ℓ wires of $\tilde{C}'$ labeled $(i, 1), \ldots, (i, \ell)$, where the parameter ℓ will be determined later. It is convenient to assume that these wires can carry ternary values from the set $\{0, 1, \$\}$. The execution of $\tilde{C}'$ will maintain the following invariant relative to an execution of C': if wire i of C' carries a value $v_i \in \{0, 1\}$, then the wires $(i, 1), \ldots, (i, \ell)$ will carry the value v_i in a random position (independently of other ℓ-tuples) and the value $\$$ in the remaining $\ell - 1$ positions. This property can be easily initialized at the inputs level by appropriately defining the input encoder of $\tilde{C}'$. Similarly, the output decoder of $\tilde{C}'$ can be easily obtained from that of C'.

It remains to describe how to emulate a gate of C' while maintaining the above invariant. Suppose that $v_i = v_{i_1} * v_{i_2}$, i.e., the value of wire i in C' is obtained by applying some commutative boolean operation '$*$' to the values of wires i_1, i_2. We replace this gate in C' with a 2ℓ-input, ℓ-output gadget in $\tilde{C}'$, which first routes the values v_{i_1}, v_{i_2} to two random but *adjacent* positions, and then combines them to form the output. One should be careful, however, to implement this computation so that even by observing intermediate values, the adversary will not be able to learn more values v_i than it is entitled to. Such an implementation for a gadget is given below.

PREPROCESSING. Let $r, r_1, \ldots, r_\ell$ be $\ell + 1$ uniformly random and independent integers from the range $[0, 2^k]$. For each $1 \leq j \leq \ell$, use the values $v_{i_1,j}, v_{i_2,j}$ (of wires (i_1, j) and (i_2, j)) to form a pair $(\text{key}_j, \text{val}_j)$ such that: (1) key_j is set to r_j if $v_{i_1,j} = v_{i_2,j} = \$$ and to r otherwise; (2) val_j is set to $\$$ if both $v_{i_1,j}, v_{i_2,j}$ are $\$$, to a bit value b if one of $v_{i_1,j}, v_{i_2,j}$ is b and the other is $\$$, and to $b_1 * b_2$ if $v_{i_1,j} = b_1$ and $v_{i_2,j} = b_2$.

SORTING. A sorting network is applied to the above ℓ-tuple of pairs using key as the sorting key. Let $(u_1, \ldots, u_\ell)$ denote the ℓ-tuple of symbols val_j sorted according to the keys key_j.

POSTPROCESSING. The jth output $v_{i,j}$ is obtained by looking at u_j, u_{j+1}, u_{j+2}: if $u_j, u_{j+1} \neq \$$ then $v_{i,j} = u_j * u_{j+1}$, if $u_j = u_{j+2} = \$$ and $u_{j+1} \neq \$$ then $v_{i,j} = u_{j+1}$, and otherwise $v_{i,j} = \$$.

Note that each such gadget can be implemented by a circuit of size $\tilde{O}(\ell k)$ and depth $\mathrm{poly}(\log \ell + \log k)$.

To complete the description of $\tilde{C}'$, we describe a (simpler) gadget replacing each random bit gate z in C'. As in the gate gadget, the random bit gadget has ℓ inputs and ℓ outputs. The jth input is a random bit z_j. A random selector $r \in [\ell]$ is used for determining which z_j will appear in the output. Specifically, the jth output is set to z_j if $r = j$ and to $\$$ otherwise. The cost of implementing this gadget is smaller than that of the gate gadget. Hence, the entire circuit $\tilde{C}'$ has size $\tilde{O}(\ell k n)$ and depth comparable to that of C' (up to polylog factors).

We now establish the relation between the worst-case privacy of $\tilde{C}'$ and the average-case privacy of C'.

Lemma 2. *Suppose that C' is p-private in the average case. Then the circuit $\tilde{C}'$, constructed with $\ell = O(t/p^4)$, is statistically t-private in the worst case.*

Proof sketch: It is convenient to make the adversary slightly stronger by assuming that it may actually probe t *logical*, rather than boolean, wires (i.e., each such wire may contain an integer, a ternary symbol, or a bit). For each compromised wire of $\tilde{C}'$, the adversary can either see some random integer r_i, a $\$$ symbol, or an actual value v_i of the ith wire of C'.[11] In the latter case, we say that v_i has been *observed*. Let S denote the set of indices i such that v_i has been observed. Note that S is a random variable, where the probability is over the execution of $\tilde{C}'$.

We will argue that for any fixed index set S_0, and for ℓ chosen as in the lemma, we have $\Pr[S_0 \subseteq S] \leq p^{|S_0|}$. Thus, an adversary attacking any fixed set of t wires in $\tilde{C}'$ is not better off than an adversary corrupting each wire of C' independently with probability p.

To make this argument, we pick a subset $S_1 \subseteq S_0$ such that: (1) $|S_1| \geq |S_0|/4$; (2) each value in S_1 is observed with probability (at most) p^4; and (3) the events of observing different values in S_1 are independent. This will make the probability of observing all values in S_1 at most $(p^4)^{|S_0|/4} = p^{|S_0|}$ as required.

We pick S_1 to be a maximal matching in the subgraph of C' induced by the wires in S_0. Since the degree of each vertex in this graph is at most 4, we have $|S_1| \geq |S_0|/4$. It remains to show that S_1 satisfies properties (2) and (3) above.

To prove (2) it suffices to show that for any *fixed* wire of $\tilde{C}'$, the probability that this wire contains a useful value (i.e., contributes to S) is $O(1/\ell)$. (Property (2) would then follow, since by taking the union over all t compromised wires of $\tilde{C}'$, the probability of

---

[11] In fact, depending on the exact implementation there may be wires of $\tilde{C}'$ containing information on two values v_i. We ignore this technicality as it does not change the analysis in any substantial way.

observing a value of C' is $O(t/\ell) \leq p^4$.) This clearly holds for input wires, by definition of the input encoder, and is maintained through all internal wires in the circuit by a symmetry argument. (In the case of gate gadgets, the argument relies on the fact that each val entry inside a sorting network contains one of the gadget's inputs, rather than some arbitrary combination of these inputs; due to the randomness of the sorting keys, the randomness of the positions of the useful entries is maintained).

It remains to argue that the independence property (3) holds. This follows from the fact that no two wires in S_1 are adjacent to a common gate in C' and from the fact that each gadget in $\tilde{C}'$ uses fresh randomness to shuffle its entries.   □

Combining Lemma 2 with Lemma 1, we have:

Theorem 3. *There exists a statistically t-private stateless transformer $(\tilde{T}, \tilde{I}, \tilde{O})$, such that $\tilde{T}(C, k)$ transforms a circuit C of size n to a circuit $\tilde{C}'$ of size $n \cdot \tilde{O}(t) \cdot k^{O(1)}$ (where k is a statistical security parameter). The depth of $\tilde{C}'$ is the same as that of C, up to polylog factors.*

Remark 1. Throughout this section, we view $k^{O(1)}$ and $\mathrm{polylog}(t)$ as being small in comparison to t, and therefore do not attempt to optimize the exact dependence on such factors. We note that all occurrences of $k^{O(1)}$ in the complexity of our constructions (e.g., in Theorem 3) can be replaced by $\mathrm{polylog}(k)$ while still satisfying our asymptotic notion of statistical security.

The above construction (and in particular the analysis of Lemma 2) crucially relies on the assumption that the adversary chooses in advance which t wires to corrupt, independently of the values it observes while invoking $\tilde{C}'$. However, for using this construction in the stateful case we need a somewhat stronger security guarantee. Indeed, since the adversary is allowed to move its t probes before each invocation based on the values it observes in previous invocations, it may gradually build more and more knowledge about the locations of useful values in $\tilde{C}'$. To get around this problem and guarantee sufficient independence between different invocations, it suffices to re-randomize each ℓ-tuple of wires representing the new content of a memory cell by applying a perfectly t-private computation of a random cyclic shift. Using our basic construction, this can be done using $\tilde{O}(\ell t^2)$ additional gates. When the size of the circuit is much larger than the number of states and t, the amortized cost per gate of this randomization step is small.

The above discussion is captured by the following theorem.

Theorem 4. *There exists a statistically t-private stateful transformer $\tilde{T}$, such that $\tilde{T}(C, k)$ maps a circuit C of size n with s memory cells to a circuit $\tilde{C}'$ of size $\tilde{O}(nt + st^3) \cdot k^{O(1)}$. The depth of $\tilde{C}'$ is the same as that of C, up to polylog factors.*

Amortizing the cost over multiple gates. The previous construction is redundant in the sense that it uses a separate gadget, of size $\Omega(t)$, for each gate in the circuit. We briefly sketch a modification of this construction which amortizes the additional cost over multiple gates, effectively eliminating the dependence on t. For the description and analysis of this construction, it is convenient to assume that the circuit is *layered* (see Section 2), and use the following modified notion of average-case security for the layered case.

Definition 2. *Let $T = T(C, k)$ be a* layered *circuit transformer producing a layered circuit C' of width w. Then, T is said to be (statistically) p-secure in the average case if $C' = T(C, k)$ is statistically secure against an adversary which corrupts a* random *subset of pw wires in each layer of C'.*

As before, we will use an average-case p-secure C' to build a worst-case t-secure $\tilde{C}'$. However, instead of representing each wire of C' by an ℓ-tuple of wires, we will now represent an entire *layer* of C' by a corresponding layer in $\tilde{C}'$ consisting of $\ell = \max(w, t/p)$ wires. These wires will contain a random permutation of the w values of C' in ℓ random positions and the symbol \$ in all other positions. Note that typically $w > t/p$, in which case there are no useless \$ entries in this list. However, the above choice of ℓ guarantees that by looking at any fixed set of t positions in the list the adversary will observe a random subset containing at most a p-fraction of the values.

Each value of C' is represented by a pair containing its index and its value. This representation naturally defines the input encoder and output decoder. It remains to show how the above representation can be maintained between subsequent layers. As before, we also need to ensure that each *intermediate* level in the computation of $\tilde{C}'$ contains a random permutation of the useful values, where the randomness of these permutations is independent for levels that are sufficiently far apart. To achieve this, we use an ℓ-input, ℓ-output gadget whose inputs represent the jth level wires in C' and whose outputs represent the $(j+1)$th level wires. The high-level idea is as before, except that we now need to *jointly* route $w/2$ pairs of wires to random adjacent positions, and then combine each pair in the right way.

Using this approach, we can obtain the following theorem:

Theorem 5. *There exists a statistically t-secure stateless transformer $(\tilde{T}, \tilde{I}, \tilde{O})$, such that $\tilde{T}(C, k)$ transforms a layered circuit C of width w and depth d to a circuit $\tilde{C}'$ of width $\tilde{O}(\max w, t) \cdot k^{O(1)}$ and depth $d \cdot \text{polylog}(w, t, k)$.*

An analogous theorem for the stateful model can be derived similarly to Theorem 4.

6 A PRG Secure against Probing Attacks

Next, we will show how to build a deterministic, stateful circuit that will produce pseudorandom output and remain secure even in the presence of probing attacks. In essence, we will be building a PRG that resists probing attacks. Because the resulting circuit is deterministic, this is helpful if true randomness is expensive.

The basic construction is as follows. Let $G : \{0, 1\}^\sigma \to \{0, 1\}^{(2t+1)\sigma + \lambda}$ be a PRG. We will build a deterministic stateful circuit $C'[s'_0]$ with $(2t+1)\sigma$ bits of internal memory, no inputs, and λ bits of output. $C'[s'_0]$ will be understood as a secure translation of the 0-input λ-output stateless randomized circuit C whose outputs are each fed by a different random-bit gate. The initial random seed s'_0 will be chosen uniformly at random, and the behavior of the circuit $C'[s'_0]$ on any one invocation is defined as follows:

1. Let $s = (s^1, \ldots, s^{2t+1})$ denote the current state of the memory cells.
2. Set $u := G(s^1) \oplus \cdots \oplus G(s^{2t+1})$. Define s', y by parsing u as $u = (s', y)$.
3. Replace the current state of the memory cells with s', and output y.

It is crucial that the circuit $C'[s_0']$ contain $2t+1$ disjoint copies of G, executing in parallel and sharing no wires or gates. Our construction is related to the method for distributed pseudorandomness generation with proactive security from [10]. For lack of space, the proof of Theorem 6 is omitted here.

Theorem 6. *If G is a secure PRG, then the stateful deterministic circuit $C'[s_0']$ defined above is a computationally t-private transformation of the circuit C defined above.*

Application: eliminating randomness gates. One application for our PRG construction is in eliminating randomness for the stateful circuit transformer of Section 4. Our basic solution for the stateless model, as described earlier, relies on the use of random bit gates within the transformed circuit $T(C)$. An appealing consequence of our probe-resistant PRG is that it allows to dispense with on-line randomness generation: an initial random seed can be coded into the initial state by T and (deterministically) "refreshed" at each invocation of the circuit.

Suppose our transformed circuit $T(C)$ uses λ random-bit gates. Let C_r be a stateless randomized circuit consisting of λ independent random-bit gates, each connected to a different output of C_r. If $C_r'[s_0']$ is any deterministic stateful circuit that is a secure translation of C_r, then we can replace the random-bit gates of $T(C)$ with the probe-resistant PRG $C_r'[s_0']$. For instance, the deterministic, stateful PRG of Theorem 6 will do the job nicely. In this way, we can derandomize C' and obtain an efficient stateful, deterministic circuit that is computationally t-private and not too much larger than the original circuit.

7 Concluding Remarks

We have developed theoretical foundations for the problem of securing hardware against side channel attacks, and initiated a systematic study of this problem within our framework. In this initial study we restricted our attention to side channels that can be modelled by *probing* attacks, i.e. whose information leakage depends on a limited number of physical wires. It would be interesting to extend our framework and results to a wider class of realistic attacks. A step in this direction is taken by Micali and Reyzin [28], who put forward a very general model for side channel attacks.

Another natural extension of the problem studied in this work is to allowing additional protection against *fault* attacks [7,26]. Similarly to our problem, solutions to this more general problem can be based on existing protocols from the MPC literature. However, even the most efficient of these (e.g., [22]) are still quite inefficient to implement on hardware. Obtaining better solutions in this setting, possibly under relaxed notions of security, remains an interesting challenge.

Acknowledgements. We wish to thank David Molnar and anonymous referees for helpful comments. We also thank an anonymous referee for suggesting the use of the brain metaphor. Amit Sahai acknowledges support from an Alfred P. Sloan Foundation Research Fellowship.

References

1. M. Ajtai , J. Komlos , E. Szemeredi. An $O(n \log n)$ sorting network. In *Proceedings of the 15th STOC*, pp. 1–9, 1983.
2. R. Anderson, M. Kuhn, "Tamper Resistance—A Cautionary Note," *USENIX E-Commerce Workshop*, USENIX Press, 1996, pp.1–11.
3. R. Anderson, M. Kuhn, "Soft Tempest: Hidden Data Transmission Using Electromagnetic Emanations," *Proc. 2nd Workshop on Information Hiding*, Springer, 1998.
4. B. Barak, O. Goldreich, R. Impagliazzo, S. Rudich, A. Sahai, S. Vadhan, and K. Yang. On the (im)possibility of obfuscating programs. CRYPTO 2001, 2001.
5. K. Batcher. Sorting Networks and their Applications. In *Proc. AFiPS Spring Joint Conference*, Vol. 32, 1988, pp. 307–314.
6. M. Ben-Or, S. Goldwasser, and A. Widgerson. Completeness theorems for non-cryptographic fault-tolerant distributed computation. In *Proc. of 20th STOC*, 1988.
7. D. Boneh, R.A. Demillo, R.J. Lipton, "On the Importance of Checking Cryptographic Protocols for Faults," *EUROCRYPT'97*, Springer-Verlag, 1997, pp.37–51.
8. R. Canetti. Security and composition of multiparty cryptographic protocols. In *J. of Cryptology*, 13(1), 2000.
9. R. Canetti, Y. Dodis, S. Halevi, E. Kushilevitz and A. Sahai. Exposure-Resilient Functions and All-or-Nothing Transforms. In *EUROCRYPT 2000*, pages 453–469.
10. R. Canetti and A. Herzberg. Maintaining Security in the Presence of Transient Faults. In *CRYPTO 1994*, pages 425–438.
11. S. Chari, C.S. Jutla, J.R. Rao, P. Rohatgi, "Towards Sound Approaches to Counteract Power-Analysis Attacks," *CRYPTO'99*, Springer-Verlag, 1999, pp.398–412.
12. D. Chaum, C. Crepeau, and I. Damgård. Multiparty unconditional secure protocols. In *Proc. of 20th STOC*, 1988.
13. R. Cramer, I. Damgård, and U. Maurer. General secure multi-party computation from any linear secret-sharing scheme. In *Proc. of EUROCRYPT '00*.
14. J.-S. Coron, L. Goubin, "On Boolean and Arithmetic Masking against Differential Power Analysis," *CHES'00*, Springer-Verlag, pp.231–237.
15. J. Daemen, V. Rijmen, "Resistance Against Implementation Attacks: A Comparative Study of the AES Proposals," *AES'99*, Mar. 1999.
16. K. Gandolfi, C. Mourtel, F. Olivier, "Electromagnetic Analysis: Concrete Results," *CHES'01*, LNCS 2162, Springer-Verlag, 2001.
17. R. Gennaro, M. O. Rabin, and T. Rabin. Simplified VSS and fast-track multiparty computations with applications to threshold cryptography. In *Proc. of 17th PODC*, 1998.
18. O. Goldreich, S. Goldwasser, and S. Micali. How to construct random functions. *JACM*, 33(4):792–807, October 1986.
19. O. Goldreich, S. Micali, and A. Wigderson. How to play any mental game (extended abstract). In *Proc. of 19th STOC*, 1987.
20. O. Goldreich and R. Ostrovsky. Software Protection and Simulation on Oblivious RAMs. *JACM* 43(3): 431–473, 1996.
21. L. Goubin, J. Patarin, "DES and Differential Power Analysis—The Duplication Method," *CHES'99*, Springer-Verlag, 1999, pp.158–172.
22. M. Hirt and U. Maurer. Robustness for free in unconditional multi-party computation. In *Proc. of CRYPTO '01*.
23. D. Kahn, *The Codebreakers*, The MacMillan Company, 1967.
24. J. Kelsey, B. Schneier, D. Wagner, "Side Channel Cryptanalysis of Product Ciphers," *ES-ORICS'98*, LNCS 1485, Springer-Verlag, 1998.

25. P. Kocher, "Timing Attacks on Implementations of Diffie-Hellman, RSA, DSS, and Other Systems," *CRYPTO'96*, Springer-Verlag, 1996, pp.104–113.
26. P. Kocher, J. Jaffe, B. Jun, "Differential Power Analysis," *CRYPTO'99*, Springer-Verlag, 1999, pp.388–397.
27. T.S. Messerges, "Securing the AES Finalists Against Power Analysis Attacks," *FSE'00*, Springer-Verlag, 2000.
28. S. Micali and L. Reyzin. A model for physically observable cryptography. Manuscript, 2003.
29. R. Ostrovsky and M. Yung. How to withstand mobile virus attacks. In *Proc. of 10th PODC*, 1991.
30. D. Page, "Theoretical Use of Cache Memory as a Cryptanalytic Side-Channel," Tech. report CSTR-02-003, Computer Science Dept., Univ. of Bristol, June 2002.
31. B. Pfitzmann, M. Schunter and M. Waidner, "Secure Reactive Systems", IBM Technical report RZ 3206 (93252), May 2000.
32. J.-J. Quisquater, D. Samyde, "Eddy current for Magnetic Analysis with Active Sensor," *Esmart 2002*, Sept. 2002.
33. J.-J. Quisquater, D. Samyde, "ElectroMagnetic Analysis (EMA): Measures and Counter-Measures for Smart Cards," *Esmart 2001*, LNCS 2140, Springer-Verlag, 2001.
34. J.R. Rao, P. Rohatgi, "EMpowering Side-Channel Attacks," IACR ePrint 2001/037.
35. US Air Force, *Air Force Systems Security Memorandum 7011—Emission Security Counter-measures Review*, May 1, 1998.
36. W. van Eck, "Electromagnetic Radiation from Video Display Units: An Eavesdropping Risk," *Computers & Security*, v.4, 1985, pp.269–286.
37. D. Wright, *Spycatcher*, Viking Penguin Inc., 1987.
38. A. C. Yao. How to generate and exchange secrets. In *Proc. of 27th FOCS*, 1986.

A   Relation with Secure Multi-party Computation

The problem studied in this paper is closely related to the problem of secure multi-party computation (MPC), introduced and first studied in [38,19,6,12] and extensively studied thereafter. We begin by explaining the relation between the problems, and then highlight some important differences.

THE MPC MODEL. In the most basic setting for secure MPC, n parties are connected by a complete network of point-to-point channels. Initially, each party holds a local input and an independent random input. The parties' goal is to evaluate some publicly known function f of their inputs while hiding their inputs from each other. To this end, they interact via a prescribed protocol. The protocol proceeds in round, where at each round each party may send a message to every other party based on its input, its random input, and messages received in previous rounds. The protocol terminates at some predetermined round, in which all parties should output the correct value of f. A protocol as above is said to be t-*private* if for any set T of at most t parties, the entire view of T (consisting of their inputs, random inputs, and received messages) reveals no more information about the other parties' inputs than what follows from their own inputs and the value of f. Note that the latter information captures what must *inevitably* be learned. To better correspond to our circuit model, it is convenient to consider a slightly modified MPC model in which each of the n inputs is initially secret-shared among the parties (say, using n out of n additive sharing), and the output produced by the protocol is also secret-shared in a similar fashion. This allows to realize a stronger and simpler privacy requirement: every collusion of t players learns nothing from their interaction with the remaining players.

RELATION TO PRIVATE CIRCUITS. To illustrate the relation between the MPC model and our circuit model, we focus on the stateless case and ignore some unimportant technicalities. First, we show that any t-private protocol corresponds to some t-private circuit computing the same function. Consider the following "hardware implementation" of an n-party protocol as above. In each round of interaction, each player's local computation is implemented by a separate sub-circuit. When the players interact, each message bit is translated into a wire connecting the corresponding sub-circuits. Note that the t-privacy of the protocol guarantees t-privacy also in the circuit model. Indeed, if an adversary can violate the circuit's privacy by probing t wires, then it could have also violated the protocol's privacy by corrupting some t players who "own" these wires.[12] The converse relation also holds. Suppose that we are given a t-private circuit computing the function f where the fan-in of each gate is at most 2. We use the circuit to define a protocol, in which each gate is owned by a distinct player. The circuit is evaluated by the players in a bottom-up fashion, starting with the encoded inputs and ending with an encoded output, where for each wire a message is sent from its source player to its destination players, and for each gate a local computation is performed by the corresponding player. It is not hard to see that if the circuit is $2t$-private, then the corresponding protocol is t-private. Indeed, a protocol-adversary corrupting t players learns content of at most $2t$ wires in the circuit.

In light of the above, one might expect to obtain the best solutions to our problem via efficient hardware implementations of state-of-the-art protocols from the MPC literature. However, this is not really the case. For instance, the BGW protocol [6] (as well as subsequent optimizations [17,13]) requires each player to evaluate a degree-t polynomial on $\Theta(t)$ points for each gate of the circuit being evaluated. Consequently, the stateless circuit transformer that can be derived from this protocol is significantly less efficient than our transformer. This state of affairs stems from some major differences in the underlying optimization goals. First, the MPC literature puts much emphasis on tolerating a constant fraction of corrupted players, whereas in our setting the number of corruptions is viewed as being independent of the number of "players" (in particular, we are willing to settle for tolerating a miniscule fraction of corruptions). Second, the MPC setting typically views the *communication complexity* and the *round complexity* as the most important resources to optimize, placing the time complexity only as a third-order optimization goal. In contrast, the main optimization criterion in our case is the *size* of a circuit, which roughly (but not exactly) corresponds to the time complexity of the underlying protocol. Finally, our main (stateful) model is quite nonstandard from the MPC point of view, as it involves extra ingredients such as a one-time trusted precomputation (via the circuit transformer), a mobile adversary (as in [29,10]), and on-line inputs and outputs (as in [31,8]). To conclude, the problem we are posing is quite different from that of implementing standard MPC protocols at the hardware level.

---

[12] Note that a wire corresponding to a message bit is owned by more than one player.

A Tweakable Enciphering Mode

Shai Halevi[1] and Phillip Rogaway[2]

[1] IBM T.J. Watson Research Center, Yorktown-Heights, NY 10598, USA
shaih@watson.ibm.com
[2] Dept. of Computer Science, University of California, Davis, CA 95616, USA, and
Dept. of Computer Science, Fac. of Science, Chiang Mai University, 50200, Thailand
rogaway@cs.ucdavis.edu, http://www.cs.ucdavis.edu/~rogaway

Abstract. We describe a block-cipher mode of operation, CMC, that turns an n-bit block cipher into a tweakable enciphering scheme that acts on strings of mn bits, where $m \geq 2$. When the underlying block cipher is secure in the sense of a strong pseudorandom permutation (PRP), our scheme is secure in the sense of tweakable, strong PRP. Such an object can be used to encipher the sectors of a disk, in-place, offering security as good as can be obtained in this setting. CMC makes a pass of CBC encryption, xors in a mask, and then makes a pass of CBC decryption; no universal hashing, nor any other non-trivial operation beyond the block-cipher calls, is employed. Besides proving the security of CMC we initiate a more general investigation of tweakable enciphering schemes, considering issues like the non-malleability of these objects.

1 Introduction

ENCIPHERING SCHEMES. Suppose you want to encrypt the contents of a disk, but the encryption is to be performed by a low-level device, such as a disk controller, that knows nothing of higher-level concepts like files and directories. The disk is partitioned into fixed-length sectors and the encrypting device is given one sector at a time, in arbitrary order, to encrypt or decrypt. The device needs to operate on sectors as they arrive, independently of the rest. Each ciphertext must have the same length as its plaintext, typically 512 bytes. When the plaintext disk sector P is put to the disk media at location T what is stored on the media should be a ciphertext $C = \mathcal{E}_K^T(P)$ that depends not only on the plaintext P and the key K, but also on the location T, which we call the *tweak*. Including the dependency on T allows that identical plaintext sectors stored at different places on the disk will have computationally unrelated ciphertexts.

The envisioned attack-model is a chosen plaintext/ciphertext attack: the adversary can learn the ciphertext C for any plaintext P and tweak T, and it can learn the plaintext P for any ciphertext C and tweak T. Informally, we want a *tweakable, strong, pseudorandom permutation* (PRP) that operates on a *wide blocksize* (like 512 bytes). We call such an object an *enciphering scheme*. We want to construct the enciphering scheme from a standard block cipher, such as AES, giving a *mode of operation*. The problem is one of current interest for

D. Boneh (Ed.): CRYPTO 2003, LNCS 2729, pp. 482–499, 2003.

standardization [11]. We seek an algorithm that is simple, and is efficient in both hardware and software.

NAOR-REINGOLD APPROACH. Naor and Reingold give an elegant approach for making a strong PRP on N bits from a block cipher on $n < N$ bits [18,17]. Their *hash–encipher–hash* paradigm involves applying to the input an *invertible blockwise-universal hash-function*, enciphering the result (say in ECB mode), and then applying yet another invertible blockwise-universal hash-function. Their work stops short of fully specifying a mode of operation, but in [17] they come closer, showing how to make the invertible blockwise-universal hash-function out of an xor-universal hash-function. So the problem, one might imagine, is simply to *instantiate* the approach [17], selecting an appropriate xor-universal hash function from the literature.

It turns out not to be so simple. Despite many attempts to construct a desirable hash function to use with the hash–encipher–hash approach, we could find no desirable realization. We wanted a hash function that was simple and more efficient, per byte, across hardware and software, than AES. The collision bound should be about 2^{-128} (degrading with the length of messages). Many techniques were explored, but nothing with the desired constellation of characteristics was ever found. We concluded that while making a wide-blocksize, strong PRP had "in principal" been reduced to a layer of block-cipher calls plus two "cheap" layers of universal hashing, the story, in practice, was that the "cheap" hashing layers would come to dominate the total cost in hardware, software, or both.

OUR CONTRIBUTIONS. Our main contribution is a simple, practical, completely-specified enciphering mode. CMC starts with a block cipher $E: \mathcal{K} \times \{0,1\}^n \to \{0,1\}^n$ and turns it into an enciphering scheme $\mathrm{CMC}[E]: \mathcal{K}' \times \mathcal{T} \times \mathcal{M} \to \mathcal{M}$ where $\mathcal{T} = \{0,1\}^n$ and $\mathcal{M}$ contains strings with any number (at least two) of n-bit blocks. See Figs. 1 and 2 for a preview. CMC stands for CBC–Mask–CBC.

CMC uses $2m + 1$ block-cipher calls. No "non-elementary" operations are used—in particular, no form of universal hashing is employed. The mode is highly symmetric: deciphering is the same as enciphering except that one uses the inverse block cipher E_K^{-1} in place of E_K. We prove that $\mathrm{CMC}[E]$ is secure, in the sense of a tweakable, strong PRP. This assumes that E itself is secure as a strong PRP. The actual results are quantitative, with the usual quadratic degradation in security.

Apart from the specific scheme, we investigate, more generally, the underlying goal. We show that being secure as a tweakable, strong, PRP implies the appropriate versions of indistinguishability [9,2] and non-malleability [8,3] under a chosen-ciphertext attack. Following Liskov, Rivest and Wagner [14], we show how tweaks can be cheaply added to the untweaked version of the primitive.

JOUX'S ATTACK. In an earlier, unpublished, manuscript we described a different version of CMC mode [19]. Although the algorithmic change between the old and new mode is small, its consequences are not: the old mode was *wrong*, as recently shown by Antoine Joux [12]. His simple and clever attack is described in Appendix A. In the same appendix we describe the bug in the proof that corresponds to the attack. This paper fixes the mode and its proof.

OTHER PRIOR WORK. Efforts to construct a block cipher with a large blocksize from one with a smaller blocksize go back to Luby and Rackoff [15], whose work can be viewed as building a $2n$-bit block cipher from an n-bit one. They also put forward the notion of a PRP and a strong ("super") PRP. The concrete-security treatment of PRPs begins with Bellare, Kilian, and Rogaway [4]. The notion of a tweakable block-cipher is due to Liskov, Rivest and Wagner [14]. Earlier work by Schroeppel describes a block cipher that was already designed to incorporate a tweak [20]. The first attempt to directly construct an nm-bit block cipher from an n-bit one is due to Zheng, Matsumoto and Imai [21], who give a Feistel-type construction. Bellare and Rogaway [5] give an enciphering mode that works on messages of varying lengths but is not a strong PRP. Another enciphering scheme that is potentially a strong PRP appears in unpublished work of Bleichenbacher and Desai [6]. Yet another suggestion we have seen [11] is forward-then-backwards PCBC mode [16]. The mode is easily broken in the sense of a strong PRP, but the possibility of a simple, two-layer, CBC-like mode helped to motivate us. A different approach for disk-sector encipherment is to build a wide-blocksize block cipher from scratch. Such attempts include BEAR, LION, and Mercy [1,7].

AFTERWARDS. Recent work by the authors has focused on providing a fully parallelizable enciphering scheme having serial efficiency comparable to that of CMC. We shall report on that work elsewhere. The full version of the current paper appears as [10].

2   Preliminaries

BASICS. A *message space* $\mathcal{M}$ is a set of strings $\mathcal{M} = \bigcup_{i \in I}\{0,1\}^i$ for some nonempty index set $I \subseteq \mathbb{N}$. A *length-preserving permutation* is a map $\pi: \mathcal{M} \to \mathcal{M}$ where $\mathcal{M}$ is a message space and π is a permutation and $|\pi(P)| = |P|$ for all $P \in \mathcal{M}$. A *tweakable enciphering scheme*, or simply an *enciphering scheme*, is a function $\mathcal{E}: \mathcal{K} \times \mathcal{T} \times \mathcal{M} \to \mathcal{M}$ where $\mathcal{K}$ (the key set) is a finite nonempty set and $\mathcal{T}$ (the tweak set) is a nonempty set and $\mathcal{M}$ is a message space and for every $K \in \mathcal{K}$ and $T \in \mathcal{T}$ we have that $\mathcal{E}(K, T, \cdot) = \mathcal{E}_K^T(\cdot)$ is a length-preserving permutation. An *untweakable enciphering scheme* is a function $\mathbb{E}: \mathcal{K} \times \mathcal{M} \to \mathcal{M}$ where $\mathcal{K}$ is a finite nonempty set and $\mathcal{M}$ is message space and $\mathbb{E}(K, \cdot) = \mathbb{E}_K(\cdot)$ is a length-preserving permutation for every $K \in \mathcal{K}$. A *block cipher* is a function $E: \mathcal{K} \times \{0,1\}^n \to \{0,1\}^n$ where $n \geq 1$ and $\mathcal{K}$ is a finite nonempty set and $E(K, \cdot) = E_K(\cdot)$ is a permutation for each $K \in \mathcal{K}$. The number n is the *blocksize*. An untweakable enciphering scheme can be regarded as a tweakable enciphering scheme with tweak set $\mathcal{T} = \{\varepsilon\}$ and a block cipher can be regarded as a tweakable enciphering scheme with tweak set $\mathcal{T} = \{\varepsilon\}$ and message space $\mathcal{M} = \{0,1\}^n$. The inverse of an enciphering scheme $\mathcal{E}$ is the enciphering scheme $\mathcal{D} = \mathcal{E}^{-1}$ where

$X = \mathcal{D}_K^T(Y)$ if and only if $\mathcal{E}_K^T(X) = Y$. An *adversary* A is a (possibly probabilistic) algorithm with access to some oracles. Oracles are written as superscripts. By convention, the running time of an algorithm includes its description size. We let $\mathrm{Time}_f(\mu)$ be a function that bounds the worst-case time to compute f on strings that total μ bits. We write $\tilde{O}(f)$ for $O(f(n)\lg(f(n)))$. Constants inside of O and $\tilde{O}$ notations are absolute constants, depending only on details of the model of computation. If X and Y are strings of possibly different lengths we let $X \overset{\leftarrow}{\oplus} Y$ be the string one gets by xoring the shorter string into the *beginning* of the longer string, leaving the rest of the longer string alone.

SECURITY NOTIONS. The definitions here are adapted from [4,14,15]. When $\mathcal{M}$ is a message space and $\mathcal{T}$ is a nonempty set we let $\mathrm{Perm}(\mathcal{M})$ denote the set of all functions $\pi\colon \mathcal{M} \to \mathcal{M}$ that are length-preserving permutations, and we let $\mathrm{Perm}^{\mathcal{T}}(\mathcal{M})$ denote the set of functions $\pi\colon \mathcal{T} \times \mathcal{M} \to \mathcal{M}$ for which $\pi(T, \cdot)$ is a length-preserving permutation for all $T \in \mathcal{T}$.

Let $\mathcal{E}\colon \mathcal{K} \times \mathcal{T} \times \mathcal{M} \to \mathcal{M}$ be an enciphering scheme and A be an adversary. We define the *advantage* of A in distinguishing $\mathcal{E}$ from a random, tweakable, length-preserving permutation and its inverse as

$$\mathbf{Adv}_{\mathcal{E}}^{\pm \widetilde{\mathrm{prp}}}(A) \overset{\mathrm{def}}{=} \Pr\left[K \overset{\$}{\leftarrow} \mathcal{K}\colon\ A^{\mathcal{E}_K(\cdot,\cdot)\,\mathcal{E}_K^{-1}(\cdot,\cdot)} \Rightarrow 1\right]$$
$$-\ \Pr\left[\pi \overset{\$}{\leftarrow} \mathrm{Perm}^{\mathcal{T}}(\mathcal{M})\colon\ A^{\pi(\cdot,\cdot)\,\pi^{-1}(\cdot,\cdot)} \Rightarrow 1\right]$$

The notation above shows, in the brackets, an experiment to the left of the colon and an event to the right of the colon. We are looking at the probability of the indicated event after performing the specified experiment. By $A \Rightarrow 1$ we mean the event that A outputs the bit 1. Often we omit writing the experiment, the oracle, or the placeholder-arguments of the oracle. The tilde above the "prp" serves as a reminder that the prp is tweakable, while the $\pm$ symbol in front of the "prp" serves as a reminder that this is the "strong" (i.e., chosen plaintext/ciphertext attack) notion of security. Thus we omit the tilde for untweakable enciphering schemes and block ciphers, and we omit the $\pm$ sign to mean that the adversary is given only the first oracle from each pair.

For each "advantage notion" $\mathbf{Adv}_{\Pi}^{\mathrm{xxx}}$ we write $\mathbf{Adv}_{\Pi}^{\mathrm{xxx}}(\mathcal{R})$ for the maximal value of $\mathbf{Adv}_{\Pi}^{\mathrm{xxx}}(A)$ over all adversaries A that use resources at most $\mathcal{R}$. Resources of interest are the running time t, the number of queries q, the total length of all queries μ (sometimes written as $\mu = n\sigma$ when μ is a multiple of some number n), and the length of the adversary's output ς. The name of an argument $(t, t', q,$ etc.) will be enough to make clear what resource it refers to.

POINTLESS QUERIES. There is no loss of generality in the definitions above to assume that regardless of responses that adversary A might receive from an arbitrary pair of oracles, it never repeats a query (T, P) to its left oracle, never repeats a query (T, C) to its right oracle, never asks its right oracle a query (T, C) if it earlier received a response of C to a query (T, P) from its left oracle, and never asks its left oracle a query (T, P) if it earlier received a response of P to a query (T, C) from its right oracle. We call such queries *pointless* because

Algorithm $\mathcal{E}_{K\widetilde{K}}^{T}(P_1 \cdots P_m)$	**Algorithm** $\mathcal{D}_{K\widetilde{K}}^{T}(C_1 \cdots C_m)$
100 $\mathbb{T} \leftarrow E_{\widetilde{K}}(T)$	200 $\mathbb{T} \leftarrow E_{\widetilde{K}}(T)$
101 $PPP_0 \leftarrow \mathbb{T}$	201 $CCC_0 \leftarrow \mathbb{T}$
102 **for** $i \leftarrow 1$ **to** m **do**	202 **for** $i \leftarrow 1$ **to** m **do**
103 $PP_i \leftarrow P_i \oplus PPP_{i-1}$	203 $CC_i \leftarrow C_i \oplus CCC_{i-1}$
104 $PPP_i \leftarrow E_K(PP_i)$	204 $CCC_i \leftarrow E_K^{-1}(CC_i)$
110 $M \leftarrow 2\,(PPP_1 \oplus PPP_m)$	210 $M \leftarrow 2\,(CCC_1 \oplus CCC_m)$
111 **for** $i \in [1 .. m]$ **do**	211 **for** $i \in [1 .. m]$ **do**
112 $CCC_i \leftarrow PPP_{m+1-i} \oplus M$	212 $PPP_i \leftarrow CCC_{m+1-i} \oplus M$
120 $CCC_0 \leftarrow 0^n$	220 $PPP_0 \leftarrow 0^n$
121 **for** $i \in [1 .. m]$ **do**	221 **for** $i \in [1 .. m]$ **do**
122 $CC_i \leftarrow E_K(CCC_i)$	222 $PP_i \leftarrow E_K^{-1}(PPP_i)$
123 $C_i \leftarrow CC_i \oplus CCC_{i-1}$	223 $P_i \leftarrow PP_i \oplus PPP_{i-1}$
130 $C_1 \leftarrow C_1 \oplus \mathbb{T}$	230 $P_1 \leftarrow P_1 \oplus \mathbb{T}$
131 **return** $C_1 \cdots C_m$	231 **return** $P_1 \cdots P_m$

Fig. 1. Enciphering (left) and deciphering (right) under $\mathcal{E} = \mathrm{CMC}[E]$, where $E\colon \mathcal{K} \times \{0,1\}^n \to \{0,1\}^n$ is a block cipher. The tweak is $T \in \{0,1\}^n$ and the plaintext is $P = P_1 \cdots P_m$ and the ciphertext is $C = C_1 \cdots C_m$.

the adversary "knows" the answer that it should receive. A query is called *valid* if it is well-formed and not pointless. A sequence of queries and their responses is valid if every query in the sequence is valid. We assume that adversaries ask only valid queries.

THE FINITE FIELD $GF(2^n)$. We may think of an n-bit string $L = L_{n-1} \ldots L_1 L_0 \in \{0,1\}^n$ in any of the following ways: as an abstract point in the finite field $GF(2^n)$; as the number in $[0..2^n - 1]$ whose n-bit binary representation is L; and as the polynomial $L(\mathsf{x}) = L_{n-1}\mathsf{x}^{n-1} + \cdots + L_1\mathsf{x} + L_0$. To add two points, $A \oplus B$, take their bitwise xor. To multiply two points we must fix an irreducible polynomial $P_n(\mathsf{x})$ having binary coefficients and degree n: say the lexicographically first polynomial among the irreducible degree-n polynomials having a minimum number of nonzero coefficients. For $n = 128$, the indicated polynomial is $P_{128}(\mathsf{x}) = \mathsf{x}^{128} + \mathsf{x}^7 + \mathsf{x}^2 + \mathsf{x} + 1$. Now multiply $A(\mathsf{x})$ and $B(\mathsf{x})$ by forming the degree $2n-2$ (or less) polynomial that is their product and taking the remainder when this polynomial is divided by $P_n(\mathsf{x})$.

Often there are simpler ways to multiply in $GF(2^n)$ than the definition above might seem to suggest. In particular, given L it is easy to "double" L. We illustrate the procedure for $n = 128$, in which case $2L = L \ll 1$ if $\mathsf{firstbit}(L) = 0$, and $2L = (L \ll 1) \oplus \mathrm{Const87}$ if $\mathsf{firstbit}(L) = 1$, where $\mathrm{Const87}$ is $0^{120}10000111$. Here $\mathsf{firstbit}(L)$ means L_{n-1} and $L \ll 1$ means $L_{n-2}L_{n-3} \cdots L_1 L_0 0$.

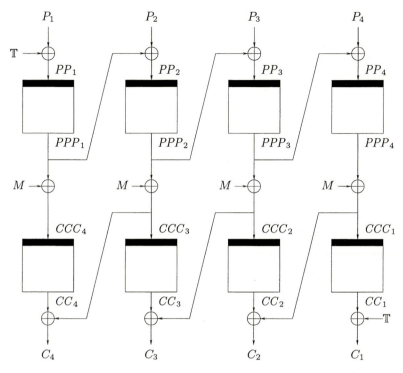

Fig. 2. Enciphering under CMC mode for a message of $m = 4$ blocks. The boxes represent E_K. We set mask $M = 2\,(PPP_1 \oplus PPP_m)$. This value can also be computed as $M = 2\,(CCC_1 \oplus CCC_m)$. We set $\mathbb{T} = E_{\widetilde{K}}(T)$ where T is the tweak.

3 Specification of CMC Mode

We construct from block cipher $E \colon \mathcal{K} \times \{0,1\}^n \to \{0,1\}^n$ a tweakable enciphering scheme that we denote by CMC-E or CMC[E]. The enciphering scheme has key space $\mathcal{K} \times \mathcal{K}$. It has tweak space $\mathcal{T} = \{0,1\}^n$. The message space $\mathcal{M} = \bigcup_{m \geq 2}\{0,1\}^{mn}$ contains any string having any number m of n-bit blocks, where $m \geq 2$. We specify in Fig. 1 both the forward direction of our construction, $\mathcal{E} = \text{CMC-}E$, and its inverse $\mathcal{D}$. An illustration of CMC mode is given in Fig. 2. In the figures, all capitalized variables except for K and $\widetilde{K}$ are n-bit strings (keys K and $\widetilde{K}$ are elements of $\mathcal{K}$). Variable names P, C, and M are meant to suggest *plaintext*, *ciphertext*, and *mask*. When we write $\mathcal{E}_K^T(P_1 \cdots P_m)$ we mean that the incoming plaintext $P = P_1 \cdots P_m$ is silently partitioned into n-bit strings $P_1, \ldots, P_m$ (and similarly when we write $\mathcal{D}_K^T(C_1 \cdots C_m)$). It is an error to provide $\mathcal{E}$ (or $\mathcal{D}$) with a plaintext (or ciphertext) that is not mn bits for some $m \geq 2$.

4 Discussion

BASIC OBSERVATIONS. Deciphering $C = \mathcal{E}^T_{K\widetilde{K}}(P)$ produces the same mask M as enciphering P because $CCC_1 \oplus CCC_m = (PPP_1 \oplus M) \oplus (PPP_m \oplus M) = PPP_1 \oplus PPP_m$. Also note that the multiply by two in computing M cannot be dispensed with; if it were, CCC_m would not depend on PPP_1 so the mode could not be a PRP.

THE CMC CORE. Consider the untweakable enciphering scheme $\mathbb{CMC}$ one gets by ignoring T and setting $\mathbb{T}$ to 0^n in Figs. 1 and 2. The CMC algorithm can then be viewed as taking $\mathbb{CMC}$ and "adding in" a tweak according to the construction $\mathrm{CMC}^T_{K\widetilde{K}}(P) = \mathbb{T} \oplus \mathbb{CMC}_K(P \oplus \mathbb{T})$ where $\mathbb{T} = E_{\widetilde{K}}(T)$. A similar approach to modifying an untweakable enciphering scheme to create a tweakable one was used by Liskov, Rivest, and Wagner [14, Theorem 2]. See Section 6.

SYMMETRY. Encryption under CMC is the same as decryption under CMC except that E_K is swapped with E_K^{-1} (apart from the computation of $\mathbb{T}$). Pictorially, this high degree of symmetry can be seen by observing that if the picture in Fig. 2 is rotated 180 degrees it is unchanged, apart from swapping letters P and C. Symmetry is a useful design heuristic in trying to achieve strong PRP security, as the goal itself provides the adversary with capabilities that are invariant with respect to replacing an enciphering scheme $\mathcal{E}$ by its inverse $\mathcal{D}$.

Notice that output blocks in CMC mode are taken in reverse order from the input blocks (meaning that $CCC_i = PPP_{m+1-i} \oplus M$ instead of $CCC_i = PPP_i \oplus M$). This was done for purposes of symmetry: if one had numbered output blocks in the "forward" direction then deciphering would be quite different from enciphering. As an added benefit, the reverse-numbering may improve the cache-interaction characteristics of CMC by improving locality of reference. That said, an application is always free to write its output according to whatever convention it wishes, and an application with limited memory may prefer to write its output as $C_m \cdots C_1$.

RE-ORIENTING THE BOTTOM LAYER. It is tempting to orient the second block-cipher layer in the opposite direction as the first, thinking that this improves symmetry. But if one were to use E_K^{-1} in the second layer then CMC would become an involution, and thus easily distinguishable from a random permutation.

LIMITATIONS. CMC has the following limitations: (1) The mode is not parallelizable. (2) The sector size must be a multiple of the blocksize. (3) In order to make due with $2m + 1$ block-cipher calls one needs $\Theta(nm)$ bits of extra memory. Alternatively, one can use $\Theta(n)$ bits of memory, but then one needs $3m + 1$ block-cipher calls and one should output the blocks in reverse order. (4) The key for CMC is longer than the key for the underlying block cipher; to keep things simple, we have done nothing to "collapse keys" for this mode. (5) Both directions of the block cipher are used to decipher, due to the one block-cipher call used for producing $\mathbb{T}$ from T.

All of the above limitations could potentially be addressed. Further limitations are inherent characteristics of the type of object that is being constructed. Namely: (a) a good PRP necessarily achieves less than semantic security: repetitions of plaintexts that share a tweak are manifest in the ciphertexts. (b) A PRP must process the entire plaintext before emitting the first bit of ciphertext (and it must process the entire ciphertext before emitting the first block of plaintext). Depending on the context, these limitations can be significant.

5   Security of CMC

The concrete security of the CMC is summarized in the following theorem. The theorem relates the advantage that an adversary has in attacking CMC-E to the advantage that an adversary can get in attacking the underlying block cipher E.

Theorem 1. [CMC security]   *Fix $n, t, q \geq 1$, $m \geq 2$, and a block cipher $E\colon \mathcal{K} \times \{0,1\}^n \to \{0,1\}^n$. Let message space $\mathcal{M} = \{0,1\}^{mn}$ and let $\sigma = mq$. Let CMC and $\mathbb{CMC}$ be the modes with the indicated message space. Then*

$$\mathbf{Adv}^{\pm\mathrm{prp}}_{\mathbb{CMC}[\mathrm{Perm}(n)]}(n\sigma) \leq \frac{5\sigma^2}{2^n} \tag{1}$$

$$\mathbf{Adv}^{\pm\widetilde{\mathrm{prp}}}_{\mathrm{CMC}[\mathrm{Perm}(n)]}(n\sigma) \leq \frac{7\sigma^2}{2^n} \tag{2}$$

$$\mathbf{Adv}^{\pm\widetilde{\mathrm{prp}}}_{\mathrm{CMC}[E]}(t, n\sigma) \leq \frac{7\sigma^2}{2^n} + 2\,\mathbf{Adv}^{\pm\mathrm{prp}}_E(t', 2\sigma) \tag{3}$$

where $t' = t + O(n\sigma)$. $\square$

Although we defined CMC and $\mathbb{CMC}$ to have message space $\bigcup_{m \geq 2}\{0,1\}^{mn}$ the theorem restricts messages to one particular length, mn bits for some m. In other words, proven security is for a fixed-input-length (FIL) cipher and not a variable-input-length (VIL) one. We believe that, in fact, security also holds in the sense of a VIL cipher, but we do not at this time provide a proof. All other results in this paper are done for arbitrary (VIL) message spaces.

The heart of Theorem 1 is Equation (1), which is sketched in Appendix C. Equation (2) follows immediately using Theorem 2, as given below. Equation (3) embodies the standard way to pass from the information-theoretic setting to the complexity-theoretic one.

Since the proof of Equation (1) is long (and only a small portion of it is included in this proceedings version), let us try to get across some basic intuition for it. Refer to Fig. 2 (but ignore the $\mathbb{T}$, as we are only considering $\mathbb{CMC}$). Suppose the adversary asks to encipher some new four-block plaintext P. Plaintext P must be different from all previous plaintexts, so it has some first block where it is different, say P_3. This will usually result in PP_3 being *new*—some value not formerly acted on by the block cipher π. This, in turn, will result in PPP_3 being nearly uniform, and this will propagate to the right, so that PPP_4 will be nearly uniform as well. The values PPP_1 and PPP_2 will usually have been different from each other, and they'll usually be different from the freshly chosen PPP_3 and PPP_4 values. Now $M = 2(PPP_1 \oplus PPP_4)$ and so M will be nearly uniform due to the presence of PPP_4. When we add M to the PPP_i values we will get a bunch of sums CCC_i that are almost always new and distinct. This in turn will cause the vector of CC_i-values to be uniform, which will cause C to be uniform. The argument for a decryption query is symmetric.

Though it is ultimately the above intuition that the proof formalizes, one must be careful, as the experience with the Joux-attack drives home [12]. One must be sure that an adversary cannot, by cutting and pasting parts of plaintexts and ciphertexts, force any nontrivial repetitions in intermediate values.

6 Transforming an Untweakable Enciphering Scheme to a Tweakable One

Let $E\colon \widetilde{\mathcal{K}} \times \{0,1\}^n \to \{0,1\}^n$ be a block cipher and let $\mathbb{E}\colon \mathcal{K} \times \mathcal{M} \to \mathcal{M}$ be an untweakable enciphering scheme where the message space $\mathcal{M}$ contains no string of length less than n bits. We construct a tweakable enciphering scheme $\mathcal{E} = \mathbb{E} \triangleleft E$ where $\mathcal{E}\colon (\mathcal{K}\times\widetilde{\mathcal{K}}) \times \{0,1\}^n \times \mathcal{M} \to \mathcal{M}$. The construction is $\mathcal{E}^T_{K\widetilde{K}}(M) = \mathbb{T} \overset{+}{\oplus} \mathbb{E}_K(M \overset{+}{\oplus} \mathbb{T})$ where $\mathbb{T} = E_{\widetilde{K}}(T)$. (Recall that $\overset{+}{\oplus}$ just means to xor in the shorter string at the beginning.) Notice that the cost of adding in the tweak is one block-cipher call and two n-bit xors, regardless of the length of the sector being enciphered or deciphered. Also notice that CMC $= \mathbb{CMC} \triangleleft E$.

The specified construction is similar to that of Liskov, Rivest and Wagner [14, Theorem 2] but, instead of a PRP E, those authors used an xor-universal hash function. One can view a secure block cipher as being "computationally" xor-universal, and try to conclude the security of the construction in that way. But we have also broadened the context to include enciphering schemes whose input is not a string of some fixed length, and so it seems better to prove the result from scratch. We show that $\mathcal{E} = \mathbb{E} \triangleleft E$ is secure (as a tweakable, strong, enciphering scheme) as long as $\mathbb{E}$ is secure (as an untweakable, strong enciphering scheme) and E is secure (as a PRP). The proof is given in the full version of this paper [10].

Theorem 2. [Adding in a tweak] *Let $E\colon \widetilde{\mathcal{K}} \times \{0,1\}^n \to \{0,1\}^n$ be a block cipher and let $\mathbb{E}\colon \mathcal{K} \times \mathcal{M} \to \mathcal{M}$ be an untweakable enciphering scheme whose message space $\mathcal{M}$ has a shortest string of $N \geq n$ bits. Then*

$$\mathbf{Adv}^{\pm\widetilde{\mathrm{prp}}}_{\mathbb{E}\triangleleft E}(t,q,\mu) \leq \frac{q^2}{2^n} + \frac{q^2}{2^N} + \mathbf{Adv}^{\pm\mathrm{prp}}_{\mathbb{E}}(t',q,\mu) + \mathbf{Adv}^{\mathrm{prp}}_{E}(t',q) \qquad (4)$$

where $t' = t + \widetilde{O}(\mu + q\mathrm{Time}_E + \mathrm{Time}_{\mathbb{E}}(\mu))$. □

7 Indistinguishability and Nonmalleability of Tweakable Enciphering Schemes

The definition we have given for the security of an enciphering scheme is simple and natural, but it is also quite far removed from any natural way to say that

Table 1. Disallowed queries. The dot refers to an arbitrary argument—all are disallowed.

When query	gets an answer of	then these queries are no longer allowed:	
$\mathcal{E}(T_0, P_0;\ T_1, P_1)$	C	$\mathcal{D}(T_0, C, \cdot, \cdot)$ $\mathcal{E}(T_0, P_0, \cdot, \cdot)$	$\mathcal{D}(\cdot, \cdot,\ T_1, C)$ $\mathcal{E}(\cdot, \cdot,\ T_1, P_1)$
$\mathcal{D}(T_0, C_0,\ T_1, C_1)$	P	$\mathcal{E}(T_0, P, \cdot, \cdot)$ $\mathcal{D}(T_0, C_0, \cdot, \cdot)$	$\mathcal{E}(\cdot, \cdot,\ T_1, P)$ $\mathcal{D}(\cdot, \cdot,\ T_1, C_1)$

an encryption scheme does what it should do. In this section we explore two notions of security that speak more directly about the privacy and integrity of an enciphering scheme. First we give a definition of *indistinguishability* and then we give a definition for the *nonmalleability*. We show that, as one would expect, security in the sense of a tweakable PRP implies both of these notions, and by tight reductions.

INDISTINGUISHABILITY. To define the indistinguishability of a tweakable enciphering scheme $\mathcal{E}\colon \mathcal{K} \times \mathcal{T} \times \mathcal{M} \to \mathcal{M}$ we adapt the left-or-right notion from [2]. We imagine the following game. At the onset of the game we select at random a key K from $\mathcal{K}$ and a bit b. The adversary is then given access to two oracles, $\mathcal{E} = \mathcal{E}_K^b$ and $\mathcal{D} = \mathcal{D}_K^b$. The attacker can query the $\mathcal{E}$-oracle with any 4-tuple $(T_0, P_0,\ T_1, P_1)$ where $T_0, T_1 \in \mathcal{T}$ and P_0 and P_1 are equal-length strings in $\mathcal{M}$. The oracle returns $\mathcal{E}_K(T_b, P_b)$. Alternatively, the adversary can query the $\mathcal{D}$ oracle with a 4-tuple $(T_0, C_0,\ T_1, C_1)$ where $T_0, T_1 \in \mathcal{T}$ and C_0 and C_1 are equal-length strings in $\mathcal{M}$. The oracle returns $\mathcal{D}_K(T_b, C_b)$ where $\mathcal{D}$ is the inverse of $\mathcal{E}$. The adversary wants to identify the bit b. We must disallow the adversary from asking queries that will allow it to win trivially. The disallowed queries are given in Table 1.

The advantage of the adversary in guessing the bit b is defined by

$$\mathbf{Adv}_{\mathcal{E}}^{\pm \widetilde{\mathrm{ind}}}(A) \stackrel{\text{def}}{=} \Pr[K \stackrel{\$}{\leftarrow} \mathcal{K}\colon\ A^{\mathcal{E}_K^1\ \mathcal{D}_K^1} \Rightarrow 1] \ -\ \Pr[K \stackrel{\$}{\leftarrow} \mathcal{K}\colon\ A^{\mathcal{E}_K^0\ \mathcal{D}_K^0} \Rightarrow 1]$$

We now show a tight equivalence between the PRP-security of a tweakable enciphering scheme and its indistinguishability. In Theorem 3 we show that PRP-security implies indistinguishability, and in Theorem 4 we show the converse. The proofs are in the full version of this paper [10].

Theorem 3. [$\pm\widetilde{\mathbf{prp}}$**-security** $\Rightarrow$ $\pm\widetilde{\mathbf{ind}}$**-security**] *Let* $\mathcal{E}\colon \mathcal{K} \times \mathcal{T} \times \mathcal{M} \to \mathcal{M}$ *be an enciphering scheme whose message space $\mathcal{M}$ consists of strings of length at least n bits. Then for any t, q, μ,*

$$\mathbf{Adv}_{\mathcal{E}}^{\pm \widetilde{\mathrm{ind}}}(t, q, 2\mu) \le 2\,\mathbf{Adv}_{\mathcal{E}}^{\pm \widetilde{\mathrm{prp}}}(t', q, \mu) + \frac{2q^2}{2^n - q}$$

where $t' = t + O(\mu)$. □

Theorem 4. [±ĩnd-security $\Rightarrow$ ±p̃rp-security] *Let* $\mathcal{E}\colon \mathcal{K} \times \mathcal{T} \times \mathcal{M} \to \mathcal{M}$ *be an enciphering scheme. Then for any* t, q, μ, *we have* $\mathbf{Adv}_{\mathcal{E}}^{\pm\widetilde{\mathrm{prp}}}(t, q, \mu) \leq \mathbf{Adv}_{\mathcal{E}}^{\pm\widetilde{\mathrm{ind}}}(t', q, 2\mu)$, *where* $t' = t + \widetilde{O}(\mu)$. □

NONMALLEABILITY. Nonmalleability is an important cryptographic goal that was first identified and investigated by Dolev, Dwork, and Naor [8]. Informally, an encryption scheme is nonmalleable if an adversary cannot modify a cipher-text C to create a ciphertext C^* where the plaintext P^* of C^* is related to the plaintext P of C. In this section we define the nonmalleability of a tweakable enciphering scheme with respect to a chosen-ciphertext attack and we show that ±p̃rp-security implies nonmalleability. The result mirrors the well-known result that indistinguishability of a probabilistic encryption scheme under a chosen-ciphertext attack implies its nonmalleability under the same kind of attack [8, 3].

Fix an enciphering scheme $\mathcal{E}\colon \mathcal{K} \times \mathcal{T} \times \mathcal{M} \to \mathcal{M}$ and an adversary A. Consider running A with two oracles: an enciphering oracle $\mathcal{E}_K(\cdot, \cdot)$ and a deciphering oracle $\mathcal{D}_K(\cdot, \cdot)$, where $\mathcal{D} = \mathcal{E}^{-1}$ and K is chosen randomly from $\mathcal{K}$. After A has made all of its oracle queries and halted, we define a number of sets:

– *Known plaintexts.* For every $T \in \mathcal{T}$ we define the $\mathcal{P}^T$ as the set of all P such that A asked $\mathcal{E}_K$ to encipher (T, P) or A asked $\mathcal{D}_K$ to decipher some (T, C) and A got back an answer of P. Thus $\mathcal{P}^T$ is the set of all plaintexts P associated to T that the adversary already "knows".

– *Known ciphertexts.* For every $T \in \mathcal{T}$ we define $\mathcal{C}^T$ as the set of all C such A asked $\mathcal{D}_K$ to decipher (T, C) or A asked $\mathcal{E}_K$ to encipher some (T, P) and A got back an answer of C. Thus $\mathcal{C}^T$ is the set of all ciphertexts C associated to T that the adversary already "knows".

– *Plausible plaintexts.* For every $T \in \mathcal{T}$ and $C \in \mathcal{M}$ we define $\mathcal{P}^T(C)$ as the singleton set $\{\mathcal{D}_K^T(C)\}$ if $C \in \mathcal{C}^T$ and as $\{0, 1\}^{|C|} \setminus \mathcal{P}^T$ otherwise. Thus $\mathcal{P}^T(C)$ is the set of all plaintexts P for which the adversary should regard it as plausible that $C = \mathcal{E}_K^T(P)$.

With enciphering scheme $\mathcal{E}\colon \mathcal{K} \times \mathcal{T} \times \mathcal{M} \to \mathcal{M}$ and adversary A still fixed, we consider the following two games, which we call games Real and Ideal. Both games begin by choosing a random key $K \stackrel{\$}{\leftarrow} \mathcal{K}$ and letting the adversary A interact with oracles $\mathcal{E}_K$ and $\mathcal{D}_K$ where $\mathcal{D} = \mathcal{E}^{-1}$. Just before termination, after the adversary has asked all the queries that it will ask, it outputs a three-tuple (T, C, f) where $T \in \mathcal{T}$ and $C \in \mathcal{M}$ and f is the encoding of a predicate $f\colon \mathcal{M} \to \{0, 1\}$ (we do not distinguish between the predicate and its encoding). Now for game Real we set $P \leftarrow \mathcal{D}_K^T(C)$ and for game Ideal we set $P \stackrel{\$}{\leftarrow} \mathcal{P}^T(C)$. Finally, we look at the event that $f(P) = 1$. Formally, we define the advantage of A, in the sense of nonmalleability under a chosen-ciphertext attack, as follows:

$$\mathbf{Adv}_{\mathcal{E}}^{\pm\widetilde{\mathrm{nm}}}(A) = \Pr[K \stackrel{\$}{\leftarrow} \mathcal{K}; (T, C, f) \stackrel{\$}{\leftarrow} A^{\mathcal{E}_K(\cdot,\cdot)\, \mathcal{D}_K(\cdot,\cdot)}; P \leftarrow \mathcal{D}_K^T(C) \colon f(P) = 1] -$$
$$\Pr[K \stackrel{\$}{\leftarrow} \mathcal{K}; (T, C, f) \stackrel{\$}{\leftarrow} A^{\mathcal{E}_K(\cdot,\cdot)\, \mathcal{D}_K(\cdot,\cdot)}; P \stackrel{\$}{\leftarrow} \mathcal{P}^T(C) \colon f(P) = 1]$$

We emphasize that in game Ideal (the second experiment) the set $\mathcal{P}^T(C)$ depends on the oracle queries asked by A and the answers returned to it (even though this is not reflected in the notation). For the resource-bounded version of $\mathbf{Adv}_{\mathcal{E}}^{\pm\widetilde{\mathrm{nm}}}$ we let the running time t include the running time to compute $f(P)$. We have the following result, the proof of which appears in the full version of this paper [10].

Theorem 5. [$\pm\widetilde{\mathrm{prp}}$-security $\Rightarrow$ $\pm\widetilde{\mathrm{nm}}$-security] *Let* $\mathcal{E}\colon \mathcal{K}\times\mathcal{T}\times\mathcal{M}\to\mathcal{M}$ *be an enciphering scheme. Then for any* t,q,μ,ς

$$\mathbf{Adv}_{\mathcal{E}}^{\pm\widetilde{\mathrm{nm}}}(t,q,\mu,\varsigma) \leq 2\,\mathbf{Adv}_{\mathcal{E}}^{\pm\widetilde{\mathrm{prp}}}(t',q+1,\mu+\varsigma)$$

$\square$

where $t' = t + \widetilde{O}(\mu+\varsigma)$.

Acknowledgments. The authors thank Jim Hughes for posing this problem to each of us and motivating our work on it. Shai thanks Hugo Krawczyk and Charanjit Jutla for discussions regarding candidates for implementing the Naor-Reingold approach and regarding the security proof for CMC. Phil thanks John Black for useful conversations on this problem, and Mihir Bellare, who promptly broke his first two-layer attempts. Phil received support from NSF grant CCR-0085961 and a gift from CISCO Systems. This work was carried out while Phil was at Chiang Mai University, Thailand.

References

1. R. Anderson and E. Biham. Two practical and provably secure block ciphers: BEAR and LION. In *Fast Software Encryption, Third International Workshop*, volume 1039 of *Lecture Notes in Computer Science*, pages 113–120, 1996. www.cs.technion.ac.il/~biham/.
2. M. Bellare, A. Desai, E. Jokipii, and P. Rogaway. A concrete security treatment of symmetric encryption: Analysis of the DES modes of operation. In *Proceedings of 38th Annual Symposium on Foundations of Computer Science (FOCS 97)*, 1997.
3. M. Bellare, A. Desai, D. Pointcheval, and P. Rogaway. Relations among notions of security for public-key encryption schemes. In H. Krawczyk, editor, *Advances in Cryptology – CRYPTO '98*, volume 1462 of *Lecture Notes in Computer Science*, pages 232–249. Springer-Verlag, 1998.
4. M. Bellare, J. Kilian, and P. Rogaway. The security of the cipher block chaining message authentication code. *Journal of Computer and System Sciences*, 61(3):362–399, 2000. www.cs.ucdavis.edu/~rogaway.
5. M. Bellare and P. Rogaway. On the construction of variable-input-length ciphers. In *Fast Software Encryption—6th International Workshop—FSE '99*, volume 1635 of *Lecture Notes in Computer Science*, pages 231–244. Springer-Verlag, 1999. www.cs.ucdavis.edu/~rogaway.
6. D. Bleichenbacher and A. Desai. A construction of a super-pseudorandom cipher. Manuscript, February 1999.
7. P. Crowley. Mercy: A fast large block cipher for disk sector encryption. In B. Schneier, editor, *Fast Software Encryption: 7th International Workshop*, volume 1978 of *Lecture Notes in Computer Science*, pages 49–63, New York, USA, Apr. 2000. Springer-Verlag. www.ciphergoth.org/crypto/mercy.

8. D. Dolev, C. Dwork, and M. Naor. Non-malleable cryptography. *SIAM Journal on Computing*, 30(2):391–437, 2000. Earlier version in STOC 91.

9. S. Goldwasser and S. Micali. Probabilistic encryption. *Journal of Computer and System Sciences*, 28:270–299, Apr. 1984.

10. S. Halevi and P. Rogaway. A tweakable enciphering mode. Manuscript, full version of this paper, May 2003. www.cs.ucdavis.edu/~rogaway.

11. J. Hughes. Chair of the IEEE Security in Storage Working Group. Working group homepage at www.siswg.org. Call for algorithms can be found at www.mail-archive.com/cryptography@wasabisystems.com/msg02102.html, May 2002.

12. A. Joux. Cryptanalysis of the EMD mode of operation. In *Advances in Cryptology – EUROCRYPT '03*, volume 2656 of *Lecture Notes in Computer Science*. Springer-Verlag, 2003.

13. J. Kilian and P. Rogaway. How to protect DES against exhaustive key search. *Journal of Cryptology*, 14(1):17–35, 2001. Earlier version in CRYPTO '96. www.cs.ucdavis.edu/~rogaway.

14. M. Liskov, R. Rivest, and D. Wagner. Tweakable block ciphers. In *Advances in Cryptology – CRYPTO '02*, Lecture Notes in Computer Science. Springer-Verlag, 2002. www.cs.berkeley.edu/~daw/.

15. M. Luby and C. Rackoff. How to construct pseudorandom permutations from pseudorandom functions. *SIAM J. of Computation*, 17(2), April 1988.

16. C. Meyer and S. Matyas. *Cryptography: A new dimension in computer security.* John Wiley and Sons, 1982.

17. M. Naor and O. Reingold. A pseudo-random encryption mode. Manuscript, available from www.wisdom.weizmann.ac.il/~naor/.

18. M. Naor and O. Reingold. On the construction of pseudo-random permutations: Luby-Rackoff revisited. *Journal of Cryptology*, 12(1):29–66, 1999. (Earlier version in STOC '97.) Available from www.wisdom.weizmann.ac.il/~naor/.

19. P. Rogaway. The EMD mode of operation (a tweaked, wide-blocksize, strong PRP). Cryptology ePrint Archive, Report 2002/148, Oct. 2002. Early (buggy) version of the CMC algorithm. http://eprint.iacr.org/.

20. R. Schroeppel. The hasty pudding cipher. AES candidate submitted to NIST. www.cs.arizona.edu/~rcs/hpc, 1999.

21. Y. Zheng, T. Matsumoto, and H. Imai. On the construction of block ciphers provably secure and not relying on any unproved hypotheses. In *Advances in Cryptology – CRYPTO '89*, volume 435 of *Lecture Notes in Computer Science*, pages 461–480. Springer-Verlag, 1989.

A   The Joux Attack

In an early, unpublished version of the current paper [19] the scheme CMC, then called EMD, worked a little bit differently: instead of computing $\mathbb{T} = \mathbb{E}_K(T)$ and xoring $\mathbb{T}$ into P_1 and C_1, we simply xored T into the mask M, setting $M = 2(PPP_1 \oplus PPP_m) \oplus T$. We claimed—incorrectly—that the scheme was secure (as a tweakable, strong PRP). Antoine Joux [12] noticed that the scheme was wrong, pointing out that it is easy to distinguish the mode and its inverse from a tweakable truly random permutation and its inverse. Below is (a slightly simplified variant of) his attack:

1. The adversary picks an arbitrary tweak T and an arbitrary 4-block plaintext $P_1P_2P_3P_4$. It encrypts $(T,\ P_1P_2P_3P_4)$, obtaining ciphertext $C_1C_2C_3C_4$, and it encrypts $(T+1,\ P_1P_2P_3P_4)$, obtaining a different ciphertext $C_1'C_2'C_3'C_4'$.
2. The adversary now decrypts $(T, C_1(C_2'+1)(C_3+1)C_4)$, obtaining plaintext $P_1''P_2''P_3''P_4''$.

If $P_1'' = P_1$ then the adversary outputs 1 (it guesses that it has a "real" enciphering oracle; otherwise, the adversary answers 0 (it knows that it has a "fake" enciphering oracle). It is easy to see that this attack has advantage of nearly 1.

What went wrong? Clearly the provided proof had a bug. The bug turns out not to be a particularly interesting one. On the 14-th page of the proof [19] begins a detailed case analysis. The case denoted X1–X5 was incorrect: two random variables are said to rarely collide, but with an appropriate choice of constants the random variables become degenerate (constants) and *always* collide. The same happens for case Y1–Y5. The current paper restructures the case analysis.

Our earlier manuscript [19] also mentioned a parallelizable mode that we called EME. Joux also provides an attack on EME, using the tweak in a manner similar to the attack on CMC. We later found that, as opposed to CMC, the EME scheme remains insecure even as an untweakable PRP. Thus one cannot repair EME simply by using a different method of incorporating the tweak.

B   A Useful Lemma — $\pm\widetilde{\mathbf{prp}}$-Security $\Leftrightarrow$ $\pm\widetilde{\mathbf{rnd}}$-Security

Before proving security for CMC, we provide a little lemma that says that a (tweakable) truly random permutation and its inverse looks very much like a pair of oracles that just return random bits (assuming you never ask pointless queries). Let $\mathcal{E}\colon \mathcal{K} \times \mathcal{T} \times \mathcal{M} \to \mathcal{M}$ be a tweaked block-cipher and let $\mathcal{D}$ be its inverse. The advantage of distinguishing $\mathcal{E}$ from random bits, $\mathbf{Adv}_{\mathcal{E}}^{\pm\widetilde{\mathrm{rnd}}}$, is

$$\mathbf{Adv}_{\mathcal{E}}^{\pm\widetilde{\mathrm{rnd}}}(A) = \Pr[K \xleftarrow{\$} \mathcal{K}:\ A^{\mathcal{E}_K(\cdot,\cdot)\,\mathcal{D}_K(\cdot,\cdot)} \Rightarrow 1] - \Pr[A^{\$(\cdot,\cdot)\,\$(\cdot,\cdot)} \Rightarrow 1]$$

where $\$(T, M)$ returns a random string of length $|M|$. We insist that A makes no pointless queries, regardless of oracle responses, and A asks no query (T, M) outside of $\mathcal{T} \times \mathcal{M}$. We extend the definition above in the usual way to its resource-bounded versions. We have the following:

Lemma 1. [$\pm\widetilde{\mathbf{prp}}$-security $\approx$ $\pm\widetilde{\mathbf{rnd}}$-security]   *Let* $\mathcal{E}\colon \mathcal{K} \times \mathcal{T} \times \mathcal{M} \to \mathcal{M}$ *be a tweaked block-cipher and let* $q \geq 1$. *Then* $|\mathbf{Adv}_{\mathcal{E}}^{\pm\widetilde{\mathrm{prp}}}(q) - \mathbf{Adv}_{\mathcal{E}}^{\pm\widetilde{\mathrm{rnd}}}(q)| \leq q(q-1)/2^{N+1}$, *where N is the length of a shortest string in* $\mathcal{M}$.

The proof, which is standard, appears in the full paper [10].

C   Sketch of Theorem 1 — Security of $\mathbb{CMC}$

Our proof of security for $\mathbb{CMC}$ is divided into two parts: (1) a game-substitution argument, reducing the analysis of $\mathbb{CMC}$ to the analysis of a simpler probabilistic game; and (2) analyzing that game. We also use Lemma 1 from above.

Initialization:

$bad \leftarrow$ false;   Domain $\leftarrow$ Range $\leftarrow \emptyset$;   **for** all $X \in \{0,1\}^n$ **do** $\pi(X) \leftarrow$ undef

Respond to the s-th adversary query as follows:

An encipher query, $\mathsf{Enc}(P_1^s \cdots P_m^s)$:

$u[s] \leftarrow$ the largest value in $[0 .. m]$ s.t. $P_1^s \cdots P_{u[s]}^s = P_1^r \cdots P_{u[s]}^r$ for some $r < s$

$PPP_0^s \leftarrow CCC_0^s \leftarrow 0^n$; **for** $i \leftarrow 1$ **to** $u[s]$ **do** $PP_i^s \leftarrow P_i^s \oplus PPP_{i-1}^s$, $PPP_i^s \leftarrow PPP_i^r$

for $i \leftarrow u[s] + 1$ **to** m **do**

$\quad PP_i^s \leftarrow P_i^s \oplus PPP_{i-1}^s$

$\quad PPP_i^s \xleftarrow{\$} \{0,1\}^n$;   **if** $PPP_i^s \in$ Range **then** $bad \leftarrow$ true, $\boxed{PPP_i^s \xleftarrow{\$} \overline{\text{Range}}}$

$\quad$ **if** $PP_i^s \in$ Domain **then** $bad \leftarrow$ true, $\boxed{PPP_i^s \leftarrow \pi(PP_i^s)}$

$\quad \pi(PP_i^s) \leftarrow PPP_i^s$, Domain $\leftarrow$ Domain $\cup \{PP_i^s\}$, Range $\leftarrow$ Range $\cup \{PPP_i^s\}$

$M^s \leftarrow 2\,(PPP_1^s \oplus PPP_m^s)$;   **for** $i \in [1 .. m]$ **do** $CCC_i^s \leftarrow PPP_{m+1-i}^s \oplus M^s$

for $i \leftarrow 1$ **to** m **do**

$\quad CC_i^s \xleftarrow{\$} \{0,1\}^n$;   **if** $CC_i^s \in$ Range **then** $bad \leftarrow$ true, $\boxed{CC_i^s \xleftarrow{\$} \overline{\text{Range}}}$

$\quad$ **if** $CCC_i^s \in$ Domain(π) **then** $bad \leftarrow$ true, $\boxed{CC_i^s \leftarrow \pi(CCC_i^s)}$

$\quad C_i^s \leftarrow CC_i^s \oplus CCC_{i-1}^s$

$\quad \pi(CCC_i^s) \leftarrow CC_i^s$, Domain $\leftarrow$ Domain $\cup \{CCC_i^s\}$, Range $\leftarrow$ Range $\cup \{CC_i^s\}$

return $C_1 \cdots C_m$

A decipher query, $\mathsf{Dec}(C_1^s \cdots C_m^s)$, is handled similarly

Fig. 3. Game CMC1 provides a perfect simulation of $\mathbb{CMC}[\mathrm{Perm}(n)]$. The boxed statements are events where we need to reset a previously chosen value.

The game-substitution sequence. Let n, m, and q all be fixed, and $\sigma = mq$. Let A be an adversary that asks q oracle queries (none pointless), each of nm bits. Our first major goal is to describe a probability space, NON2, this probability space depending on constants derived from A, and to define an event on the probability space, denoted NON2 sets bad, for which $\mathbf{Adv}_{\mathbb{CMC}[\mathrm{Perm}(n)]}^{\pm \widetilde{\mathrm{prp}}}(A) \leq 2 \cdot \Pr[\text{NON2 sets } bad] + \sigma^2/2^n$. Later we bound $\Pr[\text{NON2 sets } bad]$ and, putting that together with Lemma 1, we will get Equation (1) of Theorem 1. The rest of Theorem 1 follows easily, as explained in Section 5. Game NON2 is obtained by a game-substitution argument, as carried out in works like [13]. The goal is to simplify the rather complicated setting of A adaptively querying its oracles and to arrive at a simpler setting where there is no adversary and no interaction—just a program that flips coins and a flag bad that does or does not get set.

The various games. We describe the attack of A against $\mathbb{CMC}[\mathrm{Perm}(n)]$ as a probabilistic game in which the permutation π is chosen "on the fly", as needed to answer the queries of A. Initially, the partial function $\pi \colon \{0,1\}^n \to \{0,1\}^n$ is everywhere undefined. When we need $\pi(X)$ and π isn't yet defined at X we choose this value randomly among the available range values. When we need $\pi^{-1}(Y)$ and there is no X for which $\pi(X)$ has been set to Y we likewise choose X at random from the available domain values. As we fill in π its domain and its range thus grow. In the game we keep track of the domain and range of π by maintaining two sets, Domain and Range, that include all the points for which π is already defined. We let $\overline{\text{Domain}}$ and $\overline{\text{Range}}$ be the complement of these sets relative to $\{0,1\}^n$. The game, denoted CMC1, is shown in Fig. 3. Since game CMC1 accurately represent the attack scenario, we have that

$$\Pr[\,A^{\mathbb{E}_\pi \, \mathbb{D}_\pi} \Rightarrow 1\,] = \Pr[\,A^{\text{CMC1}} \Rightarrow 1\,] \tag{5}$$

The basic idea in the proof is that the bad events that we need to analyze are "accidental collisions", where a value that was supposed to be "new" happens to be equal to one of the values currently in the sets Domain and Range. The full proof therefore goes through a sequence of intermediate games—RND1, RND2, RND3, NON1, and NON2—that are designed to help us reason about the probability of these "accidental collisions" (and therefore the advantage an adversary can get in distinguishing CMC1 from a random permutation and its inverse).

Specifically, in game RND1 we omit all the boxed "resetting events" that immediately follow the setting of the flag *bad* (this is the usual trick under the game-substitution approach). In games RND2 and RND3 we just re-arrange the code without effecting the distribution of any of the variables in the game. In game NON1 we eliminate the interaction, essentially by letting the adversary specify not only the queries in the game, but also the answers to these queries (with some minor restrictions). Finally, in game NON2 we use the symmetry of CMC, arguing that it is sufficient to analyze only half of the "collision events" since the other half is completely symmetric. These games are designed so that:

- $\Pr[\,A^{\text{CMC1}} \Rightarrow 1\,] - \Pr[\,A^{\text{RND1}} \Rightarrow 1\,] \le \Pr[\,A^{\text{RND1}} \text{ sets } bad\,]$
- $\Pr[\,A^{\text{RND1}} \Rightarrow 1\,] = \Pr[\,A^{\text{RND2}} \Rightarrow 1\,] = \Pr[\,A^{\pm\widetilde{\text{rnd}}} \Rightarrow 1\,]$
- $\Pr[\,A^{\text{RND1}} \text{ sets } bad\,] = \Pr[\,A^{\text{RND2}} \text{ sets } bad\,] = \Pr[\,A^{\text{RND3}} \text{ sets } bad\,]$
  $\le \Pr[\,\text{NON1 sets } bad\,] + \frac{q(q-1)}{2^{n+1}} \le 2 \cdot \Pr[\,\text{NON2 sets } bad\,] + \frac{q(q-1)}{2^{n+1}}$

Combining these statements with Equation (5) and Lemma 1 we have reduced the problem of bounding the adversary's advantage to answering a question about game NON2. We now look at game NON2, which is shown in Fig. 4.

Game NON2 (the name suggests "noninteractive") depends on a fixed transcript $\tau = \langle \mathbf{ty}, \mathbf{P}, \mathbf{C} \rangle$ with $\mathbf{ty} = (\text{ty}^1, \cdots, \text{ty}^q)$, $\mathbf{P} = (\mathsf{P}^1, \cdots, \mathsf{P}^q)$, and $\mathbf{C} = (\mathsf{C}^1, \cdots, \mathsf{C}^q)$ where $\text{ty}^s \in \{\mathsf{Enc}, \mathsf{Dec}\}$ and $\mathsf{P}^s = \mathsf{P}_1^s \cdots \mathsf{P}_m^s$ and $\mathsf{C}^s = \mathsf{C}_1^s \cdots \mathsf{C}_m^s$ for $|\mathsf{P}_i^r| = |\mathsf{C}_i^r| = n$. This fixed transcript may not specify any "immediate collisions" or "pointless queries"; we call such a transcript *allowed*. Formally, saying that τ is allowed means that for all $r < s$ we have the following: if $\text{ty}^s = \mathsf{Enc}$ then (i) $\mathsf{P}^s \ne \mathsf{P}^r$ and (ii) $\mathsf{C}_1^s \ne \mathsf{C}_1^r$; while if $\text{ty}^s = \mathsf{Dec}$ then (i) $\mathsf{C}^s \ne \mathsf{C}^r$ and (ii) $\mathsf{P}_1^s \ne \mathsf{P}_1^r$. Now fix an allowed transcript τ that maximizes the probability of the flag *bad* being set. This one transcript τ is hardwired into game NON2.

ANALYSIS OF GAME NON2. It is helpful to view the multiset $\mathfrak{D}$ as a set of formal variables (rather than a multiset containing the values that these variables

$\mathfrak{D} \leftarrow \emptyset$ // Multiset
for $s \leftarrow 1$ **to** q **do**
  if $ty^s = \mathsf{Enc}$ **then**
    $u[s] \leftarrow$ largest value in $[0 \mathinner{..} m]$ s.t. $\mathsf{P}_1^s \cdots \mathsf{P}_{u[s]}^s = \mathsf{P}_1^r \cdots \mathsf{P}_{u[s]}^r$ for some $r < s$
    $PPP_0^s \leftarrow CCC_0^s \leftarrow 0^n$
    for $i \leftarrow 2$ **to** $u[s]$ **do** $PP_i^s \leftarrow \mathsf{P}_i^s \oplus PPP_{i-1}^s, \quad PPP_i^s \leftarrow PPP_i^r$
    for $i \leftarrow u[s] + 1$ **to** m **do**
      $PP_i^s \leftarrow \mathsf{P}_i^s \oplus PPP_{i-1}^s$; $\mathfrak{D} \leftarrow \mathfrak{D} \cup \{PP_i^s\}$
      $\boxed{PPP_i^s \xleftarrow{\$} \{0,1\}^n}$
    for $i \in [1 \mathinner{..} m]$ **do** $CCC_i^s \leftarrow PPP_{m+1-i}^s \oplus 2\,(PPP_1^s \oplus PPP_m^s)$
                    $\mathfrak{D} \leftarrow \mathfrak{D} \cup \{CCC_i^s\}$
  else $(ty^s = \mathsf{Dec})$
    $u[s] \leftarrow$ largest value in $[0 \mathinner{..} m]$ s.t. $\mathsf{C}_1^s \cdots \mathsf{C}_{u[s]}^s = \mathsf{C}_1^r \cdots \mathsf{C}_{u[s]}^r$ for some $r < s$
    $CCC_0^s \leftarrow PPP_0^s \leftarrow 0^n$
    for $i \leftarrow 1$ **to** $u[s]$ **do** $CC_i^s \leftarrow \mathsf{C}_i^s \oplus CCC_{i-1}^s, \quad CCC_i^s \leftarrow CCC_i^r$
    for $i \in [u[s]+1 \mathinner{..} m]$ **do** $\boxed{CCC_i^s \xleftarrow{\$} \{0,1\}^n}$; $\mathfrak{D} \leftarrow \mathfrak{D} \cup \{CCC_i^s\}$
    for $i \in [1 \mathinner{..} m]$ **do** $PPP_i^s \leftarrow CCC_{m+1-i}^s \oplus 2\,(CCC_1^s \oplus CCC_m^s)$
    for $i \in [1 \mathinner{..} m]$ **do** $PP_i^s \leftarrow \mathsf{P}_i^s \oplus PPP_{i-1}^s$; $\mathfrak{D} \leftarrow \mathfrak{D} \cup \{PP_i^s\}$
$bad \leftarrow$ (some value appears more than once in $\mathfrak{D}$)

Fig. 4. Game NON2. The boxed statements are the random choices in this game.

assume). Namely, whenever in game NON2 we set $\mathfrak{D} \leftarrow \mathfrak{D} \cup \{X\}$ for some variable X, we would think of it as setting $\mathfrak{D} \leftarrow \mathfrak{D} \cup \{\text{"}X\text{"}\}$ where "X" is the name of that formal variable. Viewed in this light, our goal now is to bound the probability that two formal variables in $\mathfrak{D}$ assume the same value in the execution of NON2. We observe that the formal variables in $\mathfrak{D}$ are uniquely determined by τ—they don't depend on the random choices made in the game NON2; specifically,

$$\mathfrak{D} = \{PP_i^s \mid \; ty^s = \mathsf{Dec}\} \; \cup \; \{PP_i^s \mid \; ty^s = \mathsf{Enc} \text{ and } i > u[s]\} \; \cup$$
$$\{CCC_i^s \mid \; ty^s = \mathsf{Enc}\} \; \cup \; \{CCC_i^s \mid \; ty^s = \mathsf{Dec} \text{ and } i > u[s]\}$$

We view the formal variables in $\mathfrak{D}$ as *ordered* according to when they are assigned a value in the execution of game NON2. This ordering too is fixed, depending only on the fixed transcript τ. The crucial claim is the following:

Claim. For any two distinct variables $X, X' \in \mathfrak{D}$ we have $\Pr[X = X'] \le 2^{-n}$ □

The proof, consisting of an exhaustive case analysis, can be found in the full version of this paper [10]. The idea is to look at the "free variables" that are directly chosen at random in game NON2 (the boxed statements in Fig. 4), and to show that the sum of any two variables in $\mathfrak{D}$ depends linearly on at least one free variable. We now show how this claim finishes the proof of Theorem 1. As there are no more than 2σ variables in $\mathfrak{D}$, we use the union bound to conclude

$\Pr[\text{NON2 sets } bad] \leq \binom{2\sigma}{2}/2^n$. Combining the results given so far we have that:

$$\mathbf{Adv}_{\mathrm{CMC[Perm}(n)]}^{\pm\widetilde{\mathrm{prp}}}(A) \leq \mathbf{Adv}_{\mathrm{CMC[Perm}(n)]}^{\pm\widetilde{\mathrm{rnd}}}(A) + q(q-1)/2^{2n+1}$$

$$\leq 2 \cdot \Pr[\text{NON2 sets } bad] + q(q-1)/2^{n+1} + q(q-1)/2^{2n+1}$$

$$\leq 2 \cdot \binom{2\sigma}{2}/2^n + q(q-1)/2^{n+1} + q(q-1)/2^{2n+1} \ \leq 5\sigma^2/2^n$$

This completes the proof, assuming the missing claim from above. ∎

A Message Authentication Code Based on Unimodular Matrix Groups

Matthew Cary[*][1] and Ramarathnam Venkatesan[2]

[1] University of Washington
cary@cs.washington.edu
[2] Microsoft Research
venkie@microsoft.com

Abstract. We present a new construction based on modular groups. A novel element of our construction is to embed each input into a sequence of matrices with determinant ± 1, the product of which yields the desired MAC. We analyze using the invertibility and the arithmetic properties of the determinants of certain types of matrices; this may be of interest in other applications. Performance results on our preliminary implementations show the speed of our MAC is competitive with recent fast MAC algorithms, achieving 0.5 Gigabytes per second on a 1.06 GHz Celeron.

Keywords: Message authentication, efficient MAC, hash functions.

1 Introduction

Algorithms to compute message authentication codes (MACs) are important in security applications, and the task of constructing them rigorously and efficiently has been a subject of many papers. An introduction may be found in [MvOV97].

MAC algorithms use a secret key K to select a function $H_K(X)$ and map an input X into a short binary string $h = H_K(X)$ of some fixed length. Then, h is encrypted using a block cipher. If the cipher acts as a random permutation, the encryptions of the hash values $h_1, ..., h_q$ of q distinct inputs $X_1, ..., X_q$ can not be distinguished from truly random outputs of the corresponding length, if the hash values $h_i = H_K(X_i)$ are distinct. Thus the collision properties of the hash function determines the security of the MAC. The main parameter of interest is the *collision probability* $\Pr_K[H_K(X) = H_K(X')]$ where X and X' are arbitrary and distinct inputs. If this probability is the inverse of the size of the range, the family of H_K is called *universal* [CW81]. This approach has enabled the construction of families (See [BHK+99,HK97,Rog99,BCK96,Sho96,JV98,Ber]) with quantifiable collision probabilities that are quite fast in practice. In this paper, we will focus on the initial mapping $X \mapsto h$ and its collision probability, and assume for simplicity that all inputs are ℓ-bit word sequences which are the same length, and can be subdivided into blocks of convenient length t evenly.

---

[*] Research done while at Microsoft Research. Partially supported by NSF grant CCR-0098066

D. Boneh (Ed.): CRYPTO 2003, LNCS 2729, pp. 500–512, 2003.

Previous Work

To motivate our construction, we recall some earlier ones. The evaluation MAC identifies an input message $X = x_1, \ldots x_t$ with a polynomial of degree t over a suitable field and computes the map $\alpha \mapsto \sum_i x_i \alpha^i$ for a random α. (See [Sho96] and [Ber] for speed-ups; the latter uses floating-point operations in a manner which may apply to our construction).

Many MAC constructions use a standard iterative rule $y_i = f_i(x_i + y_{i-1})$, where y_i are the *intermediate values* and various methods use different f_i's. In the evaluation MAC, $f_i(x) = f(x) = \alpha x$, the iteration is Horner's rule and y_m is the final value, while if one takes $f_i = f(x) = E_K(x)$ to be a block cipher, one gets the CBC MAC (see [BKR00] for an analysis). The chain and sum method in [JV98] doubles the length of the hash in a one-pass computation by outputting the pair $(y_t, \sum y_i)$. It alternates two *random* affine transformations $x \mapsto ax + b$, one for for odd i, and one for even i, each using an extra multiplier in the iteration $y_i = f(ex_i + y_{i-1})$. The invertibility of the operations allows one to combine MAC and encryption to obtain a pseudo-random permutation on X by further encrypting the intermediate values $y_1, \ldots y_{t-2}$ with an one-time pad derived from $(y_t, \sum y_i)$ using a stream cipher and encrypting $(y_t, \sum y_i)$ with a block cipher. We will also use the sum of intermediate values in our hash.

These methods work over a field where operations are typically expensive and using arithmetic modulo 2^ℓ is advantageous, as the fastest MACs indeed do. However, we lose invertibility (in the multiplicative sense) which is crucial for analysis. To wit, for $x \neq x'$, the function $f(x) = \alpha x + b$ over a field has a uniform *output differential* $f(x) - f(x') = \alpha(x - x')$ in the sense that it is uniformly distributed if α is randomly chosen. However, modulo 2^ℓ this changes sharply. If $2^{\ell-1} | (x - x')$ then $2^{\ell-1} | (y - y')$, and the output is distributed on a set of size 2 for a random odd α. Our goal is to construct reversible transformations that are suitable for MAC (and other applications) keeping the structure of the proof in the finite field case, except our equations involve coefficients from matrix groups.

UMAC [BHK$^+$99] uses the iteration $y_i = y_{i-1} + f(x_{2i}, x_{2i+1}) \mod 2^{2\ell}$ where $f(x_{2i}, x_{2i+1}) = (x_{2i} + k_{2i}) \cdot (x_{2i+1} + k_{2i+1}) \mod 2^{2\ell}$, k_i are random and all variables except y_i are ℓ-bits. This allows leveraging SIMD available on today's CPU's for media processing to hash more than a byte per cycle or gigabyte per second.

Recently Klimov and Shamir [KS] constructed an elegant family of invertible mappings (modulo 2^ℓ) that combine arithmetic and boolean operations to get non-linear maps for use in cryptographic primitives. These functions need to be randomized and modified to have suitable differential properties, in order to be used for our and similar applications.

Our Construction

PRELIMINARIES: Recall that our inputs are broken into blocks of length t words, each of size ℓ-bits. We use a sequence 2×2 matrices (A_i) with $\det(A_i) = \pm 1$

and fixed independent of the input x_i; the sequence may be periodic so that implementations can be unrolled with small code footprint. We define invertible functions $f_i(x)$ by multiplication with odd a_i, where a_i and x are ℓ bits, and the 2ℓ bit result is viewed as a vector of two ℓ bit numbers. The a_i form the secret key of the MAC and are assumed to be random. Thus $f_i(x)$ is invertible modulo $2^{2\ell}$ and can be implemented in one instruction using the usual 2ℓ-bit result of multiplication of two ℓ-bit quantities. One may be able to use floating point arithmetic as in [Ber]. All our matrix operations are over ring of integers modulo 2^ℓ.

THE ALGORITHM: We embed a given ℓ-bit input x_i into a 3×3 matrix B_i by $x_i \mapsto \left[\begin{smallmatrix} A_i & v_i \\ 0 & 0 & 1 \end{smallmatrix}\right] =: B_i$, where $v_i = f_i(x_i)$ is a vector with two elements. For each block of input, we compute the product $B = \left[\begin{smallmatrix} A & z \\ 0 & 0 & 1 \end{smallmatrix}\right]$ of these matrices B_i. The output of our hash value is the pair $(z, \sum_{i=1}^{t} v_i)$.

We show that the collision probability is nearly $2^{-2\ell}$, by using the invertibility of A_i and the arithmetic properties of the determinants of the matrices of the form $\prod_{i=j}^{k} A_i - I$ over $\mathbb{Z}$ (and not modulo 2^ℓ). We believe this new approach offers simplicity and can be helpful in other applications than MACs.

Our construction can be viewed in a more general setting. Let G be the group of integer matrices with determinant ± 1, $V = \mathbb{Z}_{2^\ell}^2$ be the additive group of 2-dimensional vectors modulo 2^ℓ and $G \ltimes V$ be their semi-direct product by the natural action of G on V. Then the above maps each x_i into $G \ltimes V$, by mapping x_i to $\big(A_i, f_i(x_i)\big)$ and takes the product of the images in $G \ltimes V$. This can be generalized into higher dimensions (see the section 6), and one can also view this hashing as a walk on the associated directed Cayley graphs.

The reader may have noticed that in the above and in UMAC the block keys are random independent sequences of words. UMAC expands a short key into a longer one by using a secure pseudo-random generator, and after the first block, needs only small amount of additional keys on a per block basis. Our algorithm is similar and a closer look security and performance at reducing this key generation may be desirable for some applications.

Our current version, implemented on an Intel Celeron, while not fully optimized, is competitive or better than known algorithms but slower than UMAC. However, we believe our novel construction is interesting in its own right, and may lend itself to other applications. Further refinement may improve the speed of our algorithm, and reduce the amount of key used. Implementations on imminent architectures such as AMD with a larger number of registers and different parallel constructs may also improve our speed.

Our algorithm suggests a *semi-universal* model for checksums for files, where MAC's may be an overkill. In section 4 we show that any two inputs that collide within a block, must differ in at least two locations. The collision probability of our MAC is much smaller if the input differs in at least three locations, which we may call 3-semi-universal hashing. Such variations and generalizations may lead to more efficient constructions. We omit the details in this version.

Section 2 describes some conventions we use in this paper. Section 3 describes the construction, which is analyzed in Section 4. We give experimental performance results in Section 5, and conclude with open problems in Section 6.

2   Conventions

Fix a modulus $m = 2^\ell$, for example, $\ell = 32$. A *word* will refer to an element of $\mathbb{Z}_m = \mathbb{Z}/m\mathbb{Z}$ and a *double word* to an element of $\mathbb{Z}_{m^2}$. Hence, words can be though of as ℓ bit integers, and double words as 2ℓ bit integers. All operations will take place over words, that is, over $\mathbb{Z}_m$, unless otherwise specified. We will take advantage of the ability of modern processors to multiply two words to produce a double word in a single instruction; this operation will be denoted $\times_*$. For $x, y \in \mathbb{Z}_m$, $x \times_* y$ will be in $\mathbb{Z}_m^2$, that is, we view the result as a two word vector.

We shall assume, if necessary by padding, the input to consist integral number of words. For simplicity we assume our input consists of b blocks each of which has a fixed *block length* of t words.

3   The Construction

Hashing One Block

We now describe construction for a map v that sends an input block $X = x_1, \ldots, x_t$ into ℓ-bit hash value $v = v(X)$. The block key consists of ℓ-bit words a_i, for $1 \le i \le t$; the same key is reused with each block. We define $f_i : \mathbb{Z}_m \to \mathbb{Z}_m^2$ by $f_i(x) = a_i \times_* x$. Our algorithm uses fixed public matrices $A_1, \ldots, A_t$. These will contain very small entries so that matrix products can be implemented very efficiently by addition and subtraction of words.

Let v_i be the column vector of two words equal to $f_i(x_i)$. Define matrices B_i, B and B_0, which have the form $\begin{bmatrix} * & * & * \\ * & * & * \\ 0 & 0 & 1 \end{bmatrix}$, where $B_0 = \begin{bmatrix} 1 & 0 & \\ 0 & 1 & z_0 \\ 0 & 0 & 1 \end{bmatrix}$, and for $i > 0$,

$$B_i := \begin{bmatrix} A_i & v_i \\ 0 & 0 & 1 \end{bmatrix}, \quad B := B_0 \cdot \prod_{i=1}^{t} B_i =: \begin{bmatrix} A & z \\ 0 & 0 & 1 \end{bmatrix} \qquad (1)$$

It is clear that B can be written as above; z is the first two components of the third column of B and A has determinant ± 1. z_0 is an initial value for the block. We also compute

$$\sigma = \sigma_0 + \sum_{i=1}^{t} v_i,$$

where σ_0 is another initial value for the block. The hash value is $v(X) = (z, \sigma)$.

Inter-block Chaining

The kth block is associated with two uniform hash functions $F_1^{(k)}$ and $F_2^{(k)}$ mapping double words to double words. We drop the superscript if the block number is clear from context. If (z', σ') is the output of a hashed block, we chain this to the next block by setting $\sigma_0 = F_2(\sigma')$ and

$$B_0 = \begin{bmatrix} 1 & 0 & \\ 0 & 1 & F_1(z') \\ 0 & 0 & 1 \end{bmatrix}$$

as the initial values for the next block. These inter-block functions may be repeated to save on key length, at some cost of security, which will be detailed in the analysis. The exact definition of these functions is not important for our applications.

Doubling the Hash Value Length

The hash value length can be doubled by performing an independent hash in parallel. We use key words b_i, $1 \le i \le t$ which is independent of the a_i, and set the functions g_i, $i \le t$ to $g(x) = b_i \times_* x$. We define $u_i = g_i(x_i)$, and as above get a map $X \mapsto u(X)$ with the hash value u using

$$C_i := \begin{bmatrix} A_i & u_i \\ 0 & 0 & 1 \end{bmatrix}, \quad C := C_0 \cdot \prod_{i=1}^{t} C_i =: \begin{bmatrix} A & w \\ 0 & 0 & 1 \end{bmatrix}, \quad C_0 := \begin{bmatrix} 1 & 0 & \\ 0 & 1 & u_0 \\ 0 & 0 & 1 \end{bmatrix}. \tag{2}$$

We also compute $\nu = \nu_0 + \sum_{i=1}^{t} u_i$. The overall hash will now be $\big(v(X), u(X)\big) = (z, \sigma, w, \nu)$.

Main Result

Theorem 1. *For $t \le 50$, if $H = (z, \sigma, w, \nu)$ and $H' = (z', \sigma', w', \nu')$ are the hash values computed from two distinct inputs, then*

$$\Pr[H = H'] \le 2^{-4\ell + 20},$$

where the probability is taken over the choice of key.

This theorem will follow directly from Lemmas 3 and 4. We note that the theorem is not optimal, in that the choice for the matrices of Lemma 4 could be improved.

4   Analysis

Collisions in a Block

We will first concentrate on the analysis of the hash of a single block, and assume that $B_0 = I$, the 3×3 identity matrix. By repeated use of the identity

$$\begin{bmatrix} A & v \\ 0\ 0 & 1 \end{bmatrix} \cdot \begin{bmatrix} B & u \\ 0\ 0 & 1 \end{bmatrix} = \begin{bmatrix} AB & Au + v \\ 0\ 0 & 1 \end{bmatrix}$$

in equation (1), we have that

$$z = v_1 + A_1 v_2 + A_1 A_2 v_3 + \cdots + A_1 A_2 \cdots A_{t-1} v_t. \tag{3}$$

For two (not necessarily distinct) input blocks, we write $X = x_1, \ldots, x_t$ and $X' = x'_1, \ldots, x'_t$ define $v'_i = f_i(x'_i)$, and define z' and σ' analogously.

We need the following technical lemma relating the distributive law of of $\times_*$ over vector subtraction. We remark that in general it is not true that $a \times_* x - a \times_* x' = a \times_* (x - x')$ and thus the operation is not linear. However, assuming $x \neq x'$, $a \times_* x - a \times_* x'$ is nearly as likely to collide with any fixed value as $a \times_* (x - x')$.

Lemma 1. *Given any fixed words $x \neq x'$ and any fixed double word $\alpha = (\alpha_1, \alpha_2)$,*

$$\Pr_a[a \times_* x - a \times_* x' = \alpha] \leq 2^{-\ell+2},$$

where the probability is taken over uniformly chosen odd words $a \in \mathbb{Z}_m$.

Proof. For this proof we let $\cdot$ denote the usual multiplication over double words. By abusing notation we write $a \cdot x = y$ for $a, x \in \mathbb{Z}_m$ and $y \in \mathbb{Z}_{m^2}$; we note also in this case there is no overflow, so that $y = ax$ as integers. The crux of this lemma is the difference between subtraction over double words as integers modulo m^2, and subtraction over two-dimensional vectors modulo m. To make this distinction explicit, for an element $x \in \mathbb{Z}_{m^2}$ we write $[x]$ as the vector corresponding x, so that $[x] \in \mathbb{Z}_m^2$. Then for double words y and z, if $[y] - [z] = (w_1, w_2)$, then $[y - z] = (w_1 - c, w_2)$, where c is either 0 and 1 depending on whether there is a carry between the low and high words or not.

Let A be the set of all odd a that cause a collision, that is, for the fixed $\alpha = (\alpha_1, \alpha_2)$, all a such that $[a \cdot x] - [a \cdot x'] = \alpha$ for x and x' as in the statement of the lemma. Then for any $a \in A$, $[a \cdot x - a \cdot x'] = (\alpha_1 - c_a, \alpha_2)$, for $c_a = 0$ or 1. Given $a, a' \in A$ with $c_a = c_{a'}$, we have $a \cdot (x - x') = a' \cdot (x - x')$ over the integers, so that as $x \neq x'$, $a = a'$. Thus A contains at most two elements, possibly one with carry 0 and possibly one with carry 1. As there are $2^{\ell-1}$ choices for odd a, the chance of choosing one in A is at most $2 \cdot 2^{-\ell+1} = 2^{-\ell+2}$, as required.   □

We now begin our analysis of the hash function proper.

Lemma 2. *If $(z, \sigma) = (z', \sigma')$ for distinct inputs X and X', then X and X' differ in at least two locations.*

Proof. Suppose not, so that $x_i = x'_i$ for all $i \neq j$, and $x_j \neq x'_j$ for some j. Then $\sigma - \sigma' = a_j \times_* x_j - a_j \times_* x'_j$. As a_j is odd and hence an invertible map from $\mathbb{Z}_m \to \mathbb{Z}_m^2$, $\sigma \neq \sigma'$, contradicting $(z, \sigma) = (z', \sigma')$.   □

We now know that colliding inputs have at least two distinct words, however we do not know which words these are. This is where computing the hash as a matrix product and sum helps us. For example, if x and y are independently distributed over $\mathbb{Z}_m$, then $2x + y$ and $2y - x$ are independently distributed as well. Note, however, that $x + y$ and $x - y$ are not independently distributed; for example, they have the same parity. The difference between these two examples is that the former arises from the matrix $\begin{bmatrix} 2 & 1 \\ -1 & 2 \end{bmatrix}$, which is invertible over $\mathbb{Z}_m$, while the the the matrix of the latter is $\begin{bmatrix} 1 & 1 \\ 1 & -1 \end{bmatrix}$ has determinant -2, and so is not invertible over $\mathbb{Z}_m$. The relationship between the two components of our hash pair, z and σ, is similar, so that if we pick our matrices carefully, z and σ will be independent.

Definition 1. *A sequence of matrices $A_1, \ldots, A_t$ is k-invertible if for any $i < j$, and Δ defined as*

$$\Delta = \det(A_i \cdots A_{j-1} - I),$$

if Δ is nonzero, and if $2^{k'} | \Delta$, then $k' \leq k$.

For any interval $\mathcal{I} = [i, j)$, the matrix $B = \prod_{\mathcal{I}} A_i - I$ formed from a k-invertible sequence of A_i is nearly invertible in the following sense. Let $\det(B) = s2^{k'}$ for odd, nonzero s and $k' \leq k$. Then $Bx = \alpha$ can be solved modulo $2^{\ell-k}$ uniquely and then there are 2^k solutions modulo 2^ℓ. Thus we need to ensure that the value k should be as small as possible.

Lemma 3. *Assume that the sequence $A_1, \ldots, A_t$ is k-invertible. Then for distinct inputs $X \neq X'$, $\Pr_{\{a_i\}}[(z, \sigma) = (z', \sigma')] \leq 2^{-2\ell+4+k}$, where $f_i(x) = a_i \times_* x$.*

Proof. Let $\delta x_i = x_i - x'_i$ and $\delta v_i = f(x_i) - f(x'_i) = a_i \times_* x_i - a_i \times_* x'_i$. By Lemma 2, we can assume that there exist $i < j$ such that $\delta x_i \neq 0$ and $\delta x_j \neq 0$. Our analysis now is in terms of matrix equations over $\mathbb{Z}_m$ involving A_i's and δv_i; the inputs x_i and x'_i are involved implicitly in a non-linear way which will by Lemma 1 will cost us a factor of 2. By fixing all a_r for $r \neq i, j$, we have that

$$\Pr_{a_i, a_j} [(z, \sigma) = (z', \sigma')] =$$
$$\Pr_{a_i, a_j} [A_1 \cdots A_{i-1} \delta v_i + A_1 \cdots A_{j-1} \delta v_j = \alpha, \delta v_i + \delta v_j = \beta], \qquad (4)$$

for appropriate fixed α and β. Rearranging (4), we have that for some fixed α', it is equivalent to

$$\Pr_{a_i, a_j} [(A_i \cdots A_{j-1} - I) \delta v_j = \alpha', \delta v_i + \delta v_j = \beta].$$

Let $B = (A_i \cdots A_{j-1} - I)$, and let $\Delta = \det B$. As the sequence $A_i, \ldots, A_{j-1}$ is k-invertible, $\Delta = s \cdot 2^{k'}$ for some odd s and $k' \leq k$. As remarked above, $B \delta v_j = \alpha'$ iff

$2^{k'} \delta v_j = \alpha^*$ in $\mathbb{Z}_m$, for some fixed α^* depending on α' and B. As from Lemma 1 $\Pr_{a_j}[\delta v_j = \gamma] \leq 2^{-\ell+2}$ for any fixed γ, $\Pr_{a_j}[2^{k'} \delta v_j = \alpha^*] \leq 2^{-\ell+2+k'} \leq 2^{-\ell+2+k}$ (recall all operations are performed over $\mathbb{Z}_m$). Finally, if the event $2^k \delta v_j = \alpha^*$ occurs, then $\Pr_{a_i}[\delta v_i + \delta v_j = \beta] \leq 2^{-\ell+2}$ (and can be possibly zero), as δv_i depends only on a_i, independently from v_j. Multiplying these probabilities gives the lemma. $\square$

Inter-block Chaining

We now consider the operation of the hash over several blocks. Let (z_k, σ_k) be the output of the kth block, so that the initial values for the $k+1$ block are $F_1^{(k)}(z_k)$ and $F_2^{(k)}(\sigma_k)$. If the keys for the pair $(F_1^{(k)}, F_2^{(k)})$ are new at each block, then the initial positions at each block are independent, using the uniformity of the F_i. Given two messages $X_1, \ldots, X_n$ and $X_1', \ldots, X_n'$, let i be the largest index of different blocks, so that $X_i \neq X_i'$ and $X_j = X_j'$ for $j > i$. Then $H(X_1, \ldots, X_n) = H(X_1', \ldots, X_n')$ iff $(z_i, \sigma_i) = (z_i', \sigma_i')$. If $H(X_1, \ldots, X_{i-1}) = H(X_1', \ldots, X_{i-1}')$, then the probability that $(z_i, \sigma_i) = (z_i', \sigma_i')$ is given in Lemma 3. Otherwise, by fixing all key bits but those for $F_r^{(i-1)}$, $r = 1, 2$, the probability that $(z_i, \sigma_i) = (z_i', \sigma_i')$ is equal to that of a collision in the $F_r^{(i-1)}$, which is smaller than that of Lemma 3. If we wish to save on key size, the $F_j^{(i)}$ can be reused. A standard union-bound shows that the bit-security of the hash decreases linearly with the frequency of reuse.

Choice of Matrices

The choice of the sequence $A_1, \ldots, A_t$ can be tailored to implementation requirements. Obviously there is a trade-off between finding k-invertible matrices for minimum k while ensuring that the matrix-vector products of the hashing algorithm can be efficiently computed. Our implementations in §5 use the families below. We caution that if the order of the matrices is changed, the determinants of interest may be identically zero!

Lemma 4. *Define the following integer matrices of determinant ± 1.*

$$A_1' = \begin{pmatrix} -1 & 1 \\ 1 & -2 \end{pmatrix}, \ A_2' = \begin{pmatrix} 2 & 1 \\ 1 & 1 \end{pmatrix}, \ and \ A_3' = \begin{pmatrix} 1 & 3 \\ 1 & 2 \end{pmatrix}.$$

We now extend this periodically into a longer sequence: $\mathcal{A}_t = (A_1, \ldots, A_t)$ where $A_{i+3s} = A_i'$. Then $\mathcal{A}_{19}$ is 4-invertible, and $\mathcal{A}_{50}$ is 6-invertible.

Proof. This can be verified by direct computation. A plot of the the the k-invertibility of $\mathcal{A}_{50}$ is shown in Figure 1. $\square$

It would be interesting to see if the noticeable structure in the plot can be exploited. We now present another family of matrices whose near-invertibility is not as good; however these have entries from $\{\pm 1, 0\}$ yielding more efficient

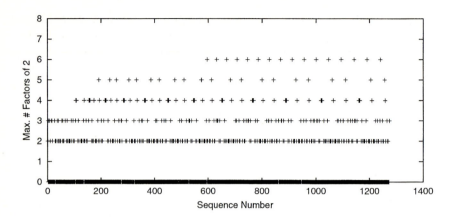

Fig. 1. The 6-invertibility of $\mathcal{A}_{50}$. The y-axis is the largest $k \geq 0$ such that $2^k | \det\left(\left(\prod_i^j A_s\right) - I\right)$, where the interval $\{i \ldots j\}$ is given by the sequence number. The determinant is nonzero in all cases.

implementations. Preliminary implementations suggest a 15% speed-up when using these simpler matrices. Of perhaps more interest, we can also show the determinants of interest are non-zero, if not nearly odd.

Lemma 5. *Define the following matrices.*

$$B_1' = \begin{pmatrix} 1 & 1 \\ 1 & 0 \end{pmatrix}, B_2' = \begin{pmatrix} -1 & -1 \\ 0 & -1 \end{pmatrix}, B_3' = \begin{pmatrix} 0 & 1 \\ 1 & 1 \end{pmatrix}, \text{ and } B_4' = \begin{pmatrix} -1 & 0 \\ -1 & -1 \end{pmatrix}.$$

Set $B_i = B'_{(i \bmod 4)+1}$ *and* $\mathcal{B}_t = (B_1, \ldots, B_t)$. *Then for any* $1 \leq i \leq j \leq t$, *if* $M = \prod_i^j B_s$, $\det(M - I) \neq 0$.

This is a necessary condition for k-invertibility, though clearly is insufficient in general. Experimentally, $\mathcal{B}_t$ is roughly $\log_{1.5} t$-invertible. For $t \sim 50$, they are not as invertible as $\mathcal{A}_{50}$, so we have not used them in our implementation. Figure 2 shows the growth of the k-invertibility of $\mathcal{B}_t$ as t is increased.

Proof. For a matrix A, we write $A \geq 0$ if each entry of A is at least 0. We write $A \leq 0$ if $-A \geq 0$, and $A \geq A'$ if $A - A' \geq 0$. We also write $|A|$ to denote the matrix whose entries are the absolute value of those of A.

In the notation of Lemma 5, note that

$$X_1 = B_1'B_2' = B_2'B_3' = \begin{pmatrix} -1 & -2 \\ -1 & -1 \end{pmatrix} \text{ and } X_2 = B_3'B_4' = B_4'B_1' = \begin{pmatrix} -1 & -1 \\ -2 & -1 \end{pmatrix}.$$

By examination we have for all $1 \leq s \leq 4$, $\det(B_s' - I) \in \{-1, 4\}$ and hence nonzero, and $\mathrm{Tr}(B_s') \in \{1, -1\}$ and at least 1 in absolute value. For $r = 1, 2$, $\det(X_r - I) = 2 \neq 0$ and $\mathrm{Tr}(X_r) = -2$. Finally, $\det(B_s'X_r - I) \in \{-4, -3, 6\}$.

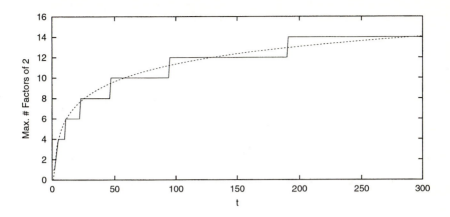

Fig. 2. The k-invertibility of $\mathcal{B}_t$ (solid line) plotted against $\log_{1.5} t$ (dashed line). Here the y-axis is the largest k such that $2^k | \det\left(\left(\prod_i^j B_s\right) - I\right)$, for all $1 \leq i \leq j \leq t$, for the specified t.

Hence we can proceed by induction, and assume $j - i > 2$. Set $M' = \prod_{s=i}^{j-2} B_s$ and fix r so that $M = M' X_r$, and by induction we can assume that $\left|\text{Tr}(M')\right| \geq 2$

Since $\det(M) = \pm 1$, $\det(M - I) = \det(M) + 1 - \text{Tr}(M)$, and $\det(M) + 1 = 0$ or 2, it will be enough to show that $|\text{Tr}(M)| > 2$. Note that $M \geq 0$ or $M \leq 0$, for $B_s = \pm 1 \cdot |B_s|$, so that $M = \pm 1 \cdot \prod_i^j |B_s|$, and $\prod |B_s| \geq 0$. As $M' \geq 0$ or $M' \leq 0$, using the same argument as for M, by examining X_r we see that $|M| \geq |M'|$.

One can label the off-diagonal elements of M' by x and y, so that

$$\text{Tr}(M) = \text{Tr}(M' X_r) = -\left(\left|\text{Tr}(M')\right| + 2|x| + |y|\right),$$

if necessary by exchanging x and y. In a similar way as showing $|M| \geq |M'|$, one can show $|M'| > 0$, so thus $\left|\text{Tr}(M)\right| \geq \left|\text{Tr}(M')\right| + 1 \geq 3$, using the inductive assumption on M'. Hence $\det(M - I) \neq 0$, as required.   □

5   Preliminary Implementation

Our hash design is mindful of operating constraints of modern processors. In particular it admits parallelization which is useful now that SIMD operations are standard on most computers. For example, the MMX instruction set standard on Intel Pentium II and later processors can operate simultaneously on 32-bit words with a throughput of 2 per cycle.

We mention some caveats. As written above, our test version is not optimized and availability of extra registers and parallelism in imminent machines must be taken in to account. The comparison numbers are to be taken as indicative

Algorithm	Security (Bits)	Peak Rate (cycles/byte)	Key Size (8 Kbyte Message)
Ours (two streams)	108	3.7	13.6 Kbits
Ours (one stream)	54	2.0	6.8 Kbits
UMAC	60	0.98	8 Kbits
SHA-1	80	12.6	512 bits

Data for other algorithms taken from [BHK$^+$99,BHK$^+$00].

Fig. 3. MAC Comparisons

of the parameters chosen in the test, and the results may depend on it. The comparison of key size generation was not closely looked at since this would mean re-coding and testing other algorithms with different parameters. The numbers presented on key sizes here may offer a different picture than what one would get if aggressive optimizations were made.

For brevity, we say that a hash or MAC has s bits of security if the collision probability (over the choice of keys) on two distinct fixed messages is $\leq 2^{-s}$. Using $\mathcal{A}_{50}$, by Lemma 3 each hash gives $2 \cdot 32 - 4 - 6 = 54$ bits of security, using 30 32-bit words of key per MAC per stream, plus the key for the inter block chaining. As two MACs are computed, the total security is 108 bits. Using MMX instructions on a 1.06 GHz Celeron, this MAC was computed at a peak rate of 3.7 cycles per byte. We have not implemented an optimized SSE2 algorithm to determine if the extended instruction set would benefit our algorithm. We also implemented the hash using a single stream, which gives 54 bits of security. This achieved a peak rate of 2.0 cycles per byte.

Our algorithm is also competitive with UMAC on the length of the generated key. To maintain the security bounds of Lemma 3, each inter-block hash needs four 32-bit words of key per hash stream. Each of our blocks then requires $50 \cdot 2$ 32-bit words of key. Thus, for an 8 Kbyte message, 42 inter-block hashes are required, for 5376 bits of key per hash stream. The total for an 8 Kbyte message and two hash streams is 13.6 Kbits of key. This compares with the current UMAC implementation [BHK$^+$00] which requires 8 Kbits of generated key to hash a message of any length to 60 bits of security.

We summarize this information with context from other algorithms in Figure 3.

6   Future Work

k-INVERTIBLE MATRICES: The proof k-invertibility of our matrix sequences is computational and analytical proofs are desirable. It is not necessary for such sequences to be periodic, as ours are. More complex families may improve the speed and the security of our hash. For example, we have found a periodic sequence of 4×4 matrices of length 80 which is 4-invertible. The larger matrices can be used to consume twice as much input per iteration, and the longer sequence length means the inter-block chaining is less frequent, improving efficiency. Preliminary implementations show this is 17% faster than the matrices of Lemma 4, and 2% faster than the matrices of Lemma 5, while providing more security than the other sequences.

OTHER HARDWARE PLATFORMS: Both our construction and UMAC benefit from the media processing instructions found on Pentium CPUs. Other platforms, such as those of AMD, or Intel's Itanium CPU, have different advantages, including larger register files. These details may change the relative performance between our MAC and UMAC, as well as motivate new directions in MAC design.

COMBINED MAC AND ENCRYPTION: Since our operations are invertible, we may try to combine authentication and encryption with stream ciphers. The idea is rather simple: use the final hash value to define a key for a stream cipher to generate a one-time pad. Instead of encrypting the input sequence x_i, one encrypts $y_i = a_i x_i + b_i$, where a_i and b_i are random key words (the first quantity is the lower half of a v_i in a step of our MAC). As before we further need to encrypt the hash value. One needs to exercise caution here: if the addition with b_i were omitted, one could still observe correlations. This would be the case if the inputs x_i end in many zeroes and RC4 is used. It would be interesting to see if such minor known or future correlations in RC4 [Gol97,Mir02,MT98] can be masked.

KEY SIZE REDUCTION: The inter-block chaining seems to contain some slack in the use of key. Almost twice as much key is used in inter-block hashing as is used for the blocks. Key reuse techniques such as a Toplitz shift (see [BHK+99]) may be able to address this problem. The use of a single pairwise independent hash may be sufficient, but our proof of that is incomplete.

References

[ALW01] Noga Alon, Alexander Lubotzky, and Avi Wigderson. Semi-direct product in groups and Zig-zag product in graphs: Connections and applications. In *FOCS 2001*, pages 630–637. IEEE, 2001.

[BCK96] Mihir Bellare, Ran Canetti, and Hugo Krawczyk. Keying hash functions for message authentication. *Lecture Notes in Computer Science*, 1109, 1996.

[Ber] D. Bernstein. Floating-point arithmetic and message authentication. draft available as http://cr.yp.to/papers/hash127.dvi.

[BHK+99] J. Black, S. Halevi, H. Krawczyk, T. Krovetz, and P. Rogaway. UMAC: Fast and secure message authentication. *Lecture Notes in Computer Science*, 1666:216–233, 1999.

[BHK+00] J. Black, S. Halevi, H. Krawczyk, T. Krovetz, and P. Rogaway. UMAC home page, 2000. URL: http://www.cs.ucdavis.edu/~rogaway/umac.

[BKR00] Mihir Bellare, Joe Kilian, and Phillip Rogaway. The security of the cipher block chaining message authentication code. *Journal of Computer and System Sciences*, 61(3):362–399, 2000.

[CW81] Carter and Wegman. New hash functions and their use in authentication and set equality. *Journal of Computer and System Sciences*, 22(3):265–279, 1981.

[Gol97]   J. Golic. Linear statistical weaknesses in alleged RC4 keystream generator. In *Advances in Cryptology — EUROCRYPT '97*, volume 1233 of *Lecture Notes in Computer Science*, pages 226–238. Springer-Verlag, 1997.

[HK97]   Shai Halevi and Hugo Krawczyk. MMH: Software message authentication in the Gbit/second rates. In *Fast Software Encryption*, pages 172–189, 1997.

[JV98]   Mariusz H. Jakubowski and Ramarathnam Venkatesan. The chain and sum primitive and its applications to MACs and stream ciphers. In *Advances in Cryptology — EUROCRYPT '98*, volume 1403 of *Lecture Notes in Computer Science*, pages 281–293. Springer-Verlag, 1998.

[KS]   Alexander Klimov and Adi Shamir. A new class of invertible mappings. Crypto 2001 Rump Session.

[Mir02]   Ilya Mironov. Not so random shuffles of RC4. In *Advances in Cryptology — CRYPTO 2002*, Lecture Notes in Computer Science. Springer-Verlag, 2002.

[MvOV97]   Alfred J. Menezes, Paul C. van Oorschot, and Scott A. Vanstone. *Handbook of Applied Cryptography.* CRC Press, 1997.

[MT98]   Serge Mister and Stafford E. Tavares. Cryptanalysis of RC4-like Ciphers In *Selected Areas in Cryptography*, 131–143, 1998.

[Rog99]   Phillip Rogaway. Bucket hashing and its application to fast message authentication. *Journal of Cryptology: the Journal of the International Association for Cryptologic Research*, 12(2):91–115, 1999.

[Sho96]   Victor Shoup. On fast and provably secure message authentication based on universal hashing. *Lecture Notes in Computer Science*, 1109, 1996.

Luby-Rackoff: 7 Rounds Are Enough for $2^{n(1-\varepsilon)}$ Security

Jacques Patarin

University of Versailles

Abstract. In [3] M. Luby and C. Rackoff have proved that 3-round random Feistel schemes are secure against all adaptive chosen plaintext attacks when the number of queries is $m \ll 2^{n/2}$. Moreover, 4-round random Feistel schemes are also secure against all adaptive chosen plaintext and chosen ciphertext attacks when $m \ll 2^{n/2}$. It was shown later that these bounds are tight for 3 and 4 rounds (see [9] or [1]).
In this paper our main results are that for every $\varepsilon > 0$, when $m \ll 2^{n(1-\varepsilon)}$:

- for 4 rounds or more, a random Feistel scheme is secure against known plaintext attacks (KPA).
- for 7 rounds or more it is secure against all adaptive chosen plaintext attacks (CPA).
- for 10 rounds or more it is secure against all adaptive chosen plaintext and chosen ciphertext attacks (CPCA).

These results achieve the optimal value of m, since it is always possible to distinguish a random Feistel cipher from a truly random permutation with $\mathcal{O}(2^n)$ queries, given sufficient computing power.
This paper solves an open problem of [1,9] and [17]. It significantly improves the results of [13] that proves the security against only $2^{\frac{3n}{4}}$ queries for 6 rounds, and the results of [6] in which the $2^{n(1-\varepsilon)}$ security is only obtained when the number of rounds tends to infinity. The proof technique used in this paper is also of independent interest and can be applied to other schemes.

An extended version of this paper is available from the author.

1 Introduction

In this paper we study the security proofs for random Feistel ciphers with k rounds, $k \in \mathbb{N}$, which is also known as "Luby-Rackoff construction with k rounds" or simply "L-R construction with k rounds" (see Section 2 for precise definitions). By definition a random Feistel cipher with k rounds, is a Feistel cipher in which the round functions $f_1, \ldots, f_k$ are independently chosen as truly random functions.

In their famous paper [3], M. Luby and C. Rackoff have shown that in an adaptive plaintext attack (CPA) with m queries to the encryption oracle, the probability to distinguish the 3-round L-R construction from a truly random

D. Boneh (Ed.): CRYPTO 2003, LNCS 2729, pp. 513–529, 2003.

permutation of $2n$ bits $\rightarrow 2n$ bits, is always $\leq m^2/2^n$. Therefore 3-round L-R constructions are secure against all chosen plaintext attacks when m is very small compared with $2^{n/2}$ (i.e. $m \ll 2^{n/2}$).

Moreover, in all adaptive chosen plaintext and chosen ciphertext attack (CPCA), the probability to distinguish the 4-round L-R construction from a truly random permutation of $2n$ bits $\rightarrow 2n$ bits, is also $\leq m^2/2^n$ (This result was mentioned in [3] and a proof published in [10]). Therefore 4-round L-R constructions are secure against CPCA when $m \ll 2^{n/2}$.
These results are valid if the adversary has unbounded computing power as long as he does only m queries.

These results, as well the results of the present paper, can be applied in two different ways:

1. Directly, using k truly random functions $f_1, \ldots, f_k$ (that requires significant storage). Then we obtain an unconditionally secure cipher, that is secure even against adversaries that are not limited in their computing power, however they have to be limited in the number of known (plaintext, ciphertext) pairs.
2. In a hybrid setting, in which instead of using k truly random functions $f_1, \ldots, f_k$, we use k pseudo-random functions. If no adversary with limited computing power can distinguish these functions from truly random functions by any existing test, a fortiori he cannot achieve worse security for the hybrid cipher, than for the ideal version with truly random functions, and all the security results will hold.

The L-R construction inspired a considerable amount of research, see [7] for a summary of existing works on this topic. One direction of research is to use less than 4 different pseudo-random functions, or to use less than 4 calls to these functions in one encryption, see [7,11,16,17]. However in these papers the proven security is still $m \ll 2^{n/2}$. In [18], the authors proved that even if the adversary has block-box access to the middle two functions of a 4 round $L-R$ construction the security proof is maintained.

Another direction of research, also followed in the present paper, is to improve the security bound $m \ll 2^{n/2}$. Then one may try to prove the security bound obtained is tight. Thus in [9] and independently in[1], it is shown that for the Luby-Rackoff theorems for 3 and 4 rounds, the bound $m \ll 2^{n/2}$ is optimal. Generic attacks exist, KPA for 3 rounds (with the notations that we will see below, just count the number of equalities $R_i \oplus S_i = R_j \oplus S_j$) and CPA for 4 rounds (take $R_i =$ constant and count the number of equalities $S_i \oplus L_i = S_j \oplus L_j$), that distinguish them from a random permutation for $m = \mathcal{O}(2^{n/2})$.

In order to improve this bound $m \ll 2^{n/2}$ we have the choice between two strategies: either to study the L-R constructions with 5 and more rounds (see for example [9,13] and the present paper), or to design new constructions. For this second strategy the best results obtained so far are in [1] and [7]. In [1] the bound $m \ll 2^n$ could be achieved for a construction "Benes" that however is not a permutation. In [7] the security of unbalanced Feistel schemes[1] is studied. A

---

[1] In [19] such unbalanced Feistel schemes are studied under the angle of linear and differential cryptanalysis.

security proof in $2^{n(1-\varepsilon)}$ is obtained, instead of $2^{n/2}$, but for much larger round functions (from $2n$ bits to ε bits, instead of n bits to n bits). This bound is basically again the birthday bound for these functions.

For the first strategy, the best security results obtained so far are in [13] and [6]. In [13] it is shown that when $m \ll 2^{\frac{3n}{4}}$ the L-R construction with 6 rounds (or more) is secure against CPCA. (In this paper, we will get $m \ll 2^{\frac{5n}{6}}$ for these conditions: 6 rounds and CPCA.) Recently in [6] it is shown that for L-R construction the security in $2^{n(1-\varepsilon)}$ can be achieved for all $\varepsilon > 0$, when the number of rounds $\to \infty$. In this paper we will show that when $m \ll 2^{n(1-\varepsilon)}$, $\varepsilon > 0$, 4 rounds are sufficient to achieve security against KPA, 7 rounds are sufficient to achieve security against CPA, and 10 rounds are sufficient for security against CPCA. Thus the number of rounds can in fact be fixed to a small value.

Thus we will solve an open problem described in [9], p. 310, as well as in [1], p. 319 and in [17], p. 149. This result also immediately improves the proven bound for one scheme of [2].

Our results are optimal with the regard of the number of queries, since an adversary with unlimited computing power can always distinguish a $k-$round L-R construction (i.e. a random Feistel cipher with k rounds) from a random permutation with $\mathcal{O}(k \cdot 2^n)$ queries and $\mathcal{O}(2^{kn2^n})$ computations by simply guessing all the round functions (this fact was already pointed out in [9] and in [14]).

Remark: It is conjectured but still unclear if 5 rounds are enough to avoid all CPCA attacks when $m \ll 2^{n(1-\varepsilon)}$. (See section 10).

In Appendix, we will summarize all the results proved so far for k rounds.

2   Notations

- $I_n = \{0, 1\}^n$ denotes the set of the 2^n binary strings of length n. $|I_n| = 2^n$.
- The set of all functions from I_n to I_n is F_n. Thus $|F_n| = 2^{n \cdot 2^n}$.
- The set of all permutations from I_n to I_n is B_n. Thus $B_n \subset F_n$, and $|B_n| = (2^n)!$
- For any $f, g \in F_n$, $f \circ g$ denotes the usual composition of functions.
- For any $a, b \in I_n$, $[a, b]$ will be the string of length $2n$ of I_{2n} which is the concatenation of a and b.
- For $a, b \in I_n$, $a \oplus b$ stands for bit by bit exclusive or of a and b.
- Let f_1 be a function of F_n. Let L, R, S and T be four n-bit strings in I_n. Then by definition

$$\Psi(f_1)[L, R] = [S, T] \overset{\text{def}}{\Longleftrightarrow} \begin{cases} S = R \\ T = L \oplus f_1(R) \end{cases}$$

- Let $f_1, f_2, \ldots, f_k$ be k functions of F_n. Then by definition:
$$\Psi^k(f_1, \ldots, f_k) = \Psi(f_k) \circ \cdots \circ \Psi(f_2) \circ \Psi(f_1).$$

The permutation function $\Psi^k(f_1, \ldots, f_k)$ is called "a Feistel scheme with k rounds" or shortly Ψ^k. When $f_1, f_2, \ldots, f_k$ are randomly and independently chosen functions in F_n, then $\Psi^k(f_1, \ldots, f_k)$ is called a "random Feistel scheme with k rounds", or a "L-R construction with k rounds".

We assume that the definitions of distinguishing circuits, and of normal and inverse (encrypting/decrypting) oracle gates are known. These standard definitions can be found in [3] and [7]. Let ϕ be a distinguishing circuit. We will denote by $\phi(F)$ it's output (1 or 0) when its oracle gates are implementing the encryption or decryption with the function F.

3   The "Coefficients H Technique"

We will formulate four theorems that we will use to prove our results. These theorems are the basis of a general proof technique, called the "coefficients H technique", that allows to prove security results for permutation generators (and thus applies for random and pseudorandom Feistel ciphers). This "coefficient H technique" was first described in [10].

Notations for This Section
In this section, $f_1, \ldots f_p$ will denote p functions of F_n, and $\Lambda(f_1, \ldots, f_p)$ is a function of F_{2n} (Λ is derived from the $f_1, \ldots f_p$).

When $[L_i, R_i], [S_i, T_i], 1 \le i \le m$, is a given sequence of $2m$ values of I_{2n}, we will denote by $H(L, R, S, T)$ or in short by H, the number if $p - tuples$ of functions $(f_1, \ldots f_p)$ such that:

$$\forall i, \ 1 \le i \le m, \ \Lambda(f_1, \ldots, f_p)[L_i, R_i] = [S_i, T_i].$$

Theorem 31 (Coefficient H technique, sufficient condition for security against KPA) *Let α and β be real numbers, $\alpha > 0$ and $\beta > 0$.*
  If :
(1) For random values $[L_i, R_i], [S_i, T_i], 1 \le i \le m$, such that $i \ne j \Rightarrow L_i \ne L_j$ or $R_i \ne R_j$, with probability $\ge 1 - \beta$ we have: $H \ge \dfrac{|F_n|^p}{2^{2nm}}(1 - \alpha)$
  Then:
(2) For all algorithm A (with no limitation in the number of computations) that takes the $[L_i, R_i], [S_i, T_i], \ 1 \le i \le m$ in input and outputs 0 or 1, we have that the expectation of $|P_1 - P_1^*|$ when the $[L_i, R_i], \ 1 \le i \le m$, are randomly chosen satisfy:

$$|E(P_1 - P_1^*)| \le \alpha + \beta.$$

With P_1 being the probability that A outputs 1 when $[S_i, T_i] = \Lambda(f_1, \ldots, f_p)[L_i, R_i]$ and when $(f_1, \ldots, f_p)$ are p independent random functions chosen in F_n.

  And with P_1^* being the probability that A outputs 1 when $[S_i, T_i] = F[L_i, R_i]$ and when F is randomly chosen in F_{2n}.

Remarks:

1. In this paper Λ will be the $L - R$ construction Ψ.
2. The condition $i \neq j \Rightarrow L_i \neq L_j$ or $R_i \neq R_j$, is in $m(m-1)/2^{2n}$.
3. Here if $\alpha + \beta$ is negligible, $\Lambda(f_1, \ldots, f_p)$ will resist to all known plaintext attacks, i.e. an attack where m cleartext/ciphertext pairs are given and when the m cleartext have random values.
4. A proof of this Theorem 31 is given in [15].
5. From this Theorem 31 we can prove that in order to attack Ψ^2 with KPA, we must have $m \geq$ about $2^{n/2}$ (see [15]).

Theorem 32 (Coefficient H technique sufficient condition for security against adaptive CPA)

Let α and β be real numbers, $\alpha > 0$ and $\beta > 0$.

Let E be a subset of I_{2n}^m such that $|E| \geq (1 - \beta) \cdot 2^{2nm}$. If :

(1) For all sequences $[L_i, R_i], 1 \leq i \leq m$, of m pairwise distinct elements of I_{2n} and for all sequences $[S_i, T_i], 1 \leq i \leq m$, of E

we have: $H \geq \dfrac{|F_n|^p}{2^{2nm}}(1 - \alpha)$

Then:

(2) For every distinguishing circuit ϕ with m oracle gates, we have :

$$\begin{cases} Adv_\phi^{PRF}(m, n) \stackrel{def}{=} |P_1 - P_1^*| \leq \alpha + \beta \\ Adv_\phi^{PRP}(m, n) \stackrel{def}{=} |P_1 - P_1^{**}| \leq \alpha + \beta + \frac{m(m-1)}{2 \cdot 2^{2n}} \end{cases}$$

With P_1 being the probability that $\phi(F) = 1$ when $F = \Lambda(f_1, \ldots, f_p)$ and when $(f_1, \ldots, f_p)$ are p independent random functions chosen in F_n.

With P_1^{**} being the probability that $\phi(F) = 1$ when F is randomly chosen in B_{2n}. And with P_1^* being the probability that $\phi(F) = 1$ when F is randomly chosen in F_{2n}.

Remarks:

1. In all this paper, "pairwise distinct elements of I_{2n}" means here that $\forall i, 1 \leq i \leq m, (L_i \neq L_j)$ <u>or</u> $(R_i \neq R_j)$.
2. Note that there is no limitation in the number of computations that the distinguishing circuit can perform, in order to analyse the m values given by its oracle gates.
3. A proof of this Theorem 32 (and more general formulations of it) can be found in [10] page 27 (for P_1^*) and pages 27 and 40 (for P_1^{**}).
4. Note that when $m \ll 2^n$ the term $\frac{m(m-1)}{2 \cdot 2^{2n}}$ is negligible and this term will not be a problem.
5. Here if $Adv^{PRP} = |P_1 - P_1^{**}|$ is negligible, $\Lambda(f_1, \ldots, f_p)$ will resist to all chosen plaintext attacks (we have only encryption gates). This includes adaptive attacks: in the distinguishing circuit the query number $i, 1 \leq i \leq m$ can depend on the results of the previous queries.

6. From this Theorem 32 (see [8], [10] or [15]), we obtain one way to prove the famous result of Luby and Rackoff: to attack Ψ^3 with CPA we must have $m \geq$ about $2^{n/2}$.

Theorem 33 (Coefficient H technique sufficient condition for security against adaptative CPCA)

Let $f_1, \ldots f_p$ be p functions in F_n, and let $\Lambda(f_1, \ldots, f_p) \in B_{2n}$. Let $\alpha > 0$.
If:
(1) For all sequences $[L_i, R_i], 1 \leq i \leq m$, *of m distinct elements of I_{2n}, and for all sequences* $[S_i, T_i], 1 \leq i \leq m$, *of m distinct elements of I_{2n}*

we have: $H \geq \dfrac{|F_n|^p}{2^{2nm}}(1 - \alpha)$

Then:
(2) For all super distinguishing circuit ϕ with m "super oracle gates" (normal/encryption or inverse/decryption gates), we have :

$$Adv_\phi^{SPRP}(m, n) \overset{def}{=} |P_1 - P_1^{**}| \leq \alpha + \frac{m(m-1)}{2 \cdot 2^{2n}}.$$

With P_1 being the probability that $\phi(F) = 1$ when $F = \Lambda(f_1, \ldots, f_p)$ and $(f_1, \ldots, f_p)$ are randomly (and independently) chosen in F_n.
And with P_1^{**} being the probability that $\phi(F) = 1$ when F is randomly chosen in B_{2n}.

Remarks:

1. This Theorem 33 can be found in [11], and in [10] p.40 where a proof is given.
2. Here if $Adv^{SPRP} = |P_1 - P_1^{**}|$ is negligible, $\Lambda(f_1, \ldots, f_p)$ will resist to all adaptive CPCA (we have both encryption and decryption oracle queries here).
3. From this Theorem 33 (see [8], [10] or [15]) we can prove that in order to attack Ψ^4 with CPCA we must have $m \geq$ about $2^{n/2}$.

Theorem 34 (Variant of Theorem 33, a bit more general)
With the same notations, let assume that

(1a) We have $H \geq \frac{|F_n|^p}{2^{2nm}}(1 - \alpha)$ for all $[L, R, S, T] \in E$, where E is a subset of I_n^{4m}.
(1b) For all super distinguishing circuit ϕ with m super oracle gates, the probability that $[L, R, S, T](\phi) \in E$ is $\geq 1 - \beta$, when ϕ acts on a random permutation f of B_{2n}. (Here $[L, R, S, T](\phi)$ denotes the successive $[S_i, T_i] = f[L_i, R_i]$ or $[L_i, R_i] = f^{-1}[S_i, T_i]$, $1 \leq i \leq m$, that will appear.)

Then (2) : $|P_1 - P_1^{**}| \leq \alpha + \beta + \frac{m(m-1)}{2 \cdot 2^{2n}}$.

Remarks:

1. This Theorem 34 can be found in [10] p. 38.
2. Theorem 33 is a special case of Theorem 34 where E is the set of all possible $[L, R, S, T]$ (with pairwise distinct $[L, R]$ and pairwise distinct $[S, T]$).
3. This Theorem 34 is sometime useful because it allows to study only cleartext/ciphertext pairs where we do not have too many equations that cannot be forced by CPCA attacks (for example like $R_i = S_i$, and unlike $L_i = R_i$).

In this paper we will use Theorem 31 for KPA on Ψ^4, Theorem 32 for CPA on Ψ^7 (and our result on Ψ^5), Theorem 33 for CPCA on Ψ^{10}, and Theorem 34 for our result for CPCA on Ψ^6.

4   An Exact Formula for H

Let $[L_i, R_i], 1 \le i \le m$ be m pairwise distinct elements of I_{2n}, and let $[S_i, T_i], 1 \le i \le m$ be some other m pairwise distinct elements of I_{2n}. We will note H the number of $(f_1, \ldots, f_k) \in F_n^k$ such that $\Psi^k(f_1, \ldots, f_k)[L_i, R_i] = [S_i, T_i]$.

This is the coefficient H that we need to apply Theorems 31, 32, 33 and 34 to k-round L-R construction Ψ^k. Fortunately it is possible to give an exact formula for H for every number of rounds k. Unfortunately when $k \ge 3$, the exact formula for H will involve a somewhat complex summation, and therefore it is not easy to use it. In this paper we will use the exact formula for H for 4 rounds. The proof of this formula (and formulas for $1, 2, 3$ rounds) can be found in [10], pages 132-136, or in [15].

An Exact Formula for H for 4 Rounds
Let P_i and Q_i, with $1 \le i \le m$, be the values such that $\Psi^2(f_1, f_2)[L_i, R_i] = [P_i, Q_i]$, i.e. the values after 2 rounds. Let $P = (P_1, \ldots P_m)$ and $Q = (Q_1, \ldots Q_m)$. Let (C) be the conditions:

$$\forall (i, j), \ 1 \le i \le m, 1 \le j \le m, \begin{cases} R_i = R_j \Rightarrow L_i \oplus P_i = L_j \oplus P_j \\ S_i = S_j \Rightarrow Q_i \oplus T_i = Q_j \oplus T_j \\ P_i = P_j \Rightarrow R_i \oplus Q_i = R_j \oplus Q_j \\ Q_i = Q_j \Rightarrow P_i \oplus S_i = P_j \oplus S_j \end{cases} \quad (C)$$

Then

$$H = \sum_{(P,Q) \text{ satisfying } (C)} \frac{|F_n|^4}{2^{4mn}} \cdot 2^{n(r+s+p+q)},$$

with p being the number of linearly independent equations of the form $P_i = P_j$, $i \ne j$, and similarly with q,r and s being the number of linearly independent equations of the form respectively $Q_i = Q_j$, $i \ne j$, $R_i = R_j$, $i \ne j$ and $S_i = S_j$, $i \ne j$.

5   A Formula for H for 4 Rounds with "Frameworks"

Most of the work in this paper is done for 4 rounds. Only at the end we will add some additional rounds to get the final results. From now on, we will use the same notations as in the formula for H for 4 rounds given in Section 4.

Definition 51 *We will call a "framework" a set $\mathcal{F}$ of equalities such that each equality of $\mathcal{F}$ is of one of the following forms: $P_i = P_j$ or $Q_i = Q_j$ with $1 \leq i < j \leq m$.*

Let (P, Q) be an element of $I_n^m \times I_n^m$.

Definition 52 *We will say that (P, Q) "satisfy" $\mathcal{F}$ if the set of all the equations of the form $P_i = P_j$ $i < j$ that are true in the sequence P, and all the equations of the form $Q_i = Q_j$ $i < j$ true in Q, is exactly $\mathcal{F}$.*
If it is so we will also say that $\mathcal{F}$ "is the framework of (P,Q)". (Each (P,Q) has one and only one framework).

Then from the exact formula given in Section 4 we have:

$$H = \sum_{\substack{\text{all frameworks} \\ \mathcal{F}}} \left[\sum_{\substack{(P,Q) \text{ satisfying} \\ (C) \text{ and } \mathcal{F}}} \frac{|F_n|^4}{2^{4mn}} \cdot 2^{n(r+s+p+q)} \right]$$

The set of conditions (C) was defined in Section 4. We observe that when $\mathcal{F}$ is fixed, from (C) we get a set of equations between the P values (and L and S values) or between the Q values (and T and R values), *i.e.* in these equations from (C), the P_i and the Q_i will never appear in the same equation.
We have:

$$H = \frac{|F_n|^4}{2^{4mn}} \sum_{\substack{\text{all frameworks} \\ \mathcal{F}}} \left[\sum_{P \text{ satisfying } (C1)} 2^{n(r+q)} \right] \cdot \left[\sum_{Q \text{ satisfying } (C2)} 2^{n(s+p)} \right]$$

With $(C1)$ and $(C2)$ being the sets of conditions defined as follows:

$$(C1) : \begin{cases} \text{The equalities } P_i = P_j, i < j \text{ that are present in } \mathcal{F}, \\ \text{and no other equalities } P_i = P_j, i < j \\ R_i = R_j \Rightarrow P_i \oplus P_j = L_i \oplus L_j \\ \text{The equalities } P_i \oplus P_j = S_i \oplus S_j \text{ for all } (i,j) \text{ such that } Q_i = Q_j \text{ is in } \mathcal{F} \end{cases}$$

$$(C2) : \begin{cases} \text{The equalities } Q_i = Q_j, i < j \text{ that are present in } \mathcal{F}, \\ \text{and no other equalities } Q_i = Q_j, i < j \\ S_i = S_j \Rightarrow Q_i \oplus Q_j = T_i \oplus T_j \\ \text{The equalities } Q_i \oplus Q_j = R_i \oplus R_j \text{ for all } (i,j) \text{ such that } P_i = P_j \text{ is in } \mathcal{F} \end{cases}$$

We have: $H = \dfrac{|F_n|^4}{2^{4mn}} \displaystyle\sum_{\substack{\text{all frameworks} \\ \mathcal{F}}} 2^{n(r+q)} \left[\text{Number of } P \text{ satisfying } (C1)\right] \cdot$

$$\cdot 2^{n(s+p)} \left[\text{Number of } Q \text{ satisfying } (C2)\right]$$

For a fixed framework $\mathcal{F}$, let:

$H_{\mathcal{F}_1} = 2^{n(r+q)}$ [Number of $(P_1, \ldots P_m)$ satisfying $(C1)$]

$H_{\mathcal{F}_2} = 2^{n(s+p)}$ [Number of $(Q_1, \ldots Q_m)$ satisfying $(C2)$]

Then: $H = \dfrac{|F_n|^4}{2^{4mn}} \displaystyle\sum_{\substack{\text{all frameworks} \\ \mathcal{F}}} H_{\mathcal{F}_1} \cdot H_{\mathcal{F}_2}$.

Remark: When $\mathcal{F}$ is fixed, in $(C1)$ we have only conditions on P and in $(C2)$ we have only conditions on Q.

6   Some Definitions on Sets of Equations and Frameworks

Definition 61 *For a fixed framework $\mathcal{F}$,*
let $J_{\mathcal{F}_1}$ = Number of $(P_1, \ldots P_m)$ such that the equalities $P_i = P_j$, $i < j$ are exactly those of $\mathcal{F}$.
let $J_{\mathcal{F}_2}$ = Number of $(Q_1, \ldots Q_m)$ such that the equalities $Q_i = Q_j$, $i < j$ are exactly those of $\mathcal{F}$.

So we have: $J_{\mathcal{F}_1} = 2^n \cdot (2^n - 1) \cdot (2^n - 2) \cdot \ldots \cdot (2^n - m + 1 + p)$
and $J_{\mathcal{F}_2} = 2^n \cdot (2^n - 1) \cdot (2^n - 2) \cdot \ldots \cdot (2^n - m + 1 + q)$

Definition 62 *Let $\mathcal{F}$ be a framework. We will say that two indices i and j, $1 \le i \le m$ and $1 \le j \le m$ are "connected in P" if the equation $P_i = P_j$ is in $\mathcal{F}$. (Similar definition for "connected in Q"). We say that i and j are connected in R if we have $R_i = R_j$ (here it does not depend on $\mathcal{F}$).*

Definition 63 *Let $\mathcal{F}$ be a framework. We will say that $\mathcal{F}$ "has a circle in R, P, Q" if there are k indices $i_1, i_2, \ldots, i_k$, with $k \ge 3$ and such that:*

1. *$i_k = i_1$ and $i_1 \ne i_2, i_2 \ne i_3, \ldots, i_{k-1} \ne i_k$.*
2. *$\forall \lambda, 1 \le \lambda \le k - 2$ we have one of the three following conditions:*

    - *i_λ and $i_{\lambda+1}$ are connected in R, and $i_{\lambda+1}$ and $i_{\lambda+2}$ are connected in P or in Q*
    - *i_λ and $i_{\lambda+1}$ are connected in P, and $i_{\lambda+1}$ and $i_{\lambda+2}$ are connected in R or in Q*
    - *i_λ and $i_{\lambda+1}$ are connected in Q, and $i_{\lambda+1}$ and $i_{\lambda+2}$ are connected in R or in P*

Examples.

- If $P_1 = P_2$ and $Q_1 = Q_2$ are in $\mathcal{F}$, then $\mathcal{F}$ has a circle in P, Q.
- If $\mathcal{F} = \{P_1 = P_2, P_2 = P_3\}$, then $\mathcal{F}$ has no circle in P, Q.

Definition 64 *Let $\mathcal{F}$ be a framework. We will say that (in $\mathcal{F}$) two indices i and j are connected by R, P, Q if there exist some indices i_1, i_2, ..., i_v such that $i = i_1$, $i_v = j$, and $\forall k$, $1 \leq k \leq v - 1$, we have either $(R_{i_k} = R_{i_{k+1}})$, or $(P_{i_k} = P_{i_{k+1}}) \in \mathcal{F}$ or $(Q_{i_k} = Q_{i_{k+1}}) \in \mathcal{F}$.*

Definition 65 *Let $\mathcal{F}$ be a framework. We will say that $\mathcal{F}$ has "no more than θ equalities in R, P, Q in the same line" if for all set of $\theta + 1$ independent equations that are either of $\mathcal{F}$ or of the form $R_i = R_j$ (with $R_i = R_j$ true), there exist two indices i and j which are not connected by R, P, Q.(Similar definition for "no more than θ equalities in S, P, Q in the same line".)*

Definition 66 *Let $\mathcal{F}$ be a framework. Let $\mathcal{F}'$ be the set of all the following equations:*

- *$P_i = P_j$ such that $P_i = P_j$ is in $\mathcal{F}$.*
- *$P_i \oplus P_j = L_i \oplus L_j$ for all $i < j$ such that $R_i = R_j$.*
- *$P_i \oplus P_j = S_i \oplus S_j$ such that $Q_i = Q_j$ is in $\mathcal{F}$.*

If from these equations of $\mathcal{F}'$ we can generate by a linear combination an equation $P_i = P_j, i \neq j$, we say that $\mathcal{F}$ has a circle in $R, P, Q, [LS]$.

We define in the same way "$\mathcal{F}$ has a circle in S, P, Q, $[RT]$" (by interchanging R and S, P and Q, and L and T).

Example. If $\mathcal{F} = \{Q_i = Q_k\}$ and we have $R_i = R_j$ and $L_i \oplus L_j = S_i \oplus S_k$ then $\mathcal{F}'$ contains $P_i \oplus P_j = L_i \oplus L_j$ and contains $P_i \oplus P_k = S_i \oplus S_k$, and then from $\mathcal{F}'$ we can generate $P_j = P_k$. Here $\mathcal{F}$ has a circle in $R, P, Q, [LS]$.

7   The Proof Strategy

We recall that from the end of Section 5, for 4 rounds we have:

$$H = \frac{|F_n|^4}{2^{4mn}} \sum_{\substack{\text{all frameworks} \\ \mathcal{F}}} H_{\mathcal{F}_1} \cdot H_{\mathcal{F}_2}.$$

We will evaluate H with this formula, in order to get the results of section 9 below. For this, the general strategy is to study this summation "framework by framework", i.e. we will compare $H_{\mathcal{F}}$ and $J_{\mathcal{F}}$ for a fixed framework $\mathcal{F}$. We will do this by using mainly four ideas:

- We will see that when $m \ll 2^n$ we can avoid all the "circles" in the equalities in the variables, and when $m^{\theta+1} \ll 2^{n\theta}$ we can avoid all the $\theta + 1$ equalities of the variables in the same line.
- We will use a property (Theorem 81 given in section 8) on sets of equations $P_i \oplus P_j = \lambda_k$.
- We will see that we can assume that the λ_k are generally random (sometime by adding 3 rounds at the beginning or at the end).
- We will need a general result of probability (Theorem 73 below).

More precisely, we will prove the following theorems.

a) Analysing Sets of Equations $P_i \oplus P_j = \lambda_k$

First we will prove Theorem 81 given in section 8. Conjecture 81 of section 8 is also of interest.

b) Avoiding "Circles" and "Long Lines"

Theorem 71 *Let $\mathcal{M}$ be the set of all frameworks $\mathcal{F}$ such that:*

1. *$\mathcal{F}$ has no circle in R, P, Q*
2. *$\mathcal{F}$ has no circle in S, P, Q*
3. *$\mathcal{F}$ has no circle in $R, P, Q, [LS]$*
4. *$\mathcal{F}$ has no circle in $S, P, Q, [RT]$*
5. *$\mathcal{F}$ has no more than θ equalities in R, P, Q in the same line*
6. *$\mathcal{F}$ has no more than θ equalities in S, P, Q in the same line*

Let M be the number of (P, Q) such that the framework $\mathcal{F}$ of (P, Q) is in $\mathcal{M}$. Then, with probability $\geq p$, M satisfies:

$$M \geq 2^{2nm} \left(1 - \mathcal{O}\left(\frac{m^2}{2^{2n}}\right) - \mathcal{O}\left(\frac{m^{\theta+1}}{2^{n\theta}}\right) \right),$$

where p is near 1 when the big O in the expression above are small, and when the R, L, S, T variables have random values, or are the output of a two rounds (or more) random Feistel scheme.

See [15] for the exact value of p. A similar result, with a small restriction on the inputs/outputs also exist if we add only one round (see [15]).

c) We Can Assume That the λ_k Are Generally Random

Theorem 72 *Let λ_k, $1 \leq k \leq a$, be some variables of I_n such that $\forall k$, $1 \leq k < a$, $\exists i, j$, such that $\lambda_k = L_i \oplus L_j$, or $\lambda_k = S_i \oplus S_j$, with no circle in the L or in the S variables that appear in the λ_k. Then if:*

(1a) The $[L_i, R_i, S_i, T_i]$ are random variables of I_n. or:

(1b) The $[S_i, T_i]$ are random variables of I_{2n} and the $[L_i, R_i]$ are obtained after a $\Psi^3(f_1, f_2, f_3)$ where f_1, f_2, f_3 are randomly chosen in F_n. or:

(1c) The $[L_i, R_i]$ are obtained after a $\Psi^3(f_1, f_2, f_3)$ and the $[S_i, T_i]$ are obtained after a $\Psi^3(g_1, g_2, g_3)$, where f_1, f_2, f_3, g_1, g_2, g_3 are randomly chosen in F_n. Then:

The probability to distinguish λ_1, λ_2, ..., λ_a from a truly random values of I_n is $\leq 1 - \mathcal{O}\left(\frac{a^2}{2^{2n}}\right)$ (with no limitation in the computing power).

Proof: See [15].

d) A General Result of Probability

Theorem 73 *Let a_i and b_i, $1 \leq i \leq N$, be N variables $a_i \geq 0$, $b_i \geq 0$, such that: $\forall i, 1 \leq i \leq N, a_i \geq b_i$ with a probability $\geq 1 - \varepsilon$.*

Then: $\forall \lambda > 0$, the probability that $\sum_{i=1}^{N} a_i \geq \left(\sum_{i=1}^{N} b_i \right)(1 - \lambda \varepsilon)$ is $\geq 1 - \frac{1}{\lambda}$.

Proof: See [15].

8   About Sets of Equations $P_i \oplus P_j = \lambda_k$

Definition 81 *Let (A) be a set of equations $P_i \oplus P_j = \lambda_k$. If by linearity from (A) we cannot generate an equation in only the λ_k, we will say that (A) has "no circle in P", or that the equations of (A) are "linearly independent in P".*

Let a be the number of equations in (A), and α be the number of variables P_i in (A). So we have parameters $\lambda_1, \lambda_2, \ldots, \lambda_a$ and $a + 1 \leq \alpha \leq 2a$.

Definition 82 *We will say that two indices i and j are "in the same block" if by linearity from the equations of (A) we can obtain $P_i \oplus P_j = $ an expression in $\lambda_1, \lambda_2, \ldots, \lambda_a$.*

Definition 83 *We will denote by ξ the maximum number of indices that are in the same block.*

Example. If $A = \{P_1 \oplus P_2 = \lambda_1,\ P_1 \oplus P_3 = \lambda_2,\ P_4 \oplus P_5 = \lambda_3\}$, here we have two blocks of indices $\{1, 2, 3\}$ and $\{4, 5\}$ and $\xi = 3$.

Definition 84 *For such a system (A), when $\lambda_1, \lambda_2, \ldots, \lambda_a$ are fixed, we will denote by h_α the number of $P_1, P_2, \ldots, P_\alpha$ solutions of (A) such that: $\forall i, j$, $i \neq j \Rightarrow P_i \neq P_j$.*
We will also denote $H_\alpha = 2^{na} h_\alpha$.

Definition 85 *We will denote by J_α the number of $P_1, P_2, \ldots, P_\alpha$ in I_n such that: $\forall i, j,\ i \neq j \Rightarrow P_i \neq P_j$.*

So $J_\alpha = 2^n \cdot (2^n - 1) \ldots (2^n - \alpha + 1)$.

Theorem 81 *Let ξ be a fixed integer, $\xi \geq 2$.*
For all set (A) of equations $P_i \oplus P_j = \lambda_k$, with no circle in P, with no more than ξ indices in the same block, with α variables P_i and a equations in (A), with $\alpha \ll 2^n$ (and also $\xi \alpha \ll 2^n$ since ξ is a fixed integer), when $\lambda_1, \lambda_2, \ldots, \lambda_a$ are randomly chosen in the subset D of I_n^a such that $H_\alpha \neq 0$, we have:
1) the average value of H_α is $\dfrac{2^{na}}{|D|} \cdot J_\alpha$ so is $\geq J_\alpha$.
2) the standard variation of H_α is $\sigma \leq J_\alpha \cdot \mathcal{O}\left(\dfrac{\alpha \sqrt{\alpha}}{2^n \sqrt{2^n}} \right)$.

Proof: See [15]
The condition $H_\alpha \neq 0$ means that for all i and j in the same block, $i \neq j$, the expression of $P_i \oplus P_j$ in $\lambda_1, \lambda_2, \ldots, \lambda_a$ is $\neq 0$. So this condition is in $1 - \mathcal{O}\left(\frac{\alpha}{2^n} \right)$.
From Bienaymé-Tchébichef Theorem, we get :

Corollary 81 *For all $\lambda > 0$, with a probability $\geq 1 - \mathcal{O}\left(\frac{1}{\lambda^2} \right) - \mathcal{O}\left(\frac{\alpha}{2^n} \right)$, we have:*

$$H_\alpha \geq J_\alpha \left(1 - \frac{\lambda \alpha \sqrt{\alpha}}{2^n \sqrt{2^n}} \right).$$

We will say that we have $H_\alpha \geq J_\alpha \left(1 - \frac{\lambda \alpha \sqrt{\alpha}}{2^n \sqrt{2^n}}\right)$ with a probability as near as 1 as we want.

Theorem 82 *Let ξ be a fixed integer, $\xi \geq 2$.*
   Let (A) be a set of equations $P_i \oplus P_j = \lambda_k$ with no circle in P, with α variables P_i, such that:

1. *$\alpha^3 \ll 2^{2n}$ (and also $\xi \alpha^3 \ll 2^{2n}$ since ξ is here a fixed integer).*
2. *We have no more than ξ indices in the same block.*
3. *The $\lambda_1, \lambda_2, \ldots, \lambda_k$ have any fixed values such that: for all i and j in the same block, $i \neq j$, the expression of $P_i \oplus P_j$ in $\lambda_1, \lambda_2, \ldots, \lambda_a$ is $\neq 0$ (i.e. by linearity from (A) we cannot generate an equation $P_i = P_j$ with $i \neq j$).*

Then we have, for sufficiently large n: $H_\alpha \geq J_\alpha$.

   Proof: See [15]

Conjecture 81 *This Theorem 82 is still true when $\alpha \ll 2^n$ (instead of $\alpha^3 \ll 2^{2n}$).*

   This conjecture 81 is not yet proved in general.

9   Results for 4, 7 and 10 Rounds in $\mathcal{O}(2^{n(1-\varepsilon)})$

From the theorems of section 7 and Theorem 91 we get the following theorems on H (see [15] for the proofs).

Theorem 91 *Let $[L_i, R_i]$, and $[S_i, T_i]$, $1 \leq i \leq m$, be random values such that the $[L_i, R_i]$ are pairwise distincts and the $[S_i, T_i]$ are pairwise distincts. Then for Ψ^4 the probability p that :*

$$H \geq \frac{|F_n|^4}{2^{2nm}} \left(1 - \mathcal{O}\left(\frac{m}{2^n}\right) - \mathcal{O}\left(\frac{m^{\theta+1}}{2^{n\theta}}\right)\right).$$

satisfy:

$$p \geq 1 - \mathcal{O}\left(\frac{m}{2^n}\right)$$

Theorem 92 *Let $[L_i, R_i]$, and $[S_i, T_i]$, $1 \leq i \leq m$, be some values such that the $[L_i, R_i]$ are pairwise distincts and the $[S_i, T_i]$ are pairwise distincts. Then for Ψ^7 we have:*
   There is a subset E of I_{2n}^m with $|E| \geq \left(1 - \mathcal{O}\left(\frac{m}{2^n}\right) - \mathcal{O}\left(\frac{m^{\theta+1}}{2^{n\theta}}\right)\right)$ such that if the $[S_i, T_i]$, $1 \leq i \leq m$ are in E we have:

$$H \geq \frac{|F_n|^7}{2^{2nm}} \left(1 - \mathcal{O}\left(\frac{m}{2^n}\right) - \mathcal{O}\left(\frac{m^{\theta+1}}{2^{n\theta}}\right)\right).$$

Theorem 93 *Let $[L_i, R_i]$, and $[S_i, T_i]$, $1 \leq i \leq m$, be some values such that the $[L_i, R_i]$ are pairwise distincts and the $[S_i, T_i]$ are pairwise distincts. Then for Ψ^{10} we have:*

$$\text{For all integer } \theta \geq 1 \quad H \geq \frac{|F_n|^{10}}{2^{2nm}} \left(1 - \mathcal{O}\left(\frac{m}{2^n}\right) - \mathcal{O}\left(\frac{m^{\theta+1}}{2^{n\theta}}\right)\right).$$

Security Results against Cryptographic Attacks

Finally our cryptographic results on 4,7 and 10 rounds are just a direct consequence of Theorem 91,92,93 and of Theorem 31, 32 and 33: this is because θ can be any integer.

Remarks:

1. In these theorems when θ is fixed, we can get explicit values for all the coefficients that appear as $\mathcal{O}()$ in our theorems. Therefore our results are not only asymptotic (when $n \to \infty$), they can also be written as explicit concrete security bounds.

2. For Ψ^4 our security results are optimal both in term of m and in term of the number of computations to be performed. With $\mathcal{O}(2^n)$ messages and $\mathcal{O}(2^n)$ computations it is indeed possible to distinguish Ψ^4 from a truly random permutation with a KPA (count the number of (i, j, k) with $R_i = R_j$ and $S_i \oplus L_i = S_j \oplus L_j$).

10 Results for 5 or 6 Rounds in $\mathcal{O}(2^{5n/6})$

Here we cannot assume that the λ_k are almost random. However, from Theorem 82 we can prove:

Theorem 101 *Ψ^5 resists all CPA when $m \ll \mathcal{O}(2^{5n/6})$. Ψ^6 resists all CPCA when $m \ll \mathcal{O}(2^{5n/6})$.*

(See [15] for the proofs. Hint: we will have $\alpha \simeq \frac{m^2}{2^n}$ and $\alpha^3 \ll 2^{2n}$, so $m \ll \mathcal{O}(2^{\frac{5n}{6}})$ will be our condition.)

Remark: If we can use Conjecture 81, then from it we can prove that Ψ^5 resists all CPA when $m \ll \mathcal{O}(2^{n(1-\varepsilon)})$ and Ψ^6 resists all CPCA when $m \ll \mathcal{O}(2^{n(1-\varepsilon)})$ since we will have to add only one or two rounds in addition of the central Ψ^4. However, Conjecture 81 is not yet proven in general.

11 Conclusion and Further Work

In this paper we were able to prove improved security bounds for random Feistel ciphers. It seems reasonable that our method can be extended, for example for 5 or 6 rounds. This method can also be used in various other directions. For example one can study Feistel schemes with a different group law than $\oplus$ (it has already been studied but only when $m \ll 2^{n/2}$). One can also study the

Feistel schemes on digits/$GF(q)$/bytes etc. instead of bits. Finally one can study cryptographic constructions of different type.

It seems particularly interesting to study dissymmetric Feistel schemes, i.e. schemes in which a round is defined as $\Psi(f_i)[L, R] = [S, T] \stackrel{\text{def}}{\Leftrightarrow} S = R$ and $T = L \oplus f_1(R)$ but with L and T having only 1 bit, and S and R having $2n - 1$ bits, and with the f_i being single Boolean functions $f_i \in I_{2n-1} \to I_1$. It seems that in such schemes the methods of the present paper should give a security proof for $m \ll 2^{2n(1-\varepsilon)}$, even against unbounded adversaries [2]. (This will improve the $2^{n(1-\varepsilon)}$ result of [7] for such schemes). For comparison, the best possible result for classical Feistel schemes with the same block size $2n$ (and achieved in the present paper) is $m \ll 2^{n(1-\varepsilon)}$ and cannot be improved in the unbounded adversary model.

In conclusion we hope that the proof techniques given in this paper will be useful in future works, on one hand in the design of cryptographic schemes with optimal proofs of security, and on the other hand to detect flaws in existing designs and suggest some new attacks.

Acknowledgement. I would like to thank the Stephan Banach Center of Warsaw where I was invited in January 2003 and where part of this work was done.

References

1. William Aiollo, Ramarathnam Venkatesan, *Foiling Birthday Attacks in Length-Doubling Transformations - Benes: A Non-Reversible Alternative to Feistel.* Eurocrypt 96, LNCS 1070, Springer, pp. 307–320.
2. John Black, Philip Rogaway, *Ciphers with Arbitrary Finite Domains*, [RS]A'2002, pp. 114–130, Springer LNCS 2271, February 2002.
3. M. Luby, C. Rackoff, *How to construct pseudorandom permutations from pseudorandom functions*, SIAM Journal on Computing, vol. 17, n. 2, pp. 373–386, April 1988.
4. U. Maurer, *A simplified and generalized treatment of Luby-Rackoff pseudorandom permutation generators*, Eurocrypt'92, Springer, pp. 239–255.
5. U. Maurer: *Indistinguishability of Random Systems*, Eurocrypt 2002, LNCS 2332, Springer, pp. 110–132.
6. U. Maurer, K. Pietrzak: *The Security of Many-Round Luby-Rackoff Pseudo-Random Permutations*, Eurocrypt 2003, May 2003, Warsaw, Poland, LNCS, Springer.
7. Moni Naor and Omer Reingold, *On the construction of pseudo-random permutations: Luby-Rackoff revisited*, Journal of Cryptology, vol 12, 1999, pp. 29-66. Extended abstract was published in: Proc. 29th Ann. ACM Symp. on Theory of Computing, 1997, pp. 189–199.
8. J. Patarin, *Pseudorandom Permutations based on the DES Scheme*, Eurocode'90, LNCS 514, Springer, pp. 193–204.

---

[2] The reason for this is that in the asymmetric Feistel scheme there are much more possible round functions.

9. J. Patarin, *New results on pseudorandom permutation generators based on the DES scheme*, Crypto'91, Springer, pp. 301–312.

10. J. Patarin, *Etude des générateurs de permutations basés sur le schéma du DES* . Ph. D. Thesis, Inria, Domaine de Voluceau, Le Chesnay, France, 1991.

11. J. Patarin, *How to construct pseudorandom and super pseudorandom permutations from one single pseudorandom function*. Eurocrypt'92, Springer, pp. 256–266.

12. J. Patarin, *Improved Security Bounds for Pseudorandom Permutations*, 4th ACM Conference on Computer and Communications Security April 2-4th 1997, Zurich, Switzerland, pp. 142–150.

13. J. Patarin, *About Feistel Schemes with Six (or More) Rounds*, in Fast Software Encryption 1998, pp. 103–121.

14. J. Patarin, *Generic Attacks on Feistel Schemes*, Asiacrypt 2001, LNCS 2248, Springer, pp. 222–238.

15. J. Patarin, *Luby-Rackoff: 7 Rounds are Enough for $2^{n(1-\varepsilon)}$ Security*. Extended version of this paper. Available from the author.

16. S. Patel, Z. Ramzan and G. Sundaram, *Toward making Luby-Rackoff ciphers optimal and practical*, FSE'99, LNCS, Springer, 1999.

17. J. Pieprzyk, *How to construct pseudorandom permutations from Single Pseudorandom Functions*, Eurocrypt'90, LNCS 473, Springer, pp. 140–150.

18. Z. Ramzan, L. Reyzin, *On the Round Security of Symmetric-Key Cryptographic Primitives*, Crypto 2000, LNCS 1880, Springer, pp. 376–393.

19. B. Schneier and J. Kelsey, *Unbalanced Feistel Networks and Block Cipher Design*, FSE'96, LNCS 1039, Springer, pp. 121-144.

Appendix: Summary of the Known Results on Ψ^k

Ψ	Ψ^2	Ψ^3	Ψ^4	Ψ^5	Ψ^6	Ψ^7	$\Psi^k, k \geq 10$
KPA	1	$\mathcal{O}(2^{\frac{n}{2}})$	$\mathcal{O}(2^{\frac{n}{2}})$	$\geq \mathcal{O}(2^{n(1-\varepsilon)})$ and $\leq \mathcal{O}(2^n)$	$\geq \mathcal{O}(2^{n(1-\varepsilon)})$ and $\leq \mathcal{O}(2^n)$	$\geq \mathcal{O}(2^{n(1-\varepsilon)})$ and $\leq \mathcal{O}(2^n)$	$\geq \mathcal{O}(2^{n(1-\varepsilon)})$ and $\leq \mathcal{O}(2^n)$	$\geq \mathcal{O}(2^{n(1-\varepsilon)})$ and $\leq \mathcal{O}(2^n)$
CPA	1	2	$\mathcal{O}(2^{\frac{n}{2}})$	$\mathcal{O}(2^{\frac{n}{2}})$	$\geq \mathcal{O}(2^{\frac{5n}{6}})$ and $\leq \mathcal{O}(2^n)$	$\geq \mathcal{O}(2^{\frac{5n}{6}})$ and $\leq \mathcal{O}(2^n)$	$\geq \mathcal{O}(2^{n(1-\varepsilon)})$ and $\leq \mathcal{O}(2^n)$	$\geq \mathcal{O}(2^{n(1-\varepsilon)})$ and $\leq \mathcal{O}(2^n)$
CPCA	1	2	3	$\mathcal{O}(2^{\frac{n}{2}})$	$\geq \mathcal{O}(2^{\frac{n}{2}})$ and $\leq \mathcal{O}(2^n)$	$\geq \mathcal{O}(2^{\frac{5n}{6}})$ and $\leq \mathcal{O}(2^n)$	$\geq \mathcal{O}(2^{\frac{n}{2}})$ and $\leq \mathcal{O}(2^n)$	$\geq \mathcal{O}(2^{n(1-\varepsilon)})$ and $\leq \mathcal{O}(2^n)$

Fig. 1. The minimum number **m** of queries needed to distinguish Ψ^i from a random permutation of B_{2n}

	Ψ	Ψ^2	Ψ^3	Ψ^4	Ψ^5	Ψ^6	Ψ^7	$\Psi^k, k \geq 10$
KPA	$\mathcal{O}(1)$	$\mathcal{O}(2^{\frac{n}{2}})$	$\mathcal{O}(2^{\frac{n}{2}})$	$\geq \mathcal{O}(2^{n(1-\varepsilon)})$ and $\leq \mathcal{O}(2^n)$	$\geq \mathcal{O}(2^{n(1-\varepsilon)})$ and $\leq \mathcal{O}(2^{\frac{7n}{4}})$	$\geq \mathcal{O}(2^{n(1-\varepsilon)})$ and $\leq \mathcal{O}(2^{2n})$	$\geq \mathcal{O}(2^{n(1-\varepsilon)})$ and $\leq \mathcal{O}(2^{2n})$	$\geq \mathcal{O}(2^{n(1-\varepsilon)})$ and $\leq \mathcal{O}(2^{2n})$
CPA	$\mathcal{O}(1)$	$\mathcal{O}(1)$	$\mathcal{O}(2^{\frac{n}{2}})$	$\mathcal{O}(2^{\frac{n}{2}})$	$\geq \mathcal{O}(2^{\frac{5n}{6}})$ and $\leq \mathcal{O}(2^{\frac{3n}{2}})$	$\geq \mathcal{O}(2^{\frac{5n}{6}})$ and $\leq \mathcal{O}(2^{2n})$	$\geq \mathcal{O}(2^{n(1-\varepsilon)})$ and $\leq \mathcal{O}(2^{2n})$	$\geq \mathcal{O}(2^{n(1-\varepsilon)})$ and $\leq \mathcal{O}(2^{2n})$
CPCA	$\mathcal{O}(1)$	$\mathcal{O}(1)$	$\mathcal{O}(1)$	$\mathcal{O}(2^{\frac{n}{2}})$	$\geq \mathcal{O}(2^{\frac{n}{2}})$ and $\leq \mathcal{O}(2^{\frac{3n}{2}})$	$\geq \mathcal{O}(2^{\frac{5n}{6}})$ and $\leq \mathcal{O}(2^{2n})$	$\geq \mathcal{O}(2^{\frac{n}{2}})$ and $\leq \mathcal{O}(2^{2n})$	$\geq \mathcal{O}(2^{n(1-\varepsilon)})$ and $\leq \mathcal{O}(2^{2n})$

Fig. 2. The minimum number λ of computations needed to distinguish Ψ^i from a random permutation of B_{2n}

Remark: The result $\lambda \leq \mathcal{O}(2^{2n})$ is obtained due to the fact that Ψ^k permutations always have an even signature. If we want to distinguish Ψ^k from random permutations with an even signature (instead of random permutations of the whole B_{2n}), or if we do not have exactly all the possible cleartext/ciphertext pairs, then we only know that (when k is even): $\lambda \leq \mathcal{O}(2^{n(k^2/2 - 4k + 8)})$, see [14].

Weak Key Authenticity and the Computational Completeness of Formal Encryption

Omer Horvitz[1] and Virgil Gligor[2]

[1] Department of Computer Science,
University of Maryland, College Park MD 20742
horvitz@cs.umd.edu
[2] Department of Electrical and Computer Engineering,
University of Maryland, College Park MD 20742
gligor@eng.umd.edu

Abstract. A significant effort has recently been made to rigorously relate the formal treatment of cryptography with the computational one. A first substantial step in this direction was taken by Abadi and Rogaway [AR02]. Considering a formal language that treats symmetric encryption, [AR02] show that an associated formal semantics is sound with respect to an associated computational semantics, under a particular, sufficient, condition on the computational encryption scheme. In this paper, we give a necessary and sufficient condition for completeness, tightly characterizing this aspect of the exposition. Our condition involves the ability to distinguish a ciphertext and the key it was encrypted with, from a ciphertext and a random key. It is shown to be strictly weaker than a previously suggested condition for completeness (confusion-freedom of Micciancio and Warinschi [MW02]), and should be of independent interest.

Keywords. Cryptography, Encryption, Authentication, Formal Reasoning, Completeness, Weak Key Authenticity.

1 Introduction

Modern cryptography has been investigated from both a formal and a computational perspective. Of the former, a typical treatment features a formal language, in which statements, representing cryptographic entities and operations, can be made. Their security properties are usually stated outside the language, captured in operations that manipulate the formal statements, or expressed with additional formal constructs. Of the latter, a typical treatment uses algorithms on strings of bits to model cryptographic operations. Security properties are defined in terms of probability and computational complexity of successful attacks.

Recently, an effort has been made to relate the two approaches, traditionally considered separately and mostly by different communities. A successful attempt holds the promise of bringing the strengths of one treatment to the other. From

D. Boneh (Ed.): CRYPTO 2003, LNCS 2729, pp. 530–547, 2003.

one direction, it is expected to quantify and highlight implicit assumptions of formal semantics. In addition, it should confirm and increase the relevance of formal proofs to concrete computational instantiations. From the other direction, the establishment of such connections may allow the application of high level formal reasoning mechanisms to the computational domain.

A first step in this direction was taken by Abadi and Rogaway [AR02]. Focusing on symmetric encryption, their work is based on a formal language that includes constructs to represent bits, keys and an encryption operation. Two semantics are defined for the language. In the first, an expression in the language is associated with a syntactic counterpart, the *pattern*, which mirrors the expression up to parts that should look unintelligible to a viewer (informally, parts that are encrypted with keys that are not recoverable from the expression). Expressions are said to be equivalent in this setting if their patterns are equal (up to key renaming). This constitutes a formal semantics. In the second definition, an expression is associated with an ensemble of distributions on strings, obtained by instantiating its encryption construct with a concrete computational encryption scheme with different security parameters. Two expressions are said to be indistinguishable in this setting if their associated ensembles are computationally indistinguishable. This constitutes a computational semantics. Under this framework, [AR02] give a soundness result: they show that under specific, sufficient, conditions on the computational encryption scheme, equivalence of expressions in the formal semantics implies their indistinguishability in the computational semantics.

Our Results. In this paper, we tightly characterize the completeness aspect of this exposition. We identify a *necessary* and *sufficient* condition on the computational encryption scheme under which indistinguishability in the computational setting implies equivalence in the formal one. For any two expressions, our condition involves the admittance of an efficient test that distinguishes a ciphertext and the key it was encrypted with, from a ciphertext and some random key, with a noticeable (i.e., non-negligible) probability, when the plaintexts are drawn from the ensembles associated with those expressions. An encryption scheme that satisfies this requirement is said to *admit weak key-authenticity tests for expressions*. The result is obtained using a new proof technique, featuring a fixpoint characterization of formal equivalence.

In the literature [MW02,AJ01], the notion of *confusion-freedom* was previously proposed as sufficient for completeness. Informally, a confusion-free encryption scheme is one in which the decryption of a ciphertext with a wrong key fails with almost certainty. The above-mentioned work suggests the use of a full-fledged *authenticated encryption scheme* [BN00,KY00] to achieve this notion. We compare confusion-freedom with a strengthened version of the notion of the admittance of weak key-authenticity tests for expressions, that involves the admittance of a single, all-purpose, weak key-authenticity test, defined purely in computational terms. Such test is referred to as a *weak key-authenticity test*. We show that the requirement that an encryption scheme admits a weak key-authenticity test is *strictly weaker* than the requirement that it be confusion-free (and certainly weaker than the requirement that it be an authenticated encryption scheme). To that effect, we present a simple encryption scheme that admits a weak key-authenticity test but is not confusion-free. The scheme thus matches our completeness criterion, but not that of [MW02]. Furthermore, it meets the soundness criterion of [AR02].

The notion of the weak key authenticity should be of independent interest. As a primitive, it relates to the absence of a weak version of key-anonymity [BB01] and which-key revealing [AR02] properties. It would be interesting to investigate ways of meeting it, other than the ones presented here, and explore its practical uses.

The paper proceeds as follows. In section 2, we revisit the formal treatment of symmetric encryption of [AR02], and give a fixpoint characterization of the "reachable parts" of expressions. In section 3, we discuss the computational treatment of symmetric encryption, and revisit the computational semantics for expressions of [AR02]. In section 4, we give our main completeness result for schemes that admit weak key-authenticity tests for expressions. In section 5, we present the strengthened version of the test and compare it with other cryptographic notions; in particular, we show that the admittance of a weak key-authenticity test is a weaker property of an encryption scheme than it being confusion-free. The proof demonstrates a method of achieving the admittance of a weak key-authenticity test.

2   Formal Treatment of Symmetric Encryption, Formal Semantics for Expressions

In this section, we revisit the formal treatment of symmetric encryption of [AR02]. That treatment consists of a formal language and a formal semantics. Our goal is to recast the definitions of [AR02] in terms that pertain closely to the tree structure of expressions in the language. In addition, we provide an alternative, fixpoint characterization of the "reachable parts" of expressions, that plays an important role in the proof of our completeness result.

2.1   A Formal Language for Symmetric Encryption

Let **Bits** be the set $\{0, 1\}$. Let **Keys** be a fixed, non-empty set of symbols, disjoint from **Bits**. The elements of **Bits** and **Keys** are referred to as bits and keys, respectively. Following the work of [AR02], we let our formal language, denoted by **Exp**, be a set of *expressions*, defined inductively as follows:

1. bits and keys are *expressions*. They are referred to as *atomic expressions*, or simply *atoms*.
2. a) If M and N are *expressions*, then so is (M, N). We say that (M, N) is *directly derived* from M and N; it is *non-atomic*.
   b) If M is an *expression* and K is a key, then $\{M\}_K$ is an *expression*. We say that $\{M\}_K$ is is *directly derived* from M; it is *non-atomic* too.

Parts 2a and 2b of the above definition are called *derivation rules*. Informally, (M, N) represents the pairing of expressions M and N; $\{M\}_K$ represents the encryption of expression M with key K.

Expressions are strings of symbols. The *length* of an expression E is the number of symbols it is comprised of (count '$\}_K$' as a single symbol), and is denoted by $|E|$. We use $E_1 = E_2$ to denote that the expressions E_1, E_2 are identical as strings of symbols.

It is important to note that every non-atomic expression can be associated with a unique rule and a unique set of expressions from which it is directly derived. Expressions in **Exp** are consequently said to be *uniquely readable*. The converse holds too. Formally, two non-atomic expressions $E_1, E_2 \in$ **Exp** are identical as strings of symbols iff either

- $E_1 = (M_1, N_1)$, $E_2 = (M_2, N_2)$ and $M_1 = M_2$, $N_1 = N_2$; or
- $E_1 = \{M_1\}_{K_1}$, $E_2 = \{M_2\}_{K_2}$ and $M_1 = M_2$, $K_1 = K_2$.

For a proof, see the full version of this paper [HG03].

The structure of an expression can be represented naturally in the form of a tree. A *derivation tree* T_E for an expression E is defined inductively as follows:

1. If E is atomic, then T_E consists of a single node, the *root*, labelled by E.
2. If E is non-atomic, then T_E consists of a single node, the *root*, labelled by E, and an ordered list of *trees* for the expressions from which E is directly derived; the sets of nodes of these trees are disjoint, and none contains the root of T_E. If $E = (M, N)$, we say that T_M and T_N are the *left* and *right subtrees* of T_E, respectively. The roots of T_M and T_N are said to be the *left* and *right children* of the root of T_E, respectively. Similarly, if $E = \{M\}_K$ then T_M is said to be the *subtree* of T_E; the root of T_M is said to be the *child* of the root of T_E.

Informally, the notion of a derivation tree resembles that of the standard parse tree; the two relate in that a node in a derivation tree is labelled with the *yield* of the corresponding node in the parse tree. We let $|T_E|$ denote the cardinality of the set of nodes of T_E.

We mention two properties of expressions and their derivation trees that are relevant to our treatment. First, two expressions are identical as strings of symbols iff their respective derivation trees are identical; to see this, apply the unique readability property of expressions and its converse inductively to the structure of the derivation trees. Second, if $|E| = n$, then T_E consists of at most n nodes; this can be shown by induction on the length of an expression.

2.2   Formal Semantics for Expressions

In defining a formal semantics for **Exp**, we seek to capture a notion of privacy, intuitively associated with the encryption operation. In particular, we would like to express our understanding that parts of expressions, representing encryptions with keys that are not recoverable from the text, are unintelligible (or unreachable) to a viewer. We would also like to capture our understanding that expressions, differing only in their unintelligible parts, "look the same" to a viewer. Just as in [AR02], we do so by mapping each expression to a syntactic counterpart, the *pattern*, which mirrors only its reachable parts. We then define equivalence of expressions in terms of their respective patterns. We state our definitions in terms of functions on derivation trees of expressions and patterns, rather than in the form of procedures on expressions and patterns, as is done in [AR02].

Let T_E be the derivation tree of an expression E, let V be its node set and $r \in V$ its root. A set $U \subseteq V$ is said to *contain the reachable nodes* of T_E if:

1. $r \in U$.
2. For all $u \in U$,
   a) if u is labelled with an expression of the form (M, N), then both the children of u in T_E (labelled M and N) are in U.
   b) if u is labelled with an expression of the form $\{M\}_K$, and there exists a $u' \in U$ labelled K, then the child of u in T_E (labelled M) is in U.

For $E \in$ **Exp** of length n, T_E consists of at most n nodes, of which there are at most 2^n subsets. It follows that the number of sets containing the set of reachable nodes of T_E is finite. Let R be the intersection of all those sets. It is easy to show that R itself contains the set of reachable nodes of T_E; it is minimal in the sense that it is contained in all such sets. We call R the *set of reachable nodes* of T_E. Informally, reachable nodes correspond to parts of an expression that should be intelligible to a viewer.

Let T_E be a derivation tree with a root r and a set of reachable nodes R. The graph induced by T_E on R must be a tree rooted at r, and not a forest (otherwise, let R' be the set of nodes in the connected component that contains r; R' is a set that contains the set of reachable nodes in T_E, contradicting the minimality of R). We call this tree the *tree of reachable nodes*, and use T_E^R to denote it.

The definition of a pattern extends that of an expression with the addition of an atomic symbol, $\square$. Informally, $\square$ will appear in parts of a pattern that correspond to unintelligible parts of the associated expression.

Let **Pat** be the set of *patterns*, defined inductively as follows:

1. bits, keys and the symbol $\square$ are *(atomic) patterns*.
2. a) If M and N are *patterns*, then (M, N) is a *(non-atomic) pattern*.
   b) If M is a *pattern* and K is a key, then $\{M\}_K$ is a *(non-atomic) pattern*.

As with expressions, we associate a pattern P with a derivation tree T_P. Two patterns are identical as strings of symbols iff their respective derivation trees are identical.

To map expressions to patterns via their respective derivation trees, we will need an appropriate notion of tree isomorphism. Let T_1, T_2 be finite, rooted, ordered trees with node sets V_1, V_2 and roots $r_1 \in V_1$, $r_2 \in V_2$, respectively. T_1, T_2 are said to be *isomorphic as rooted, ordered trees* if there exists a bijection $\varphi : V_1 \to V_2$ such that:

1. $\varphi(r_1) = r_2$.
2. For all $v \in V_1$, $(u_1, \ldots, u_k)$ are the children of v in T_1 iff $(\varphi(u_1), \ldots, \varphi(u_k))$ are the children of $\varphi(v)$ in T_2.

φ is said to be an *isomorphism* of T_1, T_2 as rooted, ordered trees.

Let T_E be the derivation tree of an expression E, V_E its node set, $R \subseteq V_E$ its set of reachable nodes. Let T_P be the derivation tree of a pattern P, V_P its node set. We say that *expression E has a pattern P* if there exists a $\varphi : R \to V_P$ such that:

1. φ is an isomorphism of T_E^R, T_P as rooted, ordered trees.
2. For all $v \in R$,
   a) if v is labelled with a bit, then $\varphi(v)$ is labelled with an identical bit.
   b) if v is labelled with a key, then $\varphi(v)$ is labelled with an identical key.
   c) if v is labelled (M, N), then $\varphi(v)$ is labelled (M', N').
   d) if v is labelled $\{M\}_K$ and there exists a $u \in R$ labelled K, then $\varphi(v)$ is labelled $\{M'\}_K$.
   e) if v is labelled $\{M\}_K$ and there does not exist a $u \in R$ labelled K, then $\varphi(v)$ is labelled $\square$.

The corresponding definition of [AR02] amounts to a walk of T_E^R and T_P that enforces the above constraints.

We note that the pattern P associated with each expression E is unique. To see this, notice that the uniqueness of T_E implies a unique set of reachable nodes R, which implies a unique T_E^R, which is mapped to a unique T_P, which, in turn, guarantees a unique P. The converse is not true, however; every pattern has infinitely many expressions that are mapped to it.

We proceed with the notion of expression equivalence. Informally, we require that the derivation trees of patterns corresponding to equivalent expressions be isomorphic up to key renaming. For $i \in \{1, 2\}$, let P_i be the pattern of expression E_i, with a derivation tree T_{P_i} over V_{P_i}. We say that E_1 is *equivalent* to E_2, and write $E_1 \cong E_2$, iff there exists a $\varphi : V_{P_1} \to V_{P_2}$ and a permutation σ on **Keys** such that:

1. φ is an isomorphism of T_{P_1}, T_{P_2} as rooted, ordered trees.
2. For all $v \in V_{P_1}$,
   a) if v is labelled with a bit, then $\varphi(v)$ is labelled with an identical bit.
   b) if v is labelled K, then $\varphi(v)$ is labelled with $\sigma(K)$.
   c) if v is labelled (M, N), then $\varphi(v)$ is labelled (M', N').
   d) if v is labelled $\{M\}_K$, then $\varphi(v)$ is labelled $\{M'\}_{\sigma(K)}$.
   e) if v is labelled $\square$, then $\varphi(v)$ is labelled $\square$.

Composing the above definitions, we obtain the following property of the equivalence relation.

Theorem 2.1. *For $i \in \{1, 2\}$, let E_i be an expression with a derivation tree T_{E_i}, a set of reachable nodes R_{E_i} and an induced tree of reachable nodes $T_{E_i}^R$. Then $E_1 \cong E_2$ iff there exist a $\varphi : R_{E_1} \to R_{E_2}$ and a permutation σ on **Keys** such that:*

1. φ is an isomorphism of $T_{E_1}^R$, $T_{E_2}^R$ as rooted, ordered trees.
2. For all $v \in R_{E_1}$,
    a) if v is labelled with a bit, then $\varphi(v)$ is labelled with an identical bit.
    b) if v is labelled K, then $\varphi(v)$ is labelled with $\sigma(K)$.
    c) if v is labelled (M, N), then $\varphi(v)$ is labelled (M', N').
    d) if v is labelled $\{M\}_K$ and there exists a $u \in R_{E_1}$ labelled K, then $\varphi(v)$ is labelled $\{M'\}_{\sigma(K)}$ and $\varphi(u)$ is labelled $\sigma(K)$.
    e) if v is labelled $\{M\}_K$ and there does not exist a $u \in R_{E_1}$ labelled K, then $\varphi(v)$ is labelled $\{M'\}_{K'}$ and there does not exist a $u' \in R_{E_2}$ labelled K'.

The proof, mostly technical, appears in the full version of this paper [HG03].

We conclude with a brief discussion of some ramifications of the formal semantics we have seen in this section. We observe that under the above definitions, the encryption operator:

- "Preserves privacy", as seen in the equivalence $\{0\}_K \cong \{1\}_K$. Informally, a ciphertext conceals the underlying plaintext.
- "Conceals plaintext repetitions", as seen in the equivalence $(\{0\}_K, \{0\}_K) \cong (\{0\}_K, \{1\}_K)$. Informally, an adversary, given two ciphertexts, cannot tell whether their underlying plaintexts are identical or not.
- "Conceals key repetitions", as seen in the equivalence $(\{0\}_{K_1}, \{1\}_{K_1}) \cong (\{0\}_{K_7}, \{1\}_{K_8})$. Informally, an adversary, given two ciphertexts, cannot tell whether they were generated with the same encryption key or not.
- "Conceals plaintext length", as seen in the equivalence $\{0\}_K \cong \{(0, (1, 0))\}_K$. Informally, the ciphertext conceals the length of the underlying plaintext.

The definitions of semantics can be modified to accommodate relaxations of the above properties. For example, the semantics can be made sensitive to different plaintext lengths, by introducing an atomic pattern symbol $\square_n$ for each size n and modifying the definition of equivalence appropriately. We stress that the results of [AR02] and ours can be modified to tolerate such changes.

2.3   A Fixpoint Characterization of the Set of Reachable Nodes

The set of reachable nodes was defined in the previous section in set-intersection terms. Here, we give an alternative characterization that plays an important role in the proof of our completeness result. We show that for an expression E, the set of reachable nodes of T_E is the least fixpoint of an associated operator, O_E. In addition, we show that this fixpoint can be achieved by an iterative application of the operator, no more than a polynomial (in the size of E) number of times. The reader is referred to the full version of this paper [HG03] for a full account.

Let S be a finite set, and let 2^S be the set of all subsets of S. A set $A \subseteq 2^S$ is said to be a *fixpoint* of $O : 2^S \to 2^S$ if $O(A) = A$; A is said to be the *least fixpoint* of O, and is denoted $\mathrm{lfp}(O)$, if A is a fixpoint of O and for all fixpoints B of O, $A \subseteq B$. The *powers* of O are defined as follows:

$$O^0 = \emptyset$$

$$O^i = O(O^{i-1}) \qquad \text{for all } i \in \mathbb{N}^+$$

Consider an expression E with a derivation tree T_E over a set of nodes V_E with a root $r_E \in V_E$. Let $O_E : 2^{V_E} \to 2^{V_E}$ be defined as follows:

$$O_E(A) = \left\{ u \in V_E \left| \begin{array}{l} \text{either:} \\ \text{(a) } u = r_E; \text{ or} \\ \text{(b) } \exists v \in A \text{ labelled } (M, N) \text{ with a left child } u \text{ in } T_E \text{ (labelled } M); \text{ or} \\ \text{(c) } \exists v \in A \text{ labelled } (M, N) \text{ with a right child } u \text{ in } T_E \text{ (labelled } N); \text{ or} \\ \text{(d) } \exists v \in A \text{ labelled } \{M\}_K \text{ with a child } u \text{ in } T_E \text{ (labelled } M) \\ \quad \text{and } \exists w \in A \text{ labelled } K. \end{array} \right. \right\}$$

We prove the following:

Theorem 2.2 (A Fixpoint Characterization of the Set of Reachable Nodes [HG03]). *Let E be an expression of length n, T_E its derivation tree over V_E, $R_E \subseteq V_E$ the set of reachable nodes. Then there exists an $i \in \mathbb{N}$, $0 \le i \le n$, such that for all $j \ge i$, $O_E^j = \mathrm{lfp}(O_E) = R_E$.*

3   Computational Treatment of Symmetric Encryption, Computational Semantics for Expressions

In this section, we describe a computational treatment of symmetric encryption: we define an encryption scheme, discuss a relevant notion of security, and review methods of achieving such a notion under standard assumptions. The discussion is similar to the one in [AR02], and may be skipped without significant damage. We then use a computational encryption scheme to define a semantics for the language of expressions of subsection 2.1, recasting the corresponding definition of [AR02] in terms of the derivation trees of expressions.

3.1   Computational Treatment of Symmetric Encryption

Let $\{0,1\}^*$ denote the set of all finite binary strings and let $|x|$ denote the length of $x \in \{0,1\}^*$.

An *encryption scheme* $\Pi = (\mathcal{K}, \mathcal{E}, \mathcal{D})$ with a security parameter $\eta \in \mathbb{N}$ consists of three polynomial-time algorithms, as follows:

- $\mathcal{K}$, the *key generation algorithm*, is a probabilistic algorithm that takes a security parameter $\eta \in \mathbb{N}$ (provided in unary—denoted by 1^η) and returns a key $k \in \{0,1\}^*$. We write $k \xleftarrow{R} \mathcal{K}(1^\eta)$, thinking of k as being drawn from the probability distribution induced by $\mathcal{K}(1^\eta)$ on $\{0,1\}^*$. When used as a set, we let $\mathcal{K}(1^\eta)$ denote the support of that distribution.
- $\mathcal{E}$, the *encryption algorithm*, is a probabilistic algorithm that takes a key $k \in \mathcal{K}(1^\eta)$ for some $\eta \in \mathbb{N}$ and a *plaintext* $x \in \{0,1\}^*$ and returns a *ciphertext* $c \in \{0,1\}^* \cup \{\bot\}$. As before, we write $c \xleftarrow{R} \mathcal{E}_k(x)$, thinking of c as being drawn from the probability distribution induced by $\mathcal{E}_k(x)$ on $\{0,1\}^*$. When used as a set, we let $\mathcal{E}_k(x)$ denote the support of that distribution.

It is common for encryption schemes to restrict the set of strings they are willing to encrypt; having the encryption algorithm return $\perp$ is intended to capture such restrictions. We make two requirements. First, we insist that for a given $\eta \in \mathbb{N}$, a plaintext $x \in \{0,1\}^*$ is either *restricted* or not, that is, for all $k \in \mathcal{K}(1^\eta)$, $\mathcal{E}_k(x) = \{\perp\}$ or for all $k \in \mathcal{K}(1^\eta)$, $\mathcal{E}_k(x) \not\ni \perp$. We use $\text{Plain}_{\Pi[\eta]}$ to denote the set of unrestricted plaintexts, for any $\eta \in \mathbb{N}$. Second, we require that for any $\eta \in \mathbb{N}$, if $x \in \{0,1\}^*$ is not restricted, then all $x' \in \{0,1\}^*$ of the same length are unrestricted.

In addition, we insist that the length of a ciphertext $c \in \mathcal{E}_k(x)$ depend only on η and $|x|$ when $k \in \mathcal{K}(1^\eta)$, for any x and η.

- $\mathcal{D}$, the *decryption algorithm*, is a deterministic algorithm that takes a key $k \in \mathcal{K}(1^\eta)$ for some $\eta \in \mathbb{N}$ and a ciphertext $c \in \{0,1\}^*$ and returns some $x \in \{0,1\}^* \cup \{\perp\}$. We write $x \leftarrow \mathcal{D}_k(c)$.

Having the decryption algorithm output $\perp$ is intended to reflect a rejection of the given ciphertext.

We require that Π be *correct*; that is, for all $\eta \in \mathbb{N}$, for all $k \in \mathcal{K}(1^\eta)$ and for all $x \in \text{Plain}_{\Pi[\eta]}$, $\mathcal{D}_k(\mathcal{E}_k(x)) = x$.

A Notion of Security. We consider a variation of the standard notion of indistinguishability under chosen-plaintext attacks (*IND-CPA* security, for short) of [GM84,BD97]. Informally, the strengthened version "conceals key repetitions" and "conceals message lengths", as discussed in subsection 2.2. This is necessary for a soundness result (see [AR02] for additional motivation).

Recall that a function $\epsilon : \mathbb{N} \to \mathbb{R}$ is *negligible* if for every constant $c \in \mathbb{N}$ there exists an η_c such that for all $\eta > \eta_c$, $\epsilon(\eta) \leq \eta^{-c}$.

Let $\Pi = (\mathcal{K}, \mathcal{E}, \mathcal{D})$ be an encryption scheme, $\eta \in \mathbb{N}$ a security parameter and A an adversary with access to two oracles (denoted $A^{(\cdot),(\cdot)}$). Define:

$$\text{Adv}^0_{\Pi[\eta]}(A) = \Pr[k, k' \xleftarrow{R} \mathcal{K}(1^\eta) : A^{\mathcal{E}_k(\cdot),\mathcal{E}_{k'}(\cdot)}(1^\eta) = 1]$$
$$- \Pr[k \xleftarrow{R} \mathcal{K}(1^\eta) : A^{\mathcal{E}_k(0),\mathcal{E}_k(0)}(1^\eta) = 1],$$

where $\mathcal{E}_k(\cdot)$ is an oracle that returns $c \xleftarrow{R} \mathcal{E}_k(m)$ on input m, and $\mathcal{E}_k(0)$ is an oracle that returns $c \xleftarrow{R} \mathcal{E}_k(0)$ on input m. We say that Π is *Type-0/IND-CPA, Key-repetition Concealing, Length Concealing* secure [AR02] if for every probabilistic, polynomial-time adversary A, $\text{Adv}^0_{\Pi[\eta]}(A)$ is negligible (as a function of η).

Pseudorandom Function Families and Achieving Type-0 Security. Given a set S, let $x \xleftarrow{R} S$ denote the sampling of x from S endowed with a uniform distribution.

Let $\eta \in \mathbb{N}$, l, L be polynomials, $\text{Func}^{l(\eta) \to L(\eta)}$ the set of all functions from $\{0,1\}^{l(\eta)}$ to $\{0,1\}^{L(\eta)}$, $F \subseteq \text{Func}^{l(\eta) \to L(\eta)}$ a family of functions indexed by $\{0,1\}^\eta$, and A an adversary with access to an oracle (denoted $A^{(\cdot)}$). Define:

$$\text{Adv}^{\text{prf}}_{F[\eta]}(A) = \Pr[k \xleftarrow{R} \{0,1\}^\eta : A^{F_k(\cdot)}(1^\eta) = 1]$$
$$- \Pr[f \xleftarrow{R} \text{Func}^{l(\eta) \to L(\eta)} : A^{f(\cdot)}(1^\eta) = 1],$$

where $F_k(\cdot)$ is an oracle that returns $F_k(x)$ on input x, and $f(\cdot)$ is an oracle that returns $f(x)$ on input x. We say that F is *pseudorandom [GG86]* if for every probabilistic, polynomial time adversary A, $\mathrm{Adv}^{\mathrm{prf}}_{F[\eta]}(A)$ is negligible (as a function of η).

Pseudorandom function families are commonly used in computational cryptography as building blocks for encryption schemes, as in the CBC and CTR modes. In [BD97], it is shown that these modes are IND-CPA secure when using an underlying pseudorandom family of functions. In [AR02], the authors describe how these results extend to achieve Type-0 security. See the mentioned references for more details.

3.2   Computational Semantics for Expressions

In this section, we define a computational semantics for the language of expressions of section 2.1. We first associate an expression with an ensemble of distributions over $\{0,1\}^*$, resulting each from the "instantiation" of the expression with a concrete computational encryption scheme with a particular security parameter. We then define expression indistinguishability in terms of the indistinguishability of associated ensembles.

Let E be an expression, T_E its derivation tree over V_E, Keys_E the set of key symbols appearing in E (that is, atomic keys and keys from derivations of the form $\{\cdot\}_K$). Let $\Pi = (\mathcal{K}, \mathcal{E}, \mathcal{D})$ be an encryption scheme with a security parameter $\eta \in \mathbb{N}$. For $x_1, \ldots, x_k \in \{0,1\}^*$ and a tag t from some finite, fixed set of tags, let $\langle x_1, \ldots, x_k, t \rangle$ denote an (arbitrary, fixed, unambiguous, polynomial-time) encoding of $x_1, \ldots, x_k, t$ as a string over $\{0,1\}^*$. Define the following procedure:

Sample$_{\Pi[\eta]}(\mathbf{E})$

1. For each $K \in \mathrm{Keys}_E$, let $\tau(K) \xleftarrow{R} \mathcal{K}(1^\eta)$.
2. Assign a *sampling label* to each $v \in V_E$, inductively, as follows:
   a) If v is labelled with a bit b, let its *sampling label* be $\langle b, \text{"bit"} \rangle$.
   b) If v is labelled with a key K, let its *sampling label* be $\langle \tau(K), \text{"key"} \rangle$.
   c) If v ia labelled (M, N), its left child in T_E has a sampling label m and its right child in T_E has a sampling label n, then let the *sampling label* of v be $\langle m, n, \text{"pair"} \rangle$ if $m, n \neq \bot$, $\bot$ otherwise.
   d) If v is labelled $\{M\}_K$ and its child in T_E has a sampling label m, then let the *sampling label* of v be $\langle \mathcal{E}_{\tau(K)}(m), \text{"ciphertext"} \rangle$ if $m \neq \bot$, $\bot$ otherwise.
3. Output the sampling label of the root of T_E.

Let $[\![E]\!]_{\Pi(\eta)}$ denote the probability distribution induced by $\mathrm{Sample}_{\Pi[\eta]}(E)$ on $\{0,1\}^* \cup \bot$; let $[\![E]\!]_\Pi$ denote the ensemble $\{[\![E]\!]_{\Pi(\eta)}\}_{\eta \in \mathbb{N}}$.

We write $x \xleftarrow{R} D$ to indicate that x is sampled from a distribution D. To make our forthcoming definitions robust, we require that Π is such that for every expression E, there exists an $\eta_E \in \mathbb{N}$ such that for all $\eta \geq \eta_E$ and $e \xleftarrow{R} [\![E]\!]_{\Pi(\eta)}$, $e \in \mathrm{Plain}_{\Pi[\eta]}$.

For $i \in \{1, 2\}$, let $D_i = \{D_i(\eta)\}_{\eta \in \mathbb{N}}$ be probability distribution ensembles, A an algorithm. Define:

$$\mathrm{Adv}^{\mathrm{ind}}_{D_1(\eta), D_2(\eta)}(A) = \Pr[x \overset{R}{\leftarrow} D_1(\eta) : A(1^\eta, x) = 1]$$
$$- \Pr[x \overset{R}{\leftarrow} D_2(\eta) : A(1^\eta, x) = 1].$$

We say that D_1, D_2 are *indistinguishable*, and write $D_1 \approx D_2$, if for every probabilistic, polynomial time algorithm A, $\mathrm{Adv}^{\mathrm{ind}}_{D_1(\eta), D_2(\eta)}(A)$ is negligible (as a function of η).

Let E_1, E_2 be expressions. We say that E_1, E_2 are *indistinguishable*, and write $E_1 \overset{\Pi}{\approx} E_2$, iff $[\![E_1]\!]_\Pi \approx [\![E_2]\!]_\Pi$.

4   Weak Key-Authenticity Tests for Expressions, Semantic Completeness

The soundness result of Abadi and Rogaway states that for acyclic expressions[1] E_1, E_2 and a Type-0 encryption scheme Π, $E_1 \cong E_2$ implies $E_1 \overset{\Pi}{\approx} E_2$. Here, we give a *necessary* and *sufficient* condition for completeness, tightly characterizing this aspect of the exposition. For any two acyclic expressions, the condition involves the admittance of an efficient test that distinguishes a ciphertext and the key it was encrypted with, from a ciphertext and some random key, with a noticeable probability, when the plaintexts are drawn from the ensembles associated with those expressions. Formally:

Definition 4.1 (Weak Key-Authenticity Test for Expressions). Let $\Pi = (\mathcal{K}, \mathcal{E}, \mathcal{D})$ be an encryption scheme with a security parameter $\eta \in \mathbb{N}$, let E_1, E_2 be acyclic expressions, A an algorithm. Define:

$$\mathrm{Adv}^{\mathrm{wka\text{-}exp}}_{\Pi[\eta], E_1, E_2}(A)$$
$$= \Pr[e \overset{R}{\leftarrow} [\![E_1]\!]_{\Pi(\eta)}; k \overset{R}{\leftarrow} \mathcal{K}(1^\eta); c \overset{R}{\leftarrow} \mathcal{E}_k(e) : A(1^\eta, c, k) = 1]$$
$$- \Pr[e \overset{R}{\leftarrow} [\![E_2]\!]_{\Pi(\eta)}; k, k' \overset{R}{\leftarrow} \mathcal{K}(1^\eta); c \overset{R}{\leftarrow} \mathcal{E}_k(e) : A(1^\eta, c, k') = 1].$$

We say that Π *admits a weak key-authenticity test for* E_1, E_2 (*WKA-EXP-* (E_1, E_2) *test*, for short), if there exists a probabilistic, polynomial-time algorithm A such that $\mathrm{Adv}^{\mathrm{wka\text{-}exp}}_{\Pi[\eta], E_1, E_2}(A)$ is non-negligible (as a function of η).

We say that Π *admits weak key-authenticity tests for expressions* (*WKA-EXP tests*, for short), if for all acyclic expressions E_1 and E_2, Π admits a weak key-authenticity test for E_1, E_2.

Our main result is the following:

Theorem 4.2 (The admittance of WKA-EXP tests is necessary and sufficient for completeness). *Let* $\Pi = (\mathcal{K}, \mathcal{E}, \mathcal{D})$ *be an encryption scheme. Then for all acyclic expressions* E_1 *and* E_2, $E_1 \overset{\Pi}{\approx} E_2$ *implies* $E_1 \cong E_2$ *iff* Π *admits weak key-authenticity tests for expressions.*

---

[1] Expressions that do not contain "encryption cycles"; see [AR02] for a formal definition.

We begin by proving the necessity part. Let E_1, E_2 be two acyclic expressions. Consider the expressions $M_1 = (\{E_1\}_K, K)$, $M_2 = (\{E_2\}_K, K')$ (without loss of generality, assume K does not occur in E_1, E_2). $M_1 \not\cong M_2$, so by the completeness assumption $M_1 \not\approx M_2$. Let B be such that $\mathrm{Adv}^{\mathrm{ind}}_{[\![M_1]\!]_{\Pi(\eta)}, [\![M_2]\!]_{\Pi(\eta)}}(B)$ is non-negligible. We use B to construct a WKA-EXP-(E_1, E_2) test A for Π. Define: $A(1^\eta, c, k) \stackrel{\mathrm{def}}{=} B(1^\eta, \langle\langle c, \text{``ciphertext''}\rangle, \langle k, \text{``key''}\rangle, \text{``pair''}\rangle)$. Now:

$$
\begin{aligned}
&\mathrm{Adv}^{\mathrm{wka\text{-}exp}}_{\Pi[\eta], E_1, E_2}(A) \\
&= \Pr[e \stackrel{R}{\leftarrow} [\![E_1]\!]_{\Pi(\eta)}; k \stackrel{R}{\leftarrow} \mathcal{K}(1^\eta); c \stackrel{R}{\leftarrow} \mathcal{E}_k(e) : A(1^\eta, c, k) = 1] \\
&\quad - \Pr[e \stackrel{R}{\leftarrow} [\![E_2]\!]_{\Pi(\eta)}; k, k' \stackrel{R}{\leftarrow} \mathcal{K}(1^\eta); c \stackrel{R}{\leftarrow} \mathcal{E}_k(e) : A(1^\eta, c, k') = 1] \\
&= \Pr \left[\begin{array}{l} e \stackrel{R}{\leftarrow} [\![E_1]\!]_{\Pi(\eta)}; \\ k \stackrel{R}{\leftarrow} \mathcal{K}(1^\eta); \\ c \stackrel{R}{\leftarrow} \mathcal{E}_k(e) \end{array} : B\left(1^\eta, \begin{array}{l} \langle\langle c, \text{``ciphertext''}\rangle, \\ \langle k, \text{``key''}\rangle, \text{``pair''}\rangle \end{array}\right) = 1 \right] \\
&\quad - \Pr \left[\begin{array}{l} e \stackrel{R}{\leftarrow} [\![E_2]\!]_{\Pi(\eta)}; \\ k, k' \stackrel{R}{\leftarrow} \mathcal{K}(1^\eta); \\ c \stackrel{R}{\leftarrow} \mathcal{E}_k(e) \end{array} : B\left(1^\eta, \begin{array}{l} \langle\langle c, \text{``ciphertext''}\rangle, \\ \langle k', \text{``key''}\rangle, \text{``pair''}\rangle \end{array}\right) = 1 \right] \\
&= \Pr[e \stackrel{R}{\leftarrow} [\![(\{E_1\}_K, K)]\!]_{\Pi(\eta)} : B(1^\eta, e) = 1] \\
&\quad - \Pr[e \stackrel{R}{\leftarrow} [\![(\{E_2\}_K, K')]\!]_{\Pi(\eta)} : B(1^\eta, e) = 1] \\
&= \Pr[e \stackrel{R}{\leftarrow} [\![M_1]\!]_{\Pi(\eta)} : B(1^\eta, e) = 1] - \Pr[e \stackrel{R}{\leftarrow} [\![M_2]\!]_{\Pi(\eta)} : B(1^\eta, e) = 1] \\
&= \mathrm{Adv}^{\mathrm{ind}}_{[\![M_1]\!]_{\Pi(\eta)}, [\![M_2]\!]_{\Pi(\eta)}}(B),
\end{aligned}
$$

where the second equality is due to the definition of A, and the third due to the definition of $\mathrm{Sample}_{\Pi[\eta]}$. It follows that A is a weak key-authenticity test for E_1, E_2, as required. This completes the necessity part of the proof.

Next, we sketch the sufficiency part; a complete proof appears in [HG03]. Assume $E_1 \not\cong E_2$. To show that $[\![E_1]\!]_{\Pi(\eta)} \not\approx [\![E_2]\!]_{\Pi(\eta)}$, we consider an algorithm that simultaneously parses its input e, a sample from either $[\![E_1]\!]_{\Pi(\eta)}$ or $[\![E_2]\!]_{\Pi(\eta)}$, and expressions E_1, E_2, attempting to construct φ, σ that bear witness to the equivalence of the expressions. By the assumption, this attempt is bound to fail. We show that upon failure, the algorithm has enough parsed information to predict the origin of the sample with a non-negligible probability of success. In some cases, the prediction depends on an application of a weak key-authenticity test for (particular, fixed) expressions to the amassed information.

Specifically, the algorithm computes the powers of the operator $O_{E_1, E_2, e}$, defined in Fig. 1 (where $S = V_{E_1} \times V_{E_2} \times \{0, 1\}^*$), as long as they satisfy the predicate TEST, also of Fig. 1. Let $i \in \mathbb{N}$. Let $V^i_{E_1} = \{v_1 \,|\, (v_1, \cdot, \cdot) \in O^i_{E_1, E_2, e}\}$, $V^i_{E_2} = \{v_2 \,|\, (\cdot, v_2, \cdot) \in O^i_{E_1, E_2, e}\}$. Let $j \in \{1, 2\}$. Let $T^{V^i_{E_j}}_{E_j}$ denote the subtree induced by $V^i_{E_j}$ on T_{E_j}. Let O_{E_j} be the operator from the fixpoint characterizations of the set of reachable nodes of T_{E_j} (see Theorem 2.2). We show that as long as $\mathrm{TEST}(O^i_{E_1, E_2, e})$ holds,

- $V^{i+1}_{E_1} = O^{i+1}_{E_1}$ and $V^{i+1}_{E_2} = O^{i+1}_{E_2}$;
- there exist φ, σ consistent with the requirements of Theorem 2.1 when restricted to $T^{V^i_{E_1}}_{E_1}$, $T^{V^i_{E_2}}_{E_2}$, $V^i_{E_1}$, and $V^i_{E_2}$ (instead of $T^R_{E_1}$, $T^R_{E_2}$, R_{E_1}, and R_{E_2}).

$O_{E_1,E_2,e}(A) =$

$$
\begin{cases}
(u_1, u_2, y) \\ \in S
\end{cases}
\left|
\begin{array}{l}
\text{either:} \\
\text{(a) } u_1 = r_{E_1},\, u_2 = r_{E_2},\, y = e;\text{ or} \\
\text{(b) } \exists (v_1, v_2, x) \in A \text{ such that:} \\
\quad v_1 \text{ is labelled } (M, N) \text{ and has a left child } u_1 \text{ in } T_{E_1}, \\
\quad v_2 \text{ is labelled } (M', N') \text{ and has a left child } u_2 \text{ in } T_{E_2} \\
\quad \text{and } x \text{ is of the form } \langle y, z, \text{``pair''}\rangle;\text{ or} \\
\text{(c) } \exists (v_1, v_2, x) \in A \text{ such that:} \\
\quad v_1 \text{ is labelled } (M, N) \text{ and has a right child } u_1 \text{ in } T_{E_1}, \\
\quad v_2 \text{ is labelled } (M', N') \text{ and has a right child } u_2 \text{ in } T_{E_2} \\
\quad \text{and } x \text{ is of the form } \langle y, z, \text{``pair''}\rangle;\text{ or} \\
\text{(d) } \exists (v_1, v_2, x) \in A \text{ and } \exists (w_1, w_2, z) \in A \text{ such that:} \\
\quad v_1 \text{ is labelled } \{M\}_K \text{ and has a child } u_1 \text{ in } T_{E_1}, \\
\quad v_2 \text{ is labelled } \{M'\}_{K'} \text{ and has a child } u_2 \text{ in } T_{E_2}, \\
\quad x \text{ is of the form } \langle c, \text{``ciphertext''}\rangle, \\
\quad w_1 \text{ is labelled } K, \\
\quad w_2 \text{ is labelled } K', \\
\quad z \text{ is of the form } \langle k, \text{``key''}\rangle \\
\quad \text{and } y = \mathcal{D}_k(c).
\end{array}
\right.
$$

$\text{TEST}(A) =$

$$
\begin{cases}
\textit{true} & \text{if for all } (v_1, v_2, x) \in A, \text{ either:} \\
& \text{(a) } v_1 \text{ is labelled with } b \in \mathbf{Bits} \text{ and } v_2 \text{ is labelled } b;\text{ or} \\
& \text{(b) } v_1 \text{ is labelled } K, v_2 \text{ is labelled } K' \text{ and for all } (u_1, u_2, y) \in A, \\
& \quad u_1 \text{ is labelled } K \text{ iff } u_2 \text{ is labelled } K';\text{ or} \\
& \text{(c) } v_1 \text{ is labelled } (M, N) \text{ and } v_2 \text{ is labelled } (M', N');\text{ or} \\
& \text{(d) } v_1 \text{ is labelled } \{M\}_K, v_2 \text{ is labelled } \{M'\}_{K'} \text{ and for all} \\
& \quad (u_1, u_2, y) \in A, u_1 \text{ is labelled } K \text{ iff } u_2 \text{ is labelled } K'. \\
\textit{false} & \text{otherwise.}
\end{cases}
$$

Fig. 1. Definitions of $O_{E_1,E_2,e} : 2^S \to 2^S$, $\text{TEST} : 2^S \to \{\text{true}, \text{false}\}$.

If TEST does not fail by the $\max(|E_1|, |E_2|)$'s power of $O_{E_1,E_2,e}$, V_{E_1}, V_{E_2} achieve the sets of reachable nodes of T_{E_1}, T_{E_2}, respectively, by the first point above, and so $E_1 \cong E_2$ by the second point, contradicting our assumption. We conclude that TEST must fail on some lower power of $O_{E_1,E_2,e}$; let $i^* \in \mathbb{N}$ be the lowest such power.

We use $O_{E_1,E_2,e}^{i^*}$ to make a prediction, based on the reason TEST fails. Here, we illustrate a case that calls for the use of a weak key-authenticity test for expressions. Assume TEST fails because there exist $(v_1, v_2, x), (u_1, u_2, y) \in O_{E_1,E_2,e}^{i^*}$ such that v_1 is labelled $\{M\}_K$, u_1 is labelled K, v_2 is labelled $\{M'\}_{K'}$, and u_2 is labelled K''. An inductive argument on the powers of our operator shows that x, y are the sampling labels of either v_1, u_1, respectively, or v_2, u_2, respectively, depending on the origin of e. Let $x = \langle c, \text{``ciphertext''}\rangle, y = \langle k, \text{``key''}\rangle$. In the first case, c is an encryption of a sample from $[\![M]\!]_{\Pi(\eta)}$ with the key k; in the second case, c is an encryption of a sample from $[\![M']\!]_{\Pi(\eta)}$ with some key, and k is a random key. The WKA-EXP-(M,M') test on c and k distinguishes these cases with a noticeable probability of success.

Finally, we show that the procedure is efficient. This completes the sketch of the sufficiency part of the proof.

5   How the Notion of Weak Key-Authenticity Relates to Other Cryptographic Notions

In this section, we strengthen the notion of the admittance of weak key authenticity tests for expressions. We consider the admittance of a *single*, all-purpose test, hereby referred to as the *weak key-authenticity test*, that distinguishes any ciphertext and the key it was encrypted with from any ciphertext and a random key, with a non-negligible probability; the test is defined in terms that are *independent* of the formal language of the preceding sections. We compare the strengthened version with the notions of confusion-freedom and authenticated encryption, previously discussed in the literature in the context of the completeness result [MW02,AJ01]. Specifically, we show that the requirement that an encryption scheme admits a weak key-authenticity test is *strictly weaker* than the requirement that it be confusion-free, as defined in the above references (which, in turn, is enough to show it is strictly weaker than authenticated encryption as well). To that effect, we present an encryption scheme that admits a weak key-authenticity test but *is not* confusion-free. The scheme we present is also Type-0. It therefore satisfies the soundness criteria of [AR02], our completeness criteria, but not the previous completeness criteria of [MW02]. The notions we present and the methods used to achieve the admittance of a weak key-authenticity test should be of independent interest.

Informally, confusion-freedom captures the ability of a decryption algorithm to distinguish a ciphertext and the key it was encrypted with from a ciphertext and a random key with almost full certainty. In contrast, the weak key-authenticity test is required to distinguish the two with merely a noticeable probability. We will separate the notions in a strong sense, pertaining directly to the gap in their required distinguishing certainties (as opposed to pertaining to the placement of the distinguisher—inside or outside the decryption algorithm).

First, we give formal definitions for the notions at hand.

Confusion-freedom is defined as it appears in the completeness result of [MW02]; our proofs can be modified to accommodate the version of [AJ01] too.

Definition 5.1 (Confusion-Freedom). Let $\Pi = (\mathcal{K}, \mathcal{E}, \mathcal{D})$ be an encryption scheme, $\eta \in \mathbb{N}$ a security parameter, and $D[\eta] = \{D_1[\eta], \dots, D_l[\eta]\}$ a series of finite sets of distributions. For $1 \leq i \leq l$, define:

$$\mathrm{Adv}^{\mathrm{cf}}_{\Pi[\eta],D[\eta],i} = \Pr[k, k' \xleftarrow{R} \mathcal{K}(1^\eta); x \xleftarrow{R} D_i[\eta] : \mathcal{D}_{k'}(\mathcal{E}_k(x)) \neq \perp].$$

We say that Π is *confusion-free* (*CF* for short) if for any $1 \leq i \leq l$, $\mathrm{Adv}^{\mathrm{cf}}_{\Pi[\eta],D[\eta],i}$ is negligible (as a function of η).

Next, we define two auxiliary notions that will enable us to focus on the the above-mentioned gap. These will provide a "middle ground" for comparing the WKA-EXP test with CF.

Definition 5.2 (Strong Key-Authenticity Test, Weak Key-Authenticity Test). Let $\Pi = (\mathcal{K}, \mathcal{E}, \mathcal{D})$ be an encryption scheme, $\eta \in \mathbb{N}$ a security parameter. Let $\mathcal{P}_1, \mathcal{P}_2$ (hereby referred to as *plaintext generators*) be probabilistic algorithms that take a security parameter η (provided in unary), and for sufficiently large η always return a $x \in \text{Plain}_{\Pi[\eta]}$; we write $x \stackrel{R}{\leftarrow} \mathcal{P}_j(1^\eta)$ for $j \in \{1, 2\}$, thinking of x as being drawn from the probability distribution induced by $\mathcal{P}_j(1^\eta)$ on $\{0,1\}^*$. Let A be an algorithm. Define:

$$\text{Adv}^{\text{tst}}_{\Pi[\eta], \mathcal{P}_1[\eta], \mathcal{P}_2[\eta]}(A)$$
$$= \Pr[x \stackrel{R}{\leftarrow} \mathcal{P}_1(1^\eta); k \stackrel{R}{\leftarrow} \mathcal{K}(1^\eta); c \stackrel{R}{\leftarrow} \mathcal{E}_k(x) : A(1^\eta, c, k) = 1]$$
$$- \Pr[x \stackrel{R}{\leftarrow} \mathcal{P}_2(1^\eta); k, k' \stackrel{R}{\leftarrow} \mathcal{K}(1^\eta); c \stackrel{R}{\leftarrow} \mathcal{E}_k(x) : A(1^\eta, c, k') = 1],$$

where tst $\in \{\text{ska}, \text{wka}\}$. We say that Π admits a *strong* (resp., *weak*) *key-authenticity test*, *SKA* (resp., *WKA*) for short, if there exists a probabilistic, polynomial-time algorithm A such that for all probabilistic, polynomial-time algorithms $\mathcal{P}_1, \mathcal{P}_2$, $\text{Adv}^{\text{ska}}_{\Pi[\eta], \mathcal{P}_1[\eta], \mathcal{P}_2[\eta]}(A)$ (resp., $\text{Adv}^{\text{wka}}_{\Pi[\eta], \mathcal{P}_1[\eta], \mathcal{P}_2[\eta]}(A)$) is negligibly close to 1 (resp., is non-negligible) as a function of η.

As for the definition of *integrity of plaintext* security (*INT-PTXT* for short), a flavor of authenticated encryption, we refer the reader to [BN00,KY00] and to [MW02].

The following diagram depicts relationships between our notions of interest.

$$
\begin{array}{ccccccc}
\text{INT-} & & & \text{Admittance} & \xrightarrow{} & \text{Admittance} & & \text{Admittance} \\
\text{PTXT} & \longrightarrow & \text{CF} \longrightarrow & \text{of a} & & \text{of a} & \longrightarrow & \text{of} \\
& & & \text{SKA test} & \xnrightarrow{} & \text{WKA test} & & \text{WKA-EXP tests}
\end{array}
$$

In the above, $A \longrightarrow B$ means that an encryption scheme that meets notion A must also meet notion B; we call such a relationship an *implication*. $A \not\longrightarrow B$ means that an encryption scheme that meets notion A does not necessarily meet notion B; we call such a relationship a *separation*.

The implications in the diagram are mostly straightforward (see [HG03]). The rest of the section is devoted to the separation of WKA from SKA. To that end, we show an encryption scheme that admits a WKA test but does not admit an SKA test. We use a standard construction based on a pseudorandom function family, with an added "weak redundancy". To simplify the exposition, we use a single, constant bit as redundancy; refer to the end of the section for a generalization.

Let F be a pseudorandom family of functions with a security parameter $\eta \in \mathbb{N}$, key domain $\{0,1\}^\eta$, domain $\{0,1\}^{l(\eta)}$ and range $\{0,1\}^{L(\eta)}$ (where l, L are polynomials in η); let ϵ be a negligible function such that $\text{Adv}^{\text{prf}}_{F[\eta]}(A) \leq \epsilon(\eta)$ for any probabilistic, polynomial-time algorithm A. We use $x_1 x_2 \cdots x_m$ to denote the individual bits of a string $x \in \{0,1\}^m$. We use $\circ$ to denote the concatenation operator on strings of bits, $\oplus$ to denote the bitwise XOR operator on strings of bits of equal length.

Define an encryption scheme $\Pi^* = (\mathcal{K}^*, \mathcal{E}^*, \mathcal{D}^*)$ with a security parameter $\eta \in \mathbb{N}$ as follows:

$$
\begin{array}{c|c|c}
\begin{array}{l}
\mathcal{K}^*(1^\eta) \\
k \stackrel{R}{\leftarrow} \{0,1\}^\eta; \\
\text{Output } k.
\end{array}
&
\begin{array}{l}
\mathcal{E}_k^*(x = x_1 x_2 \cdots x_{L(\eta)-1}) \\
r \stackrel{R}{\leftarrow} \{0,1\}^{l(\eta)}; \\
y \leftarrow (x \circ 1) \oplus F_k(r); \\
\text{Output } \langle y, r \rangle.
\end{array}
&
\begin{array}{l}
\mathcal{D}_k^*(\langle y = y_1 y_2 \cdots y_{L(\eta)}, r \rangle) \\
x' \leftarrow y \oplus F_k(r); \\
\text{Output } x_1' x_2' \cdots x_{L(\eta)-1}'.
\end{array}
\end{array}
$$

Note that $\text{Plain}_{\Pi^*[\eta]} = \{0,1\}^{L(\eta)-1}$. Also note that $\mathcal{E}^*$ and $\mathcal{D}^*$ can deduce η from k ($\eta = |k|$).

Π^* can easily be shown to be IND-CPA secure based on the pseudorandomness of F. For a proof, see [GG86], or simply think of Π^* as a degenerate version of the randomized CTR mode, and rely on [BD97]. Using the results of [AR02], it can further be shown to be Type-0.

We have that:

Theorem 5.3. Π^* *admits a WKA test.*

To see this, consider an algorithm that takes as input $\langle y, r \rangle$ and k, computes $y \oplus F_k(r)$ and outputs 1 iff the last bit of the outcome is 1. The algorithm is a WKA test for Π^* by a simple reduction to the pseudorandomness of F (see [HG03]). In addition, we have that:

Theorem 5.4. Π^* *does not admit an SKA test.*

Proof. Let A be a probabilistic algorithm that runs in time t, a function of the size of its input. Let $A(a_1, a_2, \dots; w)$ denote the outcome of running A on inputs $a_1, a_2, \dots$ and randomness w. Note that the length of w is bounded by t.

Let $\mathcal{U}$ be an algorithm that takes $\eta \in \mathbb{N}$ (in unary) as input and outputs a random, uniformly-selected element of $\{0,1\}^{L(\eta)-1}$. We have:

$$
\begin{aligned}
&\text{Adv}_{\Pi^*[\eta], \mathcal{U}[\eta], \mathcal{U}[\eta]}^{\text{ska}}(A) \\
&= \Pr\left[
\begin{array}{l}
x \stackrel{R}{\leftarrow} \{0,1\}^{L(\eta)-1}; k \stackrel{R}{\leftarrow} \{0,1\}^\eta; r \stackrel{R}{\leftarrow} \{0,1\}^{l(\eta)}; \\
w \stackrel{R}{\leftarrow} \{0,1\}^{t(\eta)}; y \leftarrow (x \circ 1) \oplus F_k(r)
\end{array}
: A(1^\eta, \langle y, r \rangle, k; w) = 1 \right] \\
&\quad - \Pr\left[
\begin{array}{l}
x \stackrel{R}{\leftarrow} \{0,1\}^{L(\eta)-1}; k, k' \stackrel{R}{\leftarrow} \{0,1\}^\eta; r \stackrel{R}{\leftarrow} \{0,1\}^{l(\eta)}; \\
w \stackrel{R}{\leftarrow} \{0,1\}^{t(\eta)}; y \leftarrow (x \circ 1) \oplus F_k(r)
\end{array}
: A(1^\eta, \langle y, r \rangle, k'; w) = 1 \right],
\end{aligned}
$$

where t is a polynomial in η.

Let S_1 and $A_1 \subseteq S_1$ denote the sample space and event, respectively, depicted by the first term above. Let S_2 and $A_2 \subseteq S_2$ be defined similarly with respect to the second term.

Let $(x_0, k_0, r_0, w_0) \in A_1$. Note that for any $k \in \{0,1\}^\eta$, if there exists an $x \in \{0,1\}^{L(\eta)-1}$ such that $(x \circ 1) \oplus F_k(r_0) = (x_0 \circ 1) \oplus F_{k_0}(r_0)$, then it must be the case that $(x, k, k_0, r_0, w_0) \in A_2$ (because in this case, A, in the second experiment, runs on the same input and randomness as in the first experiment). This happens when $x \circ 1 = (x_0 \circ 1) \oplus F_{k_0}(r_0) \oplus F_k(r_0)$, which must happen for at least $\left(\frac{1}{2} - \epsilon(\eta)\right) \cdot 2^\eta$ of the keys $k \in \{0,1\}^\eta$; otherwise, an adversary that queries its oracle on r_0, XORs the answer with $(x_0 \circ 1)$ and with $F_{k_0}(r_0)$, and outputs 1 if the last bit of the result is different than 1, 0 otherwise—breaks the pseudorandomness of F.

For a given $(x_0, k_0, r_0, w_0) \in A_1$, we've just described a way of counting at least $\left(\frac{1}{2} - \epsilon(\eta)\right) \cdot 2^\eta$ tuples in A_2. We would like to argue that for a distinct $(x_1, k_1, r_1, w_1) \in A_1$, we would be counting *different* tuples in A_2 by employing the same method. This is clear if $k_1 \neq k_0$ or $r_1 \neq r_0$ or $w_1 \neq w_0$. As for the case that $k_1 = k_0, r_1 = r_0, w_1 = w_0$, we would be double-counting a tuple iff

$$(x_0 \circ 1) \oplus F_{k_0}(r_0) \oplus F_k(r_0) = (x_1 \circ 1) \oplus F_{k_1}(r_1) \oplus F_k(r_1) = (x_1 \circ 1) \oplus F_{k_0}(r_0) \oplus F_k(r_0),$$

which happens iff $x_1 = x_0$.

We conclude that $|A_2| \geq \left(\frac{1}{2} - \epsilon(\eta)\right) \cdot 2^\eta \cdot |A_1|$. We also know that $|S_2| = 2^\eta \cdot |S_1|$. Therefore:

$$\mathrm{Adv}^{\mathrm{ska}}_{\Pi^*[\eta], \mathcal{U}[\eta], \mathcal{U}[\eta]}(A) = \frac{|A_1|}{|S_1|} - \frac{|A_2|}{|S_2|} \leq \left(\frac{1}{2} + \epsilon(\eta)\right) \cdot \frac{|A_1|}{|S_1|} \leq \frac{1}{2} + \epsilon(\eta),$$

which is *not* negligibly close to 1. $\blacksquare$

Finally, we note that our construction can be easily generalized to one that admits a WKA test with an advantage *as small as desired*, as follows. For any $c \in \mathbb{N}^+$, let Π^*_c be a variation on Π^* that adds the bit 1 with probability $\frac{1}{2} + \frac{1}{2^c}$, 0 with probability $\frac{1}{2} - \frac{1}{2^c}$, as redundancy upon encryption (instead of the fixed 1). Our proofs easily extend to show that Π^*_c admits a WKA test with advantage at least $\frac{1}{2^c} - \epsilon(\eta)$.

Acknowledgements. We thank Jonathan Katz for helpful discussions and comments. This work was supported by the Defense Advanced Research Projects Agency and managed by the U.S. Air Force Research Laboratory under contract F30602-00-2-0510; the views and conclusions contained are those of the authors and should not be interpreted as representing the official policies, either expressed or implied, of DARPA, U.S. AFRL, or the U.S. Government.

References

[AJ01]   M. Abadi, J. Jurgens. Formal Eavesdropping and its Computational Interpretation. In *Proc. of the Fourth International Symposium on Theoretical Aspects of Computer Software (TACS 2001)*, 2001.

[AR02]   M. Abadi, P. Rogaway. Reconciling Two Views of Cryptography (the computational soundness of formal encryption). In *Journal of Cryptology*, vol. 15, no. 2, pp. 103–128. (also in *Proc. of the First IFIP International Conference on Theoretical Computer Science*, LNCS vol. 1872, pp. 3–22, Springer Verlag, Berlin, August 2000.)

[BB01]   M. Bellare, A. Boldyreva, A. Desai, D. Pointcheval. Key-Privacy in Public-Key Encryption. In *Advances in Cryptology — ASIACRYPT 2001*, LNCS vol. 2248,pp. 566-582, Springer Verlag, 2001.

[BD97]   M. Bellare, A. Desai, E. Jokipii, P. Rogaway. A Concrete Security Treatment of Symmetric Encryption: Analysis of the DES Modes of Operation. In *Proceedings of the 38th Annual Symposium on Foundations of Computer Science (FOCS 97)*, 1997.

[BN00]   M. Bellare, C. Namprempre. Authenticated Encryption: Relations Among Notions and Analysis of the Generic Composition Paradigm. In *Advances in Cryptology — ASIACRYPT 2000*, LNCS vol. 1976, pp. 541–545, Springer Verlag, 2000.

[GG86]   O. Goldreich, S. Goldwasser, S. Micali. How to Construct Random Functions. In *Journal of the ACM*, vol. 33, no. 4, pp. 792–807, 1986.

[GM84]   S. Goldwasser, S. Micali. Probabilistic Encryption. In *Journal of Computer and System Sciences*, 28:270-299, April 1984.

[HG03]   O. Horvitz, V. Gligor. Weak Key Authenticity and the Computational Completeness of Formal Encryption. Full version available at
          http://www.cs.umd.edu/~horvitz, http://www.ee.umd.edu/~gligor

[KY00]   J. Katz, M. Yung. Unforgeable Encryption and Chosen Ciphertext Secure Modes of Operation. In *Proceedings of the 7th International Workshop on Fast Software Encryption (FSE 2000)*, LNCS vol. 1978, pp. 284–299, Springer Verlag, 2000.

[Ll87]   J. W. Lloyd. *Foundations of Logic Programming*. Second Edition, Springer-Verlag, 1987, section 1.5.

[LN84]   J. L. Lassez, V. L. Nguyen, E. A. Sonenberg. Fixpoint Theorems and Semantics: a Folk Tale. In *Information Processing Letters*, vol. 14, no. 3, 1982, pp. 112–116.

[MW02]   D. Micciancio, B. Warinschi. Completeness Theorems for the Abadi-Rogaway Language of Encrypted Expressions. In *Journal of Computer Security* (to appear). Also in *Proceedings of the Workshop on Issues in the Theory of Security*, 2002.

[Ta55]   A. Tarski. A Lattice-theoretical Fixpoint Theorem and its Applications. In *Pacific Journal of Mathematics*, vol. 5, pp.285–309, 1955.

Plaintext Awareness via Key Registration

Jonathan Herzog, Moses Liskov, and Silvio Micali

MIT Laboratory for Computer Science

Abstract. In this paper, we reconsider the notion of plaintext aware-ness. We present a new model for plaintext-aware encryption that is both natural and useful. We achieve plaintext-aware encryption without random oracles by using a third party. However, we do not need to trust the third party: even when the third party is dishonest, we still guar-antee security against adaptive chosen ciphertext attacks. We show a construction that achieves this definition under general assumptions. We further motivate this achievement by showing an important and natural application: giving additional real-world meaningfulness to the Dolev-Yao model.

1 Introduction

In this paper, we put forward and implement a new notion of plaintext-aware encryption that is both natural and useful.

A Beautiful But Controversial Notion. As insightfully introduced by Bel-lare and Rogaway [1] (see also [2] for refinements), an encryption scheme is *plaintext-aware* (PA) if, whenever an adversary creates a ciphertext, he must "know" its corresponding plaintext.

Despite its natural appeal, PA encryption has been somewhat controversial for two main reasons:

1. *Plaintext awareness fundamentally relies on random oracles.*
   Not only do all known implementations of PA encryption use random ora-cles, but the very definition of plaintext awareness has, so far, crucially de-pended upon them. Random oracles are fundamentally abstract constructs. Although sometimes they can be realized algorithmically, no such hope ex-ists here: traditional PA encryption requires the random oracle not only to be random, but also to be an *oracle*.
   A random oracle is in essence a trusted third party that interacts with the rest of us only in a very rigid way: if one puts a string x on a special query tape, it will write a random bit b_x on a special answer tape. This codified in-terface guarantees that even an adversary who purposely tries not to "know" what he is doing must be aware of his queries to the random oracle: after all he has to explicitly write each and every bit of x on the query tape! It is this elementary awareness that is cleverly exploited by Bellare and Rogaway to imply a much more sophisticated awareness: by looking at just the queries

D. Boneh (Ed.): CRYPTO 2003, LNCS 2729, pp. 548–564, 2003.

that an adversary makes to the random oracle during the computation of a ciphertext, one can easily deduce the underlying plaintext. The random oracle thus provides a magical "window" into the state of the encrypting algorithm, forcing it to disclose parts of its internal state.

One can hardly fault the inventors of plaintext-awareness for depending on random oracles: without any additional help plaintext awareness looks to be essentially impossible.

2. *Plaintext awareness has found no important and novel applications.*

Plaintext-awareness is so strong a property that it immediately implies security against chosen-ciphertext attacks (CCA-2 security to be exact, in the notation of [2]). In essence, if the adversary already "knows" the answer that it will receive from a decryption oracle, then the oracle gives him no additional power.

However, CCA-2 secure schemes were already known: they were constructed by Cramer and Shoup [3] under the decisional Diffie-Hellman assumption (yielding a very efficient scheme), and by Sahai [4] (improving on previous work of Naor and Yung [5]) under very general complexity assumptions. Thus, genuinely new applications of PA encryption, despite its intuitive great power, have been somewhat scarce.

Our Contributions. The main contributions of this paper are:

1. A new *definition* of PA encryption that does not use the random oracle.
2. An *implementation* of the new definition that is based on very general complexity assumptions.
3. A new and natural *application* of PA encryption that requires its full power.

To be sure, we still need to access a trusted third party, but our party is much more natural (being already used in practice) and we access it only once rather than at every encryption.

The Essence of Our Definition. Our model is very simple: encryption is available only between users who have properly registered their public keys with a *registration authority*, and plaintext-awareness is guaranteed if this authority is honest.

This third-party model has several attractive features:

– *Safety:* Only the *plaintext awareness* of our scheme depends on the honesty of the registration authority. In particular, the *security* does not. Even if the registration authority collaborated with the adversary, our scheme is guaranteed to be CCA-2 secure.
– *Naturalness:* A trusted registration authority is essentially implicit in any actual implementation of public-key encryption. Such implementations enforce a correct association between users and public keys by requiring that

users register their public key with a *certification authority*. These authorities verify the identity of the applicant and that the applicant knows the corresponding secret key.

In our system, users will have separate keys for sending and receiving messages, and our definitions only require that the sending keys key be registered. However, it is natural to require users to register their sending keys at the same time that they have their receiving keys certified, and that the certificate authority act as registering authority also.

 – *Efficiency:* As we've said, a random oracle can be thought of as a trusted third party. However, in the Bellare-Rogaway model, this trusted third party must be accessed every time that a ciphertext is generated. By contrast, in our model the (rather different) trusted party is accessed only once, and, thereafter, registered users can generate ciphertexts on their own. (To be sure, the quite general implementation that we propose is not efficient, but this inefficiency is *not* due to our model.)

The Essence of Our Implementation. Our scheme is based on those of [6,4, 5], and makes use of the following key registration process: a sender simply gives a zero-knowledge proof of knowledge of his secret sending key. Since the proof system is zero-knowledge, no registration authority (honest or dishonest) gains any information.

Following [6], we also make the encryption of a message depend on the public keys of both sender and receiver. More precisely, and giving a self-referential twist to the schemes of [5] and [4], our sender U encrypts a message for V both in V's public receiving key as well as his own public sending key — and provides a proof of having done so.

The Essence of Our Application. We apply plaintext awareness to the *Dolev-Yao* model [7], the famous alternative for cryptographic protocol analysis.[1] Unlike the more general computational models, the Dolev-Yao model has the advantage of extreme simplicity and ease of use. Although it is impossible to decide the correctness of a protocol in general, the correctness of an impressive number of specific protocols has been successfully decided by automated tools [8, 9,10].

However, these successes are qualified by their reliance on *extremely* strong assumptions. In particular, the Dolev-Yao model assumes that the adversary is not allowed to perform arbitrary computations. Instead, he is limited to selecting his actions from a small number of predetermined operations. (For example, he is prohibited from doing anything with a ciphertext except decrypting it with the right key.) These restrictions raise serious doubts about the meaningfulness of the Dolev-Yao model. After all, a real-world adversary is not required to obey them.

---

[1] It is also known as the *formal* model, due to its origins in the formal methods community.

However, we show that plaintext awareness ensures that the Dolev-Yao restrictions can be actually *enforced* in the real world. It is here that the naturalness of our model and implementation matters crucially: were our model in any way abstract or unachievable, we would simply be reducing one abstraction to another. However, since our model is concrete, we show that the Dolev-Yao adversary can be made concrete also.

2   Preliminaries

We say that an algorithm (or interactive TM) A is *history-preserving* if it "never forgets" anything. As soon as it flips a coin or receives an input or a message, A writes it on a separate history tape that is write-only and whose head always moves from left to right. The history tape's content coincides with A's internal configuration before A executes any step.

If A is an history-preserving algorithm, then if A appears more than once in a piece of GMR notation (e.g., $\Pr[\ldots; a \overset{R}{\leftarrow} A(x); \ldots; b \overset{R}{\leftarrow} A(y); \ldots : p(\cdots, a, b, \cdots)]$) then the history and state of A is preserved from the end of one "use" to the beginning of the next. The notation $h \overset{H}{\longleftarrow} A$ indicates that h is the content of the current history tape of A.

An *adversary* is an efficient history-preserving algorithm (interactive TM).

Following [11], we consider a two-party protocol as a pair, (A, B), of interactive Turing machines. By convention, A takes input (x, r_A) and B takes input (y, r_B) where x and y are arbitrary and r_A and r_B are random tapes. On these inputs, protocol (A, B) computes in a sequence of rounds, alternating between A-rounds and B-rounds. In an A-round only A is active and sends a message (i.e., a string) that will become an available input to B in the next B-round. (Likewise for B-rounds.) A computation of (A, B) ends in a B-round in which B sends the empty message and computes a private output.

If E is an execution of (A, B) on inputs (x, r_A) and (y, r_B), then the *output of A in E* (denoted $OUT_A^{A,B}(x, r_A | y, r_B)$) consists of the string z output by A in the last A-round. Similarly, $OUT_B^{A,B}(x, r_A | y, r_B)$ is the output of B in the same execution. We also define the random distribution $OUT_A^{A,B}(x, \cdot | y, \cdot)$ to be $OUT_B^{A,B}(x, r_A | y, r_B)$ where r_A and r_B are selected randomly.

We say that an execution of a protocol (A, B) has *security parameter* k if the private input of A is of the form $(1^k, x')$ and the private input of B is of the form $(1^k, y')$.

3   The Notion of Plaintext Awareness via Key Registration

3.1   Informally

Plaintext-awareness via key registration requires a significantly different definition than those of other plaintext-aware cryptosystems. We insist that not only the receiver of encrypted messages have a public key but also that the sender have a public key, registered in advance with the registration authority. In this setting, plaintext awareness means the following: the adversary can decrypt *any* ciphertext it creates, so long as the (apparent) sender has registered its sending key with the proper registration authority.

Also, we ask that plaintext-awareness hold for any key registered with the honest registration authority. However, as mentioned before, the security of our scheme should not depend on the honesty of the registration authority. The scheme should remain CCA-2 secure (i.e., the most secure possible without a trusted third party) even if the registration authority collaborates with the adversary.

A plaintext-aware encryption scheme consists of an encryption scheme (G, E, D), and a key-registration protocol (RU, RA).

Algorithm G is used for the generation of the receiver's encryption and decryption keys. In this model, E and D must also be given the public key of the sender as input.

The sender must participate in the key-registration protocol in order to generate a key. The registration is performed by having U run protocol RU on input 1^k with the registration authority running RA on input 1^k. If the registration is successful, the registration authority simply outputs the key e_s, and U should also output this key. One can think of the key as then being inserted in a public file or that U is given a certificate for e_s, but the precise mechanism of the publication is irrelevant here. What is crucial, however, is that the registration protocol be a *secure atomic operation*. That is, we can think of it as being run one user at a time, in person, and from beginning to end.[2]

It is worth noting that either RU or RA may reject in the registration protocol (presumably when the other party is dishonest), in which case we assume the output is ⊥. For ease in the definitions, we assume that if ⊥ is any input to either E or D, the output will also be ⊥.

3.2   More Formally

A *registration-based plaintext-aware encryption scheme* consists of a pair (G, E, D) and (RU, RA), where

- (G, E, D) is a public-key encryption scheme, where:
  - $G(1^k)$ produces (e_r, d_r), a key pair for the receiver, where k is a security parameter;
  - $E(m, e_r, e_s)$ produces c, where c is a ciphertext, m is the message to encrypt, and e_r and e_s are the receiver's and sender's public keys; (The ciphertext c is assumed to explicitly indicate which public keys were used in its creation.)

---

[2] Without this assumption, we would have to worry about man-in-the-middle, concurrency and other types of attacks which will obscure both the definitional and implementation aspects of our model.

- D(c, d_r, e_s) produces m, a message, where c is the ciphertext to decrypt, d_r is the receiver's private key, and e_s is the sender's public key. (If the ciphertext is invalid, the output is $\perp$.)
- $(\mathsf{RU}, \mathsf{RA})$ is a two-party protocol in which both parties should output e_s, a public key for the sender;

which satisfy the following conditions (in which ν is a negligible function):

- *Registration Completeness*: The key registration protocol between an honest registrant and an honest registration authority will almost always be successful, and the user and the authority will agree on the key.

$$\forall k$$
$$\Pr[\, r_1 \xleftarrow{R} \{0,1\}^*; r_2 \xleftarrow{R} \{0,1\}^*;$$
$$e_s \xleftarrow{R} OUT_{\mathsf{RA}}^{\mathsf{RU},\mathsf{RA}} \left(1^k, r_1 | 1^k, r_2\right);$$
$$e_s' \xleftarrow{R} OUT_{\mathsf{RU}}^{\mathsf{RU},\mathsf{RA}} \left(1^k, r_1 | 1^k, r_2\right);$$
$$e_s = e_s' \neq \perp \qquad\qquad\qquad] = 1 - \nu(k)$$

- *Encryption Completeness*: If an honest sender encrypts a message m into a ciphertext c, then the honest recipient will almost always decrypt c into m.

$$\forall k, \forall m \in \{0,1\}^k$$
$$\Pr[\, e_s \xleftarrow{R} OUT_{\mathsf{RU}}^{\mathsf{RU},\mathsf{RA}} \left(1^k, \cdot | 1^k, \cdot\right);$$
$$(e_r, d_r) \xleftarrow{R} \mathsf{G}(1^k);$$
$$c \xleftarrow{R} \mathsf{E}(m, e_r, e_s);$$
$$g \xleftarrow{R} \mathsf{D}(c, d_r, e_s):$$
$$g = m \qquad\qquad\qquad] = 1 - \nu(k)$$

- *Honest Security:* If recipient and sender are honest, the encryption is adaptively chosen-ciphertext secure even if the adversary controls the registration authority.

$$\forall \text{ oracle-calling adversaries } \mathsf{A}, \forall \text{ sufficiently large } k$$
$$\Pr[\, (d_r, e_r) \xleftarrow{R} \mathsf{G}(1^k);$$
$$e_s \xleftarrow{R} OUT_{\mathsf{RU}}^{\mathsf{RU},\mathsf{A}} \left(1^k, \cdot | 1^k, \cdot\right);$$
$$m_0, m_1 \xleftarrow{R} \mathsf{A}^{\mathsf{D}(\cdot, d_r, \cdot)}(e_r, e_s);$$
$$b \xleftarrow{R} \{0, 1\};$$
$$c \xleftarrow{R} \mathsf{E}(m_b, e_r, e_s);$$
$$g \xleftarrow{R} \mathsf{A}^{\mathsf{D}(\cdot, d_r, \cdot) - \{c\}}(c):$$
$$b = g \qquad\qquad\qquad] \leq \tfrac{1}{2} + \nu(k)$$

where

- m_0 and m_1 have the same length, and
- $\mathsf{D}(\cdot, d_r, e_s) - \{c\}$ is the oracle that returns $\mathsf{D}(c', d_r, e_s)$ if $c' \neq c$ and returns $\perp$ if $c' = c$.

Note that if $e_s = \bot$ the adversary will get only $\bot$ from the oracle, and $c = \bot$ as well, so the probability of success will be just $1/2$. Also, recall that the adversary is assumed to be history-preserving, so that it remembers every input it has ever seen.

- *Plaintext Awareness:* If the registration authority is honest and player X (either the adversary or an honest player) registers a key, then the adversary can decrypt any string it sends to an honest participant ostensibly by X:

$$\forall \text{ adversaries } \mathsf{A}, \forall \mathsf{X} \in \{\mathsf{A}, \mathsf{RU}\}, \exists \text{ efficient algorithm } \mathsf{S_X}, \forall \text{ s.l. } k$$
$$\Pr[\ (e_r, d_r) \xleftarrow{R} \mathsf{G}(1^k);$$
$$e_\mathsf{X} \xleftarrow{R} OUT_\mathsf{RA}^{\mathsf{X},\mathsf{RA}}\ (e_r, \cdot | 1^k, \cdot)\ ;$$
$$h \xleftarrow{H} \mathsf{A};$$
$$c \xleftarrow{R} \mathsf{A}^{\mathsf{D}(\cdot, d_r, \cdot)}(e_\mathsf{X}, e_r);$$
$$\mathsf{S_X}(h, c, e_r, e_\mathsf{X}) = \mathsf{D}(c, d_r, e_\mathsf{X})\ ] \geq 1 - \nu(k)$$

Remarks. Note in the definition of plaintext awareness that if $\mathsf{X} = \mathsf{RU}$, then it expects its input to be 1^k and not e_r. Hence, we assume that if RU finds input e_r that it extracts 1^k from it and proceeds as normal.

Also in the definition of plaintext-awareness, if the sender key is registered by an honest participant ($\mathsf{X} = \mathsf{RU}$) then h, the history of the adversary, will be empty.

Lastly, note that these definitions do not guarantee anonymity of the sender. That is, senders must register their keys, and so it might be that they can no longer send messages without their name attached in some way. We note three things with respect to this.

1. If plaintext-awareness is not required, a sender may simply use an unregistered key. Plaintext awareness will no longer be guaranteed, but chosen ciphertext security will still hold.
2. Each registered key does not necessarily represent a sender but rather one incarnation of a sender. Senders may register many keys in order to bolster their anonymity.
3. Lastly, we note that in our motivating application, channels will be authenticated anyway, so this is no additional loss. Indeed, authentication is almost necessary, as our definition guarantees only that a message encrypted under a registered key will be known to the party that registered that key.

We choose to regard the possibility of sender authentication as an opportunity rather than a drawback, and use it in an essential way in our implementation.

4   Implementing Plaintext Awareness via Key Registration

Our scheme uses non-interactive zero-knowledge proofs (e.g. [12,13,14,15]) in order to enhance encryption security. This approach has been pioneered by Naor and Yung [5], and greatly refined by Sahai [4]. Another fount of inspiration comes from the work of Rackoff and Simon [6] which used a very powerful registration authority (indeed, one that chooses every user's secret keys) to obtain chosen-ciphertext security.

We make use of the following three cryptographic tools:

- $(\mathsf{G}', \mathsf{E}', \mathsf{D}')$, a semantically secure cryptosystem in the sense of [16].
- (f, P, V, S), a non-malleable NIZK proof system for NP in the sense of [4], where P is the proving algorithm, V is the verification algorithm, S is the simulator, and $f(k)$ is the length of the reference string for security parameter k.
- a zero-knowledge proof of knowledge for NP, and [17,18,11].
- Authenticated channels that allow a recipient to determine if a ciphertext (c, e, e') was sent by the entity that registered the sending key e'.

The first three of the above rely only upon the existence of trapdoor permutations. The authenticated channels may introduce additional assumptions.

4.1   The Scheme $\mathcal{S}$

The scheme $\mathcal{S} = (\mathcal{G}, \mathcal{E}, \mathcal{D}, \mathcal{RU}, \mathcal{RA})$ is as follows.[3]

- $\mathcal{G}$ (receiver key generation): Generate (e_1, d_1) and (e_2, d_2) according to $\mathsf{G}'(1^k)$. Pick a random σ from $\{0,1\}^{f(k)}$. The public (receiver's) key is $e_r = (e_1, e_2, \sigma)$ and the secret key is $d_r = (d_1, d_2)$.
- $\mathcal{RU}$ and $\mathcal{RA}$: First, Generate (e_3, d_3) according to $\mathsf{G}'(1^k)$. The public (sender's) key is $e_s = e_3$. Next, we engage in a zero-knowledge proof of knowledge that the user knows d_3. If the zero-knowledge proof of knowledge terminates correctly, $\mathcal{RA}$ outputs the sender's public key, otherwise it outputs $\bot$. $\mathcal{RU}$ outputs e_s so long as the zero-knowledge proof of knowledge was not aborted, otherwise it outputs $\bot$.
- $\mathcal{E}$, on input $(m, (e_1, e_2, \sigma), (e_3))$ first computes $c_1 = \mathsf{E}'(e_1, m)$, $c_2 = \mathsf{E}'(e_2, m)$, and $c_3 = \mathsf{E}'(e_3, m)$. Here, naturally, e_1, e_2, and σ are from the receiver's public key, while e_3 is the sender's public key. Then, it computes π, a non-malleable NIZK proof that c_1, c_2, and c_3 all encrypt the same message relative to e_1, e_2, and e_3, respectively. It outputs (c_1, c_2, c_3, π).
- $\mathcal{D}$, on input $((c_1, c_2, c_3, \pi), (e_1, e_2, \sigma, d_1, d_2), (e_3))$ first determines if the ciphertext (c_1, c_2, c_3, π) was sent by the entity that registered e_3. (Authenticated channels are essential for this step.) If not, it outputs $\bot$. If so, it then determines if π is a valid proof that c_1, c_2 and c_3 are encryptions of the same message under e_1, e_2 and e_3, respectively, relative to the reference string σ. If so, it outputs $\mathsf{D}'(d_1, c_1)$. Otherwise, it outputs $\bot$.

---

[3] In these definitions, we liberally assume that any secret key contains any needed information from the associated public key.

4.2   Security of $\mathcal{S}$

$\mathcal{S}$ **Satisfies Registration Completeness.** This is natural: the registration process is a zero-knowledge protocol. By its completeness property, an honest prover will almost always be able to prove a true theorem (e_s) to an honest verifier if it possesses a witness (d_s). Since the honest registrant has access to the witness and engages in the protocol honestly, the honest registration authority will almost always accept the proof and output the public key, and the user will output the same key.

$\mathcal{S}$ **Satisfies Encryption Completeness.** This should be clear. If the sender is honest, then it produces (c_1, c_2, c_3, π) where c_1, c_2 and c_3 all contain the same plaintext m and π is an honest proof of that fact. Since the proof is honest, the recipient will almost always accept it and decrypt c_1 to receive m, the same message encrypted by the sender.

$\mathcal{S}$ **Satisfies Honest Security.** We will prove chosen-ciphertext security by the contrapositive. Suppose there is an adversary A that succeeds in an adaptive chosen ciphertext attack against an honest sender and an honest recipient. We will give two reductions, R and R', and we will prove that one of the two must break the underlying encryption scheme.

R simulates the adversary A so as to break the semantic security of the underlying encryption scheme $(\mathsf{G}', \mathsf{E}', \mathsf{D}')$. So, on input $(e, 1^k)$, R runs as follows:

1. First, we create the receiver's public key (e_1, e_2) and the sender's public key (e_3) as follows. Pick a at random from $\{1, 2\}$. Set e_{3-a} to be e and set $(e_a, d_a) \stackrel{R}{\leftarrow} \mathsf{G}'(1^k)$. Generate σ according to the simulator S for the NIZK proof system[4]. Set $(e_3, d_3) \stackrel{R}{\leftarrow} \mathsf{G}'(1^k)$.
2. Run A on input $((e_1, e_2, \sigma), (e_3))$. Whenever A asks for a decryption of (c'_1, c'_2, c'_3, π'), encrypted with sending key e', we check the correctness of π' using V. If it verifies, we decrypt c'_a using d_a and output that as the result. Otherwise we return $\perp$.
3. Eventually A will output (m_0, m_1) Output (m_0, m_1) and obtain challenge c. For notation later, let us say that m_β is the message c encrypts.
4. We then simulate the ciphertext challenge for A. Pick b at random from $\{0, 1\}$. Let $c_a \stackrel{R}{\leftarrow} \mathsf{E}'(e_a, m_b)$, and set $c_{3-a} \stackrel{R}{\leftarrow} c$. With probability $1/2$, let $c_3 \stackrel{R}{\leftarrow} \mathsf{E}'(e_3, m_b)$ and otherwise, let $c_3 \stackrel{R}{\leftarrow} \mathsf{E}'(e_3, m_{1-b})$. Fake the NIZK proof π using the simulator S.
5. Run A on input (c_1, c_2, c_3, π).
6. Again, whenever A asks for a decryption, we check the proof and decrypt using d_a.
7. Eventually A outputs an answer b'. If $b = b'$, output b'. Otherwise, output a random bit.

---

[4] Here, we assume the simulator S is history-preserving.

There are three kinds of input the adversary can get.

I. First, it is possible that c_1, c_2, and c_3 all encrypt the same message m_β. In this case, the input given to the adversary is indistinguishable from the input in the real attack the adversary succeeds in. Thus, the adversary must return β with probability $1/2 + \epsilon$, where ϵ is some non-negligible function of k.

II. Second, it may be that c_1 and c_2 both encrypt the same message m_β but c_3 encrypts $m_{1-\beta}$. Let x be such that in this case, the adversary returns β with probability x.

III. Finally, it may be that c_1 and c_2 encrypt different messages. Note that there are two subcases:

   – c_a and c_3 encrypt the same message while c_{3-a} encrypts the other, and
   – c_{3-a} and c_3 encrypt the same message while c_a encrypts the other

   These two cases are indistinguishable to the adversary. Since the adversary cannot make any proofs of false theorems, the oracle will return $\perp$ if the adversary ever makes a decryption query when c_1 and c_2 encrypt different messages. Thus, the case $a = 1$ and the case $a = 2$ give the same distribution. (See [4]. This is just like one of the main details from Sahai's proof that his scheme is CCA2-secure.)

   Let $m_{\beta'}$ be the message encrypted in c_3, and let y be such that in this case the adversary returns β'.

This reduction is parameterized by the values x and y, both of which can be chosen by the adversary. However, we will show that the only value of interest to us is x. In fact, we will show that for almost all values of x, the above reduction breaks the security of (G', E', D'). However, the reduction R will not work for certain values of x, so we give a different reduction R' and show that it does.

To begin: what is the probability that R returns the correct answer? Again, we consider two cases: when $b = \beta$ and when $b \neq \beta$:

– In the case that $b = \beta$, the adversary sees an input of type I with probability $1/2$. When the adversary sees an input of type I, R is correct with probability $\left(\frac{1}{2} + \epsilon\right) + \frac{1}{2}\left(\frac{1}{2} - \epsilon\right) = \frac{3}{4} + \frac{\epsilon}{2}$. If the adversary does not see an input of type I though $b = \beta$ then it sees an input of type II, in which case R is correct with probability $x + (1 - x)/2$ (since whenever the adversary returns β in an input of type 2, R is correct, and the rest of the time, R is correct with probability $1/2$). Thus, the total probability that R is correct when $b = \beta$ is $\frac{1}{2}\left((3/4 + \epsilon/2) + (1/2 + x/2)\right)$.

– Now let us examine the case that $b \neq \beta$. Any time this is true, we give input type III to the adversary. However, with probability $1/2$, m_b encrypted into c_3 and with probability $1/2$, m_{1-b} is encrypted into c_3. (Recall, these two cases are indistinguishable to the adversary.) Thus, the adversary returns b with probability $\frac{1}{2}y + \frac{1}{2}(1 - y)$, and otherwise returns $1 - b$. In this case, our total probability of being correct is $(y/4) + (1 - y)/4 = 1/4$, since we are only correct when the adversary returns $1 - b$, and then, only half the time.

Taking into account all cases, the probability that R is correct is

$$\frac{1}{2}\left(\frac{1}{2}(3/4 + \epsilon/2) + \frac{1}{2}(1/2 + x/2) + \frac{1}{4}\right)$$

This expression evaluates to

$$\frac{3}{16} + \frac{\epsilon}{8} + \frac{1}{8} + \frac{x}{8} + \frac{1}{8} = \frac{7}{16} + \frac{\epsilon + x}{8}.$$

Now if $\epsilon + x$ is non-negligibly different from $1/2$ then the above expression is also, and R breaks (G', E', D'). If, on the other hand, $x \approx 1/2 - \epsilon$, then we can use A to break the security of (G', E', D') directly. Let R' be the reduction that works as follows, on input e:

1. Generate (e_1, d_1) and (e_2, d_2) by running $G'(1^k)$. Generate σ according to the simulator S for the NIZK proof system. Set e_3 to e.
2. Run A on input $((e_1, e_2, \sigma), (e_3))$. Whenever A asks for a decryption query we check the correctness of the included NIZK proof using V. If it verifies, we decrypt c_1 using d_1 and output that as the result, otherwise we return $\bot$.
3. Obtain m_0, m_1 as the output of A. Output (m_0, m_1) and obtain challenge c. For notation later, let us say that m_β is the message c encrypts.
4. Pick b at random from $\{0, 1\}$. Let $c_1 \xleftarrow{R} E'(e_1, m_b)$, let $c_2 \xleftarrow{R} E'(e_2, m_b)$, and let c_3 be c.
5. Fake the NIZK proof π using the simulator S (which we assume to be history-preserving).
6. Run A on input (c_1, c_2, c_3, π). Again, whenever A asks for a decryption, we check the proof and decrypt using d_1. Eventually A outputs an answer b'. Output b'.

The proof that R' works is simple. If R' picks $b = \beta$ then A outputs b with probability $1/2 + \epsilon$. If R' picks $b \neq \beta$ then A sees input type II and so it outputs b with probability $x = 1/2 - \epsilon + \nu'$ where ν' is (positively or negatively) negligible. Thus in either case, we output β with probability at least $1/2 + \epsilon - |\nu'/2|$.

S Enjoys Plaintext Awareness.

This is fairly simple, and we show it informally. There are two cases. If $X = RU$, then e_X was registered by an honest user. Hence, when the adversary creates a ciphertext ostensibly from that honest user, it will fail to decrypt. (The use of authenticated channels will tell the receiver that it was sent by the adversary, and not the entity that registered e_X.) Hence, S_X simply outputs $\bot$ on all input.

In the other case, $X = A$, and the adversary registered e_X. We extract plaintext as follows. On input $(h, (c_1, c_2, c_3, \pi), e_r, e_X)$, we use h to rewind the adversary to the point where A engages in key registration with RA. We then use the extractor from the interactive zero knowledge proof of knowledge to find a value d, the secret key associated with e_X. We then check π; if π is invalid, we output $\bot$. Otherwise, we use d to decrypt c_3 and give the result as the answer. From the extractibility property of the proof system, d must be a secret key relative to key so this answer is correct.

However, we do need to show that the decryption under d will always be the same as the decryption under d_r. If the proof π in c is invalid, then certainly we are correct to output $\perp$. If π is valid, then by the soundness of the NIZK proof system, it must be that c_1, c_2, and c_3 all encrypt the same message, so we are still correct.

5   Plaintext Awareness and the Dolev-Yao Adversary

We conclude by considering a naturally-arising application of plaintext-awareness: the adversary of the Dolev-Yao model of cryptographic protocols [7].

The Dolev-Yao model is an alternate model of cryptographic protocol execution which grew out of the formal methods community. It differs from the standard, computational, model in two important ways:

1. The representation of messages, and
2. The ability of the adversary.

In this model, messages are not bit-strings but parse trees. The atomic elements (leaves) are considered to be abstract symbols with no internal structure, and are partitioned into three sets: names ($\mathcal{M}$), random numbers ($\mathcal{R}$) or keys ($\mathcal{K}_{Pub}$ and $\mathcal{K}_{Priv}$).[5] Compound messages are formed using two operations:

- $encrypt : \mathcal{K}_{Pub} \times \mathcal{A} \rightarrow \mathcal{A}$
- $pair : \mathcal{A} \times \mathcal{A} \rightarrow \mathcal{A}$

We denote $encrypt(K, M)$ by $\{\!|M|\!\}_K$, and denote $pair(M, N)$ by $M\,N$. We denote by $\mathcal{A}$ the set of all messages. Because messages are parse trees, every message has a unique interpretation. We assume for our purposes that the algebra contains a finite number of atomic elements, though the model itself has no such restriction.

Whereas the standard adversary is an arbitrary algorithm, the Dolev-Yao adversary is much more limited. As in the standard model, the adversary is able to know all public and predictable values. Likewise, the adversary controls the network in both models, meaning that it sees and routes all traffic between honest participants. However, when it comes to the ability of the adversary to create new messages, the two models sharply differ.

Where the standard adversary is able to create any efficiently computable bit-string, the Dolev-Yao adversary can only create new parse trees by applying to known ones a limited number of operations: pairing, separation of pairs, encryption in public keys, and decryption in known keys. Formally, the power of the Dolev-Yao adversary to create new messages is given by a set-theoretic operation:

---

[5] We will only consider the case of asymmetric encryption, though the Dolev-Yao models symmetric encryption also.

Definition 1 (Closure) *The* closure *of* S, *written* $C[S]$, *is the smallest set such that:*

1. $S \subseteq C[S]$,
2. *If* $\{\!|M|\!\}_K \in C[S]$ *and* $K^{-1} \in C[S]$, *then* $M \in C[S]$,
3. *If* $M \in C[S]$ *and* $K \in C[S]$, *then* $\{\!|M|\!\}_K \in C[S]$,
4. *If* $M\,N \in C[S]$, *then* $M \in C[S]$ *and* $N \in C[S]$, *and*
5. *If* $M \in C[S]$ *and* $N \in C[S]$, *then* $M\,N \in C[S]$.

(It is assumed that S contains all public values such as names and public keys.)

It is the central assumption of the Dolev-Yao model that the closure operation is the extent of the adversary's ability to manipulate cryptographic material:

Definition 2 (Formal Adversary) *The formal adversary is a non-deterministic process on $\mathcal{A}$ that, given a set S of messages, produces messages in* $C[S]$.

The Dolev-Yao is an attractive model in which to work. Proofs are simple and easily found, and protocol verification is easily automated. However, it is not clear how the Dolev-Yao model relates to the standard computational model. The above restriction makes the formal adversary seem fairly weak: the standard adversary can certainly calculate any value available to the formal adversary, and many more besides. Hence, security against the formal adversary seems like a fairly weak property. However, it turns out that if the underlying cryptography is plaintext aware, then the standard adversary is no more powerful than the formal adversary. That is, computational cryptography can limit the computational adversary to this closure operation.

To formalize the limitation on the formal adversary in terms of computational cryptography, we need to somehow translate the parse-tree messages of the Dolev-Yao model into bit-strings. To do this, we adapt the "encoding" operation from Abadi and Rogaway [19] from the symmetric-encryption setting to that of asymmetric encryption. In brief, the "encoding" of a message M, written $[M]_n$, depends on the parse tree of M, the security parameter, and the choice of underlying public-key encryption scheme $(\mathsf{G}, \mathsf{E}, \mathsf{D})$.[6] Recursing on the structure of M:

- If M is the nonce of an honest participant, then $[M]_n$ is a specific n-bit string, chosen at random.
- If M is a nonce of the formal adversary, then $[M]_n$ is a specific n-bit string chosen by the computational adversary
- If M is a public or private key of an honest participant, then $[M]_n$ is a specific computational key chosen at random from $G(1^n)$.
- If M is a public or private key of the formal adversary, then $[M]_n$ is a specific computational key chosen by the computational adversary
- If $M = M_1\, M_2$, then $[M]_n$ is the concatenation of $[M_1]_n$ and $[M_2]_n$.
- If $M = \{\!|M_1|\!\}_K$, then $[M]_n$ is the *distribution* on bit-strings defined by $\mathsf{E}([M_1]_n, [K]_n)$.

---

[6] Our definition of plaintext-aware encryption contains several more algorithms, which we ignore for the moment.

Now that we can relate Dolev-Yao messages and bit-strings, we can formalize the intuition of Definition 2. We will call a computational encryption scheme *ideal* if it restricts the computational adversary to the limit on the Dolev-Yao adversary:

Attempt 3. *An encryption scheme* $(\mathsf{G}, \mathsf{E}, \mathsf{D})$ *is* ideal *if*

$$\forall \mathsf{A}_{PPT}, \ \forall S \subseteq \mathcal{A}, \ \forall M \notin C\left[S\right], \ \forall \text{ polynomials } q, \ \forall \text{ sufficiently large } n :$$
$$\Pr[\ s \xleftarrow{R} [S \cup \mathcal{K}_{Pub} \cup \mathcal{K}_{Subv} \cup \mathcal{M}]_n\ ;$$
$$m \xleftarrow{R} \mathsf{A}(1^n, s) :$$
$$m \in \boldsymbol{supp}[M]_n \qquad\qquad\quad\ ] \leq \tfrac{1}{q(n)}$$

(Here, $\boldsymbol{supp}(D)$ means the support of distribution D.)

However, our results are subject to one technical limitation: S must be *acyclic*. A formal definition of an acyclic set can be found in [20]. Informally, it means that if K_1 encrypts K_2^{-1} in S K_2 encrypts K_3^{-1}, and so on, this sequence never loops back on itself.[7].

Hence, we revise the security condition:

Definition 4 *An encryption scheme* $(\mathsf{G}, \mathsf{E}, \mathsf{D})$ *is* ideal *if the adversary cannot create something outside the closure:*

$$\forall \mathsf{A}_{PPT}, \forall \text{ acyclic } S \subseteq \mathcal{A}, \forall M \notin C\left[S\right], \ \forall \text{ polynomials } q, \ \forall \text{ sufficiently large } n :$$
$$\Pr[\ s \xleftarrow{R} [S \cup \mathcal{K}_{Pub} \cup \mathcal{K}_{Subv} \cup \mathcal{M}]_n\ ;$$
$$m \xleftarrow{R} \mathsf{A}(1^n, s) :$$
$$m \in \boldsymbol{supp}[M]_n \qquad\qquad\quad\ ] \leq \tfrac{1}{q(n)}$$

This definition, it turns out, is no stronger than plaintext-awareness. Before we can prove this, however, we need to address a small technical issue. Encryption in the Dolev-Yao model depends only on the message and the receiver's public key. In our definition of plaintext-aware encryption, however, encryption uses the receiver's public receiving key and the sender's public sending key. Hence, we consider a slight variant of the Dolev-Yao model in which the formal encryption operation uses two public keys: the sender's and the receiver's. The encoding of a formal key contains both a sending portion and a receiving portion (public or private, as appropriate). The encoding of a formal encryption is then defined using, in the natural way, a computational plaintext-aware encryption scheme and the encodings of the public keys.

Theorem 5. *Any encryption scheme that achieves plaintext-awareness is also ideal, if all public keys are registered with an honest* RU.

Proof Sketch. Suppose that the encryption scheme were not ideal. Then with non-negligible probability the adversary could create an encoding m such that m is the valid encoding of an M not in $C\left[S\right]$.

---

[7] This is a reasonable assumption for most "real-world" protocols, for reasons discussed in [20]

Consider the parse tree of M. Each node in this tree is a message. Furthermore, if the adversary can create an encoding of an internal node of this tree with some probability p, then either

1. That node is in $C[S]$, or
2. The adversary can, with probability almost p, create encodings of both children.

To see this, suppose the node is not in $C[S]$ and consider its type. It is easy to separate the components of a pair. On the other hand, if the adversary creates an encryption, then plaintext-awareness tells us that there exists a simulator that can extract the plaintext (which by construction, is never $\perp$.) Hence, the adversary can create the encoding of an encryption, then run the simulator to extract the plaintext of the encryption. Also, since all public keys are known to the adversary, it can create both encryption keys. Thus, the adversary can create all children of a encryption node.

Furthermore, membership in $C[S]$ is closed up the tree: if both children are in $C[S]$, then their parent is in $C[S]$ also. Hence, if M is not in $C[S]$ then there must be one path from root to leaf in the parse tree of M where no element of the path is in $C[S]$. (If there were no such path, then M would be in $C[S]$, a contradiction.)

By recursing down the tree and making both children of every node along this path, the adversary can create an encoding of the leaf at the end of this path. Hence, the adversary can make M, the root message, with probability p, then with probability $\frac{p}{q(n)}$ (for some polynomial q) the adversary can create the encoding of some atomic message M' outside of $C[S]$.

There are two cases:

1. If M' is related to S, then it must be either as the plaintext of an encryption or as the private key of some public key used in S (or both). In this case, the adversary has broken the security of the encryption. If we assume that every party has engaged in the setup phase with every other party, then the encryption scheme is secure, leading to a contradiction.
2. If M' is not related to S, then the adversary has managed to guess a random value from input independent of that value. If there are n_1 elements of $\mathcal{R}$ then the adversary has a $\frac{n_1}{2^n}$ chance of guessing any given nonce. If there are n_2 elements of $\mathcal{K}_{Priv}$ and each key is $l(n)$ bits long, then the adversary has a $\frac{n_2}{2^{l(n)}}$ chance of guessing any private key. Since both of these are negligible, then the adversary must have a negligible change creating an encoding of M'.

Hence, $\frac{p}{q(n)}$ must be negligible, which means that p must have been negligible to begin with.   ∎

Hence, plaintext-aware encryption limits the computational adversary to the operations available to the Dolev-Yao adversary. It is unknown whether any weaker form of encryption achieves the same limitation, making this the first naturally arising application of plaintext-aware cryptography.

Acknowledgments. The authors would like to thank Ron Rivest and Nancy Lynch, under whose supervision part of this work was done. They would also like to thank the anonymous referees for their insightful comments.

References

1. Bellare, M., Rogaway, P.: Optimal asymmetric encryption– how to encrypt with RSA. In Santis, A.D., ed.: Advances in Cryptology – Eurocrypt 94 Proceedings. Volume 950 of Lecture Notes in Computer Science., Springer-Verlag (1995) 92–111
2. Bellare, M., Desai, A., Pointcheval, D., Rogaway, P.:   Relations among notions of security for public-key encryption schemes.   In Krawczyk, H., ed.: Advances in Cryptology (CRYPTO 98). Volume 1462 of Lecture Notes in Computer Science., Springer–Verlag (1998) 26–45 Full version found at http://www.cs.ucsd.edu/users/mihir/papers/relations.html.
3. Cramer, R., Shoup, V.: A practical public key cryptosystem provably secure against adaptive chosen ciphertext attack. In: Advances in Cryptology — CRYPTO 1998. Number 1462 in LNCS, Springer–Verlag (1998) 13–25
4. Sahai, A.: Non-malleable non-interactive zero knowledge and adaptive chosen-ciphertext security. In: Proceedings of 40th Annual IEEE Symposium on Foundations of Computer Science (FOCS). (1999) 543–553
5. Naor, M., Yung, M.: Public-key cryptosystems provably secure against chosen ciphertext attacks. In: 22nd Annual ACM Symposium on Theory of Computing. (1990) 427–437
6. Rackoff, C., Simon, D.: Noninteractive zero-knowledge proof of knowledge and the chosen-ciphertext attack. In: Advances in Cryptology– CRYPTO 91. Number 576 in Lecture Notes in Computer Science (1991) 433–444
7. Dolev, D., Yao, A.: On the security of public-key protocols. IEEE Transactions on Information Theory **29** (1983) 198–208
8. Lowe, G.: Breaking and fixing the Needham–Schroeder public-key protocol using FDR. In Margaria, Steffen, eds.: Tools and Algorithms for the Construction and Analysis of Systems. Volume 1055 of Lecture Notes in Computer Science. Springer–Verlag (1996) 147–166
9. Paulson, L.C.: The inductive approach to verifying cryptographic protocols. Journal of Computer Security **6** (1998) 85–128
10. Song, D.: Athena, an automatic checker for security protocol analysis. In: Proceedings of the 12th IEEE Computer Security Foundations Workshop. (1999) 192–202
11. Goldwasser, S., Micali, S., Rackoff, C.: The knowedge complexity of interactive proof systems. In: Proceedings of the 17th ACM Symposium on Theory of Computing. (1985) 291–304 Superseded by journal version.
12. Blum, M., Feldman, P., Micali, S.: Non-interactive zero knowledge proof systems and applications. In: Proceedings of the 20th Annual ACM Symposium on Theory of Computing. (1988) 103–112
13. Santis, A.D., Micali, S., Persiano, G.: Non-interactive zero-knowledge proof systems. In Pomerance, C., ed.: Proceedings Crypto '87, Springer-Verlag (1988) 52–72 Lecture Notes in Computer Science No. 293.
14. Blum, M., Santis, A.D., Micali, S., Persiano, G.: Noninteractive zero knowledge. SIAM Journal on Computing **20** (1991) 1084–1118
15. Boyar, J., Damgård, I., Peralta, R.: Short non-interactive cryptographic proofs. Journal of Cryptology: the journal of the International Association for Cryptologic Research **13** (2000) 449–472

16. Goldwasser, S., Micali, S.: Probabilistic encryption. Journal of Computer and System Sciences (1984) 270–299
17. Bellare, M., Goldreich, O.: On defining proofs of knowledge. In Brickell, E., ed.: Advances in Cryptology – Crypto 92 Proceedings. Volume 740 of Lecture Notes in Computer Science., Springer–Verlang (1992) 390–420
18. Goldreich, O., Micali, S., Wigderson, A.: Proofs that yield nothing but their validity or all languages in NP have zero-knowledge proof systems. Journal of the ACM **38** (1991) 691–729
19. Abadi, M., Rogaway, P.: Reconciling two views of cryptography (the computational soundness of formal encryption). In: IFIP International Conference on Theoretical Computer Science (IFIP TCS2000). Number 1872 in Lecture Notes in Computer Science, Springer-Verlag (2000) 3–22
20. Herzog, J.: Computational soundness for formal adversaries. Master's thesis, Massachusetts Institute of Technology (2002)

Relaxing Chosen-Ciphertext Security

Ran Canetti[1], Hugo Krawczyk[1,2], and Jesper B. Nielsen[3]*

[1] IBM T.J. Watson Research Center, PO Box 704, Yorktown Heights,
New York 10598.
canetti@watson.ibm.com.
[2] Department of Electrical Engineering, Technion, Haifa 32000, Israel,
hugo@ee.technion.ac.il.
[3] BRICS, Centre of the Danish National Research Foundation. Department of
Computer Science, University of Aarhus, DK-8000 Arhus C, Denmark.
buus@brics.dk.

Abstract. Security against adaptive chosen ciphertext attacks (or, CCA security) has been accepted as the standard requirement from encryption schemes that need to withstand active attacks. In particular, it is regarded as the appropriate security notion for encryption schemes used as components within general protocols and applications. Indeed, CCA security was shown to suffice in a large variety of contexts. However, CCA security often appears to be somewhat *too strong:* there exist encryption schemes (some of which come up naturally in practice) that are not CCA secure, but seem sufficiently secure "for most practical purposes."
We propose a relaxed variant of CCA security, called Replayable CCA (RCCA) security. RCCA security accepts as secure the non-CCA (yet arguably secure) schemes mentioned above; furthermore, it suffices for most existing applications of CCA security. We provide three formulations of RCCA security. The first one follows the spirit of semantic security and is formulated via an ideal functionality in the universally composable security framework. The other two are formulated following the indistinguishability and non-malleability approaches, respectively. We show that the three formulations are equivalent in most interesting cases.

1   Introduction

One of the main goals of cryptography is to develop mathematical notions of security that adequately capture our intuition for the security requirements from cryptographic tasks. Such notions are then used to assess the security of protocols and schemes. They also provide abstractions that, when formulated and used correctly, greatly facilitate the design and analysis of cryptographic applications.

With respect to encryption schemes, a first step was taken with the introduction of semantic security of (public key) encryption schemes in [16]. This first step is indeed a giant one, as it introduces the basic definitional approach and techniques that underlie practically all subsequent notions of security, for encryption as well as many other cryptographic primitives.

---

* Work done while at IBM T.J. Watson Research Center.

D. Boneh (Ed.): CRYPTO 2003, LNCS 2729, pp. 565–582, 2003.

However, semantic security under chosen-plaintext attacks as defined in [16] captures only the very basic requirement from an encryption scheme, namely secrecy against "passive" eavesdroppers. In contrast, when an encryption scheme is used as a component within a larger protocol or system, a much wider array of attacks against the scheme are possible. Specifically, adversaries may have control over the messages that are being encrypted, but may also have control over the ciphertexts being delivered and decrypted. This opens new ways for the attacker to infer the outcome of the decryption of some ciphertexts by observing the system (see e.g. [6,19]).

Several notions of "security against active adversaries" were proposed over the years in order to capture such often subtle security concerns. These notions include semantic security against Lunchtime Attacks (or, IND-CCA1 security), semantic security against Adaptive Chosen Ciphertext Attacks (or, IND-CCA2 security), Non-Malleability against the above attacks, and more [21,23,12,3]. In particular, CCA2 security (where the semantic-security and the non-malleability formulations are equivalent) became the "golden standard" for security of encryption schemes in a general protocol setting. Indeed, CCA2 security (or simply CCA security) was demonstrated to suffice for a number of central applications, such as authentication and key exchange [12,2,10], encrypted password authentication [17], and non-interactive message transmission [7]. In addition, in [7] CCA is shown to suffice for realizing an "ideal public-key encryption functionality" within the universally composable security (UC) framework, thus demonstrating its general composability properties.

CCA security is indeed a very strong and useful notion. But is it *necessary* for an encryption scheme to be CCA-secure in order to be adequate for use within general protocol settings? Some evidence that this may not be the case has been known all along: Take any CCA-secure encryption scheme S, and change it into a scheme S' that is identical to S except that the encryption algorithm appends a 0 to the ciphertext, and the decryption algorithm discards the last bit of the ciphertext before decrypting. It is easy to see that S' is no longer CCA-secure, since by flipping the last bit of a ciphertext one obtains a different ciphertext that decrypts to the same value as the original one, and this "slackness" is prohibited by CCA security. But it seems that this added slackness of S' is of no "real consequence" in most situations. In other words, S' appears to be just as secure as S for most practical purposes. This example may seem contrived, but it in fact turns up, in thin disguises, in a number of very natural settings. (Consider for instance an implementation of some CCA-secure scheme, where for wider interoperability the decryption algorithm accepts ciphertexts represented both in big-endian and in little-endian encodings. We mention other natural examples within.) In fact, some relaxations of CCA-security were already proposed in the literature in order to address this example and similar ones, e.g. [19,24,1,20]. However, while being good first steps, these notions were not fully justified as either sufficient or necessary for applications. (See more details within.)

We propose a new relaxed version of CCA-security, called Replayable CCA (RCCA) security. In essence, RCCA is aimed at capturing encryption schemes that are CCA secure "except that they allow anyone to generate new ciphertexts that decrypt to the same value as a given ciphertext." RCCA is strictly weaker than CCA security. In fact, it is strictly weaker than the relaxations in [19, 24,1]. The rationale behind RCCA is that as far as an attacker in a protocol setting is concerned, generating *different* ciphertexts that decrypt to the *same* plaintext as a given ciphertext has the same effect as copying (or, "replaying") the same ciphertext multiple times. Since replaying a ciphertext multiple times is unavoidable even for CCA secure encryptions, RCCA security would have "essentially the same effect" as CCA security.

To substantiate this intuition, we prove that RCCA security suffices for all of the above major applications of CCA secure encryption (authentication, key exchange, etc.). We also demonstrate that the *hybrid encryption* paradigm can be based on RCCA security rather than CCA security. (Hybrid encryption calls for encrypting a key k using an asymmetric encryption and then encrypting a long message using symmetric encryption with key k.)

It should be stressed that the above rationale holds only as long as the protocol that uses the scheme makes its decisions based on the outputs of the decryption algorithm, and does not directly compare ciphertexts. Arguably, most applications of CCA secure encryption have this property. However, in some applications it is natural and helpful to directly compare ciphertexts. For instance, consider a voting scheme in which votes are encrypted, and illegal duplicate votes are detected via direct ciphertext comparison. In such cases, the full power of CCA security is indeed used.

We provide three formulations of RCCA security. The first two are formulated via "guessing games" along the lines of CCA security. The first of these, called IND-RCCA, has the flavor of "security by indistinguishability" (or, IND-CCA2 in the terminology of [3]) with a CCA-style game that allows for plaintext replay. The second notion, called NM-RCCA, has the flavor of non-malleability in a CCA-style game that allows for plaintext replay. The third notion, called UC-RCCA, is formulated via an ideal functionality in the UC framework [7]. This ideal functionality,called $\mathcal{F}_{rpke}$, is obtained by modifying the ideal functionality $\mathcal{F}_{pke}$ in [7] to explicitly allow the environment to generate ciphertexts that decrypt to the same value as a given ciphertext. Having been formulated in the UC framework, this notion provides strong and general composability guarantees. Furthermore, in the spirit of semantic security, it provides a clear and explicit formalization of the provided security guarantee. It also explicitly demonstrates the exact sense in which RCCA weakens CCA.[1]

We show that, when applied to encryption schemes where the message domain is "large" (i.e., super-polynomial in the security parameter), the three notions are equivalent.[2] When the message domain is polynomial in size, we have

---

[1] Krohn in [20] studies various relaxations of CCA security and their respective strengths. One of these notions is essentially the same as IND-RCCA security. However, no concrete justification for this notion is provided.

[2] We say that an encryption scheme has message domain D if, for any message $m \in D$, the process of encrypting m and then decrypting the resulting ciphertext returns m.

that UC-RCCA implies NM-RCCA, and NM-RCCA implies IND-RCCA. We also show, via a separating example, that in this case IND-RCCA does *not* imply NM-RCCA. Whether NM-RCCA implies UC-RCCA for polynomial message domains remains open.

For schemes that handle large message domains, having the three equivalent formalizations allows us to enjoy the best of each one: We have the intuitive appeal and strong composability of the UC-RCCA, together with the relative simplicity of NM-RCCA and IND-RCCA. Indeed, the case of large message domains is arguably the most interesting one, since most existing encryption schemes are either directly constructed for large message domains, or can be extended to deal with large domains in a natural way. Also, most applications of public-key encryption, e.g. encrypting an identity or a key for symmetric encryption, require dealing with large message domains.

The three notions in a nutshell. Let us briefly sketch the three notions. See Section 3 for more detailed description and rationale. First recall the standard (indistinguishability based) formulation of CCA security for public-key cryptosystems. Let $S = (gen, enc, dec)$ be a public-key encryption scheme where gen is the key generation algorithm, enc is the encryption algorithm, and dec is the decryption algorithm. Informally, S is said to be CCA secure if any feasible attacker $\mathcal{A}$ succeeds in the following game with probability that is only negligibly more than one half . Algorithm gen is run to generate an encryption key e and a decryption key d. $\mathcal{A}$ is given e and access to a decryption oracle $dec(d, \cdot)$. When $\mathcal{A}$ generates a pair m_0, m_1 of messages, a bit $b \xleftarrow{R} \{0, 1\}$ is chosen and $\mathcal{A}$ is given $c = enc(e, m_b)$. From this point on, $\mathcal{A}$ may continue querying its decryption oracle, with the exception that if $\mathcal{A}$ asks to decrypt the "test ciphertext" c, then $\mathcal{A}$ receives a special symbol test instead of the decryption of c. $\mathcal{A}$ succeeds if it outputs b.

IND-RCCA is identical to CCA, with the exception that the decryption oracle answers test whenever it is asked to decrypt *any ciphertext that decrypts to either m_0 or m_1*, even if this ciphertext is different than the test ciphertext c. Indeed, in the IND-RCCA game the ability to generate new ciphertexts that decrypt to the test ciphertext does not help the adversary. (Yet, it is not immediately clear from this formulation that we did not weaken the security requirement by too much. The justification for this notion comes mainly from its equivalence with UC-RCCA, described below.)

NM-RCCA is identical to IND-RCCA, with the exception that $\mathcal{A}$ succeeds if $m_0 \neq m_1$ and it outputs *a ciphertext c' that decrypts to m_{1-b}*. Note that if we required $\mathcal{A}$ to output m_{1-b} explicitly, we would get a requirement that is only a reformulation of IND-RCCA. So the difference is in the fact that here $\mathcal{A}$ is only required to output an encryption of m_{1-b}, without necessarily being

---

Thus the larger the domain, the stronger the requirement. (Encryption schemes with large message domain should not be confused with encryption schemes that guarantee security only if the message is taken uniformly from a large domain. The latter is a weak notion of security, whereas the former is only a correctness requirement, and can be used in conjunction with any security requirement.)

able to output m_{1-b} explicitly. This requirement has a flavor of non-malleability, thus the name. (Indeed, it can be regarded as a non-malleability requirement in which the attacker is considered successful as long as the "malleability relation" it uses is not the "equality relation.")

UC-RCCA is defined via an ideal functionality, $\mathcal{F}_{rpke}$. To best understand $\mathcal{F}_{rpke}$, let us first recall the "ideal public-key encryption" functionality, $\mathcal{F}_{pke}$, from [7], that captures CCA security.[3] In fact, instead of getting into the actual mechanism of $\mathcal{F}_{pke}$ (see Section 3), let us only sketch the security guarantee it provides. Functionality $\mathcal{F}_{pke}$ captures the behavior of an "ideal encryption service." That is, $\mathcal{F}_{pke}$ provides an encryption interface that is available to all parties, and a decryption interface that is available only to one privileged party, the decryptor. When querying the encryption interface with some message m, a ciphertext c is returned. The value of c is chosen by the adversary, without any knowledge of m. This guarantees "perfect secrecy" for encrypted messages. When the decryption interface is queried with a "legitimate encryption of m" (i.e., with a string c that was the outcome of a request to encrypt m), then the returned value is m. Since there is no requirement on how "illegitimate ciphertexts," i.e. strings that were not generated using the encryption interface, are being decrypted, $\mathcal{F}_{pke}$ allows the adversary to choose the decryption values of these ciphertexts.

Functionality $\mathcal{F}_{rpke}$ is identical to $\mathcal{F}_{pke}$, except that it allows the adversary to request to decrypt "illegitimate ciphertexts" to the same value as some previously generated legitimate ciphertext. This directly captures the relaxation where the adversary is allowed to generate new ciphertexts which decrypt to the same (unknown) value as existing ciphertexts. It also demonstrates that RCCA does not weaken CCA-security beyond allowing for "plaintext replay" by the attacker.

Between RCCA security and CCA security. As sketched above, RCCA security allows anyone to modify a given ciphertext c into a different ciphertext c', as long as c and c' decrypts to the same message. One potential strengthening of RCCA security is to require that it will be possible to detect, given two ciphertexts c and c', whether one is a "modified version" of the other. (Indeed, the "endian changing" example given above has this additional property.) Here it is natural to distinguish between schemes where the detection algorithm uses the secret decryption key, and schemes where the detection can be done given only the public encryption key. We call such schemes secretly detectable RCCA (sd-RCCA) and publicly detectable RCCA (pd-RCCA), respectively.

We first observe that pd-RCCA security is essentially equivalent to the notions proposed by Krawczyk [19], Shoup [24], and Ann, Dodis and Rabin [1]. (The notions are called, respectively, loose ciphertext-unforgeability, benign malleability, and generalized CCA security, and are essentially the same.) Next we study the relations between these notions. It is easy to see that:

---

[3] In [7] it is mistakenly claimed that CCA security is a *strictly* stronger requirement than realizing $\mathcal{F}_{pke}$ for non-adaptive adversaries. However, as shown in this work, the two requirements are actually *equivalent*. The mistake in [7] and the equivalence proof were independently discovered in [18].

CCA security $\Rightarrow$ pd-RCCA security $\Rightarrow$ sd-RCCA security $\Rightarrow$ RCCA security. We show that the two leftmost implications are strict. (The first is implied by the above "endian changing" example; the second holds for schemes with message space 3 or larger, or alternatively under the DDH assumption.) Whether sd-RCCA security is equivalent to RCCA security remains open.

Finally, we provide a generic construction that turns any RCCA secure scheme (with message domain $\{0, 1\}^k$ where k is the security parameter) into a CCA scheme. The construction is quite efficient, and uses only shared-key primitives. This in essence demonstrates that the existence of an RCCA secure encryption scheme implies the existence of a CCA secure scheme without any additional computational assumptions. Also, this construction may provide an alternative way of obtaining CCA security.

Symmetric encryption. In this work we develop the RCCA notions mainly for public-key encryption. However, the notion can be adapted to the symmetric-key setting in a straightforward way. We outline this generalization in [8].

Organization. Section 2 recalls the formulation of CCA security, and establishes its equivalence with the universally composable notion of security for public-key encryption schemes (against non-adaptive adversaries) as defined in [7]. Section 3 presents the three variants of RCCA security and establishes the relationships among them. Section 4 studies the detectable variants of RCCA security. It also shows how to turn any RCCA secure scheme into a CCA secure one. Section 5 demonstrates several central applications where RCCA security can be used instead of CCA security. Most proofs are omitted from this extended abstract. They can be found in [8].

2   Prolog: On CCA Security

Before introducing our RCCA definitions, we recall the formulation of CCA security. We also demonstrate that the notion of secure public-key encryption in the universally composable framework [7] is *equivalent* to CCA security. (In particular, this corrects the erroneous claim from [7] that the UC characterization is *strictly weaker* than CCA security. The mistake in [7] and the equivalence proof were discovered independently in [18].) This equivalence sets the stage for the presentation of RCCA. In particular, by comparing the UC formalizations of CCA security and of RCCA security it is easier to see that the technical relaxation from CCA to RCCA coincides with the intuition behind the later notion as described above, namely, that "replayable CCA" is identical to CCA except for the added ability of the attacker to generate new ciphertexts that decrypt to the same plaintexts as previously seen ciphertexts. We start by establishing the basic formal setting for public-key encryption schemes.

2.1   CCA Secure Encryption Schemes

Public-key encryption schemes.   Throughout the paper we consider (public key) encryption schemes as triples of probabilistic polynomial-time algorithms $S = (gen, enc, dec)$ together with an ensemble of finite domains (sets) $\mathcal{D} =$

$\{D_k\}_{k\in\mathbb{N}}, D_k \subset \{0,1\}^*$. Algorithm *gen*, on input 1^k (k is a security parameter), generates a pair of keys (e,d). The encryption and decryption algorithms, *enc* and *dec*, satisfy that if $(e,d) = gen(1^k)$, then for any message $m \in D_k$ we have $dec_d(enc_e(m)) = m$ except with negligible probability. The range of the decryption function may include a special symbol invalid $\notin \mathcal{D}_k, \forall k$.

The CCA Game

The game proceeds as follows, given an encryption scheme $S = (gen, enc, dec)$, an adversary F, and value k for the security parameter.

Key generation: Run $(e,d) \leftarrow gen(1^k)$, and give e to F.

First decryption stage: When F queries (ciphertext,c), compute $m = dec_d(c)$ and give m to F.

Encryption stage: When F queries (test messages,m_0, m_1), $m_0, m_1 \in D_k$, and $m_0 \neq m_1$, compute $c^* = enc_e(m_b)$ where $b \xleftarrow{R} \{0,1\}$, and give c^* to F. (This step is performed only once.)

Second decryption stage: When F queries (ciphertext,c) after c^* is defined, proceed as follows. If $c = c^*$ then give test to F.[a] Otherwise, compute $m = dec_d(c)$ and give m to F.

Guessing stage: When F outputs (guess,b'), the outcome of the game is determined as follows. If $b' = b$ then F wins the game. Otherwise, F loses the game.

---

[a] The symbol test is a reserved symbol, which is different from all possible outputs of *dec*.

Fig. 1. The CCA Game.

CCA security. We recall the definition of CCA security (or IND-CCA2) for public key encryption schemes. See [23,12,3]. Let $S = (gen, enc, dec)$ be an encryption scheme over domain $\mathcal{D} = \{D_k\}_{k\in\mathbb{N}}$. This definition, presented next, is based on the CCA game described in Figure 1.[4]

Definition 1. *An encryption scheme S is said to be* CCA-secure *if any polynomial-time adversary F wins the IND-CCA game of Figure 1 with probability that is at most negligibly more than one half.*

2.2   Equivalence of CCA and UC Security of Encryption Schemes

A UC characterization of CCA security. Here we assume some familiarity of the reader with the UC framework. See [8] for a quick review.   Within the

---

[4] The explicit requirement in this game that $m_0 \neq m_1$ is immaterial for the definition of CCA and the later definition of IND-RCCA (in which choosing $m_0 = m_1$ is of no benefit to the attacker), but will be substantial in our definition of NM-RCCA. Thus, for the sake of uniformity we present all our definitions using the explicit requirement $m_0 \neq m_1$.

UC framework, public-key encryption is defined via the public-key encryption functionality from [7], denoted $\mathcal{F}_{pke}$ and presented in Figure 2. Functionality $\mathcal{F}_{pke}$ is intended at capturing the functionality of a public-key encryption scheme as a *tool* to be used within other protocols. In particular, $\mathcal{F}_{pke}$ is written in a way that allows realizations that consist of three non-interactive algorithms without any communication. (The three algorithms correspond to the key generation, encryption, and decryption algorithms in the traditional definitions.) All the communication is left to the higher-level protocols that use $\mathcal{F}_{pke}$.

Referring to Figure 2, we note that id serves as a unique identifier for an instance of functionality $\mathcal{F}_{pke}$ (this is needed in a general protocol setting when this functionality can be composed with other components, or even with other instantiations of $\mathcal{F}_{pke}$). The "public key value" e has no particular meaning in the ideal scenario beyond serving as an identifier for the public key related to this instance of the functionality, and can be chosen arbitrarily by the attacker. Also, in the ideal setting ciphertexts serve as identifiers or tags with no particular relation to the encrypted messages (and as such are also chosen by the adversary). Yet, rule 1 of the decryption operation guarantees that "legitimate ciphertexts", i.e. those produced and recorded by the functionality under an Encrypt request, are decrypted correctly and the resultant plaintexts remain unknown to the adversary. In contrast, ciphertexts that were not legitimately generated can be decrypted in any way chosen by the ideal-process adversary (yet, since the attacker obtains no information on legitimately encrypted messages, illegitimate ciphertexts will be decrypted to values that are independent from the legitimately encrypted messages.) Note that illegitimate ciphertexts can be decrypted to different values in different activations. This provision allows the decryption algorithm to be non-deterministic with respect to ciphertexts that were not legitimately generated.

$\mathcal{F}_{pke}$ is parameterized by $\mathcal{D} = \{D_k\}_{k\in\mathbf{N}}$, the ensemble of domains of the messages to be encrypted. Given security parameter k, $\mathcal{F}_{pke}$ encrypts messages in domain D_k.

Remarks. In [7] $\mathcal{F}_{pke}$ is slightly different. Specifically, when invoked to encrypt a message m, the functionality there is instructed to hand $|m|$, the length of m, to the adversary. Here we do not hand $|m|$ to the adversary. This means that a protocol that realizes $\mathcal{F}_{pke}$ may not reveal any information on $|m|$, beyond the fact that $m \in \mathcal{D}_k$. (Indeed, knowing that $m \in D_k$ may by itself reveal some information on the length of m.)

$\mathcal{F}_{pke}$ *captures CCA security.* We show the equivalence between the notion of security induced by functionality $\mathcal{F}_{pke}$ and the notion of CCA security. First, recall the following natural transformation from an encryption scheme S to a protocol π_S that is geared towards realizing $\mathcal{F}_{pke}$.

1. When activated, within some P_i and with input (KeyGen, id), run algorithm *gen*, output the encryption key e and record the decryption key d.
2. When activated, within some party P_j and with input (Encrypt, id, e', m), return $enc_{e'}(m, r)$ for a randomly chosen r. (Note that it does not necessarily hold that $e' = e$.)

Functionality $\mathcal{F}_{pke}$

$\mathcal{F}_{pke}$ proceeds as follows, when parameterized by message domain ensemble $\mathcal{D} = \{D_k\}_{k\in\mathbf{N}}$ and security parameter k, and interacting with an adversary $\mathcal{S}$, and parties $P_1, ..., P_n$. (Recall that $\mathcal{S}$ can be either an ideal-process adversary or an adversary in the hybrid model.)

Key Generation: Upon receiving a value (KeyGen, id) from some party P_i, do:
1. Hand (KeyGen, id) to the adversary.
2. Receive a value e from the adversary, and hand e to P_i.
3. If this is the first activation then record the value e.

Encryption: Upon receiving from some party P_j a value (Encrypt, id, e', m) proceed as follows:
1. If $m \notin D_k$ then return an error message to P_j.
2. If $m \in D_k$ then hand (Encrypt, id, e', P_j) to the adversary. (If $e' \neq e$ or e is not yet defined then hand also the entire value m to the adversary.)
3. Receive a tag c from the adversary and hand c to P_j. If $e' = e$ then record the pair (c, m). (If the tag c already appears in a previously recorded pair then return an error message to P_j.)

Decryption: Upon receiving a value (Decrypt, id, c) from P_i (and P_i only), proceed as follows:
1. If there is a recorded pair (c, m) then hand m to P_i.
2. Otherwise, hand the value (Decrypt, id, c) to the adversary. When receiving a value m from the adversary, hand m to P_i.

Fig. 2. The public-key encryption functionality, $\mathcal{F}_{pke}$

3. When activated, within P_i and with input (Decrypt, id, c), return $dec_d(c)$.

We show:

Theorem 1. *Let $S = (gen, enc, dec)$ be an encryption scheme over domain $\mathcal{D}$. Then S is CCA-secure if and only if π_S securely realizes $\mathcal{F}_{pke}$ with respect to domain $\mathcal{D}$ and non-adaptive adversaries.*

3   Replayable CCA (RCCA) Security

This section presents the three notions of security for public-key encryption sketched in the Introduction, all aimed at capturing the intuition that "the adversary should gain nothing from seeing a legitimately generated ciphertext, except for the ability to generate new ciphertexts that decrypt to the same value as the given ciphertext". Following the presentation of the three notions, we demonstrate their equivalence for encryption schemes with super-polynomial message domains, and present separating examples for polynomial message domains.

3.1   UC-RCCA: Functionality $\mathcal{F}_{rpke}$

The UC-based formulation of RCCA security is obtained by modifying the $\mathcal{F}_{pke}$ functionality from Figure 2, as to explicitly allow the adversary, together with the environment, to generate ciphertexts that decrypt to the same values as legitimately generated ciphertexts. Specifically, the new functionality, $\mathcal{F}_{rpke}$, modifies step 2 of the **Decryption** stage in Figure 2 in which the decrypting party asks to decrypt a ciphertext that was not legally generated by the functionality. In this case, $\mathcal{F}_{rpke}$ allows the adversary to fix the decrypted value to be the same as a previously encrypted value (without letting the adversary know what this value is). Thus, functionality $\mathcal{F}_{rpke}$ is defined identically to $\mathcal{F}_{pke}$ from Figure 2 except that step 2 of the **Decryption** stage is re-defined as follows:

Decryption: Upon receiving a value $(\texttt{Decrypt}, id, c)$ from P_i (and P_i only), proceed as follows:
1. If there is a recorded pair (c, m) then hand m to P_i.
2. Otherwise, hand the value $(\texttt{Decrypt}, id, c)$ to the adversary, and receive a value (α, v) from the adversary. If $\alpha =$'plaintext' then hand v to P_i. If $\alpha=$'ciphertext' then find a stored pair (c', m) such that $c' = v$, and hand m to P_i. (If no such c' is found then halt.)

In the following definition we use the transformation from an encryption scheme S into a protocol π_S as described in the previous section.

Definition 2. *Let S be an encryption scheme. We say that S is UC-RCCA secure if protocol π_S securely realizes $\mathcal{F}_{rpke}$ with respect to non-adaptive adversaries.*

Remark: We stress that protocol π_S is only required to realize $\mathcal{F}_{rpke}$ with respect to *non-adaptive* adversaries. Realizing $\mathcal{F}_{rpke}$ with respect to *adaptive* adversaries is a considerably stronger requirement. Indeed, using the techniques of [22], it can be shown that no protocol π_S can realize $\mathcal{F}_{rpke}$ with respect to adaptive adversaries, for any scheme S. Realizing $\mathcal{F}_{rpke}$ with respect to adaptive adversaries requires additional mechanisms, such as forward secure encryption or highly interactive solutions based on non-committing encryption.

3.2   NM-RCCA and IND-RCCA: The RCCA Games

This section presents two notions of security for encryption schemes, that are formulated via relaxed versions of the CCA game (See Figure 1 in Section 2), and demonstrates the equivalence of these notions to the UC-RCCA formulation for encryption schemes with super-polynomial domains. The two notions are called IND-RCCA and NM-RCCA, and are defined via the IND-RCCA game and the NM-RCCA game, respectively.

The IND-RCCA game (see Figure 3) differs from the CCA game in one point: When the adversary generates a ($\texttt{decrypt}$, c) request, the answer is $\texttt{test}$ whenever c decrypts to either m_0 or m_1. Roughly speaking, this captures the intuition that "the ability to generate different ciphertexts that decrypt to the same values as a given ciphertext should not help the adversary to win the game." (As we demonstrate below, in the case of large message spaces, this intuition is supported by the equivalence between IND-RCCA and UC-RCCA.)

The NM-RCCA game is identical to the IND-RCCA game (Figure 3), with the exception that the guessing stage is defined as follows:

The IND-RCCA game

The game proceeds as follows, given an encryption scheme $S = (gen, enc, dec)$ over a domain ensemble $\mathcal{D}$, an adversary F, and value k for the security parameter.

Key generation: Run $(e, d) \leftarrow gen(1^k)$, and give e to F.

First decryption stage: When F queries (`ciphertext`,c), compute $m = dec_d(c)$ and give m to F.

Encryption stage: When F queries (`test messages`,m_0, m_1), with $m_0, m_1 \in D_k$, and $m_0 \neq m_1$, compute $c^* = enc_e(m_b)$ where $b \stackrel{R}{\leftarrow} \{0, 1\}$, and give c^* to F. (This step is performed only once.)

Second decryption stage: When F queries (`ciphertext`,c) after c^* is defined, compute $m = dec_d(c)$. If $m \in \{m_0, m_1\}$ then give `test` to F. Otherwise, give m to F.

Guessing stage: When F outputs (`guess`,b'), the outcome of the game is determined as follows. If $b' = b$ then F wins the game. Otherwise, F loses the game.

Fig. 3. The IND-RCCA game

Guessing stage for NM-RCCA: When F outputs (`guess`,c), the outcome of the game is determined as follows. Compute $m = dec_d(c)$; if $m = m_{1-b}$ then F wins the game. Otherwise, F loses the game.

In order to consider F successful we require it to output an encryption of m_{1-b}. Changing this requirement to explicitly output m_{1-b} would result in a reformulation of IND-RCCA. Thus the difference from IND-RCCA is in the fact that F is only required to output an encryption of m_{1-b}, without necessarily being able to explicitly output m_{1-b} (or, equivalently, b). As we demonstrate below, this added strength relative to IND-CCA is significant for small message domains. The formulation of the attacker's goal in the NM-RCCA definition follows the non-malleability approach (and hence the name); see the discussion below.

Definition 3. *An encryption scheme S is said to be* IND-RCCA *secure (resp.,* NM-RCCA *secure) if any polynomial-time adversary F wins the IND-RCCA game of Figure 3 (resp., the NM-RCCA game) with probability that is at most negligibly more than one half.*

Discussion. The formulation of NM-RCCA is syntactically different than the usual formulation of definitions of non-malleability. We thus provide an intuitive explanation as to why NM-RCCA indeed captures the non-malleability requirement, in spite of its different formalization. Roughly speaking, an encryption scheme is called *non-malleable* [12,3] if it is infeasible for an adversary to output ciphertexts which decrypt to plaintexts that satisfy some (non-trivial) relation with the plaintext encrypted under a given "challenge ciphertext" c^*. (A bit more precisely, the attacker is not given the plaintext encrypted under c^* but she may choose the probability distribution under which this plaintext is taken.

Thus, "trivial relations" are those that hold for randomly chosen elements from this distribution.) In the case of non-malleability under chosen-ciphertext attacks (NM-CCA) the only restriction on the attacker is that it is not allowed to include the ciphertext c^* as one of its output ciphertexts (otherwise, the attacker could always output c^* and satisfy the "equality" relation.)

In our formulation of NM-RCCA we use the above non-malleability approach to capture our intuition behind the "replayable CCA" notion. The idea is to relax NM-CCA so that there is only one form of malleability *allowed* to the attacker: outputting a ciphertext that decrypts to *the same* plaintext as c^*. In other words, the attacker is considered successful as long as it uses any relation *other than the "equality" relation*. Now, if we carry this idea to the case where the probability distribution $\mathcal{P}$, from which the plaintext to be encrypted as c^* is selected, is of the special form $\mathcal{P} = [\{m_0, m_1\}, Prob(m_0) = Prob(m_1) = 1/2]$, with m_0, m_1 chosen by the attacker, then we obtain our "non-malleability game" NM-RCCA. Beyond this intuition and relationship to general non-malleability, the main source of confidence for this definition comes from its equivalence (at least over super-polynomial domains) with the UC-RCCA notion which captures in a more explicit way the "intuitive semantics" of RCCA.

3.3   Equivalence for Large Message Domains

Theorem 2. *Let S be an encryption scheme whose domain ensemble $\mathcal{D}$ is super-polynomial in size. Then the following three conditions are equivalent: (I) S is UC-RCCA secure; (II) S is NM-RCCA secure; (III) S is IND-RCCA secure.*

3.4   Polynomial Message Domains

In Theorem 2 it was proved that for super-polynomial domain ensembles, the notions UC-RCCA, NM-RCCA and IND-RCCA are equivalent. Here we show that this premise is necessary. As mentioned in the proof of Theorem 2 it holds for all domain ensembles that UC-RCCA implies NM-RCCA and that NM-RCCA implies IND-RCCA. A minimal assumption for separating is therefore that there exist IND-RCCA secure encryption schemes in the first place. We do not know whether UC-RCCA and NM-RCCA are equivalent for polynomial domain ensembles.

Theorem 3. *For all polynomial-size domain ensembles $\mathcal{D}$, if there exists an IND-RCCA secure encryption scheme with domain ensemble $\mathcal{D}$, then there exists an IND-RCCA secure encryption scheme with domain ensemble $\mathcal{D}$ which is not NM-RCCA secure.*

For the proof, let $S = (gen, enc, dec)$ be an IND-RCCA secure encryption scheme with polynomial-size domain ensemble $\mathcal{D} = \{D_k\}$. Consider the encryption scheme $S' = (gen', enc', dec')$, where $gen' = gen$, $enc'_e(m) = enc(m)_e enc_e(n)$ for $n \in_R D_k \setminus \{m\}$, and $dec'_d(c_1, c_2) = dec_d(c_1)$. We claim that S' is IND-RCCA secure but not NM-RCCA secure. See details in [8].

4   Between RCCA and CCA Security: Detectable RCCA

In this section we investigate the relations between RCCA security and CCA security, and introduce the notion(s) of "detectable RCCA". In particular, we establish the relationship between RCCA security and the relaxation of CCA security presented in [19,24,1]. These results include a (strict) separation between these notions, and consequently between RCCA and CCA security. In particular, this demonstrates that there exist encryption schemes that are not secure in the sense of the definitions from [19,24,1] and yet are sufficiently secure for most practical applications of CCA secure encryption. We complement these findings by showing (Section 4.3) how to construct a CCA-secure scheme (with any domain size) from any RCCA-secure scheme whose domain size is exponential in the security parameter. This transformation uses symmetric encryption and message authentication only, thus demonstrating that once RCCA security is obtained for large enough message spaces, CCA security can be obtained with moderate overhead (and without further assumptions).

Remark. Due to the equivalence between the notions of IND-RCCA, NM-RCCA and UC-RCCA security for super-polynomial domain ensembles (Theorem 2), we will usually refer to these notions under the generic term of RCCA security, and assume, for simplicity, super-polynomial domains (except if otherwise stated).

4.1   Detectable RCCA

The first and obvious fact to observe regarding the relation between RCCA and CCA security is that the former is strictly weaker than the latter. Indeed, a simple inspection of the definitions of CCA and RCCA security shows that any scheme that is CCA-secure is also RCCA-secure (under any of the definitions of RCCA security from Section 3). On the other hand, there are simple examples of encryption schemes that are RCCA but not CCA secure. One such example was mentioned in the introduction in which a CCA-secure scheme is modified by instructing the (modified) encryption to append a '0' bit to each ciphertext, and defining the (modified) decryption algorithm to ignore this bit. It is easy to see that the obtained scheme is not CCA but it does satisfy our definition(s) of RCCA security. Other examples exist. Specifically, consider the usual practice of allowing encryption schemes to add arbitrary padding to ciphertexts and later discard this padding before performing decryption. (This padding is usually required in order to align the length of ciphertexts to a prescribed length-boundary – e.g., to a multiple of 4 bytes.) Other examples include encryption schemes that naturally allow for more than one representation of ciphertexts, such as the endianess example in the introduction, or the example in [24] related to dual point representations in elliptic-curve cryptosystems.

All these examples have the property that given a certain ciphertext c, anyone can easily produce a different ciphertext c' that decrypts to the same plaintext (e.g., by changing the endianess representation or modifying the padding). Also common to these examples is the fact that if someone (say, the attacker) indeed modifies a ciphertext c into a ciphertext c' in one of the above ways then c and c' satisfy a relation that is easy to test with the sole knowledge of the public key. This fact (and the realization that these "syntactic deficiencies" do not seem to effect the actual security of these encryption schemes when used in many applications) has motivated the introduction of the relaxations of CCA security presented in [19,24,1] and mentioned in the introduction. Essentially, all these notions allow for "replay" of plaintexts by modifying the ciphertext, but restrict the allowed modifications to be efficiently detectable given the public key. RCCA further relaxes this requirement by allowing *any form* of ciphertext modification that do not change the plaintext, without insisting in the ability to detect such a replay. (Indeed, as argued in the introduction and demonstrated in Section 5, RCCA security is sufficient for many applications that use CCA secure encryption.)

A natural question that arises from this discussion is whether RCCA security is truly more relaxed than the notions considered in [19,24,1], namely, is there an RCCA-secure scheme for which the modification of ciphertexts is not "publicly detectable"? Here we provide a positive answer to this question. We start by formalizing some of the notions discussed above.

Definition 4. *Let $S = (gen, enc, dec)$ be be an encryption scheme.*

1. *We say that a family of binary relations $\equiv_e$ (indexed by the public keys of S) on ciphertext pairs is a* compatible *relation for S if for all key-pairs (e, d) of S we have:*
   a) *For any two ciphertexts c, c', if $c \equiv_e c'$ then $dec_d(c) = dec_d(c')$, except with negligible probability over the random choices of algorithm dec.*
   b) *For any plaintext m in the domain of S, if c and c' are two ciphertexts obtained as independent encryptions of m (i.e., two applications of algorithm enc on m using independent random bits), then $c \equiv_e c'$ only with negligible probability.*
2. *We say that a relation family as above is* publicly computable *(resp.* secretly computable*) if for all key pairs (e, d) and ciphertext pairs (c, c') it can be determined whether $c \equiv_e c'$ using a probabilistic polynomial time algorithm taking inputs (e, c, c') (resp. (e, d, c, c')).*
3. *We say that S is* publicly-detectable replayable-CCA (pd-RCCA) *if there exists a compatible and publicly computable relation family $\equiv_e$ such that S is secure according to the standard definition of CCA with the following modification to the CCA game from Figure 1: if, after receiving the challenge ciphertext c^*, the adversary queries the decryption oracle on a ciphertext c such that $c \equiv_e c^*$ then the decryption oracle returns* test.
   Similarly, we say that S is secretly-detectable replayable-CCA (sd-RCCA) *if the above holds for a secretly computable relation family $\equiv_e$.*

The reader can verify that the notion of pd-RCCA is essentially equivalent to the notions of loose ciphertext-unforgeability, benign malleability, and generalized CCA security, presented, respectively, in [19,24,1]. The term "compatible relation" is adapted from [24] where it is defined without item 1(b). Indeed, in the case of "public detectability" this requirement is redundant as it follows from the semantic security of the encryption scheme. In contrast, it is significant in the case of "secret detectability"; without it this notion is trivially equivalent to RCCA. More significantly, requirement 1(b) captures the intuition that "detectability" is useful as long as it can tell apart legitimately generated ciphertexts from those that are created by "mauling" a given (legitimate) ciphertext. We expand on this aspect in [8].

4.2   Relations among Notions of Detectable RCCA

Here we investigate the relations among the different flavors of detectable RCCA security, and between these and CCA security. We show:

Theorem 4. *CCA $\Rightarrow$ pd-RCCA $\Rightarrow$ sd-RCCA $\Rightarrow$ RCCA.   If RCCA-secure encryption schemes with super-polynomial message space exist then the first two implications are strict.*

It is easy to see that any encryption scheme that is CCA secure is also pd-RCCA (simply consider the equality relation as a compatible relation). Also, immediate from the definition we get that pd-RCCA implies sd-RCCA. On the other hand, as discussed above, any CCA-secure scheme can be transformed into a pd-RCCA scheme that is not CCA by appending to the ciphertext a "dummy bit" that is ignored by the decryption operation. Therefore, we get a strict separation between CCA and pd-RCCA security.   In [8] we provide examples that separate between sd-RCCA and pd-RCCA.

Remark: The above results leave the open question of whether one can separate between sd-RCCA and RCCA. An interesting related question is whether there exist RCCA-secure encryption schemes, where anyone can "randomize" a given ciphertext c to another ciphertext c' so that c' looks like an "honestly generated ciphertext", even when given the decryption key. In particular, c and c' should not be "linkable" in any way. We call such encryption schemes *randomizable.*

On the other hand, given an RCCA encryption scheme that can encrypt long messages (say, $O(k)$-bit messages where k is the security parameter), one can easily obtain an sd-RCCA scheme by appending a fresh random (and sufficiently long) tag to each message before encryption. Replay of ciphertexts can then be privately detected by decrypting and comparing the received tag to previously received tags.

4.3   From RCCA Security to CCA Security

This section demonstrates that the existence of an RCCA secure public-key encryption scheme with large message domain implies the existence of a CCA secure public-key encryption scheme. To be precise, by large we will mean that the encryption scheme can encrypt messages of length k, where k is the security parameter.

The construction consists of two steps. First we recall that the existence of any secure encryption scheme implies the existence of a CCA secure *symmetric* encryption scheme. We then show how to combine an RCCA secure public-key encryption scheme with large domain and a CCA secure symmetric encryption scheme to obtain a CCA secure public-key encryption scheme. The first step is very inefficient, whereas the second step results in an efficient encryption scheme if the RCCA secure public-key encryption scheme and the CCA secure symmetric encryption scheme are both efficient.

For the first step, if an RCCA secure public-key encryption scheme (gen, enc, dec) exists, then a one-way function exists, e.g. the function $(e, d) = gen(r)$, where r is the random bits used in generating (e, d). Now, the existence of a one-way function through a series of well-known reductions implies the existence of a CCA secure symmetric encryption scheme (E, D) encrypting unbounded length messages.[5] To prepare for the second step, let $l(k)$ denote the key-length of (E, D) as a function of the security parameter k and consider the public-key encryption scheme $(\overline{gen}, enc, dec)$ given by

$$\overline{gen}(1^k) = gen(1^{\max(k, l(k))}) \ .$$

Clearly $(\overline{gen}, enc, dec)$ is RCCA secure if (gen, enc, dec) is RCCA secure. Furthermore, $(\overline{gen}, enc, dec)$ can encrypt messages of length $l(k)$. In the following we will therefore assume that we have access to a CCA secure symmetric encryption scheme (E, D) with key-length l and an RCCA secure public-key encryption scheme (gen, enc, dec) capable of encrypting messages of length l.

Consider then the public-key encryption scheme $(gen, \overline{enc}, \overline{dec})$ given by

$$\overline{enc}_e(m) = [K \xleftarrow{R} \{0,1\}^{l(k)}; c_1 = enc_e(K); c_2 = E_K(c_1 \| m) : (c_1, c_2)] \ ,$$

$$\overline{dec}_d(c_1, c_2) = [K \leftarrow dec_d(c_1); c_1' \| m = D_K(c_2); \text{if } c_1' \neq c_1 \text{ then } m \leftarrow \texttt{invalid} : m] \ .$$

This construction of $(gen, \overline{enc}, \overline{dec})$ resembles the usual extension of a CCA secure public-key encryption scheme with a CCA secure symmetric encryption scheme for doing hybrid encryption. The only difference is the encryption of c_1 under the symmetric key. This encryption of c_1 functions as a MAC which protects against 'mauling' of c_1. Indeed, if one is not interested in hybrid encryption but only in obtaining CCA security, the encryption of m could be done as $c_1 = enc_e(K \| m); c_2 = mac_K(c_1)$, where mac is a strong message authentication code. The proof would follow the same lines as the proof of the following theorem.

Theorem 5. *If (E, D) is a CCA secure symmetric encryption scheme with key-length l and (gen, enc, dec) is an RCCA secure public-key encryption scheme capable of encrypting messages of length l, then the public-key encryption scheme $(gen, \overline{enc}, \overline{dec})$ is CCA secure.*

---

[5] I.e., the encryption scheme itself does not contain any bound on the message-length. However, under attack by a given adversary F the length of the messages encrypted is of course bounded by a polynomial as F is required to be PPT.

5   Using RCCA Security

This section demonstrates the power of RCCA security by proving its sufficiency for several core applications of public key encryption. Prior proofs for the security of these applications relied on the CCA security (or, in some cases, on the pd-RCCA security) of the underlying encryption schemes.

We first demonstrate that RCCA security suffices for achieving secure communication channels, in exactly the same way that CCA does. That is, we show that the natural protocol of [7] for realizing secure channels given access to $\mathcal{F}_{pke}$ remains secure even if $\mathcal{F}_{pke}$ is replaced with $\mathcal{F}_{rpke}$. Next, we consider the simple key-exchange protocol of [10] based on any CCA-secure encryption scheme. We show that this protocol remains secure even when the underlying encryption is RCCA. A similar result is demonstrated with respect to the password-based key exchange protocol of Halevi and Krawczyk [17]. Finally, we demonstrate that the *hybrid encryption* paradigm can be based on RCCA security rather than CCA security. (Hybrid encryption calls for encrypting a key k using an asymmetric encryption and then encrypting a long message using symmetric encryption with key k.) For lack of space, these results are deferred to [8].

As shown by these results, RCCA security is adequate for most typical encryption applications. Yet, it should be stressed that RCCA cannot be considered as a "drop-in" replacement for *all* applications of CCA. As pointed out in the introduction, if any such application makes decisions based on the ciphertext strings themselves (e..g compares them), rather than just using the ciphertexts as inputs to the decryption algorithm, then replacing CCA with RCCA may not be secure. It is indeed unusual that such examination of the ciphertext strings is performed by applications, yet this cannot be discounted. A simple example is a voting scheme where votes are encrypted and illegitimate duplicate votes are detected via ciphertext comparison.

References

1. JH An, Y. Dodis, and T. Rabin, "On the Security of Joint Signature and Encryption", in *Eurocrypt '02*, pages 83–107, 2002. LNCS No. 2332.
2. M. Bellare, R. Canetti and H. Krawczyk, "A modular approach to the design and analysis of authentication and key-exchange protocols", *30th STOC*, 1998.
3. M. Bellare, A. Desai, D. Pointcheval, and P. Rogaway, "Relations Among Notions of Security for Public-Key Encryption Schemes", *Advances in Cryptology – CRYPTO'98 Proceedings*, Lecture Notes in Computer Science Vol. 1462, H. Krawczyk, ed., Springer-Verlag, 1998.
4. M. Bellare and C. Namprempre, "Authenticated encryption: Relations among notions and analysis of the generic composition paradigm", *Advances in Cryptology – ASIACRYPT'00 Proceedings*, Lecture Notes in Computer Science Vol. 1976, T. Okamoto, ed., Springer-Verlag, 2000.
5. Bellovin, S. M. and Merritt, M., "Encrypted key exchange: Password- based protocols secure against dictionary attacks", In *Proceedings of the IEEE. Computer Society Symposium on Research in Security and Privacy* 1992, pp. 72–84.

6. Bleichenbacher, D., "Chosen Ciphertext Attacks against Protocols Based on RSA Encryption Standard PKCS #1", *Advances in Cryptology – CRYPTO'98 Proceedings*, Lecture Notes in Computer Science Vol. 1462, H. Krawczyk, ed., Springer-Verlag, 1998, pp. 1–12.
7. R. Canetti, "Universally Composable Security: A new paradigm for cryptographic protocols", http://eprint.iacr.org/2000/067. Extended Abstract appears in *42nd FOCS*, 2001.
8. R. Canetti, H. Krawczyk and J. Nielsen, "Relaxing Chosen Ciphertext Security," available online at http://eprint.iacr.org, 2003.
9. R. Canetti and S. Goldwasser, "A practical threshold cryptosystem resilient against adaptive chosen ciphertext attacks", *Eurocrypt'99*, 1999.
10. Canetti, R., and Krawczyk, H., "Analysis of Key-Exchange Protocols and Their Use for Building Secure Channels", *Eurocrypt 01*, 2001. Full version in: *Cryptology ePrint Archive* (http://eprint.iacr.org/), Report 2001/040.
11. R. Cramer and V. Shoup, "A practical public-key cryptosystem provably secure against adaptive chosen ciphertext attack", *Advances in Cryptology – CRYPTO'98 Proceedings*, Lecture Notes in Computer Science Vol. 1462, H. Krawczyk, ed., Springer-Verlag, 1998.
12. D. Dolev, C. Dwork and M. Naor, Non-malleable cryptography, *SIAM. J. Computing,* Vol. 30, No. 2, 2000, pp. 391-437. Preliminary version in *23rd Symposium on Theory of Computing (STOC)*, ACM, 1991.
13. T. ElGamal, A Public-Key cryptosystem and a Signature Scheme based on Discrete Logarithms, *IEEE Transactions*, Vol. IT-31, No. 4, 1985, pp. 469–472.
14. O. Goldreich, "Foundations of Cryptography: Basic Tools", Cambridge Press, 2001.
15. O. Goldreich, S. Goldwasser and S. Micali, "How to construct random functions," *Journal of the ACM,* Vol. 33, No. 4, 210–217, (1986).
16. S. Goldwasser and S. Micali, Probabilistic encryption, *JCSS,* Vol. 28, No 2, 1984.
17. S. Halevi, and H. Krawczyk, "Public-Key Cryptography and Password Protocols", *ACM Transactions on Information and System Security*, Vol. 2, No. 3, August 1999, pp. 230–268.
18. Dennis Hofheinz and Joern Mueller-Quade and Rainer Steinwandt, "On Modeling IND-CCA Security in Cryptographic Protocols", http://eprint.iacr.org/2003/024. 2003.
19. H. Krawczyk, "The order of encryption and authentication for protecting communications (Or: how secure is SSL?)", Crypto 2001. http://eprint.iacr.org/2001/045
20. M. Krohn, "On the definitions of cryptographic security: Chosen-Ciphertext attack revisited," Senior Thesis, Harvard U., 1999.
21. M. Naor and M. Yung, "Public key cryptosystems provably secure against chosen ciphertext attacks". *Proceedings of the 22nd Annual ACM Symposium on Theory of Computing*, 1990.
22. Jesper B. Nielsen, " Separating random oracle proofs from complexity theoretic proofs: The non-committing encryption case", in M. Yung, editor, *Advances in Cryptology - Crypto 2002*, pages 111–126,Lecture Notes in Computer Science Volume 2442.
23. C. Rackoff and D. Simon, "Non-interactive zero-knowledge proof of knowledge and chosen ciphertext attack", *CRYPTO '91*, 1991.
24. V. Shoup, "A Proposal for an ISO Standard for Public Key Encryption", Crypto Eprint archive entry 2001:112, http://eprint.iacr.org, 2001.
25. A. Sahai, "Non malleable, non-interactive zero knowledge and adaptive chosen ciphertext security", FOCS 99.

Password Interception in a SSL/TLS Channel

Brice Canvel[1], Alain Hiltgen[2], Serge Vaudenay[1], and Martin Vuagnoux[3]

[1] Swiss Federal Institute of Technology (EPFL) – LASEC
http://lasecwww.epfl.ch
[2] UBS AG
alain.hiltgen@ubs.com
[3] EPFL – SSC, and Ilion
http://www.ilionsecurity.ch

Abstract. Simple password authentication is often used e.g. from an email software application to a remote IMAP server. This is frequently done in a protected peer-to-peer tunnel, e.g. by SSL/TLS.

At Eurocrypt'02, Vaudenay presented vulnerabilities in padding schemes used for block ciphers in CBC mode. He used a side channel, namely error information in the padding verification. This attack was not possible against SSL/TLS due to both unavailability of the side channel (errors are encrypted) and premature abortion of the session in case of errors. In this paper we extend the attack and optimize it. We show it is actually applicable against latest and most popular implementations of SSL/TLS (at the time this paper was written) for password interception.

We demonstrate that a password for an IMAP account can be intercepted when the attacker is not too far from the server in less than an hour in a typical setting.

We conclude that these versions of the SSL/TLS implementations are not secure when used with block ciphers in CBC mode and propose ways to strengthen them. We also propose to update the standard protocol.

1 Introduction

1.1 CBC-PAD in Secured Channels

Peer-to-peer secure channels can be established by the TLS protocol [9]. It consists in, first, negotiating a cipher suite and security parameters, second, exchanging secret keys. Then messages are first authenticated with a Message Authentication Code (MAC), then encrypted with a symmetric cipher. Block ciphers, e.g. the Triple DES (3DES) [3] are frequently used in Cipher Block Chaining (CBC) mode [4] with padding. Let b be the block length in characters (e.g. $b = 8$ for DES).

Let MES be the message to be sent. First we append the MAC of MES to MES. We obtain MES|MAC. Then we pad MES|MAC with a padding PAD such that MES|MAC|PAD|LEN is of length a multiple of b where LEN is a single byte whose value ℓ is the length of PAD in bytes. PAD is required by the TLS specifications to consist of ℓ bytes equal to ℓ. Then MES|MAC|PAD|LEN is cut

D. Boneh (Ed.): CRYPTO 2003, LNCS 2729, pp. 583–599, 2003.

into a block sequence $x_1, x_2, \ldots, x_n$ (each x_i has a length of b), then encrypted in CBC mode, i.e. transformed into $y_1, y_2, \ldots, y_n$ with

$$y_i = \mathrm{ENC}(y_{i-1} \oplus x_i)$$

where ENC denotes the block cipher. (We do not discuss about the initial vector y_0 which can be either a part of the secret key, or a random value sent with the ciphertext, or a fixed value.)

When $y_1, y_2, \ldots, y_n$ is received, it is first decrypted back into $x_1, x_2, \ldots, x_n$. Then we look at the last byte LEN, call ℓ its value, and separate the padding PAD of length ℓ and LEN from the plaintext. It is required that PAD should be checked to consist of bytes all equal to ℓ. If this is not the case, a padding error is generated. Otherwise the MAC is extracted then checked. If the MAC is not valid, a MAC error is generated. Otherwise the cleartext MES is extracted and processed.

In TLS, fatal errors such as incorrect padding or bad MAC errors simply abort the session. It should also be outlined that error messages are sent through the same channel, i.e. they are MACed, padded, then encrypted before being sent.

A typical application of TLS is when an email application connects to a remote IMAP server [8]. For this, the application (client) simply sends the user name and password through the secured channel, i.e. the message MES includes the password in clear.

1.2   Side Channel Attack against CBC-PAD

In 2002, Vaudenay [17] presented an attack which enables the decryption of blocks provided that error messages are available (as a side channel attack) and sessions do not abort. We thus assume that we can send a ciphertext to the server and get the answer which is either an error or an acknowledgment. We modelize this as an oracle $\mathcal{O}$. When the answer is a padding error message (decryption_failed), we say that $\mathcal{O}$ answers 0. Otherwise (bad_record_mac), the oracle $\mathcal{O}$ answers 1.

Let y be the ciphertext block to decrypt. The purpose of the attack is to find the block x such that $y = \mathrm{ENC}(x)$. Following [17], and as depicted on Fig. 1, we first transform the oracle $\mathcal{O}$ into an oracle **Check1**(y, u) which checks whether "the $\mathrm{ENC}^{-1}(y)$ block ends with the byte sequence u" or not. We then use this oracle in **DecryptByte1**(y, s) in order to decrypt a new character in $\mathrm{ENC}^{-1}(y)$ from the known tail s of x. We then use this process in **DecryptBlock1**(y) in order to decrypt a full block y.

The attack of [17] works against WTLS [2]. It does not work against TLS for two reasons. First of all, as soon as a padding or MAC error occurs the session is broken and needs to restart with a freshly exchanged key. As pointed out in [17], the attack could have still worked in order to decrypt only the rightmost byte with a probability of success of 2^{-8}. It can also be adapted in order to test if x ends with a given pattern. (In [17] the oracle in the corresponding

DecryptBlock1(y)
1: **for** $i = 1$ to b **do**
2:    $c_i \leftarrow$**DecryptByte1**$(y, c_{i-1}|\ldots|c_1)$
3: **end for**
4: **return** $c_b|\ldots|c_1$

DecryptByte1(y, s)
1: **for all** possible values of byte c **do**
2:    **if** **Check1**$(y, c|s) = 1$ **then**
3:       **return** c
4:    **end if**
5: **end for**

Check1(y, u)
1: let i be the length of u
2: let L be a random string of length $b - i$
3: let $R = (i-1)|(i-1)|\ldots|(i-1)$ of length i
4: $r \leftarrow L|(R \oplus u)$
5: build the fake ciphertext $r|y$ to be sent to the oracle
6: **return** $\mathcal{O}(r|y)$

Fig. 1. Side Channel Attack against CBC-PAD.

attacks is called a "bomb oracle".) This does not work either against TLS for another reason: because error messages are not available to the adversary (they are indeed encrypted and indistinguishable). In order to make them "even less distinguishable", standard implementations of the TLS protocol now use the same error message for both types or errors (as specified for SSL) in order to protect against this type of attack [13].

Other studies investigated attacks against paddings, like Black-Urtubia [6] which considered other modes than the CBC mode, and Paterson-Yau[1] which considered the CBC mode of ISO/IEC 10116 [1].

1.3   Structure of This Paper

In this paper we first explain in Section 2 how to distinguish between the two types of errors by using another side channel attack based on timing discrepancies. We perform experimental analysis and optimize the attack. Section 3 then explains how to push the attack forward in several broken TLS sessions. We analyze the attack. Section 4 optimizes the attack by performing dictionary attacks against password authentication. We finally describe an experimental attack (Section 5), discuss about practical consequences (Section 6), and conclude.

---

[1] Personal communication.

2   Timing Attack

2.1   Attack Principles

In order to get access to the error type which is not directly available, we try to deduce it from a side channel by performing a timing attack [12]. Instead of getting 0 or 1 depending on the error type, we now have an oracle which outputs the timing answer T of the server. The principle of the attack is as follows: in order to check if the padding is correct, the server only needs to perform simple operations on the very end of the ciphertext. When the padding is correct, the server further needs to perform cryptographic operations throughout the whole ciphertext in order to check the MAC, and this may take more time. We use the discrepancy between the time it takes to perform the two types of operations in order to get the answer from the oracle.

We increase the discrepancy of the two types of errors by enlarging the ciphertext: the longer the ciphertext, the longer the MAC verification. (The MAC verification time increases linearly with the length of MES.) Hence we replace the $r|y$ fake ciphertext in **DecryptByte1** by $f|r|y$ where f is a random block sequence of the longest acceptable length (i.e. $2^{14} + 2048$ bytes in TLS).[2]

Formally we let D_W (resp. D_R) be the distribution of the timing answer from the server when there is a padding error (resp. a MAC error). We let μ_W (resp. μ_R) be its expected value. We will use the following approximation.

Conjecture 1. We approximate D_R and D_W by normal distributions with the same standard deviation σ and expected values μ_R and μ_W respectively. We assume w.l.o.g. that $\mu_W < \mu_R$.

We further make the query to the oracle several times in order to make a statistical analysis. We use a predicate ACCEPT in order to decide whether or not the error type is a padding or a MAC.

Provided that the adversary is close to the server, the time measurement may be influenced by a little noise. An unexpectedly long answer may occur due to other protocol issues. These answers are ignored in the experiment. In practice we ignore times which are greater that a given threshold B.

On Fig. 2 is the updated algorithm. It uses a **DecryptBlock2** algorithm which is similar to **DecryptBlock1**. Note that **Check2** may miss the right byte, so **DecryptByte2** needs to repeat the loop until the byte is found.

2.2   Experiment

We made a statistical analysis of the answer time for the two types of errors. The distributions D_W and D_R can be seen on the graph of Fig. 3. The expected values

---

[2] This trick applies assuming that CBC decryption and MAC verification are not done at the same time.

DecryptByte2(y, s)
1: **repeat**
2:    **for all** possible values of byte c **do**
3:       **if Check2**$(y, c|s) = 1$ **then**
4:          **return** c
5:       **end if**
6:    **end for**
7: **until** byte is found

Check2(y, u)
1: make r in order to test u as in **Check1**
2: build the fake ciphertext $f|r|y$ to be sent to the oracle
   (f is the longest possible random block sequence)
4: query the oracle n times and get $T_1, \ldots, T_n$
   (answers which are larger than B are ignored)
6: **return** ACCEPT$(T_1, \ldots, T_n)$

Fig. 2. Regular Timing Attack.

μ_R and μ_W and the standard deviations σ_R and σ_W for the two distributions are as follows:

$$\mu_R \approx 23.63 \quad \mu_W \approx 21.57$$
$$\sigma_R \approx 1.48 \quad \sigma_W \approx 1.86.$$

We take $\sigma = \sigma_W \approx 1.86$. From the graph, we can clearly see that the two distributions are distinguishable. The following section formalizes this distinguishability. Note that these values were obtained on a LAN where a firewall was present between the attacker and the server, so the attacker was not directly connected to the server.[3]

2.3   Analysis of the Best ACCEPT Predicate

The ACCEPT predicate is used in order to decide whether the distribution of the answers is D_R (the predicate should be true) or D_W (the predicate should be false). The predicate introduces two types of wrong information. We let ε_+ (resp. ε_-) be the probability of bad decision when the distribution is D_W (resp. D_R). The ε_+ and ε_- probabilities can be interpreted as the probabilities of false positives and false negatives of a character correctness test. The optimal tradeoff between ε_+ and ε_- is achieved by the ACCEPT predicate which is given by the Neyman-Pearson lemma:

$$\text{ACCEPT} : \frac{f_R(T_1)}{f_W(T_1)} \times \ldots \times \frac{f_R(T_n)}{f_W(T_n)} > \tau$$

---

[3] More precisely, the route between the attacker and the server included two switches and a firewall (a PC running Linux).

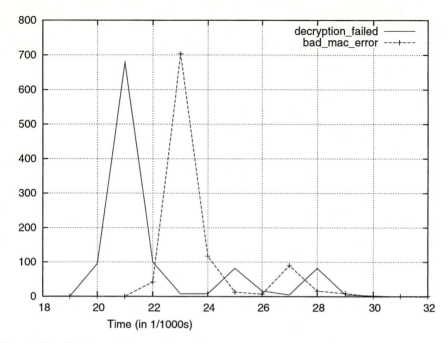

Fig. 3. Distribution of the number of `decryption_failed` and `bad_mac_error` error messages with respect to time.

with f_R and f_W the density functions of D_R and D_W respectively and a given threshold τ. Depending on τ we trade ε_+ against ε_-.

With the approximation by a normal distribution the ACCEPT test can be written

$$\prod_{i=1}^{n} \frac{e^{-\frac{(T_i - \mu_R)^2}{2\sigma^2}}}{e^{-\frac{(T_i - \mu_W)^2}{2\sigma^2}}} > \tau$$

which is equivalent to

$$\frac{T_1 + \ldots + T_n}{n} > \frac{\mu_R + \mu_W}{2} + \frac{\sigma^2 \log \tau}{n(\mu_R - \mu_W)}$$

so we equivalently can change the definition of τ and consider

$$\text{ACCEPT} : \frac{T_1 + \ldots + T_n}{n} > \tau'$$

as the ACCEPT test choice. With this we obtain

$$\varepsilon_- = \varphi\left(\frac{\tau' - \mu_R}{\sigma}\sqrt{n}\right)$$

$$\varepsilon_+ = \varphi\left(-\frac{\tau' - \mu_W}{\sigma}\sqrt{n}\right)$$

where

$$\varphi(x) = \frac{1}{\sqrt{2\pi}} \int_{-\infty}^{x} e^{-\frac{t^2}{2}} dt.$$

2.4   Using Sequential Decision Rules

On Fig. 4 is a more general algorithm skeleton. Basically, we collect timing samples T_j until some STOP predicate decides that there are enough of these for the ACCEPT predicate to decide. We use **DecryptByte3** and **DecryptBlock3** algorithms which are similar to **DecryptByte2** and **DecryptBlock2**.

> **Check3**(y, u)
> 1: make r in order to test u as in **Check1**
> 2: build the fake ciphertext $f|r|y$ to be sent to the oracle as in **Ckeck2**
> 3: $j \leftarrow 0$
> 4: **repeat**
> 5: $j \leftarrow j + 1$
> 6:   query the oracle and get T_j
>   (a T_j larger than B is ignored and the query is repeated)
> 8: **until** STOP$(T_1, \dots, T_j)$
> 9: **return** ACCEPT$(T_1, \dots, T_j)$

Fig. 4. Timing Attack with a Sequential Distinguisher.

By using the theory of hypothesis testing with sequential distinguishers (see Junod [11]), we obtain that the most efficient algorithms are obtained with the following STOP and ACCEPT tests.

$$\text{STOP} : \frac{f_R(T_1)}{f_W(T_1)} \times \dots \times \frac{f_R(T_j)}{f_W(T_j)} \notin [\tau_-, \tau_+]$$

$$\text{ACCEPT} : \frac{f_R(T_1)}{f_W(T_1)} \times \dots \times \frac{f_R(T_j)}{f_W(T_j)} > \tau_+$$

where τ_- and τ_+ are two given thresholds. Assuming Conjecture 1, this is equivalent to

$$\text{STOP} : T_1 + \dots + T_j - j\frac{\mu_R + \mu_W}{2} \notin [\tau'_-, \tau'_+] \tag{1}$$

$$\text{ACCEPT} : T_1 + \dots + T_j - j\frac{\mu_R + \mu_W}{2} > \tau'_+ \tag{2}$$

where

$$\tau'_+ = \frac{\sigma^2}{\mu_R - \mu_W} \log \tau_+ \tag{3}$$

$$\tau'_- = \frac{\sigma^2}{\mu_R - \mu_W} \log \tau_-. \tag{4}$$

Thanks to the Wald Approximation, we can freely select ε_+ and ε_-, compute the corresponding τ_+ and τ_- by

$$\tau_+ \approx \frac{1 - \varepsilon_-}{\varepsilon_+}$$

$$\tau_- \approx \frac{\varepsilon_-}{1 - \varepsilon_+}$$

and deduce the expected number J of samples (i.e. the j iterations) until STOP holds and ACCEPT takes the right decision by

$$J_W \approx -\frac{2\sigma^2}{(\mu_R - \mu_W)^2} \log \tau_-$$

$$J_R \approx \frac{2\sigma^2}{(\mu_R - \mu_W)^2} \log \tau_+$$

when the character to be tested is wrong or right respectively.[4]

3   Multi-session Attack

3.1   Attack Strategy

Since sessions are broken as soon as there is an error, the attacks from previous sections do not work. We now assume that each TLS session includes a critical plaintext block x which is always the same (e.g. a password) and that we intercept the corresponding ciphertext block $y = \text{ENC}(x \oplus y')$. (Here y' is the previous ciphertext block following the CBC mode.) The target x is constant in every session, but y and y' depend on the session. The full attack is depicted on Fig. 5. Here the **Check4** oracle no longer relies on some y since this block is changed in every session. The **Check4**(u) is called in order to check whether "the x plaintext block ends with the byte sequence u" or not. The plaintext block x is equal to $\text{ENC}^{-1}(y) \oplus y'$ for some current key, and some current ciphertext blocks y and y'. We assume that the oracle can get y and y'.

3.2   Analysis

Let C be the average complexity of **DecryptBlock4**. Let Z denote the set of all possible byte values. Let p be the probability of success of **DecryptBlock4**. Let p_i be the success probability of **DecryptByte4**(s) assuming that s is the right tail of length $i - 1$. We have $p = p_1 \cdots p_b$.

In order to simplify our analysis, we assume that the target block is uniformly distributed in Z^b so that step 1 of **DecryptByte4** can be ignored. We further consider a weaker algorithm in which the outer repeat/until loop of **DecryptByte4** is removed (i.e. we consider that the attack fails as soon as a STOP predicate is satisfied but the ACCEPT predicate takes a bad decision).

---

[4] For a mathematical treatment on these results we refer to Siegmund [15].

DecryptBlock4
1: **for** $i = 1$ to b **do**
2:    $c_i \leftarrow$ **DecryptByte4**$(c_{i-1}|\ldots|c_1)$
3: **end for**
4: **return** $c_b|\ldots|c_1$

DecryptByte4(s)
1: sort all possible c characters in order of decreasing likelihood.
2: **repeat**
3:    **for all** possible values of character c **do**
4:       **if** **Check4**$(c|s) = 1$ **then**
5:          **return** c
6:       **end if**
7:    **end for**
8: **until** byte is found

Check4(u)
1: $j \leftarrow 0$
2: **repeat**
3:    $j \leftarrow j + 1$
4:    wait for a new session and get the current y and y' blocks
5:    let i be the length of u
6:    let L be a random string of length $b - i$
7:    let $R = (i-1)|(i-1)|\ldots|(i-1)$ of length i
8:    $r \leftarrow (L|(R \oplus u)) \oplus y'$
9:    build the fake ciphertext $f|r|y$ to be sent to the oracle
      (f is the longest possible random block sequence)
11:   query the oracle and get T_j
      (if it is larger than B then go back to Step 4)
13: **until** STOP$(T_1,\ldots,T_j)$
14: **return** ACCEPT$(T_1,\ldots,T_j)$

Fig. 5. Password Interception inside SSL/TLS.

Here we assume that all characters of the unknown block x are independent and uniformly distributed in the alphabet Z. Thus all p_is are equal. We have

$$p_1 = \sum_{i=1}^{|Z|} \frac{1}{|Z|}(1 - \varepsilon_+)^{i-1}(1 - \varepsilon_-)$$
$$= \frac{1 - (1 - \varepsilon_+)^{|Z|}}{|Z|\varepsilon_+}(1 - \varepsilon_-)$$
$$\approx (1 - \varepsilon_+)^{\frac{|Z|-1}{2}}(1 - \varepsilon_-)$$

when $\varepsilon_+ \ll \frac{1}{|Z|}$ thus

$$p \approx (1 - \varepsilon_+)^{b\frac{|Z|-1}{2}}(1 - \varepsilon_-)^b.$$

Note that $p \approx 1$ when $\varepsilon_+ \ll \frac{1}{b|Z|}$ and $\varepsilon_- \ll \frac{1}{b}$. Assuming that the algorithm succeeds, the average number or iterations per byte is

$$\sum_{i=1}^{|Z|} \frac{1}{|Z|}\left((i-1)J_W + J_R\right) = \frac{|Z|-1}{2}J_W + J_R$$

so the average complexity per block is

$$C = b\frac{|Z|-1}{2}J_W + bJ_R.$$

Numerical examples will be given in a more general context in Section 4.3.

4   Password Interception with Dictionary Attack

4.1   Attack Description

We now use the *a priori* distribution of x in the previous attack in order to decrease the complexity. For instance, if x is a password corresponding to an IMAP authentication, we perform a kind of dictionary attack on x. We assume that we have precomputed a dictionary of all possible x blocks with the corresponding probability of occurrence. We use it in the first step of **DecryptByte4** in order to sort the c candidates.

4.2   Analysis

We consider a list of possible blocks $c_b \dots c_1$. We let $\Pr[c_b \dots c_1]$ be the occurrence probability of a plaintext block. We also let $\Pr[c_i \dots c_1]$ be the sum $\Pr[c_b \dots c_1]$ for all possible $c_b \dots c_{i+1}$.

We arrange the dictionary of all blocks into a search tree. The root is connected to many subtrees, each corresponding to a c_1 character. Each subtree corresponding to a c_1 character is connected to many sub-subtrees, each corresponding to a c_2 character... We label each node of the tree by a $c_i \dots c_1$ string. We assume that the list of subtrees of any node $c_i \dots c_1$ is sorted in decreasing order of values of $\Pr[c_{i+1}c_i \dots c_1]$. We let $N(c_{i+1} \dots c_1)$ be the rank of the $c_{i+1} \dots c_1$ subtree of the node $c_i \dots c_1$ in the list.

We let C_i be the average number to trials for finding c_i (if the attack succeeds) and $C_0 = C_1 + \dots + C_b$.

We have

$$C_i = \sum_{c_i,\dots,c_1} \Pr[c_i,\dots,c_1]N(c_i \dots c_1)$$

so

$$C_0 = \sum_{i=1}^{b} \sum_{c_i,\dots,c_1} \Pr[c_i,\dots,c_1]N(c_i \dots c_1).$$

Note that $C_0 = b\frac{|Z|-1}{2} + b$ when $c_b \ldots c_1$ is uniformly distributed in Z^b so this generalizes the analysis from Section 3.2.

The expected complexity in case of success is $(C_0 - b)J_W + bJ_R$ thus approximately

$$C \approx \frac{2\sigma^2}{(\mu_R - \mu_W)^2} \left((C_0 - b) \log \frac{1}{\varepsilon_-} + b \log \frac{1}{\varepsilon_+} \right)$$

$$p \approx (1 - \varepsilon_+)^{C_0 - b} (1 - \varepsilon_-)^b$$

when ε_- and ε_+ are small against b^{-1} and C_0^{-1} respectively. The problem is to select ε_- and ε_+ in order to maximize p and minimize C. Computations shows that this is the case when

$$\left(\frac{\delta C}{\delta \varepsilon_+} \right) / \left(\frac{\delta \log p}{\delta \varepsilon_+} \right) = \left(\frac{\delta C}{\delta \varepsilon_-} \right) / \left(\frac{\delta \log p}{\delta \varepsilon_-} \right)$$

hence,

$$\frac{b/\varepsilon_+}{(C_0 - b)/(1 - \varepsilon_+)} = \frac{(C_0 - b)/\varepsilon_-}{b/(1 - \varepsilon_-)}.$$

With the assumption that $\varepsilon_+ \ll 1$ and $\varepsilon_- \ll 1$, we deduce

$$\frac{\varepsilon_+}{\varepsilon_-} \approx \frac{b^2}{(C_0 - b)^2}$$

thus

$$\varepsilon_+ = \frac{b^2}{t}$$

$$\varepsilon_- = \frac{(C_0 - b)^2}{t}$$

for some parameter t. With the same approximation we obtain

$$p \approx \exp \left(-\frac{C_0(C_0 - b)b}{t} \right)$$

$$\varepsilon_+ \approx \frac{b}{C_0(C_0 - b)} \log \frac{1}{p}$$

$$\varepsilon_- \approx \frac{C_0 - b}{C_0 b} \log \frac{1}{p}$$

and finally, we deduce C in terms of the success probability p as well as τ_- and τ_+:

$$C \approx \frac{2\sigma^2}{(\mu_R - \mu_W)^2} \left(C_0 \log C_0 - (C_0 - 2b) \log \frac{C_0 - b}{b} - C_0 \log \log \frac{1}{p} \right) \quad (5)$$

$$\log \tau_+ \approx \log C_0 + \log(C_0 - b) - \log b - \log \log \frac{1}{p} \quad (6)$$

$$\log \tau_- \approx \log(C_0 - b) - \log C_0 - \log b + \log \log \frac{1}{p}. \quad (7)$$

Note that $C = O(-\log\log\frac{1}{p})$. If $p \sim 1$ we have $\log\frac{1}{p} \sim 1-p$ so $p = 1 - e^{-\Omega(C)}$. The failure probability decreases exponentially with the complexity.

4.3   Numerical Example

We have used dictionary [5] from which we have selected only words of size $b = 8$ characters (i.e. 8 bytes), giving a total word count of $712'786$ words and ordered it as described the previous section. For this dictionary, we have calculated that $C_0 = 31$ and then implemented algorithm **DecryptBlock4** and confirmed this result. Note that $C_0 = 31$ is a quite remarkable result since the best search rule for finding a password out of a dictionary of $D = 712'786$ words consists of $\lceil \log_2 D \rceil = 20$ binary questions, so the overhead is only of 11 questions.

Example 1. Table 1 shows complexity C and values $\log \tau_-$ and $\log \tau_+$ for a given success probability p in the case of the dictionary [5] being used and also for uniform distributions with $|Z| = 256$, 128, and 64.

The complexity for $p = 50\%$ is 166 with the dictionary attack and 4239 for a fully random block. Similar computation with the **DecryptBlock2** strategy shows that sequential distinguishers lead to a improvement factor between 4 and 5. This illustrates the power of this technique.

5   Implementation of the Attack

In this section we describe how the **DecryptBlock4** was implemented in practice against an IMAP email server.

5.1   Setup

The multi-session attack has been implemented using the Outlook Express 6.x client from Microsoft under Windows XP and an IMAP Rev 4 server[5]. Outlook sends the login and password to the IMAP server using the following format:

XXXX␣LOGIN␣"username"␣"password"<0x0d><0x0a>

Here XXXX are four random digits which are incremented each time Outlook connects to the server.

An interesting feature of Outlook is that (by default) it checks for messages automatically every 5 minutes and also that it requires an authentication for each folder created on the IMAP user account, i.e. we have a bunch of free sessions every 5 minutes. For instance, with five folders (in, out, trash, read, and draft), we obtain 60 sessions every hour. If Outlook is now configured to check emails every minute, the fastest attack of Table 1 with 166 sessions requires half an hour. Outlook notices that some protocol errors occur but this does not seem to bother it at all.

---

[5] http://www.washington.edu/imap/

Table 1. Calculated complexity C and threshold values $\log \tau_-$ and $\log \tau_+$ for algorithm **DecryptBlock4** given success probability p with $\mu_R = 23.63$, $\mu_W = 21.57$, $\sigma = 1.86$ and heuristic value $B = 32.93$ for dictionary [5] and uniform distributions in Z^b.

Dictionary, $C_0 = 31$

p	0.5	0.6	0.7	0.8	0.9	0.99
C	166	181	199	223	261	380
$\log \tau_-$	-2.74	-3.05	-3.41	-3.88	-4.62	-6.98
$\log \tau_+$	4.86	5.16	5.52	5.99	6.74	9.09

Uniform distribution, $|Z| = 256$, $C_0 = 1028$

p	0.5	0.6	0.7	0.8	0.9	0.99
C	4239	4750	5353	6139	7397	11335
$\log \tau_-$	-2.45	-2.76	-3.12	-3.59	-4.34	-6.69
$\log \tau_+$	12.15	12.46	12.81	13.28	14.03	16.38

Uniform distribution, $|Z| = 128$, $C_0 = 516$

p	0.5	0.6	0.7	0.8	0.9	0.99
C	2179	2346	2738	3132	3764	5741
$\log \tau_-$	-2.46	-2.77	-3.12	-3.60	-4.35	-6.70
$\log \tau_+$	10.76	11.07	11.43	11.90	12.65	15.00

Uniform distribution, $|Z| = 64$, $C_0 = 260$

p	0.5	0.6	0.7	0.8	0.9	0.99
C	1140	1269	1421	1620	1938	2934
$\log \tau_-$	-2.48	-2.78	-3.14	-3.61	-4.36	-6.71
$\log \tau_+$	9.38	9.68	10.04	10.51	11.26	13.61

The TLS tunneling between the IMAP server and Outlook Express was implemented using stunnel v3.22[6].

The attack is a man-in-the-middle type attack where connection requests to the IMAP server from the Outlook client are redirected to the attacker's machine using DNS spoofing [16] where the attacker intercepts the authentication messages and attempts to decrypt it using **DecryptBlock4**.

Note that the attack is performed on a Local Area Network.

5.2   Problems and Notes

Two main problems arose when implementing the multi-session attack using the above setup. Firstly, Outlook uses the RC4_MD5 algorithm by default[7] despite the fact that [9] and [14] suggest that 3DES_EDE_CBC_SHA should be supported by default. Hence, we had to force the IMAP server to only offer block ciphers in CBC mode.

---

[6] http://www.stunnel.org
[7] Some other applications like stunnel use the CBC mode by default.

The second problem comes from the format of the authentication message. It can be the case that the last bytes of the password belong to the first block of the MAC of the message. So it sometimes happens that the last few bytes of the password cannot be decrypted using the multi-session attack described in this paper. For example, assume that the user name has four characters so that we have the following for an 8-byte block cipher:

`|0021␣LOG|IN␣"name|"␣"passw|ord"<0x0d><0x0a><HMAC1><HMAC2>|`

then it will not be possible to decrypt the last three characters from the password.

As explained in [13], a countermeasure against the CBC-PAD problem [17] has been implemented in OpenSSL 0.9.6d and following versions so that only the `bad_mac_error` error message is sent when an incorrect padding or an incorrect MAC are detected. However, during our experiments, we have seen that the timing differences still exist and are even easier to identify with this version than previous ones.

6   Discussion

Obviously, the attack works if the following conditions are met.

1. A critical piece of information is repeatedly encrypted at a predictable place.
2. A block cipher in CBC mode is chosen.
3. The attacker can sit in the middle and perform active attacks.
4. The attacker can distinguish time differences between two types of errors.

In this paper we focused on the password access control in the IMAP protocol. We can also consider the basic authentication in HTTP [10] which is also used for access control. This means that we can consider intercepting the password for accessing to an Intranet server (e.g. the web server of the program committee of the Crypto'03 Conference!). The attack would work in the same way provided that the above conditions are met, and in particular when the clients of program committee members send their passwords more than a hundred times and do not care about errors.

We can also consider other critical information than just passwords which are sent by the clients to the server. We can consider decrypting a particular constant block of plaintext of a private URL which is retrieved from a server by many clients or by a single one many times. For instance, we can try to decrypt data about the bank account of a client in electronic banking systems. Systems whose security fully rely only on the SSL/TLS protocol would certainly face to security threats. Fortunately, systems that we are aware of use additional security means. They use challenge-response authentication instead of password authentication. They also block accesses in case of multiple failures.

Fixing the problem is quite simple (it was actually done in OpenSSL versions older than 0.9.6i).[8] One can simply try to make error responses time-invariant

---

[8] See http://www.openssl.org

by simulating a MAC verification even when there is a padding error. One can additionally add some random noise in the time delay. Obviously, session errors should be audited. Note that the problem should also be fixed on the client side which will take much more time than fixing it on the server side. In the meantime, servers should become more sensitive to the types or errors issued by our attack.

Some people claimed that the problem we pointed out in this paper is related to an implementation mistake for SSL/TLS. We can however argue that the problem was raised in [17] in 2002, that an update was made, and that an error was still present in current versions one year after. Since there are so many possible mistakes in implementing the protocol we can reasonably claim that this is a protocol problem rather than an implementation one.

For future versions of the TLS protocol we recommend to invert the MAC and padding processes: the sender first pad the plaintext then MAC it, so that the receiver can first check the MAC then check the padding if the MAC is valid. This would thwart any active attack in which messages which are received are not authentic.

7   Conclusion

In this paper we have derived a multi-session variant of the attack of [17] in order to show that it is possible to attack SSL/TLS in the case when the message that is being encrypted remains the same during each session. This is the case, for example, when an email client such as Outlook Express connects to an IMAP server. We have detailed the attack and described the setup we have used in order to perform it.

One problem we have encountered is that the error messages sent in SSL/TLS are encrypted and it is not possible to easily differentiate which is being sent by the client or the server. A solution to this problem is to look at timings between errors messages. We have shown that when using sequential distinguishers, we can efficiently intercept and decrypt a password for an IMAP account in less than an hour. In doing so, we have also shown that the post-[17] version of OpenSSL [13] is not secure when used with block ciphers in CBC mode. Hopefully, this will have been easily fixed by the time the present paper is published.

Interestingly, we have run one of the first timing attacks through a network and not directly on the attacked device. Another timing attack (against RSA) was run in parallel by Brumley and Boneh [7].

Our attack also illustrates how the theory of sequential distinguishers [11] can be used in order to optimize practical attacks.

Acknowledgments. We owe Pascal Junod useful discussions about the timing attack, the idea to fill the message with f, and some helpful details about sequential tests. We thank Gildas Avoine for useful comments. We would also like to thank Bodo Möller for his immediate feedback and the OpenSSL community for caring about our attack in real time. We thank the media for there very positive interest in our results. We also received threats from several companies involved in security which seems to mean that they cared about our results.

The work presented in this paper was supported (in part) by the National Competence Center in Research on Mobile Information and Communication Systems (NCCR-MICS), a center supported by the Swiss National Science Foundation under grant number 5005-67322.

References

1. *ISO/IEC 10116*, Information Processing — Modes of Operation for an n-bit Block Cipher Algorithm. International Organization for Standardization, Geneva, Switzerland, 1991.
2. Wireless Transport Layer Security. Wireless Application Protocol WAP-261-WTLS-20010406-a. Wireless Application Protocol Forum, 2001. http://www.wapforum.org/
3. *FIPS 46-3*, Data Encryption Standard (DES). U.S. Department of Commerce — National Institute of Standards and Technology. *Federal Information Processing Standard Publication 46-3*, 1999.
4. *FIPS 81*, DES Modes of Operation. U.S. Department of Commerce — National Bureau of Standards, National Technical Information Service, Springfield, Virginia. Federal Information Processing Standards 81, 1980.
5. English Word List Elcomsoft Co. Ltd. http://www.elcomsoft.com
6. J. Black, H. Urtubia. Side-Channel Attacks on Symmetric Encryption Schemes: The Case for Authenticated Encryption. In *Proceedings of the 11th Usenix UNIX Security Symposium*, San Francisco, California, USA, USENIX, 2002.
7. D. Brumley, D. Boneh. Remote Timing Attacks are Practical. To appear in *Proceedings of the 12th Usenix UNIX Security Symposium*, USENIX, 2003.
8. M. Crispin. Internet Message Access Protocol - Version 4. RFC 1730, standard tracks, University of Washington, 1994.
9. T. Dierks, C. Allen. The TLS Protocol Version 1.0. RFC 2246, standard tracks, the Internet Society, 1999.
10. J. Franks, P. Hallam-Baker, J. Hostetler, S. Lawrence, P. Leach, A. Luotonen, L. Stewart. HTTP Authentication: Basic and Digest Access Authentication. Internet standard. RFC 2617, the Internet Society, 1999.
11. P. Junod. On the Optimality of Linear, Differential and Sequential Distinguishers. In *Advances in Cryptology EUROCRYPT'03*, Warsaw, Poland, Lectures Notes in Computer Science 2656, pp. 17–32, Springer-Verlag, 2003.
12. P. Kocher. Timing Attacks on Implementations of Diffie-Hellman, RSA, DSS, and other Systems. In *Advances in Cryptology CRYPTO'96*, Santa Barbara, California, U.S.A., Lectures Notes in Computer Science 1109, pp. 104–113, Springer-Verlag, 1996.
13. B. Möller. Security of CBC Ciphersuites in SSL/TLS: Problems and Countermeasures. 2002. http://www.openssl.org/~bodo/tls-cbc.txt
14. C. Newman Using TLS with IMAP, POP3 and ACAP. RFC 2595, standard tracks, the Internet Society, 1999.
15. D. Siegmund. *Sequential Analysis — Tests and Confidence Intervals*, Springer-Verlag, 1985.

16. M. Ricca. The Denver Projet - A Combination of ARP and DNS Spoofing. Ecole Polytechnique Fédérale de Lausanne, LASEC, Semester Project, 2002. http://lasecwww.epfl.ch
17. S. Vaudenay. Security Flaws Induced by CBC Padding — Applications to SSL, IPSEC, WTLS... In *Advances in Cryptology EUROCRYPT'02*, Amsterdam, Netherland, Lectures Notes in Computer Science 2332, pp. 534–545, Springer-Verlag, 2002.
18. M. Vuagnoux. CBC PAD Attack against IMAP over TLS. omen. Ecole Polytechnique Fédérale de Lausanne, LASEC, Semester Project, 2003. http://omen.vuagnoux.com

Instant Ciphertext-Only Cryptanalysis of GSM Encrypted Communication

Elad Barkan[1], Eli Biham[1], and Nathan Keller[2]

[1] Computer Science Department
Technion – Israel Institute of Technology
Haifa 32000, Israel
{barkan,biham}@cs.technion.ac.il
http://tx.technion.ac.il/~barkan/,
http://www.cs.technion.ac.il/~biham/
[2] Department of Mathematics
Technion – Israel Institute of Technology
Haifa 32000, Israel
nkeller@tx.technion.ac.il

Abstract. In this paper we present a very practical ciphertext-only cryptanalysis of GSM encrypted communication, and various active attacks on the GSM protocols. These attacks can even break into GSM networks that use "unbreakable" ciphers. We describe a ciphertext-only attack on A5/2 that requires a few dozen milliseconds of encrypted off-the-air cellular conversation and finds the correct key in less than a second on a personal computer. We then extend this attack to a (more complex) ciphertext-only attack on A5/1. We describe new attacks on the protocols of networks that use A5/1, A5/3, or even GPRS. These attacks are based on security flaws of the GSM protocols, and work whenever the mobile phone supports A5/2. We emphasize that these attacks are on the protocols, and are thus applicable whenever the cellular phone supports a weak cipher, for instance they are also applicable using the cryptanalysis of A5/1. Unlike previous attacks on GSM that require unrealistic information, like long known plaintext periods, our attacks are very practical and do not require any knowledge of the content of the conversation. These attacks allow attackers to tap conversations and decrypt them either in real-time, or at any later time. We also show active attacks, such as call hijacking, altering of data messages and call theft.

1 Introduction

GSM is the most widely used cellular technology. By December 2002, more than 787.5 million GSM customers in over 191 countries formed approximately 71% of the total digital wireless market. GPRS (General Packet Radio Service) is a new service for GSM networks that offer 'always-on', higher capacity, Internet-based content and packet-based data services. It enables services such as color Internet browsing, e-mail on the move, powerful visual communications, multimedia messages and location-based services.

D. Boneh (Ed.): CRYPTO 2003, LNCS 2729, pp. 600–616, 2003.

GSM incorporates security mechanisms. Network operators and their customers rely on these mechanisms for the privacy of their calls and for the integrity of the cellular network. The security mechanisms protect the network by authenticating customers to the network, and provide privacy for the customers by encrypting the conversations while transmitted over the air.

There are three main types of cryptographic algorithms used in GSM: A5 is a stream-cipher used for encryption, A3 is an authentication algorithm and A8 is the key agreement algorithm. The design of A3 and A8 is not specified in the specifications of GSM, only the external interface of these algorithms is specified. The exact design of the algorithm can be selected by the operators independently. However, many operators used the example, called *COMP128*, given in the GSM memorandum of understanding (MoU). Although never officially published, its description was found by Briceno, Goldberg, and Wagner [6]. They have performed cryptanalysis of COMP128 [7], allowing to find the shared (master) key of the mobile phone and the network, thus allowing cloning. The description of A5 is part of the specifications of GSM, but it was never made public. There are two currently used versions of A5: A5/1 is the "strong" export-limited version, and A5/2 is the "weak" version that has no export limitations. The exact design of both A5/1 and A5/2 was reverse engineered by Briceno [5] from an actual GSM telephone in 1999 and checked against known test-vectors. An additional new version, which is standardized but not yet used in GSM networks is A5/3. It was recently chosen, and is based on the block-cipher KASUMI. We note that a similar construction based on KASUMI is also used in third generation networks (3GPP) [1], on which we make no claims in this paper.

A5/1 was initially cryptanalyzed by Golic [14], and later by: Biryukov, Shamir and Wagner [4], Biham and Dunkelman [2], and recently by Ekdahl and Johansson [11].

After A5/2 was reverse engineered, it was immediately cryptanalyzed by Goldberg, Wagner and Green [13]. Their attack is a known plaintext attack that requires the difference in the plaintext of two GSM frames, which are exactly 2^{11} frames apart (about 6 seconds apart). The average time complexity of this attack is approximately 2^{16} dot products of 114-bit vectors.[1] A later work by Petrović and Fúster-Sabater [17] suggests to treat the initial internal state of the cipher as variables, write every output bit of A5/2 as a quadratic function of these variables, and linearize the quadratic terms. They showed that the output of A5/2 can be predicted with extremely high probability after a few hundreds of known output bits. However, this attack does not discover the initial key of A5/2. Thus, it is not possible to use this attack as a building block for more advanced attacks, like those presented in Section 5. This attack's time complexity is proportional to 2^{17} Gauss eliminations of matrices of size of about 400×719.[2] This latter attack actually uses overdefined systems of quadratic

---

[1] We observe that this attack [13] is not applicable (or fails) in about half of the cases, since in the first frame it requires that after the initialization of the cipher the 11th bit of R4 is zero.

[2] The paper [17] does not clearly state the complexity. The above figure is our estimate.

equations during its computations, however, there is no need to solve the equations to perform cryptanalysis — the equations are only used to predict the output of future frames.

Solving overdefined systems of quadratic equations — as a method of cryptanalysis drew significant attention in the literature. This method has been applied initially by Kipnis and Shamir to the HFE public key cryptosystem in [15], later improved by Courtois, Klimov, Patarin, and Shamir in [9]. Further work include: Courtois and Pieprzyk's cryptanalysis of block ciphers [10]. The method has also been applied to stream ciphers, see Courtois work on Toyocrypt [8]. The complexity of most of these methods seems difficult to evaluate.

In this paper we show how to mount a ciphertext-only attack on A5/2. This attack requires a few dozen milliseconds of encrypted data, and its time complexity is about 2^{16} dot products. In simulations we made, our attack found the key in less than a second on a personal computer. We show that the attack we propose on A5/2 can be leveraged to mount an active attack even on GSM networks that use A5/1, A5/3, or GPRS networks, thus, realizing a real-time active attack on GSM networks, without any prior required knowledge. The full attack is composed of three main steps:

1. The first step is a very efficient known plaintext attack on A5/2 that recovers the initial key. This first attack is algebraic in nature. It takes advantage of the low algebraic order of the A5/2 output function. We represent the output of A5/2 as a quadratic multivariate function in the initial state of the registers. Then, we construct an overdefined system of quadratic equations that expresses the key-stream generation process and we solve the equations.
2. The second step is improving the known plaintext attack to a ciphertext-only attack on A5/2. We observe that GSM employs Error-Correction codes before encryption. We show how to use this observation to adapt the attack to a ciphertext-only attack on A5/2.
3. The third step is leveraging of an attack on A5/2 to an active attack on A5/1, A5/3, or GPRS-enabled GSM networks. We observe that due to the GSM security modules interface design, the key that is used in A5/2 is the same as in A5/1, A5/3, and GPRS. We show how to mount an active attack on any GSM network.

We then show how to mount a passive ciphertext-only attack on networks that employ A5/1. It is basically a time/memory/data tradeoff attack. There are many choices for the parameters of the attack, four of which are given in Table 1. This attack on A5/1 can be similarly leveraged to active attacks on the protocols of GSM, but the complexity is higher than the A5/2 case.

This paper is organized as follows: In Section 2 we describe A5/2, and the way it is used, and give some background on GSM security. We present our new known plaintext attack in Section 3. In Section 4 we improve our attack to a ciphertext-only attack. In Section 5 we show how to leverage the ciphertext-only attack on A5/2 to an active attack on any GSM network. We then describe the passive ciphertext-only attack on A5/1 in Section 6. We discuss the implications of the attacks under several attack scenarios in Section 7. Section 8 summarizes the paper.

2   Description of A5/2 and GSM Security Background

In this section we describe the internal structure of A5/2 and the way it is used. A5/2 consists of 4 maximal-length LFSRs: R1, R2, R3, and R4. These registers are of length 19-bit, 22-bit, 23-bit, and 17-bit respectively. Each register has taps and a feedback function. Their irreducible polynomials are: $x^{19} \oplus x^5 \oplus x^2 \oplus x \oplus 1$, $x^{22} \oplus x \oplus 1$, $x^{23} \oplus x^{15} \oplus x^2 \oplus x \oplus 1$, and $x^{17} \oplus x^5 \oplus 1$, respectively. For the representation of the registers we adopt the notation of [2,4,5,17], in which the bits in the register are given in reverse order, i.e., x^i corresponds to a tap with index $len - i - 1$, where len is the register size. For example, when R4 is clocked, the XOR of $R4[17 - 0 - 1 = 16]$ and $R4[17 - 5 - 1 = 11]$ is computed. Then, the register is shifted by one bit to the right, and the value of the result of the XOR is placed in $R4[0]$.

At each step of A5/2 R1, R2 and R3 are clocked according to a clocking mechanism that we describe later. Then, R4 is clocked. After the clocking is performed, one output bit is ready at the output of A5/2. The output bit is a non-linear function of the internal state of R1, R2, and R3.

After the initialization 99 bits[3] of output are discarded, and the following 228 bits of output are used as the output key-stream.

Denote the i'th bit of the 64-bit session-key K_c by $K_c[i]$, the i'th bit of register j by $Rj[i]$, and the i'th bit of the 22-bit publicly known frame number by $f[i]$.

The initialization of the internal state with K_c and the frame number is done in the following way:

- Set all LFSRs to 0 ($R1 = R2 = R3 = R4 = 0$).
- For $i := 0$ to 63 do
  1. Clock all 4 LFSRs.
  2. $R1[0] \leftarrow R1[0] \oplus K_c[i]$
  3. $R2[0] \leftarrow R2[0] \oplus K_c[i]$
  4. $R3[0] \leftarrow R3[0] \oplus K_c[i]$
  5. $R4[0] \leftarrow R4[0] \oplus K_c[i]$
- For $i := 0$ to 21 do
  1. Clock all 4 LFSRs.
  2. $R1[0] \leftarrow R1[0] \oplus f[i]$
  3. $R2[0] \leftarrow R2[0] \oplus f[i]$
  4. $R3[0] \leftarrow R3[0] \oplus f[i]$
  5. $R4[0] \leftarrow R4[0] \oplus f[i]$

The key-stream generation is as follows:

---

[3] Some references state that A5/2 discards 100 bits of output, and that the output is used with a one-bit delay. This is equivalent to stating that it discards 99 bits of output, and that the output is used without delay.

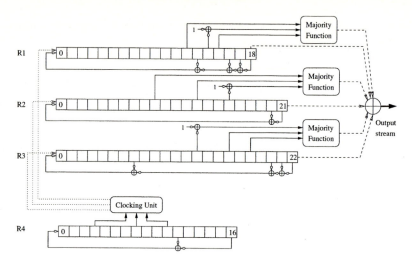

Fig. 1. The A5/2 internal structure

1. Initialize the internal state with K_c and frame number.
2. Force the bits R1[15], R2[16], R3[18], R4[10] to be 1.
3. Run A5/2 for 99 clocks and ignore the output.
4. Run A5/2 for 228 clocks and use the output as key-stream.

After the first clocking is performed the first output bit is ready at the output of A5/2. In Figure 1 we show the internal structure of A5/2. The clocking mechanism works as follows: R4 controls the clocking of R1, R2, and R3. When clocking of R1, R2, and R3 is to be performed, bits R4[3], R4[7], and R4[10] are the input of the clocking unit. The clocking unit performs a majority function on the bits. R1 is clocked if and only if R4[10] agrees with the majority. R2 is clocked if and only if R4[3] agrees with the majority. R3 is clocked if and only if R4[7] agrees with the majority. After these clockings, R4 is clocked.

Once the clocking is performed, an output bit is ready. The output bit is computed as follows: in each register the majority of two bits and the complement of a third bit is computed; the results of all the majorities and the rightmost bit from each register are XORed to form the output (see Figure 1). Note that the majority function is quadratic in its input: $maj(a, b, c) = a \cdot b \oplus b \cdot c \oplus c \cdot a$.

A5/2 is built on a somewhat similar framework of A5/1. The feedback functions of R1, R2 and R3 are the same as A5/1's feedback functions. The initialization process of A5/2 is also somewhat similar to that of A5/1. The difference is that A5/2 also initializes R4, and that one bit in each register is forced to be 1 after initialization . Then A5/2 discards 99 bits of output while A5/1 discards 100 bits of output. The clocking mechanism is the same, but the input bits to the clocking mechanism are from R4 in the case of A5/2, while in A5/1 they are from R1, R2, and R3. The designers meant to use similar building blocks to save hardware in the mobile phone [16].

This algorithm outputs 228 bits of key-stream. The first block of 114 bits is used as a key-stream to encrypt the link from the network to the customer, and the second block of 114 bits is used to encrypt the link from the customer to the network. Encryption is performed as a simple XOR of the message with the key-stream.

Although A5 is a stream cipher, it is used to encrypt 114-bit "blocks", called *frames*. The frames are sequentially numbered (modulo 2^{22}) by a *TDMA frame number*. The frame number f that is used in the initialization of a A5 frame is actually a fixed bit permutation of the TDMA frame number. In the rest of this paper we ignore the existence of this permutation, since it does not affect our analysis.

2.1   GSM Security Background: A3/A8 and GPRS

In this section we give a more detailed description on the usage and specification of A3 and A8. This and more can be found in [12].

A3 provides authentication of the mobile phone to the network, and A8 is used for session-key agreement. The security of these algorithms is based on a user-specific secret key K_i that is common to the mobile phone and the network. The GSM specifications do not specify the length of K_i, thus it is left for the choice of the operator, but usually it is a 128-bit key. Authentication of the customers to the network is performed using the A3 authentication algorithm as follows: The network challenges the customer with a 128-bit randomly chosen value $RAND$. The customer computes a 32-bit long response $SRES = A3(K_i, RAND)$, and sends $SRES$ to the network, which can then check its validity.

The session key K_c is obtained using A8 as follows: $K_c = A8(K_i, RAND)$. Note that A8 and A3 are always invoked together and with the same parameters. In most implementations, they are one algorithm with two outputs, $SRES$ and K_c. Therefore, they are usually referred to as A3/A8.

The security in GPRS is based on the same mechanisms as of GSM. However, GPRS uses a different encryption key to encrypt its traffic. To authenticate a customer, the same A3/A8 algorithm is used with the same K_i, but with a different $RAND$. The resulting K_c is used to encrypt the GPRS traffic. We refer to this key as $GPRS\text{-}K_c$, to differentiate it from K_c which is used to encrypt the GSM voice-traffic. Similarly we refer to the $SRES$ and $RAND$ as $GPRS\text{-}SRES$ and $GPRS\text{-}RAND$ to differentiate them from their GSM-voice counterparts. The GPRS cipher is referred to as GPRS-A5, or GPRS Encryption Algorithm (GEA). There are currently three versions of the algorithm: GEA1, GEA2, and GEA3 (which is actually A5/3).

3   A Known Plaintext Attack on A5/2

In this section we present a new known plaintext attack (known key-stream attack) on A5/2. Namely, given a key-stream, divided to frames, and the respective frame numbers, the attack recovers the session key.

Goldberg, Wagner and Green presented the first attack [13] on A5/2. The time complexity of this attack is very low. However, it requires the knowledge of the XOR of plaintexts in two frames that are 2^{11} frames apart. Their attack shows that the cipher is quite weak, yet it might prove difficult to implement such an attack in practice. The problem is knowing the exact XOR of plaintexts in two frames that are 6 seconds apart. Another aspect is the elapsed time from the beginning of the attack to its completion. Their attack takes at least 6 seconds, because it takes 6 seconds to complete the reception of the data. Our attack might look as if it requires more information, however, it works with only a few milliseconds of data. We improve our attack in Section 4 to a ciphertext-only attack that requires only a few dozen milliseconds of encrypted unknown data. Therefore, our attack is very easy to implement in practice. We have simulated our known plaintext attack on a personal computer, and verified the results. This simulation recovers the key in less than a second. The computation time and memory complexity of this attack are similar to Goldberg, Wagner and Green's attack. The known plaintext attack of Petrović and Fúster-Sabater [17] has similar data requirements as our attack, however, it does not recover the initial key.

Knowing the initial internal state of R1, R2, R3, R4, and the initial frame number, the session key can be retrieved using simple algebraic operations. This is mainly because the initialization process is linear in the session key and the initial frame number. Therefore, in the attack we focus on revealing the initial internal state of the registers.

Let $k_f, k_{f+1}, k_{f+2}, ...$ be the output of A5/2 divided to frames. Note that each k_j is the output key-stream for a whole frame, i.e., each k_j is 114-bit long.[4] Let $f, f+1, f+2, ...$ be the frame numbers associated with these frames. We denote the i'th bit of the key-stream at frame j by $k_j[i]$. The initial internal state of register Ri at frame j (after the initialization but before the 99 clockings) is denoted by Ri_j.

Assume we know the initial state $R4_f$ of R4 at the first frame. An important observation is that R4 controls the clockings of the other registers. Since we know $R4_f$, then for each output bit we know the exact number of times that a register is clocked to produce that output bit. Each register has a linear feedback, therefore, once given the number of times a register is clocked, we can express every bit of its internal state as a linear combination of bits of the original internal state.

The output of A5/2 is an XOR of the last bits of R1, R2, and R3, and three majority functions of bits of R1, R2, R3 (see Figure 1 for the exact details). Therefore, the resulting function is quadratic with variables which are the bits in the initial state of these registers. We take advantage of this low algebraic degree of the output. The goal in the next paragraphs is to express every bit of the whole output of the cipher (consisting of several frames) as a quadratic

---

[4] Note that this notation is somewhat imprecise, since the output is actually 228 bits, where the first half is used to encrypt the network-to-mobile link, and the second half is used to encrypt the mobile-to-network link.

multivariate function in the initial state. Then, we construct an overdefined system of quadratic equations which expresses the key-stream generation process and finally we solve it.

Given a frame number f, there is an algebraic description of every output bit. We perform linearization to the quadratic terms in this algebraic description. We observe that each majority function operates on bits of a single register. Therefore, we have quadratic terms consisting of pairs of variables of the same register only. Taking into account that one bit in each register is set to 1, R1 contributes 18 linear variables and all their $\frac{17*18}{2} = 153$ products. In the same way R2 contributes $21 + \frac{21*20}{2} = 21 + 210$ variables and R3 contributes $22 + \frac{22*21}{2} = 22 + 231$ variables. So far we have $18 + 153 + 21 + 210 + 22 + 231 = 655$ variables after linearization. Together with constant 1 we have a set of 656 variables. We denote the set of these 656 variables by V_f. Of these variables, $18 + 21 + 22 = 61$ variables form the full initial state of R1, R2, and R3.

Every output bit we have adds a linear equation with variables from V_f. A frame consists of 114 bits. Therefore, we get 114 equations from each frame. The solution of the equation system reveals the value of the variables in V_f, and among them the linear variables that directly describe the initial internal state of R1, R2, and R3. However, we do not have enough equations at this stage to efficiently solve the system.

The main observation is that given the variables in V_f defined on frame f, we can describe the bits of any other frame in linear terms of the variables in the set V_f. When moving to the next frame, the frame number is incremented by 1 and the internal state is re-initialized. We assume we know the value of $R4_f$. Due to the initialization method, where the frame number is XORed bit by bit into the registers (see Section 2), we know the value of $R4_{f+1}$. Since we do not know $R1_f$, $R2_f$, and $R3_f$, we do not know the value of $R1_{f+1}$, $R2_{f+1}$, and $R3_{f+1}$, but we know the XOR-differences between $R1_f, R2_f, R3_f$ and $R1_{f+1}, R2_{f+1}, R3_{f+1}$ respectively. We define the set of variables that describe their state and the linearization of these variables as V_{f+1}, in the same way as we did with the first frame to create the set V_f. Due to the initialization method, for each register Ri we know the difference between Ri_{f+1} and Ri_f. Thus, we can describe the variables in the set V_{f+1} as linear combinations of the variables from V_f. We emphasize that even the quadratic terms can be represented in this way. To see this, we assume that $a_{f+1} \cdot b_{f+1}$ is a quadratic term in V_{f+1}, and $a_f \cdot b_f$ is a quadratic term in V_f. We know the differences $d_a = a_{f+1} \oplus a_f$ and $d_b = b_{f+1} \oplus b_f$. Therefore, $a_{f+1} \cdot b_{f+1} = (a_f \oplus d_a) \cdot (b_f \oplus d_b) = a_f \cdot b_f \oplus a_f \cdot d_b \oplus b_f \cdot d_a \oplus d_a \cdot d_b$. Since d_b and d_a are known constants, this equation is linear in the variables in V_f. This fact allows us to use output bits of the second frame in order to get additional linear equations in the variables of V_f. In a similar way, we describe the variables in any set V_i as linear combinations of the variables from V_f. In total, we get an equation system of the form: $S \cdot V_f = k$, where S is the system's matrix, and k is the concatenation of k_f, k_{f+1}, etc.

It is clear that once we obtain 656 linearly independent equations the system can be easily solved using Gauss elimination. However, it is practically very difficult to collect 656 linearly independent equations. This is an effect of the frequent re-initializations, and the low order of the majority function. We note that we do not actually need to solve all the variables, i.e., it suffices to solve the linear variables of the system, since the other variables are defined as their products. We have tested experimentally and found that after we sequentially obtain about 450 equations, the original linear variables in V_f can be solved using Gauss elimination.[5]

We can summarize this attack as follows: we try all the 2^{16} possible values for $R4_f$, and for each such value we solve the linearized system of equations that describe the output. The solution of the equations gives us a suggestion for the internal state of R1, R2, and R3. Together with R4, we have a suggestion for the full internal state. Most of the $2^{16} - 1$ wrong states are identified due to inconsistencies in the Gauss elimination. If more than one consistent internal state remains, these suggestions are verified by trial encryptions.

The time complexity of the attack is as follows: There are 2^{16} possible guesses of the value of $R4_f$. We should multiply this figure by the time it takes to solve a linear binary system of 656 variables, which is about $656^3 \approx 2^{28}$ XOR operations. Thus, the total complexity is about 2^{44} bit-XOR operations. When performed on a 32-bit machine, the complexity is 2^{39} register-XOR operations.

An implementation of this algorithm on our Linux 800MHz PIII personal computer finds the internal state within about 40 minutes, and requires relatively small amount of memory (holding the linearized system in memory requires 656^2 bits $\approx 54KB$).

3.1   Optimization of the Known Plaintext Attack on A5/2

In an optimized implementation that will be described in detail in the full version of this paper, the average time complexity can be further reduced to about 2^{28} bit-XOR operations (less than 1 second on our personal computer). The memory complexity rises to about $2^{27.8}$ bytes (less than 250MBs). The optimization requires a pre-computation step whose time complexity is about 2^{46} bit-XOR operations (about 160 minutes on our personal computer). The data complexity is slightly higher, and still in the range of a few dozen milliseconds of data. The optimization is based on the observation that for every candidate for the value of $R4_f$ the system of equations contains linearly dependent rows. In the pre-computation stage we consider all the possible values of $R4_f$, for each such value we compute the system of equations. We find which rows in the equation system are linearly dependent by performing a Gauss elimination. In the real-time phase of the attack we filter wrong values for $R4_f$ by checking if the linear dependencies that we found in the pre-computation step hold on the key-stream

---

[5] In case the data available for the attacker is scarce, there are additional methods that can be used to reduce the number of required equations. For example, whenever a value of a linear variable x_i is discovered, any quadratic variable of the form $x_i \cdot x_j$ can be simplified to 0 or x_j depending whether $x_i = 0$ or $x_i = 1$, respectively. The XL algorithm [9] can also be used in cases of scarce available data.

bits. This kind of filtering requires two dot products on average for each wrong value of $R4_f$. Once we have a suggestion for the correct value of $R4_f$, the correct key is found using the methods of Section 3.

Note that when using this optimized attack some compromise is needed. Since four known plaintext frames are required, we must know the XOR-differences: $f \oplus (f+1)$, $f \oplus (f+2)$, $f \oplus (f+3)$ in advance, before we know the exact value f. These XOR-differences are required in order to express the frames' key-stream bits as linear terms over the set V_f, and to compute the system of equations. The problematic element is the addition operation, for example, $f+1$ can result in a carry that would propagate through f, thus not allowing the calculation of the XOR-difference in advance. Therefore, we require that f has a specific bit set to 0. This requirement prevents a carry from propagating beyond the specific bit. We require the two last bits in f have a fixed value, and perform the pre-computation for each of the four combinations of the last two bits in f. This requirement is the source of the data complexity being slightly higher, and it also causes a factor four increase in the memory complexity and the pre-computation time complexity.

4   An Instant Ciphertext-Only Attack on A5/2

In this section we convert the attack of Section 3 on A5/2 to a ciphertext-only attack. We observe that error-correction codes are employed in GSM before encryption. Thus, the plaintext has a highly structured redundancy.

There are several kinds of error-correction methods that are used in GSM, and different error-correction schemes are used for different data channels. We focus on control channels, and specifically on the error-correction codes of the Slow Associated Control Channel (SACCH). We note that this error-correction code is the one used during the initialization of a conversation. Therefore, it suffices to focus on this code. Using this error-correction code we mount a ciphertext-only attack that recovers the key. However, we stress that the ideas of our attack can be applied to other error-correction codes as well.

In the SACCH, the message to be coded with error-correction codes has a fixed size of 184 bits. The result is a 456-bit long message. The 456 bits of the message are then interleaved, and divided to four frames. These frames are then encrypted and transmitted.

The coding operation and the interleaving operation can be modeled together as one 456×184 matrix over $GF(2)$, which we denote by G. The message to be coded is regarded as a 184-bit binary vector, P. The result of the coding-interleaving operation is: $M = G \cdot P$. The resulting vector M consists of 4 frames. In the encryption process each frame is XORed with the output key-stream of $A5/2$ for the respective frame.

Since G is a 456×184 binary matrix, there are $456 - 184 = 272$ equations that describe the kernel of the inverse transformation (and the dimension of the kernel is not larger than 272 due to the properties of the matrix G). In other words, for any vector M, $M = G \cdot P$, there are 272 linearly independent equations on its elements. Let K_G be a matrix that describes these 272 linear equations, i.e., $K_G \cdot M = 0$ for any such M.

We denote the output sequence bits of A5/2 for a duration of 4 frames by $k = k_j||k_{j+1}||k_{j+2}||k_{j+3}$, where $||$ is the concatenation operator. The ciphertext C is computed by $C = M \oplus k$. We use the same 272 equations on C, namely:

$$K_G \cdot C = K_G \cdot (M \oplus k) = K_G \cdot M \oplus K_G \cdot k = 0 \oplus K_G \cdot k = K_G \cdot k.$$

Since the ciphertext C is known, we actually get linear equations over elements of k. Note that the equations we get are independent of P — they depend only on k. We substitute each bit in k with its description as linear terms over V_f (see Section 3), and thus get equations on variables of V_f. Each 456-bit coding block, provides 272 equations. The rest of the details of the attack and its time complexity are similar to the case in the previous section, but in this attack we know linear combinations of the key-stream, and therefore, the corresponding equations are the respective linear combinations of the equations: Let $S \cdot V_f = k$ be the system of equations from Section 3, where S is the system's matrix. In the ciphertext-only attack we multiply this system by K_G as follows: $(K_G \cdot S) \cdot V_f = (K_G \cdot k)$. K_G is a fixed known matrix that depends only on the coding-interleaving matrix G, S is the system's matrix which is different for each value of $R4_f$ (and for different XOR-differences of $f \oplus (f+1)$, $f \oplus (f+2)$, $f \oplus (f+3)$, etc.). Thus, in the optimized attack, all the possible matrices $K_G \cdot S$ are computed during the pre-computation, and for each such matrix we find linear dependencies of rows by a Gauss elimination. In the real-time phase of the attack we filter wrong values of $R4_f$ by checking if the linear dependencies that we found in the pre-computation step hold on the bits of $K_G \cdot k$.

Note that while four frames of data suffice to launch the attack in Section 3, in the ciphertext-only attack we need eight frames, since from each encrypted frame we get only about half the information compared to the known plaintext attack. The time complexity of the attack is the same as the attack in Section 3.1.

We now analyze the time and memory complexity of this ciphertext-only attack while using the optimized attack of Section 3.1. In Section 3.1 we restrict the values of the three least significant bits of the frame number f, since we need four frames of data. The attack in this section requires eight frames of data, therefore, we restrict the four least significant bits of the frame number f. This restriction doubles the memory complexity compared to the optimized attack of Section 3.1, and it also doubles the pre-computation complexity.

We summarize the complexity of the ciphertext-only attack (using the optimized implementation) as follows: the average time complexity of this ciphertext-only attack is approximately 2^{16} dot products, the memory complexity is about $2^{28.8}$ bytes (less than 500MBs), and the pre-computation time complexity is about 2^{47} bit-XORs. Our implementation on a personal computer (taking advantage of our machine's 32-bit XOR) recovers K_c in less than a second, and it takes about 320 minutes (less than 5.5 hours) for the one-time pre-computation.

We also successfully enhance the attack of Goldberg, Wagner, and Green and the attack of Petrović and Fúster-Sabater to a ciphertext-only attack using our methods. The details of these attacks will appear in the full version of this paper.

5   Leveraging the Attacks to Any GSM Network

The attack shown in Section 4 assumes that the encryption algorithm is A5/2. Using the attack it is easy to recover K_c in real-time from a few dozen milliseconds of ciphertext. We ask the question what happens when the encryption algorithm is not A5/2, but rather is A5/1 or the newly chosen A5/3. The surprising answer is that using weakness of the protocols almost the same attack applies. Also, a variant of the attack works on GPRS networks. All that is needed for the new attack to succeed is that the mobile phone supports A5/2 for voice conversations. The vast majority of handsets support A5/2 encryption (to allow encryption while roaming to networks that use only A5/2).

The following three attacks retrieve the encryption key that the network uses when A5/1 or A5/3 is employed. In the first attack the key is discovered by a man-in-the-middle attack on the victim customer. In this attack, the attacker plays two roles. He impersonates the network to the customer and impersonates the customer to the network. In the second attack, the attacker needs to change bits (flip bits) in the conversation of the mobile and the network (this attack can also be performed as a man-in-the-middle attack). In the third one the attacker impersonates the network for a short radio-session with the mobile. We note that these kind of attacks are relatively very easy to mount in a cellular environment.

This man-in-the-middle attack is performed as follows: when authentication is performed (in the initialization of a conversation), the network sends an authentication request to the attacker, and the attacker sends it to the victim. The victim computes $SRES$, and returns it to the attacker, which sends it back to the network. Now the attacker is "authenticated" to the network. Next, the network asks the customer to start encrypting with A5/1 or A5/3. In our attack, since the attacker impersonates the customer, the network actually asks the attacker to start encrypting with A5/1 or A5/3. The attacker does not have the key, yet, and therefore, is not able to start the encryption. The attacker needs the key before he is asked to use it. To achieve it, the attacker asks the victim to encrypt with A5/2 just after the victim returned the $SRES$, and before the attacker returns the authentication information to the network. This request sounds to the victim as a legitimate request, since the victim sees the attacker as the network. Then, the attacker employs cryptanalysis of A5/2 to retrieve the encryption key of the A5/2 that is used by the victim. Only then, the attacker sends the authentication information to the network. The key only depends on $RAND$, that means that the key recovered through the A5/2 attack is the same key to be used when A5/1 is used or even when 64-bit A5/3 is used! Now the attacker can encrypt/decrypt with A5/1 or A5/3 using this key.

Some readers may suspect that the network may identify this attack, by identifying a small delay in the time it takes to the authentication procedure to complete. However, the GSM standard allows 12 seconds for the mobile phone to complete his authentication calculations and to return an answer, while the delay incurred by this attack is less than a second.

A second possible attack, which can be relatively easily spotted (and prevented) by the network, is a class-mark attack. During initialization of conversation, the mobile phone sends his ciphering capabilities to the network (this information is called class-mark). Most mobile phones currently support A5/1, A5/2, and A5/0 (no encryption), but this may change from phone to phone, and can change in the future. The attacker changes (for example using a man-in-the-middle attack) the class-mark information that the mobile phone sends, in a way that the network thinks that the mobile phone can only support A5/2, and A5/0. The network then defaults to A5/2, and thus allowing the attacker to listen to the conversation. This attack takes advantage of the following protocol flaw: the class-mark information is not protected.

Many networks initiate the authentication procedure rarely, and use the key created in the last authentication. An attacker can discover this key by impersonating the network to the victim mobile phone. Then the attacker initiates a radio-session with the victim, and asks the victim mobile phone to start encrypting using A5/2. The attacker performs the attack, recovers the key, and ends the radio session. The owner of the mobile phone and the network have no indication of the attack.

The leveraging in the first and last attacks relies on the fact that the same key is loaded to A5/2 and A5/1 and even to 64-bit A5/3 (in case A5/3 is used in GSM, according to GSM standards). Thus, discovering the key for A5/2 reveals the key for A5/1 and 64-bit A5/3. We note that although A5/3 can be used with key lengths of 64-128 bits, the GSM standard allows the use of only 64-bit A5/3.

A similar attack can be performed on GPRS. The attacker listens to the *GPRS-RAND* sent by the network to the customer. The attacker can impersonate the voice network, initiate radio session with the customer and start authentication procedure using the *GPRS-RAND* value that he intercepted, as the GSM-voice *RAND*. The result is that K_c equals *GPRS-K_c*. The attacker asks the customer to encrypt with A5/2, recovers K_c, and ends the radio-session. The attacker can now decrypt/encrypt the customer's GPRS traffic using the recovered K_c. Alternatively, the attacker can record the customer's traffic, and perform the impersonation at any later time in order to retrieve the *GPRS-K_c* with which the recorded data can be decrypted. If A5/2 is not supported by the phone, but A5/1 is supported, then the above attacks against A5/3 and GPRS can be performed using the (more complex) cryptanalysis of A5/1 instead of A5/2 as a building block.

6 Passive Ciphertext-Only Cryptanalysis of GSM-A5/1 Encrypted Communication

In this section we discuss possible ciphertext-only attacks on GSM communications that are encrypted using A5/1. Unlike Section 5, this section discusses passive attacks, i.e., these attacks do not transmit any radio signals. Our starting point for the passive ciphertext-only attacks is that error-correction-codes are employed before encryption, as discussed in Section 4.

Table 1. Four points on the time/memory/data tradeoff curve

Available data D	M	Number of 200GBs disk	P=N/D	Number of PCs to complete preprocessing in one year	T	Number of PCs to complete attack in real-time
2^{12} (≈ 5 min)	2^{38}	≈ 22	2^{52}	140	2^{28}	1
$2^{6.7}$ (≈ 8 sec)	2^{41}	≈ 176	$2^{57.3}$	5000	$2^{32.6}$	1000
$2^{6.7}$ (≈ 8 sec)	2^{42}	≈ 350	$2^{57.3}$	5000	$2^{30.6}$	200
2^{14} (≈ 20 min)	2^{35}	≈ 3	2^{50}	35	2^{30}	1

We apply a time/memory/data tradeoff, similar to the one presented by Biryukov, Shamir and Wagner [4], and further discussed and generalized by Biryukov and Shamir [3]. Their original attack requires about two second of known plaintext, and takes a few minutes to execute on a personal computer. It also requires a preprocessing time of about 2^{48} and memory of four 73GByte disks.

In our attack, we are given a block of four encrypted frames. We use the methods of Section 4 to compute $K_G \cdot k$, which is 272 bits long. These 272 bits depend only on the output of A5/1 for four consecutive frames. We call these output bits the *coded-stream*. Let's assume that we know that the frame number of the first frame of these four frames is divisible by four without remainder — this assumption limits us to use only a quarter of the GSM data stream on average. We can view the whole process as a function from the internal state of A5/1 to the coded-stream. Let $h : \{0,1\}^{64} \to \{0,1\}^{272}$ be the function that takes an internal state of A5/1 after initialization, and outputs the coded-stream. Inverting h reveals the internal state, and breaks the cipher. Note that the output of h is calculated from the key-stream of four frames, while its input is the internal state after the initialization of A5/1 at the first frame. The assumption we make on the frame numbers facilitates the computation of the initial internal states at the other three frames from the initial internal state at the first frame. It suffices to look at only 64 bits of the output of h, therefore, we can assume that $h : \{0,1\}^{64} \to \{0,1\}^{64}$. We apply Biryukov and Shamir's time/memory/data tradeoff attack with sampling [3] and adopt their notations: the number of possible states is $N = 2^{64}$, the number of four-frame blocks that we allow is denoted by D, T is the number of applications of h during the real-time phase of the attack, M is the number of memory rows (in our case each row is about 16-byte long), and the tradeoff curve is: $TM^2D^2 = N^2$, $D^2 \leq T \leq N$. The preprocessing complexity $P = N/D$ is the number of applications of h during preprocessing. The attack performs about $\sqrt{T}$ memory accesses. We assume that h can be applied 2^{20} times every second on a personal computer. We list four points on the tradeoff curve in Table 1.

A more detailed description about the passive attack will appear in the full version of this paper.

7   Possible Attack Scenarios

The attacks presented in this paper can be used in several scenarios. In this section we present four of them: call wire-tapping, call hijacking, altering of data messages (sms), and call theft — dynamic cloning.

7.1   Call Wire-Tapping

The most naive scenario that one might anticipate is eavesdropping conversations. Communications encrypted using GSM can be decrypted and eavesdropped by an attacker, once the attacker has the encryption key. Both voice conversations and data, for example SMS messages, can be wire-tapped. Video and picture messages that are sent over GPRS can also be tapped, etc. These attacks are described in Section 5.

In another possible wire-tapping attack the attacker records the encrypted conversation. The attacker must make sure that he knows the $RAND$ value that created the used key (the $RAND$ is sent unencrypted). At a later time, when it is convenient for the attacker, the attacker impersonates the network to the victim. Then the attacker initiates a GSM radio-session, asks the victim to perform authentication with the above $RAND$, and recovers the session key used in the recorded conversation. Once the attacker has the key he simply decrypts the conversation and can listen to its contents. Note that an attacker can record many conversations, and with subsequent later attacks recover all the keys. This attack has the advantage of transmitting only in the time that is convenient for the attacker. Possibly even years after the recording of the conversation, or when the victim is in another country, or in a convenient place for the attacker.

7.2   Call Hijacking

While a GSM network can perform authentication at the initiation of the call, encryption is the means of GSM for preventing impersonation at later stages of the conversation. The underlying assumption is that an imposter do not have K_c, and thus cannot conduct encrypted communications. We show how to obtain encryption keys. Once an attacker has the encryption keys, he can cut the victim off the conversation, and impersonate the victim to the other party. Therefore, hijacking the conversation after authentication is possible. Hijacking can occur during early call-setup, even before the victim's phone begins to ring. The operator can hardly suspect there is an attack. The only clue of an attack is a moment of some increased electro-magnetic interference.

Another way to hijack incoming calls is to mount a kind of a man-in-the-middle attack, but instead of forwarding the call to the victim, the attacker receives the call.

7.3   Altering of Data Messages (SMS)

Once a call has been hijacked, the attacker decides on the content. The attacker can listen to the contents of a message being sent by the victim, and send his own version. The attacker can stop the message, or send his own SMS message. This compromises the integrity of GSM traffic.

7.4   Call Theft – Dynamic Cloning

GSM was believed to be secure against call theft, due to authentication proce-dures of A3/A8 (at least for operators that use a strong primitive for A3/A8 rather then COMP128).

However, due to the mentioned weaknesses, an attacker can make outgoing calls on the expense of a victim. When the network asks for authentication, a man-in-the-middle attack similar to the one described in Section 5 can be ap-plied: the attacker initiates an outgoing call to the cellular network in parallel to a radio session to a victim. When the network asks the attacker for authenti-cation, the attacker asks the victim for authentication, and relays the resulting authentication back to the network. The attacker can also recover K_c as de-scribed in Section 5. Now the attacker can close the session with the victim, and continue the outgoing call to the network. This attack is hardly detectable by the network, as the network views it as normal access. The victim's phone does not ring, and the victim has no indication that he is a victim. At least until his monthly bill arrives.

8   Summary

In this paper we present new methods for attacking GSM and GPRS encryption and security protocols. The described attacks are easy to apply, and do not require knowledge of the conversation. We stress that GSM operators should replace the cryptographic algorithms and protocols as early as possible, or switch to the more secure third generation cellular system.

In GSM, even GSM networks using the new A5/3 succumb to our attack. We suggest to change the way A5/3 is integrated to protect the networks from such attacks. A possible correction is to make the keys used in A5/1 and A5/2 unrelated to the keys that are used in A5/3. The integration of GPRS suffers from similar flaws that should be taken into consideration.

We would like to emphasize that our ciphertext-only attack is made possible by the fact that the error-correction codes are employed before the encryption. In the case of GSM, the addition of such a structured redundancy before encryption is performed crucially reduces the system's security.

Acknowledgments. We are grateful to Orr Dunkelman for his great help and various comments on early versions of this paper, and to Adi Shamir for his advice and useful remarks. We would like to thank David Wagner for providing us with information on his group's attack on A5/2. We also acknowledge the anonymous referees for their important comments.

References

1. *The 3rd Generation Partnership Project (3GPP)*, http://www.3gpp.org/.
2. Eli Biham, Orr Dunkelman, *Cryptanalysis of the A5/1 GSM Stream Cipher*, Progress in Cryptology, proceedings of Indocrypt'00, Lecture Notes in Computer Science 1977, Springer-Verlag, pp. 43–51, 2000.
3. Alex Biryukov, Adi Shamir, *Cryptanalytic Time/Memory/Data Tradeoffs for Stream Ciphers*, Advances in Cryptology, proceedings of Asiacrypt'00, Lecture Notes in Computer Science 1976, Springer-Verlag, pp. 1–13, 2000.
4. Alex Biryukov, Adi Shamir, David Wagner, *Real Time Cryptanalysis of A5/1 on a PC*, Advances in Cryptology, proceedings of Fast Software Encryption'00, Lecture Notes in Computer Science 1978, Springer-Verlag, pp. 1–18, 2001.
5. Marc Briceno, Ian Goldberg, David Wagner, *A pedagogical implementation of the GSM A5/1 and A5/2 "voice privacy" encryption algorithms*, http://cryptome.org/gsm-a512.htm (originally on www.scard.org), 1999.
6. Marc Briceno, Ian Goldberg, David Wagner, *An implementation of the GSM A3A8 algorithm*, http://www.iol.ie/~kooltek/a3a8.txt, 1998.
7. Marc Briceno, Ian Goldberg, David Wagner, *GSM Cloning*, http://www.isaac.cs.berkeley.edu/isaac/gsm-faq.html, 1998.
8. Nicolas Courtois, *Higher Order Correlation Attacks,XL Algorithm and Cryptanalysis of Toyocrypt*, proceedings of ICISC'02, Lecture Notes in Computer Science 2587, Springer-Verlag, pp. 182–199, 2003.
9. Nicolas Courtois, Alexander Klimov, Jacques Patarin, Adi Shamir, *Efficient Algorithms for Solving Overdefined Systems of Multivariate Polynomial Equations*, Advances in Cryptology, proceedings of Eurocrypt'00, Lecture Notes in Computer Science 1807, Springer-Verlag, pp. 392–407, 2000.
10. Nicolas Courtois, Josef Pieprzyk, *Cryptanalysis of Block Ciphers with Overdefined Systems of Equations*, Advances in Cryptology, proceedings of Asiacrypt'02, Lecture Notes in Computer Science 2501, Springer-Verlag, pp. 267–287, 2002.
11. Patrik Ekdahl, Thomas Johansson, *Another Attack on A5/1*, to be published in IEEE Transactions on Information Theory, http://www.it.lth.se/patrik/publications.html, 2002.
12. European Telecommunications Standards Institute (ETSI), *Digital cellular telecommunications system (Phase 2+); Security related network functions*, TS 100 929 (GSM 03.20), http:/www.etsi.org.
13. Ian Goldberg, David Wagner, Lucky Green, *The (Real-Time) Cryptanalysis of A5/2*, presented at the Rump Session of Crypto'99, 1999.
14. Jovan Golic, *Cryptanalysis of Alleged A5 Stream Cipher*, Advances in Cryptology, proceedings of Eurocrypt'97, LNCS 1233, pp.239–255, Springer-Verlag,1997.
15. Aviad Kipnis, Adi Shamir, *Cryptanalysis of the HFE Public Key Cryptosystem by Relinearization*, Advances in Cryptology, proceedings of Crypto'99, Lecture Notes in Computer Science 1666, Springer-Verlag, pp. 19–30, 1999.
16. Security Algorithms Group of Experts (SAGE), *Report on the specification and evaluation of the GSM cipher algorithm A5/2*, http://cryptome.org/espy/ETR278e01p.pdf, 1996.
17. Slobodan Petrović, Amparo Fúster-Sabater, *Cryptanalysis of the A5/2 Algorithm*, Cryptology ePrint Archive, Report 2000/052, http://eprint.iacr.org, 2000.

Making a Faster Cryptanalytic Time-Memory Trade-Off

Philippe Oechslin

Laboratoire de Securité et de Cryptographie (LASEC)
Ecole Polytechnique Fédérale de Lausanne
Faculté I&C, 1015 Lausanne, Switzerland
philippe.oechslin@epfl.ch

Abstract. In 1980 Martin Hellman described a cryptanalytic time-memory trade-off which reduces the time of cryptanalysis by using precalculated data stored in memory. This technique was improved by Rivest before 1982 with the introduction of distinguished points which drastically reduces the number of memory lookups during cryptanalysis. This improved technique has been studied extensively but no new optimisations have been published ever since. We propose a new way of precalculating the data which reduces by two the number of calculations needed during cryptanalysis. Moreover, since the method does not make use of distinguished points, it reduces the overhead due to the variable chain length, which again significantly reduces the number of calculations. As an example we have implemented an attack on MS-Windows password hashes. Using 1.4GB of data (two CD-ROMs) we can crack 99.9% of all alphanumerical passwords hashes (2^{37}) in 13.6 seconds whereas it takes 101 seconds with the current approach using distinguished points. We show that the gain could be even much higher depending on the parameters used.

Keywords: Time-memory trade-off, cryptanalysis, precomputation, fixed plaintext

1 Introduction

Cryptanalytic attacks based on exhaustive search need a lot of computing power or a lot of time to complete. When the same attack has to be carried out multiple times, it may be possible to execute the exhaustive search in advance and store all results in memory. Once this precomputation is done, the attack can be carried out almost instantly. Alas, this method is not practicable because of the large amount of memory needed. In [4] Hellman introduced a method to trade memory against attack time. For a cryptosystem having N keys, this method can recover a key in $N^{2/3}$ operations using $N^{2/3}$ words of memory. The typical application of this method is the recovery of a key when the plaintext and the ciphertext are known. One domain where this applies is in poorly designed data encryption system where an attacker can guess the first few bytes of data (e.g.

D. Boneh (Ed.): CRYPTO 2003, LNCS 2729, pp. 617–630, 2003.

"#include <stdio.h>"). Another domain are password hashes. Many popular operating systems generate password hashes by encrypting a fixed plaintext with the user's password as key and store the result as the password hash. Again, if the password hashing scheme is poorly designed, the plaintext and the encryption method will be the same for all passwords. In that case, the password hashes can be calculated in advance and can be subjected to a time-memory trade-off.

The time-memory trade-off (with or without our improvement) is a probabilistic method. Success is not guaranteed and the success rate depends on the time and memory allocated for cryptanalysis.

1.1   The Original Method

Given a fixed plaintext P_0 and the corresponding ciphertext C_0, the method tries to find the key $k \in N$ which was used to encipher the plaintext using the cipher S. We thus have:

$$C_0 = S_k(P_0)$$

We try to generate all possible ciphertexts in advance by enciphering the plaintext with all N possible keys. The ciphertexts are organised in chains whereby only the first and the last element of a chain is stored in memory. Storing only the first and last element of a chain is the operation that yields the trade-off (saving memory at the cost of cryptanalysis time). The chains are created using a *reduction function* R which creates a key from a cipher text. The cipher text is longer that the key, hence the reduction. By successively applying the cipher S and the reduction function R we can thus create chains of alternating keys and ciphertexts.

$$k_i \xrightarrow{S_{k_i}(P_0)} C_i \xrightarrow{R(C_i)} k_{i+1}$$

The succession of $R(S_k(P_0))$ is written $f(k)$ and generates a key from a key which leads to chains of keys:

$$k_i \xrightarrow{f} k_{i+1} \xrightarrow{f} k_{i+2} \rightarrow ...$$

m chains of length t are created and their first and last elements are stored in a table. Given a ciphertext C we can try to find out if the key used to generate C is among the ones used to generate the table. To do so, we generate a chain of keys starting with $R(C)$ and up to the length t. If C was indeed obtained with a key used while creating the table then we will eventually generate the key that matches the last key of the corresponding chain. That last key has been stored in memory together with the first key of the chain. Using the first key of the chain the whole chain can be regenerated and in particular the key that comes just before $R(C)$. This is the key that was used to generate C, which is the key we are looking for.

Unfortunately there is a chance that chains starting at different keys collide and merge. This is due to the fact that the function R is an arbitrary reduction of the space of ciphertexts into the space of keys. The larger a table is, the higher is the probability that a new chain merges with a previous one. Each merge reduces the number of distinct keys which are actually covered by a table. The chance of finding a key by using a table of m rows of t keys is given in the original paper [4] and is the following:

$$P_{table} \geq \frac{1}{N} \sum_{i=1}^{m} \sum_{j=0}^{t-1} \left(1 - \frac{it}{N}\right)^{j+1} \tag{1}$$

The efficiency of a single table rapidly decreases with its size. To obtain a high probability of success it is better to generate multiple tables using a different reduction function for each table. The probability of success using ℓ tables is then given by:

$$P_{success} \geq 1 - \left(1 - \frac{1}{N} \sum_{i=1}^{m} \sum_{j=0}^{t-1} \left(1 - \frac{it}{N}\right)^{j+1}\right)^{\ell} \tag{2}$$

Chains of different tables can collide but will not merge since different reduction functions are applied in different tables.

False alarms. When searching for a key in a table, finding a matching endpoint does not imply that the key is in the table. Indeed, the key may be part of a chain which has the same endpoint but is not in the table. In that case generating the chain from the saved starting point does not yield the key, which is referred to as a false alarm. False alarms also occur when a key is in a chain that is part of the table but which merges with other chains of the table. In that case several starting points correspond to the same endpoint and several chains may have to be generated until the key is finally found.

1.2   Existing Work

In [2] Rivest suggests to use distinguished points as endpoints for the chains. Distinguished points are points for which a simple criteria holds true (e.g. the first ten bits of a key are zero). All endpoints stored in memory are distinguished points. When given a first ciphertext, we can generate a chain of keys until we find a distinguished point and only then look it up in the memory. This greatly reduces the number of memory lookups. All following publications use this optimisation.

[6] describes how to optimise the table parameters t, m and ℓ to minimise the total cost of the method based on the costs of memory and of processing engines. [5] shows that the parameters of the tables can be adjusted such as to increase the probability of success, without increasing the need for memory or the cryptanalysis time. This is actually a trade-off between precomputation time and success rate. However, the success rate cannot be arbitrarily increased.

Borst notes in [1] that distinguished points also have the following two advantages:

- They allow for loop detection. If a distinguished point is not found after enumerating a given number of keys (say, multiple times their average occurrence), then the chain can be suspected to contain a loop and be abandoned. The result is that all chains in the table are free of loops.
- Merges can easily be detected since two merging chains will have the same endpoint (the next distinguished point after the merge). As the endpoints have to be sorted anyway the merges are discovered without additional cost. [1] suggest that it is thus easy to generate collision free tables without significant overhead. Merging chains are simply thrown away and additional chains are generated to replace them. Generating merge free tables is yet another trade-off, namely a reduction of memory at the cost of extra precomputation.

Finally [7] notes that all calculations used in previous papers are based on Hellman's original method and that the results may be different when using distinguished points due to the variation of chain length. They present a detailed analysis which is backed up by simulation in a purpose-built FPGA.

A variant of Hellman's trade-off is presented by Fiat and Noar in [3]. Although this trade-off is less efficient, it can be rigorously analysed and can provably invert any type of function.

2   Results of the Original Method

2.1   Bounds and Parameters

There are three parameters that can be adjusted in the time-memory trade-off: the length of the chains t , the number of chains per table m and the number of tables produced ℓ.

These parameters can be adjusted to satisfy the bounds on memory M, cryptanalysis time T and success rate $P_{success}$. The bound on success rate is given by equation 2. The bound on memory M is given by the number of chains per table m, the number of tables ℓ and the amount of memory m_0 needed to store a starting point and an endpoint (8 bytes in our experiments). The bound in time T is given by the average length of the chains t, the number of tables ℓ and the rate $\frac{1}{t_0}$ at which the plaintext can be enciphered (700'000/s in our case). This bound corresponds to the worst case where all tables have to be searched but it does not take into account the time spent on false alarms.

$$M = m \times \ell \times m_0 \qquad\qquad T = t \times \ell \times t_0$$

Figure 1 illustrates the bounds for the problem of cracking alphanumerical windows passwords (complexity of 2^{37}). The surface on the top-left graph is the bound on memory. Solutions satisfying the bound on memory lie below this surface. The surface on the bottom-left graph is the bound on time and solutions also have to be below that surface to satisfy the bound. The graph on the right side shows the bound on success probability of 99.9% and the combination of the two previous bounds. To satisfy all three bounds, the parameters of the solution must lie below the protruding surface in the centre of the graph (time

M < 1.4GB

Success > 0.999, min(M <1.4GB, T < 220)

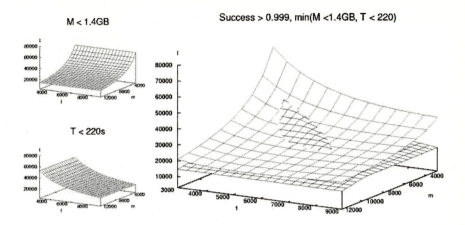

T < 220s

Fig. 1. Solution space for a success probability of 99.9%, a memory size of 1.4GB and a maximum of 220 seconds in our sample problem.

and memory constraints) and above the other surface (success rate constraint). This figure nicely illustrates the content of [5], namely that the success rate can be improved without using more memory or more time: all the points on the ridge in the centre of the graph satisfy both the bound on cryptanalysis time and memory but some of them are further away from the bound of success rate than others. Thus the success rate can be optimised while keeping the same amount of data and cryptanalysis time, which is the result of [5]. We can even go one step further than the authors and state that the optimal point must lie on the ridge where the bounds on time and memory meet, which runs along $\frac{t}{m} = \frac{T}{M}$. This reduces the search for the optimal solution by one dimension.

3   A New Table Structure with Better Results

The main limitation of the original scheme is the fact that when two chains collide in a single table they merge. We propose a new type of chains which can collide within the same table without merging.

   We call our chains rainbow chains. They use a successive reduction function for each point in the chain. They start with reduction function 1 and end with reduction function $t-1$. Thus if two chains collide, they merge only if the collision appears at the same position in both chains. If the collision does not appear at the same position, both chains will continue with a different reduction function and will thus not merge. For chains of length t, if a collision occurs, the chance of it being a merge is thus only $\frac{1}{t}$. The probability of success within a single table of size $m \times t$ is given by:

$$P_{table} = 1 - \prod_{i=1}^{t}(1 - \frac{m_i}{N})$$ (3)

where $m_1 = m$ and $m_{n+1} = N\left(1 - e^{-\frac{m_n}{N}}\right)$

The derivation of the success probability is given in the appendix. It is interesting to note that the success probability of rainbow tables can be directly compared to that of classical tables. Indeed the success probability of t classical tables of size $m \times t$ is approximately equal to that of a single rainbow table of size $mt \times t$. In both cases the tables cover mt^2 keys with t different reduction functions. For each point a collision within a set of mt keys (a single classical table or a column in the rainbow table) results in a merge, whereas collisions with the remaining keys are not merges. The relation between t tables of size $m \times t$ and a rainbow table is shown in Figure 2. The probability of success are compared in Figure 3. Note that the axes have been relabeled to create the same scale as with the classical case in Figure 1. Rainbow tables seem to have a slightly better probability of success but this may just be due to the fact that the success rate calculated in the former case is the exact expectation of the probability where as in the latter case it is a lower bound.

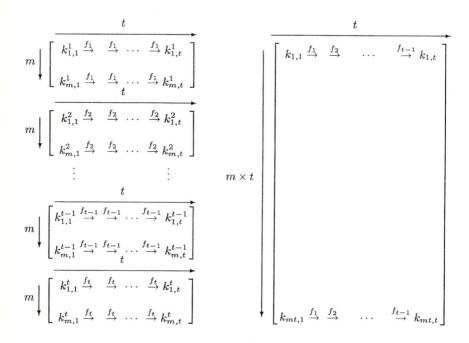

Fig. 2. t classic tables of size $m \times t$ on the left and one rainbow table of size $mt \times t$ on the right. In both cases merges can occur within a group of mt keys and a collision can occur with the remaining $m(t-1)$ keys. It takes half as many operations to look up a key in a rainbow table than in t classic tables.

To lookup a key in a rainbow table we proceed in the following manner: First we apply R_{n-1} to the ciphertext and look up the result in the endpoints of the table. If we find the endpoint we know how to rebuild the chain using the

Success > 0.999 and min(Memory <1.4GB, Time < 110)

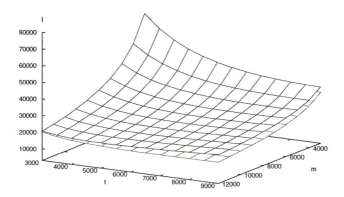

Fig. 3. Comparison of the success rate of classical tables and rainbow tables. The upper surface represents the constraint of 99.9% success with classical tables, the lower surface is the same constraint for rainbow tables. For rainbow tables the scale has been adjusted to allow a direct comparison of both types of tables $m \rightarrow \frac{m'}{t}, \ell \rightarrow \frac{\ell'}{t}$

corresponding starting point. If we don't find the endpoint, we try if we find it by applying R_{n-2}, f_{n-1} to see if the key was in the second last column of the table. Then we try to apply $R_{n-3}, f_{n-2}, f_{n-1}$, and so forth. The total number of calculations we have to make is thus $\frac{t(t-1)}{2}$. This is half as much as with the classical method. Indeed, we need t^2 calculations to search the corresponding t tables of size $m \times t$.

Rainbow chains share some advantages of chains ending in distinguished points without suffering of their limitations:

- The number of table look-ups is reduced by a factor of t compared to Hellman's original method.
- Merges of rainbow chains result in identical endpoints and are thus detectable, as with distinguished points. Rainbow chains can thus be used to generate merge-free tables. Note that in this case, the tables are not collision free.
- Rainbow chains have no loops, since each reduction function appears only once. This is better than loop detection and rejection as described before, because we don't spend time on following and then rejecting loops and the coverage of our chains is not reduced because of loops than can not be covered.
- Rainbow chains have a constant length whereas chains ending in distinguished points have a variable length. As we shall see in Section 4.1 this reduces the number of false alarms and the extra work due to false alarms.

624   P. Oechslin

This effect can be much more important that the factor of two gained by the structure of the table.

4   Experimental Results

We have chosen cracking of MS Windows passwords as an example because it has a real-world significance and can be carried out on any standard workstation. The password hash we try to crack is the LanManager hash which is still supported by all versions of MS Windows for backward compatibility. The hash is generated by cutting a 14 characters password into two chunks of seven characters. In each chunk, lower case characters are turned to upper case and then the chunk is used as a key to encrypt a fixed plain-text with DES. This yields two 8 byte hashes which are concatenated to form the 16 byte LanManager hash. Each halves of the LanManager hash can thus be attacked separately and passwords of up to 14 alphanumerical generate only 2^{37} different 8 byte hashes (rather than 2^{83} 16 byte hashes).

Based on Figure 1 we have chosen the parameters for classic tables to be $t_c = 4666, m_c = 8192$ and for rainbow tables to be $t_r = 4666, m_r = t_c \times m_c = 38'223'872$. We have generated 4666 classic tables and one rainbow table and measured their success rate by cracking 500 random passwords on a standard workstation (P4 1.5GHz, 500MB RAM). The results are given in the table below:

Table 1. Measured coverage for classic tables with distinguished points and for rainbow tables, after cracking of 500 password hashes

	classic with DP	rainbow
t, m, ℓ	4666, 8192, 4666	4666, 38'223'872, 1
predicted coverage	75.5%	77.5%
measured coverage	75.8%	78.8%

This experiment clearly shows that rainbow tables can achieve the same success rate with the same amount of data as classical tables. Knowing this, it is now interesting to compare the cryptanalysis time of both methods since rainbow tables should be twice as fast. In Table 2 we compare the mean cryptanalysis time, the mean number of hash operations per cryptanalysis and the mean number of false alarms per cryptanalysis.

What we see from table 2 is that our method is actually about 7 times faster than the original method. Indeed, each cryptanalysis incurs an average of 9.3M hash calculations with the improved method whereas the original method incurs 67.2M calculations. A factor of two is explained by the structure of the tables. The remaining speed-up is caused by the fact that there are more false alarms with distinguished points (2.8 times more in average) and that these false alarms generate more work. Both effects are due to the fact that with distinguished points, the length of the chains is not constant.

4.1   The Importance of Being Constant

Fatal attraction: Variations in chain length introduce variations in merge probability. Within a given set of chains (e.g. one table) the longer chains will have more chances to merge with other chains than the short ones. Thus the merges will create larger trees of longer chains and smaller trees of shorter chains. This has a doubly negative effect when false alarms occur. False alarm will more probably happen with large trees because there are more possibilities to merge into a large tree than into a small one. A single merge into a large tree creates more false alarms since the tree contains more chains and all chains have to be generated to confirm the false alarm. Thus false alarms will not only tend to happen with longer chains, they will also tend to happen in larger sets.

Larger overhead: Additionally to the attraction effect of longer chains, the number of calculations needed to confirm a false alarm on a variable length chains is larger than with constant length chains. When the length of a chain is not known the whole chain has to be regenerated to confirm the false alarm. With constant length chains we can count the number of calculations done to reach the end of a chain and then know exactly at what position to expect the key. We thus only have to generate a fraction of a chain to confirm the false alarm. Moreover, with rainbow chains, false alarms will occur more often when we look at the longer chains (i.e. starting at the columns more to the left of a table). Fortunately, this is also where the part of the chain that has to be generated to confirm the false alarms is the shortest.

Both these effects can be seen in Table 2 by looking at the number of endpoints found, the number of false alarms and the number of calculations per false alarm, in case of failure. With distinguished points each matching point generates about 4 false alarms and the mean length of the chains generated is about 9600. With rainbow chains there are only about 2.5 false alarms per endpoint found and only 1500 keys generated per false alarm.

The fact that longer chains yield more merges has been noted in [7] without mentioning that it increases the probability and overhead of false alarms. As a result, the authors propose to only use chains which are within a certain range of length. This reduces the problems due to the variation of length but it also reduces the coverage that can be achieved with one reduction function and increases the precalculation effort.

4.2   Increasing the Gain Even Further

We have calculated the expected gain over classical tables by considering the worst case where a key has to be searched in all columns of a rainbow table and without counting the false alarms. While a rainbow table is searched from the amount of calculation increases quadraticly from 1 to $\frac{t^2-1}{2}$, whereas in classical tables it increases linearly to t^2. If the key is found early, the gain may thus be much higher (up to a factor of t). This additional gain is partly set off by the fact

Table 2. Statistics for classic tables with distinguished points and for rainbow tables

	classic with DP	rainbow	ratio
t, m, ℓ	4666, 8192, 4666	4666, 38'223'872, 1	1
	mean cryptanalysis time		
to success	68.9s	9.37s	7.4
to failure	181.0s	26.0s	7.0
average	96.1s	12.9s	7.4
	mean nbr of hash calculations		
to success	48.3M	6.77M	7.1
to failure	126M	18.9M	6.7
average	67.2M	9.34M	7.2
	mean nbr of searches		
to success	1779	2136	0.83
to failure	4666	4666	1
average	2477	2673	0.93
	mean nbr of matching endpoints found		
to success	1034	620	1.7
to failure	2713	2020	1.3
average	1440	917	1.6
	mean nbr of false alarms		
to success	4157	1492	2.8
to failure	10913	5166	2.1
average	5792	2271	2.6
	mean nbr of hash calculations per false alarms		
to success	9622	3030	3.2
to failure	9557	1551	6.2
average	9607	2540	3.8

that in rainbow tables, false alarms that occur in the beginning of the search, even if rarer, are the ones that generate the most overhead. Still, it should be possible to construct a (possibly pathological) case where rainbow tables have an arbitrary large gain over classical tables. One way of doing it is to require a success rate very close to 100% and a large t. The examples in the literature often use a success rate of up to 80% with $N^{1/3}$ tables of order of $N^{1/3}$ chains of $N^{1/3}$ points. Such a configuration can be replaced with a single rainbow table of order of $N^{2/3}$ rows of $N^{1/3}$ keys. For some applications a success rate of 80% may be sufficient, especially if there are several samples of ciphertext available and we need to recover just any key. In our example of password recovery we are often interested in only one particular password (e.g. the administrator's password). In that case we would rather have a near perfect success rate. High success rates lead to configurations where the number of tables is several times larger than the length of the chains. Thus we end up having several rainbow tables (5 in our example). Using a high success rate yields a case were we typically will find the key early and we only rarely have to search all rows of all tables. To benefit from this fact we have to make sure that we do not search the five rainbow tables sequentially but that we first look up the last column of each table and then only move to the second last column of each table. Using this procedure we reach a gain of 12 when using five tables to reach 99.9% success rate compared to the gain of 7 we had with a single table and 78% success rate. More details are given in the next section.

4.3   Cracking Windows Passwords in Seconds

After having noticed that rainbow chains perform much better than classical ones, we have created a larger set of tables to achieve our goal of 99.9% success rate. The measurements on the first table show that we would need 4.45 tables of 38223872 lines and 4666 columns. We have chosen to generate 5 tables of 35'000'000 lines in order to have an integer number of tables and to respect the memory constraint of $1.4GB$. On the other hand we have generated 23'330 tables of 4666 columns and 7501 lines. The results are given in Table 3. We have cracked 500 passwords, with 100% success in both cases.

Table 3. Cryptanalysis statistics with a set of tables yielding a success rate of 99.9%. From the middle column we see that rainbow tables need 12 times less calculations. The gain in cryptanalysis time is only 1.5 times better due to disk accesses. On a workstation with 500MB of RAM a better gain in time (7.5) can be achieved by restricting the search to one rainbow table at a time (rainbow sequential).

	classic with DP	rainbow	ratio	rainbow sequential	ratio
t, m, ℓ	4666, 7501, 23330	4666, 35M, 5	1	4666, 35M, 5	1
cryptanalysis time	101.4s	66.3	1.5	13.6s	**7.5**
hash calculations	90.3M	7.4M	**12**	11.8M	7.6
false alarms (fa)	7598	1311	5.8	2773	2.7
hashes per fa	9568	4321	2.2	3080	3.1
effort spent on fa	80%	76%	1.1	72%	1.1
success rate	100%	100%	1	100%	1

From table 3 we see that rainbow tables need 12 times less calculations than classical tables with distinguished points. Unfortunately the gain in time is only a factor of 1.5. This is because we have to randomly access 1.4GB of data on a workstation that has 500MB of RAM. In the previous measurements with a single table, the table would stay in the filesystem cache, which is not possible with five tables. Instead of upgrading the workstation to 1.5GB of RAM we chose to implement an approach where we search in each rainbow table sequentially. This allows us to illustrate the discussion from the end of the previous section. When we search the key in all tables simultaneously rather than sequentially, we work with shorter chains and thus generate less work (7.4M operations rather than 11.8M). Shorter chains also mean that we have less false alarms (1311 per key cracked, rather than 2773). But short chains also mean that calculations needed to confirm a false alarm are higher (4321 against 3080). It is interesting to note that in all cases, the calculations due to false alarms make about 75% of the cryptanalysis effort.

Looking at the generic parameters of the trade-off we also note that the precalculation of the tables has needed an effort about 10 times higher than calculating a full dictionary. The large effort is due to the probabilistic nature of the method and it could be reduced to three times a full dictionary if we would accept 90% success rate rather that than 99.9%.

5   An Outlook at Perfect Tables

Rainbow tables and classic tables with distinguished points both have the property that merging chains can be detected because of their identical endpoints. Since the tables have to be sorted by endpoint anyway, it seems very promising to create perfect tables by removing all chains that merge with chains that are already in the table. In the case of distinguished points we can even choose to retain the longest chain of a set of merging chains to maximise the coverage of the table. The success rate of rainbow tables and tables with distinguished points are easy to calculate, at least if we assume that chains with distinguished points have a average length of t. In that case it is straight forward to see that a rainbow table of size $mt \times t$ has the same success rate than t tables of size $m \times t$. Indeed, in the former case we have t rows of mt distinct keys where in the latter case we have t tables containing mt distinct keys each.

Ideally we would want to construct a single perfect table that covers the complete domain of N keys. The challenge about perfect tables is to predict how many non-merging chains of length t it is possible to generate. For rainbow chains this can be calculated in the same way as we calculate the success rate for non-perfect tables. Since we evaluate the number of distinct points in each column of the table, we need only look at the number of distinct points in the last column to know how many distinct chains there will be.

$$\hat{P}_{table} = 1 - e^{-t\frac{m_t}{N}} \quad \text{where} \quad m_1 = N \quad \text{and} \quad m_{n+1} = N\left(1 - e^{-\frac{m_n}{N}}\right) \quad (4)$$

For chains delimited by distinguished points, this calculation is far more complex. Because of the fatal attraction described above, the longer chains will be merged into large trees. Thus when eliminating merging chains we will eliminate more longer chains than shorter ones. A single experiment with 16 million chains of length 4666 shows that after elimination of all merges (by keeping the longest chain), only 2% of the chains remain and their average length has decreased from 4666 to 386! To keep an average length of 4666 we have to eliminate 96% of the remaining chains to retain only the longest 4% (14060) of them.

The precalculation effort involved in generating maximum size perfect tables is prohibitive (Nt). To be implementable a solution would use a set of tables which are smaller than the largest possible perfect tables.

More advanced analysis of perfect tables is the focus of our current effort. We conjecture that because of the limited number of available non-merging chains, it might actually be more efficient to use near-perfect tables.

6   Conclusions

We have introduced a new way of generating precomputed data in Hellman's original cryptanalytic time-memory trade-off. Our optimisation has the same property as the use of distinguished points, namely that it reduces the number of table look-ups by a factor which is equal to the length of the chains. For an equivalent success rate our method reduces the number of calculations needed for cryptanalysis by a factor of two against the original method and by an even more important factor (12 in our experiment) against distinguished points. We have shown that the reason for this extra gain is the variable length of chains that are delimited by distinguished points which results in more false alarms and more overhead per false alarm. We conjecture that with different parameters (e.g. a higher success rate) the gain could be even much larger than the factor of 12 found in our experiment. These facts make our method a very attractive replacement for the original method improved with distinguished points.

The fact that our method yields chains that have a constant length also greatly simplifies the analysis of the method as compared to variable length chains using distinguished points. It also avoids the extra precalculation effort which occurs when variable length chains have to be discarded because they have an inappropriate length or contain a loop. Constant length could even prove to be advantageous for hardware implementations.

Finally our experiment has demonstrated that the time-memory trade-off allows anybody owning a modern personal computer to break cryptographic systems which were believed to be secure when implemented years ago and which are still in use today. This goes to demonstrate the importance of phasing out old cryptographic systems when better systems exist to replace them. In particular, since memory has the same importance as processing speed for this type of attack, typical workstations benefit doubly from the progress of technology.

Acknowledgements. The author wishes to thank Maxime Mueller for implementing a first version of the experiment.

References

1. J. Borst, B. Preneel, and J. Vandewalle. On time-memory tradeoff between exhaustive key search and table precomputation. In P. H. N. de With and M. van der Schaar-Mitrea, editors, *19th Symp. on Information Theory in the Benelux*, pages 111–118, Veldhoven (NL), 28-29 1998. Werkgemeenschap Informatie- en Communicatietheorie, Enschede (NL).
2. D.E. Denning. *Cryptography and Data Security*, page 100. Addison-Wesley, 1982.
3. Amos Fiat and Moni Naor. Rigorous time/space tradeoffs for inverting functions. In *STOC 1991*, pages 534–541, 1991.
4. M. E. Hellman. A cryptanalytic time-memory trade off. *IEEE Transactions on Information Theory*, IT-26:401–406, 1980.
5. Kim and Matsumoto. Achieving higher success probability in time-memory tradeoff cryptanalysis without increasing memory size. *TIEICE: IEICE Transactions on Communications/Electronics/Information and Systems*, 1999.

6. Koji KUSUDA and Tsutomu MATSUMOTO. Optimization of time-memory trade-off cryptanalysis and its application to DES, FEAL-32, and skipjack. *IEICE Transactions on Fundamentals*, E79-A(1):35–48, January 1996.
7. F.X. Standaert, G. Rouvroy, J.J. Quisquater, and J.D. Legat. A time-memory tradeoff using distinguished points: New analysis & FPGA results. In *proceedings of CHES 2002*, pages 596–611. Springer Verlag, 2002.

Appendix

The success rate of a single rainbow table can be calculated by looking at each column of the table and treating it as a classical occupancy problem. We start with $m_1 = m$ distinct keys in the first column. In the second column the m_1 keys are randomly distributed over the keyspace of size N, generating m_2 distinct keys:

$$m_2 = N(1 - \left(1 - \frac{1}{N}\right)^{m_1}) \approx N\left(1 - e^{-\frac{m_1}{N}}\right)$$

Each column i has m_i distinct keys. The success rate of the table is thus:

$$P \;=\; 1 - \prod_{i=1}^{t}(1 - \frac{m_i}{N})$$

where

$$m_1 = m \quad , \quad m_{n+1} = N\left(1 - e^{-\frac{m_n}{N}}\right)$$

The result is not in a closed form and has to be calculated numerically. This is no disadvantage against the success rate of classical tables since the large number of terms in the sum of that equation requires a numerical interpolation.

The same approach can be used to calculate the number of non-merging chains that can be generated. Since merging chains are recognised by their identical endpoint, the number of distinct keys in the last column m_t is the number of non-merging chains. The maximum number of chains can be reached when choosing every single key in the key space N as a starting point.

$$m_1 = N \, , \, m_{n+1} = N\left(1 - e^{-\frac{m_n}{N}}\right)$$

The success probability of a table with the maximum number of non-merging chains is:

$$\hat{P} = 1 - (1 - \frac{m_t}{N})^t \approx 1 - e^{-t\frac{m_t}{N}}$$

Note that the effort to build such a table is Nt.

Author Index

Lecture Notes in Computer Science

For information about Vols. 1–2665
please contact your bookseller or Springer-Verlag

Vol. 2708: R. Reed, J. Reed (Eds.), SDL 2003: System Design. Proceedings, 2003. XI, 405 pages. 2003.

Vol. 2709: T. Windeatt, F. Roli (Eds.), Multiple Classifier Systems. Proceedings, 2003. X, 406 pages. 2003.

Vol. 2710: Z. Ésik, Z, Fülöp (Eds.), Developments in Language Theory. Proceedings, 2003. XI, 437 pages. 2003.

Vol. 2711: T.D. Nielsen, N.L. Zhang (Eds.), Symbolic and Quantitative Approaches to Reasoning with Uncertainty. Proceedings, 2003. XII, 608 pages. 2003. (Subseries LNAI).

Vol. 2712: A. James, B. Lings, M. Younas (Eds.), New Horizons in Information Management. Proceedings, 2003. XII, 281 pages. 2003.

Vol. 2713: C.-W. Chung, C.-K. Kim, W. Kim, T.-W. Ling, K.-H. Song (Eds.), Web and Communication Technologies and Internet-Related Social Issues – HSI 2003. Proceedings, 2003. XXII, 773 pages. 2003.

Vol. 2714: O. Kaynak, E. Alpaydin, E. Oja, L. Xu (Eds.), Artificial Neural Networks and Neural Information Processing – ICANN/ICONIP 2003. Proceedings, 2003. XXII, 1188 pages. 2003.

Vol. 2715: T. Bilgiç, B. De Baets, O. Kaynak (Eds.), Fuzzy Sets and Systems – IFSA 2003. Proceedings, 2003. XV, 735 pages. 2003. (Subseries LNAI).

Vol. 2716: M.J. Voss (Ed.), OpenMP Shared Memory Parallel Programming. Proceedings, 2003. VIII, 271 pages. 2003.

Vol. 2718: P. W. H. Chung, C. Hinde, M. Ali (Eds.), Developments in Applied Artificial Intelligence. Proceedings, 2003. XIV, 817 pages. 2003. (Subseries LNAI).

Vol. 2719: J.C.M. Baeten, J.K. Lenstra, J. Parrow, G.J. Woeginger (Eds.), Automata, Languages and Programming. Proceedings, 2003. XVIII, 1199 pages. 2003.

Vol. 2720: M. Marques Freire, P. Lorenz, M.M.-O. Lee (Eds.), High-Speed Networks and Multimedia Communications. Proceedings, 2003. XIII, 582 pages. 2003.

Vol. 2721: N.J. Mamede, J. Baptista, I. Trancoso, M. das Graças Volpe Nunes (Eds.), Computational Processing of the Portuguese Language. Proceedings, 2003. XIV, 268 pages. 2003. (Subseries LNAI).

Vol. 2722: J.M. Cueva Lovelle, B.M. González Rodríguez, L. Joyanes Aguilar, J.E. Labra Gayo, M. del Puerto Paule Ruiz (Eds.), Web Engineering. Proceedings, 2003. XIX, 554 pages. 2003.

Vol. 2723: E. Cantú-Paz, J.A. Foster, K. Deb, L.D. Davis, R. Roy, U.-M. O'Reilly, H.-G. Beyer, R. Standish, G. Kendall, S. Wilson, M. Harman, J. Wegener, D. Dasgupta, M.A. Potter, A.C. Schultz, K.A. Dowsland, N. Jonoska, J. Miller (Eds.), Genetic and Evolutionary Computation – GECCO 2003. Proceedings, Part I. 2003. XLVII, 1252 pages. 2003.

Vol. 2724: E. Cantú-Paz, J.A. Foster, K. Deb, L.D. Davis, R. Roy, U.-M. O'Reilly, H.-G. Beyer, R. Standish, G. Kendall, S. Wilson, M. Harman, J. Wegener, D. Dasgupta, M.A. Potter, A.C. Schultz, K.A. Dowsland, N. Jonoska, J. Miller (Eds.), Genetic and Evolutionary Computation – GECCO 2003. Proceedings, Part II. 2003. XLVII, 1274 pages. 2003.

Vol. 2725: W.A. Hunt, Jr., F. Somenzi (Eds.), Computer Aided Verification. Proceedings, 2003. XII, 462 pages. 2003.

Vol. 2726: E. Hancock, M. Vento (Eds.), Graph Based Representations in Pattern Recognition. Proceedings, 2003. VIII, 271 pages. 2003.

Vol. 2727: R. Safavi-Naini, J. Seberry (Eds.), Information Security and Privacy. Proceedings, 2003. XII, 534 pages. 2003.

Vol. 2728: E.M. Bakker, T.S. Huang, M.S. Lew, N. Sebe, X.S. Zhou (Eds.), Image and Video Retrieval. Proceedings, 2003. XIII, 512 pages. 2003.

Vol. 2729: D. Boneh (Ed.), Advances in Cryptology – CRYPTO 2003. Proceedings, 2003. XII, 631 pages. 2003.

Vol. 2731: C.S. Calude, M.J. Dinneen, V. Vajnovszki (Eds.), Discrete Mathematics and Theoretical Computer Science. Proceedings, 2003. VIII, 301 pages. 2003.

Vol. 2732: C. Taylor, J.A. Noble (Eds.), Information Processing in Medical Imaging. Proceedings, 2003. XVI, 698 pages. 2003.

Vol. 2733: A. Butz, A. Krüger, P. Olivier (Eds.), Smart Graphics. Proceedings, 2003. XI, 261 pages. 2003.

Vol. 2734: P. Perner, A. Rosenfeld (Eds.), Machine Learning and Data Mining in Pattern Recognition. Proceedings, 2003. XII, 440 pages. 2003. (Subseries LNAI).

Vol. 2741: F. Baader (Ed.), Automated Deduction – CADE-19. Proceedings, 2003. XII, 503 pages. 2003. (Subseries LNAI).

Vol. 2742: R. N. Wright (Ed.), Financial Cryptography. Proceedings, 2003. VIII, 321 pages. 2003.

Vol. 2743: L. Cardelli (Ed.), ECOOP 2003 – Object-Oriented Programming. Proceedings, 2003. X, 501 pages. 2003.

Vol. 2745: M. Guo, L.T. Yang (Eds.), Parallel and Distributed Processing and Applications. Proceedings, 2003. XII, 450 pages. 2003.

Vol. 2746: A. de Moor, W. Lex, B. Ganter (Eds.), Conceptual Structures for Knowledge Creation and Communication. Proceedings, 2003. XI, 405 pages. 2003. (Subseries LNAI).

Vol. 2748: F. Dehne, J.-R. Sack, M. Smid (Eds.), Algorithms and Data Structures. Proceedings, 2003. XII, 522 pages. 2003.

Vol. 2749: J. Bigun, T. Gustavsson (Eds.), Image Analysis. Proceedings, 2003. XXII, 1174 pages. 2003.

Vol. 2750: T. Hadzilacos, Y. Manolopoulos, J.F. Roddick, Y. Theodoridis (Eds.), Advances in Spatial and Temporal Databases. Proceedings, 2003. XIII, 525 pages. 2003.

Vol. 2751: A. Lingas, B.J. Nilsson (Eds.), Fundamentals of Computation Theory. Proceedings, 2003. XII, 433 pages. 2003.

Vol. 2752: G.A. Kaminka, P.U. Lima, R. Rojas (Eds.), RoboCup 2002: Robot Soccer World Cup VI. XVI, 498 pages. 2003. (Subseries LNAI).

Vol. 2753: F. Maurer, D. Wells (Eds.), Extreme Programming and Agile Methods – XP/Agile Universe 2003. Proceedings, 2003. XI, 215 pages. 2003.

Vol. 2758: D. Basin, B. Wolff (Eds.), Theorem Proving in Higher Order Logics. Proceedings, 2003. X, 367 pages. 2003.

Vol. 2759: O.H. Ibarra, Z. Dang (Eds.), Implementation and Application of Automata. Proceedings, 2003. XI, 312 pages. 2003.

Vol. 2762: G. Dong, C. Tang, W. Wang (Eds.), Advances in Web-Age Information Management. Proceedings, 2003. XIII, 512 pages. 2003.